GREEK AND ROMAN TECHNOLOGY

In this new edition of *Greek and Roman Technology*, the authors translate and annotate key passages from ancient texts to provide a history and analysis of the origins and development of technology in the classical world.

Sherwood and Nikolic, with Humphrey and Oleson, provide a comprehensive and accessible collection of rich and varied sources to illustrate and elucidate the beginnings of technology. Among the topics covered are energy, basic mechanical devices, hydraulic engineering, household industry, medicine and health, transport and trade, and military technology. This fully revised Sourcebook collects more than 1,300 passages from over 200 ancient sources and a diverse range of literary genres, such as the encyclopaedic *Natural History* of Pliny the Elder, the poetry of Homer and Hesiod, the philosophies of Plato, Aristotle, and Lucretius, the agricultural treatises of Varro, Columella, and Cato, the military texts of Philo of Byzantium and Aeneas Tacticus, as well as the medical texts of Galen, Celsus, and the Hippocratic Corpus. Almost 100 line-drawings, indexes of authors and subjects, introductions outlining the general significance of the evidence, notes to explain the specific details, and current bibliographies are included.

This new and revised edition of *Greek and Roman Technology* will remain an important and vital resource for students of technology in the ancient world, as well as those studying the impact of technological change on classical society.

Andrew N. Sherwood is associate professor in the School of Languages and Literatures at the University of Guelph, Canada.

Milorad Nikolic is associate professor in the Department of Classics at Memorial University of Newfoundland in St. John's, Canada.

John W. Humphrey is retired from the Department of Classics and Religion at the University of Calgary, Canada.

John P. Oleson is retired from the Department of Greek and Roman Studies at the University of Victoria, Canada.

ROUTLEDGE SOURCEBOOKS FOR THE ANCIENT WORLD

Titles include:

THE HISTORIANS OF ANCIENT ROME
An Anthology of the Major Writings, 3rd Edition
Ronald Mellor

WOMEN IN THE ANCIENT NEAR EAST
A Sourcebook
Edited by Mark Chavalas

POMPEII AND HERCULANEUM
A Sourcebook, 2nd Edition
Alison E. Cooley and M.G.L. Cooley

ANCIENT ROME
Social and Historical Documents from the Early Republic to the Death of Augustus, 2nd Edition
Matthew Dillon and Lynda Garland

PAGANS AND CHRISTIANS IN LATE ANTIQUITY
A Sourcebook, 2nd Edition
A.D. Lee

GREEK AND ROMAN TECHNOLOGY
A Sourcebook of Translated Greek and Roman Texts, 2nd Edition
Andrew N. Sherwood, Milorad Nikolic, John W. Humphrey, and John P. Oleson

www.routledge.com/Routledge-Sourcebooks-for-the-Ancient-World/book-series/RSAW

GREEK AND ROMAN TECHNOLOGY

A Sourcebook of Translated Greek and Roman Texts

Second Edition

*Andrew N. Sherwood,
Milorad Nikolic,
John W. Humphrey,
and John P. Oleson*

LONDON AND NEW YORK

Second edition published 2020
by Routledge
2 Park Square, Milton Park, Abingdon, Oxon OX14 4RN

and by Routledge
52 Vanderbilt Avenue, New York, NY 10017

Routledge is an imprint of the Taylor & Francis Group, an informa business

© 2020 Andrew N. Sherwood, Milorad Nikolic, John W. Humphrey, and John P. Oleson

The right of Andrew N. Sherwood, Milorad Nikolic, John W. Humphrey, and John P. Oleson to be identified as authors of this work has been asserted by them in accordance with sections 77 and 78 of the Copyright, Designs and Patents Act 1988.

All rights reserved. No part of this book may be reprinted or reproduced or utilised in any form or by any electronic, mechanical, or other means, now known or hereafter invented, including photocopying and recording, or in any information storage or retrieval system, without permission in writing from the publishers.

Trademark notice: Product or corporate names may be trademarks or registered trademarks, and are used only for identification and explanation without intent to infringe.

First edition published by Routledge 1997

British Library Cataloguing-in-Publication Data
A catalogue record for this book is available from the British Library

Library of Congress Cataloging-in-Publication Data
Names: Sherwood, Andrew N. (Andrew Neil), 1955– author. | Oleson, John Peter, author. | Nikolic, Milorad, 1968– author. | Humphrey, John William, 1946– author.
Title: Greek and Roman technology : a sourcebook annotated translations of Greek and Latin texts and documents / Andrew N. Sherwood, Milorad Nikolic, John P. Oleson, John W. Humphrey.
Description: Second edition. | Abingdon, Oxon ; New York, NY : Routledge, 2019. | Series: Routledge sourcebooks for the ancient world | Includes bibliographical references and indexes.
Identifiers: LCCN 2019016909 (print) | LCCN 2019017300 (ebook) | ISBN 9781315682181 (ebook) | ISBN 9781317402411 (web pdf) | ISBN 9781317402398 (mobi/kindle) | ISBN 9781317402404 (epub) | ISBN 9781138927902 (hbk :
alk. paper) | ISBN 9781138927896 (pbk : alk. paper) |
ISBN 9781315682181 (ebk)
Subjects: LCSH: Technology–Greece–History–Sources. | Technology–Rome–History–Sources.
Classification: LCC T16 (ebook) | LCC T16 .S44 2019 (print) | DDC 609.38–dc23
LC record available at https://lccn.loc.gov/2019016909

ISBN: 978-1-138-92790-2 (hbk)
ISBN: 978-1-138-92789-6 (pbk)
ISBN: 978-1-315-68218-1 (ebk)

Typeset in Times New Roman
by Wearset Ltd, Boldon, Tyne and Wear

TO

DEVON
NANCY AND LAURA
BENJAMIN, SOPHIA, WILLIAM, ROSE
MAC, ALLISON, MIKE, JOHN, ANGUS†,
RYAN, ALANNA, DAN, NEIL
SURI AND LINCOLN
SUSAN

WITH LOVE AND GRATITUDE

CONTENTS

List of illustrations ix
Preface to the second edition xii
Extract from the preface and acknowledgements of the first edition xv
List of abbreviations xviii

Introduction   1

1 The rise of humans and human technology   12

2 Sources of energy and basic mechanical devices   34

3 Agriculture   88

4 Food processing   158

5 Mining and quarrying   185

6 Metallurgy   221

7 Sculpture   252

8 Construction engineering   266

9 Hydraulic engineering   335

10 Household crafts, health and well-being, and workshop production   388

CONTENTS

11	Transport and trade	508
12	Record-keeping	621
13	Military technology	664
14	Attitudes towards labour, innovation, and technology	709
	Indexes	732

ILLUSTRATIONS

2.1	Wind-powered organ	38
2.2	The *Aeölipile*	39
2.3	Water-powered whistling wheel	41
2.4	Earliest pictorial representation of a crank drive (third century AC).	45
2.5	The *Barylkos*	66
2.6	Screwcutter	69
2.7	Bronze force pump from Bolsena, second century AC	74
2.8	Singing bird automaton	77
2.9	Whistling bird and owl automaton	78
2.10	Automatic trumpeting doorbell	79
2.11	Coin-operated holy-water dispenser	80
2.12	Lamp with self-trimming wick	81
2.13	Self-opening temple doors	82
4.1	Donkey mill in Pompeii	161
4.2	Relief decoration on the Tomb of Eurysaces, the baker, showing the individual steps in breadmaking, Rome, first century BC	164
4.3	Lever press	166
4.4	*Trapetum* olive press	166
5.1	Pinax with open-pit miners, sixth century BC	185
5.2	The arrangement and operation of mining shafts	186
5.3	Supporting pillar in an underground quarry	194
5.4	Hushing	203
5.5	Catching gold dust in a cloth	206
5.6	Transport of a marble block on a sled	210
5.7	Splitting a rock with wedges	212
6.1	Milling the ore	226
6.2	Washing the ore	227
6.3	Forging the bloom	237
7.1 a, b	Vase painting of bronze sculptors at foundry	254
7.2	Carving a sarcophagus	259

7.3	Blacksmithing *putti*	260
7.4	A metal workshop	260
8.1	Clever ways to move heavy marble pieces	280
8.2	*Opus reticulatum*	282
8.3	Reconstructed centering of the Pantheon	287
8.4	*Dioptra*	290
8.5	*Chorobates*	291
8.6	Raising the obelisk of Theodosius	301
8.7	Stages in shaping a column drum	304
8.8	A-frame hoist with capstan	306
8.9	Tomb of the Haterii	306
8.10	Reconstruction of *Insula Dianae* at Ostia	316
8.11	Wall painting with bricklayers	316
8.12	Hypocaust system	321
9.1	Waterwheels in Rio Tinto Mine, Spain	336
9.2	Egyptian wall painting with *shaduf*	362
9.3	Wheel with compartmented body (axle not shown)	363
9.4	Wheel with compartmented rim (detail of compartments)	364
9.5	Pair of compartmented wheels (axles and frame not shown)	365
9.6	Tread-wheel driven bucket chain	366
9.7	Paddle-wheel driven bucket chain	367
9.8	Design for a Vitruvian water screw	370
9.9	Philo of Byzantium's force pump	372
9.10	Force pump with swivelling nozzle	374
10.1	Vase painting with blacksmith	391
10.2	Roman relief with knife seller	393
10.3	Vase painting with vertical loom	424
10.4	Pinax showing a potter at the kiln	432
10.5	*Diatretum* cup	440
10.6	Strigil, 21 cm	453
10.7	Shears/scissors, 10 cm	455
10.8	Bleeding cup, 15 cm	464
10.9	Relief of a box of scalpels with flanking bleeding cups, 43 × 33 cm	465
10.10a	Scalpel, 8.7 cm	466
10.10b	Scalpel, 14 cm	466
10.10c	Probe, 18 cm	467
10.10d	Spatula probe, 18 cm	467
10.10e	Spatula probe, 15.5 cm	467
10.10f	*Ligula*, 18.4 cm	468
10.10g	Bifurcated probe, 15 cm	468
10.10h	*Vulsellum*, 5.8 cm	469
10.10i	Uvula forceps, 20.2 cm	469
10.10j	Bone forceps, 21 cm	469

10.10k	Bone lever, 15.5 cm	470
10.11	Bladder sound (same shape and size as catheter, but solid not hollow), 15 cm	471
10.12	Vaginal speculum, 23 cm	471
10.13	Thong drill	473
10.14	Syringe	482
10.15	Vase painting showing slave hauling furniture	493
10.16	Firing of bricks in a kiln	496
11.1	Roman road construction	511
11.2	Typical street in Pompeii with wheel ruts	518
11.3	Vase painting with drowning sailors	547
11.4	Mortise and tenon joint	551
11.5	Reconstruction of prefabricated *caissons* being loaded with concrete off Caesarea	580
11.6	Towing a river boat on the Durance River	591
11.7	Ox-hide ingots, *c.*1200–1050 BC	592
11.8	Athenian silver tetradrachm, Athena and owl	595
11.9	Striking a coin	595
12.1	*Kleroterion*: Aristotle's voting machine	632
12.2	*Pinakion*: voting ballot, Athens, fourth century BC	632
12.3	Ballot disks	635
12.4	Roman coin with voting procedure	636
12.5	Ostraca	637
13.1	*Testudo* from Trajan's Column	677
13.2	Greek fire after 12th century illustration from the Madrid Skylitzes	687
13.3	*Cheiroballistra*	693
13.4	*Carroballista* from Trajan's Column	703

PREFACE TO THE SECOND EDITION

The first edition, published in 1998 and itself a product of 20 years of diligent accumulation of material, has been used successfully for 20 years as a resource and textbook. Even though the past appears immutable from a modern point of view, it is anything but – at least not in the perception of researchers as scholarly curiosity, acumen, and initiative bring to light new information. Since the publication of the first edition, for example, the use of a crank was confirmed in a Roman context and the date of Archimedes' screw was revised. On a more personal level, mid-career tenured faculty members (i.e. two of the original authors) have retired, and undergraduate students (specifically one of the current authors) have become mid-career tenured faculty members: plenty of change to warrant a new edition. The publisher canvassed users who had indicated their desire for a new edition to increase the usefulness of the book as a textbook, and collected numerous helpful suggestions. Many of these contradicted one another. Hence the second edition would ideally contain more material and more images, while having a smaller physical format and lower price. We did the best we could to respond to the suggestions and to include the results of scholarship from the past two decades. Updated chapter bibliographies reflect the advances made in the study of ancient technology over the past two decades years. We have retained most of the bibliography items from the first edition, as they give the reader a glimpse into the progression of the discipline – a kind of meta-analysis of the study of the history of technology, as it were. One conspicuous difference from 20 years ago is the big increase in the number of female scholars who are making groundbreaking contributions to the field. May it forever continue in this fashion. In general, the volume of excellent publications has virtually exploded, so that it is hardly possible for one or two individuals to be aware of every relevant article that has appeared in journals around the world. Most items newly included in the chapter bibliographies are, therefore, edited collections and important monographs, each with its own bibliography that will, we hope, guide interested readers to more detailed publications in their individual field of interest. We call it, to use a phrase from urban water distribution, a dendritic system. It is almost inevitable to miss publications, and we apologise to those authors whose work we have overlooked in spite of our best efforts.

PREFACE TO THE SECOND EDITION

The addition of a considerable number of line-drawings, based on ancient depictions and renderings of excavated material and sites enhances the new edition. Changes and additions to most of the sections, and some new units respond to readers' wishes for more material. We are aware that these additions and changes are merely steps on the way to improvement, and we look forward to holding in our hands one day a third edition that may be prepared by our own current students another two decades from now.

The authors wish to thank a great many colleagues, friends, and assistants for their suggestions and encouragements, but the list has to be selective for considerations of space. First, we have benefitted from all the reviewers of the first edition and from anonymous reviewers of the proposed second edition, but we gratefully acknowledge the many observations, suggestions, and corrections of Professor Houston in particular. We also would like to thank Carol Benson for her "disappointment" that sculpture was not represented in the first version, and have tried to alleviate that distress. Erik de Bruijn kindly provided a mass of texts regarding ancient divers, from which we have included a sampling in several chapters. Many others, Clare Barker in particular, have offered new texts for consideration and, more surprisingly, convincing arguments to retain texts that we planned to remove since they were "not technological enough". Their basic arguments that those texts provided important and interesting evidence for the cultural impact of the technology and were significant for students were instrumental in staying the implementation of our deletion-technology for many texts. Several scholars, yet again, have been very generous with their time to help us with difficult texts. Professors D.A. Campbell, P. O'Cleirigh, J. Yardley, L.J. Sanders, H. Westra, M. Cropp, B. Levett, and K. Simonsen have all been very patient and helpful. Two scholars have also generously provided some imagery for the book. Professor Grewe has produced a wonderful reconstruction of a water-powered saw from Hierapolis [**fig. 2.4**] and architect Christopher Brandon has rendered the activity surrounding the construction of the harbour at Caesarea Maritima [**fig. 11.5**] so vividly that, for two of us, it brings back fond memories of our own work below the surface there many years ago.

Two final groups remain to be thanked. First our student assistants Theresa Ainsworth and Rose Conlin who drew almost all of the line-drawings that appear in the volume. Their patience and willingness to put up with suggestion after suggestion by all four of us is quite remarkable, and their very fine and clear images greatly enhance the new edition. Thank you both for saving us the embarrassment of our own stick-figures. Jennifer Zantingh, a third student assistant, also deserves much thanks for taking on many of the most tedious and repetitive corrections and other necessary tasks when we reorganised the new edition. Jenn, thank you so much since, yes, we could have done this ourselves, but we may have sought out a *xylospongium* in the process (**9.47**).

Finally, we would like to thank the staff at Routledge for their continual encouragement and patience. This has been a very long process and we have encountered some extremely difficult personal situations that have led to more

PREFACE TO THE SECOND EDITION

delays and setbacks than one could possibly have imagined, yet Routledge has been behind us the entire time. Amy Davis-Poynter has helped us considerably in the early stages, but Elizabeth Risch has been a rock of perseverance and understanding for us throughout the entire process. We cannot really express our gratitude sufficiently to you Lizzi – thank you so much!

Selections from *Documents in Mycenaean Greek* are reproduced with the permission of Cambridge University Press and John Chadwick (passages **6.27**, **8.7**, **10.32**, **10.125**, **10.126**, and **12.3**).

Andrew N. Sherwood, Milo Nikolic

EXTRACT FROM THE PREFACE AND ACKNOWLEDGEMENTS OF THE FIRST EDITION

This book is the product of a long gestation. In 1976, Humphrey prepared a *Sourcebook of Ancient Technology* based on a short collection of ancient texts, for use in his course on ancient technology at the University of Calgary. In 1978, Oleson altered and slightly expanded this informal text for use in his own course on ancient technology at the University of Victoria. Both of us agreed that a much larger sourcebook with new translations and appropriate commentary and bibliography was desirable, and in 1979 we agreed on the framework for the present Sourcebook and began the work of compiling citations and preparing new translations. We decided to focus mainly on Greek and Roman technology, with glances at the Bronze Age technologies of the Aegean, Egypt, the Levant, and Mesopotamia where these had a marked effect on Graeco-Roman technology, or on the imaginations of Greek and Roman writers concerned with technology. We also decided to place more emphasis on longer texts by Greek and Roman authors, which describe activities surrounding the technologies than on texts simply describing the materials or the results. In consequence, we have included fewer papyri and inscriptions than would otherwise be the case. Because of an embarrassment of riches, we have also had to omit hundreds of passages that are relevant to the topic – although many appear as simple citations. Even so, this book has grown beyond the size originally envisioned by the authors and the publisher; we are grateful to Dr. Stoneman of Routledge for his patience and understanding. In the end, we have translated more than 750 passages by 148 ancient authors (not counting individual inscriptions and papyri). The text is intended in particular for university-level courses on Greek and Roman technology and related subjects, but scholars should find the collection useful, and the general reader will find much of the material fascinating.

Because of other commitments to excavation and publication, heavy teaching loads, and – in particular – major administrative responsibilities, progress was slow. In 1992 the two original authors decided to seek a third collaborator in order to speed up completion of the project, and Sherwood, now at McGill University, agreed to take on approximately one-third of the text. In the end, the responsibility for translation and introductory material was apportioned as follows. Humphrey: Chapter 3, Agriculture; Chapter 4, Food Processing;

EXTRACT FROM THE PREFACE AND ACKNOWLEDGEMENTS

Chapter 10C, Standards of Trade; Chapter 11B, Writing and Literacy; Chapter 12, Military Technology; Subject index; Oleson: Chapter 2, Sources of Energy; Chapter 5, Mining and Quarrying; Chapter 8, Hydraulic Engineering; Chapter 9A, Metalworking; Chapter 9B, Woodworking; Chapter 9D, Ceramics and Glass; Chapter 10B, Navigation; Chapter 13, Attitudes; Bibliographies; Citation Index; Sherwood: Chapter 1, Rise of Humans; Chapter 6, Metallurgy; Chapter 7, Construction Engineering; Chapter 9C, Textiles and Leather; Chapter 9E, Applied Chemistry; Chapter 9F, Large-Scale Production; Chapter 10A, Land Transport; Chapter 11A, Time-Keeping; Subject index. All three authors contributed to the Introduction.

We have done our best to smooth over differences in style, approach to introductory material, and citation of references. Since the texts incorporate references to hundreds of names, some cited in their Greek form, others in Latinised form, the problem of spelling was a major one. In the end, we have transliterated most names in the spelling closest to that used by the ancient author cited, but we have preferred Latinised forms where students might have found the Greek version confusing. Given the scope of the project, it is unlikely that we have achieved total consistency. In the translated passages, parentheses surround Greek or Latin words reproduced by the translator to illustrate the context or to supply something omitted by the ancient author. Square brackets surround editorial comment or expansion. All Greek words are transliterated in Latin letters. Cross-references to passages in the Sourcebook are printed in bold. We have appended bibliographies to the end of each chapter. After some discussion, we decided to include a selection of the most recent or particularly useful books and articles in all European languages, with a preference for English titles where there was duplication. Within a chapter, citation of a modern work listed in the bibliography for that chapter is by author and date only; full publication information is given for other titles.

We have included only a relatively small number of illustrations, all line-drawings of machines or gadgets that seemed too complicated to document with verbal descriptions alone. Many other passages would have benefited from illustration, but in the interests of economy, we decided to abandon the idea of comprehensive visual documentation. If the present volume is well received, we might consider preparation of a companion volume devoted to the visual documentation of ancient technology.

Many of our colleagues have given assistance in one way or another over the long history of this project. We would like to extend special thanks to David A. Campbell, Ann Dusing, Harry Edinger, Lyn Rae, James Russell, Robert Todd, and Hector Williams. Research assistants for Oleson and Humphrey provided careful, much appreciated help with collating material and checking references: Vicky Karas, Terresa Lewis, Brian Woolcock, Cindy Nimchuck, Alison Cummings, and Benjamin Garstad. Over the years, we have received financial assistance from the Social Sciences and Humanities Research Council of Canada, the University of Victoria Committee on Research and Travel, the University of

EXTRACT FROM THE PREFACE AND ACKNOWLEDGEMENTS

Victoria Work-Study Programme, the University of Calgary Research Grants Committee, and the Arts and Science Faculty of Concordia University (with special thanks to P.H. Bird and C. Vallejo). We are very grateful for all this assistance.

John W. Humphrey, John P. Oleson, Andrew N. Sherwood

ABBREVIATIONS

CIG   Boeckh, August, *Corpus Inscriptionum Graecarum* (Berlin, 1828–1877).
CIL   *Corpus Inscriptionum Latinarum* (Berlin, 1863–).
DMG   Ventris Michael, and John Chadwick, *Documents in Mycenaean Greek.* 2nd edn (Cambridge, 1973).
ESAR   Frank *et al.*, eds., *Economic Survey of Ancient Rome* (Baltimore, 1933–1940).
IG   *Inscriptiones Graecae* (Berlin, 1873–).
IGR   Cagnat *et al.*, eds., *Inscriptiones Graecae ad res Romanas pertinentes* (Paris, 1906–).
ILS   Dessau, Hermann, *Inscriptiones Latinae Selectae* (Berlin, 1892–1916).
OGIS   Dittenberger, Wilhelm, ed., *Orientis Graeci Inscriptiones Selectae* (Leipzig, 1903–1905).
SEG   *Supplementum Epigraphicum Graecum* (Leiden, 1923–).
SIG   Dittenberger, Wilhelm, *Sylloge Inscriptionum Graecarum.* 3rd edn (Leipzig, 1915–1924).

INTRODUCTION

A SOCIETY AND TECHNOLOGY IN ANTIQUITY

At its most basic level technology can be regarded as the attempt by humans to control and master the natural environment, changing it into a more hospitable, if artificial, one. Unlike pure science, which involves a theoretical understanding of the environment that may involve research but often lacks any immediate implementation, technology is the process by which humans accomplish this change. A varied body of knowledge, composed of recipes and practical skills as well as the abstract knowledge of inventions and designs, is utilised to alleviate perceived problems; in the process various devices and machines are manufactured to aid in the conversion of the environment. It is clear that technological innovations have improved human existence considerably by providing the basics of life (better food, shelter, clothing, defence, and transport), and few aspects of life escape their influence. One goal of this book is the assembly in English translation of the most important passages in Greek and Latin literature illustrative of the role of technology in Greek and Roman society. This Introduction attempts to present some of the themes and sources of this topic.

Throughout human history, attempts have been made to discover the impetus that prompted technological innovation. Ancient and modern theories identify numerous stimuli. Advanced societies realise they had once been simpler and often examine the process by which their standard of living had improved. When historical sources are absent, logic, emotion, and religion become rational sources of explanation. The assumption of a very primitive existence for early humans, a time in which society was too simple to create its own technological inventions, resulted in the belief that advances were made by divine gifts/interventions, chance, or natural occurrence: fire was given by Prometheus; lightning created fire, which then melted naturally occurring ores by chance; trees provided food, shelter, and clothing, which humans merely had to harvest or put to use.

A more sympathetic view of humankind suggested that humans might see natural examples and then imitate them by artificial means to alter their environment: observation of a tree leaning across a river prompts people to produce bridges; an acanthus plant inspires the creation of the Corinthian capital. The next step was obvious – people invented/manufactured technologies as necessity prompted: the need for shelter compelled them to create them from the materials at hand; lack of appropriate materials or labour forced humankind to create or discover alternatives; military disadvantages drove them to invent new and better weapons; desire for wealth stimulated crafts and trades. In certain cases, the lack of need or lack of demand for an invention might result in the failure to put a machine to practical use, retaining it for its value as a novelty. Many of Hero's and Ctesibius' devices fall into this category. In other cases, a conservative audience might actively resist innovations that could have led to significant changes in society: the intellectual elite often scorned applied science, and others were content with the status quo as long as it functioned. On the other hand, progressive societies or individuals could radically stimulate technology by encouraging or rewarding productive inventors: Hieron II, ruler of Syracuse, funded a cluster of inventions by Archimedes.

But it is humankind's ingenuity that most often has been given credit for technological advances. Accounts of individual genius stretch from legendary/mythical inventors, like Daedalus and Palamedes, to historical figures, like Archimedes and Hero. Humankind's visionary ideal of a better world stimulated many advances that improved civilisation. Even the philosophical tenets of the Greeks, that a rational order governed the world, led to theoretical examination of the environment, which in turn did stimulate technology. The human spirit thrived on the challenge to provide a better life for humankind, generating a competitive side that hastened improvements via technological advance. Generally, the competitive motive was positive, but occasionally jealousy hindered technological developments as people worked through the process of trial and error in their march to improve their environment. And surely human curiosity – our fascination with machines, materials, and power – was responsible for much of the impetus towards technological advance.

Many of humankind's most important technological discoveries, which were subsequently used to further or to create other advances, occurred long before the Greek and Roman civilisations existed. The use and control of fire is recognised as humankind's most important advance, crucial in the struggle against the cold and wild animals, and crucial for a wider selection of foods. Agricultural advances, like domestication of crops and animals, once humankind's nomadic existence began to end, led to civilisation and permanent settlements that produced food surpluses, which in turn eventually resulted in specialisation of labour that promoted more technological changes. Humans began to produce textiles, which replaced animal skins for clothing, and to decorate their shelters, and for the first time had leisure for "unproductive" activities. The discovery and use of the wheel for pottery and transport is often cited as one of the greatest inventions, since it is ubiquitous today, but in antiquity its importance was probably much less. Instead, the use of domesticated animals and the discovery and use of water transport were most important to those who made their way through and around the rough terrain of the Mediterranean lands. The discovery of ores, their smelting, and alloying also began at an early date, but new processes, metals, and alloys were discovered during the classical period. Writing, a major technological advance that had a profound effect upon human culture in the Bronze Age, had to be re-introduced to Greece during the eighth century BC.

The appearance of similar technologies in diverse geographical areas occasionally may have been the result of simultaneous creation, but for the most part technologies were dispersed and developed as different peoples came into contact. Not only might the lack of raw materials in an area lead to a pattern of exchange that could transfer a technology that formed the need for the exchange (one component for an alloy, for example), but also such contact might lead to other incidental technological transfers (a new form of transport or method of construction). Such trade could be conducted on a peaceful level, but need/desire might also produce dispersal of technologies in a more aggressive manner, by conquest.

Although the Greeks often are characterised as theoretical thinkers without interest in practical application and the Romans as interested only in practical matters and not in theory, both characterisations are exaggerated. Obviously, the Greeks created and put into use many technological innovations, and the Romans spent time on theoretical observations. A different perception of the world and of humankind's position in it, however, did exist between the two cultures, at least at the level of the elite. The Greeks were interested in theory, science, and philosophy as they tried to determine and understand the rational order of the universe. Technological and practical advances may have been regarded as less "worthy" than theoretical ones by many Greek thinkers, but Greece did produce important inventors throughout her history – Eupalinus, Archimedes, Ctesibius, and Hero – some of whom were merely better than their contemporaries at putting their discoveries into practice.

The Romans, on the other hand, were superb at building upon the discoveries of others. Their society thrived on organisation and improving its effectiveness. One might think of Rome's military forces and the production of direct, paved roads on a scale hitherto unattempted, largely to make the army more effective. Again and again Roman writers like Cato, Columella, Frontinus, and Pliny the Elder stress the practicality of their culture. Yet they were also creative when the need arose. Military innovations gave them an empire; the development and implementation of the arch, cement, and kiln-baked brick created engineering marvels that inspired subsequent cultures and that stand to this day.

Technology without doubt improved the life of peoples in antiquity as it continues to do today. Arrangements such as the division of labour potentially meant better and faster production. The standard of living improved as nature was harnessed or "conquered". Food was better, more varied, and more abundant; shelters were stronger, larger, and more comfortable; leisure and luxuries became available; one even could work, play, or eat at night instead of merely sleep. Life had improved as humankind began to dominate and mould its environment.

INTRODUCTION

Yet at the same time technology brought disadvantages. The division of labour brought divisions within society, arrogance, and slavery. Military superiority fed greed and aggression, bringing widespread conquest, slavery, and death. The sense of community was gradually weakened by technological advance and continues to weaken to this day as technology allows society to fragment even further and creates greater gulfs between peoples. Present-day concerns regarding deforestation and pollution as the cost of technological advance were also voiced in antiquity, but little was done to stop the disasters.

Certainly the benefits of technology far outweigh its negative aspects both today and in antiquity: without it we would have no civilisation. Yet it has always been a double-edged sword in the hands of humankind. Pliny's thoughts (**14.20**) regarding iron, which serves as the best and worst example of technological progress, say much about technology in the hands of humans. The good it has done and continues to do is incalculable, but the harm has been and can be devastating and shameful.

There are four major sources of information about technologies from various periods of antiquity: contemporary pictures such as wall-paintings and decorated pottery; the comparison of less developed societies in the modern world with those of antiquity, from the early agricultural villages of the Neolithic to the specialised holdings of Classical times; ancient written records such as histories and technical reports; and the excavation of ancient sites. Of these, the two last are of greatest importance for the student of Greek and Roman technology, though ancient depictions and the relatively underdeveloped techniques of some parts of the eastern Mediterranean are often useful supplements. This book, of course, concentrates on the literary sources. Many general surveys of ancient technology are available, however, and most of these make significant use of the archaeological and ethnographic evidence. A few standard surveys are listed here.

B LITERARY SOURCES FOR ANCIENT TECHNOLOGY

Written records are available from the Bronze Age onwards, but it is not until the classical period of Greece and Rome that the works are particularly valuable for a study of ancient technology. Selections from over 150 ancient authors appear in the following pages; of these, the following groupings (of which some overlap) represent the most important.

Encyclopaedists. Although a Greek concept initially, the gathering of knowledge into a collected body was best carried out by the practical Romans. Most important for technology is the *Natural History* of Pliny the Elder (first century AC), an uncritical treatment in 37 books of a multitude of subjects: physics, geography, ethnology, human physiology, zoology, botany, architecture and art, stones, and metallurgy. Other important compilers include Aulus Gellius (second century AC) and etymologists like Varro (first century BC) and Isidore (sixth to seventh centuries AC).

Topical Monographs or Treatises. Of the many treatises written about specific topics, often by inventors or builders, very little survives beyond author and title. In a few cases, like Ctesibius, later writers sometimes provide a bit more information. We are fortunate to have a few complete or almost complete handbooks: the *Mechanical Problems*, originating in the school of Aristotle, concerned with the basic machines, as well as more complex ones; portions of Philo of Byzantium's (late third century BC) work on catapults and other weapons, and on hydraulic machinery survive as an important source of information; and work by Hero of Alexandria (mid-first century AC) that provides important information regarding devices using air and water pressure, gearing, and making of screw threads. Best known of the monographs, however, are several by Roman writers: Vitruvius' *On Architecture*, a late first century BC work, which examines construction, materials, time recording, water supply, and machines in ten books; Frontinus' late first-century AC book *On the Aqueducts of Rome*, and the more derivative military work, *On Stratagems*. This last topic is supplemented by a military treatise by Vegetius (fourth century AC).

Agricultural Writers. Since farming was the basis of the ancient economy, and since it was considered one of the few honourable occupations, it is not surprising that we have three classical Roman monographs on the subject: written by Cato (second century BC), Varro (first century BC), and Columella (first century AC), they treat cultivation, livestock, farm equipment, and so on. Poets, like Hesiod (eighth century BC) and Vergil (first century BC), also provide important agricultural information (see Poets).

Historians. Although technology is rarely an important focus for ancient historians, their narratives often provide detailed information about ancient technologies. The Greek Herodotus (fifth century BC), for example, gives details on the building of the Egyptian pyramids; Julius Caesar provides many useful descriptions of military technology in his accounts of his own campaigns in the first century BC; and Josephus (first century AC) provides accounts of military technology and harbour construction.

Geographers. Travellers like Strabo (early first century AC) and Pausanias (mid-second century AC) have left us accounts of their journeys in the Mediterranean during the Roman Empire, often providing interesting descriptions of engineering feats, famous structures, and artistic marvels, and recording legends and myths regarding early inventors.

Poets. Homer's and Hesiod's works, our earliest Greek literature (eighth century BC), not only provide evidence for agricultural life in the archaic period, a harsh existence, but give information regarding seafaring, smithing, and fanciful robots among others. The later poets, both Greek and Latin, continue to provide unexpected bits of evidence for farming (Vergil), textiles (Aristophanes; fifth century BC), tradespeople (Plautus; early second century BC), and so on. Satirists like Horace (first century BC) and Juvenal (first century AC) relate information regarding construction practices and transport in their descriptions of life.

Philosophers. Thinkers like the Pre-Socratics (seventh to fifth centuries BC), Plato (fourth century BC), Aristotle (fourth century BC), and Lucretius (first century BC) all offer ideas concerning the origin of mankind and the rise of the species via technological and other means. Some, like Plato, also devised ideal societies promoting theories, like specialised labour, that would improve the community.

Biographers. Although often filled with intrigues and fiction, sources such as Suetonius (second century AC), Plutarch (second century AC), and the *Historia Augusta* (fourth century AC) sometimes provide insights into organisation, technological wonders, and marvellous discoveries; accounts of engineering policies and rivalries also appear.

Authors of Letters. Best known are the collections of Cicero's letters (first century BC) and those of Pliny the Younger (first century AC), sometimes official in character, but more often providing day-to-day accounts of their lives. Among the information they provide appear details regarding engineering projects (private and public), residences, and furnishings.

Inscriptions and Papyri. Although our book of sources is largely compiled from the works of individual writers, two other written sources require brief mention: inscriptions and papyri. Both groups have thousands of records that touch on every aspect of Greek and Roman life. The class of inscriptions includes a specialised category, the Linear B tablets of the Bronze Age, which give us our earliest written evidence for Greek technology, meagre as it is. From the Classical periods, however, inscriptions provide much evidence for building construction (quarrying, transport, construction, workforce, decoration), water supply, mining, metallurgy, and food production. Most of the papyri documents written on a form of paper derived from the papyrus plant – come from the dry soil of Egypt and thus involve a bias for that region. Nevertheless, from their vast numbers important information regarding chemistry, textile manufacture, metallurgy, agriculture, and water management survives that is often applicable in other parts of the Mediterranean.

Greek and Roman writers must be read with caution; the ancients often show a surprising ignorance of or scorn for their own technologies, and the evidence they provide is often indirect. Their approach may often be superficial or unscientific, but the documents are fascinating and are often our only source of detailed information about certain technologies. An excellent reference source for biographical and other details about particular classical authors is S. Hornblower, A.S. Spawforth, eds., *Oxford Classical Dictionary*. 4th edn Oxford, UK: Oxford University Press, 2012. For ease of reference, and to put the authors translated in this volume into some general chronological context, the Index of Passages Translated includes a brief characterisation and date for all the authors for whom translated passages appear in the book.

INTRODUCTION

C ARCHAEOLOGY AS A SOURCE FOR THE HISTORY OF TECHNOLOGY

The value of archaeology is two-fold: it allows us to date material within a site and so to form some conclusions about the chronological development of human cultures; and it allows us to obtain and study artefacts in their original form.

During excavation, earth is removed over the entire surface of a trench in very shallow layers. In this way the archaeologist can determine any changes in the colour of the soil that might indicate, for example, a destruction level or an ancient floor. These superimposed layers of earth and deposits, laid down in order by both human occupation and nature, are called strata; they are one of the archaeologist's most useful tools, since they allow him or her to arrange in chronological order finds within the various levels. It is almost always true that, where one deposit overlies another, the upper deposit must have accumulated later than the lower one (given no subsequent disturbance). Hence, any objects discovered in one level – pottery fragments, coins, and so on – must antedate objects found in the level above and postdate objects from the levels below. Notice, however, that stratigraphy can supply only a relative date for a find in comparison with other finds: it cannot give an absolute date of xxx BC or yyyy years ago.

For this kind of absolute dating the archaeologist relies upon datable small finds like coins that bear the year of a ruler's reign, or potsherds that are known from other sites to be of a shape and decoration of a specific period, or building materials and styles similar to dated ones from analogous sites. This kind of cross-dating between excavations is one of the most important and difficult skills of an archaeologist: he or she must be as familiar with other sites from the same area and period as with his or her own, and must have access to a good library of recent archaeological publications.

While coins generally give the most precise date to a stratum, potsherds are far more common. Fortunately, ceramic shapes and styles changed frequently in antiquity, so details like the shape of the lip, the angle of the handle, or the design painted on the belly can often date a pot to within a decade of its manufacture. Other more advanced methods of dating material – the use of carbon-14 or potassium-argon, for example – are either inappropriate or too expensive to be employed in any but the most important cases on classical sites.

For artefacts recovered and then examined by the archaeologist, problems beyond dating are numerous. For example, the object's material and its environment (moisture, pressure, acidity, and temperature) in the earth affect durability and state of preservation. Iron is durable when cared for but rusts away under poor conditions; woods and textiles fare even worse although they can survive very well under arid, wet, or oxygen-free conditions; pottery is fragile and easily broken, but does not disintegrate easily (hence its importance to the archaeologist). As a result, most recovered artefacts are damaged and incomplete. Various techniques are employed to clean and stabilise them but often the

INTRODUCTION

archaeologist is forced to turn to examples from other sites or to ancient images and written descriptions to obtain a more complete understanding of the artefact and the technology used to produce it.

Such situations have produced collaborative efforts between the archaeologist, historian, and technician and have sometimes led to modern experiments to reproduce the working conditions, classical techniques, and working models of the ancients. Experiments with pottery, glass, earthwork fortifications, catapults, and even the construction of a full-scale model of a trireme, the most famous fighting ship of antiquity, have provided modern scholars with a clearer, if still imperfect, understanding of ancient technologies.

Although this book focuses upon textual rather than archaeological evidence, numerous line-drawings, usually derived from recovered artefacts, have been incorporated into the volume in an attempt to provide some visual references to the literary descriptions.

Works on the history, theory, and practice of archaeological excavations are numerous, and their number has swollen significantly over the past two decades. Anyone not familiar with the methodology and terminology of archaeological excavation can find good surveys in Green 2009, Hester *et al.* 2009, Renfrew and Bahn 2016.

D ANCIENT WEIGHTS, MEASURES, AND COINAGE

The following Greek and Latin terms are used frequently in the text (where they appear in italicised form unless translated) rather than the English or metric equivalent, to give readers a more accurate sense of the measurements that appear in the original. It should be noted that, especially for the Greek examples that follow, there are frequent local variations – for example (see **11.117**).

Weights

Greek		Roman	
1 *obolos* [obol]	0.7–1.1 g	1 *uncia* [ounce]	27.3 g
2 *drachmai* =1 *stater*	8.6–12.6 g	12 *unciae* =1 *libra* [pound]	327 g
6 *oboloi* =1 *drachma*	4.3–6.3 g		
100 *drachmai* =1 *mina*	431–630 g		
50 *stateres* =1 *mina*	431–630 g		
60 *minai* =1 *talanton* [talent]	26–38 kg		

INTRODUCTION

Length

Greek		Roman	
1 *daktylos* [finger]	19 mm	1 *uncia* [inch]	25 mm
4 *daktyloi* = 1 *palaiste* [palm]	75 mm		
16 *daktyloi* = 1 *pous* [foot]	0.305 m	12 *unciae* = 1 *pes* [foot]	0.296 m
24 *daktyloi* = 1 *pechus* [cubit]	0.45 m		
2.5 *podes* [feet] = 1 *bema* [pace]	0.75 m	5 *pedes* [feet] = 1 *passus* [pace]	1.48 m
6 *podes* = 1 *orguia* [fathom]	1.80 m		
100 *podes* = 1 *plethron*	30 m		
600 *podes* = 1 *stadion*	180 m	125 *passus* = 1 *stadium*	180 m
		1,000 *passus* = *mille (passus)*	1.48 km

Area

Greek		Roman	
1 *plethron*	100 feet by 100 feet	1 *iugerum*	120 feet by 240 feet

Liquid volume

Greek		Roman	
1 *kyathos* [cup]	35–40 ml	1 *sextarius*	546 ml
6 *kyathoi* = 1 *kotyle*	210–240 ml	6 *sextarii* = 1 *congius*	3.3 litres
144 *kotylai* = 1 *amphora*	25–35 litres	8 *congii* = 1 *amphora*	26 litres

Dry volume

Greek		Roman	
1 *kyathos*	40 ml	1 *sextarius*	546 ml
6 *kyathoi* = 1 *kotyle*	240 ml	16 *sextarii* = 1 *modius*	8.7 litres
192 *kotylai* = 1 *medimnos*	46 litres		

Money

Greek	Roman
1 *obolos* [obol]	1 *as*
6 *oboloi* = 1 *drachma*	2.5 *asses* = 1 *sestertius*
6,000 *drachmai* = 1 *talanton* [talent]	4 *sestertii* = 1 *denarius*
	25 *denarii* = 1 *aureus*

INTRODUCTION

Bibliography

Society and technology in antiquity

Blank, Horst, *Einführung in das Privatleben der Griechen und Römer*. Darmstadt: Wissenschaftliche Buchgesellschaft, 1976.

Blümner, Hugo, *Technologie und Terminologie der Gewerbe und Künste bei Griechen und Römern*. 5 Vols. Leipzig: Teubner, 1875–1887.

Cotterell, Brian and Johan Kamminga, *Mechanics of Pre-Industrial Technology*. Cambridge: Cambridge University Press, 1990.

Cuomo, Serafina, *Technology and Culture in Greek and Roman Antiquity*. Cambridge: Cambridge University Press, 2007.

Daumas, Maurice, ed., *A History of Technology and Invention*, I: *The Origins of Technological Civilization*. New York: Crown Publishers, 1969.

De Camp, L. Sprague, *The Ancient Engineers*. New York: Doubleday, 1963.

Feldhaus, Franz Maria, *Die Technik der Vorzeit, der geschichtlichen Zeit und der Naturvölker*. Leipzig, Berlin: Wilhelm Engelmann, 1914.

Feldhaus, Franz Maria, *Die Technik der Antike und des Mittelalters*. Potsdam: Athenaion, 1931 [Repr. Hildesheim: Olms, 1971. Foreword and Bibliography by Horst Callies].

Forbes, Robert J., *Studies in Ancient Technology*. 9 Vols. 2nd edn. Leiden: Brill, 1964–1972.

Gille, Bertrand, *Histoire des techniques: Technique et civilisations, Technique et science*, ed. Bertrand Gille, with contributions by André Fell, Jean Parent, Bertrand Quemada, and François Russo. Paris: Gallimard, 1978.

Gille, Bertrand, *Les mécaniciens grecs. La naissance de la technologie*. Paris: Éditions du Seuil, 1980.

Greene, Kevin, "Historiography and Theoretical Approaches". Pp. 62–90 in John P. Oleson, ed., *Oxford Handbook of Engineering and Technology in the Classical World*. Oxford: Oxford University Press, 2008.

Hodges, Henry, *Technology in the Ancient World*. New York: Alfred A. Knopf, 1970.

Hodges, Henry, *Artifacts. An Introduction to Early Materials and Technology*. 2nd edn. London: John Baker, 1976.

Humphrey, John W., *Ancient Technology*. Westport: Greenwood Press, 2006.

Irby, Georgia L., ed., *A Companion to Science, Technology, and Medicine in Ancient Greece and Rome*. Chichester: Wiley-Blackwell, 2016.

Landels, John G., *Engineering in the Ancient World*, revised edition. Berkeley: University of California Press, 2000.

Moorey, Peter R.S., *Ancient Mesopotamian Materials and Industries: The Archaeological Evidence*. Oxford: Clarendon Press, 1994.

Oleson, John P., ed., *Bronze Age, Greek, and Roman Technology. A Select, Annotated Bibliography*. New York: Garland, 1986.

Oleson, John P., ed., *The Oxford Handbook of Engineering and Technology in the Classical World*. Oxford: Oxford University Press, 2008.

Rihll, Tracey E., *Technology and Society in the Ancient Greek and Roman Worlds*. Washington: American Historical Association/Society for the History of Technology, 2013.

Roebuck, Carl, *The Muses at Work. Arts, Crafts, and Professions in Ancient Greece and Rome*. Cambridge: MIT Press, 1969.

Schneider, Helmuth, *Einführung in die antike Technikgeschichte*. Darmstadt: Wissenschaftliche Buchgesellschaft, 1992.

INTRODUCTION

Schneider, Helmuth, *Geschichte der antiken Technik.* Munich: C.H. Beck, 2007.
Singer, Charles, Eric J. Holmyard, Alfred R. Hall, and Trevor I. Williams, eds., *A History of Technology*, Vol. I: *From Early Times to Fall of Ancient Empires c.500 B.C.* Oxford: Clarendon Press, 1954.
Singer, Charles, Eric J. Holmyard, Alfred R. Hall, and Trevor I. Williams, eds., *A History of Technology, Vol. II: The Mediterranean Civilizations and the Middle Ages.* Oxford: Clarendon Press, 1957.
White, Kenneth D., *Greek and Roman Technology.* Ithaca: Cornell University Press, 1984.

Archaeology as a source for the history of technology

Alcock, Susan E., and Robin Osbourne, eds., *Classical Archaeology*, 2nd edn. Hoboken: Wiley-Blackwell, 2012.
Barrett, John, Michael J. Boyd, and Richard Hodges. *From Stonehenge to Mycenae: The Challenges of Archaeological Interpretation.* London: Bloomsbury Academic, 2019.
Bowens, Amanda, ed., *Archaeology Underwater: The NAS Guide to Principles and Practice.* 2nd edn. Chichester: Blackwell Publishing, 2009.
Cohen, Getzel, and Martha S. Joukowsky, eds., *Breaking Ground: Pioneering Women Archaeologists.* Ann Arbor: University of Michigan Press, 2004.
Green, Jeremy N., *Maritime Archaeology: A Technical Handbook* 2nd edn. New York: Routledge, 2009.
Hester, Thomas R., Harry J. Shafer, and Kenneth L. Feder, *Field Methods in Archaeology*, 7th edn. New York: Routledge, 2009.
Renfrew, Colin, and Paul Balm, *Archaeology: Theories, Methods, and Practice.* 7th edn. London: Thames & Hudson, 2016.
Shanks, Michael, and Christopher Tilley, *Re-Constructing Archaeology*, 2nd edn. New York: Routledge, 2000.
Trigger, Bruce G., *A History of Archaeological Thought*, 2nd edn. Cambridge: Cambridge University Press, 2006.
Ulrich, Roger, "Representations of Technical Processes". Pp. 35–61 in John P. Oleson, ed., *Oxford Handbook of Engineering and Technology in the Classical World.* Oxford: Oxford University Press, 2008.

1

THE RISE OF HUMANS AND HUMAN TECHNOLOGY

The early history of humankind and its technological development is necessarily recorded in terms of vague memories and suppositions. The rise of humans is sometimes entirely cloaked in mythological tales based occasionally upon parallel conditions in contemporary societies. More advanced cultures had only to compare their own state with that of the primitive peoples with whom they had contact to comprehend that they themselves had once been a simple society. This development of primitive peoples into more complex civilisations was explained by a variety of methods.

Legend and myth are often used to explain the advance of technologies through the direct intervention or gift of divinities. Many immortal benefactors are cited, but Prometheus remains the most beneficent god of all towards humans (**1.6**). In the historical period, the bestowal of "gifts" by more advanced cultures upon less civilised peoples might be regarded as the practical realisation of the earlier myths (**1.8–9**).

Nature and natural phenomena (**1.10–13**) represent the second agency by which humans obtained their technological advances. Nature is virtually personified in some of the sources, making "her" godlike in her gifts. For the most part, however, these passages indicate a more scientific analysis of the advancement of the human race, attributing major steps in development not to human ability but to random elements outside their control.

The third major explanation for the rise of civilisation might be considered a theory of natural evolution. Humanity, because it is a unique species with unique abilities, is forced by necessity or driven by its intelligence and desires to improve its lot.

In all three types of interpretation, humans are often regarded as initially helpless, little more than beasts. But with their unique skills they are able not only to survive but to advance against all odds. These steps towards civilisation involve a wide variety of special qualities that are listed with reasonable consistency by our sources: thought, use of hands, speech, laws, writing, and the use of fire. Sometimes the most significant stages of progress are set out in a single passage; sometimes only one or two traits are emphasised (**1.14–16**). The ability to use fire must in many respects be regarded as the physical "tool" of primary importance in humankind's early technological development (Harari 2016, Dodds 1973).

EARLY MYTHICAL AND HISTORICAL CIVILISATIONS

Many myths and legends exist that tell of earlier peoples and lands in which humans did not have to work for their sustenance or who were culturally advanced. Some stories record single civilisations that had been destroyed, others tell of successive generations and their decline from an initial "Golden Age" to ages of toil. The Golden Age represented a paradise when humans mingled with the gods and the age to which later generations looked back with longing. Its loss meant the loss of a simple yet blissful existence of early civilisation; the achievements of material culture with the accompanying moral sins of mortals were not adequate compensation.

1.1 The five ages of humankind

Hesiod explains how humans have fallen to the state where they have to work for their livelihood; sad indeed, considering that gods and humans had a common origin. The Five Ages of Humankind are, with the exception of the Age of Heroes, named after metals – each Age having some of the qualities of its namesake metal. But not all ages make use of their associated metals: not until the Bronze Age does metallurgy come into the saga of human development.

Hesiod, *Works and Days* 107–178

I beg you to consider seriously that gods and mortals are born from the same source. First the immortal gods dwelling in Olympian homes made the golden race of people who lived in the time when Cronus ruled the heavens. They lived like the gods, carefree in heart and free from labour and misery; all good things were theirs: grain-giving earth spontaneously bore her copious and ungrudging fruit, and in pleasant peace they lived off their lands with much abundance.... Then the immortal gods, dwelling in Olympian homes, made the second race, the silver one, much worse than the previous, unlike the golden in either thought or appearance... Then Father Zeus made the third race, the bronze race of mortals, not at all like the silver race, from ash trees, terrible and mighty; they loved the wretched works of Ares [war] and acts of arrogance.... Their armour and weapons were bronze, bronze their houses, and with bronze tools they worked: dark iron did not yet exist. And overcome by their own hands they went into the dank and dark house of cold Hades, leaving no name.... Then Zeus, the son of Cronus, made another race, the fourth on the bountiful earth, better and more just, the divine race of heroes who are called demi-gods, the race before ours on the boundless earth. Some were destroyed by grim war and terrible battle [at Troy and Thebes]; to others Father Zeus, the son of Cronus, gave the gift of a home and means of living and settled them at the end of the earth apart from everyone. And they live free from worry on the Islands of the Blessed along the shore of deep-swirling Oceanus. Fortunate are these heroes, since the grain-giving earth produces a honey-sweet harvest three times a year for them. Oh, that I were not living among the fifth race, but had either died before or been born afterwards. For now is the iron race, when humans never will cease from labour and sorrow by day and from suffering at night, since the gods will give only grievous concerns.... And Zeus will destroy this race of mortals too.

1.2 The "uncivilised" Cyclopes

The Cyclopes, a monstrous race of lawless and unsocial giants, were visited by Odysseus during his long return voyage from Troy. Although he regards them as backward and without "modern" technology, their life is surprisingly easy compared to his own technologically advanced culture. This is a result of chance, however, rather than divine intervention. The principal point of the passage is that lack of technology and need have hindered the Cyclopes, who are unable and probably uninterested in gaining access to a nearby fertile island, leaving its potential unrealised.

Homer, *Odyssey* 9.105–131

Heavy at heart, we sailed from the land of the Lotus Eaters and came to the land of the Cyclopes, an overbearing and lawless race, who, relying on the immortal gods, neither sow with their hands nor plough. Everything grows without sowing or ploughing: wheat and barley and even the vines with their grape clusters ripe for wine, and the rain of Zeus fosters them. They possess neither counselling assemblies nor laws, but live in hollow caves on the peaks of towering mountains. Each one governs his children and wives, paying no attention to any other.

A flat and wooded island stretches out beyond the harbour, neither close nor far from the land of the Cyclopes. On it live countless wild goats since neither the comings and goings of people frighten them nor do the hunters come there who endure woodland hardships chasing across the mountainous peaks. Neither domesticated flocks nor ploughed lands possess it, but unsown and unploughed it is destitute of humans all its days, and nourishes the bleating goats. For the Cyclopes possess no scarlet-prowed ships, nor do they have shipwrights among them who might build well-oared ships, which would accomplish all their needs by going to cities, just as people often cross the sea in ships and visit one another; such would have made the island good to live in. For the island is not barren in any sense, but would bear everything in season.

1.3 Hard-hearted Jupiter makes humans work for their survival

Vergil compares the hard existence of contemporary society with an earlier and more pleasant life under Saturn. The struggle for survival has resulted in the creation and development of human technology.

THE RISE OF HUMANS AND HUMAN TECHNOLOGY

Vergil, *Georgics* 1.121–146

Jupiter, the Father himself, has willed that the path of cultivation should not be easy, and he first awakened the fields through agricultural skill, sharpening mortal minds with cares and not allowing his realms to lie sluggish in heavy lethargy. Before Jupiter, during Saturn's reign, no peasants subdued the lands; it was unlawful to even mark or to divide the field with boundaries. Humans made gain for the common good, and Earth herself bore everything more freely when no one demanded them. Jupiter added evil venom to black serpents and commanded wolves to plunder and the sea to stir. He shook honey from the leaves and took away fire, and he stopped up the wine running in streams everywhere so that practice through consideration might gradually hammer out a variety of arts and seek out the grain stalk in furrows and strike the hidden fire from the veins of flint. Then, for the first time, the rivers felt the hollowed alders; then the sailor counted and named the stars: Pleiades, Hyades, and Arctos, the illustrious daughter of Lycaon. Then humans discovered how to trap animals with snares, to deceive with birdlime, and to circle great glades with dogs. And now one assails the broad stream with the casting net seeking the depths, another draws his watery drag-net in the sea. Then came unyielding iron and shrill-toothed saw blade – for the first peoples used to rend the splitting wood with wedges; then came the variety of arts.

1.4 The hardy race of humans weakened by technology

Greek and Roman writers clearly recognised the importance of technology as an ingredient of an advanced culture, while the lack of technological progress was considered a sign indicating the absence of true civilisation. Lucretius here describes a situation, in which early humans were supported by the earth without the use of any technology except primitive weapons. In the passage, early people are seen as much stronger and living in an environment of great danger. But with the growth of advanced technologies humankind became softer and life easier: a direct contradiction of Hesiod's Iron Age (**1.1**). Earlier in this passage, Lucretius (5.837–859) had expressed the concepts of natural evolution and survival of the fittest.

Lucretius, *On the Nature of Things* 5.925–1025

And the human race [in a primitive state] was at that time by far more hardy on land, as was fitting, since the hard earth had made it: built up inside with larger and more solid bones, fitted with strong sinews throughout the flesh, and not easily overcome by heat or cold or novel food or any defect of body. Humans drew out their lives in the manner of the rambling wild animals for many *lustra* [cycles] of the sun rolling through the sky. No steady guide for the curved plough existed, nor did anyone know how to work the fields with iron nor to dig new shoots into the ground nor to cut off the old branches from the high trees with a sickle. What the sun and rain had given, what the earth had created by her own accord, that gift sufficed to content their hearts.... Not yet did they know how to treat things with fire nor to use skins and to clothe their bodies with the pelts of wild animals; but they inhabited the woods and forests and mountain caves, and they concealed their rough bodies among the undergrowth when forced to escape the lashing of the winds and rains.... Confident in the wonderful power of their hands and feet, they used to hunt the woodland haunts of wild animals with stone missiles and with great, heavy clubs, overpowering many, and avoiding few from their places of ambush. And like the bristly boars, when overcome by night, they surrendered their wild, naked bodies to the earth, rolling leaves and boughs around themselves.... [Sometimes they had to flee in terror when wild animals troubled their rest; humankind was more likely to die if attacked by wild animals, since no medical knowledge existed.].... But at that time a single day did not send many thousands of men led under military standards to destruction, nor did the rough waters of the sea dash men and ships on the rocks; at that time the wicked skill of navigation lay hidden. Instead it was lack of food that sent weak bodies to death; now, to the contrary, abundance of everything destroys mortals.... After they had procured huts and skins and fire, then, for the first time, the human race began to grow soft. For fire took care that their shivering bodies could no longer endure the cold under the vault of the sky.... Then also neighbours began to join in friendships among themselves, eager neither to do harm nor be harmed. They entrusted their children and women to one another and indicated with stuttering voice and gesture that it was right for everyone to pity the feeble. It was not possible, however, to produce harmony among everyone, but a good and large part of them piously maintained the agreement, or else even then the human race would have been wholly destroyed, and begetting would not have been able to prolong the generations up to the present.

1.5 An advanced civilisation reduced to a primitive existence

At the other end of the scale, earlier civilisations were believed to have existed that partially owed their advanced state to developed technology. With the collapse of technology and civilisation (whether due to nature or war), the peoples either disappeared, or reverted to a more primitive condition. In the following passage, Plato describes the fall of such a civilisation and its subsequent level of existence: an unnamed Athenian and a Cretan, Clinias, discuss early cultures and the result of lost civilisations.

Plato, *Laws* 3.677a–679b

ATHENIAN: "So do you think there is any truth in the ancient stories?"
CLINIAS: "What stories?"
A: "That there have been many destructions of mortals by floods and plagues and many other things so that only a small portion of the human race has survived".
C: "Certainly, everyone would think that credible".
A: "Come then, let us consider one of the many catastrophes, the one that occurred once through the flood [of Deucalion]".

THE RISE OF HUMANS AND HUMAN TECHNOLOGY

C: "What are we to consider about it?"
A: "That those escaping the destruction then must have been some mountain herdsmen, tiny sparks of the human race preserved on the mountain peaks".
C: "Clearly".
A: "And by necessity these kinds of men must be unskilled in the arts in general and in those devices men use against each other in the cities for the purposes of greed and rivalry and all the other knavish tricks, which they contrive against each other".
C: "It is probable".
A: "Shall we assume that the cities located in the plains and near the sea were utterly destroyed at that time?"
C: "Let us assume".
A: "And so shall we say all tools were destroyed and that everything important in the arts and inventions that they may have had, whether pertaining to politics or other skills, all these perished at that time? For if these things had remained the whole time just as they are now arranged, my good man, how could anything new ever have been discovered".

The dialogue continues, discussing how and by whom certain skills were recovered and the human condition following the destruction of cities and the loss of technology.

A: "Were they not happy to see one another since there were very few round about at that time, since the ways of passage, by which they might cross by land and by sea to each other, were almost all destroyed along with the arts as the story tells us? Thus to mingle with one another was not very possible, I imagine. For iron and bronze and all the minerals in the confusion [of the flood] vanished so that it was difficult to extract all these, and as a result there was a scarceness of felled timber. For even if some tools happened to be on the mountains, these soon were worn out and disappeared and were not to be found again until the art of metal-working was rediscovered".
C: "How could they?"
A: "How many generations do we think had passed before this happened?"
C: "Clearly very many".
A: "And so all the arts needing iron and bronze and all such metals must have remained secret the whole period and even longer?"
C: "How else?"
A: "Moreover civil strife and war disappeared because of many reasons at that time".

The Athenian speaker concludes that this primitive state without metals had many advantages.

C: "How so?"
A: "In the first place they were quite content and friendly towards each other on account of their isolation, secondly there was no fighting over their food.

For there was no scarcity of pasturage, except perhaps at the outset for some, which for the most part was what humans lived on at that time: in no way were they lacking milk and meat since they were able to obtain excellent and plentiful foods by hunting. And they were well equipped with clothing and coverlets and houses and cooking pots and other pots; for moulding and weaving are skills that don't need iron; and god gave these two skills to humankind to supply them with everything so that whenever the human race should come into distress, it might have the means for sprouting up and increasing".

The Athenian concludes that without gold and silver, humans were neither poor nor rich, and thus there were no rivalries or deceptions; instead, this was a race of simple but noble people. This scenario is also found in the more famous account of Atlantis (Plato, *Timaeus* 20e–26e): the peoples of Atlantis disappeared, and an earlier, advanced Athenian people survived on a more primitive level.

THE RISE OF CIVILISATION: GODS, NATURE, AND HUMANKIND

The concept of divine help as the source for all human achievements was widespread. Prometheus was often regarded as the greatest benefactor of human development, but other divinities were also credited with providing specific gifts; the theme is that a superior being aids an inferior one, an idea anchored in reality; advanced and technologically superior cultures did help inferior ones. The unique positions of the Athenians and Romans in their respective empires led to an inflated opinion of their cultures. Their confident belief that they were superior to other peoples is revealed in the last two passages, where they replace the role of the gods by improving the lives of more primitive peoples through the instrument of technology.

1.6 Prometheus: humankind's greatest benefactor

Although the verses of Aeschylus provide the most elegant description of Prometheus's gifts to the human race, other authors also describe his benevolence. Plato (*Protagoras* 320c–322d) describes his theft of technologies (especially fire) from the Olympian deities and then enumerates the acquisition of elements that lead to civilisation: worship, speech, invention, socialisation for protection, then a grant of civic order by Zeus, and finally specialisation of labour. In contrast to Aeschylus, Plato attributes only fire and arts in general to the actions of Prometheus, crediting the rest to human inventiveness.

Aeschylus, *Prometheus Bound* 442–506

What I, Prometheus, did for mortals in their misery, hear now. At first mindless, I gave them mind and reason. What I say is not to criticise mortals, but to show you how all my gifts to them were guided by goodwill. At first, they had eyes, but sight was meaningless; they heard but did not listen. They passed their long lives in utter confusion like dreamy images. They had no knowledge of well-built houses warmed in the sun, nor the working of timber, but lived like crawling ants in the ground in deep, sunless caverns. Nor did they have a fixed sign to mark off winter or flowery spring or fruitful summer; their every act was without knowledge until I came. I showed them the risings and settings of the stars, hard to interpret till now. I invented for them also numbering, the supreme skill, and how to set words in writing to remember all things, the inventive mother of the Muses. I was the first to harness beasts under the yoke with a trace or saddle as a slave, to take the man's place under the heaviest burdens; put the horse to the chariot, made him obey the rein, and be an ornament to wealth and greatness. No one before me discovered the sailor's wagon, the flax-winged craft that roam the seas. Such tools and skills I discovered for humans.... [Other benefits are recounted here.].... So much for prophecy. But as for those benefits to humans that lay hidden in the earth, the bronze, iron, silver, and gold – who else before me could claim to have found them first? No one, I know well, unless he wishes to sound like a fool. Learn the whole matter in a brief phrase: all arts possessed by mortals come from Prometheus.

1.7 Other divine benefactors

Although Prometheus is often regarded as humankind's greatest benefactor, other divinities were credited with specific gifts to mortals. Euripides (*Suppliants* 201–213) attributes human progress to an unnamed god; Hesiod (*Works and Days* 47–105), after relating the theft of fire by Prometheus, states that Athena bestowed the skill of weaving on Pandora (cf. *Theogony* 561–616 and **10.28**). Pliny (*Natural History* 7.191–215) provides a list of divinities, heroes, and historical characters who introduce technological advances, and in the *Homeric Hymn to Hephaestus* (20), Hephaestus and Athena elevate humans from living in caves like beasts to skilled workers living in comfort. This belief is supported by Athenian festivals celebrating the gift of fire by Hephaestus.

Harpocration s.v. *Lampas*

Torch (*Lampas*): Lysias in his speech against Euphemos [writes that] the Athenians stage three festivals of the torch: during the Panathenaia, the Hephaistia, and the Prometheia, as Polemon says in his work about the paintings in the Propylaia. But Istros in his history of Attica, having described how in the festival of the Apatouria some of the Athenians, clothed in the most beautiful robes, having taken burning torches from the hearth, sing hymns while sacrificing to Hephaistos, a remembrance well understood of the need of fire by the one who instructed others [in its use].

1.8 The Athenians claim credit for the progress of civilisation

After fawning over Athens, Isocrates turns to the benefactions that Athens has bestowed upon less fortunate peoples. He attributes to the favour of Demeter the gifts of agriculture and the celebration of the Mysteries, which elevated the Athenians from a life like that of beasts, and then gives all the credit to the Athenians for the advance of civilisation in Greece.

Isocrates, *Panegyricus* 29–40

... Our city was not only dear to the gods, but also so generous to all mortals that, having gained possession of these wonderful things, it did not begrudge them to other peoples, but shared with everyone all that had been received. And, in short, even now every year we reveal the Mysteries and our city has instructed the whole human race in the uses, methods, and benefits coming from cultivation.... If we leave all this aside and examine things from the beginning, we shall find that those peoples first appearing on the earth did not immediately lead the kind of life that we now enjoy, but little by little reached it by their own joint efforts.... This was but the beginning of our benefactions, to discover for those in need the sustenance, which humans must have to live a well-ordered life in other respects. But believing that life limited to subsistence alone was not worth living, the Athenian city gave such careful consideration to the remaining desires that none of the good things enjoyed by mortals now, and which we owe not to the gods but to each other, not one is unconnected with our city, in fact most are her creation.... [Laws and civilisation are first developed in Athens.].... As for the arts and skills, both those useful for the necessities of life and those created for pleasure, they were either invented or tested by our city, who then passed them to the rest of the human race to use.

The widespread and persistent belief of the Athenian importance to the progress of Greek civilisation is supported by other passages. Diodorus of Sicily (*History* 13.26.3) has a Syracusan victor recommend mercy for Athenian prisoners captured in 413 BC on the basis of their benefactions to the human race. An inscription from the late second century BC (Dittenberger, *SIG* 704 lines 11–22; Vol. 2 p. 324 n. 12 for further references) states that the Athenian people gave to the world laws, agriculture, civilisation, and admitted some people to the Mysteries.

1.9 The subtle process of urbanisation

Less advanced peoples were lured into accepting the rule of Rome through the introduction and adoption of the insidious comforts of civilisation that were often provided by their mighty rulers. The use of developed technologies to enhance the lifestyles of these conquered peoples was not always a gradual or subtle process, as indicated by the aggressive tactics of Agricola recorded by Tacitus. Tertullian sums up the spread of civilisation in North Africa in the early third century AC.

Dio Cassius, *History* 56.18.1–3

The following events took place in Germany during this time [AD 9]. The Romans were occupying parts of it, ... and while their soldiers were spending the winter there, cities were being founded. The barbarians were gradually switching to Roman practices, adopting the markets as one of their own customs, and holding peaceful assemblies. They had not, however, given up their ancestral traditions, their innate character, their independent lifestyle, nor forgotten that their power was derived from military strength. Nevertheless, as long as they were unlearning these things gradually and under the eyes of the Roman garrisons, they were not upset by their changing lifestyle and were becoming different without noticing it.

Tacitus, *Agricola* 21

The following winter [AD 79] was spent in very beneficial consultations. For in order that the scattered and barbaric Britons, a people ready for war, might be accustomed to pleasure by means of peace and relaxation, Agricola, by praising the enthusiastic and scolding the lazy, urged on individuals and assisted communities to construct temples, market places, and homes.... Gradually the Britons yielded to the enticing vices: the covered porticoes, the baths, and the elegance of banquets. And this condition was called "civilisation" among simple Britons, although it was part of their slavery.

Tertullian, *De Anima* 30.3

Now all places are accessible, all are known, all are full of commerce. Most charming farms have consigned to oblivion the infamous wastes, ploughed fields have vanquished the forests ... [domestication and reclamation of useless land are described], ... and where once were hardly cottages, there are now large cities. No longer are [solitary] islands dreaded, nor their rocky shores feared; everywhere are houses, everywhere are people, everywhere is the stable government, and everywhere is civilised life.

Compare Aelius Aristeides (*To Rome* 36–39), who paints a fawning and enthusiastic description of the benefits of Roman urban life.

1.10 Humankind's humble origin

Human progress was not always attributed to the gods, but sometimes was believed to be a result of human response to nature. In addition, individual elements are often cited as the principal factors contributing to the development of civilisation.

Plutarch, *Moralia* 8.8.730e

Those descended from Hellenus of old also sacrificed to Poseidon as patriarch, believing, as do the Syrians, that humans came into being from the moist element. And so, they revere the fish as being of the same race and raised together with humankind.

1.11 Trees: the supreme gift of nature to humans

When other natural riches were as yet unknown, people obtained all their needs from the forests: food, shelter, clothing. This simple life was virtually forgotten with the rise of luxury (cf. **3.40**).

Pliny, *Natural History* 12.1–2

For a long time, the riches of the earth were hidden, and trees and forests were thought to be the greatest gift given to mortals. These first provided food, made caves more comfortable with leaves, and clothed bodies with their bark. Even now there are peoples living thus. We are amazed more and more that from these beginnings, humans have proceeded to quarry marbles in the mountains, to chase to China for clothing, to search the depths of the Red Sea for the pearl and the bowels of the earth for the emerald....

1.12 Imitation of nature: the inspiration of technology

The two Philostrati describe paintings, occasionally recounting simplified explanations for the source of human inventiveness. The first passage describes a landscape with a river and herders on a bridge. The "clever construction" of the bridge by the forces of nature possibly provided a visual model for artificial bridges (cf. **8.41**).

Philostratus, *Imagines* 1.9 (308.23–35)

The painter has thrown a bridge of date palms across the river for a very clever reason. For knowing that the date palms are said to be male and female, and having heard about their marriage that the male takes the female, embracing her with branches and stretching out upon her, he has painted one on one bank and the other on the other bank. Thus, the male palm falls in love, leans down and springs over the river; but unable to reach the female tree, still separated, he lies down and slavishly bridges the water. And he is safe for crossing because of the roughness of his bark.

The Younger Philostratus, *Imagines* 3.1 ("Hunters") (395.18–20)

Nature (*physis*) is wholly sufficient in whatever she desires and needs nothing from art; in fact, nature is the origin of the arts themselves.

1.13 Imitation of nature leads to agriculture

Lucretius, *On the Nature of Things* 5.1361–1369

Nature herself was the mother who first brought forth the model of sowing and the beginning of grafting: berries and acorns, having fallen from the trees, sent out swarms of sprouts on the ground beneath in proper season. From these examples, humans learnt to graft shoots into branches and to plant new seedlings in the earth throughout the fields. Then they tried one method after another to cultivate their dear little plot and saw wild fruits grow tame in the ground with tender treatment and careful cultivation.

1.14 Human technologies and nature

Vitruvius, *On Architecture* 10.1.4–6

All mechanisms are created by nature and founded on the revolution of the universe, our guide and teacher. For example, let us first contemplate and examine the continuous motion of the sun, moon, and the five planets. Unless they revolved by natural means, we would not have had alternating light [and dark], nor would fruits ripen. Thus, when our ancestors had realised that this was so, they took their examples from nature, and by imitating natural examples they were borne onward by divine truths which they adapted to their way of living. As a result, they discovered that some things were more easily done with machines (*machina*) and their revolutions, some others with instruments (*organum*). Thus, they took care to improve gradually by their learning all those things which they believed useful for research, for the arts, and for established traditions.

Let us first consider an invention from necessity, such as clothing, how, with instrumental arrangements of the loom, the combination of warp and woof not only protects bodies by covering, but also adds an elegant apparel. Indeed, we should not have had an abundance of food if yokes and ploughs had not been invented for oxen and other draught animals. If there had been no provision of windlasses, press-beams, and levers for presses, neither shining oil nor the fruit of the vine would we have had for our enjoyment, and their transport would not be possible except for the invention of contrivances: carts or wagons for land and ships for sea. Indeed, the discovery of a means to test weights by balances and scales has delivered our life from fraud by means of just practices. Countless numbers of machines also exist about which it is not necessary to speak since they are at hand every day: mills, blacksmiths' bellows, wagons, two-wheeled vehicles, turning lathes, and other things that are commonly suited to general use.

1.15 Climate: an important factor in progress

Aristotle, *Politics* 7.6.1 (1327b)

The nations in cold regions and those around Europe are full of heart but lacking in intellect and art, on which account they proceed rather free-spirited, but without political government and unable to rule neighbouring people. The nations of Asia, however, are intelligent and skilful in character, but without spirit, with the result that they are continuously dominated and enslaved. But the Greek nation, just as it occupies the middle region, partakes of both characteristics: it is both full of heart and intelligent. On account of this, it continues free, has the best government, and has the potential to rule all the human race if it gains political unity.

Similar sentiments regarding location, climate, and intelligence are provided by Plato (*Timaeus* 22d–e, 24c–d) and Vitruvius (*On Architecture* 6.1.9–12). The latter suggests Italy as the perfect location.

1.16 The origin of fire

Not believing in the Prometheus story, Lucretius looks to nature for the proper explanation of how humans obtained fire, our most important technological breakthrough. Vitruvius recounts the same theory, then describes the outcome of the initial discovery.

Lucretius, *On the Nature of Things* 5.1091–1104

In case you are perhaps silently wondering at this point, lightning first brought down fire to the earth for mortals, and from it all the brilliance of flames was spread about. We see many things glittering once they have been implanted with the celestial flames, when the blast from the sky has given up its heat to them. Yet, when a many-branched, swaying tree tosses about, pounded by the winds, and lies upon the branches of another tree, fire is squeezed out through the great force of the rubbing branches. Sometimes the fiery heat of the flames flashes out while the branches and trunks rub upon each other in turn. Either of these causes could have given fire to mortals. Then the sun taught them to cook food and to soften it with the heat of flame, since they saw many things become ripe, overcome by the lashings and heat of his rays in the fields.

Vitruvius, *On Architecture* 2.1.1–2

In antiquity humans were born like wild animals in the woods and caves and groves and lived their lives eating wild food. Later, in a place thickly crowded with trees that were tossed about by storms and winds, fire flared up when the branches rubbed upon each other. Terrified by the raging flames, the humans living nearby fled the area. After it subsided, they drew nearer to the heat of the fire and realised that their bodies enjoyed a great advantage. They threw on logs to maintain the fire and brought others before it, showing them by signs the advantages they had received. In that assembly of humans, when sounds were gasped out in various utterances, from daily usage they decided upon the names of things as they happened to come up; then, by indicating things more often in use and by chance, they began to speak similarly about events and created conversation among themselves. As a result of the invention of fire, therefore, concourse among people developed, deliberation, and social intercourse, and groups of people continued to gather in one place. They also possessed by nature a gift beyond all other animals: they walked not stooped down but erect and viewed the magnificence of the world and stars. Moreover, they could easily handle whatever they wished with their hands and fingers.

Vitruvius continues with the advances in housing (**8.1**) and the advantages of imitation and rivalry for improving themselves. Lucretius (**6.1**) relates the power of fire and its role in the discovery of metals and metallurgy.

THE RISE OF HUMANS AND HUMAN TECHNOLOGY

1.17 Humankind's unique nature: the ability to learn

Diodorus of Sicily (*History* 1.8.1–6) regards speech as one of the most important advantages of human beings. Like Vitruvius (**1.14**), he regards socialisation of humans as the first step towards civilisation, since it led to common signs and terms that developed into speech. In addition, Diodorus observes that individual social groups would have created different terms, a conclusion reached by observing the diverse languages of his own day. The passage, however, makes it clear that the ability to learn step-by-step is our greatest advantage.

Diodorus of Sicily, *History* 1.8.7–9

Gradually, humans learnt by experience to take refuge in caves during the winter and to store the fruits that could be preserved. When they had become acquainted with fire and other useful things, little by little the arts and other things capable of improving the common life of mortals were discovered. For, in general, necessity itself became their teacher, dutifully providing instruction in every matter to a creature well endowed by nature and having as co-workers for everything, hands and speech, and shrewdness of mind.

1.18 Hands, the most important human attribute

Anaxagoras, Fragment 59.A.102 (Diels-Kranz)

The human race is the wisest of all living creatures because it has hands.

Aristotle, *On the Soul* 3.8 (432a)

The hand is the tool that makes and uses tools.

Galen, *On the Usefulness of the Parts of the Body* 1.2–3

[Galen has just reviewed the natures and strengths of various animals] But to man – for he is an intelligent animal and, alone of those on earth, godlike – in place of all defensive weapons together, she gave hands, tools necessary for all arts and crafts, and as useful in peace as in war. Accordingly, there was no need of a horn as a natural armament, since at need he could grasp in his hands a weapon better than a horn whenever he wished; for certainly, a sword and a spear are greater weapons and more ready to maim than a horn ... [horns and hooves are effective only in close quarters, while javelins and darts are superior since they are also effective at a distance].... Indeed, man is not naked, not unarmed, not easily routed, not barefoot, and, whenever he wishes, he has his breastplate of iron (a product harder to damage than all types of skins), all sorts of footwear, weapons, and clothing of all sorts. Nor is the breastplate his only protection, for he has houses, the city walls, and the tower [which could not be made if his hands had defensive weapons growing from them]. With these hands, man weaves a cloak and fabricates hunting-nets, fish-traps and fish-nets, and fine bird-nets, so that he rules not only over animals upon the earth, but also over those in the sea and the air. Such is the hand of man as an instrument of might. Yet, man is also both a peaceful and civil animal, and with his hands he writes his laws, raises altars and statues to the gods, and makes the ship and flute and lyre and knife and fire-tongs and all the other tools of the arts, and he leaves behind him commentaries on the theories of them in his writings.... Thus, man is the most intelligent of the animals and, thus, hands are the tools suitable for an intelligent animal ... [intelligence is the key to success, while] ... hands are a tool, like the lyre of the musician and the tongs of the smith....

1.19 The unique nature of the human species

People did not only imitate nature but were compelled by it to survive and improve. From very humble beginnings, humankind was able to survive and conquer the adverse environment thrust upon it by nature.

Pliny, *Natural History* 7, praef. 1–4

The first place is rightly to be given to humans, for whose sake great nature seems to have created all other things. Yet, in return for such great benefactions she demands a cruel price, so that one does not have enough evidence to decide whether she makes a better parent or worse stepmother for mortals. Before everything else, humans alone of all animals she clothes in borrowed resources ... only a human on the day of birth does she cast down naked onto the bare ground, immediately to cry and wail ... and thus when successfully born, the child lies crying with feet and hands [useless, as if] bound, the creature who is to control all others.... Alas the folly of those people thinking that from these beginnings, they were born to lofty spirit! ... All the rest of the animals know their own natural abilities, some use their agility, others swift flight, others swim; man knows nothing unless taught: not to speak, not to walk, not to eat, and in short, he knows nothing by natural instinct other than to cry....

1.20 The relation of skill and mechanics to nature

[Aristotle], *Mechanical Problems* 847a.10–25

Some things, whose cause and origin are unknown, happening in accordance with nature are marvelled at, while other things happen contrary to nature, things which originate through skill (*technē*) for the benefit of humans. For in many instances, nature creates measures acting in opposition to our advantage; for nature always acts in the same fashion and simply, but our advantage changes in many ways. So, when it is necessary to produce something contrary to nature, the ensuing difficulty creates a need for skill to resolve the problem. As a result, we call that part of skill which helps with such difficulties, a device (*mechanē*). For as the poet Antiphon wrote, and this is true: "We overcome with skill the

THE RISE OF HUMANS AND HUMAN TECHNOLOGY

things with which we are conquered by nature". Such things, as when the lesser overcome the greater, and a small weight moves a heavy weight, and all similar devices, which we label mechanical problems.

The gods are often credited with providing a spark of inspiration, but civilisation is a result of unique human abilities.

Xenophanes, Fragment 18 (Edmonds = 16 Diehl)

Not from the outset did the gods teach all to mortals; but in due course humans have discovered improvements by research.

Compare Euripides (*Suppliants* 201–215) where all of human progress is credited to the gift of reason by an unnamed god.

1.21 The limits of human technology

Xenophon defends Socrates, his teachings, and beliefs against charges of impiety. He explains the counsels of Socrates with matter-of-fact examples.

Xenophon, *Memorabilia* 1.1.7–8

Those who intended to govern a house or a city, Socrates said, needed the help of the art of divination; for the craft of carpenter, smith, farmer or ruler, or the ability to delve into these crafts, or arithmetic, or economics, or generalship, all such things could be learnt and understood by the intelligence of man. But the greatest of these things the gods, he said, kept back for themselves, and those things were not evident to humans.

1.22 The accomplishments of humankind

Eventually it is not the gods but humans that are held totally responsible for the advancement of the race. Sophocles is, in some respects, offering a response to Aeschylus (**1.6**) who had credited Prometheus with all cultural progress. Here Sophocles sets out the abilities and skills of the human race.

Sophocles, *Antigone* 332–372

There are many wonders, but nothing is more wonderful than humankind. This creature crosses over the grey sea in the face of the wintry south wind and ploughs its way through the roaring billows. And they vex the indestructible and inexhaustible Earth, the eldest of the gods, with their ploughs and the race of horses going up and down, turning over the earth year after year. And after snaring the race of nimble birds and the host of fierce wild beasts and the maritime creatures of the sea in the nets' meshy folds, this most skilful being carries them off. By his devices he masters the rustic wild animals and makes obedient to the bit the shaggy-maned, mountain-ranging horse and the unwearying mountain bull, yoking them about the neck. He has learnt speech and lofty thought and public speaking, and to flee the arrows of the storm and the barbs of inhospitable frost in the open air. Always inventive, he never meets the future unprepared. Only from death has he not created an escape, but he has developed cures for unrelenting diseases; skilful beyond all hope are the devices of his art. And sometimes he glides towards evil, at other times towards good. When he honours the laws of the land and the justice sworn to the gods, his city stands high; but he is a man without a city whomsoever the daring spirit leads to consort with wickedness. Let him not share my hearth, let his thought not be mine, the man who does these things.

For a similar sentiment emphasising the human spirit, compare Lucretius, *On the Nature of Things* 1.62–79. As might be expected, immortal and legendary characters are sometimes assigned a variety of inventions, since the true inventors' names have been lost, while historical individuals are generally credited with only one or two technological discoveries (Pliny, *Natural History* 7.191–215 = **13.1**).

1.23 Palamedes: a man of many talents

Sophocles, *Nauplius* Fragment 432 (Radt)

This man Palamedes devised a fortification wall for the Argive army; the inventions of weights, numbers, and measures; invented the battle array and interpretation of heavenly signs. He was also the first to find how to count from one to ten, and in turn from ten to 50 and to 1,000; he brought to light the fiery beacons of the army and revealed things not known before. He discovered the periods and revolutions of the stars, the watches of the night – trusty signs.

1.24 Reason and memory as humankind's greatest attribute

Lucretius recognises that the progress of humankind can be deduced only by reasoning, since record-keeping is too recent an invention.

Lucretius, *On the Nature of Things* 5.1448–1457

Ships and cultivation of the field, fortifications, laws, arms, roads, clothing, and all the rest of these types of things, prizes, and all the allurements of life right from the top to bottom, poetry, pictures, and skilfully constructed, polished statues: the use and testing of an active, developing mind taught all these things gradually and carefully. Thus, eventually time draws forth every single thing into focus, and reason raises it into the realm of light. They saw one thing after another become clear in their minds until they attained the height of perfection in the arts.

THE RISE OF HUMANS AND HUMAN TECHNOLOGY

PROGRESS AND ITS MOTIVATING FACTORS

A variety of elements were regarded as important for pushing humankind along the road of progress to a civilised species. Most often need or necessity is regarded as a vital element, but competition, greed, pleasure, and luxury are also provided as motivations.

1.25 Survival of the fittest

On Ancient Medicine 3.5–32 (Jones[1])

... Sheer necessity caused humans to search for and to invent the art of medicine ... I think that originally we used the same type of nourishment [as wild animals]. Our present type of life, I believe, has developed through discoveries and inventions over a long period of time. The sufferings of humans were many and terrible as a result of the strong and savage diet when they lived on raw and simple foods of strong qualities, ... and it is reasonable to assume that most humans were of a weaker constitution and perished, while the stronger put up a longer resistance.... Then they produced bread from wheat, after steeping, winnowing, grinding and sifting, kneading, and baking [see **fig. 4.2**]; and cake from barley. Experimenting with many other foods in this manner, they boiled, cooked, and mixed, combining the strong and simple with weaker components, adapting everything to their human constitution and strength.

Compare Diodorus (**1.17**).

1.26 The advantages of specialised labour in the city

Plato was especially intrigued by the concept of justice and the make-up of a city. In order to define justice, he looked first for principles that make a state just. Here he begins by building the social structure of society, based on the driving force of human needs. Socialisation allows the development of specialised labour, an important factor in the advancement of civilisation, since all humans do not have the same innate qualities and function best as a social unit (cf. Plato, *Timaeus* 24a–b). Socrates is speaking to Adeimantus.

Plato, *Republic* 2.369b-370b

"Well then", I said, "I think that a city comes into being because of the fact that each of us is not independent of others, but lacking in many things. Or do you think a city is founded from some other origin?"

"No other", he said.

"So then, needing so many things, one man calls in someone for one thing and another man summons someone else for another thing. Thus, many people gathering together as helpers and associates, live in one place; on this settlement we bestow the name city. Isn't that so?"

"By all means".

"So, one man exchanges with another what he has to exchange, or to take, thinking it is to his own advantage".

"Certainly".

"Come then", said I, "let us build in theory a city from the beginning. As it seems, our needs create it".

"Indeed".

"The first and greatest of our needs is the provision of food to stay alive".

"Certainly".

"The second is housing and the third is clothing and such things".

"True".

"So, tell me", I said, "how will our city be strong enough to provide all these things? Will there be one man to be a farmer, another a builder, and a third a weaver? And shall we add to them a shoemaker or some other physician to attend to the body?"

"By all means".

"The smallest city, then, will consist of four or five individuals".

"So it seems".

"What of this, then? Is each one of these to contribute work into a common pool for all? Should the one farmer provide food for four and spend four times the time and labour on the production of food, and share it with the others? ... [Yes, since] it occurred to me that each person born is not like another, but that each person's innate differences fit that person to one task".

Plato continues, creating larger cities with examples of further specialisation in arts and crafts as a result of need; on this, see also Plato, *Protagoras* 322d–e. For a concise statement of the innate ability of a person and the fitness to one task compare *Republic* 4.433d. Aristotle (**1.27**) discusses a similar theory but sees knowledge, not experience, as the key to specialisation.

1.27 Distinction between artist and craftsman

In the opinion of Aristotle, the difference between artist and artisan rested in two principal distinctions: first, theoretical knowledge versus that obtained through habitual use, and second, abstract versus utilitarian use.

Aristotle, *Metaphysics* 1.1.11–17 (981a-982a)

We think that the master craftsmen (*architectonoi*) in every field are more honourable and know more and are wiser than the artisans (*cheirotechnai*), because they understand the reasons for the things done, while the artisans do things, just like some inanimate objects, without knowing how to do the things they do.... Artisans accomplish their work through custom. Thus, the master craftsmen are wiser not because they are practical, but because they have understanding and know the causes.... As more and more arts (*technai*) were discovered, some pertaining to necessities and some to pastimes, the inventors of the latter were always considered wiser than the inventors of the former, because their knowledge was not oriented towards utility.

1.28 Pleasure and utility: the ultimate goals of technology

Polybius is discussing the growth of Roman power and concludes that domination of other lands is not the ultimate objective. He compares the superficial reasons for actions in a variety of spheres and concludes that pleasure, good, and utility are the true objectives for every action.

Polybius, *Histories* 3.4.10–11

A sane person does not make war upon neighbours merely for the sake of defeating enemies, nor sails on the open sea only for the pleasure of crossing it, and indeed no one undertakes crafts (*empeiria*) and arts (*technai*) for the sake of knowledge. All people do everything for the sake of the subsequent pleasures or good or usefulness.

Plato (*Republic* 2.372e–373e) states the negative side: desire for luxuries results in vices and eventually leads to war.

1.29 Rivalry and envy drive technological advance

Eris [Rivalry and Envy] has both a good and a bad side, and both aspects stimulate technological progress as humankind strives to obtain or better the possessions of a neighbour.

Hesiod, *Works and Days* 20–26

Eris stirs even the lazy man to work. For a man who sees another man rich because he hastens to plough and to plant and who is putting his house in good order, that man becomes eager to work. And one neighbour vies with another, pressing on after wealth. This is the Eris who is good to mortals. And potter competes with potter, craftsman with craftsman, beggar with beggar, and singer with singer.

1.30 Multiple explanations for the rise of civilisation

After first describing the primitive condition of early human life in terms similar to Vitruvius (**1.14**) and Diodorus (**1.17**), Moschion lists advances made by means of technology. In the second passage Philostratus presents a more pragmatic opinion of the various theories.

Moschion, Fragment 6 (Nauck[2])

Justice was of no account, and violence shared the throne of Zeus. But when time, creator and sustainer of all things, fashioned a change in mortal life – whether through the care of Prometheus or from necessity or by long experience, putting forth nature itself as teacher – then was found the sacred gift of Demeter, the nourishment of cultivated grain, and Bacchus' sweet spring. The earth, once barren, was ploughed by oxen under yoke, towering cities rose, people constructed sheltering homes and turned their lives from savage to civilised ways.

Philostratus, *Imagines* 1.1 (294.5–12)

For the person wishing to devise cleverly, the discovery of painting comes from the gods – observe that the *Horai* [Seasons] paint the meadows on the earth and the displays in the sky – but for the person seeking the origin of art, imitation is the eldest invention and the most related to nature; the wise invented it, calling it now painting, now plastic art.

1.31 A philosopher's view of the fate of civilisation

Seneca offers his opinion of the end of humankind, and it differs little from the prediction of Hesiod (**1.1**) eight centuries earlier.

Seneca, *Moral Epistles* 71.15

All the human race, whatever is and whatever will be, is condemned to death. All the cities that ever have held dominion over the world and that have been the great decorations of empires not their own, someday people will ask where they were. And they will be destroyed by various types of destructions: wars will ruin some, idleness and the kind of peace turned to sloth will consume others, and by luxury, a deadly thing to people with great wealth. All these fertile plains will be blotted out of sight by a sudden overflowing of the sea, or the slippage of the settling earth will sweep them suddenly into the abyss.

Note

1 From William H.S. Jones, *Philosophy and Medicine in Ancient Greece*. Baltimore, 1946.

Bibliography

Blundell, Sue, *The Origins of Civilization in Greek and Roman Thought*. London: Croom Helm, 1986.
Cole, Thomas, *Democritus and the Sources of Greek Anthropology*. American Philological Association, Philological Monographs, 25. Cleveland: American Philological Association, 1967.
Diamond, Jared, *Guns, Germs, and Steel*. New York: W.W. Norton, 1999.
Dodds, Eric R., *The Ancient Concept of Progress and Other Essays on Greek Literature and Belief*. Oxford: Oxford University Press, 1973.
Guthrie, William K.C., *A History of Greek Philosophy*, Vol. 3: *The Fifth-Century Enlightenment*. Cambridge: Cambridge University Press, 1969. Pp. 55–134.
Harari, Yuval N., *Sapiens: A Brief History of Humankind*. New York: Penguin Random House, 2016.

Hodges, Henry, *Artifacts. An Introduction to Early Materials and Technology*. 2nd edn. London: John Baker, 1976.
Hodges, Henry, *Technology in the Ancient World*. New York: Knopf, 1970.
Jaynes, Julian, *The Origin of Consciousness in the Breakdown of the Bicameral Mind*. Boston: Houghton Mifflin, 1976.
Jones, William H.S., *Philosophy and Medicine in Ancient Greece*. Baltimore: The Johns Hopkins University Press, 1946.
Lovejoy, Arthur O., and George Boas, *Primitivism and Related Ideas in Antiquity*. Baltimore: The Johns Hopkins University Press, 1935.
Mund-Dopchie, Monique, "La notion de progrès chez les grecs. Mise au point préliminaire". *Les Études Classiques* 51 (1983) 201–218.
Phillips, E.D., "The Greek Vision of Prehistory". *Antiquity* 38 (1964) 171–178.
Snell, Bruno, *The Discovery of the Mind: The Greek Origins of European Thought*. Cambridge, MA: Harvard University Press, 1953.

2

SOURCES OF ENERGY AND BASIC MECHANICAL DEVICES

Greek and Roman conceptions about available sources of energy and the basic principles of mechanical devices to which some forms of energy could be applied are wide ranging and interrelated. Section A, on sources of energy and prime movers, includes passages concerned with both animate and inanimate sources of power, along with energy conversion and basic fuels. The material has been organised as follows: under inanimate sources appear solar and geothermal energy, and magnetic forces, followed by steam, wind, and water. Animate sources of energy include animals and humans. Topics associated with energy conversion include fire, fuels, lighting, heating, and cooling. Section B collects passages concerned with the basic mechanical devices known in antiquity, including discussions of the five basic machines and a few elementary applications. Discussion of the actual mechanical devices developed to serve specific technologies, however, such as presses, cranes, wagons, catapults, and pumps, appears in the chapters appropriate to their primary applications. Some of the mechanical principles were put to work as well in imaginative gadgets devised, and apparently sometimes constructed and put to use, by Hellenistic and Roman "scientists", perhaps better termed research engineers or mechanicians. Section C presents texts that deal with these relatively small mechanical contrivances (or imaginary contrivances), most of which are remarkable more for the complexity or sophistication of their mechanical design than for any spectacular application of power or practical effect. Although few of these gadgets bear any relation to practical technology, they are gathered here as an indication of the mechanical basis for innovation in ancient machine-building. This last section should be compared with the passages in Chapter 14 that deal with attitudes towards experiment and innovation (Wikander 2008, Wilson 2008, Drachman 1973).

A SOURCES OF ENERGY AND PRIME MOVERS
SOLAR AND GEOTHERMAL ENERGY

Sunlight is a resource available in abundance in the Mediterranean area during most months of the year, and the Greeks and Romans were well aware that it was the source of virtually all energy on earth. The light and warmth of the sun not only

SOURCES OF ENERGY AND BASIC MECHANICAL DEVICES

allowed the production, curing, and drying of agricultural crops (see Chapter 3), but it could also be put to work lighting or warming structures, and even kindling fires. Obvious sources of geothermal energy were much less common, but were much appreciated where practical use could be made of them (Wikander 2008).

2.1 The sun as the primary source of energy

Theophrastus, *Concerning Fire* 5

And again, if the Sun is by nature a certain form of fire, it will be a great deal different from the fire we know, being a first principle and reaching all things. For light comes from the Sun, also the generating of warmth in animals and plants. Still more, the force of this earthy, burning fire arises from it, for many people think that in kindling a fire they are capturing rays from the Sun.

2.2 Passive solar heating in Greek houses

The hot summers typical of the Mediterranean world made it important to design structures that would produce shade during the hot months. With proper orientation and roof design, the same structure could obstruct the noonday sun in summer, yet allow it in to heat the structure in winter. The Romans also made use of solar heating; see **8.69**.

Xenophon, *Memorabilia* 3.8.8–10

"So, if one intends to have just the right sort of house, must one work to make it as pleasant to inhabit and as useful as possible?"

When this was agreed, "Is it not, then, pleasant to have a house that is cool in the summer and warm in the winter?"

After they agreed with this too, "Now the winter sun shines into the porticoes of houses that face south, but the summer sun follows a path high overhead, above the roof, which provides shade. If this, then, is a good arrangement, we should build the south-facing side of the house higher, so that the winter sun might not be blocked out, the side that opens northward lower, that it might not be exposed to the north winds. To sum up, the dwelling in which the owner can find a pleasant refuge at every season and keep his belongings securely is presumably both the most pleasant and most beautiful".

2.3 Roman legal rulings on solar energy

Roman law formalised some of the rights of property owners interested in protecting their access to sunlight.

Ulpian in *Digest* 8.2.17

If a party plants a tree in such a way that it interferes with the sunlight, it must justly be said that the party has done it contrary to the easement placed on the property, for the tree brings it about that less of the sky can be seen. If, however, the tree that is planted does not interfere at all with the general brightness, but does block the direct sunlight – if it blocks it from a spot where it was not wanted, it can be said to do nothing contrary to the easements. But if it blocks the sunlight from a true sun-room or terrace, it must be said that it was done contrary to the easement placed on the property, because it creates shade in a place for which direct sun was essential. In the contrary case, if a party takes down a building or some tree limbs and in consequence a spot formerly shady is suddenly in the direct sunlight, he does not do this contrary to an easement, for the easement obliged him not to interfere with the light, but in this case, he does not interfere with the light – he makes more light than is suitable.

2.4 Kindling fires by focusing sunlight

Several passages in Greek and Roman writers indicate that individuals interested in science, at least, knew how to kindle fire by focusing sunlight by means of curved mirrors, glass spheres, or possibly intentionally ground lenses. The absence of any general application of this principle to daily life probably resulted from the difficulty and expense of producing the optics, and the availability of relatively convenient alternatives (see **2.22**). Nevertheless, it is at least possible that the burning-glass mentioned by Aristophanes was available for sale rather than for some special use in the shop.

Aristophanes, *Clouds* 771–773

Strepsiades: "Now then, have you seen in the druggists' shops that stone, beautiful, transparent, by which they kindle fire?" Socrates: "You mean the [burning] Glass?"

Pliny, *Natural History* 2.239

… since also concave mirrors [or possibly "plano-concave lenses"], turned towards the rays of the sun, set things on fire more easily than any other source of heat (cf. **2.22**).

2.5 Geothermal energy used in a Roman bath

Since the Mediterranean area is a seismically active region, hot springs are relatively common. At several sites the Greeks and Romans directed this natural hot water and steam into their baths, saving heating costs while also benefiting from the high mineral content. Dio Cassius refers to the region around Baiae on the Bay of Naples. For another comment on the Baiae baths, see **9.44**.

Dio Cassius, *Roman History* 48.51.1–2

These mountains, which are right next to the lagoons, have springs that emit much fire mixed with water. Neither is found by itself (that is, neither fire nor

SOURCES OF ENERGY AND BASIC MECHANICAL DEVICES

cold water appears), but from their mixing the water is heated, and the fire is moistened. The water runs down the lower slopes to the sea and into reservoirs. As vapour rises from the water, the inhabitants conduct it into upper rooms through pipes and use it there for steam baths; for it becomes drier the higher it rises above the earth and the water. Expensive equipment has been set up to make use of water and steam, very serviceable both for the needs of daily life and for cures.

2.6 A magnetic field in a Hellenistic temple

Although the Greeks and Romans were well aware of magnetic forces, it is likely that the story of Timochares' project is only a fanciful application.

Pliny, *Natural History* 34.148

The architect Timochares started out using lodestone to build the vaulted roof of the Temple of Arsinoe at Alexandria, so that the iron cult image in it might seem to hang in mid-air. His own death interfered, and that of King Ptolemy II [286–247 BC], who had ordered the work carried out in his sister's honour.

WIND, STEAM, AND WATER-POWER

Although the energy of the wind was harnessed for propelling ships by means of sails as early as the fourth millennium BC, the following passage (**2.7**) by the first-century AC author Hero is the only surviving evidence that the Greeks or Romans conceived of other applications of wind power. Even so, the use envisioned – driving the air pump for a water organ – seems neither practical nor particularly useful. The neglect of this source of energy is all the more puzzling given the very nautical-looking triangular sails typical of medieval and early modern windmills in the eastern Mediterranean. The forces exerted by steam were widely understood (**2.9**), but this knowledge was of little practical use, given the scarcity of fuel in most of the Mediterranean world and the relatively small scale of ironworking, particularly regarding potential boiler plate. Water-power, in contrast, was widely used from the second or third century BC onwards, at first for raising water in compartmented wheels with paddles, and later principally for grinding grain (**2.10–2.16**).

2.7 The only ancient reference to the windmill

The translation has been adapted from Woodcroft (1851) 108.

Hero, *Pneumatics* 1.43

The construction of an organ from which, when the wind blows, the sound of a flute shall be produced [**fig. 2.1**]. Let *a* be the pipes, *b c* the transverse tube

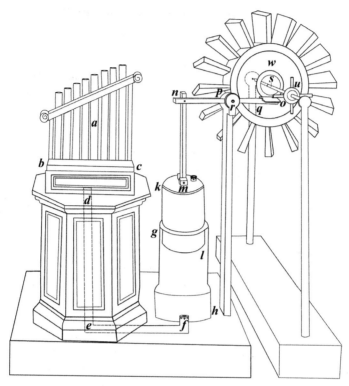

Figure 2.1 Wind-powered organ.

Source: Adapted from Schmidt, 1899, fig. 44.

communicating with them, *d e* the vertical tube, and *e w* another transverse tube leading from *d e* into a cylinder *h g*, the inner surface of which is made smooth to accommodate a piston. Into this cylinder fit the piston *k l*, which can slip into it easily. To the piston attach a rod, *m n*, and to this another, *n q*, working on the rod *p r*. At **v** let there be a pin moving readily, and to the extremity *q* fasten a small plate, *q o*, near which a rod, *s*, is to be placed, moving on iron pivots placed in a frame, which is capable of being moved. To the rod attach two small discs, *u* and *w*, of which *u* is furnished with pegs placed close to the plate *q o*, and *w* with broad vanes like the so-called windmills (*anemouria*). When all these vanes, pushed by the wind, drive the wheel *w* around, the rod *s* will be driven around, so that the wheel *u* and the pegs attached to it will strike the plate *q o* at intervals and raise the piston; when each peg passes on, the piston, descending, will force out the air in the cylinder *h g* into the tubes and pipes, and produce the sound. It is possible to move the frame holding rod *s* towards the prevailing wind, so that its revolutions may be more rapid and uniform.

SOURCES OF ENERGY AND BASIC MECHANICAL DEVICES

2.8 Is Ixion's wheel powered by wind?

Vergil, *Georgics* 4.484

Indeed, even the House of Death and deepest Tartarus were dumbstruck, and the Eumenides, their hair woven through with blue snakes. And Cerberus himself was silent with gaping triple maw, and even Ixion's wheel stood still in the wind.

2.9 The *Aeölipile*: a primitive reaction turbine

In contrast to wind power, the application of steam power to machinery – also attested by Hero alone – had no future in the ancient Mediterranean. The expense and relatively low thermal value of the typical fuels – wood, charcoal, chaff, and dung – restricted its development, a drawback also impeding the essential co-technology of large-scale forging of iron. Translation adapted from Woodcroft (1851) 108.

Hero, *Pneumatics* 2.11

Place a cauldron over a fire: a ball shall revolve on a pivot [fig. 2.2]. A fire is lit under a cauldron, *a b*, containing water, and covered at the mouth by the lid *c d*. The bent tube *e f g* communicates with this, the end of the tube being fitted into a hollow ball, *h k*. Opposite to the extremity *c* place a pivot, *l m*, resting on the lid *c d*, and let the ball have two bent pipes, communicating with it at the opposite extremities of a diameter, and bent in opposite directions, the bends

Figure 2.2 The *Aeölipile*.
Source: Adapted from Schmidt 1899, figs. 55 and 55a.

SOURCES OF ENERGY AND BASIC MECHANICAL DEVICES

being at right angles and perpendicular to the line *g l*. As the cauldron becomes hot, it will be found that the steam, entering the ball through *e f g*, passes out through the bent tubes towards the lid, and causes the ball to revolve, as in the case of the dancing figures.

2.10 The earliest surviving reference to a water-powered wheel

The development of techniques for harnessing the force of running or falling water was of great importance to the Roman world, although it never replaced human labour to the extent seen in medieval Europe. Unfortunately, the origins of this technology are obscure, but our knowledge of the sources used by Philo of Byzantium and by Vitruvius suggest that the technology should be connected with the research institute at Alexandria, perhaps with the mid-third century BC scientist Ctesibius. Some of the earliest applications can be seen in devices like the water-powered whistling wheel described in the following passage. This model was designed to display to the royal patron, a Ptolemaic king, the principles involved. The water-powered paddle wheel was used in irrigation machinery from the third century BC onwards, but was applied to tasks such as grinding grain only in the late Republic.

Philo of Byzantium, *Pneumatics* 61[1]

Construction of a fine water-powered whistling wheel [fig. 2.3]. Prepare a wheel of wood or copper having a certain depth, similar to the wheels used for irrigation [in the Arabic text, *hannana*]. Its diameter is two cubits; it is labelled *c*. The wheel has some curved panels turning with it, marked *k l*. Towards the outer circumference of the wheel, on the exterior face marked *s*, these panels are closed. On the outside they have openings like those of the irrigating wheels (*hannana*) that do not have paddles, that is, on one of the sides of the closed part that we have mentioned. On the other side it has a square enclosure forming some empty compartments capable of receiving the water. The apertures opening on these empty compartments are labelled *f*. The space within *s*, labelled *v*, is enclosed. It has some open apertures labelled *o*. The central part of the wheel ought to have a diameter equal to one-third the diameter of the wheel.

The wheel is mounted solidly on strong posts. When you have made this, which must be well prepared, place the wheel in a vessel filled up to the line labelled *e g*. Above, there is a channel that pours water into the square openings labelled *f*. The wheel is symmetrical and equal in weight on every side. When only one side is filled with water, it outweighs the other, and the wheel turns. As it does so, the empty openings receive water, that is, the square openings labelled *f*. When this side of the water has been thrust down in the water and turns underneath, air is captured. And when the water flows in and reaches the hollow compartments, the air that was in them whistles, and that which was carried beneath the water is expelled with a sound. While this one whistles, another section descends, and the same thing occurs, so that as one stops whistling, another begins.

SOURCES OF ENERGY AND BASIC MECHANICAL DEVICES

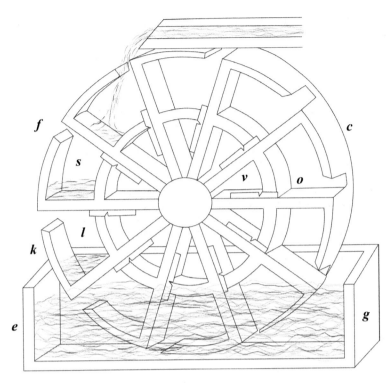

Figure 2.3 Water-powered whistling wheel.
Source: Adapted from Beck, 1911, fig. 21.

When these compartments designed to receive the water are lifted above it, the water leaves them, and the peripheral compartments rise empty. They outweigh the compartments where the sound is produced because they are larger and farther away from the centre ... [The chapter ends with instructions on how to tune the notes of the wheel and the design of a spigot for regulating the water supply.]

2.11 Early references to the waterwheel

Neither of the following passages gives an explicit description of the design of the waterwheel involved, but the undershot design, in which the lower circumference of the wheel is allowed to rest in a flowing stream and pushed along by its movement, is the most likely. This design would have been the easiest to adapt to the river-bank environment mentioned by Lucretius. The use of the term "scoops" (*austra*) instead of "paddles" (*pinnae*) may indicate that Lucretius envisaged a paddle wheel with a compartmented rim designed to raise water, rather than a paddle wheel driving a mill. The context of the passage in Strabo suggests that the watermill at Cabeira in Pontus dates to the reign of Mithridates VI (120–63 BC). This attribution is not certain, but Strabo was a native of the region and would have had knowledge of such things. He uses the term *hydraletes*, which becomes the typical Greek term for this device.

SOURCES OF ENERGY AND BASIC MECHANICAL DEVICES

Lucretius, *On the Nature of Things* 5.517

(The sphere of heaven is driven around in a circular motion …) as we see rivers turn wheels with their scoops.

Strabo, *Geography* 12.3.30

At Cabeira was built the palace of Mithridates and the watermill, along with the zoological garden, the nearby hunting grounds, and the mines.

2.12 Poetic eulogies to the benefits of water-power

This anonymous poem, a poetic description of a water-powered water-lifting wheel, constitutes a valuable expression of an individual's delight that a self-regulating device powered by an inanimate prime mover has taken on what was previously an arduous household task. The *antlia*, a term derived from the Greek verb *antlein*, "to bail or lift water", is a wheel with peripheral paddles and a compartmented rim for lifting the water, possibly the same device alluded to in **2.11a**. There is a full description of this device in Vitruvius (**9.28**).

Latin Anthology 284

About an *Antlia*.

It pours out and scoops up water, discharging on high the stream it carries, and it drinks up a river only to disgorge it. A marvellous achievement! It carries water and is carried along by water. In this fashion a stream is pushed up by a stream, and a new device scoops up the old-fashioned fluid.

Antipater expresses the same sentiments concerning the use of water-power to take over another labour-intensive task of ancient households, the production of flour. This source of power is said to bring back the Golden Age, when humans were fed by the produce of the earth without having to work. The design of the watermill is not clear; it may have been either a geared mill of the Vitruvian type (see **2.13**), or a simpler, "Norse" mill, in which the paddles project like spokes from a hub at the bottom of an axle that drives the runner stone directly. For a discussion of these designs, see Wikander 2004, 2000, 1985, 1979.

Antipater of Thessalonica, *Greek Anthology* 9.418

Rest your mill-turning hands, maidens who grind! Sleep on even when the cock's crow announces dawn, for Demeter has reassigned to the water nymphs the chores your hands performed. They leap against the very edge of the wheel, making the axle spin, which, with its revolving cogs [or "twisted paddles" or "spokes"] turns the heavy pair of porous millstones from Nisyros. We once again have a taste of the old way of life, if we learn to feast on the produce of Demeter [i.e. grain] without toil.

2.13 The earliest description of the geared watermill

This passage is a valuable testimony to the design of the geared watermill, which Vitruvius describes in the context of water-driven water-lifting wheels that use the same source of power. He describes

SOURCES OF ENERGY AND BASIC MECHANICAL DEVICES

an angle-gear drive, in which a cog-wheel mounted vertically on the same axle as the paddle wheel meshes at right angles with a cog-wheel mounted horizontally on the axle that drives the runner stone. The final sentence is fascinating evidence for the early use of a self-governing grain-feeding device (probably operated by a stick vibrated by the upper stone) and a dough-kneading device apparently worked off the same gear system. Since millers in the Roman world usually baked and sold bread as well, the kneading machine was a useful addition.

Vitruvius, *On Architecture* 10.5.1–2

Wheels of the same design as those described above (see **9.28**) can also be set up in rivers. Around the circumference are fixed paddles, which, as they are struck by the force of the river, move along and cause the wheel to turn. And in this manner drawing up the water in compartments and carrying it to the top without the use of labourers for treading, the wheels are turned by the force of the river itself and provide what is needed.

Watermills (*hydraletae*) are turned by the same system. With them everything is the same except that a toothed drum [cog-wheel] is set on one end of the axle. This drum, oriented upright on its edge, is turned along with the paddle wheel. A smaller drum, oriented horizontally and toothed in the same manner, is placed near this larger drum and meshes with it [there seems to be a gap in the text at this point]. Thus, the teeth of that drum that is set on the axle, by driving the teeth of the horizontal drum, of necessity brings about the revolution of the millstones. In this machine a hanging hopper supplies grain to the millstones, and by the same revolution the dough is kneaded.

2.14 Urban watermills in Byzantine Rome

Although only a few installations have been identified in archaeological remains, most watermills in large Roman cities that had an extensive aqueduct system were often powered by the flow of aqueduct water or of waste water from public baths. Such an arrangement did not pollute the water, and the aqueduct channel provided a readymade sluice for a single wheel or a whole series. An installation for 16 mills built on the Arles aqueduct has been excavated at Barbegal, and Procopius writes about a similar installation on the Janiculum hill in Rome, remains of which also have been identified. Mills have also been found in the basement of the Baths of Caracalla, driven by run-off water, and at Ephesus.[2] Mills such as these in large public baths provided flour for the bakeries that served hungry bathers. The bakeries may even have made use of waste heat from the furnaces serving the baths, rounding out a nicely efficient complex of technologies.

Palladius, *On Agriculture* 1.41

If there is sufficient water, mills should also take up the waste water from the baths, so that by the installation of watermills in such a location grain might be ground without animal or human labour.

Procopius, *History of the Wars* 5.19.8–9

Opposite this level area, on the other side of the Tiber river, there happened to be a certain high ridge [Janiculum hill], where for a long time all the city's mills have been constructed, since a vast amount of water is brought to the edge of the ridge in an aqueduct and runs down the slope with great force. For this reason, the Romans of old decided to build a wall around the ridge and the river bank in front of it, so that an enemy force might never be able to destroy the mills and, crossing the river, attack the city wall with ease.

2.15 Urban boat-mills solve a difficult problem

The cutting of the aqueducts by the Goths during their siege of Rome in AD 537 created a food crisis by stopping the working of the grain mills. The Byzantine general Belisarius solved the problem by setting watermills on boats, which could adapt themselves to the fluctuating depth of the Tiber and did not require extensive construction on the vulnerable river bank. Boat-mills remained at work in the city through the nineteenth century.

Procopius, *History of the Wars* 5.19.19–22

After the aqueducts had been broken open, as I have stated, their discharge no longer drove the mills. The Romans could not operate them with any sort of animal at all, since while under siege they lacked all fodder – in fact, they could scarcely care for the horses, which were essential to them. Belisarius devised the following solution. Just downstream from the bridge I mentioned, which was attached to the circuit wall, he let out ropes from either river bank, stretched as tightly as possible. He fastened two boats to them side by side, two feet apart, at the point where the current comes along in full force below the arch of the bridge; he placed two mills on each of the boats and installed between them the device, which customarily drives watermills. He tied on other boats downstream, all fastened one after another to those below them, and he put the driving machinery between them in the same manner, for a great distance. With the current pushing them in this way, all the wheels, rotating one after another independently, drove the mills attached to each, and they ground sufficient flour for the city.

2.16 Water-powered saws

Although water-power was a splendid resource in many areas of the Roman world, it seems to have been applied only seldom to tasks other than grinding grain or lifting irrigation water. Vitruvius (2.13) alludes to a dough-kneading device attached to his grain mill, but otherwise we have only the following passages, both from the fourth century AC, which describe some kind of water-powered saws to cut revetment slabs from stone. A recently discovered relief from Hierapolis in Phrygia (modern Turkey), dated to the third century AC, shows a similar machine, including details of the transmission mechanism, such as gears, crank, and connecting rod [**fig. 2.4**]. Aristotle, *Mechanical Problems* 35.3 notes the harsh sound produced by a standard hand-held, reciprocating saw cutting stone.

SOURCES OF ENERGY AND BASIC MECHANICAL DEVICES

Ausonius, *Moselle* 359–64

The rushing Celbis [Kyll] and the Erubris [Ruwer], renowned for its stone, hasten to mix their tributary waters with you [the Moselle] as soon as they can – Celbis, noted for its excellent fish; Erubris, which, as it drives millstones in rapid rotation and draws shrieking saws through the smooth stone, hears incessant noise from either bank.

Gregory of Nyssa, *On Ecclesiastes* 3.656A (Migne)

Laconian and Thessalian stone and stone from Carystus (S. Euboea) is cut into slabs with iron. The mines on the Nile and those in Numidia are sought out; Phrygian stone, too, is carried off with such effort, where the hue of porphyry covers the whiteness of marble. It becomes a manifold luxury to greedy eyes, and a multifarious spread of colour is painted on white. O how great are the efforts to acquire these things, how many the machines, cutting the material with water and iron, others labouring with human hands by night and by day, finishing the cut; nor do such things suffice for those who labour around the foolish world, but even the purity of glass is defiled with gilding and dye so that this, too, may add to the luxury that pleases the eyes.

Figure 2.4 Earliest pictorial representation of a crank drive on a water-powered twin stone saw from Hierapolis (third century AC).

Source: Illustration following the interpretation by Klaus Grewe of the Ammianos Relief (Drawing by Hajo Lauffer, used by permission).

ANIMAL AND HUMAN POWER

Most ancient authors who wrote about energy and movement usually assumed that the task described would be executed by means of animal or human power, with water and wind power taking distant second and third places. Human power could be adapted more easily to a variety of tasks than that of animals, which usually required elaborate harnesses, intermediary machinery, and supervisors. Nevertheless, animals were put to work at most tasks that required significant force over a long period of time.

2.17 An ox-powered paddle-wheel boat

The same angle-gear system that was used with the waterwheel to drive grain mills was also applied to the harnessing of animal power for the task of lifting water. A pair of oxen was yoked to a horizontal cog-wheel that turned as they trod a circular walkway, meshing with a vertical cog-wheel mounted on the same axle as the water-lifting device (see **9.30**). The imaginative Byzantine bureaucrat who wrote the following passage around AD 370 hoped to use the same device in a warship to drive a pair of paddles mounted next to the hull. The proposal may have been stimulated by a shortage of trained oarsmen.

Anonymous, *On Matters of War* 17.1–3

Specifications for a warship. Animal power, enhanced by the resources of the human intellect, drives a galley suited for naval warfare – the size of which and the limits of human strength prevent its propulsion in any way by manpower – with easy passage wherever it might be needed. In the belly or hold of the ship, pairs of oxen yoked to machinery turn wheels mounted on the sides of the hull; paddles projecting radially from the circumference of the wheels do their work by a certain marvellous accomplishment of human skill, striking the water like oars as these wheels begin to turn, and thus moving the ship. And yet this galley, through its weight and by means of the machinery operating within it, enters a battle with such rumbling strength that it easily crushes and shatters all hostile galleys that approach it.

2.18 Human- and animal-driven grain mills

The main application of animal power, other than pulling carts and ploughs, was for the grinding of grain with large grain mills carved from hard, porous varieties of volcanic stone. The most common type, now termed the "Pompeian mill" from the site where it was first identified, has an hour-glass shaped runner stone about one metre high to which the harness frame was fitted. This stone was held on a spindle just above a tapering bedding stone; as the runner stone was rotating, grain held in its upper section trickled slowly down through a central hole into the lower section, where the upper stone worked against the lower. The resulting flour was caught in a bin built around the lower stone. It is uncertain when or where this device was invented, but it probably became popular first in late third-century BC Rome and the cities of Campania, where the large urban populations were forced to purchase bread from baking establishments – which also did their own milling. Several Greek writers of the fourth century BC refer to the upper stone of a hand mill as the "mule stone" (Xenophon, *Anabasis* 1.5.5, Aristotle, *Problems* 35.3), but this term does not imply knowledge of the Pompeian mill.

SOURCES OF ENERGY AND BASIC MECHANICAL DEVICES

The allusion in Plautus seems to be the earliest reference to an ass-driven mill (c.200 BC?). In the second passage, Libanus pretends his master is an ass and threatens him with work in the mill. There is, however, no evidence that humans were ever routinely set to work turning these heavy stones. Augustine's letter proves the case, since he complains that respectable men were reduced to the condition – not of slaves – but of animals. The emphasis on the ridiculous lengths to which the miser in the *Latin Anthology* poem has gone to save the rent of an ass reinforces the interpretation that grinding grain was a job for animals rather than men. Slave labour or cheap hired labour was also used to work smaller reciprocating or circular hand mills. Cato (*On Farming* 10.4, 11.4) explicitly distinguishes between ass-driven mills (*mola asinaria*) and the reciprocating push mill (*mola trusatilis*). The confiscation of draught animals and shortages of fodder caused shortages of bread in Rome on several occasions (see **2.15**; [see **figs. 4.1 and 4.2**]).

Aulus Gellius, *Attic Nights* 3.3.14

Now Varro and many others have reported that Plautus wrote ... (several of his comedies) while working in a bakery, when – after having lost in trade all the money he had earned on the stage – he returned to Rome destitute. To support himself he hired himself out to work for a baker, driving the mills which are called *trusatiles* ["push mills"].

Plautus, *Comedy of Asses* 707–709

Libanus: "By Hercules, you will never get off today by praying! I'll drive you up a hill with my riding spurs, then hand you over to the millers to do some running under their goads".

St. Augustine, *Letters* 185.4.15

Certain family men of good birth and well educated ... were carried off scarcely alive or were harnessed to a mill and forced with blows to turn it, as if they were the lowest form of draught animal.

Latin Anthology 103

On a Man Who Mills His Own Grain.
   Since you could hire at small cost an ass trained to turn the rounded millstone properly, why are you a penny pincher, my friend, and why hold yourself in such low regard that you want to put your own neck beneath the hard yoke? I beg you, give up the circular track! You could relax and have the benefit of a nice white quarter-loaf from the miller's hands. For in grinding grain yourself you endure the troubles Ceres bore when searching for her daughter!

Suetonius, *Caligula* 39.1

... to transport [the furniture from the palace] he seized the carts let for hire and the animals in the bakeries, so that there was even a shortage of bread at Rome....

2.19 Cheap labour in a Roman milling establishment

This passage provides one of the best depictions of the sad complement of mules and human slaves being worked to death in a Roman mill-and-bakery establishment in the mid-second century AC. The narrator is a human turned by magic into a mule and suffering various misfortunes before being restored to his proper form. Compare another passage in Apuleius quoted below (**4.3**).

Apuleius, *The Golden Ass* 9.10–13

By chance, a baker from the next village passed by who had purchased a large quantity of wheat. He bought me as well and led me off with a heavy load to the mill he ran, along a steep road made difficult by sharp stones and roots of every sort.

    There, a throng of mules continually went round and round turning a number of mills, not just by day, but truly, kept at it even by night, they made flour through the small hours at the turning mills that never stopped. But my new master generously provided me with a fine stall, in case the first taste of servitude might put me off, and he gave me a holiday that first day and supplied my manger abundantly with fodder. This delightful rest and fattening did not last very long, for early the next day I was harnessed to a mill that seemed the biggest of all, and with blinders on I was set walking on the curved path of a circular track, so that by keeping on the same restricted circuit and retracing my steps again and again I might keep wandering around the set course. I did not, however, so far forget my human wisdom and foresight as to let myself be a ready learner, but even though I had often seen similar machines turned while I was a man, I nevertheless stood stock-still in feigned puzzlement, as if I did not know what was wanted of me. For I thought that they would put me to some other, lighter task, or else put me out to pasture without anything to do, as an ass useless or unsuited for milling. But I played my trick in vain, to my own loss, for a group of the slaves armed with sticks gathered around me, even now all unawares because of my blinders, and all at once at a signal they came at me together with a shout landing their blows. They so confused me with the clamour that there and then I discarded all my plans, put all my weight skilfully against the grass rope and followed the course at a trot; my sudden change in attitude roused a laugh in the whole group.

    When the best part of the day had passed and I was all but played out, they released me from the rope harness and the machine and tethered me at my manger. Although I was dreadfully tired, much in need of a restorative rest, and just about dead with hunger, nevertheless I was struck by my usual curiosity and somewhat worried, and leaving aside the fodder which was there in plenty I took a certain pleasure in examining the organisation of that unlikable shop. Great gods! What unfortunate creatures the slaves there were, their bodies black and blue all over, their backs marked by the lash, and covered – rather than clothed – in torn rags, some with only a strip of cloth to hide their private parts, everyone's nakedness visible through the holes in their tunics, foreheads branded and heads

SOURCES OF ENERGY AND BASIC MECHANICAL DEVICES

half shaved, ankles in shackles, complexions sallow and spoiled, and eyelids all in sores from the darkness, smoke, and soot, in consequence nearsighted, and like boxers who fight sprinkled with dust, dappled white and black with ashes and flour. And how should I speak of my companions, the other draught animals – such as they were, old mules and worn-out geldings. Around the manger they ducked their heads and ate the heaps of straw, their necks furrowed and eaten away by stinking sores, their nostrils racked by a constant cough, chests ulcerated by the constant rubbing of the harness, flanks laid bare to the bone by incessant beatings, their hoofs splayed out to enormous dimensions by the constant turning, and their coats ragged with old age and mange.

2.20 Convict labour at water-lifting devices

Roman engineers assumed the availability of human labour to drive water-lifting devices and other machines that were not suited to the application of water or animal power (see **9.28**, **9.31**). Such work usually entailed treading the rim of a wheel with compartments to lift water into baths, irrigation ditches, or out of mines (see **9.32–34**), a task so difficult and degrading that it was usually reserved for slaves, convicts, or poor farmers. Artemidorus provides some dream some dream analyses that relate to this topic.

Artemidorus, *Interpretation of Dreams* 1.48

I know of a certain man who dreamt that his feet alone were walking while all the rest of his body was at rest, not advancing even a short distance but at the same time stirring. It turned out that he was condemned to the waterwheel (*antlia*), for it is the lot of those who work the waterwheel to take great strides as if walking, but always to remain in the same place. Still another man dreamt that water flowed from his feet, and he also came to be condemned to the waterwheel as a criminal, and in this way, water did flow from his feet.

ENERGY CONVERSION, FIRE, AND FUELS

Fire was most likely first used by humans to provide convenient light and heat, but it soon must have become obvious that fire could be used to speed up a variety of chemical reactions as well, from the cooking of food to the refining of metals. The Graeco-Roman understanding of fuels, combustion, and the thermal yield was thorough and practical, although not scientific in character.

2.21 The special nature of fire

Not understanding fire to be the result of a chemical reaction, the Greeks made much of the special place of fire in the physical world, perhaps best summarised here by Theophrastus, the philosopher-scientist who succeeded Aristotle as head of the Lyceum.

Theophrastus, *Concerning Fire* 1–2

Of the elemental substances, fire has by nature the most special powers. Air, water, and earth undergo only physical changes into one another; none is generated of and by itself. But fire has the natural capacity to generate and to destroy itself – the smaller fire to generate the bigger, the bigger to destroy the smaller. Further, most methods for generating it require force, as it were, such as the striking of solids like stones, or friction and compression, as with fire drills and all substances that are in process, such as those that catch fire and melt, … and whatever other methods we have observed, whether occurring above, on, or below the earth. Most of these seem to be the result of force.

But if this is not the case, at least the following is obvious, that fire has many modes of generation, none of which occurs in the other elements; nor do the others have a completely unique method, but among all of them, as I have stated, there is a certain natural exchange and generation. For this reason, we are not capable of creating any of them – for when we dig for water we do not create it, but we guide it into the open, collecting what was dissipated – but fire comes into existence not by one method, but by many.

2.22 Methods for making fire

While most Greek and Roman households kept a fire alive on the hearth and banked it overnight, new fires had to be kindled from time to time. In such cases, an isolated household or individual or a military group in the field would use flint and iron or a fire drill (a bow drill working on a plank). The fire drill and the use of flint on flint were methods probably in use as early as the Mesolithic period, and the only improvements brought into play in the Graeco-Roman period were the use of an iron striker and a greater sophistication in the selection of woods for the drill. As Theophrastus notes, sparks are struck from flint directly into the tinder; with the fire drill, tinder is laid around the drill point only after it has begun to smoulder. Within a house or in an urban neighbourhood, sticks, the ends of which had been dipped in sulphur, could be used as matches to carry a flame from one point to another. The match seller in Martial's poem is collecting broken glass for recycling. The final passage by Theophrastus provides evidence for the use of reflected sunlight to kindle fires (cf. **2.4**).

Theophrastus, *Concerning Fire* 63

"Why are fire sticks made of wood, although sparks are not generated by wood, and less often from stones, which do generate sparks". This statement is not true, since from many stones fire is generated more effectively and quickly. But even if that were not the case, one must suppose that the cause is as follows, that wood has fuel close at hand, matter susceptible to fire – for if fuel is added, the fire does not catch any faster owing to its weakness – but stone, being very dry, does not have this fuel. For this reason, they guide the spark struck from stone straight to the tinder.

SOURCES OF ENERGY AND BASIC MECHANICAL DEVICES

Theophrastus, *Enquiry into Plants* 5.9.6–7

Fire drills are made from many kinds of wood, but the best, according to Menestor's report, from ivy, for that catches fire most quickly and strongly. It is also said that a very good fire drill is made from what some call smoke-wood; this is a tree like the vine or the wild vine, and one which also climbs up trees. One should make the fire board from this wood, the drill stick from bay; the moving and stationary parts should not be made of the same wood, but woods quite different by nature, the one of an active, the other of a passive character.... The wood of the buckthorn is also good and makes a useful fire board, for in addition to being dry and free of sap, wood for the board should have an open texture, so that the friction might have some effect. The drill stick should be of a very resistant wood, for which reason that of the bay is best; being resistant to wear, it does the job through its biting quality.

Lucretius, *On the Nature of Things* 6.160–163

Lightning occurs also when clouds collide and hammer out many seeds of fire, as if stone or steel should strike a stone, for then also a light flashes and scatters the bright sparks of fire.

Pliny, *Natural History* 16.207–208

Other hot woods are mulberry, laurel, ivy, and all those from which fire starters (*igniaria*) are made.

Experience in the camps of military scouts and of shepherds has found this out, since there is not always a stone on the spot for striking a fire. For this reason, wood is rubbed on wood and by the friction generates fire, which is transferred to some dry tinder-fungus or leaves catching most readily. Nothing is better than ivy wood for the drilling plank, and laurel for the drill stick; one of the wild vines, which climbs up a tree, like ivy, has also found favour, but not the *lambrusca*.

Martial, *Epigrams* 1.41.2–5

You think you are sophisticated, Caecilius, but you are not – believe me. What, then, are you? An average guy, like the sidewalk vendor from Trastevere who exchanges pale sulphur matches for broken glass....

Theophrastus, *Concerning Fire* 73

That light from the sun kindles by reflection from smooth surfaces that are unpierced ... but is not kindled from the fire [faulty reasoning is supplied]. Fire is kindled from crystal and glass, copper, and silver having been worked in a certain way....

SOURCES OF ENERGY AND BASIC MECHANICAL DEVICES

2.23 Keeping the home fire going

Despite the variety of methods to create fire, considerable attention was paid to avoid the process – it was much simpler to maintain or to 'get a light from' existing fires. The Vestal Virgins' duty to maintain the Eternal Flame of Rome (Plutarch, *Tiberius Gracchus* 15.4) is a famous example of the first practice. Prometheus' theft of fire is a mythical reflection of the second practice, of which we have many historical examples, including contests of torch races (Aristotle, *Constitution of Athens* 57.1). Boxes, lamps, and lanterns are a natural progression.

Hesiod, *Works and Days* 50–52

Zeus hid fire; but the noble son of Iapetus stole it back again for men from all-wise Zeus in a hollow fennel-stalk, to escape the notice of thunder-delighting Zeus.

Plutarch, *Alexander* 24. 7–8

[Alexander and a few men] had to spend a night of darkness and extreme cold on a rough area. When he spied many fires of the enemy burning scattered about in the distance, ... he ran to the nearest of the burning fires, and slaying the two barbarians sitting around the fire with his knife and seizing a fire-brand, carrying off the prize he came to his men, who kindled a huge fire that immediately frightened off the enemy.

Pausanias, *Description of Greece* 1.30.2

In the Academy is an altar of Prometheus from which they run to the city holding burning torches. The contest is that while running to preserve the still burning torch. Nothing remains of victory for the first runner once the torch goes out, but for the second runner it remains. But if his torch also does not burn, then the third man is the victor.

Festus, *On the Significance of Words* 9.78 (Mueller p. 105)

Ignitabulum: a little container for [holding/producing?] fire.[3]

2.24 An early headlight and other lamps

Plutarch discusses the reforms of Solon and his restrictions placed upon the citizens. Many are against ostentatious display by women, but he includes this odd note. Herodotus describes a simple lamp used in Egypt. See **2.31** for others.

Plutarch, *Solon* 21.4

They were not to travel by night unless they rode safely by wagon with a lamp lighting their way forwards.

SOURCES OF ENERGY AND BASIC MECHANICAL DEVICES

Herodotus, *Histories* 2.62.1

Whenever they are assembled at the city Saïs, on the night of the sacrifice, they all light many lamps in the open air in a circle around their houses. These lamps are saucers full of salt and oil, on which the lamp-wick floats, and they burn all night. And the name given to this festival is the Feast of Lamps.

2.25 How to put fires out

Theory entered into the discussion of extinguishing as well as creating fire. For a Roman force pump used as a fire extinguisher, see **9.35**.

Theophrastus, *Concerning Fire* 59

The greater extinguishing power of liquids results in particular from their penetrating to the source, as has been said above, concerning vinegar. The effect is even more marked if a viscous substance is mixed with it. For the vinegar is able to pass through it and penetrate, while the viscous substance spreads an oily layer on the surface. For this reason, they say the best extinguisher is a mixture of vinegar and egg white, for the former is penetrating and the latter viscous, and that this is particularly helpful against fires caused by siege engines.

Pliny, *Natural History* 29.11

And lest something be overlooked regarding the worth of eggs, [know that] the white from them, mixed with quicklime, glues together broken fragments of glass; indeed, so great is its potency, that wood coated with egg will not catch fire, and not even treated clothing will burn.

Aeneas Tacticus, *Treatise on the Defense of a Besieged City* 34.1

If the enemy tries to set anything aflame with a powerful preparation of fire, it is necessary to quench it with vinegar, for then it cannot be reignited. Even better to smear it beforehand with birdlime [a preparation of mistletoe-berry], for this does not catch fire.[4]

2.26 The many uses of fire

In contrast to Theophrastus' philosophical appreciation of fire in **2.21**, Pliny concentrates on the range of its practical applications. He does not understand, of course, that its utility in manufacturing processes stems from the fact that heat speeds up most chemical reactions.

Pliny, *Natural History* 36.200–201

Now that we have gone through everything that depends on the human genius for making art reproduce nature, it occurs to me how remarkable it is that almost nothing is brought to a finished state without fire. It receives a sandy ore and, depending on the source, melts it into glass, silver, cinnabar, various types of lead, pigments, or medicinal substances. By means of fire ore is smelted into copper, iron is produced and tempered, and gold is purified; limestone roasted in a fire supplies the binding agent for the aggregate used for construction in concrete. There are other substances that benefit from being subjected to fire several times, and a single material yields one by-product at the first firing, another at the second, and yet another at the third. For example, charcoal itself begins to take on strength when set on fire and smothered, and, when believed to have perished, it becomes more vigorous. Fire is an immense, unruly element of the natural world, about which there is doubt whether it destroys more things, or brings more into existence.

2.27 Charcoal, the most important ancient fuel

Charcoal, made by carbonising wood in a fire from which a free supply of air is excluded, was, aside from wood and straw, the most popular fuel in antiquity because it could produce a lasting, smokeless, high heat from fuel of relatively little bulk or weight. Theophrastus provides a compendium of information on the selection of wood for making charcoal and the different types of charcoal favoured for specific applications. The exact identification of the various species he mentions is not always certain. The heap of wood to be made into charcoal had to be very carefully laid, then covered with earth. Pitch could be produced by a similar process (see **10.22**).

Theophrastus, *Enquiry into Plants* 5.9.1–4, 6

We must next state and try to determine the properties of each type of wood with regards to making fire. The best charcoal is made from close-grained woods, such as holm oak or arbutus, for these are the most compact and, as a result, last the longest and are the strongest. For this reason, charcoal from this tree is used in silver mines for the first smelting. The worst woods are the oaks, for they produce the most ash. The wood of older trees is worse than that of younger ones, and for the same reason that of very old trees is particularly bad; it is very dry and sparks while it burns, but wood for charcoal ought to contain sap.

The best trees for making charcoal are those in their prime, which have been pruned, for these have the proper balance of close grain, ash content, and moisture. Better charcoal comes from trees in a sunny, dry, and north-facing position than from those in a shaded, swampy, south-facing location. But if a damper wood is used, let it be close-grained, for close-grained wood has more sap. And in every case wood that is close-grained – whether by nature or on account of its drier location – is better for the same reason, no matter what the species. One variety of charcoal is useful for one purpose, another for something else. For some purposes soft charcoal is sought, as for example in the iron works, where they use charcoal from the sweet chestnut for iron that has already been smelted, while at the silver mines they use charcoal from pine.

Craft applications also make use of these varieties. Smiths prefer charcoal from fir rather than from oak, although it is not as strong, for it can be blown up into flames more easily and is less apt to smoulder. In general, the flame from this type of wood is hotter, as is that from any woods that are porous, light, and dry, while close-grained and green woods produce a sluggish and dull flame. Brushwood produces the hottest flame of all, but charcoal cannot be made from it since it does not have enough substance.

They cut and seek out for the charcoal heap straight, smooth pieces of wood, for they must be laid as close together as possible for the process of smouldering. When they have covered the reduction heap, they kindle it section by section, poking holes in the covering with poles....

For the crafts requiring furnaces, and for those that do not, one wood is useful for one, another for another. Fig and olive wood are the best for starting fires – fig because it is tough and open textured, so that it easily catches the flame and does not let it pass, olive because it is close-grained and oily.

Pliny, *Natural History* 16.23 repeats much of this information, but one elaboration is worth recording.

Pliny, *Natural History* 16.23

The broad-leaved oak is ... useful as charcoal only in the shops of bronze workers, since it dies down the moment there is an interruption in the forced-air supply and too often needs to be rekindled; besides, no other wood sparks as much.

2.28 Getting the best heat out of charcoal

Theophrastus naturally was aware that breaking fuel up and subjecting it to a draft would assist combustion, but he also notes that compressed charcoal – something like modern charcoal briquettes – would provide a particularly high heat.

Theophrastus, *Concerning Fire* 28–29, 37

... But charcoal and wood cannot burn without a source of air, since they are earthy solids....

For this reason, charcoal sometimes is broken up and, after being piled in a heap, subjected to a draft. For fire is generated from small contributions, somewhat like springs of water are; therefore, the fire drill does the same thing by friction, they add light tinder to the wood and blow on it. Charcoal itself burns better and more quickly in a draught, as in forges....

... According to their type, those crafts that require, as it were, a softening or melting or a certain dissolution into small particles seek a mild and soft heat, while those that require a more forceful operation, as in metalworking, seek a very forceful heat. For this reason, metalworkers choose the earthiest and most dense charcoals, and sometimes even compress them for the sake of a strong heat, and they use bellows besides. In this way the heat is stronger and more focused, as the draught assists the combustion.

2.29 Coal and peat: uncommon fuels in the Mediterranean world

Because of geological history, topography, and climate, there are few deposits of coal or peat in the Mediterranean world. Several surface deposits of lignite – an inferior, soft coal – were known in the Peloponnesus in the third century BC, but, as far as we can tell, no extensive use was made of it for metalworking. Pliny clearly was puzzled by the character of peat in his description of the miserable life of the Chauci, who lived on the shore of the North Sea near the mouth of the Elbe.

Theophrastus, *On Stones* 16

Among the stones that are mined on account of their utility is that which is called simply *anthrakes* [here, lignite] It is earthy in character but catches fire and burns just like wood charcoal (*anthrakes*). It is found in Liguria, in the same places as amber, and in Elis by the road heading through the mountains to Olympia. Smiths make use of it.

Pliny, *Natural History* 16.4

The Chauci weave ropes from sedge and swamp reeds for setting nets to catch fish, and scooping up mud with their hands, they dry it more by the wind than by the sun. They use earth as a fuel to warm their food and thus their bodies, frozen by the north wind.

2.30 Petroleum by-products and their applications

Various natural sources of liquid and solidified petroleum were known around the Mediterranean world, but the most renowned were near Babylon, in the still famous oil fields of Iraq (see **5.29**). The substance was used for a variety of applications in its solid and liquid forms, including as a fuel; the particularly flammable variety may have been a naturally occurring light petroleum distillate similar to gasoline. The high flammability of *naphtha*, as it was called, was so startling that Pliny, at least, thought it could have no practical application. It is interesting that in nineteenth-century Pennsylvania gasoline was dumped as an unwanted and dangerous by-product of the production of kerosene. On the use of petroleum as a fuel in lamps, see also Pliny, *Natural History* 35.179. Although the source is a late one, it is nevertheless interesting that in the fifteenth century George Codinus reports (*On the Origins of Constantinople* 14) that Septimius Severus built a bath in Constantinople fuelled by "Median fire", which must have been some form of petroleum.

Strabo, *Geography* 16.1.15

Babylonia also produces a great amount of petroleum (*asphaltos*), concerning which Eratosthenes states that the liquid variety, which they call *naphtha*, originates near Susis, but the dry variety, which can be solidified, near Babylonia, and that there is a spring of this material near the Euphrates ... and large clods of it form, which are suitable [as a form of mortar] for building with baked brick.... The dry variety is said to be particularly useful for building, but they also state that boats are woven [from reeds] and made water tight by being smeared with asphalt. It seems that the liquid variety, which they call *naphtha*, has an amazing nature, for if *naphtha* is brought close to a flame, the flame leaps to it, and if you smear something with the substance and bring it near a flame, it bursts into flames. It is impossible to extinguish such a fire with water [for it burns all the more], except with a great amount, but it may be smothered and extinguished with mud, vinegar, alum, and birdlime.... Poseidonius says of the springs of *naphtha* in Babylonia that some produce white *naphtha*, ... while others produce black, that is, liquid asphalt, which they use instead of olive oil to fuel lamps.

Plutarch, *Life of Alexander* 35.1–4

As Alexander passed through all of Babylonia, which came under his control quickly, he was particularly struck by the chasm from which fire continually poured, as from a spring, and at the stream of *naphtha*, which on account of its quantity pooled up not far from the chasm. In other respects this *naphtha* is quite similar to petroleum (*asphaltos*) except that it is so easily affected by fire that, before the flame touches it, it catches fire from the very glow around the flame and often sets on fire the air in between.

In order to demonstrate its nature and power, the locals sprinkled the street leading to Alexander's quarters with a small amount of the substance, then – standing at the end of the street – touched their torches to the moistened area, for it was getting dark. Once the first spots had caught fire, without any perceptible delay and quick as a thought the flame darted to the other end, and the whole street was in flames.

Pliny, *Natural History* 35.179

There are some who also include naphtha ... among the varieties of petroleum (*bitumen*), but truly, its volatile nature and flammability are far removed from any practical application.

2.31 Lamps and their management

Artificial illumination in antiquity was cumbersome, mediocre in quality, and expensive, depending on fires, torches, reed bundles soaked in wax or pitch, or – most often – on lamps fuelled with poor-quality olive oil. Most activities were simply carried out by day, under natural light, but lamps would be used for late-night study or recreation. Street lighting was extremely rare (cf. **8.73**), so anyone out at night for a legitimate purpose would carry a lamp. The two vexatious aspects of lighting by an oil lamp were the need to keep filling the lamp, and to keep adjusting the wick. Hero describes a lamp with self-adjusting wick (**2.56**).

Aristophanes, *Ecclesiazusae* 1–16

Oh, radiant eye of the wheel-thrown lamp, well hung on this conspicuous spot, we will recount your ancestry and your fate: turned on a wheel by a potter's strength, with your nozzle you take on the radiant duties of the sun. Arouse the agreed-upon signal of your flame. For to you alone we have revealed the plan – quite reasonably to you, since you are even at hand in our bedrooms when we are busy with lovemaking, and no one puts your watchful eye outside while bodies are entwined. You alone cast light on our most secret places and help us singe off unwanted hair. And when we secretly open up the storerooms full of grain and wine, you stand beside us, and though you know all these things, you don't tell a soul.

Pausanias, *Description of Greece* 1.26.6–7

Callimachus made a golden lamp for the goddess [Athena in the Erechtheum on the Acropolis in Athens]. They fill the lamp with oil and wait until the same day the following year [to refill it], and that amount of oil is sufficient to keep the lamp alight night and day during the interval. The wick in the lamp is made of Carpasian flax [asbestos?], which is the only form of flax not consumed by fire. A bronze palm tree above the lamp extends as far as the roof and draws off the smoke.

Theophrastus, *Concerning Fire* 21–22, 57

A flame goes out under an excessive draught, for it is smothered....

For this reason, when some sort of shelter is placed around a lamp it is less likely to be extinguished; the bronze windscreens they make nowadays, which they use for the lanterns, for the most part prevent the flame from going out in all but the most extraordinary breeze....

Why are persons who take lamps into their mouths and extinguish them not burned?

Juvenal, *Satires* 3.283–288

... although burning with wine and the impetuosity of youth, [the night-time mugger] steers clear of the man whose crimson cloak and long train of attendants give warning, along with the quantity of torches and bronze lantern. But me, for whom the moon is the usual guide, or the flickering light of a fat-soaked reed which I husband and trim – me he scorns.

2.32 Ice and refrigeration

Both the Greeks and the Romans knew the use of snow for cooling drinks, and the practice of keeping or importing snow for this purpose during the hot months became a source of indignation among social philosophers. It does not seem that refrigeration was used very commonly for solid food stuffs, more easily preserved by salting or drying.

Athenaeus, *Philosophers at Dinner* 3.124c–f

Chares of Mytilene in his *History of Alexander* has told how one may preserve snow. Recounting the siege of the city Petra in India, he asserts that Alexander had 30 cold pits excavated, which he filled with snow and covered with oak boughs, for in this way snow will last a long time.

Strattis also says in his play *Keeping Cool* that they used to chill wine in order to drink it rather cold: "No one would choose to drink wine hot, but instead much rather drink it chilled in the well and mixed with snow".... Protagorides, in the second book of his *Comic Histories*, recounts the voyage of King Antiochus down the Nile and says something about the devices they used to chill water, in these words: "By day they expose the water to the sun; at night, straining off the thick sediment, they expose it to the air in ceramic water jars on the highest parts of the dwelling, and through the entire night two slaves wet down the vessel walls with water. At dawn, taking the jars downstairs they take off the sediment again, making the water clear and as healthful as possible, and place the jars in heaps of chaff and make use of it thus, without need of snow or anything else".

Pliny, *Natural History* 19.55

Water too is divided, and the very elements of nature are given distinctions by the power of money. Some drink snow water, others ice water, and human gluttony transforms the curse of mountain regions into delight. Coolness is stored up for the hot season, and stratagems are devised to keep snow cold in months that are foreign to it. Some boil their water, then immediately chill it. Nothing in the natural order of things pleases humans.

Seneca, *Speculations about Nature* 4B.13.3

... we have discovered how to compress snow so that it might last until summer and to keep it in a cold place against the hot time of the year.

B BASIC MACHINES

The first three of the five simple machines – lever, roller or wheel, wedge, pulley, and screw – were known throughout the Mediterranean world from the Early Bronze Age. The pulley appeared in the seventh century BC, and the screw seems to be an invention of the early fourth century BC or the Hellenistic period (see **2.35**). The gear wheel appeared at the same time. By combining these elements, individuals created devices that made a wide variety of tasks easier to accomplish, or less time consuming (Wilson 2008).

2.33 The subdivisions and study of mechanical engineering

It is typical of ancient handbooks to take a sweeping view of the fields of theoretical study necessary for successful practice of the technology involved, and the portion of Pappus's book that deals with the mathematical and geometrical basis of machine construction and engineering is no exception. Compare Vitruvius' comments on the training needed by a Roman architect (**8.37**). The second section gives a good summary of the technologies that required mechanical devices, both small and large.

Pappus, *Mathematical Collection* 8.1–2

The science of mechanics, my son Hermodorus, is useful for many applications in daily life. In addition, it is found worthy of great favour among the philosophers and is eagerly pursued by all mathematicians, for it has very nearly primary relevance to the science of nature, which concerns the material of the elements making up the universe. Being the general, theoretical consideration of inertia and mass and of movement through space, it not only examines the causes of objects moved according to their nature, but also causes objects to leave their own position against opposing forces, contrary to their nature, devising it through the theories suggested to it by matter itself. The mechanicians (*mechanikoi*) of Hero's school state that part of the science of mechanics is theoretical, part is practical. The theoretical part consists of geometry, arithmetic, astronomy, and physics, while the practical part consists of metalworking, construction, carpentry, painting, and the manual practice of these arts. They say, then, that he will be the best inventor and master builder of mechanical contrivances who from his boyhood has been involved with the fields of knowledge mentioned above and has received training in the above-mentioned crafts, and having a nature inclined towards them. But since it is not possible for the same person to grasp so many fields of knowledge and at the same time learn the above-mentioned crafts, they advise the individual who desires to undertake projects involving mechanics to use the particular crafts of which he himself has a command for the purposes suitable to each.

Of all the mechanical arts the most indispensable with respect to human needs are the following (the mechanical considered before the architectural): the art of the "conjurers" (*manganarioi*), termed mechanicians (*mechanikoi*) by the ancients. By means of machines (*mechanai*) they lift great weights to a height, moving them with little force, contrary to nature. Also, the art of those who construct the siege machines necessary for war (*organopoioi*), they too are termed mechanicians. They design catapults that shoot stone or iron missiles and other objects of this type a great distance. Additional to these is the art of those properly called machine-builders (*mechanopoioi*). Water is quite easily raised from a great depth by means of the water-lifting machines that they devise. The ancients also termed the gadget-designers (*thaumasiourgoi*) mechanicians. Some of them practice their art on the principles of air pressure, like Hero in his *Pneumatika*, while others utilise sinews and cords to imitate the movements of living creatures, like Hero in his *Automata* and *Balances;* still others depend upon bodies floating on water, like Archimedes in his *On Floating Bodies*, or on water clocks, which seem to be connected with the theory of sundials, like Hero in his *On Water Clocks*. Finally, they also termed mechanicians the sphere makers, who construct models of the heavens based on the even and circular motion of water.

2.34 The difference between "machines" and "devices"

Vitruvius' handbook on architecture depends heavily, and often not very perceptively, on a variety of Hellenistic Greek handbooks of architecture and related technologies. In consequence, Greek vocabulary, examples, and cultural ideals appear frequently.

Vitruvius, *On Architecture* 10. 1. 1–4

A machine is a built, interrelated system having an aptitude for moving weights. A machine is set in motion systematically by means of the revolution of circles, which the Greeks call *cyclice cinesis*. There is, then, one type of machine for climbing, which in Greek is named *acrobaticon*, another associated with air pressure, which they term *pneumaticon*, a third concerned with hoisting, which the Greeks call *baru ison*. The type of machine for climbing was set up in such a way that when beams had been erected and braces tied on, one could safely climb up to examine military engines. But the devices utilising air pressure were so arranged that when air was forced out under pressure, beats and notes were produced by instruments.

The type concerned with hoisting, however, is set up so that loads can be lifted by the machines and set in place at a height. The climbing system boasts not of science but of boldness; it comprises chains and the support of stays. But the machine put into motion by the power of air pressure attains elegant effects by subtlety of design. Hoisting machines, however, have more frequent and impressive opportunities for practical application, and the greatest aptitude, when operated with care.

Some of these machines are set in motion mechanically, others like devices. This seems to be the difference between machines (*machinae*) and devices (*organa*), that machines are driven by several workmen as by a greater force producing its effects, like catapults, or the beams of wine, or oil presses. Devices, however, carry out their task by the careful touch of one worker, such as winding up a light dart thrower [*scorpio*, cf. Vitruvius, 10.10] or a screw press [*anisocycla:* literally "screws" or "springs"]. Both devices and machine systems are necessary for practical applications, and without them nothing can be done easily.

Now all machinery has been modelled on Nature and set up with the revolution of the universe as instructor and teacher. For unless we first could observe and examine the unbroken succession of sun, moon, and the five planets as well,

SOURCES OF ENERGY AND BASIC MECHANICAL DEVICES

unless these were set in motion by Nature's skilful system, we would not know their light in its succession, nor the maturing of the harvest. So, when our ancestors noticed that this was the case, they took their models from Nature and imitating her, led on by divine help, they developed the amenities of life. And so, they made some things easier by means of machines and their cyclical motions, others by devices, and in this way, they took care to augment little by little with precepts what they had noticed in their studies, crafts, and practices to be of practical use.

2.35 The five simple machines

The Greeks and Romans realised that complex machines depended on very simple mechanical principles embodied in a few basic forms. According to Pappus, this knowledge went back to Philo in the third century BC and Hero in the first century AC.

Pappus, *Mathematical Collection* 8.52

There are five machines by the use of which a given weight is moved by a given force, and we will undertake to give the forms, the applications, and the names of these machines. Now both Hero and Philo have shown that these machines, though they are considerably different from one another in their external form, are reducible to a single principle. The names are as follows: wheel and axle, lever, system of pulleys, wedge, and, finally, the so-called endless screw.

2.36 Speculation on the mechanics of a pulley

The simple, single pulley, a very useful device for changing the direction of a force, seems to have been invented in the eighth century BC. The compound pulley, which allowed the enormous multiplication of a force, probably followed not long after, perhaps spawned by the construction of early Greek stone temples.

[Aristotle], *Mechanical Problems* 18.853a–b

Why is it that if one sets up two pulleys in two sheave blocks that assist each other but are oriented in opposite directions, and one passes a rope around them in a circle with one end fastened to one of the blocks and the rest passing under and over the pulleys, if the free end of the rope is pulled, great weights can be lifted, even if the drawing force is small? Is it because the same weight is lifted by a lesser force if a lever is used? The pulley acts in the same manner as the lever, so that even one person lifts a weight more easily, and with one haul by hand lifts a much heavier weight. And two pulleys will lift more than twice as much. For the second draws less than it would if drawing by itself, since the rope passes over the other pulley, for that makes the weight still less. And in this manner, if one runs the rope through even more pulleys, a great effect is accomplished by a few men: for example, if a weight of four *minai* is carried by the first,

SOURCES OF ENERGY AND BASIC MECHANICAL DEVICES

much less is carried by the last. In building construction, they easily move great weights, for they shift them from one pulley to the next, and back from that to windlasses and levers. This is the same as setting up many pulleys.

2.37 A practical demonstration of mechanical advantage (and possible allusion to the crank)

It has been suggested by Drachmann (1973) that this passage conceals the clearest allusion in the ancient world to a complex machine driven by a crank, a very useful device for turning linear motion into circular motion and *vice versa*. He asserts that only a crank would have allowed Archimedes to turn with a single hand – as the text seems to suggest – the geared device winding up the compound pulleys. The principle of the crank was used in the hand-driven quern with a vertical handle that was present in most rural households from the third century BC onwards and found its large-scale application in a water-powered machine at least by the third century AC (cf. **2.16**).

Plutarch, *Marcellus* 14.7–9

But even Archimedes, a relative and friend of King Hieron, wrote that with a given force it was possible to move a given weight, and becoming audacious, as they say, at the strength of his proof, he stated that if there were another world and he went over to it, he could move this one. Hieron expressed astonishment and asked him to make a practical demonstration of his proposition and display some great weight moved by a small force. Archimedes chose a merchantman of the royal fleet with three sails that had been pulled up on shore with great trouble by a large work force, and he put on board a large number of people and the usual cargo. He sat down at a distance and, without effort but gently moving with his hand a certain drive mechanism for a compound pulley, he drew the boat towards him smoothly and without a lurch, just as if it were running through the sea. The king, then, astounded by this act and realising the power of his art, persuaded Archimedes to build for him defensive and offensive war machines for every type of siege warfare.

2.38 The mechanics of a steelyard

The steelyard was a weighing device that depended on the careful application of the principle of levers and lever arms. Sometime in the Hellenistic period it replaced the Bronze Age balance pans for weighing most materials because it allowed the use of a small, convenient sliding weight to determine the weight of much heavier objects. The object to be weighed was suspended from a hook close to the fulcrum, while the sliding counterweight was moved along a marked scale on the opposite, longer end of the bar until it counterbalanced the other.

[Aristotle], *Mechanical Problems* 20.853b–854a, 1.849b–850a

How do steelyards weigh heavy pieces of meat with a small counterweight, when the whole is only half the beam? For only the pan hangs where the weight is suspended, and towards the other end there is only the steelyard. Is it because the steelyard happens to be at once both a balance and a lever? For it is a balance in that each of the suspension cords becomes the centre of the steelyard. Now at one end it has a pan, but at the other end the spherical weight that weighs down the beam, just as if one were to put the other pan and the weight at the tip of the steelyard. For it is obvious that it draws just the same weight when lying in the other pan. But in order that the one beam might serve as many beams, just so many suspension cords are attached to such a beam. In each case the part on the side towards the spherical weight constitutes half the weight, and the balance is equal when the cords are spaced one from another, so that a measurement is made of how much weight the object placed in the pan draws down. And when the beam is level, one can calculate on the basis of which suspension cord is used how much weight the pan holds, just as was said. In general terms, then, this is a balance, having one pan in which the weight sits, and another consisting of the weight of the steelyard. The steelyard at the other end, therefore, is the weight. Such being the case, it is like many beams, just as many as the number of suspension cords. The cord nearer the pan and the weight placed there always balance the greater weight, because the whole steelyard is an inverted lever (for each cord attached above it is the fulcrum, but the load is what is in the pan) and the longer the length of the lever arm from the fulcrum, the easier it moves them. But in this case, it makes a balance and balances the weight of the steelyard on the side of the spherical weight.

And by this principle the merchants who sell purple dye contrive balance beams for cheating, placing the suspension cord off centre, or pouring lead into one part of the beam, or utilising the portion of a tree towards the root or where there is a knot for the side which they wish to sink. For the portion of a tree constituted by the root is heavier, and a knot is a sort of root.

2.39 The mechanics of a nutcracker

Like the steelyard, the nutcracker is a device that embodies a straightforward, effective application of one of the five simple machines, and as a result it too remains in use today essentially unaltered.

[Aristotle], *Mechanical Problems* 22.854a

Why are nuts easily cracked without a blow by means of the implement which they make for cracking them, even though there is no violent, forceful movement? Furthermore, one could crack them more readily by compressing them with a hard and heavy implement than with a light one made of wood. Is it because the nut is compressed on both sides by two levers, and heavy bodies are easily split by the lever? For this implement is composed of two levers having the same fulcrum, the point where they are joined....

SOURCES OF ENERGY AND BASIC MECHANICAL DEVICES

2.40 The *Barylkos*: a gear system for lifting heavy weights

The Greeks and Romans found it more difficult to multiply forces with the use of gears than with pulleys or levers, since a much higher degree of precision is involved in preparing the gear wheels. In large devices, such as the geared watermill, the problem was not critical because the wooden crown, star, or cage gears used depended on the meshing of wooden pegs and did not require precision. Likewise, in small astronomical models or calculators such as the Antikythera Mechanism or Archimedes's Planetarium (see **2.45**), the absence of high loads or rapid rotation made it possible to use triangular gear teeth. This type of gearing, however, will not work in the type of device Hero describes here. As a result, it seems likely that he is simply working out the theoretical potential of a geared system. The appearance of a crank handle in the drawing has no basis in the text and is probably incorrect.

Hero, *Dioptra* 37

To move a given weight with a given force, by an arrangement of toothed wheels. Prepare a framework like a chest [**fig. 2.5**]. In the long parallel sides let there be fixed axles, parallel to each other, arranged at intervals so that the toothed wheels fixed on them are adjacent to one another and mesh, as we are going to show. Let the above-mentioned chest be *abcd* and in it let there be fixed the axle *ef*, as has been stated, capable of being turned easily. Let there be fixed to it the toothed wheel *gh* having a diameter, let us say, five times the diameter of the axle *ef*. And for the sake of an example, let the weight to be moved be 1,000 talents, the motivating force, five talents – that is, the man or youth activating it is to be capable of lifting five talents by himself, without a machine. Now if the ropes fastened to the weight pass through some [opening] in wall *ab* and are wound around the axle *ef* (when the wheel *gh* turns,) the ropes fastened to the weight are taken up and move the load. In order to move the wheel *gh* there [must be a force] of more than 200 talents, because it was given that the diameter of the wheel is five times the diameter of the axle. This result was shown in our explanation of the five powers. However, we do not have a force of 200 talents, but only of five. Let there be, therefore, another axle *kl* mounted [parallel] to *ef*, having fastened on it the toothed wheel *mn* with teeth like those of wheel *gh*, so that the teeth of both wheels can mesh. On the same axle *kl* fasten the wheel *q* also having a diameter five times the diameter of wheel *mn*. So, as a result of this arrangement, anyone who wishes to move the weight through the wheel e will have to have a force of 40 talents, since one-fifth of 200 talents is 40. In turn, therefore, another toothed [wheel *pr*] must be placed adjacent [to the toothed wheel *q*] [and there should be] fixed to the toothed wheel *pr* another toothed wheel [*st*], likewise having a diameter five times the diameter of wheel *pr*. The power of the wheel *st* is equivalent to eight talents, but our allotted force has been given as five talents. So, in the same manner, let another toothed wheel *yw* be placed adjacent to the toothed wheel *st*. And to the axle of this wheel *yw* let there be fixed the toothed wheel *xy*, and let the ratio of its diameter to the diameter of the wheel *cw* be the same as that between these eight talents and the five talents of the given force.

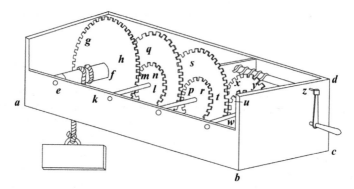

Figure 2.5 The *Barylkos*.
Source: Adapted from Schöne, 1903, fig. 115.

And when these arrangements have been made, if we conceive of the chest ***abcd*** raised on high, and if we fasten the weight to the axle ***ef*** and the pulling force to the wheel *xy*, neither of them will sink down, even if the axles turn easily and the wheels mesh smoothly, but the weight will balance the force, as if on a balance beam. If we add another small weight to one of them, the end to which the weight has been added will sink and go downward, so that if the weight of one *mina* should happen to be added to the five talents force, it will overcome and draw the weight.

But instead of adding to it, let there be installed next to the wheel [*xy*] a screw having a helix, which fits the teeth of the wheel, turning easily on tenons set into round holes. Let one of these tenons project through the wall of the box at ***cd***, [the wall adjacent] to the screw. Let the projecting end be squared off and hold the handle (*cheirolabe*) *s*, by gripping which and turning it, someone will turn the screw and the wheel *xy* and likewise the wheel *cw* fastened to it. By this means it will also turn the adjacent wheel *st* and the wheel *pr* fixed to it, and the adjacent wheel *q* and the wheel *mn* fixed to it, and the adjacent wheel *gh*, so that the axle ***ef*** fastened to it, around which are taken up the ropes from the weight, will move the load. For it is obvious that it will move it by the addition of the handle to the previous force, for the handle describes a circle greater than the perimeter of the screw, and it has been shown that greater circles conquer the smaller when they turn around the same centre.

2.41 Machine screws and nuts in bone-setting devices

The reduction of a badly fractured long bone requires both significant force and careful placement of the bones. Frames were developed for this purpose in the early imperial period, making use of harnesses tightened by screws. The screw press became common in the first century AC as well, for pressing olives or finishing woollen cloth (see **4.12**). These applications were made possible only by the invention of the nut in which the screw works, a complementary device that may have appeared only in the mid-first century AC. Because of the great difficulty of carving accurate threads inside a nut (see **2.42**), the early designs and rural applications usually incorporate only projecting pegs working in the screw channels.

Oribasius, *Compendium of Medicine* 49.4.52–58, 5.1–5, 7–9

Since for the most part the tortoises are moved by screws, I will explain next the preparation and operation of the screws. Indeed, screws have been employed in devices for moving other mechanical parts, such as drums and tortoises, so that their tension might be gentle and not cause laceration. Some screws are square, others lenticular. These designs are named after the shape of the screw thread, not the shape of the shaft, for every screw is round in section, turned on a lathe to a cylindrical shape. But the screws are distinguished from each other by the screw thread, by which one class is termed square, the other lenticular. A square screw is one which has threads square in section, both the furrow and the peak. (The screws of the device of Andreas are of this type.) Lenticular screws are those, which have furrows narrow at the bottom and wide at the top, and the peaks wide at the bottom but tapered off towards the peak, like a section across a lentil. And so, from its shape such a screw is called lenticular. Being of such a sort, the square and the lenticular screws are fit for moving different mechanical devices: the square screws move tortoises, the lenticular chiefly drums, but sometimes tortoises as well, in union with the so-called female screws.

How a square screw moves a tortoise. Let it be constructed in this manner, like the device in the tortoise of Andreas. Each of them is bored through, and the screw is put through the hole. But within the hole, along its wall, let an iron or copper tenon be driven into the tortoise. This tenon is called the "tooth". Now this tooth in the tortoise is engaged with the thread of the screw. It happens, then, that the tenon, the so-called "small cog", gripped by a turning of the screw either way, and held in the hollow screw thread around the screw, moves the tortoise. Some of these square screws are single, others double. The single screw is one which is cut with one screw thread and moves one tortoise. The double screw is the sort which is cut with two screw threads and moves two tortoises. The screw in the great frame device of Andreas is of such a sort. For from its mid-point out, between the cross-pieces, the wooden screw is cut with screw threads turning in opposite directions, so that, according to which direction the screw is turned, the tortoises either move from the middle towards the cross-pieces, or draw together from the cross-pieces to the middle spot.... The lenticular screw moves a drum, for the threads of the screw engage the cogs of the drum and turn the mechanism. This same screw, turned by a peg [*epitonion;* a "crank"?] or windlass and, engaging the teeth of the drum, moves the mechanism. The same screw, then, sometimes moves the tortoise, but not by means of a tenon like the square screw. Rather, it is gripped by the so-called female screw (*pericochlion*) which is a part of the very structure of the tortoise. For the hole in the tortoise that receives the screw is cut all around with screw threads corresponding to those of the lenticular screw, so that the raised threads of the lenticular screw are held in the hollow threads of the female screw and receive into their own hollow threads the raised threads of the female screw. Accordingly, it happens that as the screw is turned, since the screw threads are engaged, the tortoise is moved now up and now down.

2.42 How to cut threads in a nut

The procedure Hero outlines for cutting a nut, or female screw, is much the same as that used by machinists until recently. Hero's *Mechanics* is preserved in complete form only in an Arabic version. Translation adapted from a translation of the Arabic text by Drachmann (1963: 135–137) [**fig. 2.6**].

Hero, *Mechanics* 3.21

As for the female screw, it is made in this way: we take a piece of hard wood the length of which is more than twice the length of the female screw, and its thickness the same as that of the female screw, and we make on one part, on half the length of the piece of wood, a screw in the way we have already described, and the depth of the screw-turns on it should be like the depth of the screw-turns which we want to turn in this female screw. And we turn from the other part as much as the thickness of the screw-turns, so that it becomes like a peg of equal thickness. And we draw two diameters on the base of the piece of wood, and we divide each of them into three equal parts. And we draw from one of the two points a line at right angles to the diameter. Then we draw from the two ends of this line at right angles to this diameter, on the whole length of the peg, two lines at right angles; and this is easy for us to do, if we place this peg along a straight board and scratch it with the scribe [?] until we reach the screw-furrow. Then we use with great care a fine saw, till we have sawed down to the screw-furrow; then we cut off this third that was marked on the peg. And we cut out in the remaining two-thirds, in their middle, a groove like a canal on the whole length, and its size should be half the thickness of the remaining part. Then we take an iron rod and sharpen it according to the screw-turnings; then we fit it into the peg with the canal in it. Then we make its end come out into the screw-turns, after we have fastened the two pieces together with a strong fastening, so that the two are fixed to one another and cannot come apart at all. Then we take a small wedge and insert it into the canal-like groove and knock it until the iron rod comes out and lies between the two parts. When we have done this, we fit the screw into a piece of wood into which there has been bored a hole that corresponds exactly to the thickness of the screw. Then we bore in the sides of this wide hole small holes side by side, and we fit into them small, oblique, round pegs and drive them in until they engage the screw-furrow. Then we take the plank in which we want to make the female screw, and we bore in it a hole the size of the screw-peg; and we make a joint between this plank and the plank into which we have fitted the screw by strong cross-pieces, which we fasten very solidly. Then we insert the peg that carries the wedge into the hole that is in the

SOURCES OF ENERGY AND BASIC MECHANICAL DEVICES

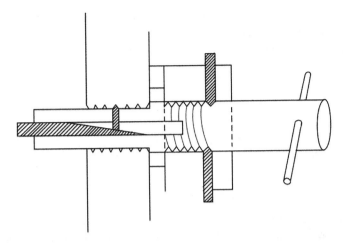

Figure 2.6 Screwcutter.

Source: Adapted from Drachmann, 1963, figs. 53 a and b.

plank in which we want to cut the female screw, and we bore on the upper end of the screw holes in which we place handles. And we turn it till it comes into the plank, and we keep on turning it up and down, and we serve the wedge with blows again and again, until we have cut out the female screw with the furrow we wanted. And so, we have made the female screw.

2.43 Structural nuts and bolts in the Jerusalem temple

Although the wooden bolt and nut were used in several types of Roman presses, their small metal equivalents – so important to modern machines – have been found only in the occasional piece of jewellery or parade helmet. Problems with quality of metal and accuracy of machining may be the reason. This passage from Josephus seems to describe large, structural tie-rod bolts in the tabernacle of the Temple in Jerusalem. Because of its context, however, the appearance of this device here may be more a reflection of the jeweller's art than of structural or mechanical engineering. The rods were only 2.6 m long.

Josephus, *Jewish Antiquities* 3.120–121

Each of the pillars had gold rings attached to its outer face, fixed to them, as it were, by roots, arranged in a line next to one another, touching edges. Through these passed gilded rods, each five cubits long, as tie bars for the pillars. The head of each rod passed into the next by means of a cleverly made socket crafted in the form of a screw. Along the back wall there was one bar passing through all the pillars, into which were set at a right angle the last of the rods from both of the two longer walls. They were held by these sockets, the male fitting the female. This, then, kept the tabernacle from being swayed by the wind or any other cause and was intended to maintain it unmoved, in absolute rest.

C MECHANICAL GADGETS

Since both small, intricate gadgets and complex mechanical devices were not only rare in the ancient world but also easily damaged and vulnerable to complete loss from the archaeological record, our knowledge of them depends almost entirely on the descriptions preserved in Greek and Latin literature. Many of these devices suggest a sophisticated knowledge of mechanics and an extensive repertoire of devices for transferring energy and changing the direction and speed of motion (Wikander 2008).

2.44 An allusion to devices using friction gears

Since it is very difficult to cut gears accurately by hand, early discussions of the theory of motion and some early gadgets involve friction gears, wheels that contact each other edge to edge to transmit motion, working by friction rather than the meshing of teeth. The gadget maker's goal of astonishing the viewer is already explicit in this work by a follower of Aristotle.

[Aristotle], *Mechanical Problems* Preface, 848a.20–38

Because the circumference of a circular object moves in opposite directions and one end of a diameter, point *A*, moves forwards while the other, point *B*, moves backwards, some people devise it that from one movement many wheels are driven simultaneously in opposite directions – as when they make little wheels of bronze and iron, which they dedicate in temples. For if there is a second wheel with diameter *CD* touching wheel *AB*, when the diameter of *AB* is moved forwards, that of *CD* is moved backwards, if the diameter is moved around a single point. That is, the wheel with diameter *CD* is moved in the opposite direction from that with diameter *AB*. And again, for the same reason, it will move the next in the series, *EF*, in the direction opposite its own. And in the same way, if there are more wheels, they will act the same when only one is moved. Seizing upon this innate characteristic of the circular body, then, craftsmen prepare a device whose first principle they conceal, so that only the wonder of the machine is apparent, the cause hidden.

2.45 The mechanical planetarium of Archimedes

Archimedes' Planetarium, a model that reproduced the movement of the heavenly spheres and the celestial bodies held in them – as was thought at the time – required the use of very precise gearing set up in a complex gear chain. The teeth of the small metal gears probably were equilateral triangles in shape, similar to those seen on the few computing devices that have survived from antiquity (see Field, Hill, and Wright 1985). Such gears are sufficient for the transmission of small amounts of force at slow speed.

Cicero, *Republic* 1.14.21–22

Although I had very often heard this celestial globe spoken of, because of Archimedes' renown, the sight of the globe itself did not greatly impress me. For that other celestial globe, also built by Archimedes, which Marcellus had deposited in the temple of Virtue, was lovelier and more famous among the people. But after Gallus began to explain in a very knowledgeable way the nature of this device, I decided that there was more genius in that Sicilian than human nature seems able to encompass. For Gallus said that other sphere, solid and without a hollow interior, was an old invention: Thales of Miletus had first fashioned it on a lathe, and afterwards Eudoxus of Cnidus, a student, so it was said, of Plato, had marked it with the constellations and stars that are fixed in the heavens.... And he said that this present type of sphere, in which are contained the movements of the sun and moon and of those five stars which are called "wanderers" [planets] or, so to speak, "roamers", could not have been marked out on that solid sphere, and the invention of Archimedes deserved admiration on this account, that he had discovered how to reproduce the unequal velocities and different orbits in a single turning mechanism. When Gallus set this globe in motion, it came about that the moon was as many revolutions behind the sun on that bronze instrument as in the heavens themselves, and therefore there was that same eclipse of the sun in that sphere, and the moon then met that point, which is the earth's shadow, when the sun ... from the region.... [The manuscripts break off at this point.]

Claudian, *Shorter Poems* 51

When Jupiter saw the heavens in a small glass sphere, he laughed and addressed the gods in such words as these: "Has the power of mankind's careful striving come this far? Is my handiwork now mocked in a fragile sphere? Look! By his skill an old man of Syracuse has copied the laws of the heavens, nature's reliability, and the ordinances of the gods. An animate force shut within attends the different stars and moves along the living mechanism with its regulated motions. A pretend zodiac runs through its own private year, and the simulated moon waxes with the new month. Bold diligence now rejoices to set its own heaven turning and regulates the stars through human intellect...."

2.46 The mechanism of a water organ

The water organ is an example of a quasi-practical complex mechanical device that grew out of the researches of Hellenistic mechanicians and was relatively common – or at least not rare – in the Roman world. The organ itself works on the same principles as a modern wind organ. It is termed a water organ because the wind reservoir was a large cauldron inverted in a container of water, the pressure of the atmosphere on the exposed water surface acting like a constant-force spring to prevent variations in the air pressure of the wind reaching the pipes. The pressurised air itself was produced by assistants working bellows or piston pumps (cf. Hero's idea for a windmill to drive the pump: **2.7**). For another description of the organ mechanism, see Hero, *Pneumatica* 1.42.

Vitruvius, *On Architecture* 10.8.1–6

As far as water organs go, then, I will not leave out touching as briefly and accurately as possible on their principles and supplying a written description. A rectangular chest made of bronze is set up on a base made of wood. Uprights are erected on the base to the right and left, framed like a ladder, enclosing bronze cylinders. Sliding pistons turned accurately on a lathe and provided with iron piston rods are linked to levers by pivoting joints. The pistons are wrapped with fleecy hides. Then holes about three digits in diameter are made in the upper surfaces of the cylinders. Bronze dolphins are mounted on pivoting joints near these holes, holding in their mouths chains from which valve flaps are suspended, hanging freely just inside the openings in the cylinders.

Within the chest, where the water is held, there is a cover like an upside-down funnel beneath which three blocks about three digits high are put to level the interval between the rim of the air vessel and the bottom. A little box is mounted above the neck of the funnel to carry the head of the machine, which in Greek is called *kanon mousikos*. There are pipes along its length: four if it is tetrachord, six if hexachord, eight if octochord.

In each of the pipes is enclosed a single stop cock fitted with an iron handle. When these handles are turned, passages are opened from the box into the pipes. Corresponding to these passages, however, the *kanon* has openings from the pipes, oriented transversely. These are in the upper panel, which in Greek is called *pinax*. Bars called *plinthides* are fixed between the panel and the *kanon* to obstruct these openings, and perforated in the same manner. They are rubbed with oil, so that they can easily be pushed in and spring back into place again, their movement to and fro closing some bores and opening others. These bars have iron pulls attached to them and linked with keys, so that touching the keys brings about in succession the movement of the bars. Rings are soldered on to the holes above the panel through which air escapes from the pipes, and the lower ends of all the organ pipes flanged into them. Then from the cylinders a series of pipes runs to the neck of the funnel cover and connects with the openings that are in the little box. In these openings are mounted flap valves turned on a lathe that, when the box receives air, obstruct the openings and do not allow it to come out again.

Thus, when the levers are raised, the piston rods pull the pistons down to the bottom of the cylinders, and the dolphins set in their pivots let the valve flaps open inward, filling the interior of the cylinders with strong and quick strokes. Then, closing the upper openings with the valve flaps by the force of the pumping, the pistons push into the pipes the air closed up there. It rushes through these pipes into the cover and through its neck into the box. The densely packed air, compressed by the vigorous movement of the levers, flows through the openings of the stop cocks and fills the pipes with air. And so, when the keys are touched, they push in and pull back the bars in succession, closing some of the holes and opening others, and by the art of music produce ringing notes in numerous different measures.

I have tried as best as I could to give a clear written description of an obscure matter. This account, however, is not easy, nor is it readily understood by everyone, but only by those who have experience in this type of thing. But if someone should understand it insufficiently from my description, when he comes to a first-hand knowledge of the device, he will immediately discover that everything has been put down carefully and accurately.

2.47 The force pump of Ctesibius and some other gadgets

The force pump, which works by pushing water along through the action of pistons working in cylinders, was a practical device that grew out of Ctesibius' experiments with pistons and cylinders working on air and water – based at least in part on the work of his teacher Philo. For the most part, however, his discoveries were embodied in the banquet-table devices Vitruvius alludes to at the end of this discussion, a kind of "publication" intended to impress his royal patron. **Fig. 2.7** illustrates a second-century AC Roman pump very similar to the pump Vitruvius describes. For other descriptions and illustrations of force pumps, see **9.34–35**.

Vitruvius, *On Architecture* 10.7.1–4

Now follows a discussion of the Ctesibian machine that raises water to a height. It is made of bronze. Its lower parts consist of two identical cylinders situated slightly apart, with pipes that converge in a fork, meeting at a central collecting chamber. In this chamber are made circular flap valves with fine joints, positioned above the upper openings of the pipes; the valves, by obstructing the openings, do not allow what has been pushed by the air into the chamber to escape back. A covering similar to an inverted funnel is fitted over the chamber and is held in place by a wedge forced through a lug so that the water pressure may not blow it off. An upright pipe (called a "trumpet") is attached to it above. The cylinders, however, have circular flap valves positioned above the intake ports in their bases, just below the lower openings of the pipes.

Then pistons smoothed on a lathe and rubbed down with lubricant are fitted into the cylinders from above. When the pistons enclosed in the cylinders are worked with connecting rods and levers, they will push along the air-water mixture present in them (since the flap valves obstruct the intake pipes), and by blowing and pushing force the water through the pipe openings into the collecting chamber. The funnel receives the water from the chamber, and pneumatic pressure forces it out upwards through the pipe. In this way, water is supplied from a low spot for spouting when a reservoir has been put in position.

And this is not the only choice device of Ctesibius, but there are many others of varied design driven by water pressure. Pneumatic pressure can be seen to provide effects that imitate nature, such as singing of blackbirds and mechanical acrobats, and little figures that drink and move, and other things that delight the senses by pleasing the eye and catching the ear.

Of these devices I have chosen those which I judged to be especially useful and necessary, and in the previous book I decided to discuss clocks, in this book,

SOURCES OF ENERGY AND BASIC MECHANICAL DEVICES

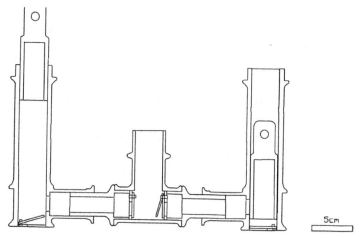

Figure 2.7 Bronze force pump from Bolsena, second century AC.
Source: Drawing by Chris Mundigler, used by permission.

devices for pumping water. The remaining devices, which are not of practical use, but are meant simply to give pleasure, can be found in Ctesibius' own commentaries by those with a special interest in their subtlety.

REAL AND IMAGINED AUTOMATA

Once tools had become relatively common, the Greeks succumbed to the universal human dream of tools that move themselves at a task, or machines that can manage complex tasks by themselves. From the time of Homer onward this impulse is expressed in literature, and in the Hellenistic period, experts in machinery created devices that seemed to be able to work, or at least execute diverting movements, on their own. Although seemingly trivial, such devices expressed the skill and research interests of their creators.

2.48 Domestic robots on Olympus

Self-motivating machines were a vision reflected in the very earliest Greek literature, where Homer, for example, has the smithy god Hephaestus create robots to serve the gods.

Homer, *Iliad* 18.369–379

Silver-footed Thetis came to the starry, everlasting house of Hephaestus, renowned among the immortals, a house of bronze, which the club-footed god himself had made. She found him sweating as he busied himself in haste among his bellows, for he was making tripods, 20 in all, to stand along the wall of the well-built house. He put golden wheels beneath the base of each so that of their own accord they might enter the assembly of the gods and again go back to his house, a marvel to see. They were nearly finished, but he had not yet attached the cleverly devised ring handles. He was preparing these and forging the rivets.

2.49 The secret of an early Greek robot

Daedalus, a name, which means "cunning craftsman", was supposed to have been the earliest human sculptor (of the eighth or seventh century BC). His creations were said to have been so life-like and close to human scale that they could actually move.

Aristotle, *On the Soul* 1.3.406b

Some also say that the soul moves the body in which it is present just as it moves itself; for example, Democritus, speaking like Philippus the comic dramatist. He says that Daedalus made his "Wooden Aphrodite" move by pouring mercury in it.

2.50 One of Archytas' inventions: a flying bird

Flight was another goal of early technological thought and dreaming, but it was never realised in antiquity. Aulus Gellius clearly does not believe the story he reports at third hand about the early fourth-century BC mathematician and philosopher Archytas.

Aulus Gellius, *Attic Nights* 10.12.9–10

But the bird, which Archytas the Pythagorean is reported to have designed and constructed certainly ought to seem no less astonishing, and not so fruitless. For not only many of the noble Greeks, but also the philosopher Favorinus, an avid student of old records, have written with the greatest assurance that, by certain principles and a knowledge of mechanics, Archytas made a wooden model of a dove that flew. Evidently it was balanced nicely by weights and propelled by compressed air concealed within it. Concerning a story so difficult to believe, I prefer, by Hercules, to put down Favorinus' own words: "Archytas of Tarentum, being also a mechanical engineer, made a flying dove of wood. Whenever it alighted, it did not take off again. For until this …" [The text breaks off here.]

2.51 Entertaining automata in Hellenistic royal parades

The Hellenistic kings were very taken by showy processions, much like modern holiday parades, where "floats" were exhibited on which stories from classical mythology or allegorical scenes were executed by living performers or by automata. The ingenuity and mechanical knowledge of many Hellenistic engineers was put into play designing these impressive *tableaux*, and smaller devices used on the banquet tables of the kings. Athenaeus, quoting a Hellenistic historian, describes one such automaton in an elaborate parade put on in Alexandria by Ptolemy Philadelphus (283/2–246 BC), and Polybius a similar, but more graphic, device that preceded a procession of Demetrius of Phaleron at Athens (*c*.317–307 BC).

SOURCES OF ENERGY AND BASIC MECHANICAL DEVICES

Athenaeus, *Philosophers at Dinner* 5.198e–f

After these women, a four-wheeled wagon, eight cubits wide, was pulled along by 60 men. On it sat an image of Nysa eight cubits high, dressed in a yellow cloak with gold spangles and wrapped in a Laconian cloak. This image stood up automatically without anyone touching it, poured milk from a gold libation cup, and sat back down again. The image held in its left hand a Bacchic wand bound with fillets.

Polybius, *Histories* 12.13.11

... (and Demochares tells us) that a mechanical snail moving by itself preceded his procession, leaving a trail of slime.

2.52 An automated manikin of Caesar's body

Appian, *Civil Wars* 2.20.147

While they were in this state, and close to fighting, someone lifted up above the funeral couch an image of Caesar himself made of wax; the corpse itself was not visible, since it lay on its back on the couch. A mechanism turned the image around in all directions, and the 23 wounds were visible all over the body and up the face....

2.53 Singing bird automaton

Hero's *Pneumatics*, a study of devices that illustrate principles of water and air pressure, borrows heavily from similar works by the Hellenistic authors Philo and Ctesibius, some of them now lost. It preserves descriptions of the mechanisms that worked automata similar to those described in the preceding passages. Devices that made sounds, opened doors, or dispensed goods appear in the passages that follow. Translations adapted from Woodcroft (1851) 29–33, 37, 52–53, 57–58.

Hero, *Pneumatics* 1.15–16

Vessels may be made such that, when water is poured into them, the song of the blackcap warbler, or a whistling sound, is produced. They are constructed in the following way. Let *a b c d* be a hollow, air-tight pedestal [**fig. 2.8**]; through the top, *a d*, let a funnel, *e f*, be introduced and soldered into the surface, its tube approaching only so near to the bottom as to leave a passage for the water. Let *g h k* be a small pipe, such as will emit sound communicating with the pedestal and likewise soldered into *a d*. Its extremity *k*, which is curved, must dip into water contained in a small vessel placed nearby at *l*. If water is poured in through the funnel *e f*, the result will be that the air, being driven out, passes through the pipe *g h k* and emits a sound. When the extremity of the pipe dips into water, a bubbling sound is heard, and the note of the warbler is produced: if no water is near, there will be only whistling. These sounds are produced through pipes, but

Figure 2.8 Singing bird automaton.
Source: Adapted from Schmidt, 1899, fig. 16.

the quality of the sounds will vary as the pipes are more or less narrow, or longer or shorter, and as a larger or smaller portion of the pipe is immersed in the water. And so, by this means the distinct notes of many birds can be produced.

The figures of several different birds are arranged near a fountain, or in a cave, or in any place where there is running water. Near them sits an owl, which, apparently of her own accord, turns at one time towards the birds, and then again away from them. When the owl looks away the birds sing, and when she looks at them, they are mute: this may be repeated frequently. It is built in the following manner. Let *a* be a stream perpetually running [**fig. 2.9**]. Underneath place an air-tight vessel, *b c d e*, provided with an enclosed *diabetes* or bent siphon *f g*, and having inserted in it a funnel, *h k*, between the extremity of whose tube and the bottom of the vessel a passage is left for the water. Let the funnel be provided with several smaller pipes, as described before, at *l*. It will be found that, while *b c d e* is being filled with water, the air that is driven out will produce the notes of birds; and as the water is being drawn off through the siphon *g f* after the vessel has filled, the birds will be mute.

We are now to describe the contrivance by which the owl is enabled to turn herself towards the birds, or away from them, as we have said. Let a rod *n q* turned in a lathe rest on any support *m*: around this rod, let a tube *o p* be fitted, so as to move freely about it, and having attached to it the disc top *r s*, on which the owl is to be securely fixed. Round the tube *o p* let a chain pass, the two extremities of which, *t u*, *w x*, wind off in opposite directions and are attached, by means of two pulleys, the one, *t u*, to a weight suspended at *y*, and *w x* to an empty vessel, *z*, which lies beneath the siphon or enclosed *diabetes f g*. It will be found that while the vessel *b c d e* is emptying, the liquid pouring into vessel *z* causes the tube *o p* to revolve, and the owl with it, so as to face the birds: but when *b c d e* is empty, the vessel *z* empties itself in the same manner by means of an enclosed or bent siphon contained within it. Then the weight *y* again

Figure 2.9 Whistling bird and owl automaton.
Source: Adapted from Schmidt, 1899, fig. 17.

preponderating, it causes the owl to turn away just at the time when the vessel *b c d e* is being filled again and the notes once more issue from the birds.

2.54 Automatic trumpeting doorbell for a temple

Hero, *Pneumatics* 1. 16–17

[The sound of trumpets can be produced] in the manner just described. Insert into a carefully sealed vessel the tube of a funnel, reaching nearly to the bottom and soldered into the surface of the vessel [**fig. 2.10**]; and, by its side, let there be a trumpet provided both with a mouthpiece and bell, the mouthpiece of which is inserted into the upper surface of the vessel. If water is poured through the funnel, it will be found that the air contained in the vessel, as it is being driven out through the mouthpiece, will produce the sound of a trumpet.

The sound of a trumpet may be produced at the opening of the doors of a temple in the following way. Behind the door, let there be a vessel *a b c d* containing water. In this, invert a narrow-necked air chamber, that is, an inverted vessel with a narrow mouth, *f*. Let a trumpet, *h k*, provided with bell and mouthpiece, communicate with it at its lower extremity. [Let the rod *l m* run] parallel with the tube of the trumpet and attached to it, fastened at its lower end to the

SOURCES OF ENERGY AND BASIC MECHANICAL DEVICES

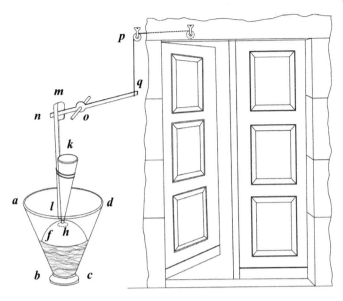

Figure 2.10 Automatic trumpeting doorbell.
Source: Adapted from Schmidt, 1899, fig. 18.

vessel *f*, and having at the other end a loop, *m*. Let the rod *n q* pass through this loop, thus supporting the vessel *f* at a sufficient height above the water. The rod *n q* must turn on pivot *o*, and a chain or cord attached to its extremity *q* must be passed through a pulley-wheel *p* and fastened to the rear of the door. When the door is drawn open, the cord will be pulled and draw the extremity *q* of the rod upward, so that the rod *q* no longer supports the loop *m*. And when the loop (in consequence) changes its position, the vessel *f* will descend into the water, and give forth the sound of a trumpet by the expulsion of the air contained in it through the mouthpiece and bell.

2.55 Automatic holy-water dispenser

Hero, *Pneumatics* 1.21

If a five-*drachma* piece is thrown into certain libation vessels, water will flow out for purifying ablutions. Let *a b c d* be a libation vessel or money box, having *a* as its mouth [**fig. 2.11**]. In the container let there be a vessel *f g h k* containing water, and a small box *l*, from which a pipe *l m* exits the chest. Near the vessel place a vertical rod *n q*, about which turns a lever *o p*, widening at *o* into the plate *r* parallel to the bottom of the vessel. From the extremity *p* is suspended a lid *s*, which fits into the box *l*, so that no water can flow through the tube *l m*. This lid, however, must be heavier than the plate *r*, but lighter than the plate and coin combined. When the coin is thrown through the mouth *a*, it will fall upon

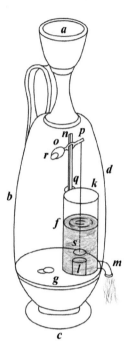

Figure 2.11 Coin-operated holy-water dispenser.
Source: Adapted from Schmidt, 1899, fig. 22.

the plate *r* and, outweighing it, will tilt the lever *o p* and raise the lid of the box so that the water will flow. But when the coin slips off, the lid will descend and close the box, so that the discharge stops.

2.56 Self-trimming lamp

Hero, *Pneumatics* 1.34

To construct a self-trimming lamp. Let *a b c* be a lamp, through the mouth of which is inserted an iron rod *d e* capable of sliding freely along the point *e*, and let the wick be wound loosely about the pin [**fig. 2.12**]. Adjacent to it, place a toothed wheel *f*, moving freely about an axle. Its teeth are in contact with the iron pin so that as the wheel revolves the wick may be pushed on by the teeth. Let the opening for the oil be of considerable width, and when the oil is poured in let a small float-vessel *g* rise with it. To this is attached a vertical toothed bar *h*, the teeth of which mesh with those on the wheel. It will be found that, as the oil is consumed, the float-vessel sinks and (by means of the teeth on the bar) causes wheel *f* to revolve, and in this way the wick is pushed out.

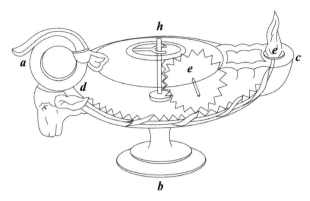

Figure 2.12 Lamp with self-trimming wick.
Source: Adapted from Schmidt, 1899, fig. 35.

2.57 Automatic door opener for a temple
Hero, *Pneumatics* 1.38

The construction of a small temple whose doors shall open by themselves when a fire is lighted and close again when it is extinguished. Let the proposed temple stand on a pedestal *a b c d*, on which sits a small altar *e d* [**fig. 2.13**]. Through the altar insert a tube *g f*, one of the openings of which (*f*) is within the altar, and the other (*g*) opens in a sphere *h*, reaching nearly to its centre. The tube must be soldered into the sphere, into which a bent siphon *k l n* is also placed. Let the hinges of the doors be extended downwards and turn freely on pivots within the base *a b c d*, and from the hinges let two interconnected chains be attached through a pulley to a hollow vessel *n q*, which is suspended from them. Two other interconnected chains, wound around the hinges in an opposite direction from the first pair, are attached through a pulley to a lead weight which shuts the doors when it descends. Let the outer portion of the siphon *k l n* lead into the suspended vessel, and through a hole *p*, which must be carefully closed afterwards, pour enough water into the sphere to fill half of it.

It will be found that, when the fire has grown hot, the air in the altar will become heated and expand to fill a larger space, and, passing through the tube *g f* into the sphere, it will drive out the liquid contained there through the siphon *k l n* into the suspended vessel. This, descending with the weight, will pull the chains and open the doors. Again, when the fire is extinguished, the air will be drawn out of the sphere as it contracts and the bent siphon (the extremity of which will be immersed in the water in the suspended vessel) will draw up the liquid in the vessel in order to fill the void left by the condensing vapour. As the vessel is lightened, the suspended weight will overbalance it and shut the doors. Some use liquid mercury in place of water, as it is heavier than water and easily caused to evaporate by fire.

Figure 2.13 Self-opening temple doors.
Source: Adapted from Schmidt, 1899, fig. 39.

Notes

1 Bernard Carra de Vaux, "Le livre des appareils pneumatiques et des machines hydrauliques de Philon de Byzance d'après les versions arabes d'Oxford et de Constantinople". *Académie des Inscriptions et des Belles Lettres: Notice et extraits de mss. de la Bibliothèque nationale, Paris* 38 (1903) 27–235; see pp. 177–179. The figure keys have been changed to correspond to the figures used here.
2 Schiøler and Wikander (1983).
3 The definition could apply to either an ember-box or fire-starting kit.
4 Philo (*Parasceuastica et poliorcetica* 99.25) recommends sponges, sheepskin, and algae in nets soaked in water or vinegar.

Bibliography

Energy, heating, lighting

Black, Ernest W., "Hypocaust Heating in Domestic Rooms in Roman Britain". *Oxford Journal of Archaeology* 4 (1985) 77–92.
Brodner, Erika, "Untersuchungen an frühen Hypokaustenanlagen". *Technikgeschichte* 43 (1976) 249–267.

Butti, Ken, and John Perlin, *A Golden Thread: 2500 Years of Solar Architecture and Technology.* New York: Van Nostrand Reinhold, 1980.
Davidson, H.R. Ellis, "The Secret Weapon of Byzantium". *Byzantinische Zeitschrift* 66 (1973) 61–74.
Forbes, Robert J., "Bitumen and Petroleum in Antiquity". Pp. 1–124 in Robert J. Forbes, *Studies in Ancient Technology*, Vol. 1. 2nd edn. Leiden: Brill, 1964.
Forbes, Robert J., "Heat and Heating", "Refrigeration", "Light", in Robert J. Forbes, *Studies in Ancient Technology*, Vol. 6. 2nd edn. Leiden: Brill, 1966.
Halleux, R., "Problèmes de l'énergie dans le monde ancien". *Les Études Classiques* 45 (1977) 49–61.
Halydon, J., and M. Byrn, "A Possible Solution to the Problem of Greek Fire". *Byzantinische Zeitschrift* 70 (1977) 91–99.
Jantzen, Ulf, and Renate Tölle, "Beleuchtungsgerät". Pp. 83–98 in Siegfried Laser, ed., *Hausrat*. Archaeologica Homerica II, P. Göttingen: Vanderhoeck & Ruprecht, 1968.
Keyser, Paul T., "The Purpose of the Parthian Galvanic Cells: A First-Century A.D. Electric Battery Used for Analgesia". *Journal of Near Eastern Studies* 52 (1993) 81–98.
Loeschcke, Siegfried, "Antike Laternen und Lichthäuschen". *Bonner Jahrbücher* 118 (1909) 370–430.
Morgan, Morris H., "De ignis eliciendi modis apud antiquos". *Harvard Studies in Classical Philology* 1 (1890) 13–64.
Partington, James Riddick, *A History of Greek Fire and Gunpowder.* Cambridge: Heffer & Sons, 1960.
Pászthory, Emmerich, "Über das 'Griechische Feuer'". *Antike Welt* 17.2 (1986) 27–37.
Pease, Arthur S., "Scintillae". *Classical Philology* 34 (1939) 148.
Perlzweig, Judith, *Lamps from the Athenian Agora.* Princeton: American School of Classical Studies at Athens, 1963.
Pernot, L., "Pour une préhistoire de l'électricité: la Grèce antique et l'ambre jaune". *Bulletin d'histoire d'électricité* 2 (1983) 19–30.
Radl, A., *Der Magnetstein in der Antike. Quellen und Zusammenhänge.* Stuttgart: 1988.
Radt, Wolfgang. "Lampen und Beleuchtung in der Antike". *Antike Welt* 17 (1986) 40–58.
Rehder, J.E., *Mastery and Uses of Fire in Antiquity.* Montreal: McGill-Queens Press, 2000.
Sandklef, Albert, and Dagmar Selling. "The Heating of Classical Thermae". *Opuscula Romana* 11 (1976) 123–125. 2 figs.
Séguin, André, "Recherches sur le pétrole dans l'antiquité". *Revue des Questions Historiques* 250 (1936) 3–41.
Séguin, André, "Étude sur le pétrol dans l'antiquité grècque et latine. Gisements de pétrole connus des latins et des grecs". *Revue des Questions Historiques* 261 (1938) 36–70.
Turcan-Deleani, M. "*Frigus amabile*". Pp. 691–696 in Marcel Renard and Robert Schilling, eds., *Hommages à Jean Bayet.* Collection Latomus, 70. Brussels: Latomus, 1964.
Webster, Graham, "A Note on the Use of Coal in Roman Britain". *Antiquaries Journal* 35 (1955) 199–216. 1 fig.
Webster, Graham, and Theodore A. Wertime, "The Furnace versus the Goat: The Pyrotechnic Industries and Mediterranean Deforestation in Antiquity". *Journal of Field Archaeology* 10 (1983) 445–452.
Wikander, Örjan, "Sources of Energy and Exploitation of Power". Pp. 136–157 in John P. Oleson, ed., *Oxford Handbook of Engineering and Technology in the Classical World.* Oxford: Oxford University Press, 2008.

Power and prime movers

Ad, Adi, Uzi Ad, ʿAbd al-Salam Saʿid, and Rafael Frankel, "Water-mills with Pompeian-type millstones at Nahal Tanninim". *Israel Exploration Journal* 55 (2005) 156–171.

Amouretti, Marie-Claire, "L'Attelage dans l'antiquité. Le prestige d'une erreur scientifique". *Annales: Économies, Sociétés, Civilisations* 1 (1991) 219–232.

Bennett, Richard, and John Elton, *History of Corn Milling*, Vol. 1: *Handstones, Slave, and Cattle Mills*; Vol. 2: *Watermills and Windmills*. London: Simpkin, Marshall & Company, 1898–1904.

Blaine, Bradford B., "The Enigmatic Water-Mill". Pp. 163–176 in Bert S. Hall, and Delna C. West, eds., *On Pre-Modern Technology and Science. A Volume of Studies in Honor of Lynn White, Jr.*, Malibu: Undina Publications, 1976.

Clutton-Brock, Juliet, *Horse Power: A History of the Horse and the Donkey in Human Societies*. Cambridge, MA: Harvard University Press, 1992.

Drachmann, Aage G., "Heron's Windmill". *Centaurus* 7 (1961) 145–151.

Forbes, Robert J. "Power". Pp. 80–130 in Robert J. Forbes, *Studies in Ancient Technology*, Vol. 2. 2nd edn. Leiden: Brill, 1965.

Grewe, Klaus, "Recent Developments in Aqueduct Research". *Mouseion Supplement 2: Festschrift in Honour of Professor John P. Oleson*. Forthcoming (2019).

Keyser, P. "Classics and Technology: A Re-evaluation of Heron's First Century A.D. Steam Engine", in Sarah U. Wisseman, and Wendell S. Williams, eds., *Ancient Technologies and Archaeological Materials*. London: Gordon & Breach, 1994.

Lewis, Michael J.T., "The Greeks and the Early Windmill". *History of Technology* 15 (1993) 141–189.

Lewis, Michael J.T., "Theoretical Hydraulics, Automata, and Water Clocks". Pp. 343–369 in Örjan Wikander, ed., *Handbook of Ancient Water Technology*. Leiden: Brill, 2000.

Lucas, Adam, *Wind, Water, Work: Ancient and Medieval Milling Technology*. Leiden: Brill, 2006.

Lo Cascio, Elio, and Paolo Malanima, "Mechanical Energy and Water Power in Europe: A Long Stability?" Pp. 201–208 in Ella Hermon, ed., *Vers une gestion intégrée de l'eau dans l'empire romain*. Rome: L'Erma di Bretschneider, 2008.

Ludwig, Karl-Heinz, "Die technikgeschichtlichen Zweifel an der 'Mosella' des Ausonius sind unbegründet". *Technikgeschichte* 48 (1981) 131–134.

Malanima, Paolo, "Energy Consumption in the Roman World". Pp. 13–36 in William V. Harris, ed., *The Ancient Mediterranean Environment Between Science and History*. Leiden: Brill, 2013.

Mangartz, Fritz, *Die byzantinische Steinsäge von Ephesos: Baubefund, Rekonstruktion, Architekturteile*. Mainz: Verlag des Römisch-Germanischen Zentralmuseums, 2010.

Moritz, Ludwig Alfred, *Grain-Mills and Flour in Classical Antiquity*. Oxford: Oxford University Press, 1955.

Polge, Henri, "L'amélioration de l'attelage a-t-elle réellement fait reculer le servage?" *Journal des Savants* (1967) 5–42.

Raepsaet, Georges, "Attelages antiques dans le Nord de la Gaule: les systèmes de traction par équidés". *Trierer Zeitschrift* 45 (1982) 215–273.

Ritti, Tullia, Klaus Grewe, and Paul Kessener, "A Relief of a Water-Powered Stone Saw Mill on a Sarcophagus at Hierapolis and its Implications". *Journal of Roman Archaeology* 20 (2007) 139–163.

Roos, Paavo, "For the Fiftieth Anniversary of the Excavation of the Water Mill at Barbegal: A Correction of a Long-Lived Mistake". *Revue Archéologique* (1986) 327–333.

Schiøler, Thorkild, and Örjan Wikander, "A Roman Water-Mill in the Baths of Caracalla". *Opuscula Romana* 14 (1983) 47–64.

Seigne, J., "A Sixth-Century Water-Powered Sawmill at Jerash". *Annual of the Department of Antiquities of Jordan* 46 (2002) 205–213.

Sellin, Robert H.J., "The Large Roman Water Mill at Barbegal (France)". *History of Technology* 8 (1983) 91–109.

Simms, D.L., "Water-Driven Saws, Ausonius, and the Authenticity of the Mosella". *Technology and Culture* 24 (1983) 635–643.

Sleeswyk, André W., "Hand-Cranking in Egyptian Antiquity". *History of Technology* 6 (1981) 23–37.

Smith, Norman A.F., "The Origins of Water Power: A Problem of Evidence and Expectations". *Transactions of the Newcomen Society* 55 (1985) 67–84.

Spruytte, Jean, *Early Harness Systems. Contribution to the History of the Horse.* London: A. Allen, 1983.

Vigneron, Paul, *Le cheval dans l'antiquité gréco-romaine (des guerres médiques aux grandes invasions). Contribution à l'histoire des techniques.* Nancy: Université de Nancy, 1968.

Wefers, S., ed., *Die Mühlenkaskade von Ephesos. Technikgeschichtliche Studien zur Versorgung einer spätantiken bis frühbyzantinischen Stadt*, Monographien des Römisch-Germanischen Zentralmuseums Mainz, Band 118, in Kooperation mit dem Österreichischen Archäologischen Institut, Institut für Kulturgeschichte der Antike der Österreichischen Akademie der Wissenschaften. Mainz: Verlag des Römisch-Germanischen Zentralmuseums, 2015.

Wikander, Örjan, "Water-Mills in Ancient Rome". *Opuscula Romana* 12 (1979) 36.

Wikander, Örjan, "The Use of Water-Power in Classical Antiquity". *Opuscula Romana* 13 (1981) 91–104.

Wikander, Örjan, *Exploitation of Water-Power or Technological Stagnation. A Reappraisal of the Productive Forces in the Roman Empire.* Lund: Royal Society of Letters, 1984.

Wikander, Örjan, "Mill-Channels, Weirs and Ponds. The Environment of Ancient Water Mills". *Opuscula Romana* 15 (1985) 149–154.

Wikander, Örjan, "Archaeological Evidence for Early Water-Mills-an Interim Report". *History of Technology* 10 (1985) 151–179.

Wikander, Örjan, "Ausonius' Saw-Mills Once More". *Opuscula Romana* 17 (1989) 185–190.

Wikander, Örjan, "The Water-Mill". Pp. 371–400 in Örjan Wikander, ed., *Handbook of Ancient Water Technology.* Leiden: Brill, 2000.

Wikander, Örjan, "Industrial Applications of Water-Power". Pp. 401–410 in Örjan Wikander, ed., *Handbook of Ancient Water Technology.* Leiden: Brill, 2000.

Wikander, Örjan, "Invention, Technology Transfer, Breakthrough – The Ancient Water-Mill as an Example". Pp. 127–133 in F. Minonzio, ed., *Problemi di macchinismo in ambito romano.* Archeologia dell'Italia Settentrionale 8. Como: Commune di Como, 2004.

Wikander, Örjan, "Sources of Energy and Exploitation of Power". Pp. 136–157 in John P. Oleson, ed., *Oxford Handbook of Engineering and Technology in the Classical World.* Oxford: Oxford University Press, 2008.

Wikander, Örjan, "The Water-Mills on the Janiculum". *Memoirs of the American Academy in Rome* 45 (2000) 219–246.

Wikander, Örjan, "Water-Mills at Amida: Ammianus Marcellinus 18.8.11". *Classical Quarterly* 51 (2001) 231–236.

Wikander, Örjan, "Machines, Power, and the Ancient Economy". *Journal of Roman Studies* 92 (2002) 1–32.

Wilson, Andrew, "Water-Power in North Africa and the Development of the Horizontal Water-Wheel". *Journal of Roman Archaeology* 8 (1995) 498–510.

Machines and gadgets

Amouretti, Marie-Claire, Georges Comet, Claude Ney, and Jean-Louis Paillet, "À propos du pressoir à huile: de l'archéologie industrielle à l'histoire". *Mélanges de l'École Française de Rome* 96 (1984) 379–421.

Apel, Willi, "Early History of the Organ". *Speculum* 23 (1948) 191–216.

Brumbaugh, Robert S., *Ancient Greek Gadgets and Machines*. New York: Crowell, 1966.

Cotterell, Brian, and Johan Kamminga, *Mechanics of Pre-Industrial Technology: An Introduction to the Mechanics of Ancient and Traditional Material Culture*. New York: Cambridge University Press, 1989.

Drachmann, Aage G., *Ancient Oil Mills and Presses*. Copenhagen: Levin & Munksgaard, 1932.

Drachmann, Aage G., "Heron's Screwcutter". *Journal of Hellenic Studies* 56 (1936) 72–77.

Drachmann, Aage G., *The Mechanical Technology of Greek and Roman Antiquity: A Study of the Literary Sources*. Copenhagen: Munksgaard, 1963.

Drachmann, Aage G., "The Crank in Graeco-Roman Antiquity". Pp. 33–51 in Teich Mikulás and Robert Young, eds., *Changing Perspectives in the History of Science*. London: Heinemann, 1973.

Field, J.V., D.R. Hill, and M.T. Wright, "Byzantine and Arabic Mathematical Gearing". *Annals of Science* 42 (1985) 87–138.

Fleury, Philippe, *La méchanique de Vitruve*. Zürich: Olms, 1991.

Foley, Bernard, Werner Soedel, John Turner, Brian Wilhoite, "The Origin of Gearing". *History of Technology* 7 (1982) 101–129.

Gaitzsch, Wolfgang, "Die 'römische' Schraube aus dem Kastell von Niederbieber". *Bonner Jahrbücher* 183 (1983) 595–602.

Gille, Bertrand, "Machines". Pp. 629–658 in Charles Singer, Eric J. Holmyard, Alfred R. Hall, and Trevor I. Williams, eds., *A History of Technology. Vol. II*. Oxford: Oxford University Press, 1956.

Hesberg, H.V., "Mechanische kunstwerke und ihre bedeutung für die höfische kunst des frühen hellenismus". *Marburger Winckelmann-Programm* (1987) 47–72.

Kiechle, Franz, "Zur Verwendung der Schraube in der Antike". *Technikgeschichte* 34 (1967) 14–22.

King, Henry C., and John R. Millburn, *Geared to the Stars. The Evolution of Planetariums, Orreries, and Astronomical Clocks*. Toronto: University of Toronto Press, 1978.

Lobell, Jarrett A., "The Antikythera Mechanism". *Archaeology* 60.2 (2007) 42–45.

Mattingly, David J., "Paintings, Presses and Perfume Production at Pompeii". *Oxford Journal of Archaeology* 9 (1990) 71–90.

Mayor, Adrienne, *Gods and Robots*. Princeton: Princeton University Press, 2018.

Moritz, Ludwig Alfred, *Grain-Mills and Flour in Classical Antiquity*. Oxford: Clarendon Press, 1958.

Mutz, Alfred, "Römische Bronzegewinde". *Technikgeschichte* 36 (1969) 161–167.

Perrot, Jean, *The Organ from its Invention in the Hellenistic Period to the End of the Thirteenth Century*. London: Oxford University Press, 1971.

Price, Derek J. de Solla, "Automata and the Origins of Mechanism and Mechanistic Philosophy". *Technology and Culture* 5 (1962) 9–23.

Price, Derek J. de Solla, *Gears from the Greeks. The Antikythera Mechanism – A Calendar Computer from c.80 B.C.* Transactions of the American Philosophical Society, 64. 7. Philadelphia: American Philosophical Society, 1974.

Ring, James, "Windows, Baths, and Solar Energy in the Roman Empire". *American Journal of Archaeology* 100 (1996) 717–724.

Ringst, Emil, and P. Thielscher, "Catos Keltern und Kollergänge. Ein Beitrag zur Geschichte von Öl und Wein". *Bonner Jahrbücher* 154 (1954) 32–93; 157 (1957) 53–126.

Truitt, E.R., *Medieval Robots*. Philadelphia: University of Pennsylvania Press, 2015.

Usher, Abbott Payson, *A History of Mechanical Inventions*. Rev. edn. Cambridge, MA: Harvard University Press, 1966.

Whitney, Elspeth, *Paradise Restored: The Mechanical Arts from Antiquity through the Thirteenth Century*. Transactions of the American Philosophical Society, 80.1. Philadelphia: American Philosophical Society, 1990.

Wikander, Örjan, "Gadgets and Scientific Instruments". Pp. 785–799 in John P. Oleson, ed., *Oxford Handbook of Engineering and Technology in the Classical World*. Oxford: Oxford University Press, 2008.

Wilson, Andrew I., "Machines in Greek and Roman Technology". Pp. 337–367 in John P. Oleson, ed., *Oxford Handbook of Engineering and Technology in the Classical World*. Oxford: Oxford University Press, 2008.

Woodcroft, Bennet, *The Pneumatics of Hero of Alexandria*. London: Taylor, Walton, & Maberly, 1851.

Wright, Michael T., "A Planetarium Display for the Antikythera Mechanism". *Horological Journal* 144 (2002) 169–173.

Wright, Michael T., "The Antikythera Mechanism and the Early History of the Moon-Phase Display". *Antiquarian Horology* 29.3 (2006) 319–329.

3

AGRICULTURE

Humankind's desire for a permanent and assured food supply led early hunter-gatherers to begin cultivating plants and domesticating animals. Of cultivated crops in the Mediterranean, cereals were the earliest, of the greatest economic importance, and more than any other development made settled life possible. Domestication of animals, which involved the subjugation of one species to suit the needs of another – humans – surely followed primitive cultivation, since a supply of fodder is always necessary for animals no longer hunting in the wild.

The evolution of agricultural techniques affected other aspects of technology as well: domestication of animals, for example, made transport easier; sowing, harvesting, and grinding demanded new agricultural tools, sometimes based on new mechanical devices; and the increasing surplus of food prompted an expansion of trade in both raw materials and luxuries to satisfy the needs and desires of the burgeoning populations in urban centres, themselves formed to administer to the increasingly complex social and technological arrangements of an agrarian economy.

Still, despite our modern tendency to equate "civilisation" with the characteristics of large urban areas – an opinion we derive from the ancients themselves – we must keep in mind that, until well into the Industrial Revolution, the great majority of the inhabitants of the Mediterranean world were dependent on agriculture for their livelihood, either directly as farmers or absentee landlords, or indirectly in the transport, processing, and distribution of foodstuffs. It is, then, not surprising to discover that our ancient sources give us more information about agricultural techniques than about any other aspect of their technologies.

Yet even in a field of such basic interest to a Greek or Roman peasant, the texts that have come down to us were not for the most part directed at the small farmer who cultivated a plot of five or ten hectares, but at the wealthy estate owner who relied on the labour of hired hands or slaves to support his own comfortable life in the city. The majority of selections that follow are of this sort, though our aristocratic authors are undoubtedly better informed about the proper cultivation of the soil than about most other technological processes.

It is also worth noting in advance that most of our information comes from Roman rather than Greek writers, particularly from the agricultural handbooks of Cato, Varro, and Columella, from the encyclopaedia of Pliny the Elder, and – more modestly – from the letters of Pliny's nephew, who owned a number of productive estates. Still, with some notable exceptions, farming procedures changed remarkably little between the time of Hesiod in the Archaic Age and the late Roman Empire of Palladius.

We begin with a survey of some real and imaginary farms and rural estates (**3.1–4**), then examine the criteria for a productive farmstead, including location, equipment, and labour (**3.5–14**) and the desirable soils, irrigation techniques, and fertilisers (**3.16–27**). There follows a survey of the common crops and their cultivation techniques: grains and legumes (**3.28–37**), vegetables (**3.38–39**), and the vine and olive (**3.40–53**). The section on animal husbandry (**3.54–67**) explains the maintenance and breeding of most of the important domesticated species, from cattle and donkeys to geese and bees to fish; and a short collection of passages on hunting, fishing, and sponge diving (**3.68–80**) serves to remind us that in antiquity humankind still relied on untamed sources of food and supplies. (Thibodeau 2016, Halstead 2014)

SOME FARMS AND ESTATES

The first selection gives a general impression of the variety of crops, livestock, and occupations on farms in antiquity. This description of the estate of a wealthy Athenian outside the city can be compared to the delights of rural life from the inexperienced view of an urban Roman provided by Martial, *Epigrams* 3.58.

3.1 A rich Attic farm

The plaintiff, here addressing the court, is seeking to be relieved of certain financial obligations placed on the wealthiest citizens of Athens; to be successful, he must have those obligations transferred to another citizen who is demonstrably wealthier than he. The rich farmer Phaenippus, his victim in this procedure, has evidently tried to camouflage the prosperity of his holdings by claiming that they were mortgaged and by removing some of the produce.

Demosthenes 42, *Against Phaenippus* 5–7, 20

After citing Phaenippus I took some relatives and friends and set out for Kytheron where his boundary estate is located. First, I led them around the farm, which is more than 40 *stadia* in circumference, and pointed out to them as my witnesses, and in the presence of Phaenippus, that there was not a single pillar on the property to indicate that it was mortgaged.... Afterwards I asked him where the grain was that had been threshed – for by the gods and goddesses, men of the jury, there were two threshing-floors there, each of them just under a *plethron* in diameter. His answer to me was that some of the grain had been sold and the rest was laid up in the storehouse. In the end – to make a long story short – I stationed some men inside to keep a watchful eye and, by Jupiter, I strictly forbade the donkey-drivers to carry the logs off the estate. For on top of his other assets, men of the jury, Phaenippus has this profitable source of income: six donkeys working throughout the year to transport wood, for which he receives more than 12 *drachmai* a day.... You, Phaenippus, are a rich man, as is to be expected of one who today sells barley from his farm at 18 *drachmai* and wine at 12 *drachmai*, while producing more than 1,000 *medimni* of grain and over 800 *amphorae* of wine.

3.2 Early Roman agriculture

In the late Republic and Empire criticisms are directed against city-dwellers whose desire for a place in the country contributed in part to the decline of the old rural peasantry, those simple and solid folk, whose hard work and honest principles had been the foundation of Rome's greatness, as Pliny observes.

Pliny, *Natural History* 18.5–21

This discussion is, quite naturally, about the countryside and rustic practices; but it is on these that life relies and that the greatest esteem was bestowed by our ancestors. At the very start Romulus established the Arval Brethren or "Priests of the Fields" and named himself one of the 12.... In those times two *iugera* satisfied the Roman citizenry, and no one was assigned a larger portion. I dare say none of the Emperor Nero's lackeys would recently have been content with a flower-garden of that size!

The area that could be ploughed in a day by a single yoke of oxen (*iugum*) was called a *iugerum*.... The most magnificent gift for which generals and heroic citizens were eligible was the largest area of land that one man could plough around in a single day, together with a *quartarius* or *hemina* of emmer wheat as a contribution from each of the citizens. Even our original surnames came from agriculture: Pilumnus from the man who invented the *pilum* or pestle used in mills, Piso from our word for "pounding", Fabius [bean], Lentulus [lentil], and Cicero [chickpea] from the crops each grew best.... Bad farming was judged a crime under the jurisdiction of the censors, and conversely (according to Cato) to commend a man by calling him a good farmer and a good tiller of the soil was thought to be the highest praise of all. Hence our word "affluent" (*locupletes*) literally means "those full of space", that is, "of land". The word for "money", *pecunia*, was derived from *pecus* meaning "cattle", and still to this day all sources of state income are called "pastures" in the censors' ledgers because for a long time these were the only thing subject to taxation....

In those days the fields were cultivated by the hands of our very generals.... These days, on the other hand, those same tasks are carried out by chained feet, by convict hands, by branded faces.... And we are amazed that chain-gangs are not as productive as generals!

AGRICULTURE

3.3 The spread of Italian *latifundia*

As the next two passages indicate, small, independent farms in Italy were being replaced from the third century BC onwards by large tracts of land, called *latifundia*, assembled by absentee landlords and worked largely by slave labour. Many Roman writers fastened on these large estates as the cause of Rome's increase in unemployment and decline in moral standards. In part they were right, but they seem to have exaggerated the amount of land encompassed by these *latifundia*: much of Italy and the provinces was still made up of small, privately owned plots worked by a family and perhaps a few retainers.

Appian, *Civil War* 1.1.7

As the Romans conquered Italy bit by bit, they would take possession of a large part of the land and build cities there or settle colonists from their own number in existing towns: they conceived of these as substitutes for forts. And each time they acquired territory in war they would immediately distribute the cultivated land to colonists, or sell it or rent it out. But as for the land that lay fallow as a result of the war – and this amounted to the greater part – they did not yet have the time to divide it up and so issued a statement in the meantime that those who wished could work it in return for a tax on the annual produce amounting to a tenth of the grain and a fifth of the fruit; and they established taxes for those who bred hoofed animals, whether cattle or sheep. Their purpose in doing this was to increase the population of Italian stock, which they perceived as particularly sturdy, and so to ensure staunch allies.

But things turned out quite the opposite, since the wealthy took possession of much of the land that had not been formally allotted, and with time felt confident that it would never be taken away from them. They also acquired neighbouring lots and any adjacent farms of the poor, who sometimes were persuaded to sell and sometimes forced. In this way the rich would cultivate large tracts of land in place of individual farms, and would use slaves as field-hands and herdsmen, to avoid having their free labourers conscripted for military service. At the same time, they made a handsome profit from the ownership of slaves who produced many children, their numbers increasing out of harm's way thanks to their immunity from service in the army. For these reasons the powerful grew ever more wealthy, and the race of slaves increased throughout the land; while the Italians, ground down by poverty, taxation, and military service, suffered a decline in population. And even if they won a respite from these evils, they were still reduced to idleness since the land was held by the rich, and slaves rather than free men were used in the fields.

Pliny, *Natural History* 18.35

A principal concern for our ancestors was to keep the size of a farm within reasonable limits, feeling as they did that it was preferable to sow less soil and plough it better (an opinion that I find shared by Vergil). If the truth be known, *latifundia* have destroyed Italy, and they are now destroying the provinces as well: six proprietors owned half of Africa when the Emperor Nero had them put to death. And we must not deprive Gnaeus Pompey of this sign of his greatness, that he never bought a field that bordered on his own holdings. Mago began his text of agricultural instruction with a preface in which he stated his opinion that a man who has bought a farm must sell his house in town: a severe opinion that does no good for the general situation, but does at least point out Mago's insistence on constant supervision.[1]

3.4 An estate of the younger Pliny

Pliny the Younger, nephew and adopted son of Pliny the encyclopaedist, was a prominent lawyer and public servant in Rome of the late first and early second centuries AC. His wealth, like that of most senators, was based on extensive agricultural holdings throughout Italy. Indeed, since the time of the Republic Roman law had encouraged the acquisition of large estates by aristocratic absentee owners, who were barred from investing in commercial enterprises and obliged to own a certain amount of land in Italy itself. If Pliny's comments here are typical of his senatorial colleagues, their customary urban delight in the rustic life was matched by their commitment to good agricultural techniques that would ensure a reasonable profit from their substantial investments. See also *Letters* 5.6.

Pliny, *Letters* 3.19 (to Calvisius Rufus)

I am following my usual custom of asking your advice in a matter of property. The estate that lies next to my land and even infringes on it is up for sale. Many things attract me about it, but I am put off by other considerations no less important.

The appeal of combining the two estates is the first attraction. Then there is the possibility – no less expedient than agreeable – of visiting the two together on a single journey, of running them under a single agent and with practically the same staff, and of maintaining and furnishing a single farmhouse (so long as the other is kept in good order). In my calculations I have included the cost of furnishings and the expense of household servants, gardeners, carpenters, and even of hunting equipment; for it makes a considerable difference whether you collect this in one place or spread it among several.

Conversely, I feel it might be imprudent to subject such a large investment to the same risks of weather and accident; it seems safer to be exposed to the uncertainties of Fortune in a variety of different properties. The enjoyment in changing one's environment and in travelling about from place to place is also to be considered.

At the moment my main concern is this: the fields are fertile, rich, and well-watered; they comprise open meadows, vineyards, and woodlands that yield enough timber for a steady if modest income. Unfortunately, the fertility of the land is being exhausted by inept tenants. Quite often the previous owners sold off the equipment they had pledged as security for rent, thus reducing their arrears temporarily but draining their vitality over the long run; and their debts grew again as their resources withered. They will have to be provided with slaves, not fettered ones (I never use them myself, nor does anyone there) but decent slaves, and that will add to the expense. [Pliny finishes with cost and financing].

THE FARMSTEAD

3.5 Choosing a good farm

Some of Pliny's concerns in the previous passage had already been expressed almost three centuries earlier (though with the absence of elegance and altruism) by Cato the Elder, a model of traditional Roman conservatism. Of particular interest is his last paragraph in this selection, where he gives a preferential list of the various types of farms: grain ranks remarkably low, probably because it was already becoming less profitable as the recently acquired provinces in the western Mediterranean were being exploited as "bread-baskets" for Italy.

Cato, *On Agriculture* 1

When you are thinking of purchasing an estate, remember three things: do not buy one just because you want it, spare no effort in examining it, and do not be satisfied with touring it just once, for a good piece of property will please you more every time you go to see it. Pay attention to how well the neighbours keep their farms: in a good district they are bound to be well kept. When you enter the farmstead, cast your eyes around, so you can find your own way out.

The place should have a good climate, one that is not liable to be destructive; the soil should be good and naturally vital. If you can find such a one, the estate should lie at the foot of a hill, facing south; and the locality should be healthful.

There should be a supply of day-labourers and a good source of water. The neighbourhood should include one of the following four things: a thriving town, the sea, a navigable river, or a good and busy road. The place should be among farms that do not often change owners; and those who have sold their estates there should be sorry to have done so. It should be well equipped with buildings; be careful not to be too quick to dismiss the technique of others, since you will do better to buy from an owner who is a good farmer *and* a good builder.

When you come to the farmhouse, ensure that there is a good supply of vats for the pressing-room and lots of storage jars; if there are not many, you will know that the yield is in direct proportion. The farm need not be elaborately equipped: it is the good location that matters. Make sure that it is fitted out as economically as possible, and that the land is not overpriced. Bear in mind that a farm is like a man: however productive it is, there will not be much left if it is also extravagant.

If you ask me what kind of estate I favour, my answer is: 100 *iugera* of land in the best location. And of all the types of fields, a vineyard comes first, provided it produces good wine in quantity; second, a well-watered garden; third, a willow grove; fourth, an olive orchard; fifth, meadow land; sixth, a grain field; seventh, a stand of timber; eighth, an orchard; and ninth, an acorn grove.

3.6 The buildings on a large estate

Cato's characteristic brevity leaves unanswered many important questions about the most advantageous location and arrangement of an estate, for which we depend on Varro in the first century BC (*On Agriculture* 1.11–16) and on the following passage from Columella, a century later. As Cato observed in **3.5**, the structures on a farm are of prime importance; here Columella outlines the principles to be observed in planning the farmhouse and barns, the storage sheds, and outbuildings (Rossiter 1978).

Columella, *On Agriculture* 1.6

The configuration of the farmstead and the number of its rooms should conform to the size of the whole enclosure and should consist of three parts: the owner's villa, the farmhouse, and the rooms devoted to the produce....

The farmhouse and stables

In the farmhouse will be located a kitchen with a high ceiling designed to protect the beams from the danger of fire, and roomy enough to give the hands a convenient place to while away the time regardless of the season. The rooms for unchained slaves will best be built with a southern exposure; for the chained slaves, an underground cell, as healthful as possible and illuminated by several windows – as long as they are narrow ones and far enough off the ground to be out of arm's reach.

The stables for the livestock will be such as not to be affected by either cold or heat. For the draught animals there should be a double set of stalls for winter and summer; for the rest of the livestock that belongs within the farmstead there should be paddocks surrounded by high walls to prevent attacks by wild beasts, and with a roofed part where the animals can rest in winter and an open part for the summer.... Ox stalls will have to be ten feet wide, nine at the least, a size that affords room for the animal to recline and for the oxherd to go about his tasks. The appropriate height for the cribs is one that allows the ox or other beast of burden to feed standing up without inconvenience.

Lodgings for the overseer should be placed next to the door and those for the steward above it, so they can both scrutinise anyone entering and leaving, and so the steward can keep a close eye on the overseer. Near these quarters should be the storeroom for all the farm's apparatus, and inside this a locked room for the iron tools. The oxherds and shepherds should have chambers located near their animals to make it easy for them to get out to care for the beasts. Notwithstanding the above, all the hands ought to be housed as near one another as possible to avoid straining the steward's diligence as he makes his rounds of the various locales; and at the same time this allows the hands to see for themselves the activity and inactivity of each of their comrades.

The Rooms Used for Produce

The area for handling the produce is divided into rooms for olive oil, for presses, for wine, and for making *defrutum*, into lofts for storing hay and chaff, and into storerooms and granaries. They are arranged so that the ground-floor rooms hold liquid products like wine or olive oil destined for the market, while dry products – grain, hay, leaves, chaff, and other fodders – are stored in the wooden lofts. As already noted, the granaries should have ladders for access and a few slits for ventilation on the north side (this direction is the coldest and least humid, two aspects that help extend the life of stored grain). It is for the same reason that the wine room is placed on the ground floor, but it must be at a good distance from the baths, oven, manure pit, and the other filthy areas that give off a rank smell; it should not be close even to water tanks and springs, which release a dampness that spoils wine....

The storerooms for olive oil, and particularly the press-rooms, should be warm, since any liquid is more fluid in the heat and congeals in severe cold; and when olive oil freezes (not a common occurrence) it turns rancid. Still, what we want is not fires or open flames whose smoke and soot spoil the flavour of oil, but rather the natural heat that is a consequence of location and orientation. So, the pressing-room should have a southern exposure that eliminates the need to use fires and torches during the pressing....

Outbuildings

Next, the following essentials will be ranged around the farmstead. First, an oven and a mill, of a size required by the projected number of hands; and at least two ponds, one for the use of geese and livestock, the other for soaking lupins, withes, and branches of elm, the other items suited to our needs.

Then two manure pits, one to take the fresh dung and keep it for a year, the other from which the old manure is removed. Both should have sides sloped at a gentle incline, as in fishponds, an embankment, and a floor of rammed earth to prevent leakage. For it is especially important to retain the strength of the compost by avoiding evaporation, and to soak it constantly in liquid; this causes any seeds of thorns and grasses mixed in with the straw or chaff to decompose, so they will not be taken out to the fields and re-establish weeds among the standing crop. Hence experienced farmers cover with branches all the waste they have removed from the sheepfolds and stables, to keep the beating sun from drying out or burning it.

If suitable, the threshing floor should be laid where it can be surveyed by the owner or at least by his steward. It is best paved with hard stone since the grain is threshed out quickly when the ground does not give under the blows of hoofs and threshing sledges, and the winnowed grain is cleaner and free from the pebbles and lumps of earth that are usually produced from an earthen floor.

A rain shelter should be attached to the threshing floor, where half-threshed grain can be brought for protection from a sudden downpour. This is particularly necessary in Italy with its changeable weather, though it would serve no useful purpose in certain areas overseas where it does not rain during the summer.

Finally, the orchards and gardens should be surrounded by fencing and be located in a spot nearby where all the faecal muck and the watery fluid pressed from olives can flow; for vegetables and trees alike thrive on this sort of nutrient.

3.7 Equipment for an olive orchard and a vineyard

The two lists that follow give some indication of the complexity of agricultural techniques even in Cato's day, and of the remarkable specialisation of equipment (though the specific function of some is now a mystery) (White 1975).

Cato, *On Agriculture* 10–11

For an olive orchard of 240 *iugera*:

Workers, a total of 13: 1 overseer, 1 housekeeper, 5 labourers, 3 ploughmen, 1 donkey-driver, 1 swineherd, 1 shepherd

Livestock: 3 pairs of oxen, 3 donkeys fitted with packsaddles to carry manure, 1 donkey for the mill, 100 sheep

Pressing equipment: 5 fully equipped sets of oil presses, 1 bronze vessel (capacity 30 *quadrantalia*) with 1 bronze lid, 3 iron hooks, 3 water pitchers, 2 funnels, 1 bronze vessel (capacity 5 *quadrantalia*) with 1 bronze lid, 3 hooks, 1 small basin, 2 amphorae for oil, 1 pitcher (capacity 50 —), 3 ladles, 1 water bucket, 1 shallow basin, 1 small pot, 1 slop pail, 1 small platter, 1 chamber[?]-pot, 1 watering pot, 1 ladle, 1 candelabrum, 1 *sextarius*-measure

Equipment for draught animals: 3 four-wheeled carts, 6 ploughs with shares, 3 yokes fitted with leather harnesses, 6 sets of trappings for oxen, 1 harrow, 4 hurdles for carrying manure, 3 manure hampers, 3 *semuniciae*[?], 3 saddle cloths for donkeys

Iron tools: 8 iron forks, 8 hoes, 4 spades, 5 shovels, 2 four-pronged drag hoes, 8 mowing scythes, 5 reaping sickles, 5 pruning hooks, 3 axes, 3 wedges, 1 mortar for grain, 2 tongs, 1 oven rake, 2 braziers

Containers: 100 storage jars for oil, 12 receiving vats, 10 jars for storing grape pulp, 10 jars for holding the lees, 10 wine jars, 20 grain jars, 1 vat for steeping lupins, 10 medium-sized storage jars, 1 washtub, 1 bathtub, 2 water basins, separate covers for jars (both large and medium)

Milling equipment and furniture: 1 donkey mill, 1 pushing mill, 1 Spanish mill, 3 mill rests[?], 1 table with a slab top, 2 bronze pans, 2 tables, 3 long benches, 1 bedroom stool, 3 low stools, 4 chairs, 2 arm-chairs, 1 bed in the bedroom, 4 beds with cord mattresses, 3 beds, 1 wooden mortar, 1 fuller's mortar, 1 loom for making cloth, 2 mortars, 4 pestles (1 each for beans, emmer, seeds, and cracking nuts), 1 *modius* measure, 1 *semimodius* measure, 8 stuffed mattresses, 8 coverlets, 16 cushions, 10 blankets, 3 table-napkins, 6 patchwork cloaks for young male slaves.

For a vineyard of 100 *iugera*:

Workers, a total of 16: 1 overseer, 1 housekeeper, 10 labourers, 1 ploughman, 1 donkey-driver, 1 hand for the willow grove, 1 swineherd

Livestock: 2 oxen, 2 draught donkeys, 1 donkey for the mill

Pressing equipment and containers: 3 fully equipped presses, jars to hold five vintages (totalling 800 *cullei*), 20 jars for storing grape pulp, separate covers and lids for the jars, 6 grass-covered pitchers, 4 grass-covered amphorae, 2 funnels, 3 wicker strainers, 3 strainers for removing the "flower" or fungus from wine, 10 pitchers for the must.

Equipment for draught animals: 2 carts, 2 ploughs, 1 yoke for the cart, 1 yoke for carrying wine jars(?), 1 donkey yoke....

The rest of Cato's vineyard list – vessels and furniture, iron tools, and miscellaneous items – is similar to the equipment for an olive orchard.

3.8 An inventory of agricultural tools

It is revealing to compare the preceding lists of Cato with the more compendious but descriptive inventory that follows, written more than 500 years later in the twilight of the Roman Empire.

Palladius, *A Study of Agriculture* 1.42

We should obtain the following tools indispensable on a farm: ploughs of the basic design or, if the area is level enough, ploughs with mould-boards whose more elevated furrow-ridges can raise the crops above the winter water table; two-pronged hoes; hatchets; pruning hooks for use on trees and vines; sickles and scythes; mattocks; "wolves" or hafted pruning saws, short ones and the larger size up to three feet in length, handy in tight spots for cutting back trees or vines, jobs that cannot be done with a regular saw; digging sticks for inserting shoots in ground that has been turned over with a dibble; vine-dresser's knives that are crescent-shaped and sharpened on the back edge; small, curved pruning knives for easier pruning of dry or projecting twigs from young trees; small, toothed sickles with short handles commonly used for cutting out ferns; small pruning saws; spades with foot-bars; grubbing hoes for going after bramble bushes; axes with a single or double head; hoes with one or two prongs, or with an adze on the opposite end; drag hoes; branding irons, and iron tools for castration, for shearing sheep, and for veterinary surgery; and hooded tunics made of skins, and leggings and gloves also of skin, appropriate for wear in woods or brambles for both farming and hunting.

3.9 The best markets for buying farm equipment

Despite the injunction of all agricultural writers (and particularly of Cato himself), it was not possible for farms to be entirely self-sufficient. As a partial shopping list of materials that could not be grown or manufactured on the estate, consider this chapter from Cato.

Cato, *On Agriculture* 135.1–2

Tunics, togas, blankets, quilts, wooden shoes: *Rome*. Hoods, iron tools, sickles, long-handled spades, mattocks, axes, harnesses, bridle bits and small chains: *Cales* and *Minturnae*. Spades: ... *Venafrum*. Carts and threshing sledges: *Suessa* and in *Lucania*. Jars and vats: *Alba* and *Rome*. Rooftiles: *Venafrum*. *Roman*-style ploughs good for heavy soil. *Campanian* ones for volcanic soil. *Roman* yokes will be the best; so, too, the ploughshares that can be slipped into place. Olive mills: *Pompeii* and *Nola* (at Rufrius' lot). Keys and door bolts: *Rome*. Water buckets, urns for olive oil, water pitchers, wine urns, and other bronze vessels: *Capua* and *Nola*. Campanian baskets are useful: *Capua*. Hoisting ropes and anything made of broom: *Capua*. Roman baskets: *Suessa* and *Casinum*.

3.10 The three classes of farm equipment

Inanimate objects were not the only required "implements" on an estate, as this passage baldly reveals: human labourers, whether free or slave, are here considered by Varro among the categories of farm equipment. While Cato's attitude to his slaves might seem to modern readers more than just callous (see **3.11–12**), there are in this and other selections (e.g. **3.3**) some sparks of the humanitarian feeling we associate more with Romans of the Empire. Of interest in this passage, too, is Varro's criticism of Cato's estimates of manpower and equipment already encountered in **3.7**: over precise, perhaps, but one of the rare opportunities we have to appreciate an analysis of one surviving author by another.

AGRICULTURE

Varro, *On Agriculture* 1.17–22

Now I shall speak of the actual cultivation of the land. Some divide this topic into two parts: men, and the aids men use without which they could not farm. Others use three classes: implements that are articulate (the slaves), those that are semi-articulate (the cattle), and the inarticulate (the carts).

The Articulate: Farmhands, Particularly Slaves

All fields are worked by men, whether they be slaves, free men, or both. Free men include those who themselves are farmers, like the majority of the poor together with their offspring; farmhands working for pay, when the more demanding tasks like the vintage or harvest are conducted with the assistance of hired free labour; and those whom our people call "debt-workers" and are still a common sight in Asia, Egypt, and Illyricum. My advice about this group as a whole is this: it is better to use hired labour than slaves to cultivate unpleasant lands and, even in healthy areas, for the more demanding tasks on the farm such as the storing of the vintage or harvest. Cassius describes as follows the kind of men these hands should be: you should obtain labourers who can endure hard work, who are not less than 22 years old, and who are ready to learn agricultural techniques....

Slaves must be neither timid nor brash.... Do not obtain a large number from the same ethnic origin, for this more than anything usually leads to private hatreds. Make your foremen more eager by rewarding them, and see to it that they have some property of their own and fellow-slaves as mistresses[2] to bear them sons; this will make them more stable and more attached to the farm.... And as for your workers who surpass their colleagues, ... their devotion to work increases when they are treated generously in a variety of situations: food, for example, or more clothing, or a holiday from work, or receiving permission to pasture a small herd of their own on the farm. The result is that, when they are given some disagreeable orders or are punished in some way, the consolation afforded by these benefits will restore their feelings of devotion and goodwill towards their master.

Cato has in mind two criteria for the size of the workforce: the precise extent of the farm, and the type of crop. So, when he writes about olive orchards and vineyards, he gives two formulas. In the first he prescribes how an olive orchard of 240 *iugera* should be fitted out, listing the following 13 slaves as necessary for an estate of this size: an overseer, a housekeeper, five labourers, three ploughmen, a donkey-driver, a swineherd, and a shepherd. The second formula he gives for a vineyard of 100 *iugera*, listing 15 required slaves: an overseer, a housekeeper, ten labourers, a ploughman, a donkey-driver, and a swineherd. Yet Saserna[3] notes that one man is enough for every eight *iugera*, and that he ought to be able to work over that much land in 45 days. (Though a single *iugerum* would take only four days' work, Saserna says that he allowed 13 spare days to account for sickness, bad weather, inactivity, and carelessness.)

Neither of these writers has left us a standard of measurement well enough defined. If Cato wanted to do this, he should have put it in such a way that we would add or subtract from his number in proportion to the size of the farm, be it larger or smaller. What is more, he should have listed the overseer and housekeeper as separate from the body of slaves; for if you were to cultivate less than 240 *iugera* of olive trees you would still not be able to have fewer than one overseer, just as if you had an estate twice as large or even more you would hardly need two or three of them. For the most part it is only the number of labourers and ploughmen that should be increased in proportion to the size of larger farms, and then only if all the land is alike. But if the terrain varies so much that it cannot all be ploughed – as is the case with uneven and steeply sloping ground – then there is less need for a large number of oxen and ploughmen. I ignore here the fact that the module of 240 *iugera* put forward by Cato is neither a common nor an accepted unit (the standard unit being a *centuria* of 200 *iugera*); but if we deduct from his figure of 240 a sixth (that is, 40 *iugera*), I fail to see how I should deduct from his model a sixth of the 13 slaves as well.... As for what he says about the need for 15 slaves on a 100-*iugera* vineyard: if someone has a *centuria* that is half vineyard and half olive orchard, then it follows that he would have two overseers and two housekeepers, which is ridiculous.

So, a different calculation must be used to determine the number of slaves in each trade, and here Saserna is more to be recommended. He says that to work one *iugerum* occupies a single labourer for four days. On the other hand, while this may be adequate on Saserna's farm in Gaul, it does not follow that the same is true for a farm in the mountains of Liguria. So in the end you will arrive at the most appropriate figure for the size of the workforce and the quantity of other equipment you should provide if you consider carefully three criteria: the type and size of farms in the vicinity, the number of labourers that work each one, and how the addition or subtraction of a certain number of workers would improve or impair your own cultivation.

The Semi-Articulate: Animals

As for the next class of equipment, the one I have called "semi-articulate", Saserna writes that two yokes of oxen are sufficient for 200 *iugera* of ploughland, and Cato suggests three pairs for an olive orchard of 240 *iugera*. So, it works out that, if Saserna is accurate, one yoke is needed for every 100 *iugera*; or, if Cato is right, one for every 80. To my mind neither of these ratios is appropriate for all farms, and each will suit some. Some land is easier or harder; some land can be broken up only with great exertion from the oxen, often breaking the plough beam in the process and leaving the share in the earth. There are, then, three examples that we should follow for each farm individually until we have gained experience: the practice of the former owner, the practice of the neighbours, and a certain amount of trial and error....

The Inarticulate Equipment

With regard to the last group, the "inarticulate" equipment, in which are included baskets, jars, and so on, this is what I suggest. Whatever can be grown on the farm or made by the staff should never be purchased. This generally includes articles made from withies and wood from the countryside, like panniers, baskets, threshing sledges, winnowing fans, and rakes; so, too, items such as ropes, cords, and mats when made of hemp, flax, reeds, palm leaves, and rushes. The cost of buying articles unobtainable on the farm will not eat away at your profit if you purchase them with an eye to their usefulness rather than for display. This is even more true if you do your shopping primarily in a place that is close at hand and where well-made items can be picked up for a very low price.

The different varieties and quantities of this equipment are determined by the size of the farm: the larger the farm, the more equipment is required. Consequently Cato, when postulating a farm of a certain size and type, writes that an owner who cultivates 240 *iugera* of olive orchard should fit it out with five sets of oil presses, which he inventories item by item.... In the same way he gives a second prescription for equipping a vineyard: if it is 100 *iugera* in size, he writes, it should have three fully equipped presses, enough jars with covers to hold 800 *cullei*, 20 more for holding grapes, 20 for grain, and other articles of this sort. In fact, other writers suggest lower numbers, but I suspect Cato set the number of *cullei* as high as he did so the wine would not have to be sold off every year (for aged wines are worth more than new ones, and the same wine can bring a higher price at one time than at another).

3.11 Upkeep of farmhands

The usual lot of agricultural workers in the Republic was an unenviable one, a life of drudgery and dangers. But as Varro pointed out above, the best labourers were those with skills, and they represented a substantial investment that had to be kept in good working condition. Consider the pragmatic Cato's prescription for rationing.

Cato, *On Agriculture* 56–59

Rations of food for the farmhands:

> Workers: 4 *modii* of wheat in the winter, 4.5 *modii* in summer.
> Overseer, housekeeper, foreman, and herdsman: 3 *modii*.
> Fettered slaves: 4 *librae* of bread in the winter, 5 *librae* from when they begin to trench the vines until they start consuming figs, then back to 4 *librae*.

AGRICULTURE

Wine for the farmhands:

> For the first three months after the vintage, let them drink the wine made from grape skins; in the fourth month, a *hemina* of wine a day (that is, 2.5 *congii* for the month); from the fifth to eighth month, a *sextarius* a day (that is, 5 *congii* a month); from the ninth to twelfth month, 3 *heminae* a day (that is, an *amphora* a month).
> For the festivals of the Saturnalia and Compitalia, increase the ration by 3.5 *congii* a person. This gives a yearly total of 7 *quadrantalia* of wine per person.
> For the fettered slaves, increase the ration in proportion to the amount of work they do; for them, 10 *quadrantalia* of wine apiece is not an excessive annual intake.

Condiments for the farmhands:

> First, store up as many windfall olives as you can, and later the ripe olives that will yield only a small amount of oil; be frugal with them so they last as long as possible.
> When the olives have been consumed, give the hands fish brine and vinegar.
> Allot 1 *sextarius* of olive oil per person each month.
> One *modius* of salt per person each year is sufficient.

Clothing for the farmhands:

> A tunic weighing 3.5 *librae* and a woollen blanket every second year. When you issue the tunic or blanket, retrieve the old one first: it can be used for quilts.
> You should issue a decent pair of wooden shoes every second year.

3.12 Farm management for the absentee owner

While the day-to-day running of these estates was usually under the supervision of a steward, all our authorities agree on the importance of regular visits from the landowner. And lest we think of these owners as urban dilettantes ignorant of agricultural technology, the following passage reveals how thoroughly familiar with their farm they must be, and how comprehensive their knowledge of its techniques and productivity.

Cato, *On Agriculture* 2

When the proprietor has arrived at the farmhouse and has paid his respects to the protective deity of the household, he should tour the farm that same day if he can; if not then, at least on the day following. Once he has ascertained how the farm has been tended, what projects have been carried out and what still remain, he should summon his overseer the next day and ask him what work has been done and what is left to be done; whether the tasks have been completed on time; whether the remaining work can be finished; and what is the yield of wine and grain and all the other products. From this information he should then make up an account of the number of workmen and the number of working days. If the results of the labour are not evident, and the overseer claims that *he* has been diligent but the slaves have not been healthy, the weather has been poor, slaves have escaped, there were public works to take care of – when he has made these and 1,000 other excuses, refer him to your calculation of work done and number of hands. If the weather has been wet, point out the sort of tasks that could have been carried out while it was raining: washing the storage jars and coating them with pitch, cleaning the farmhouse, transporting the grain, carrying the manure outside and starting a dung-heap, cleaning the seed, repairing ropes and twining new ones; and suggest that the members of the household should have been mending their own patchwork cloaks and hoods. And, for festival days: cleaning out old ditches, paving the highway, cutting back thorn bushes, digging up the garden, clearing out the meadow, bundling up twigs, weeding out brambles, grinding emmer, and generally tidying things up. When the slaves were ill, shorter rations should have been issued.

Once these matters have been patiently identified, you should see to it that the leftover tasks are carried out. Settle the various accounts – the financial statement, the grain account, the fodder purchases, the wine and oil accounts – what has been sold, what collected, what is the balance outstanding, what is there that could be sold (when a satisfactory guarantee can be given in this regard, it should be accepted, and the balance due clearly set out).

Anything lacking for the year should be acquired; anything superfluous should be sold off. Necessary contracts should be awarded; orders issued – and put down in writing – about the work he wants done and what he wants farmed out. He should inspect his livestock and hold an auction to sell his olive oil (if he gets the right price), his wine, and his surplus grain. Timeworn oxen, cattle, and sheep with slight blemishes, wool, hides, and obsolete wagon, iron tools not worth saving, a slave past his prime, another slave prone to illness, and any other surplus items: all these, too, should be disposed of at auction. In short: the head of a household should have a fondness for selling, not buying.

3.13 Tenant farmers or slave workers?

Our Roman authors describe at some length the positive and negative aspects of employing slave workers or tenants on an estate (cf. **3.10**). Pliny, *Letters* 9.37 provides evidence for failure to collect rent from tenants and this passage reveal some of the labour difficulties encountered by absentee owners who rely upon careless and dishonest slaves.

Columella, *On Agriculture* 1.7.6–7

On distant estates not easily reached by the owner, it would be more acceptable to have every kind of productive land under free tenants than under the stewardship of slaves. This is especially true of land used for grain, which a tenant can damage less than vineyards or trees, while slaves can cause it serious harm in a variety of ways: they rent out the oxen; they are negligent in feeding them and the other livestock, they take no care in ploughing the soil, they enter into the accounts considerably more seed than they actually sowed, they fail to nurture the proper growth of what they have planted, and, once they have brought the grain to the threshing floor, they reduce the yield day by day through deceptive or careless threshing. They steal the grain themselves and fail to protect it from other thieves, and in making out the accounts they misrepresent the amount stored away. So, it happens that both the manager and the farmhands are at fault, and quite often the farm acquires a bad reputation. So, my advice is that a farm of this sort (if, as I said, the owner cannot be on hand) should be rented out.

3.14 The importance of proper tools and labour

This inspiring anecdote related by Pliny serves as an antidote to the sometimes cynical attitude of his fellow Romans: the key to agricultural success lies in good tools, a content labour force, and hard work.

Pliny, *Natural History* 18.39–43

What, then, will be the most advantageous way to farm? In the words of the oracle, I suppose: "From bad to good".... Then there are all those other oracular responses: "Not Good For Much is the farmer who buys what his farm could give him; Bad is the head of a household who does during the day what he could do at night, except in inclement weather; Worse is he who does on working days what he should do on holidays; and Worst Of All is the one who spends a sunny day working in his house instead of in his field". I must indulge myself by introducing here one example from antiquity, which incidentally illustrates the custom of pleading even agricultural questions before the Popular Assembly, as well as how those men of old used to conduct their defence.

Considerable prejudice was directed against Gaius Furius Chresimus, a former slave, because he got a far greater yield from what was little more than a tiny piece of ground than his neighbours did from much larger farms, as if he were using sorcery to lure away the produce of others. For this he was indicted by the curule aedile, Spurius Albinus. Afraid of being condemned when the time arrived for the tribes to cast their votes, he took into the Forum all his agricultural equipment and brought along his farmhands, a robust group that was ... well cared for and well dressed. There were the splendidly made iron tools, the substantial mattocks, the heavy ploughshares, and the well-fed oxen. And Chresimus pointed out: "These, citizens, are my magic charms. What I am not in a position to show you or bring into court are the tasks I do at night or early in the morning, and the sweat of my brow". The verdict was unanimous: he was acquitted.

AGRICULTURE

Make no mistake about it: agriculture relies on the amount of work expended. That is why our ancestors once said that the most productive thing to spread over the fields was the owner's gaze.

3.15 An agricultural calendar

Much of what is described in subsequent selections of this chapter is neatly summarised in this Roman agricultural calendar, found in the sixteenth century. It is more an almanac than this abridged version would suggest, including the number of days, celestial activity, festivals, and so on; only its seasonal activities are included here, applicable to agriculture in central Italy.

CIL 6.2305 (= ILS 8745)

January:	Stakes are sharpened; willow and reeds are cut.
February:	Wheat fields are hoed; the upper parts of vines are tended; reeds are burnt.
March:	The vines are propped (their ground having been prepared by digging and trenching) and pruned; spring wheat is planted.
April:	The sheep are ceremonially purified.
May:	The wheat fields are weeded; the sheep are shorn and the wool is washed; the young oxen are broken in; vetch is cut for fodder; the wheat fields are ceremonially purified.
June:	Mowing of the hay; the soil around the vines is broken up.
July:	Harvesting of barley and beans.
August:	Stakes are prepared; harvesting of wheat and other grains; the stubble is burnt.
September:	The wine-casks are covered with pitch; fruits are picked and the soil around the trees is dug up.
October:	The vintage.
November:	Sowing of wheat and barley; holes are dug and trees planted.
December:	The vines are manured; sowing of beans; felling of timber; the olives are harvested and sold.

SOIL AND FERTILITY

3.16 Soil types

An analysis of the qualities of a particular soil is indispensable in determining what crops can best be grown in it. Though of no great concern to Cato in the second century BC (it is briefly discussed in his *On Agriculture* 6, 35), the topic received extensive treatment from Varro (*On Agriculture* 1.9), Pliny (*Natural History* 17.25–41), and Columella, whose passage here alludes to the three major characteristics of soil – texture, density, and moisture – observed not by any scientific testing but by a long experience.

105

Columella, *On Agriculture* 2.2.1–7

According to agricultural experts, there are three varieties of terrain: the flat, the hilly, and the mountainous. They especially recommended a flat field lying on a slight incline (rather than one on a perfectly level and horizontal plane); a hilly field that rises gently and gradually; and a mountainous one that is not high and rugged but covered with woods and grasses. What is more, under each of these classes of terrain are collected six qualities of soil types – fat or thin, loose or dense, wet or dry – which, when combined and interchanged one with another, produce a great variety of fields ... I should point out the following generalisation about all of the earth's produce: more things do better on flat than on hilly ground, and in fat rather than thin soil. We have come to no conclusion about the preference for dry or wet soils, since the number of crops that thrive in each is almost infinite; but any one of these does better in loosened than in dense soil....

Thus, the greatest profits come from a field that is fat and crumbly, since it demands the least effort for the highest return, and the effort it does demand involves no great amount of labour and expense; so, we are justified in calling this type of soil the most outstanding. Next to it comes the dense and fat, since its considerable productivity repays the farmer's expense and effort. Third place goes to the wet field, since it can produce a profit at no expense.... Soil that is at the same time dry, dense, and thin is rated as the worst of all: it is hard to work, when it is worked it gives no return, and when left untouched it is not much good as meadow or grazing land.

3.17 Techniques for determining soil types

This selection – admittedly from a poetic work but hardly less scientific than the observations of Pliny (*Natural History* 17.38–39: "It is certainly true that the best soil will be the one that has the flavour of perfumes ...") – gives a good indication of the simple methods of analysis used by the ancients. But this is hardly surprising, if we consider their ignorance of the chemical testing of properties.

Vergil, *Georgics* 2.226–258

Now I shall explain how you should be able to recognise each type. Suppose you ask whether the soil is loose or unusually dense, the denser being well disposed to Ceres the grain goddess, while all the loosest soils favour the vines of Bacchus. First search out a spot and have a pit sunk deep in solid ground, then put all the earth back in it and tramp it down on top. If it runs short, the soil will be light and more suitable for cattle and fruitful vines; but if it proves overmuch for its own cavity and there is earth left over once the pit is filled, the soil is dense, so look out for recalcitrant clods and stiff ridges, and break your land with a powerful team.

Salty earth, on the other hand, and the type described as bitter – a troublesome soil for crops, the kind that ploughing does not temper, where the vine degenerates and the fruit trees lose their character – this soil gives the following indication. Take down from the smoky rafters your densely woven wicker baskets and wine strainers, pack them to the top with that unwholesome soil, and add water drawn fresh from the spring. The water will all force its way out, of course, and pass through the wicker in thick drops. Its flavour is obvious, and will serve as proof: the mouths of those who taste it will pucker at the bitter sensation.

Likewise, what follows is the only test for discovering which soil is fat: it never crumbles when tossed from hand to hand but grows soft in handling and sticks to the fingers like pitch. Damp soil encourages taller grass and by its nature is excessively luxuriant; not for me this overly fruitful earth that proves too strong when the ears begin to form. Its own weight explicitly betrays soil that is heavy; so, too, with the light variety; and it is easy for the eye to divine what soil is black or any other colour. But to detect the destructive cold soil is difficult: only pines and poisonous yews or black ivies sometimes give a clue.

3.18 Drainage

While a scarcity of water at the right time and in the right place, particularly in the eastern Mediterranean, prompted the development of hydraulic technology (see **3.19–21** and **9.7–12**), an excess of moisture from natural runoff and high water-tables demanded a converse solution. Indeed, in Italy drainage ditches were more a part of the agricultural environment than were irrigation channels.

Cato, *On Agriculture* 155

It is necessary to drain water from the land throughout the winter, and to keep the hillside drainage ditches free from dirt. The greatest danger from water occurs at the beginning of autumn, when there is dust about. The moment it starts to rain the household staff must sally forth with shovels and hoes, breach the ditches, channel the water down onto the roads, and make sure that it runs off. While it is raining you should make a circuit of the farmstead and mark with charcoal everywhere there is a leak, with a view to replacing the tile when the downpour is over. During the growing season, should there be water standing anywhere in the grain fields – among the crops or in the seed bed or ditches – it ought to be drained off, and anything obstructing the flow should be opened up and removed.

Directions for digging ditches are given by Pliny (*Natural History* 18.47) and Columella (*On Agriculture* 2.2.9–10).

3.19 Field irrigation in Italy

From the two brief comments of Pliny that follow, together with the scarcity of ancient references to rural irrigation systems,[4] we can conclude that, in Italy at least – and with the exception of meadows, vineyards, and gardens – crops were watered by Nature, not by farmers (which accounts for Pliny's apparent surprise in the second passage here). For the water-lifting devices mentioned, see also **9.26–33**.

Pliny, *Natural History* 19.60

There is no doubt that the gardens should adjoin the farmhouse, and above all that they should be kept irrigated by a passing stream, if there happens to be one. But if not, they should be irrigated from a well by means of a pulley or force pumps or the bailing action of a *shaduf*.

Pliny, *Natural History* 17.250

In the Italian territory of Sulmo, in the Fabian district, they irrigate even the ploughed land.

3.20 Irrigation and drought farming

The contrast between the natural watering of crops by generally reliable rainfall and spring runoff on the one hand, and artificial irrigation systems that manipulated the natural course of events on the other, was nowhere more evident than in the Near East, where topography and annual floods determined which method was used to water fields and, not incidentally, the relative productivity of the land.

Deuteronomy 11:8–11

You shall therefore keep all the commandments, which I command you this day, that you may be strong and go in and possess the land, which you are going over to possess; and that you may live long in the land, which the Lord swore to your fathers to give to them and to their descendants, a land flowing with milk and honey. For the land, which you are going in to take possession of is not like the land of Egypt where you came from, where you sowed your seed and watered it with your foot like a vegetable garden. But the land, which you are going over to possess is a land of hills and valleys, which drinks water by the rain of heaven.

3.21 Natural irrigation in Egypt

All our examples of more extensive agricultural irrigation systems come from the more arid regions of the Near East, in particular the long-established civilisations of Egypt and Mesopotamia, where the manipulation of the annual floods of the Nile and the Tigris-Euphrates rivers created an agricultural base sufficient to support the first urban societies of the Mediterranean world. It is ironic that intensive agriculture was developed first in areas known as much for their general aridity as their fertility and where the rivers, when they did rise above their banks, often spread their waters over the surrounding plain in excessive quantities or at the wrong time of year to benefit the crops. Though we speak of "natural" irrigation along the flood plains of the Tigris-Euphrates and Nile river systems, from the early Bronze Age to the Roman Empire the agricultural prosperity of this Fertile Crescent depended equally on human efforts to control, direct, and assist the annual floods of the great rivers. It was this need for intervention that gave birth to the technology of hydraulics and prompted, as a necessary adjunct to the nilometer (see **9.7**), the invention of writing and record keeping.

Pliny, *Natural History* 5.57–58; 18.167–170

The Nile river begins to rise at the first new moon after the summer solstice, by gradual degrees as the sun passes through Cancer. It reaches its crest when the sun is in Leo and, in Virgo, subsides at the same rate as it rose. As Herodotus relates, it is on the hundredth day, when the sun is in Libra, that it withdraws entirely within its banks.[5] For kings or prefects to sail on the river as it rises has been pronounced contrary to divine law. The amounts of its rise are determined with calibrated marks in water shafts: a rise of 24 feet is just right; if any less, the waters do not irrigate all the fields and there is no time for sowing because the earth is still thirsty; if any more, the floods delay work by receding too slowly and waste the time for sowing since the ground is sodden. The province takes account of both extremes: with a rise of 18 feet it perceives a famine; even with 19.5 feet it goes hungry; 21 feet brings optimism, 22.5 feet confidence, and 24 feet euphoria. The greatest rise to date was one of 27 feet in Claudius' reign; the least was 7.5 feet in the year the Battle of Pharsalus was fought, as if the Nile by some freak occurrence was trying to keep remote from the murder of Pompey the Great.[6] When the flood peaks, the dams are opened and the water let loose; as soon as the waters retreat from each parcel of land, sowing begins....

The seeds are ploughed under once they have been scattered over the mud left by the river as it recedes. This is done at the beginning of November, after the area has been weeded by a small gang (a procedure they call *botanismos*). The rest visit the fields only with a sickle a little before the first of April, and the harvest is brought in during May.... A similar method is used at Seleucia in Babylonia, where the Euphrates and Tigris Rivers overflow their banks; but there the fertility is greater since the extent of the inundation is regulated by hand.

3.22 The productivity of Mesopotamia

Pliny's final observation above alludes to the Mesopotamian system of "perennial" irrigation that required control of the annual flood with dikes and sluice gates, the storage of water until it was needed on the fields, and the mechanical lifting of it from the rivers and reservoirs into irrigation channels (for details, see **9.8**). This complex and labour-intensive system was made necessary because the rivers flooded prematurely for proper irrigation and, if the Tigris and Euphrates crested simultaneously, inundation would be excessive without some human interference; but if the rivers were controlled properly, the productivity of the adjacent land was remarkable.

Herodotus, *Histories* 1.193.1–4

The land of the Assyrians receives little rainfall, enough to fatten the roots of the grain. But the standing crop is watered from the river, which brings it to ripeness and causes the grain to mature. This is done not as in Egypt, where the river by itself overflows its banks into the field; here there is manual irrigation with the use of *shadufs* or swipes. For the whole countryside of Babylonia, like that of Egypt, is partitioned by canals, the largest of which is navigable: it extends southeast from the Euphrates to another river, the Tigris, on whose banks the city of Ninos stood.

Of all the lands we know, this is the best by far for growing grain. It makes no attempt at all to produce other growths like the fig or vine or olive. But it is so productive of grain that it usually yields 200-fold, and as much as 300-fold in the best harvests. The blades of wheat and barley there easily measure four fingers across. As for the height of millet and sesame plants, I shall not say what I know; for I am quite conscious of the deep scepticism with which those who have had no experience of the Babylonian countryside view what I have already said about its grain.

3.23 Manure and compost

To replenish nutrients in the soil that were constantly being diminished by over-cultivation, the ancients relied on natural fertilisers and on fallowing and crop rotation. For the former, their usual sources were animal dung and plant compost. See also Cato, *On Agriculture* 5, 29–30, 36–37, 50; Varro, *On Agriculture* 1.38, 3.7.5; and Columella, *On Agriculture* 2.14–15 – Pliny refers to all of these passages here.

Pliny, *Natural History* 17.50–57; 18.193–194

There are a great many varieties of manure, and it has a long history. As early as Homer we find an old man of royal blood using it to fertilise his fields – and with his own hands.[7] Augeas, a king in Greece, is generally given credit for having thought it up, and Hercules for having made it popular in Italy, though the Italians have immortalised their king Stercus, son of Faunus, for having discovered it.[8]

Varro assigns first place to thrush droppings gathered from aviaries.... Columella ranks the droppings from dovecotes first and from the poultry sheds second; he dismisses out of hand the excrement from water birds. The other authorities universally recommend the by-product from men's meals as one of the best substances for this purpose: some of them single out human fluids steeped with hair from tanneries, others the liquid on its own but with even more water added now than when the wine is first drunk (there is, you see, more toxicity that has to be modified after man adds to the natural poison of the wine). These are matters of dispute.... Next, they recommend pig manure, Columella alone disapproving of it. Some promote the dung of any quadruped that eats clover, others the leavings from dovecotes. Next is the dung of goats, then of sheep, then of cattle, and finally of draught animals.... Certain people actually put the dung of draught animals ahead of that of cattle, of sheep before goats', and of donkeys at the top because they chew very slowly. Experience, however, pronounces against these rankings. Still, everyone does agree that nothing works better than a crop of lupine dug in with a plough or mattocks before it forms pods, or handfuls of cut lupine used as mulch around the roots of trees and vines. And where there are no animals, the principle is to use the straw itself or even fern as manure.

To quote Cato: "Substances from which you can make manure: litter from the stables, lupine, chaff, beanstalks, and the leaves of oak and holm oak. From the standing grain, root out dwarf elder and hemlock, and the high grass and sedge from around your willow groves; the latter you can use as litter for your sheep, though for oxen use rotting leaves". Also: "If your vine is spindly, burn up its stalks and plough the results under". Also: "Pasture your sheep on land where you are planning to sow grain".

He goes on to say that some crops by themselves add nutrients to the earth: "Grain, lupines, beans, and vetch all fertilise wheat fields". Likewise, he notes the opposite: "Chickpea – because it is removed root and all, and because it is salty – and barley, fenugreek, bitter vetch, all these and anything else pulled out with its roots cause grain fields to wither. Do not plant a grain field with anything that has a pit or stone". Vergil believes that grain is withered also by flax, oats, and poppies.[9]

Our authorities suggest that manure pits be located in the open air and in a hollow so as to collect the moisture, and be covered with straw to prevent the sun from drying them out. A stake of oak is to be driven into the pit to keep snakes from breeding there.

It is particularly efficacious to spread the manure on the ground while the wind is blowing from the west and in the dark of the moon. Most people erroneously understand this to mean that it should be done when the west wind *begins* to blow and only in February, despite the fact that most crops need fertilising in other months as well. Whatever time is decided on for the operation, care must be taken that it is carried out when the wind blows from due west, the moon is waning, and there is no threat of rain. It is quite remarkable how a precaution like this increases manure's effectiveness as a fertiliser....

If planning to sow something in the autumn, a farmer should heap manure on the land in September, in any case after it has rained. But if a spring sowing is planned, he should spread the manure during the winter. The proper coverage is 18 cart-loads per *iugerum*, and it must be spread before you plough.... A field that is not fertilised becomes cold, while one that is overly fertilised is scorched; so better to do it frequently than overdo it. It is logical that, the warmer the soil is, the less manure is to be given it.

3.24 Human fertiliser

Marius' devastating victory over the Teutones and Ambrones at Aquae Sextiae, near Massilia, in 102 BC resulted in so many deaths that the fields were imbued with long-lasting fertility. Elsewhere, as at Flanders, human remains are often credited as the reason for the fertility of recurrent battlefields. Ground human bone, in particular, has a long history of use as fertiliser, and modern society is now examining this type of option to replace burial or cremation for environmental reasons.

Plutarch, *Marius* 21.3

Other writers do not agree about the division of the spoils, nor about of the number of the dead [100,000 to 200,000 enemy dead]. But nevertheless, they say that the people of Massalia enclosed their vineyards with the bones of the fallen, and that the soil, once the bodies had dissipated in it and the rains had fallen throughout the winter upon it, grew so rich and, having become so full with the thickness of the putrefied matter pressing into it, that the soil brought forth a surpassingly large harvest for many years, and that this provides the proof for the saying of Archilochus that the fields are fattened by such a course of events.

3.25 Tasks on the farm: composting to repairing pots

Composting was an endless and valuable task on the farm, but many other duties were regularly performed when there was time. Not all equipment could be manufactured on the farm (**3.9**) but as much as possible of the equipment would be repaired. Cato provides a sample of the tasks for a rainy day, providing two ways to repair damaged pots.

Cato, *On Agriculture* 39.1–2

When the weather is poor and work cannot be done, carry out the dung into the compost pit. Properly clean out the ox stalls, the sheep pens, the poultry-yard and the farmhouse. Mend the *dolia* [large, wide-mouthed pots] with lead, or cinch them with thoroughly dried oak bands. If you have mended or cinched them well, inserted plaster-cement [into the cracks] and smeared pitch abundantly on the rims, you can make any *dolium* can become a wine pot. Make the plaster-cement for the *dolium* in this way: use one pound of wax, one pound of resin, and two-thirds of a pound of sulphur. Put all these ingredients into a new pot then add to it crushed gypsum enough so that it becomes the density of plaster. Then mend the *dolium* with it. When you have mended it, to make it all the same colour, mix together two parts of raw Cretan earth [chalk or white clay] and a third part of lime. Fashion small bricks, fire in the oven, crush and apply it.

3.26 Fallowing and crop rotation

With the small amount of arable land available in the ancient Mediterranean, over-cultivation was a constant danger for any permanent urban area dependent on an extensive agricultural base, since the semi-nomadic tradition of packing up the tents and moving to new land when the old was exhausted was no longer possible. Hence the development of fallowing – leaving a field to "rest" for a season, perhaps as pasture to be naturally fertilised by the grazing herds – and of crop rotation – the alternation of a cereal crop with a leguminous one to replace lost nitrogen in the soil – that circumvented the unproductive season of a field lying fallow.

Pindar, *Nemean Odes* 6.10–12

... the fruitful wheat fields, which by turns supplied men with ample sustenance from the plains at one time, at another gained strength by taking a rest....

Vergil, *Georgics* 1.71–83

At the same time, you will allow your harvested fields to lie fallow by turns and the idle plain to grow hard by inactivity; or you will sow yellow grain at another season in the same field that earlier provided you with pulse blessed with nodding pod, or with the produce of delicate vetch and the fragile stalks and rustling foliage of the bitter lupine. For a crop of flax burns the earth, oats burn it, and so do poppies, imbued with the sleep of forgetfulness. But the work becomes light with the changing of crops, as long as you are not loath to drench the parched soil with rich manure and to scatter sooty ashes over your exhausted fields. In this way, too, a change of crops relieves the land under cultivation, while the untilled earth shows its gratitude.

See, too, Pliny, *Natural History* 18.187, 191 ("Some people allow the sowing of wheat only in a field that has lain fallow the preceding year....").

3.27 Pesticides

Without the advantages (and attendant dangers) of artificial chemical pesticides, farmers in antiquity depended instead on natural substances that included the dregs of oil pressing (*amurca*) and that were often combined with a large dose of gullible superstition. For other craft uses of *amurca*, see **10.67**.

Cato, *On Agriculture* 92 and 95

To protect your grain from damage by weevils and mice, make a mudpack by adding a little chaff to *amurca*; let it marinate nicely and knead well; smear the whole granary with a thick coating of the mud, then sprinkle *amurca* everywhere you have put the mud. When this has dried, store your grain there, keeping it cool, and the weevil will do no damage....

To keep the caterpillar off your vines: reserve some *amurca*, purify it well, and pour two *congii* into a bronze cauldron. Then simmer it over a low flame, stirring frequently with a stick until it has the consistency of honey. Then take a third *sextarius* of bitumen and a quarter of sulphur, pulverise them separately in a mortar, and crumble them a pinch at a time into the hot *amurca* while stirring. Bring to the boil again uncovered (if you boil it covered it will flare up when the bitumen and sulphur are added). When it reaches the consistency of birdlime, let it cool down. Smear this on your vine around the top and under the branches, and the caterpillar will stay away.

Columella, *On Agriculture* 2.9.9

Some farmers use the skin of a hyena to clothe a three-*modii* sowing basket in which the seeds are stored for a short time before being broadcast; they are convinced that anything sown in this way is bound to flourish.

AGRICULTURE

CROPS AND CULTIVATION

3.28 The labour required for cultivation

Pliny (*Natural History* 18.48) states: "two principal varieties of crops exist: the grains, such as wheat and barley; and the legumes, like the bean and chickpea. The distinction between them is too well known to require an explanation". Columella adds fodder crops (*On Agriculture* 2.6–7) before his more extensive discussion of each of these categories (*On Agriculture* 2.8–9 [cereals], 2.10.1–4 [legumes], and 2.10.24–25 [fodder crops]). The following chart, reconstructed from Columella, gives a fair estimate of the number of workdays required for each stage of cultivating a *iugerum* of each major crop. The figures may be compared to the more general calculations given in **3.7–10**.

Columella, *On Agriculture* 2.12.1–7

A calculation of the number of days' work from planting the crop to bringing it in for threshing:

Tasks

Crop	Quantity per *iugerum*	Workdays required for each task*						Total
		I	II	III	IV	V	VI	
wheat	4–5 *modii*	4	1	2	1	1	1.5	10.5
soft wheat	5 *modii*	4	1	2	1	1	1.5	10.5
emmer	9–10 *modii*	4	1	2	1	1	1.5	10.5
barley	5 *modii*	3	1	1.5			1	6.5
beans	4–6 *modii*	1–2**	1.5	1.5	1	1***	1	7–8
vetch	6–7 *modii*	1–2**	1				1	[3-]4
bitter vetch	5 *modii*	2	1	1		1	1	6
fenugreek	6–7 *modii*	2					1	[3]
cowpeas	4 *modii*	2	1				1	[4]
chickling vetch	4 *modii*	3	1			1	1	6
lentil	1.5 *modii*	3	1	2		1	1	8
lupine	10 *modii*	1	1				1	[3]
millets	4 *sextarii*	4	3	3			varies	
chickpea	3 *modii*	4	2	1		1	3	11
flax	8–10 *modii*	4	3	1			3	11
sesame	6 *sextarii*	3	4	4	2		2	15
hemp					varies			
clover		X****	2	1			1	[4+]

Notes
* I ploughing and planting, II harrowing, III first hoeing, IV second hoeing, V weeding, and VI harvesting.
** the lower figure for annually tilled ground, the higher for old fallow ground.
*** third hoeing.
**** raked in; no workdays listed.

From this summary of man-days it is reckoned that a farm of 200 *iugera* can be worked by two yoke of oxen, two ploughmen, and six labourers, so long as it is un-treed; if set with trees, the same area can be cultivated properly enough with the addition of three men by Saserna's account.

3.29 Construction of a plough

Neither of these passages is easy to understand, even for those familiar with the ancient languages. And both have caused a considerable amount of scholarly debate.[10]

Hesiod, *Works and Days* 427–436

Cut a quantity of bent timbers, and carry home a plough beam whenever you find one; look for it on the mountain or in the field, one of holm oak since it is the strongest for oxen to plough with whenever Athena's servant has fixed it to the share beam and fastened it with pegs close to the yoke beam. Prepare two ploughs, one all of a piece and the other compound, and work on them at home; this is far preferable, since you can fit the second to your oxen should you break the first. Yoke beams of laurel or elm are the soundest, a share beam of oak, a plough beam of holm oak.

Vergil, *Georgics* 1.160–175

We must speak as well of the tools available to the hardy peasants, without which the crops could not be sown and would not grow. First the ploughshare and the heavy oak of the curved plough, and the slowly rolling carts of the Eleusinian Mother,[11] and the threshing sledges and drag hoes and rakes of excessive weight.... Remember to make these ready long in advance and store them away, if the glory of the divine countryside awaits you as your reward.

Even before it is removed from the forest an elm is bent by great force and shaped for the plough beam, taking on the form of a curved plough. To this are fitted, first a pole extending eight feet from the stock, then a pair of "ears", and a bifurcated share beam. A lightweight linden tree has already been cut down for the yoke, and a lofty beech as the plough handle to turn the bottom of the device from behind. The hard wood is hung above the hearth to be seasoned by smoke.

3.30 Ploughshares

For centuries after its invention in the Bronze Age, the ploughshare usually consisted of little more than a hoe affixed to a pole. But with the Roman conquest and subsequent economic development of the western Mediterranean, the great variety of soil types (and especially the heavy soils of central Europe) demanded modifications to the traditional form. The most sophisticated design, the Gallic wheeled plough, seems not to have been adopted elsewhere prior to the medieval period.

AGRICULTURE

Pliny, *Natural History* 18.171–173

There are several kinds of ploughshares.

The *culter* or knife is fitted in front of the share beam to cut the earth before it is broken up and to mark the tracks for subsequent furrows with its grooves, which the horizontal share is to bite into while ploughing.

The second kind is the common pole with a beaked end.

The third share, designed for light soil, does not extend the length of the share beam but has a small spike at the front end.

In the fourth type, this spike is broader and sharper, wedge-shaped with a pointed end; it uses this single blade to cleave the soil and to cut off the roots of weeds with its sharp edges.

A recent invention in Raetian Gaul is the kind they call the *plaumoratum*, a share of the fourth type to which are added a pair of small wheels; the spike in this case has the shape of a spade. They use it for sowing only in land that has been tilled or is almost fallow. The width of the share turns back the sod, seed is scattered immediately, and toothed harrows are dragged over the top. Fields sown in this way require no hoeing, but the ploughing takes two or three teams (40 *iugera* of easy soil, 30 of difficult is a reasonable estimate for one team of oxen to work in a year).

3.31 Ploughing

Columella, following Hesiod (*Works and Days* 436–447), indicates that proper ploughing depended as much on the qualities of the ploughman and his team of oxen as on the design of plough and share.

Columella, *On Agriculture* 1.9.2–3; 2.2.22–28

For the ploughman a natural intelligence, though necessary, is not enough unless the dimensions of his voice and body make him an object of fear to his animals. Still, he should moderate his strength with forbearance, since he ought to be a source more of fear than of cruelty; in this way the oxen will obey his commands and last longer, since they are not exhausted by the simultaneous hardships of work and flogging.... We shall make the tallest labourers our ploughmen, both for the reason I have just mentioned and because in farming the taller man is least fatigued by his work: while ploughing he stands almost erect, supporting himself on the plough handle....

It is suitable to have working oxen yoked close together for a number of reasons: the appearance of their gait is improved since they do not stoop and their heads are elevated, there is less chafing of their necks, and the yoke rests more comfortably on their shoulders. This is, in fact, the most acceptable design of yoke. The kind employed in some of the provinces – a yoke tied onto the horns – has been pretty well rejected by all writers of agricultural handbooks, and deservedly so: animals can make a greater effort using neck and chest rather than horns, so bringing to bear the whole massive weight of their body. The horn yoke, on the other hand, tortures the animals by drawing their heads back and forcing their faces up; and the ploughshare they can pull is very light so they barely scratch the surface of the earth....

It is harmful to the animal to plough a furrow longer than 120 feet, since beyond this limit he becomes disproportionately tired. When he arrives at the turn the driver should push the yoke forward onto the foreparts and hold the team back to give their necks a chance to cool down. (The necks are prone to inflammation from constant chafing, which leads to swelling and finally ulcerous sores.)

Other information on ploughing can be found in Pliny, *Natural History* 18.177–179.

3.32 Seeds and sowing

In many parts of the Mediterranean the climate allowed two sowings a year, the primary one in the autumn with a spring sowing of wheat in favoured areas.[12] Timing and location were obviously important for planting: Pliny (*Natural History* 18.201–205) provides information about sowing seasons in different regions while Varro (*On Agriculture* 1.44) discusses yields according to location. For sowing itself, broadcasting by hand was the technique most commonly used; the efficient seed drill depicted in Mesopotamian sculpture seems not to have found its way westward. The following passage applies primarily to grains; for the cultivation of leguminous plants, see Columella, *On Agriculture* 2.10.1–24; Pliny, *Natural History* 18.117–148.

Pliny, *Natural History* 18.195–200

The best seed is from last year; two-year-old seed is not as good, three-year-old is quite poor, and anything older is sterile; the germinal power of everything has its limits.... The seed that settles to the bottom on the threshing floor should be kept for sowing: since the best is the heaviest, there is no more efficient way of picking it out....

To scatter the seed evenly also requires a certain skill: hand and foot should keep the same rhythm, coordinated with the right foot.... The general rule is to sow between four and six *modii* per *iugerum*, 20 per cent less or more according to the nature of the soil.

THE HARVEST

3.33 Harvesting

In early summer grain was reaped by using a balanced sickle with an iron blade. The one-handed and long-handled scythes – more like a sickle on a pole than our two-handed design – were used almost exclusively for haying (see **3.37**); and the mechanical harvester described below by Palladius and Pliny was unsuitable for Mediterranean terrain.

Columella, *On Agriculture* 2.20.1–3

The grain should be harvested quickly, as soon as it is ripe and before it is scorched by summer heat waves, which are particularly harsh at the rising of the dog-star.[13] Delay is costly, first because it allows birds and other animals to pilfer the crop, and also because the kernels and even the ears soon fall off once the stalks and beards dry up. What is more, if it is hit by windstorms or hurricanes, more than half ends up on the ground....

There are as well several methods of harvesting. Many farmers cut the stalk in the middle with a spitted sickle, either bill-shaped or toothed; many collect the ear by itself, some using a pair of flat boards, others a comb – quite an easy procedure in a thin crop but very difficult in a thick one.

Varro, *On Agriculture* 1.50

There are three ways to harvest grain. In the first, practised in Umbria, they use a sickle to cut the straw close to the ground and then they lay each sheaf on the ground as they cut it; when they have made up a good number of sheaves, they go through them again and one by one cut the ears from the stalks. The ears they toss into a basket and send to the threshing floor; the straw they leave in the field, where it is stacked. In the second method of harvesting, used in Picenum, they employ a curved wooden stick with a small iron saw at the end. As this tool gathers in a bundle of ears it cuts them off and leaves the stalks standing in the field, to be cut close to the ground at a later time. In the third method, employed in the neighbourhood of Rome and most other places, they grasp the top of the stalk in the left hand and cut it halfway down; ... the stalk below the hand is still attached to the ground and will be cut later, while the part attached to the ear is carried off in baskets to the threshing floor....

On land that presents no difficulties the harvesting of one *iugerum* is considered just about what should be expected as a day's work for one man, whose duties include carrying off the baskets of reaped ears to the threshing floor.

Palladius, *A Study of Agriculture* 7.2.2–4

The following shortcut is used for harvesting in the more level area of the Gallic provinces; manual labour apart, the device lessens the time spent in the whole harvesting operation, and this with the effort of a single ox. They build a cart, carried on two small wheels and with a square upper part made of planks that slope outwards from the bottom, thus providing more space at the top. The height of the planks at the front of this container is lower, and here are set up a great many teeth curving inward at the top and spaced along a row at the height of the ears of grain. At the rear of the cart two very small yoke beams are attached like the poles of a litter; to these an ox is fastened with yoke and chains, his head facing the cart – a docile ox by all means, the sort not to exceed the pace set by the driver. When he begins to drive the cart through the standing crops, each ear is caught by the teeth and gathered into the container, while the straw is cropped and left behind. The ox driver, walking at the rear, makes frequent adjustments by raising or lowering it. In this way the whole harvest is completed in a few hours, after a few passes in either direction. This device is functional in open or level fields, and in places where the straw is considered worthless.[14]

3.34 Building a threshing floor

We have already met descriptions of the threshing floor in **3.6**. Here Varro gives directions for its construction and outlines some variations.

Varro, *On Agriculture* 1.51

The threshing floor should be located in the field and on a slight rise, so the wind can blow over it. Its size should correspond to the size of the harvest, and above all it should be round and slightly elevated in the centre: in this way, if it should rain, the water will not stand but will be able to flow off the floor by the shortest possible route (in a circle every line from the centre to the circumference is the shortest). The earth should be beaten solid, particularly if it is clay, to prevent cracks forming in the heat and providing fissures where the kernels can hide, become damp, and open the gates to mice and ants. Consequently, it is usual to sprinkle the floor with *amurca*, which is poisonous to weeds, ants, and moles. To make their threshing floor solid, some farmers strengthen it with stone or even pave it over, and a few actually roof their floors (the Bagienni, for example, because rainstorms are a common occurrence there at this time of year). In hot climes where the threshing floor is open to the sky, a shady bower should be erected close by, to which the labourers can retreat in the noon-day heat.

Further details can be found in Cato's *On Agriculture* (91) and Columella's *On Agriculture* (2.19).

3.35 Threshing and winnowing

On both Greek and Roman farms, the grain was separated from the harvested ear either by being beaten with sticks or flails, or by being trodden out by draught animals driven around the floor in a circle; both techniques still to be seen in some parts of the eastern Mediterranean. The Romans extended the use of machines adopted from other cultures: the rough-bottomed sledge of the Greeks and the Carthaginian cart fitted with spiked rollers. Fans, shovels, and sieves used in winnowing were common to all agricultural communities.

Varro, *On Agriculture* 1.52

A yoke of oxen and a sledge are used by some farmers to accomplish this [separating the kernels from the ears]. The sledge is constructed in one of two ways: a board with a rough surface of stones or pieces of iron and loaded down by the weight of the driver or some heavy object is dragged along by a team of oxen and severs the kernel from the ear; or the driver sits on toothed axles between small wheels and drives the oxen that pull it: they call this a Punic cart, and it is used in eastern Spain among other places. Elsewhere threshing is done by herding draught animals onto the floor and keeping them moving with goads; their hoofs then separate the grain from the ears.

The threshed grain should then be tossed from the ground during a gentle breeze, with winnowing scoops or shovels. In this way the lightest part, called the chaff and husks, is fanned outside the threshing floor and the grain, because of its weight, falls into the basket clean.

Columella, *On Agriculture* 2.20.5

Grain mixed with chaff is separated by the wind. For this operation a west wind is considered best because it blows gently and steadily throughout the summer months; still, to wait for it is the sign of an indifferent farmer since often while we bide our time a fierce storm catches us off guard. So, after it has been threshed the grain should be heaped on the threshing floor in such a way that it can be winnowed in any breath of wind. On the other hand, if from all directions the air is still for several days in a row, the grain should be cleaned in sifting baskets: this avoids the danger of losing the whole year's work should a devastating storm follow the protracted stillness of the winds.

Palladius (*On Agriculture* 7.1) mentions the use of an old column drum for threshing, possibly a sign of the decreasing prosperity of his age.

3.36 Storage of grain

Varro nicely illustrates the wide variations in agricultural techniques practised in the diverse regions of the Roman Empire. Several other storage methods are described by Pliny (*Natural History* 18.301–303) and Columella (*On Agriculture* 1.6.9–17). In the second passage, Varro provides a warning to those who use underground storage (Rickman 1971).

Varro, *On Agriculture* 1.57

Wheat should be stored in elevated granaries ventilated from the east and north, sheltered from any damp breeze wafting in from the neighbourhood. The walls and floor should be coated with a plaster made from marble, or at least with clay tempered with chaff and *amurca*, a combination that controls mice and worms while making the grain sounder and firmer. Certain farmers sprinkle even their wheat with *amurca*, adding a *quadrantal* to 1,000 *modii* or so.... Some – for example, in Cappadocia and Thrace – store their grain in subterranean caves, which they call *siri*. Others use wells, as in the regions of eastern Spain from New Carthage to Osca: they spread straw over the floor of these wells and make sure that no dampness or air can reach the grain until it is brought out to be used (for no air means no weevils). Wheat stored in these conditions lasts up to 50 years, and millet for more than a century. Some farmers – certain ones in eastern Spain and Apulia are examples – build granaries in the field and elevated above ground level, thus allowing cool ventilation not just through the side windows but even from the ground below.

Varro, *On Agriculture* 1.63

Those who keep their grain under the earth in those [pits], which they call a *sirus*, ought to remove the grain considerably after the pits were opened, because it is dangerous to enter into them when recently opened, to the point that some people have been suffocated while doing so.

3.37 Haying

Since all farms required animals for labour even when they were not raised for profit, fodder crops were essential to ancient husbandry. Cato, for example, ranks meadow land above even a field of grain (**3.5**), and meadows were irrigated more regularly than most other fields.

Pliny, *Natural History* 18.258–263

Meadows are mown about 1 June. Their cultivation is not at all difficult for farmers and involves a minimum of expense. The following points need to be made.

Land that is fertile or damp or well-watered ought to be left as meadow and irrigated by rainwater or from a public channel.... It is time to mow when the grass starts to lose its bloom and to acquire strength; and it must be done before the grass begins to wither. In Cato's words, "Do not mow your hay too late: cut it before its seed ripens".[15] Some farmers water the day before, though if there is no way to irrigate it is advisable to mow when the nights are wet with dew. In some parts of Italy haying follows the harvest.

The operation was more costly in earlier days, when the only whetstones known were from Crete and other overseas provinces; and since these required olive oil for invigorating the blade of a scythe, the labourer who was mowing used to walk along with a horn for the oil strapped to his leg. Italy has given us whetstones that can be used with water for keeping the iron sharp instead of a file, but the water quickly turns it green with corrosion.

As for the scythes, there are two designs. The Italian one is shorter and can be manipulated even among thorn bushes, while those used on the *latifundia* in the provinces of Gaul are larger and actually shorten the work by cutting the grass halfway up and passing over the shorter stalks. The Italian mower uses only his right hand to cut with. One man can be expected to mow a *iugerum* in a day and to bind 200 sheaves of four pounds each.

The mown hay should be turned towards the sun but should not be stacked until it is dry; if this advice is not followed to the letter, it is inevitable that the stacks will give out a kind of gaseous exhalation in the morning and soon after will be ignited by the sun and burn up.

Once it has been mown, the meadow should be irrigated again to allow a second mowing, this time of autumn hay, the so-called late variety.

For additional information on haying and irrigated meadows, see Cato, *On Agriculture* 9; Varro, *On Agriculture* 1.31.5, 37.5; and particularly Columella, *On Agriculture* 2.16–18.

HORTICULTURE
3.38 Locating and working a kitchen garden

Here, Pliny provides practical advice on the proper location and arrangement of the Roman *hortus*, a kitchen garden for vegetables and herbs. For a charming poem describing a poor peasant's breakfast preparations and the rustic simplicity of an idealised garden with a huge variety of vegetables, so admired by urban Romans of the Augustan period, see [Vergil], *Moretum* 61–83 (Jashemski 1979).

Pliny, *Natural History* 19.60

There is no doubt that the gardens should adjoin the farmhouse, and above all that they should be kept irrigated by a passing stream, if there happens to be one. But if not, they should be irrigated from a well by means of a pulley or force pumps or the bailing action of a *shaduf*.[16] The soil should be broken up two weeks after the onset of the west wind, while preparations are being made for the autumn, and a second time before the winter solstice.[17] Eight man-days are sufficient for working over a garden of one *iugerum*, including the mixing of dung with the earth to a depth of three feet, dividing the garden into plots, and separating off each plot with rounded banks of earth that has been scooped out to leave furrows in which a man can walk and water can gush.

3.39 Cold-frames for cucumbers

With the exception of irrigation, methods used to cultivate vegetables were the same as those for most other crops, including soil treatment and hoeing. Rarely do we find devices especially designed for horticulture, so the following invention to extend the growing season (described also by Pliny in *Natural History* 19.64) is of special interest.

Columella, *On Agriculture* 11.3.52–3

It is even possible – if the effort is worth it – to put small wheels under the larger containers, which allows them to be taken outside and brought back in again with less strain. Whether or not this device is used, the containers will have to be covered with transparent mica or gypsum so that even on cold but clear days they can be brought out into the sun without suffering damage. This was how Tiberius Caesar was supplied with cucumbers almost every day of the year.

AGRICULTURE

ARBORICULTURE

The produce of arboriculture – grapes, olives, and fruit are the main categories – formed as significant a part of the ancient diet as cereal crops, legumes, and vegetables. In addition, many trees and plants were cultivated for "industrial" purposes like basketry, paper, containers, and fuel.

3.40 An imaginary Greek plantation

On his long journey home from Troy, Homer's hero Odysseus finds himself shipwrecked on the coast of Phaeacia, an imaginary land of great bounty. The poet's description of the wonderful garden plantation there is among our earliest documentary evidence of horticulture and arboriculture, and reveals much of the agricultural diversity of archaic Greece as well as tidbits of agricultural technology.

Homer, *Odyssey* 7.112–133

Outside the palace courtyard near the gates is a large garden, four days' worth of ploughing in size and bordered on both sides by a hedge. In that place grow tall trees with luxuriant foliage, pears and pomegranates and apple trees with shining fruit, and sweet figs and olives in bloom. The fruit of these trees never dies and never fails throughout the year, winter or summer; some is always being formed and some brought to maturity by the breath of the westerly wind. Pear ripens on pear, apple on apple, one cluster of grapes after another, fig after fig. There, too, a fruitful vineyard is planted out: on level ground in the distance an area, desiccated by the sun's heat, for drying the grapes; here the vintage being gathered by some, there being pressed out by others; on one side in the foreground, unripe grapes casting their blossoms, on the other, those beginning to turn red. There, too, beyond the last row of vines, are neatly ordered garden beds planted with varieties of every kind and perpetually in bloom. Within it are two springs, one channelled off all through the garden while the other flows from the opposite side beneath the threshold of the courtyard up to the lofty palace, where the townsfolk draw from it their water. Such radiant gifts had the gods bestowed on Alcinous' palace.

3.41 Trees and their uses

Both Pliny and Columella (*On Agriculture* 3.1.1–2) are aware of the antiquity of our dependence on trees for survival, long before the domestication of grains. Yet the actual cultivation of trees began long after that of cereal crops, in part because the interval between planting and bearing could amount to decades.

Pliny, *Natural History* 12.1–4

For a long time, the favours of the Earth remained hidden, and her greatest gift to man was thought to be the trees and forests. From these he first obtained his food, cushioned his cave with their foliage, and clothed himself with their bark. Even now there are races that live in this fashion.... Trees were once the shrines of deities, and even today unsophisticated rustic places follow a venerable ritual in dedicating a towering tree to a god.... Afterwards trees soothed man with juices more seductive than grain – the oil of the olive to refresh his limbs, draughts of wine to restore his strength – in short, such a variety of delicacies offered spontaneously by the seasons, foods that still make up the second course of our meals.

3.42 The importance of vineyards

Viticulture was a popular business in antiquity for a number of reasons: the vine's tolerance of a variety of climates, profits that often exceeded those from cereal crops (Columella, *On Agriculture* 3.3.2–10 and Pliny, *Natural History* 14.47–52), and a market guaranteed by the universal consumption of wine. Not surprisingly greater profit from vines induced widespread conversion of grain fields into vineyards (Suetonius, *Domitian* 7.2, 14.2).

Columella, *On Agriculture* 3.1.3–4

The vine we justly rank before the other varieties of plants – the trees and shrubs – not just for the delightfulness of its fruit but for the ease with which it responds to man's attentions in almost every region and latitude of the world except the icy cold and scorching hot. It flourishes as well in plains as on hillsides; it matters not whether the soil is dense and crumbly (or even thin at times), fat or lean, dry or damp; and besides, the vine more than any other plant can tolerate the two extremes of climate, whether in a zone susceptible to frosts or to heat and storms.

3.43 Planting vines

Ancient vintners were as selective as their modern counterparts in choosing a vine best suited to the local conditions and the desired product. For a catalogue of some of the dozens of varieties cultivated in antiquity, read Pliny, *Natural History* 14.25–43; and on the importance of varieties, soil, and aspect, see Columella, *On Agriculture* 3.1.8–10, 2.1–6, 11.1–4; *On Trees* 3.5–7. The stages in establishing a vineyard have changed little over almost three millennia: propagation by taking cuttings from established vines and cultivating them in a nursery for three years or so, until ready for transplanting to the new field.

Columella, *On Trees* 2.1–3.5

In February or early March, once the trenching [of the nursery] is complete, select your shoots. The best of these are chosen from "marked" vines: the farmer who is determined to create good nurseries notes the vines that have produced large, sound grapes at maturity, and about the time of the vintage marks them with red earth that has been mixed with vinegar to prevent it being washed off by the rain.... Choose a plant whose grapes are large and thin-skinned, with few small pits, and sweet tasting. The best shoots are considered to come from the lower parts of the stem, the second best from the upper side branches, the third-class from the top of the vine. These last take root very quickly and are quite productive, but they do fade equally as fast....

Then, in soil that has been thoroughly trenched and manured, set the shoot upright in such a way that at least four of its nodes or "gems" are beneath the surface; it will suffice to space the plants a foot apart. When they have taken root, trim back the foliage at regular intervals to keep them from supporting more shoots than they should.

For details on the selection and treatment of shoots, see Vergil, *Georgics* 2.345–361; Pliny, *Natural History* 17.156–158; Columella, *On Agriculture* 3.13.6–13.

3.44 Propping and training vines

Few vines in antiquity grew satisfactorily without being supported on a sturdy pole or trellis, or being "wedded" to a tree when cultivated in a sort of plantation called by the Romans an *arbustum*.

Pliny, *Natural History* 17.164–166, 199–202; 14.10–12

The management of the nursery is now followed by the arrangement of the vineyards. There are five categories of vineyards: those in which the vine stands upright without support, vines trained to a stake without a cross-piece, or to a stake with a single cross-piece, or to a four-sided trellis....

Long tradition will prove that noble wines are produced only from vines that are wedded to trees; and so salutary a criterion is elevation that wines of greater distinction come from the grapes at the top of the trees, while those at the base give a larger yield.... If the land is to be ploughed, the proper spacing between the trees is 40 feet to the front and rear and 20 feet on the sides; if it is not to be ploughed, 20 feet in all directions. It is common to train ten vines to each tree; fewer than three and the grower gets a bad name....

Everywhere vines overtop elm trees. There is a story about an ambassador of King Pyrrhus named Cineas, who was amazed at the height the vines reached in Aricia and made a witty joke about the rather astringent flavour of the wine produced from them: it was only right, he observed, that the mother of this wine should be hanged from a cross as high as that!

See further on this topic Cato, *On Agriculture* 33.

3.45 Grafting vines

To propagate hybrid vines combining desirable characteristics from both parents, the ancients developed forms of grafting scion to stock that are still used today.

Pliny, *Natural History* 17.115–117

Cato gives three methods of grafting a vine.[18] His first prescription is to cut the vine back and split it through the pith, then to insert shoots that have been sharpened as described before, mating their pith to that of the vine. He advises a second method if the vines grow close together: to shave down on an angle the side of each vine that faces another and then bind them together with the piths mated. His third procedure is to bore a hole into the vine on a slant down to the pith, to insert two-foot slips and bind the graft into position with the scions vertical, and then cover it with a mud of rubbed earth. Our age has refined this last technique by using a Gallic auger that makes a hole without singeing the vine (since any scorching weakens it), and by selecting a slip that is beginning to bud and leaving no more than two eyes projecting from the graft....

The time for grafting vines has been fixed between the autumn equinox and the appearance of the first buds. Domesticated plants are grafted onto the roots of wild ones, which are characteristically thicker; if they are grafted directly onto the wild vines they revert to their wild condition.... Light dews are favoured when propagation is by inoculation.[19]

See, too, Columella, *On Trees* 8.

3.46 The vintage

The gathering of the vintage had to be undertaken at precisely the right moment and as speedily as possible, so most farms would import a casual labour force to assist.

Varro, *On Agriculture* 1.54.2

During the grape harvest the scrupulous farmer not only *gathers* his grapes – they are the ones to be used for drinking – but also *selects* the best from them, the ones for eating. So, the gathered grapes are carried off to the vintage-tank and from there find their way to an empty storage jar. The select grapes, on the other hand, are put into a separate basket [for one of three kinds of treatment]: they are transferred to small pots that in turn are packed tightly in large, wide-mouthed jars filled with the refuse from the grape pressing; some are put into an amphora sealed with pitch and lowered into the farm's pond; while still others are hoisted up to the meat locker.

For further details on treading, pressing, and winemaking, see **4.10–16**.

3.47 Pruning

The last seasonal task for the *vinitor* was to prune back the vines to ensure robust growth the following year. The techniques were more complex than this short passage suggests, as a glance at Columella's lengthy descriptions of pruning will show (*On Agriculture* 4.7–10, 23–28; *On Trees* 5).

Pliny, *Natural History* 17.191–193

Pruning is undertaken right after the vintage, when the warmth of the weather is favourable; but even with warm weather, regard for the natural course of things demands that it never be done before 20 December.... Given favourable weather, the earlier vines are pruned the more wood they put forth, and the later the pruning the fuller the yield. As a result, it will be advantageous to prune spindly vines earlier and robust ones last.

Each cut should be made on a slant facing the ground, making it easy for rain to drip off. To leave as slight a scar as possible, a pruning knife with a very sharp blade allows a very smooth cut. Always make the cut between two buds to avoid harming the eyes of the part cut back.... Should a spindly vine not have suitable sprouts, it is best to cut it right back to the ground and have it put out new shoots.

3.48 The spread of olive culture

Though grown on Crete even before the arrival of the Minoans, olive trees were not widely cultivated until the Archaic period because of the more pressing need for grain and the long interval between planting and a productive yield. By the sixth century BC Athens was an exporter of oil, and in the third century BC olive culture became common in Italy, as the comparative prices quoted here by Pliny confirm.

Pliny, *Natural History* 15.1

About 314 BC Theophrastus, one of the best-known Greek authors, affirmed that the olive grew only within 40 miles of the sea, and according to Fenestella it did not exist at all in Italy, Spain, or Africa in 581 BC, during the reign of Tarquinius Priscus. Now it has spread even across the Alps and into the hinterlands of the Gallic and Spanish provinces. In 249 BC 12 *librae* of olive oil went for ten *asses*; and later Marcus Seius, the curule aedile of 74 BC, made it available all year to the Roman people at the charge of a single *as* for every ten *librae*. This would be less of a surprise to anyone who knows that in Gnaeus Pompey's third consulship, 22 years later, Italy was exporting oil to the provinces. Hesiod, too, who believed that agriculture should be one of life's fundamental studies, observed that no one who had planted an olive tree had actually picked any fruit from it – that was how slow the operation was in those days. But today olive trees bear even in nurseries, and olives are picked from them the year after they are transplanted.

3.49 The cultivation of olive groves

Pliny's comment that olive culture had improved dramatically since Hesiod's day is well illustrated by the techniques described here by Columella.

Columella, *On Agriculture* 5.8–9

The olive occupies first place among all trees, and it is far more economical than any other plant. Although it normally bears fruit every other year rather than annually, it is still exceptionally well regarded because only modest cultivation is needed to support it and virtually no expense is involved when it is not bearing....

Four-foot holes are prepared for the plants the previous year; but if there is not time for this before the trees are to be planted, by putting straw and twigs in the holes and setting them alight, the fire will do the job of the sun and frost in mellowing the holes. In rich soil that is used for grain as well, the trees should be spaced 60 feet apart in one direction and 40 in the other....

Once the olive grove is established and ready for bearing, divide it into two sections, each to be draped in fruit in alternate years (it is a fact that the olive tree is not productive two years in succession). When the field beneath has not been sown, the tree is sending out stalks; when the field is filled with crops, the tree is bearing its fruit. So, an olive grove divided in this way delivers an equal return year by year.

3.50 Harvesting olives

Varro, *On Agriculture* 1.55

As far as the olive grove is concerned, you should pick rather than beat down those olives you can reach from the ground or from ladders, because an olive that has been bruised shrivels and does not give as much oil. Of the hand-picked ones, those handled with bare fingers are better than those handled with leather gloves, the stiffness of which not only crushes the berry but even strips the bark off branches, leaving them exposed to the frost. The olives out of reach should be beaten down, but with a cane rather than a pole since the heavier blow is an invitation to the tree doctor. The beater should not strike a direct blow, for an olive hit in this way often rips the shoot from the branch along with it; and when this happens, the harvest is lost for years to come. This is not the least reason why they say that an olive grove in alternate years is unproductive or bears a smaller crop.

The processing of harvested olives into oil (or for consumption at table) is described in **4.17–4.19**.

3.51 Preparations for an orchard

For a description of the most commonly cultivated fruit and nut trees of antiquity, see Pliny, *Natural History* 15.35–105, who lists the following harvests from them: pine nuts, quinces, peaches, pomegranates, plums (12 varieties, including the damson from Syrian Damascus), apples, pears, figs, medlars, sorb-apples, walnuts, hazelnuts, almonds, chestnuts, carobs, mulberries, strawberries, elderberries, cherries, and cornelberries.

Cato, *On Agriculture* 48

Make a nursery for your fruit trees in the same way as for your olive trees. Plant each variety of slip on its own.

When planting cypress seed, for example, turn over the ground with a spade. Sow at the beginning of spring. Make ridges five feet wide, to which you add fine manure, hoe it in, and break up the clods. Level off the ridge, making a shallow depression on the top. Then sow the seed as thick as flax and cover with sieved earth to the depth of a finger's width. Level off the soil using a board or your feet, drive forked stakes into the perimeter, insert poles into them, and hang from the poles vine canes or the sort of wicker frames used for drying figs, to keep off the cold and the sun. Make it so that a man can walk beneath. Hoe frequently. The moment weeds begin to grow, root them out; for if you pull out a weed when it has grown hardy, you will pull out the cypresses with it.

Use the same method for sowing and covering the seeds of pear and apple; plant pine nuts, too, in the same way....

Columella, *On Trees* 18

Before establishing your orchard, enclose the area you want to use with a wall or ditch, to keep men as well as animals from getting in anywhere except through the gate, at least until the seedlings come to maturity: they are ruined forever if the heads are too often broken off by hand or nibbled away by cattle. To arrange your trees according to variety is more advantageous, especially to keep the weak ones from being stifled by the stronger, since trees are not equal in strength or size and do not grow at the same rate. The same soil that is fit for vines is also suitable for trees.

Cf. Columella, *On Agriculture* 5.10.1–9; Pliny, *Natural History* 17.65–78, 88–94; for transplanting trees, Columella, *On Trees* 19; Cato, *On Agriculture* 28.

3.52 Grafting trees

Pliny, *Natural History* 17.101

It was Accident that taught us grafting – a different teacher from Nature and, if I may say so, a busier one. It happened like this. An assiduous farmer who was building a fence around his house to protect it set his fence posts on a footing of ivy wood to keep them from rotting. But the posts were gripped by the bite of the still vital ivy and revivified themselves from the other's life force: the wood of the tree seemed to be playing the part of the earth.

Cato, *On Agriculture* 40

Graft olives, figs, pears, and apples as follows. Cut back the branch you intend to graft and angle it slightly to allow water to run off; be careful when you make the cut not to tear away the bark. Get hold of a stake that will not bend and sharpen the end of it; then split a Greek willow. Combine clay or chalk with a little sand and cattle dung; knead the mixture thoroughly to make it as glutinous as possible. Take your split willow and use it to tie around the cut branch to prevent the bark from splintering. This done, force the sharpened stake between the bark and the woody part to a depth of two fingers. Then take the shoot of the variety you want to graft on and sharpen its end at a slant for a distance of two fingers. Take out the dry stick that you forced in and replace it with the shoot you want to graft. Line up bark with bark, and force the shoot in up to the end of the slant. Do the same for a second, third, and fourth graft, inserting as many varieties as you like. Wind on more Greek willow and then daub the muck you have kneaded onto the branch until it is three fingers thick. Cover it all over with the plant called "ox-tongue" to keep water from penetrating to the bark should it rain. Bind the ox-tongue with bark to keep it from falling off. Then wrap straw around and tie it into place to prevent frost damage.

Pliny, *Natural History* 17.120; 15.57

Next to the waterfalls at Tivoli I have seen a tree grafted in all the various ways I have described, loaded with every kind of fruit: nuts on one branch, berries on another, grapes in one place, elsewhere pears, figs, pomegranates, and different kinds of apples. But that tree had no long life....

That humankind has tried everything comes as no surprise when we read in Vergil of an arbutus tree grafted with nuts, a plane tree with apples, and an elm with cherries.[20] No further refinements are possible; certainly, no new fruit has been discovered for a long time now. Nor is it in keeping with religious principles to adulterate everything by grafting: thorn bushes, for example, ought not to be grafted since they cannot easily be purified when struck by lightning, and people swear that the number of varieties grafted onto them is the same as the number of thunderbolts that strike them in a single blow.

Descriptions of grafting abound in ancient literature. See, as a small sample, Pliny, *Natural History* 17.58–64, 96–138; Varro, *On Agriculture* 1.40.5–41.3; Columella, *On Agriculture* 5.11.1–15; *On Trees* 26–27; Lucretius 5.1361ff; and Vergil, *Georgics* 2.22–82.

3.53 The many uses of the palm

Trees and plants were harvested not only for food but for materials useful in a variety of applications: timber for construction of buildings and boats (**10.13–15, 11.66**), bark and gourds for containers (**10.14**), charcoal for fuel (**2.27**), pitch and tar for use as sealants (**10.22**), pliable stalks for basketry and matting (**10.16**), and papyrus as a writing surface (**12.30**). The utilitarian palm – like the buffalo for the natives of North America – was particularly useful. It provided Near Eastern peoples with their three main necessities: food, clothing, and shelter.

AGRICULTURE

Pliny, *Natural History* 13.28–44

There are a great many varieties of palm, the first being no larger than a bush.... The others are tall and rounded, with projections or rings set close together in their bark like steps, making them easy for the eastern peoples to climb: they tie a plaited loop around themselves and the tree, and the rope and the man scramble up together with surprising agility. All the leaves are at the top of the tree, and so is the fruit, though it does not hang among the foliage as on other trees but shares the characteristics of both grapes and orchard fruit, hanging in a cluster but each on a stem of its own. The leaves have edges as sharp as a knife and fold down the middle, which at first suggested hinged writing tablets; these days they are split apart for making ropes, woven wickerwork, and parasols ... Assyria and all of Persia use the sterile variety of palm for timber and the more elegant products of woodworking.... The wood of the palm makes an intense and slow-burning charcoal.... Among the most famous varieties are the date palms, which yield copious amounts of food and also of juice. These are the palms used to produce those extraordinary eastern wines that cause headaches.

ANIMAL HUSBANDRY

The technology of animal husbandry (and agriculture) goes beyond mechanical devices. Calendar-dates, weather and numerous other elements are all considered in order to produce the best possible results, whether animal or plant. Thus, the inclusion of some passages that, at first glance, seem only peripheral to technology: classifications, herd size, health, growth rates, breeding, economic return, and others (Kitchell 2016, Kron 2014).

3.54 Farming vs. ranching

Columella, *On Agriculture* 6, Preface 1–3

I realise that some experienced farmers have rejected animal husbandry and have consistently shown contempt for the herdsman's role as being harmful to their own profession. And I admit that there is some reason for this attitude, insofar as the object of the farmer is the antithesis of the herdsman's: pleasure for the one comes from land every inch of which has been cleared and tilled, for the other from land that is fallow and covered in grass; the farmer hopes for reward from the earth, the rancher from his beasts, and for that reason the ploughman abhors the same green growth that the herdsman prays for.

Still, in the face of these discordant aspirations there is yet a certain compatible relationship between the two. In the first place, there is usually more advantage in using a farm's fodder to feed one's own herds rather than someone else's; and second, the luxuriant growth of the earth's produce results from abundant applications of manure, and manure is a product of the herds. Nor for that matter is there any region where grain is the only thing to grow, and where it is not cultivated with the help of animals as much as of men.... It is for these reasons that I follow the recommendation of ancient Romans in insisting that our familiarity with animal husbandry be as thorough as our understanding of agriculture.

For similar observations, see Columella, *On Agriculture* 1.3.12, and Varro, *On Agriculture* 2, Preface 4–6.

3.55 The divisions of animal husbandry

Varro presents a "scientific classification" of animals that clearly illustrates the study of animals to increase their return on investment.

Varro, *On Agriculture* 2.1.11–24

There is accordingly a science of acquiring and pasturing herds of animals so that they yield the greatest returns of which they are capable – hence the origin of our word for "money", since the herd (*pecus*) is the source of all wealth (*pecunia*). There are nine divisions of this science, grouped into three sets of each:

I The Smaller Animals:
   1 Sheep
   2 Goats
   3 Pigs
II The Larger Animals:
   1 Oxen
   2 Donkeys
   3 Horses
III The Animals acquired not for the profit they might yield but because they either assist in or result from animal husbandry:
   1 Mules
   2 Dogs
   3 Herdsmen

Now each of these nine major divisions of animal husbandry comprises at least nine subdivisions, of which four are requisites for acquiring a herd, another four for maintaining it, and the ninth is common to both aspects. Thus, there are at least 81 subdivisions of the whole topic, every one of them essential and significant.

   Under the topic of acquiring a good herd, the first requisite is to know at what age it is expedient to acquire and keep each kind of animal.... The second subdivision in this first group of four is a knowledge of the appropriate morphology of each breed, an aspect that has a significant effect on profit.... The third subdivision is the investigation of breed.... The fourth subdivision concerns the law of purchase, how legal procedure should play a part in the procurement of each herd....

Once you have bought your herd, attention must be given to the second group of four subdivisions: pasturing, breeding, nurturing, and health. The first subject is pasturing, for which there are three considerations: in what location should you preferably pasture each group of animals, when should you put them to pasture, and how.... The second subdivision concerns breeding, which I define as the phase from conception to birth, the two parameters of pregnancy. So, the first consideration is controlled mating; ... the second consideration here is gestation, since animals differ in the time they carry their young; ... and the third topic comprises the aspects of nurturing to which one must be alert, including the number of days the young should feed from their mother.... The fourth subdivision concerns health, a complex topic but an indispensable one since a sickly herd is a ruinous herd whose ill health is frequently disastrous for its owner; ... within this topic there are three divisions: consideration must be given to the cause of each malady, to its symptoms, and to the proper course of treatment....

There remains my ninth subdivision – the size of a herd – which is common to both its acquisition and maintenance. For it is unavoidable that anyone putting together a herd should decide on its extent ... in order to prevent either a shortage or an excess of pasture land and an attendant loss of profit. What is more, he must understand how many females capable of breeding he should have in his flock, how many males, how many young of each sex, and how many rejects must be culled from it.

3.56 Cattle and oxen: training and treatment

In antiquity cattle were raised almost exclusively as draught animals rather than as a source of food. Beef and veal are seldom included – and then only briefly – in any ancient description of diet, and cow's milk was generally neglected in favour of dairy products from sheep and goats (see **3.59**). Most oxen not yoked to a plough or cart were served up to the gods (as part of the Roman *suovetaurilia*, a triple sacrifice of a pig, sheep, and bull); the leftovers might then be distributed to the poor for consumption. For types and breeding see Varro (*On Agriculture* 2.5.7–11, 18) and Pliny (*Natural History* 8.176–178; Ikeguchi 2017 for meat diet).

TRAINING OXEN

Vergil, *Georgics* 3.156–165

After calving all attention is transferred to the young. No time is lost in branding them with the signs that mark the breed, and in keeping apart those selected for breeding purposes, for sacrifice, and for ploughing and turning over a field lumpy with broken clods. The rest of the oxen graze on the fresh grasses. The ones you will mould to the practice of agricultural pursuits you should foster while still young bullocks, setting them on the path of discipline while they still have the docile spirits of youth and are of an age open to change.

Columella, *On Agriculture* 6.2.1–7

What follows is the commonly accepted method for taming those animals that are captured wild from the herd. First of all is the preparation of a spacious stable in which the trainer can move about easily and from which he can safely escape…. If your cattle are quiet and relaxed, you will be able to lead them outside before dark on the same day you tied them up, and train them to walk in an orderly way and without apprehension for a full mile. Once you have led them back home, hitch them tightly enough to the stocks that they cannot turn their heads. While the oxen are thus restrained it is finally time to approach them, not from behind or the side, but head-on, calmly, and with comforting words to accustom them to seeing you approach. Then scratch their noses to teach them the smell of a man. It is advisable to put them on more familiar terms with the man who will be driving them by stroking the lengths of their backs and sprinkling them with undiluted wine; and to place your hand on their bellies and under their thighs to forestall any alarm later when they are touched like that (as well as to remove the ticks that usually cling to the thighs). For this procedure the trainer should stand to one side to keep from being kicked. Next, spread the animal's jaws apart, draw out his tongue, and rub his whole mouth and palate with salt; put down his throat one-pound balls of flour moistened with salted fat drippings; and use a funnel to pour wine, a *sextarius* at a time, into his gullet. With enticements like these they usually grow tame in three days and can be yoked on the fourth.

CONSTRUCTION OF ENCLOSURE TO TREAT CATTLE

Columella, *On Agriculture* 6.19.1–3

After addressing the use of simple tools like pipes and reeds to introduce oils, vinegar, and smoke into the animals to rid them of leeches, Columella describes the type of enclosure that might be needed to handle large animals.

A device, in which one can contain beasts of burden and oxen and treat them, must also be constructed so that those men applying the remedies may have easier access to the animal, and so that the quadrupeds, during the actual doctoring, may not reject the remedies by their resistance. The form of the device is as follows: a section of ground nine feet long and two and half feet wide in front and four feet wide at the back is floored with oak planks. In this space four seven-foot-high upright posts are placed on the two flanks. Moreover, they are affixed in the four corners and are all secured to each other with six cross-poles, just like a log frame, so that the animals can be led in from the back, which is broader, just like into a stall, but cannot get out any other side since the guiding poles check them. And on the two front posts a solidly-constructed yoke is placed, to which beasts of burden are fastened with halters or the horns of oxen are bound fast. And here it is possible to construct shackles so that once the head has been inserted, bars may descend through holes so that the neck is gripped tightly. The rest of the body, bound and stretched out, is secured by the cross-poles and, now immobile, is available to the examination of the practitioner.

3.57 Donkeys

Varro, *On Agriculture* 2.6.2–3

There are two varieties of asses: one is the wild ass, called an onager, which is prevalent in Phrygia and Lycaonia. The other is the domesticated donkey, the only species found in Italy. The onager lends itself to breeding because it is easy to tame and, once tamed, never reverts to the wild.

Columella, *On Agriculture* 7.1

Since we are now going to speak of the less important animals, the one to lead off is the donkey from the district of Arcadia, an inexpensive and common beast that most agricultural writers claim deserves special consideration when one buys and raises beasts of burden. And here they are not mistaken, for the donkey can be kept in country even devoid of pasture and is happy to feed on a modicum of whatever sort of fodder it finds, from leaves and bramble thorns to a bundle of twigs when offered – in fact, it quite thrives on chaff, which is available in great quantities almost everywhere.

Then, too, it holds up very well when neglected by an inconsiderate master and is remarkably tolerant of blows and deprivations. It is for these reasons that it is slower to tire than any other beast of burden; since it endures hard work and hunger so very well, it is seldom affected by ill health. The countless essential tasks it performs are quite out of proportion to the modest care that the animal requires: it can pull a light plough to break up the kind of soft soil found in Baetica and throughout Libya, and it can pull wagons with loads that are by no means trifling. Often, too, as our most celebrated poet recalls, "… the driver loads his slow donkey's sides with cheap fruits, and returns from town bringing with him a mill cut from stone or a chunk of black pitch".[21] These days this beast will almost inevitably be found grinding grain at rotary mills.

Compare Varro, *On Agriculture* 2.6, and **2.18–19**.

3.58 Breeding mules

Breeding of equines receives quite a bit of attention in the sources, but generally not very much mechanical technological information is provided. For horses much is made of the uses of different breeds with noble racing horses and those used for getting mules being valued much more highly than the common horse (Columella, *On Agriculture* 6.27–29; Varro, *On Agriculture* 2.7; Pliny, *Natural History* 8.167–170). The next two passages provide information about why the mule was admired and a method for breeding them.

Pliny, *Natural History* 8.167, 171–173

The work a donkey does in ploughing is undoubtedly beneficial as well, but the animal is especially useful in breeding mules....

A mule is born in the thirteenth month after a jackass mates with a mare. Its physical strength makes it ideally suited for hard work [cf. **8.56**]. For breeding them they choose mares between four and ten years old; and since it is also claimed that neither species will mate with the other unless as a foal it was suckled by the other species, they secretly exchange the foals in the darkness, giving the young donkeys to the mares and the colts to the jennies to be suckled. It is true that a stallion and a jenny can produce a mule, but it is intractable and impossibly hard to arouse.... It has been observed that among all classes of animals, offspring of two distinct species form a third species that displays the characteristics of neither parent and is itself incapable of reproducing; hence mules are sterile. While cases of mules breeding are not infrequently to be found in our Annual Records of Public Events, they are classified as prophetic signs.

It is tempting to see this device that accommodated inter-species coupling as the origin of Daedalus' mythical contraption to enable Minos' wife Pasiphaë to mate with a bull and produce the Minotaur (see, e.g. Diodorus, *History* 4.77.1–4).

Columella, *On Agriculture* 6.37.10–11

A place is specially built for the breeding of mules, which farmers call The Device. Two parallel walls are erected down a gentle incline, with the space between them narrow to prevent the mare from struggling or turning away from the jackass as he mounts her. There is an entrance at either end, the one lower down the slope fitted with bars to which the mare is harnessed. Here she stands facing the bottom of the hill, her angle of incline affording a better entry for the donkey's sperm and making it easier for the smaller quadruped to mount her body from the higher ground.

Compare Varro, *On Agriculture* 2.8.

3.59 Sheep

Sheep and goats were significant elements of ancient farming from the Bronze Age onwards: Homer, for example, describes the flocks of Odysseus (*Odyssey* 14.96–104) and of Polyphemus (**4.22**).

Columella, *On Agriculture* 7.2.1–2

The sheep takes second place after the donkey, though it comes in first if you consider the extent of its usefulness. For it is primarily the sheep that protects us from the fury of the cold and unselfishly offers its own coverings for our bodies. It is the sheep, too, that not only appeases the hungry farmers with cheese and milk in abundance but adorns the tables of discriminating diners with an assortment of tasty dishes. It is a fact that the sheep provides sustenance to certain races that have no grain, which is why most of the natives of Africa and the Getae are called "Milk-drinkers".

Varro, *On Agriculture* 2.10.10–11; 11.1, 5–9

The proper number of herdsmen varies from farm to farm, some setting it lower and others higher. I have assigned one shepherd to every 80 of my shaggy-haired sheep, while Atticus makes it 1 to 100. If the flocks are large – some farmers number their sheep in the thousands – you can reduce the number of men more easily than you can with smaller flocks like Atticus' and mine. I have 700 head and you, I believe, had 800, but still with 10 per cent of them rams as is my practice....

Now for a few words as promised about the supplementary profit that can be made out of flocks from milking and shearing. Of all the liquids in our diet, milk is the most nourishing: sheep's milk first, then goat's milk. As a laxative the best is mare's milk, then donkey's milk, cow's milk, and finally goat's milk....

Shearing is done between the spring equinox and the summer solstice, once the sheep have begun to sweat (hence freshly shorn wool is called "juicy"). Sheep fresh from a clipping are smeared that very day with wine and olive oil, sometimes mixed with white wax and pig's fat; and if the sheep is normally "jacketed",[22] the inside of the skin coats is smeared with the same concoction before they are refitted. Any wounds inflicted on the animals during shearing are treated with liquid pitch....

Wool that has been clipped and bundled is called a "tuft" by some and a "tug" by others, both expressions indicating that plucking of wool was invented before shearing. Even these days some farmers continue to pluck their sheep, which they keep from eating for three days beforehand since the roots of the wool are less painfully tenacious when the sheep are faint from hunger.

For more information on the uses and varieties of sheep, see Columella, *On Agriculture* 7.2–5; Pliny, *Natural History* 8.187–199; Varro, *On Agriculture* 2.2. On specific topics: observations on shepherds can be found in Columella 7.3.26, on shearing in Columella 7.4.7–8, and for advice on the raising of sheep, see Varro 2.2.7–20 and Columella 7.3.8–25.

3.60 Goats

Varro, *On Agriculture* 2.3.6–10

My advice about pasturing is this: since goats are very sensitive to the cold, the herd had better be stabled so they face in the direction of the winter sunrise. Their pens – and this is true of all stables – should be paved in stone or tile to keep out the damp and mud. When they must spend the night outside, their enclosures should have the same orientation and be spread with boughs to prevent them from getting filthy....

As for breeding, at the end of autumn the billies are separated from the herd at pasture and driven into their pens, the same procedure as followed for rams.[23] A pregnant nanny gives birth in the spring, four months later, and her kids are set loose at three months to join the herd.

What can I say about the health of creatures that are never healthy? ... About 50 head is considered a suitably large herd, for which the experience of Gaberius, a Roman knight, is taken as proof. This fellow, who owned an estate of 1,000 *iugera* near Rome, heard from some goatherd or other who brought ten animals into the city that each one of them brought him in a *denarius* for every day spent on them. So Gaberius assembled a herd of 1,000 head in the hope that he would make 1,000 *denarii* a day from his estate. What a mistake! He soon lost all of them to disease. It is true, though, that among the Sallentini and in the neighbourhood of Casinum they do pasture herds of up to 100. There is pretty much the same lack of conformity in the proportion of males to females: some, myself included, provide one billy for every ten nannies; for others, like Menas, the number is one to 15; and there are even some, like Murrius, for whom it is one to 20.

See also Columella, *On Agriculture* 7.6.5, and 7.6.9 on goatherds; Pliny, *Natural History* 8.200–204.

3.61 Pigs

In antiquity swine were easy to raise, put on weight rapidly, graced sacrificial altars and "fed" both divine and mortal consumers. Unlike many other large food mammals, the pig litter was substantial (eight to ten)[24] which, combined with their growth rate, led to pork being favoured over beef for consumption.

Pliny, *Natural History* 8.207–209

Sows that are excessively fat experience a shortage of milk and produce fewer piglets in their first litter. The species is fond of rolling in the mud. Their tail is curly, and it has even been observed that it is easier to sacrifice them on the altar when their tail curves to the right rather than the left. They can be fattened in 60 days but do even better if starved for three days before the gorging process begins.

Of all animals the pig is the most slothful.... But there is a well-known incident about some stolen pigs that, when they recognised the voice of their swineherd, they crowded to one side of the ship they were on, capsized and sank it, and swam back to shore. And what is more, the lead pigs actually learn how to find the weekly market and get back home again; and wild boars know how to cover their tracks on marshy ground and to ease their escape by urinating.

Sows are spayed just as camels are: after two days of fasting they are strung up by their front haunches and their womb cut out, a procedure that speeds up the fattening process....[25]

Varro, *On Agriculture* 2.4.11–12

The Spaniard Atilius, who is quite reliable and well informed in a variety of fields, used to tell the story of a pig that was slaughtered in Lusitania in western Spain. A piece of the animal with two ribs, sent to the senator Lucius Volumnius, weighed 23 *librae* and measured 15 inches from skin to bone. I have personal experience of an equally astonishing incident in Arcadia: I recall going to see a sow that was so fat that not only was she incapable of standing up, but a mouse had actually eaten a hole in her flesh, where it built its nest and bore its young.

Varro, *On Agriculture* 2.4.7–8, 22

For breeding, the boars should be isolated for two months in advance. The best time for letting them mate is from the rise of the west wind to the spring equinox, to produce a litter four months later, in the summer when the earth is well provided with food.... While mating they are driven into muddy swamps and sloughs where they can wallow in the mud and refresh themselves the way men do at the baths. Once the sows have all conceived, the boars are again isolated. A boar begins to couple at eight months and can perform reasonably until three years old; thereafter he is on the decline until he reaches the butcher, the appointed mediator between pork and populace....

With regards to the quantity of animals, ten boars for 100 sows is considered satisfactory, though some breeders reduce this ratio. There is no standard size for a herd; while I think 100 is a sensible number, a few keep larger herds that can reach 150 pigs; some split the herd in two, others make it even larger.

3.62 Poultry, the henhouse, and eggs

Varro, *On Agriculture* 3.9.1–7

There are three varieties of fowl: the barnyard, the wild, and the African or guinea fowl.... Anyone aiming at a complete poultry house should obviously acquire all three types, but particularly the barnyard fowl....

If you are intending to raise 200 birds, you should appropriate for them an enclosed area and within it build two large, interconnected coops with an eastern exposure, each about ten feet long, half as wide, and a little less than that in height. Each should have a window three feet wide by four high and made of withes spaced wide enough to let in ample light but not so separated as to permit entry to any of the hen's traditional enemies. A door between the two houses should afford access for the *gallinarius*, or keeper. Poles should be laid across the interior of the coops close enough together to give each hen a perch, and their nests are to be sequestered along the wall opposite each row of perches. In front of the henhouses ... there is to be a fenced yard that the hens can occupy during the daytime and where they can roll about in the dust.

Varro, *On Agriculture* 3.9.11–12

The keeper should make regular rounds every few days and turn the eggs over to ensure that they are evenly warmed. They say you can tell whether or not eggs are full and fertilised by dropping them into water: the empty egg floats, the full one drops to the bottom. Farmers who try to find out the same thing by shaking eggs are making a mistake, for they disturb the vital blood vessels inside. The same people claim that, if you hold an egg up to the light and it is translucent, then it is lifeless.

Pliny, *Natural History* 10.152–153

Chicks emerge on the nineteenth day in the summer and on the twenty-fifth in winter. Eggs die if it thunders during the brooding period, and they spoil if the cry of a hawk is heard. As a remedy against the effects of thunder, an iron nail or earth from a plough is placed under the straw on which the eggs rest. Nature hatches some eggs on her own without a brooding hen, as happens on the dung heaps of Egypt. There is a good story about a particular drunk in Syracuse who was in the habit of imbibing for as long as it took eggs covered with earth to hatch.

In his section 154, Pliny describes how to incubate chicken eggs artificially. For more details on the henhouse, see Columella, *On Agriculture* 8.3.1–7; on eggs and chicks, Columella, *On Agriculture* 8.5.7–16; and Varro, *On Agriculture* 3.9.13–16.

3.63 Geese: pâté and feathers

Always fascinated by the preternatural, Pliny begins his description of geese with anecdotes about their unusual – and remarkably human – behaviour: the sacred geese on Rome's Capitoline Hill warning of a night attack by the Gauls, geese that fell in love with young boys or girls, and another that was the constant companion of a philosopher. He continues, with typical Roman practicality.

Pliny, *Natural History* 10.52–54

Our people are more perceptive: they recognise geese by the fine flavour of their liver, which grows to a considerable size when the animals are stuffed, and even if removed increases if soaked in milk mixed with honey. There is good reason to enquire who made such a felicitous discovery; ... still, there is no doubt that it was Messalinus Cotta, son of Messala the orator, who discovered how to roast the soles of their feet and to blend them into a pâté with the combs taken from cocks....

The plumage of white geese yields a second source of income. In some locations they are plucked twice a year, the feather coating growing back again. The feathers closest to the body are softer, and the ones from Germany are particularly sought after: the geese there, called *gantae*, are smaller but a shining white and fetch five *denarii* for a pound of their feathers. It is for this reason that prefects of the auxiliary troops are not infrequently charged with having removed whole cohorts from their sentry duty and sent them out as bird-catchers. And luxury has reached such a point that these days even the necks of men cannot bear to be without goose-feather bedclothes.

Further details on geese can be found in Varro, *On Agriculture* 3.10; and Columella, *On Agriculture* 8.13–14.

3.64 Ducks

Columella, *On Agriculture* 8.15.1–6

The maintenance of a yard for ducks is not unlike that for geese, though it is more expensive. Ducks, various teal, coots, and similar birds that poke about in pools and swamps are all reared in captivity.

The level area selected for a duck yard is protected by an enclosure wall 15 feet high and covered over with lattice or wide-meshed netting to prevent the domestic birds from flying out and eagles or hawks from flying in. And to keep cats and ferrets from crawling through, the entire wall is smoothed off inside and out by the application of plaster. In the middle of the duck yard a pond is excavated to a depth of two feet.... Around the pond the banks should be covered with grass for a distance of 20 feet on all sides, and beyond this border, at the base of the wall, are stone boxes 1 foot square and finished with plaster, where the birds can build their nests.... Then no time should be lost in excavating a continuous channel along the ground to carry their daily feed mixed with water, which is what this kind of bird eats.

See Varro, *On Agriculture* 3.11.

3.65 An aviary

Varro, *On Agriculture* 3.4.2–5.6

There are two kinds of aviary: one for amusement ... and the other built for profit, the kind of coop that poulterers have in the city or rural areas ... Lucullus claimed that the "combination" aviary he built in Tusculum formed a third type, since under one roof he had an aviary and a dining room where he could dine in luxury while watching some birds dished up cooked and others in captivity flitting about the windows. But the scheme was not very successful: the birds fluttering around the windows hardly compensated for the offensive smell that filled the nostrils.

As for the aviary built for profit, ... it is a vaulted building or a peristyle roof with tiles or netting, a large structure in which several thousand thrushes or blackbirds can be kept.... It should have only enough openings for light to allow the birds to see the location of their perches, food, and water. Smooth plaster should be applied around the doors and windows to keep out mice and other vermin....

When the time comes for appropriate birds to be culled from the aviary, they should be confined in a small coop called a *seclusiorum*, next door to the larger aviary but better lighted. When the desired number has been shut up there, the breeder kills them all. The reason for doing this privately and separately from the others is to avoid any despondency on the part of the survivors leading to premature – and unprofitable – death.

3.66 Beekeeping

Since honey was the primary sweetening agent in antiquity, the potential profit from beekeeping was great: before this selection Varro has described a Tuscan farm of only one *iugerum* whose honey production brought in 10,000 *sestertii* annually.

Varro, *On Agriculture* 3.16.12–17

Herewith instructions for building apiaries.... First, they should if at all possible be near the farmhouse where there are no reverberating echoes (a sound thought to cause flight by their king bees). The climate should be moderate, not overly hot in summer and with some sunshine in winter. The hive should definitely face the winter sunrise and be near areas with a ready food supply and clean water. If food does not grow there naturally, the apiarist should sow the particular kind of crops that bees sniff out: rose, thyme, balm, poppy, bean, lentil, pea, basil, rush, alfalfa, and particularly clover, which is very good for them when they are not strong....

Some build circular hives of twigs, others use wood and bark, or a hollow tree, or clay, or even the stalk of a giant fennel to make squared hives about 3 feet long by a foot deep, though when there are too few bees to fill the hive this last measurement is reduced to prevent them being depressed by a vast, empty space.... Tiny holes are cut in the middle of the hive on both sides to give the bees access, and covers are fitted to the ends to allow the keepers to remove the comb.

See Pliny, *Natural History* 21.73–86 for more specific suggestions, including how to move beehives on boats and mules when the bees' local food fails.

Vergil, *Georgics* 4.228–235

Whenever you unseal their noble home to take away the honey hoarded in their treasuries, first moisten and rinse your mouth with a draught of water,[26] and in your hand stretch out the probing fumes. Their anger knows no limits, and when vexed they breathe poison into their bites and leave invisible stings implanted in the veins, laying down their lives in the wound. Twice they collect the heavy produce, and there are two seasons of harvest: first when Taygete the Pleiad shows the earth her handsome face, ... and again when she falls disconsolate into the wintry waves.[27]

Columella, *On Agriculture* 9.15.5–16.1

We take resinous sap or dry dung mixed with burning coals in a clay pot and introduce the smoke into the hive after removing the back part of it.... The bees cannot bear the fumes and at once retreat to the front and sometimes right outside.... All the combs should not be removed at one time: at the first harvest, while there is still plenty of food in the countryside, a fifth of the combs should be left untouched, and a third of them at the second harvest, with the threat of winter in the air....

Then the whole group of combs should be collected in the spot where you are planning to make the honey. Here the openings in walls and windows must be sealed over to prevent access by the bees, who track down their honey as if it were lost wealth and devour it if they find it.... The honey should be made each day from whatever combs you have gathered, while they are still warm. A wicker basket or a loosely woven bag of fine twigs like the inverted cone used for filtering wine is hung up in a dark place, and the honeycombs are piled into it one by one.... Then, when the honey has melted and flowed down into the tub placed underneath, it is transferred into clay jars that are left uncovered for a few days while the stock is still frothing like new wine and has to be regularly skimmed with a ladle....

Though its monetary value is slight, we should still make some mention of wax as a by-product indispensable for a number of purposes. What remains of the combs after they have been pressed is first washed carefully in fresh water and then placed in a bronze pot; water is added and the combs are then melted over heat. Once this is done, the wax is poured off and strained through straw or rushes. The melting procedure is repeated, water is added, and the wax is poured off into moulds of the desired shape.

Compare Varro, *On Agriculture* 3.16.32–38; and Pliny, *Natural History* 11.44–45, for another account of harvesting honey.

3.67 Pisciculture: fish farming

Seafood played a significant role in the Graeco-Roman diet as attested by descriptions in texts, depictions in mosaics, paintings, and other abundant archaeological remains. Literary and archaeological evidence also provides substantial evidence of the Roman obsession to nurture fish not only for food, but for ostentatious display by constructing elaborate fish farms for favoured types of fish. During the late Republic and Empire, complex structures of all sizes were created to control water flow, salinity, and fish movement, in order to raise the fish. This expensive undertaking and the excessive enjoyment of these magnificent artificial environments is often criticised in the sources (Kron 2008, Higginbotham 1997).

AGRICULTURE

Varro, *On Agriculture* 3.17.2–4

There are two kinds of fishponds (*piscinae*), the fresh and the salt. The one is open to common folk and not without profit, where the Nymphs attend to the water for our domestic fish. But the seawater ponds of the nobility, for which Neptune attends to the fish as well as the water, appeal more to the eye than to the purse, and empty the pouch of the master rather than fill it. For first of all they are built at great cost, second, they are stocked at great cost, and third they are maintained at great cost [examples of specific costs follow as well as descriptions of elaborate *piscinae*]. For just as Pausias and the other painters ... have large boxes with compartments, where they have their different coloured pigments, so these people have compartmented ponds, where they have disparate varieties of fish kept separate ... [these fish are often not eaten but kept for visual use, almost as pets].

Columella, *On Agriculture* 8.17.1–3, 6

Columella (8.16.1–8.17.16) provides a lengthy account of fish farming, history, types of fish and their care, and the tanks. After criticism, he acknowledges the economic value of such farms where land is less fertile, then explains that different species of fish require different environments and discusses specific types of constructed tanks.

We rank as the best pond, without doubt, the one which is situated so that the incoming wave of the sea expels the water of the previous wave and does not allow stale water to remains within the enclosure. For that pond is most like the sea if it is moved by the winds and is constantly renewed and is not able to become warm, since it keeps rolling up a cold flow of water from the bottom to the upper layer. That pond is either cut in the rock, which only rarely occurs, or is constructed with *opus signinum*[28] on the shore. But however it is constructed, if it is kept cold by the whirlpool constantly flowing in, it also should contain cavities close to the bottom, some of them simple and straight to which the "scaly flocks" may withdraw, others twisted back into a spiral and not overly wide, in which the lampreys may lurk.... If the nature of the place permits, it is fitting that passageways be provided on every side of the fishpond. For the old layer of water is carried away more easily when an outlet lies open opposite from whatever side the wave presses in. [discussion of depths and submarine passages and recesses for shelter follow with the caution that water flow must be maintained].... Assuredly it will be proper to remember that bronze gratings with tiny holes should be fixed in front of the channels through which the fishpond pours out its waters, and by which the flight of the fish is impeded. And if the space allows, place throughout the parts of the pond rocks from the seashore, especially those which are covered with bunches of seaweed, so that it will not be foreign to the fish ... [Columella concludes with design details for specific types of fish and weather conditions, and feeding instructions].

AGRICULTURE

HUNTING AND FISHING

In antiquity, most hunting and fishing was done, not for sport, but to obtain food, though our literary sources inevitably tend to suggest otherwise. Because of its popularity as a sport among the upper classes, hunting figures prominently in ancient literature: Homer describes his heroes pursuing lions and boars (e.g. *Iliad* 18.573–586; *Odyssey* 19.429–446); Xenophon wrote a treatise on the subject, the *Cynegeticus*, which gives instructions chiefly for the capture of wild hares. The hunt is a common motif in the personal poetry of the Greeks and Romans, and the Romans transformed this rustic exigency into an urban pleasure with their elaborate productions of wild-beast fights (*venationes*) in the arena.

Fishing was an equally common pursuit, though because it was not a part of an aristocratic male's upbringing, our written sources are less full on the subject. Wall-paintings from Bronze Age Thera and Etruscan tombs depict young fishermen, and bucolic poets like Theocritus use the poor fisherman as a symbol of the simple pastoral life so admired by Hellenistic city-dwellers. But fish was an important element in the diet of all ancient peoples living on the shores of the Mediterranean,[29] which suggests an extensive commercial fishery with sizable fleets and special techniques for large-scale harvesting of fish; yet we have almost no literary evidence for this.

3.68 The hunter's tools

After giving the mythological origins of hunting, which included Centaurs, Perseus and Medusa, and other deities and heroes, Oppian provides a list of hunters' equipment (cf. **10.61** for nets).

Oppian, *Hunting* 1.147–157

And these are his weapons ...: hunter-nets and well-twisted, flexible twigs and long-tapered-net and hayes [a large net] and forked-props and grievous, slip-knot fetters, a three-barbed spear, a broad-headed lance, *harpalagos* [hunting implement] and light poles, and swift well-plumed arrows, swords, and heavy axes and a hare-slaying trident, curved [sticks?] with noose and leaden-bound crooks, esparto-bound cord and a well-plaited foot-trap, and ropes and stakes for nets and the much-meshed drag-net.

3.69 Hunting hare and boar

As an example of ancient hunting techniques, consider the following two passages that describe, from quite different perspectives, the pursuit of wild game: the first is a selection from Xenophon's detailed monograph on capturing hares, the second a typical anecdote from the bookish Pliny. The elaborate preparations outlined by Xenophon suggest no great sense of sport in the chase of a small and harmless animal, while Pliny's studied nonchalance belies the serious dangers of the boar-hunt.

AGRICULTURE

Xenophon, *On Hunting* 1.1, 18; 6.5–8; 11.1

Game and hounds are blessings from the gods, from Apollo and Artemis.... So, I advise our youth not to disdain hunting or any other element of their education. For it is through these pursuits that men become expert in war and other matters, from which necessarily evolves proficiency in thought, speech, and action....

The fellow who takes care of the nets should set out for the hunt wearing light clothing. He should set up the nets where paths converge – paths that are rough, on an incline, narrow, and shaded – and at creeks, gullies, and constantly fed streams, since it is here that game takes refuge; ... and the narrow lanes leading up to and through these places should be kept clear. These preparations should be made at daybreak, not earlier, so that if the line of nets is near where the beating will dislodge the game, the hares will not be frightened by hearing the noise close by.... He should plant at an angle the sticks used to prop the nets, so they can bear the strain of the animals' impact; and to the tops of the stakes he should attach nets of equal length and, by inserting equally spaced props, raise the pouch of the net higher in the centre. To the cord that runs around the [bottom of the] mesh he should tie a long, heavy stone to keep the net in place once the hare is inside. Finally, he should make sure that his line of netting is long and high enough to keep the hare from hopping over it....

Lions, leopards, lynxes, panthers, bears, and other wild animals of this sort are captured in foreign lands, some around Mts. Pangaion and Cissus east of Macedon, some on Olympus in Mysia and on Pindus, and some on Nysa beyond Syria and in other mountain ranges that can sustain such beasts. Because of the rugged terrain in the mountains, they are sometimes captured by drugging them with aconite, which the hunters mix with the animals' favourite food and lay out around water courses and other places where they gather.[30] Others are taken by armed horsemen who cut them off as they descend into the plains at night – but this is a dangerous way to capture them. At times men dig large holes, round and deep, leaving a column of earth in the middle; towards evening they place on the column a goat that they have hobbled, and then encircle the [edge of the] pit with a hedge of brushwood, leaving no gaps, so that the wild animals will not be able to see what is on the other side – so when they hear the bleating at night, they run around the edge, cannot find an opening, leap over the hedge, and are trapped.

Pliny, *Letters* 1.6 (to Tacitus)

You'll laugh, and well you may! I – the Pliny you know – have captured three wild boars, and especially fine ones, too. "You?" you ask. Yes, I – and without having had to depart at all from my usual laziness and inaction. I was sitting beside the hunting nets; I had no thrusting spear or javelin at hand, only my stylus and writing tablets; I was thinking over something and noting it down, so that even if I went away emptyhanded, I would at least have a full notebook.

You shouldn't look down on intellectual activity like this: it's amazing how the mind is stimulated by physical exercise. Being surrounded by forest, alone, amidst the silence that is a part of hunting: all these are important stimulants for thought. So, the next time you go hunting, you should follow my advice: take along your writing tablets as well as your basket of bread and flask of wine. You'll find that Minerva wanders the mountainsides as much as Diana.[31]

3.70 Methods to hunt and capture lions

Literary evidence indicates that lions were present in southeast Europe until the last century BC, and later were prized animals to be captured for entertainment. Oppian provides several methods to hunt and capture lions. Here the Libyan method is described once the den has been found and the lion's track to the river identified.

Oppian, *Hunting* 4.85–211

There indeed they dig out a walled-in pit, wide and very large; but in mid-trench they construct a great pillar, steep and high, from which they suspend high aloft a newborn lamb, dragging it from its mother having just given birth. Outside the pit they wreath around a circular wall, made dense with boulders one after another, so that the lion, when approaching, may not perceive the deceptive chasm ... [The bleating lamb draws the hungry lion who] ... suddenly springs over the wall and the wide, crowned chasm receives him unknowing, so that he comes into the depths of the unexpected pit ... [The lion wears itself out running and trying to escape].... And observing from their far-seeing outlook, the hunters rush forward, and having bound a plaited, skilfully wrought cage with well-cut straps, they send it down, having concealed some roasted bait of meat. Then the lion, immediately thinking to flee from the pit and rejoicing, springs in; no further return home is at hand [Oppian then describes other more physical methods: driving the lion into nets or armoured men ringing the lion and tiring it until it can be bound].

3.71 The importation of wild animals for the arena

The capture of wild animals, as indicated by Xenophon (**3.69**) has a long history, but by the Roman period, substantial demand for these beasts, primarily for display in arenas throughout the Empire, meant a great increase of effort to procure large varieties and numbers of them.

Historia Augusta, The Three Gordians 33.1

In Gordianus' time [AD 248] there were at Rome: 32 elephants, ... ten elk, ten tigers, 60 lions (tamed), 30 leopards (tamed), ten *belbi* (that is, hyenas), 1,000 pairs of gladiators belonging to the imperial treasury, six hippopotami, one rhinoceros, ten wild lions, ten giraffes, 20 wild asses, 40 wild horses, and innumerable other wild animals of various species. At the Secular Games Philip either dedicated or killed every one of them.

3.72 Poor fishermen

Theocritus' poem, while romanticising the simple and difficult life of fishermen, does provide us with a useful list of their angling equipment. Oppian provides a more prosaic inventory below.

Theocritus, *Idyll* 21.6–14

Two old stalkers of fish lay side by side on a bed of dried, mossy seaweed, bolstered by the leafy wall of their wattled hut. Near them lay the implements of their labours: their small baskets, their rods and fish-hooks, their weedy nets and horsehair lines, fish-traps and bow-nets of woven rushes, fishing lines, a pair of oars, and an aged dory resting on its props. Beneath their heads was a humble seaman's cloak, and thick coats were their coverlets. For these two fishermen, these were all their resources, this was their wealth.

3.73 The fisherman's tools and techniques

Oppian has just described the best attributes for a fisherman, then has discussed the best seasons and winds for fishing. Here he describes a variety of nets and other equipment needed to catch the cunning and deceitful fish by employing one of four methods (cf. **3.68** and **10.61** for nets).

Oppian, *Fishing* 3.72–91

Fishermen have developed four techniques of hunting sea prey. (1) Some are glad at heart with fish-hooks, and some of these men go hunting, having fastened a well-twisted horsehair line to long reed poles. Likewise others, having attached a flaxen line, cast it out from their hands, another takes delight in vertical lines or glories in horsehair lines with many hooks. (2) To arrange fish-nets is of more concern to others: some of them are called casting-nets, and some are called creels, small round oyster-nets, round fish-nets, and large drag-nets. They call others cover-nets, and, along with the large drag-nets are the ground-nets and spherical-nets and the curved fishing [trawling?] nets. Countless are the types of nimble flaxen nets with their treacherously wrought bosom-like hollows. (3) Again, others have their minds set more upon weels [conical baskets that trap animals once they enter the wide mouth], which gladden their masters slumbering at ease, and great gain follows from little toil. (4) Others thrust at the fish with the long, pronged trident from land or from a ship, as they desire.

3.74 Angling on the Moselle

Modern sport fishers will recognise the distressing scene described here in a long fourth-century AC poem celebrating the beauties of one of Europe's great rivers.

Ausonius, *The Moselle* 240–249

Now wherever the bank affords easy access to the river, a destructive horde probes all the depths for fish that are – what a pity! – not well protected by the river's secret places. This fellow well out in the middle of the stream drags his dripping nets and sweeps up whole schools of fish caught in their knotted mesh. Another, where the river glides along with gentle current, draws in his seine kept afloat by a string of corks. That one on the rocks leans out over the stream below and, bending the tip of his supple rod back in an arc, casts hooks baited with lethal delicacies.

Ausonius goes on to describe, in piteous detail, the final, gasping struggles of the fish that is landed.

3.75 On fly-fishing

From the previous selection we know that fly-fishing was practised at least in late antiquity. In fact, it was a much older technique than that. As the following passage reveals, by the early third century AC anglers had already progressed from using live bait to an artificial fly.

Aelian, *On the Nature of Animals* 15.1

This is the means of catching fish that I have heard is used in Macedonia, ... where the fish feed off native [horse]flies that flit over the river.... The fishermen know about this, but they don't actually use these insects as bait, because any human hand that touches them destroys their natural appearance: their wings wither away and the fish will not eat them or even approach them, using some mysterious sense to recognise those that have been caught. So, with a cleverness typical of anglers, the Macedonians outwit the fish by using the following ploy. They wrap crimson wool around each fish-hook and attach to the wool a pair of feathers from a cock's comb that resemble the colour of wax.... So, they drop this decoy, and the fish swims up to it, drawn by the exciting colour, assuming from the attractive vision that it is about to enjoy a wonderful feast. He opens his mouth wide, gets entangled with the hook, and enjoys a bitter feast – he's hooked.

SPONGE DIVING: TOOLS, TECHNIQUES AND DANGERS

Sponge was utilised in a variety of ways in antiquity and demand was great, but so were the risks and dangers for an activity that was brutally draining, both mentally and physically (Voultsiadou 2007).

3.76 Light: danger and necessity

Working under water presented difficulties for the sponge diver: on the one hand light skin "shone" and presented an inviting target for predators, on the other hand the degraded light made for difficult working conditions gathering the sponge. The following passage provide some evidence of efforts to improve both conditions (cf. **11.108**).

Aelian, *On the Nature of Animals* 15.11

There is also a small fish which humans called *galê*.... And since the species of these fish is carnivorous, all men spending their lives fishing and pursuing the deepest recesses blacken their feet and the palms of their hands, trying to obscure the light from them. For truly the limbs of men, particularly because they are shining bright in the water, are an attraction for these fish.

3.77 Calming the water with oil for better visibility

The practice of oiling the surface of the water "to calm" the water is a long one and continues today. The thin layer of oil on the water leads to energy dissipation, which prevents energy production into waves. The calmer water then reduces the effect of "wind-grip" on the surface of the water. In the first passage hunting "pearl-ichor" is being described but the diver is equipped like the sponge diver (cf. **11.108**).

Philostratus, *Life of Apollonius* 3.57

[When hunting for pearl-ichor in India, the inhabitants] watch for a still sea, or they themselves smooth the sea, and this they accomplish with an influx of oil; and then a man plunges down on the hunt for the oyster. He is in other regards equipped just like those shearing off the sponges from the rocks, although he has a rectangular iron block and an alabaster case of myrrh. [The myrrh is bait, the iron block has holes to capture the ichor, which flows out from the oyster when stabbed].

Plutarch, *Moralia* 950b–c

Plutarch has discussed the effects of air and water concerning white vs. black and light vs. dark. He now turns to the impact of oil on these opposites and its calming properties. Pliny (*Natural History* 2.106.234) wrongly assigns Plutarch's two uses of oil to sponge divers.

[Oil,] ... being sprinkled over the waves makes calmness upon the sea.... And peculiarly to oil, it furnishes light and clarity in the depths of the sea.... For not only for those passing the night at the surface, but also for the sponge divers below, the oil being spurted out of the mouth in the sea surrenders up a light in the water.

3.78 Attempts to resolve diving issues

[Aristotle] *Problems* 32.3 and 5 (960b15 and 21)

Why do divers bind sponges about their ears? Is it so that the sea in its violence does not rupture the ears? ... Why do the divers for sponges slit their ears and nostrils? Is it so that one may breathe more freely (cf. **11.106**)?

3.79 The sponge diver: cares, cautions, and death

Oppian, *Fishing* 5.612–74

Oppian provides one of the best texts about sponge divers in antiquity. Here he sets out the risks to the divers and their efforts to counter them in order to survive the dangers of the undertaking. Free divers did not have a great amount of time to carry out their exploration and recovery of the sponges, perhaps four to five minutes at depths of 30 or more metres. In order to descend quickly they used stone or lead weights tied to ropes for recovery, just as the divers themselves were tethered to help draw them up quickly when they were finished.

I declare there to be no other contest worse nor any endeavour more woeful for men than the task of the sponge-cutters. Indeed, these men, when they prepare for their undertaking, first take care to consume less-filling food and drink and calm themselves with sleep unbecoming to mere fishermen. [Like a singer preparing for a contest] so do they fervently take watchful care that their breath may hold out safe and sound when they journey down into the depths and that they may be refreshed from their earlier toil. But when they strive to accomplish their fearsome task, making their vows to the blessed gods, who rule the deep sea, they pray that they protect them from all hurt from the scourge of sea monsters and that they encounter no outrage of the sea.... A man is girded with a long rope above his waist and, using both hands, taking hold he grasps a heavy mass of cast lead in one hand and in his right hand he extends a sharp billhook; at the same time, he remains vigilant of the white oil held between the jaws of his mouth. Standing upon the prow he examines the swell of the sea, pondering his heavy task and the unfathomable water. [Comrades stir him to action].... But once he plucks up the courage in his heart, he leaps into the churning waters and as he springs the hurling of the heavy grey lead drags him down. Then, coming to the bottom, he spits out the oil, which shines strongly and the light mingles with the water, just like a torch showing its light through the darkness of the night. Approaching the rocks, he sees the sponges, which grow on the lowest ledges, fixed together on the rocks; and experience tells him that breath is in them [i.e. alive], even as the other things as many as grow upon the far-sounding rocks. Instantly rushing upon them, the sickle with his stout hand cuts, just like a reaper of sponges, nor is he about to tarry, but quickly yanks the plaited rope, signalling to his companions to pull him up swiftly. For hateful blood from the sponges drifts along straightway and swirls around the man; and too often the cloying current clogs his nostrils, and the heavy blood chokes the man. Because of this, swift as thought, he is pulled to the surface. And seeing him free from the sea, one would both rejoice and grieve pitying him in his collapse, with limbs numb with fear and heart-grieving weariness; his body is virtually paralysed. Often having leapt into the deep waters of the sea and come upon his most loathsome and harsh prey, he comes up no more, wretched man, having encountered some huge, monstrous beast of prey. Shaking repeatedly the link with his comrades, he exhorts them to pull him up again. And his shipmates and the mighty sea monster tear away at his half-dead body, pitiful to see, still longing for ship and companions. And grieving, they hastily quit those waters and their baneful struggle, and come back down to dry land, standing about weeping over the remains of their unhappy companion.

See Pliny, *Natural History* 9.70.151–153 for the dangers of dog-fish (sharks) and other terrors for the sponge diver.

3.80 Rousting an octopus

In antiquity just as today, divers hunted the octopus for food. Extracting them from their dens could be difficult since time was short, so techniques were developed to speed the process.

Athenaeus, *Philosophers at Dinner* 7.102 (316f.)

It is said that should one introduce salt into its lair, the octopus will immediately come out. Further, it is observed that when it flees in fear it changes colours and texture and assimilates those of the places in which it hides.

A SOBERING EPILOGUE

3.81 Humans and the Italian landscape

Lucretius here comments on the effect of human interference with the natural environment. He has just described (with some irony?) how farmers learnt to cultivate plants by observing and imitating their behaviour in the wild (a passage found at **1.13**); now he contemplates the spread of agriculture.

Lucretius, *On the Nature of Things* 5.1370–1378

Day by day people drove the forests to retreat further up the mountainside and to surrender the land below to their tillage, that they could own meadows, pools and canals, wheat fields and prosperous vineyards, and that a silver-green tract of olive trees could run between [their plots] to separate them, spilling over hills and valleys and plains. So today you see the whole landscape divided up with charming diversity: people embellish it by planting trees with sweet fruit, and keep it under control by filling it in with plantations that proliferate.

3.82 Three examples of ecological disaster

Pliny, *Natural History* 8.217–218

In Spain the animals they call rabbits are actually hares. They are prolific beyond measure and cause famine in the Balearic Islands by raiding the crops.... It is a fact that the islands asked Augustus (this was before he became a god) for military intervention against the spread of these hares.

AGRICULTURE

Pliny, *Natural History* 11.104–106

A plague of locusts is considered a sign of divine anger. Quite large ones are found, and in flight they make such a noise with their wings that they are thought to be birds. They block out the sun, causing whole nations to gaze skywards in fear that they might swarm over their lands ... Italy is plagued by locusts that come mainly from Africa, and the fear of famine has often prompted the Romans to turn to the Sibylline Books for prescriptions against them.[32] In the neighbourhood of Cyrene there is even a law that prescribes three campaigns against them every year: their eggs are squashed in the first, then the grubs, and finally the adult locusts; and anyone who does not actively participate is given the same punishment as a military deserter. On the island of Lemnos, too, they have prescribed a fixed number of locusts that each man has to kill and bring in to the magistrates. For the same purpose they also raise jackdaws that wipe out the locusts in a flying frontal assault. And people in Syria are under military orders to kill them. This pest spreads through all these areas of the world, but to the Parthians even locusts are a welcome part of their diet.

Pliny, *Natural History* 19.38–39

Next to be described is the *laserpicium*, which the Greeks call *silphium*, a plant of outstanding significance that was discovered in the province of Cyrenaica. Its juice, called laser, is an important source of nourishment and medication, and worth its weight in silver *denarii*. But it has not been found in Cyrenaica for many years now because the contractors who rent the fields for pasture, realising where their greater profit lies, wipe out the *silphium* by grazing flocks on it. A single stock is all that has been found there in my memory, and it was dispatched to the Emperor Nero. If ever a flock does happen upon a sprout of it, they give themselves away after eating it – a sheep by falling asleep on the spot, and a goat by a fit of sneezing.

Notes

1 Cf. Varro, *On Agriculture* 1.17.10:

> There are more than fifty authors whose Greek works contain scattered comments on a variety of [agricultural] topics, ... but their reputations are overshadowed by that of Mago the Carthaginian, who assembled these diffuse subjects in a 28-book compendium written in Punic.

Columella, who calls Mago "The father of husbandry", hints at the importance of this treatise when he mentions that it was translated first into Greek, and then into Latin by decree of the Roman Senate (*On Agriculture* 1.110–113). Neither the original nor the translations have survived.

2 In Roman society slaves were not allowed to marry, but were encouraged to cohabit, as much for economic as practical reasons, since their offspring, the *vernae*, were the property of their master.

3 See Pliny, *Natural History* 17.199: "The procedure of planting vineyards with trees has been condemned by the Sasernas, both father and son, but commended by Scrofa. These three are the earliest agricultural writers after Cato and are eminent experts".
4 For one example, notice the reference to a "public channel" for irrigation in **3.37**. There are, too, a number of extant laws from both Greece and Rome designed to discourage interruption or contamination of a natural watercourse to the detriment of a neighbouring farm.
5 See Herodotus, *Histories* 2.19–27, an attempt to determine the reasons for the Nile's annual flood.
6 After his defeat at Pharsalus by Caesar in 48 BC, Pompey fled to Egypt where he was murdered as he stepped from his ship onto the beach.
7 See Homer, *Odyssey* 24.225–231.
8 This is a typical etiological myth devised by the ancients to account for the origin of the word: the Latin *stercus* = "dung or turd".
9 See *Georgics* 1.77.
10 Robert Aitken, "Virgil's Plough", *Journal of Roman Studies* 46 (1956) 97–106; Fairfax Harrison, "The Crooked Plough", *Classical Journal* 11 (1916) 323–332.
11 The Greek goddess Demeter, identified by the Romans with their Ceres, protective divinity of the grain.
12 Longer calendars for other produce are well represented in the sources (e.g. **3.15** and Columella, *On Agriculture* 1.44 for vegetables, herbs and garnishes).
13 Sirius first appears on 26 July: see Columella, *On Agriculture* 11.2.53.
14 The same machine is briefly described by Pliny (*Natural History* 18.296):

> On the *latifundia* in the Gallic provinces huge frames with teeth set along the edge and with a pair of wheels are driven through the crop by a draught animal yoked to the rear; this tears off the ears, which then fall into the frame.

15 See Cato, *On Agriculture* 53.
16 See **9.27–31**.
17 See, too, Columella, *On Agriculture* 11.3.8–13, who places trenching in November for planting in the spring, and in May for an autumn planting.
18 See Cato, *On Agriculture* 41.2–4.
19 Inoculation was described by Pliny earlier in this book (17.100):

> Inoculation involves opening up an "eye" in the tree by cutting the bark with a metal punch similar to that used by a shoemaker, and then inserting in it a scion taken from another tree by using the same tool.

20 See Vergil, *Georgics* 2.67–72. Pliny's memory of the passage is faulty: Vergil has acorns, not cherries, growing on the elm.
21 Vergil, *Georgics* 1.273–275.
22 The term is explained by Varro (*On Agriculture* 2.2.18):

> "Skin-clad" sheep, like those of Tarentum and Attica, are covered with hides because of the quality of their wool, to prevent any staining of the fleece that makes it difficult to dye it properly, wash it, or scour it.

See **10.33**

23 Such controlled mating of goats suggested by Varro was necessary for reasons better explained by Columella (*On Agriculture* 7.6.3):

> The billy goat is suitable for breeding when he is seven months old, since his passion knows no bounds: even while still sucking he will try to mount his mother. So it is inevitable that he ages rapidly before the age of six, having

exhausted himself with premature lust while still in the prime of youth, and by the time he is five is considered less than suitable for impregnating nannies.

24 For Aeneas, the white sow with her 30 piglets was certainly a wondrous, and long-awaited, encounter (Vergil, *Aeneid* 3.390–391, 8.42–45, 81–83).
25 Columella (*On Agriculture* 7.9.4–5) expresses bewilderment that sows are thus prevented from breeding, but adds that boars, too, were made plumper by castration (for which see Columella 7.11).
26 This is presumably intended to cleanse the apiarist's breath, since bees are sensitive to the slightest smell. An alternative translation is equally plausible: "Dampen your face with water" as protection against sparks from the smoking torch.
27 The Pleiades rise about the time of the summer solstice and set half a year later.
28 A Roman hydraulic concrete composed of lime, sand, crushed terra cotta, and small stones.
29 Note, for example, the variety and cost of seafood in Diocletian's *Price Edict* (**11.136**).
30 Oppian (*Cynegetica* 4.316) describes using wine to drug and capture leopards.
31 Minerva (Athena) was the goddess of wisdom and the creative arts, Diana (Artemis) patron of the hunt.
32 The Sibylline Books, a collection of oracular utterances by a prophetic priestess of Apollo at Cumae in Italy, were frequently consulted by state officials for remedies against natural disasters.

Bibliography

Agriculture

André, Jacques, *Les Noms des plantes dans la Rome antique*. Paris: Belles Lettres, 1985.

Burford, Alison, *Land and Labour in the Greek World*. Baltimore: Johns Hopkins University Press, 1993.

Dalby, Andrew, *Geoponika: Farm Work: A Modern Translation of the Roman and Byzantine Farming Handbook*. Totnes, Devon: Prospect Books, 2011.

Frayn, Joan M., *Subsistence Farming in Roman Italy*. London: Centaur Press, 1979.

Fussell, George E., *The Classical Tradition in West European Farming*. Newton Abbot: David & Charles, 1972.

Halstead, Paul, "Traditional and Ancient Rural Economy in Mediterranean Europe: *plus ça change*?" *Journal of Hellenic Studies* 107 (1987) 77–87.

Halstead, Paul, *Two Oxen Ahead: Pre-Mechanized Farming in the Mediterranean*. Malden, MA; Oxford; Chichester: Wiley Blackwell, 2014.

Heurgon, Jacques, "L'Agronome Carthaginois Magon et ses traducteurs en Latin et en Grec". *Comptes Rendus de l'Académie des Inscriptions et des Belles-Lettres* (1976) 441–456.

Isager, Signe, and Jens Skydsaard, *Ancient Greek Agriculture: An Introduction*. London: Routledge, 1992.

Jashemski, Wilhelmina F., *The Gardens of Pompeii, Herculaneum, and the Villas Destroyed by Vesuvius*. New Rochelle: Caratzas, 1979.

Lewit, T., *Agricultural Production in the Roman Economy, A.D. 200–400*. British Archaeological Reports, International Series, Suppl. 568. Oxford: British Archaeological Reports, 1991.

Margaritis, Evi, and Martin K. Jones, "Greek and Roman Agriculture". Pp. 158–174 in John P. Oleson, ed., *The Oxford Handbook of Engineering and Technology in the Classical World*. Oxford: Oxford University Press, 2008.

Précheur-Canonge, Thérèse, *La vie rurale en Afrique d'après les mosaïques*. Paris: Presses Universitaires de France, 1961.

Rickman, Geoffrey E., *Roman Granaries and Store Buildings*. Cambridge: Cambridge University Press, 1971.

Rossiter, Jeremy J., *Roman Farm Buildings in Italy*. Oxford: British Archaeological Reports, 1978.

Spurr, M.S., *Arable Cultivation in Roman Italy, c.200 B.C.—A.D. 100*. London: Society for the Promotion of Roman Studies, 1986.

Thibodeau, Philip, "Greek and Roman Agriculture". Pp. 519–532 in Georgia L. Irby, ed., *A Companion to Science, Technology, and Medicine in Ancient Greece and Rome, Vol. II*. Chichester: Wiley-Blackwell, 2016.

Thompson, Dorothy B., and R.E. Griswold, *Garden Lore of Ancient Athens*. Princeton: American School of Classical Studies at Athens, 1963.

Wells, Berit, ed., *Agriculture in Ancient Greece*. Gothenburg: Åström, 1992.

White, Kenneth D., *Roman Farming*. Ithaca: Cornell University Press, 1970.

White, Kenneth D., *Bibliography of Roman Agriculture*. Reading: University of Reading, 1970.

White, Kenneth D., *Farm Equipment of the Roman World*. Cambridge: Cambridge University Press, 1975.

White, Kenneth D., *Country Life in Classical Times*. London: Paul Elek, 1977.

Animal husbandry, hunting, fishing

Anderson, John K., *Ancient Greek Horsemanship*. Berkeley and Los Angeles: University of California Press, 1961.

Campbell, Gordon L., ed., *The Oxford Handbook of Animals in Classical Thought and Life*. Oxford: Oxford University Press, 2014.

Crane, Eva, *The Archaeology of Beekeeping*. London: Duckworth, 1983.

David, Simon J.M., *The Archaeology of Animals*. New Haven: Yale University Press, 1987.

Forbes, Robert J., "The Coming of the Camel". Pp. 193–213 in Robert J. Forbes, *Studies in Ancient Technology*, Vol. 2. 2nd edn. Leiden: Brill, 1965.

Fraser, Henry M., *Beekeeping in Antiquity*. 2nd edn. London: University of London Press, 1951.

Frayn, Joan M., *Sheep-Rearing and the Wool Trade in Italy during the Roman Period*. Bristol: Cairns, 1984.

Froehner, Reinhard, *Kulturgeschichte der Tierheilkunde*, Bd. 1: *Tierkrankheiten, Heilbestrebungen, Tierärzte im Altertum*. Konstanz: Terra, 1952.

Georgoudi, Stella, *Des chevaux et des boeufs dans le monde grec*. Paris: Daedalus, 1990.

Ghigi, Alessandro, *Poultry Farming as Described by the Writers of Ancient Rome (Cato, Varro, Columella, and Palladius)*. Milan: R. Bertieri, 1939.

Higginbotham, James A., *Piscinae: Artificial Fishponds in Roman Italy*. Chapel Hill and London: University of North Carolina Press, 1997.

Ikeguchi, Mamoru, "Beef in Roman Italy". *Journal of Roman Archaeology* 30 (2017) 7–37.

Jennison, George, *Animals for Show and Pleasure in Ancient Rome*. Manchester: University of Manchester Press, 1937.

Keller, Otto, *Die antike Tierwelt*. 2 Vols. Leipzig: Engelmann, 1909–1913.

Kitchell, Kenneth F. Jr., "Animal Husbandry". Pp. 533–549 in Georgia L. Irby, ed., *A Companion to Science, Technology, and Medicine in Ancient Greece and Rome, Vol. II.* Chichester: Wiley-Blackwell, 2016.

Kron, Geoffrey, "Animal Husbandry, Hunting, Fishing, and Fish Production". Pp. 175–224 in John P. Oleson, ed., *The Oxford Handbook of Engineering and Technology in the Classical World.* Oxford: Oxford University Press, 2008.

Kron, Geoffrey, "Animal Husbandry", in Gordon L. Campbell (2014) 109–135.

Lazenby, Francis D., "Greek and Roman Household Pets". *Classical Journal* 44 (1949) 245–252, 299–307.

Meadon, Richard H., and Hans-Peter Uerpmann, eds., *Equids in the Ancient World.* Wiesbaden: Teichert, 1989.

Merlen, Rene H.A., *De Canibus. Dog and Hound in Antiquity.* London: J.A. Allen, 1971.

Metzger, Henri, "Les images de faune aquatique et marine dans l'art grec". *Revue des Études Grecques* 103 (1990) 673–683.

Radcliffe, William, *Fishing from the Earliest Times.* 2nd edn. London: Murray, 1928.

Toynbee, Jocelyn M.C., *Animals in Roman Life and Art.* Ithaca: Cornell University Press, 1973.

Vigneron, Paul, *Le cheval dans l'antiquité greco-romaine.* Nancy: Université de Nancy, 1968.

Voultsiadou, Eleni, "Sponges: An Historical Survey of their Knowledge in Greek Antiquity". *Journal of the Marine Biological Association of the United Kingdom.* 87 (2007) 1757–1763.

4

FOOD PROCESSING

Together with shelter and clothing, food is traditionally seen as one of humankind's essential requirements. While the consumption of food is a common theme in almost all ancient literary genres, its preparation is less fully described. Even lengthy accounts of extravagant banquets inevitably ignore the elaborate preparations that took place ahead of time, and in any case, they reflect the experience of only the upper social strata. Petronius' wonderful description of Trimalchio's dinner (Petronius, *Satyricon* 26–78) bears no closer relationship to the average Roman's evening meal than Apicius' recipe for stuffed flamingo. So, we must be especially careful in making generalisations about diet and culinary technology from the imbalanced evidence we are given by the ancients themselves.

Between its harvesting in the countryside and its appearance on a table in Athens or Rome, most food had passed through many hands. Almost all the agricultural products described in the previous chapter were subjected to further processing to make them edible and to prevent them from spoiling quickly. Three of the most important foods of antiquity – flour, olive oil, and wine – all needed fairly complicated preparatory treatment, for which ever more efficient and specialised machines were developed, so that originally simple tasks of processing food once undertaken at home or on the farm gradually came to be assumed by skilled professionals.

In early times, for example, grain was either pounded into flour by a mortar and pestle (**4.1**), or ground in a simple saddle-quern, both of which could be managed by any householder. Even the superior efficiency of the rotary quern – originally a pair of circular stones, the upper with a vertical handle for turning it (**4.2**) – benefited the domestic baker, until it was improved and enlarged into the enormous "hourglass" design of rotary mills powered by donkeys or slaves and familiar from Roman sites throughout the Mediterranean (**4.3**; see also **2.12**). The great quantity of flour produced from these machines was then made into a gruel, the staple of many ancient diets, or baked into bread (**4.4–8**) to feed the urban dwellers who were as ignorant as modern consumers about the origin and treatment of their food.

Similar machines were developed for extracting the liquid from olives and grapes. Beginning with simple devices like a flat stone rotated over a trough full of olives, the technology for crushing and pressing evolved into one of the most mechanically advanced in antiquity, imaginatively employing the principles of rotary motion, the lever, the wedge, and the screw to make the tasks easier and more efficient with olive mills like the *trapetum* and beam-presses whose force was exerted by windlasses or screws (**4.10–12, 17, 19**).

Other food products, too, underwent preliminary processing: combs into honey (**3.66**), milk into cheese (**4.22**), grain into beer (**4.20**), wine into sauces (**4.24**). It all seems, in fact, quite contemporary, except for the problem of preserving food: here, lacking modern techniques of refrigeration and canning, the ancients relied, where they could, on the simple techniques of smoking and salting, though we can assume that, like contemporary societies without rapid transport links to warmer climates, they depended much more on seasonally available foods than most of us do today.

For reasons of space, there is little here that treats the ancient diet specifically, and no recipes from Apicius to illustrate the culinary technology of the ancients. Some information about the former can be culled from other selections in this volume, like the list of foodstuffs in Diocletian's *Price Edict* (**11.136**); for the latter, prospective Roman chefs are directed to one of the several good translations of Apicius by scholars who have tested the recipes in their own kitchens (Paulas 2016, Curtis 2008).

MILLS AND BREAD

4.1 Methods of milling

Pliny, *Natural History* 18.97–98

Not all grains are easy to mill, for which Etruria can serve as obvious proof: here, the ear of emmer wheat is first roasted, then crushed with a pestle fitted with an iron tip in a mortar whose inner surface has toothed ridges radiating from the centre; if the millers put their weight behind their pounding, the result is that they can shatter the iron tip while breaking up the grains. The greater part of Italy uses a plain pestle, as well as water wheels and the millstone.

As for the precise method of milling, I shall quote the opinion of Mago, who instructs that wheat should first be soaked in a fair bit of water, then hulled, dried in the sun, and worked on again in a mortar. The same technique can be used for barley, except that 20 *sextarii* of grain should be moistened with two *sextarii* of water. Lentils, he says, should first be roasted and then lightly milled together with bran or with the addition of a bit of unbaked brick and half a *modius* of sand for every 20 *sextarii* of lentils. Vetch similarly. Sesame should be steeped in warm water, spread out [to dry], then rubbed vigorously and immersed in cold water so the chaff can float to the surface, and again spread out in the sun on a linen sheet: unless this is done with all speed, it turns pale yellow and begins to go mouldy.

4.2 Operation of a hand mill

Hand mills are first mentioned by Homer, who describes 12 female servants toiling over the mills in Odysseus' palace, producing coarse-ground barley groats and finer wheat flour (*Odyssey* 20.106–111; cf. 7.104). Those mills may have been simple querns like the ones commonly found in Bronze Age sites on Crete, for example: the hand-held upper stone, shaped like the inverted hull of a boat, was moved back and forth over the grain placed on a larger, flat lower stone. More efficient than this reciprocal action was the rotary motion of the hand mill affectionately described here as part of the early-morning activities of the rustic Roman farmer Simylus.

[Vergil], *Moretum* 16–29, 39–51

A meagre heap of grain was poured upon the ground, from which he helps himself to as much as his measure would hold, amounting to 16 pounds in weight. He leaves his storeroom and takes his position beside the mill, placing his trusty lamp on a small shelf firmly fixed to the wall for just such a purpose. Then from his clothing he frees his two arms and, first putting on an apron of hairy goatskin, he sweeps the stones and hollow of the mill with a brush made of tail. He mobilises his two hands for the task, allotting a job to each: his left is given to feeding the mill, his right to the work of turning and driving the unceasing spin of the wheel, while the left from time to times helps out her weary sister by taking her turn – and the grain passes through, braised by the stone's swift strokes....

Once his work of turning has made up the proper amount, he transfers handfuls of the bruised meal from the mill to a sieve and shakes it. The husks remain behind in the sieve, while the flour filters through the holes and falls out clean and pure. At once he lays it out on a smooth table, pours warm water over it, mixes together the meal and liquid and forms a ball of it, and kneads it by hand until it is firm, now and again sprinkling the mound with salt. Once he has tamed the dough, he smoothes it off and with his palms presses it out into a circle and impresses lines to divide it into four equal pieces. He then places it on the hearth, where his wife has already cleaned off a suitable spot, covers it with tiles, and heaps up the fire on top.

4.3 A donkey mill

The earliest reference to a rotary mill whose upper stone was apparently turned by animal power – the predecessor of the well-known hourglass mills found in bakers' shops in Ostia, Pompeii, [**fig. 4.1**] and many of the provincial cities of the Roman Empire – is in a passage of Aristotle (*Problems* 964b.38) that describes sensations: among the various things that cause people to shudder are the tearing of cloth, a saw being sharpened, and the movable upper millstone grinding directly against the fixed lower stone, with no grain between – and the word he uses for the rotating millstone is the Greek for "donkey", the beast that by this time must have been a common enough sight in bakeries to have given its name to the stone it turned. Such mills were a necessity for the large urban populations of the Roman Empire, the date of the next passage. The narrator has been miraculously transformed into a donkey and is made to work the mill. Compare another passage from Apuleius quoted above (**2.19**).

FOOD PROCESSING

Apuleius, *The Golden Ass* 7.15

For without a moment's hesitation his mercenary wife – to my mind a completely unscrupulous woman – had me yoked to a grinding mill and, by beating me continuously with a leafy stick, she supplied bread for herself and her family out of my hide. Not content with exhausting me for her own food, she would make a profit from my going round and round by having me grind her neighbours' grain as well. And in return for all these labours I was distressed not to be given even the prescribed diet. For my own barley that I broke up and ground down during my circuits around that very millstone she used to sell to farmers nearby, while for me – after a day spent working at that toilsome machine – she would serve up an evening meal of dirty, unsifted grain husks, made gritty by its liberal dose of stone fragments.

It is clear from Pliny (*Natural History* 18.112) that forced labour in the mills was not restricted to donkeys: "Gruel is made from emmer.... Its grain is pounded in a wooden mortar to prevent it from being ground to a powder by the hardness of stone. The pestle, as is well known, is worked by the effort of chained convicts". A famous passage from Vitruvius (*On Architecture*) 10.5 (**2.13**) shows that water-driven grain mills were known in the first century BC, but they were not common until late antiquity (see also **2.14**).

Figure 4.1 Donkey mill in Pompeii.

Source: Adapted from Adam, 1994, figs. 734–737 and White, 1984, figs. 54–55.

4.4 Types of wheat and their yield

The principal types of cereal crops used for food in the ancient Mediterranean were wheat and barley. Of the wheat, considered here by Pliny, the most important species were called by the Romans *far* and *triticum*: the former (*triticum dicoccum*=emmer or spelt), described by Pliny in *Natural History* 18.83–84, was the earlier of the two, a coarse grain that was difficult to thresh, its husk needing to be removed by pounding;[1] the latter became the common type of wheat, its varieties called by Pliny "hard wheat" (*triticum durum*) or "soft" and "common wheat" (*triticum vulgare*) (Foxhall and Forbes 1982, Rickman 1980).

Pliny, *Natural History* 18.86–90, 92

From soft wheat comes the finest bread and the most famous baked goods. In Italy, top place goes to a blend of grains from Campania and Pisa: the former are reddish, the latter whiter and heavier, like chalk. A reasonable yield from the Campanian grain that they term "emasculated" is four *sextarii* of fine flour from a *modius* (or five *sextarii* from a *modius* of common grain that has not been "emasculated"), together with half a *modius* of choice flour, four *sextarii* of ration-bread (called "seconds"), and four *sextarii* of bran....

A *modius* of flour from Celtic soft wheat yields 20 pounds of bread, from the Italian variety two or three pounds more, in the case of bread baked in a metal pan; for oven-baked loaves they add two pounds for either variety.

The best flour made from hard wheat is from Africa. A reasonable yield is half a *modius* from a *modius*, and five *sextarii* of fine flour (... they use this in the production of bronze and paper), as well as four *sextarii* each of second-quality flour and bran. A *modius* of flour from hard wheat yields 22 pounds of bread, or 16 pounds if the *modius* is choice flour.

The price for this in an average market is 40 *asses* for a *modius* of flour, 48 *asses* for flour from hard wheat, and 80 *asses* for flour from the "emasculated" soft grain.[2]

The sweetest bread comes from *arinca*, a denser grain than emmer with a larger and heavier ear. Rarely does a *modius* of this grain fail to yield 16 pounds [of bread]. In Greece it is difficult to thresh, which is why Homer,[3] who calls it *olyra*, says it is fed to the draught animals; but the same grain in Egypt is easy [to thresh] and gives a high yield.

4.5 The manufacture of leaven

In early antiquity bread was always simple and unleavened, as we read in Genesis 18.6: "And Abraham hastened into the tent unto Sarah, and said, 'Make ready quickly three measures of fine meal, knead it, and make cakes upon the hearth'". Even after the discovery of leavening agents like those described below, this traditional type of bread remained popular.

Pliny, *Natural History* 18.102–104

The principal use of millet is in making leaven: mixed with unfermented wine and kneaded, it will last a year. A similar leaven is made from the best and most powdery bran of the wheat itself, which is marinated for three days in unfermented white wine, then kneaded and left to dry in the sun. In breadmaking, they first dissolve tablets made of this leaven, add emmer flour, and bring the mixture to the boil; this is then mixed with the flour to produce, in their opinion, the best bread. The Greeks have established a workable proportion of eight ounces of leaven for every two half-*modii* of flour.

Though these kinds of leaven can be made only during grape harvests, you can at any time you like make a leavening agent out of water and barley: two-pound loaves of this mixture are baked in ash and charcoal on a hot hearth or in an earthenware dish until they turn reddish-brown, after which they are sealed in containers until they turn sour; when soaked in water they become leaven....[4]

These days leaven is made from the flour itself, which is kneaded before the addition of salt. It can be boiled down into a kind of mush, and then left until it turns sour, though in general they do not bother with this simmering process, but rather use some dough left over from the day before. Two things are equally obvious: that fermentation occurs naturally from sourness, and that people fed on leavened bread have weaker bodies, as you might expect since our ancestors attributed special wholesomeness to whichever wheat was heaviest.

4.6 Kinds of bread

Pliny, *Natural History* 18.105–106

It seems unnecessary to describe the various kinds of bread itself – sometimes named after the side-dishes served with it (like oyster bread), at other times after its special delicacy (like cake bread), or even at times after the speed with which it can be made (like hasty bread). It can also be called after the way it is baked (like oven bread, pan bread, or bread made in a clay baker), and quite recently there was even a type imported from Parthia called water bread, ... since water is used in drawing it out into a thin consistency full of holes like a sponge.

The highest quality is derived from the excellence of the wheat and the fineness of the sieve. Some knead the dough with the addition of eggs or milk, and there are even peoples who use butter, when the absence of war allows them to transfer their attention instead to the various styles of pastry-making.

Cato, *On Agriculture* 74

Recipe for kneaded bread: Wash thoroughly your hands and a bowl. Put the flour in the bowl, add water a little at a time, and knead well. When you have kneaded it thoroughly, form it into a loaf and bake it under an earthenware lid.

4.7 White bread or dark?

From the lines of satirists of the early Roman Empire, we can deduce that, although the healthfulness of wholewheat bread was recognised (Petronius, *Satyricon* 66.2), those who could afford it preferred the more refined white loaves as a sign of social superiority. Juvenal's text is a lament for the poor but honest client, humiliated by being given inedible food while his patron enjoys the finest.

FOOD PROCESSING

Juvenal, *Satires* 5.67–75

Notice the grumbling with which another has offered you some bread that you can hardly break in two, and mouldy bits of hardened dough, the sort of stuff to get your back teeth going while resisting every bite. At the same time a soft and snowy white loaf moulded from the finest flour is reserved for the master of the household. Be sure to keep your paws off it and maintain the necessary respect for the bread pan. But imagine yourself giving in just a little to temptation – and there's somebody there to make you put it back. "You're a presumptuous table companion! You really must fill yourself up from your *proper* breadbasket and learn the colour of your bread".

4.8 Professional bakers appear in Rome

In both Greece and Rome, the grinding of grain and baking of bread were traditionally done at home until urban development in both cultures made it possible (and more efficient) to support professionals, who normally functioned as both millers and bakers: in Athens this occurred in the fifth century BC, and in Rome, as we learn from this passage, around 170 BC. This is a particularly fine example of Pliny's scholarly sleuthing, as he looks for corroborating evidence in datable sources. Plautus' *Aulularia* was produced sometime before the playwright's probable death in 184 BC.

A frieze from the first century BC that decorates the Tomb of Eurysaces the baker in Rome [**fig. 4.2**], shows the individual steps involved in bread production in the large-scale establishment owned by the tomb's owner during his lifetime. It depicts the grinding of grain, the mechanical kneading of dough, the shaping of the dough into loaves, and the baking and weighing of bread (Petersen 2003).

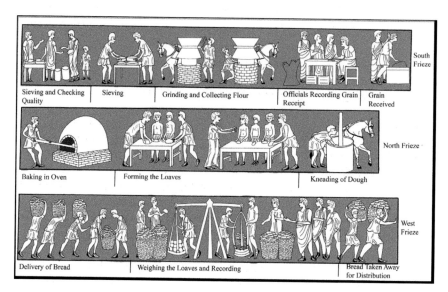

Figure 4.2 Relief decoration on the Tomb of Eurysaces, the baker, showing the individual steps in breadmaking, Rome, first century BC.

Source: Composite of reliefs *in situ*, Rome, Italy.

164

Pliny, *Natural History* 18.107–108

There were no bakers in Rome before the war with Perseus, more than 580 years after the city was founded. The citizens used to make their own bread, a task that belonged particularly to the women, as it still does nowadays for most races. Plautus already mentions the Greek word for bakers in the play he titled *Aulularia*,[5] which has prompted considerable controversy among the learned about whether this line can be attributed to the poet himself. It is, in fact, confirmed by a sentence in Ateius Capito, to the effect that it was customary in his day for cooks to bake bread for the grander households, and that only those who ground (*piso*) emmer were called bakers (*pistores*); and rather than having cooks on their permanent staff, people would hire them from the market.

4.9 Rice

Just as grain was the dietary staple of the ancient Mediterranean, so rice served the same function in Asia (and maize in pre-Columbian America). Though known to the Greeks and Romans (see, too, Strabo *Geography* 15.1.13; Horace, *Satires* 2.3.155), it seems never to have made its way westwards.

Pliny, *Natural History* 18.71

The inhabitants of India are especially fond of rice, from which they make the sort of drink for which the rest of the world uses barley. The leaves of rice are fleshy, similar to the leek's but broader. It grows to the height of a cubit, with a purple blossom and a root that is round like a stone.

PRESSES, WINE, AND OIL

To produce both grape juice and olive oil, it was necessary first to crush the fruit, then squeeze out the liquid. The first stage was simple in the case of grapes: the fruit was gently trodden by naked feet, producing only a small amount of liquid (*mustum lixivium*) that was reserved for special uses (see **4.24**). For the second stage – extraction of the juice or must – a variety of presses common to both wine and olive production were invented in the classical period, most notably those that created pressure through levers or screws [**fig. 4.3**]. Before the olive could be pressed its pulp had to be bruised, but without cracking the bitter pit, a difficult procedure for which a small number of special mills like the *trapetum* [**fig. 4.4**] were designed. This type of mill also allowed the *amurca*, or bitter watery liquid, which was heavier than the oil, to be drawn off from bottom of the vat and the best oil to be extracted even before the pulp was pressed for the poorer quality oil (Frankel *et al.* 2016, Amouretti *et al.* 1984, 1993, Rossiter 1981).

FOOD PROCESSING

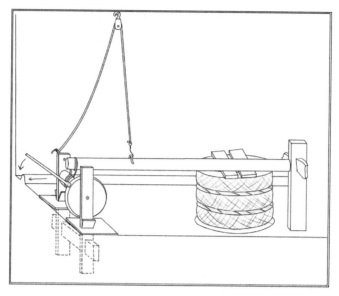

Figure 4.3 Lever press.
Source: Adapted from Adam, 1994, fig. 724.

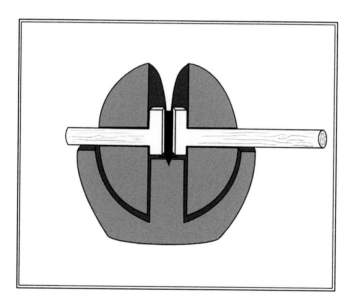

Figure 4.4 Trapetum olive press.
Source: Adapted from Forbes, 1965.3, fig. 24, Adam, 1994, figs. 726–727, and White, 1984, fig. 63.

4.10 Pressing the grapes for wine

Varro, *On Agriculture* 1.54.3

As for the grapes that have been trodden, their stalks should be put under the press together with their skins to squeeze into the same vat whatever must they still contain. When there is no more flow from under the press, some people trim off the edges and press again; they call this second pressing *circumsicium* [literally, "cut around"] and keep separate what it produces because it has taken on the taste of the iron knife. The pressed skins of the grapes are collected in large, wide-mouthed jars, to which water is added: this is called *lora* because the grapes have been washed [*lota*], and is a winter substitute for the labourers' allotment of wine.

The first part of this passage appears in **3.46**. For vineyard work leading up to the gathering of the grapes, traditionally undertaken in early autumn, see **3.44–45**.

4.11 Constructing and equipping a winery

Palladius, *On Agriculture* 1.18

We should have a winery that faces north, is cool or as dim as possible, and at a distance from the baths, stables, oven, dung hills, cisterns, standing water, and other places that give off a terrible stench. It should be furnished with enough equipment to match the harvest and should be laid out in a way that, like the plan of a basilica,[6] it has its treading floor built at a higher elevation with access by (normally) three or four steps rising between a pair of tanks inserted one on each side to catch the wine. From these tanks, built channels or clay pipes run around the outside walls and pour the wine, flowing through tributary channels, into large basins placed along its edge. If there is a rather large vintage, the intervening space can be assigned to barrels, which we can keep from blocking the passageways by placing them on higher stands above the tanks, separated by a rather wide space between them to allow a passage for the attendant as the need arises. But if we decide on their own space for barrels, it can be solidly built with slightly raised curbs and a tiled pavement, like a treading floor, so that even if a cask has been broached by mistake, the leaking wine will not be lost but can be caught in a vat placed underneath.

4.12 Using various styles of grape press

In Palladius' winery the juice was extracted from the grapes by the age-old technique of treading. Yet, to judge from the agricultural handbooks of Cato, Varro, and Columella, most establishments were fitted out with presses that could extract more of the residual juice.[7]

Pliny, *Natural History* 18.317

A reasonable basis for calculation is that a single pressing should fill 20 *cullei*, for which – together with vats for 20 *iugera* [of vineyards] – a single lever-press is enough. Some use individual presses, though it is more expedient to use a pair, however large the single presses. Length is what counts with these beams, not their thickness: the long ones press better.

Our forebears used to pull down the beams using ropes and straps and by levers;[8] but within the last 100 years the Greek system was adopted, in which the grooves of a vertical screw run in a spiral. Four turning-spikes are fitted to the screw by some, while in other examples the screw lifts up boxes of stones with it – a design that is particularly recommended.[9] Within the past 20 years we have discovered how to use small presses and a less spacious pressing shed: a shorter vertical screw projects downward onto the middle of the pressing discs, which exert a continuous downward pressure on the grape sacks set beneath; a heavy stone weight is placed on top of the apparatus [adding to the force of the screw].

4.13 Protecting wine during fermentation

Macrobius, *Saturnalia* 7.12.13–16

Avienus speaks again: "Hesiod says that, when you get to the middle of a jug of wine, that is the part that should be hoarded, while you can fill up by consuming the rest – indicating clearly that the best wine is found in the middle of the bottle. But it has been proved by experience that, in the case of olive oil, the best floats on top, while for honey the best is at the bottom. My question, then, is this: why are the best oil, wine, and honey thought to come from the top, the middle, and the bottom respectively?"

Disarius responds without hesitation: "The honey that is best is heavier than the rest; in a jar of honey, the portion at the bottom is certainly the heaviest, and for this reason is more valuable than the honey floating on top. In wine bottles, on the other hand, contact with the sediment renders the bottom portion not only cloudy but inferior in flavour, while the top part is spoiled by its proximity to the air, contact with which spoils it. It is for this reason that farmers, not happy with laying down their wine jars indoors, bury them and strengthen them by applying a coat of something to the outside, preventing as much as possible the wine's contact with the air, whose harmful effects are so obvious that wine is scarcely safe in a full bottle, let alone when exposed to the air. Once you drink from the bottle and create a space for contact with the air, the rest that remains is all contaminated as well. The further the middle part of the bottle is separated from the two extremes, the more removed it is from harm, as neither disturbed nor spoiled".

4.14 Pitch and other additives to wine

The ancients frequently added various supplements to their wines, believing them to improve the flavour as well as to help in preservation: spices, drugs, and aromatic herbs; pitch, turpentine, and even seawater. Modern vintners are somewhat more discriminating, adding little more than sulphites to stabilise and preserve their products. Readers fond of Greek *retsina* will especially enjoy the following passage (and might want to read more about the various ways of using liquid pitch to preserve wine in Columella, *On Agriculture* 12.22–24), while those who are not will be happy to echo Pliny's final comment here.

Pliny, *Natural History* 14.124, 126–130

The method of preserving wine involves the sprinkling of pitch on the must when it begins to ferment, which is over within nine days at most. In this way the wine is tinged with the scent of the pitch and a bit of its sharp flavour....

As for the other additives to wine, such attention is paid to them that some people add doses of ash and elsewhere gypsum; ... but ash made from the branches of vines or from oak is the additive of choice. Even seawater is prescribed for this same purpose, as long as it is obtained from the open ocean by the spring equinox and thereafter kept in storage or at least is collected at the solstice on a night with a north wind blowing, or – if collected at the time of the vintage – first reduced by boiling.

The pitch especially recommended in Italy for wine-storage vessels comes from the region of Bruttium and is made from the resin of the spruce tree.... The more assiduous vintners mix in black mastic [a mineral pitch], which is produced in Pontus and looks like bitumen, as well as the root of iris and olive oil. It has been found that wine from vessels treated with wax turns sour; and that it is less harmful to transfer wine into vessels that once contained vinegar than into ones used for sweet wine or mead.

Cato recommends that wine be "conditioned" – that is his word – with the addition to each *culleus* of a fortieth part of lye[10] boiled with *defrutum*, or a pound and a half of salt, and from time to time crushed marble. He mentions sulphur as well, but resin he puts at the very end. As the wine begins to age, he recommends topping up the whole batch with the must he calls *tortivum*, which I understand to be the juice from the final pressing.

And we know that colours are added, like some kind of stain, to tint the wine, and that this makes it more full-bodied. We use so many poisons to force our wine to taste the way we want, and yet we are surprised that it is noxious!

4.15 Storage of wine

Pliny, *Natural History* 14.133–135

Climate makes a great difference even for wine that has already been processed. In the region of the Alps they store it in wooden kegs that they surround with hoops, and in very cold weather they use fires to keep off the chill. Though rarely reported, it has been observed from time to time that these kegs burst open and the wine is left standing in frozen blocks – it is like a miracle, since wine does not naturally freeze but otherwise is only made sluggish by the cold. People in milder zones store their wine in jars that they bury in the earth either completely or only partially covering the belly, to protect them from the elements; other places use overhanging roofs to keep off the weather....

The shapes [of the storage jars] are also important: those with pot bellies and wide mouths are less suitable.... They must not be filled to the top, and the space left should be smeared with a mixture of *passum* or *defrutum*, and saffron or iris pounded together with boiled-down new wine. The caps of the jars should be treated similarly, with the addition of mastic or pitch from Bruttium.

The opening of wine jars in winter is viewed with general disfavour, except on a clear day, and not when the south wind is blowing or the moon is full.

4.16 Varieties of wine

Pliny describes at length the various wines from different centres of production (Italian wines alone are treated in *Natural History* 14.59–72). In the following two passages we read of noble wines of exceptional age and of a cheap wine substitute made from the lees. Athenaeus, *Philosophers at Dinner* 1.26a, sets out reasons for preferring old wine to young.

Pliny, *Natural History* 14.54–55

... and Pramnian wine – the same wine whose praises Homer sang – is still highly esteemed; it is produced around Smyrna, near the sanctuary of the Mother of the Gods. There was no special distinction to any particular variety of the other wines, though the year when Lucius Opimius was consul (the same year the tribune Gaius Gracchus was assassinated for inciting the common people by his factional discord) was famous for the quality of every type of wine: that year [121 BC], the sun did its work, and the weather was bright and warm for what they call the "cooking" of the grape. Those wines have kept until now, almost 200 years later, though (as is natural for wines of a mature age) they are reduced to the consistency of unrefined honey; ... still, they can be used in small amounts as a treatment to improve other wines.

Pliny, *Natural History* 14.86

Leftovers taken from the grape pressing and steeped in water make a drink – called "seconds" by the Greeks, *lora* by Cato and the rest of us – that we cannot properly term wine but is still listed among the wines drunk by the working class. It comes in three varieties; ... but none of them is good for more than a year.

4.17 Preparing olives for processing

For orchard work preliminary to the olive harvest, see **3.48–50**.

Varro, *On Agriculture* 1.55.4–6

The olive follows the same two paths to the farmstead as the grape: one part for solid food, the other to flow forth as liquid to anoint the body both internally and externally. So it is that the olive follows its master into the baths and gymnasium!

The share from which oil is made is usually piled on the wooden floor, one heap each day, where it can turn slightly mushy.[11] And each pile is sent in order via the earthenware casks and the olive jars to the *trapeta*, which are olive mills made from hard stone with roughened surfaces.[12] Once the olives are picked, if they remain too long in their piles, they grow soft from the heat and their oil goes rancid. So, if you cannot process them at the proper time, you should expose them to air by stirring them up in their piles.

4.18 Preserving olives

Columella, *On Agriculture* 12.50.1–3

The cold of winter follows, when the olive harvest (like the vintage) again demands the attention of the overseer's wife.... So, when your olives have already turned black but are not yet quite ripe, it is time to pick them by hand when the weather is fair, to sieve them and separate out those that will seem spotted or spoiled or of inferior size. Then, to every *modius* of olives add three *heminae* of pure salt, pour the lot into wicker baskets, and spread lots more salt over the top to cover the olives. Let them sweat thoroughly for 30 days, allowing all the lees to drip through [the wickerwork]. Next, pour them into a trough and with a clean sponge wipe off the salt to prevent it from penetrating; then put them in a jar and top up the vessel with must that has been reduced by boiling, placing on top a plug of dried fennel to keep the olives down. Most people mix three parts (some only two) of reduced must or of honey with one part vinegar, and season their olives with this decoction.

4.19 The production of olive oil

The making of oil from olives required the crushing of the fruit in a *trapetum* to break its skin and the pressing out of its liquid in a press similar to that used for grapes.

Pliny, *Natural History* 15.5

It takes greater skill to master the production of oil than of wine, since the oils that come from a single olive tree are not in fact the same. The immature olive gives the very first oil, before it has begun to ripen, and its flavour is quite extraordinary. And it is a fact that the first stream of this oil to come from the press is the richest, the quality decreasing with each successive issue, whether the olives are pressed in wicker baskets or – a recent invention – with the mush enclosed between thin strips of wood.

Columella, *On Agriculture* 12.52.1, 6–7, 10–11

The middle of the olive harvest usually falls at the beginning of December (oil made before this time, called "summer oil" is bitter), and it is round about this month that the green oil is pressed, and afterwards the ripe oil....

Rotary mills (*molae*) are more expedient for making olive oil than *trapeta*.[13] They allow for very easy operation, since they can be lowered or raised according to the size of berries to avoid spoiling the flavour of the oil by crushing the pit. The *trapetum* in turn does more work more easily than tramping with clogs on the olives in a vat. There is also an instrument like a vertical threshing sledge, called a *tudicula*, which accomplishes its purpose with no real trouble except that it frequently breaks down and becomes jammed if you throw in a few too many berries. Local conditions and custom determine the choice of machines I have described; but the rotary mill works best, and then the *trapetum*....

Once the olives have been carefully cleaned, they must be taken immediately to the press, put into new woven bags while still whole,[14] and placed under the press so that as much as possible can be pressed out in a short space of time. Then, once the rinds have been opened and softened and two *sextarii* of pure salt have been added for each *modius*, the pulp will have to be pressed out either with slats, if that is the local custom, or at least with new woven bags. Then the fellow with the ladle must immediately empty what has flowed down first into the round basin (which is better than a square one made of lead or a two-chambered concrete one) and pour it into earthenware vats made ready for this purpose.

There should be three ranks of vats in the oil cellar, one to receive the first-quality oil (that is, from the first pressing), another for the second-quality, and a third for the third. For it is extremely important not to mix the first two issues, let alone the third with the first; the flavour of the first pressing is better by far, because it has not felt as much the force of the press and has flowed like grape juice. When the oil has stood for a little in the first vats [of each rank], the ladler will have to strain it off into the second row of vats, and then into subsequent ones right up to the last. For the more often it is aerated by being transferred [from one vat to another], rather like being agitated, the clearer it becomes and it loses its dregs.[15]

Pliny, *Natural History* 15.23

To press more than 100 *modii* is not recommended. This is called a pressing [or batch], ... and it is normal for three pressings to be carried out in a day and night by four men at a time using a double basin [beneath the press].

OTHER FOODS AND DRINKS

4.20 Other alcoholic beverages

In areas of the Mediterranean where grapes were not grown, people turned to local products for making alcoholic drinks: fermented dates, for example (Xenophon, *Anabasis* 2.3.14–16 describes the arrival of the Greek soldiers in a village where they found palm wine, a sour, boiled date drink and the edible but headache-producing crown of the date palm; cf. **3.53**), or honey mead, or beer from fermented grains, a drink probably as old as agriculture itself.

Plutarch, *Moralia* 4.6 (672b)

The Jews used honey as wine[16] before the appearance of vineyards. Even up to the present those non-Greeks who do not make wine drink mead, whose sweetness they reduce with the addition of bitter roots that have something of a wine flavour.

Diodorus of Sicily, *History* 5.26.2–3

Extreme cold can ruin the otherwise temperate climate of Gaul, which cannot produce wine or olive oil. So, the Gauls who find themselves without these fruits prepare a beverage from barley, which they call beer.[17] They also soak their honeycombs in water and use the resulting infusion as a drink. But they have an excessive addiction to wine and saturate themselves with the vintages imported by traders, which they drink neat.

Pliny, *Natural History* 14.149

People who live in the west have their own form of intoxicating drink. There are different techniques used in the various districts of Gaul and Spain, and the drink goes by different names, but the basic principle is the same: grain is steeped in water. We have already learned from our Spanish provinces that these beverages even age well. And Egypt, too, has invented for its own consumption similar grain-based drinks – intoxication is constantly with us in every corner of the world![18] – and these liquids they actually drink straight, not adding any water to make them less potent, as we do with wine. God knows, we used to think that Egypt produced grain for bread; but men's vices are wonderfully resourceful, and a way has been found to make even water intoxicating!

4.21 Water

A pure and constant supply of water was uncommon in many areas of the Mediterranean, which explains in part the prevalence of beverages whose alcoholic content made them safer to drink. Given the agricultural exploitation of many rivers upstream from their urban functions, we can assume that the following experience of Horace with fouled water was not unusual. Still, despite such widespread contamination and the absence of almost any fresh water in the arid coastal regions of the eastern Mediterranean, Aristotle's principle of distillation seems never to have been practically applied.

Horace, *Satires* 1.5.1–8

I left the metropolis of Rome, accompanied by Heliodorus, easily the most learned of the Greeks. I put up at an unpretentious inn in Aricia; then on to Forum Appii, full of sailors and money-hungry innkeepers. We were lazy and divided this leg in two, though more serious travellers than we would do it in one; but the Appian Way is easier to bear if you take it slowly. It was here that I declare war on my stomach because of the water, which was pretty foul, and wait with no great patience as my fellow travellers have dinner.

Aristotle, *Meteorologica* 2.3 (358b)

I can say from experience that salt water becomes fresh water when it is vaporised, and the vapour, when it condenses again, does not turn back into salt water.

Without artificial means of refrigeration, the ancients could rely only on naturally occurring snow and ice to cool their food and drinks (cf. **2.32**). But the expense and impracticalities of preserving and transporting water in its solid state made it unavailable to any but the wealthiest – hence the common literary complaint that the use of these natural refrigerants not only was decadent, but violated the very laws of nature.

Pliny, *Natural History* 19.55

Varieties of water are differentiated as well: the power of money has separated the very elements of nature into classes. Some people drink snow, others ice, turning what is an unpleasant condition of the mountains into an agreeable experience for their appetite. Water is preserved frozen for the hot seasons and a way is being sought to keep snow cold for the months when it does not occur naturally. Some bring to a winter temperature even the water they have previously boiled. It is a fact that what pleases nature is unacceptable to humans.[19]

4.22 Cheesemaking

In antiquity as now, cheese made a contribution to the Mediterranean diet in the general absence of fresh milk and meat. The first passage, from the imaginary recollections of Odysseus, describes the mythical home of the Cyclops Polyphemus, whose processing of sheep and goats' milk reflects techniques from the late Bronze and Archaic Ages; the selection from Columella adds details but reveals how little that tradition of cheesemaking changed over the centuries.

Homer, *Odyssey* 9.218–223, 237–239, 244–249

We went into the cave and gazed in wonder at each thing in turn. There were crates full of cheeses, and pens crowded with lambs and kids. They were confined separately [by age], the spring lambs, summer lambs, and newborns each by themselves. All the handmade vessels were overflowing with whey, the pails and bowls into which he milked them.... He drove his plump herds into the broad grotto ... at least, all those that he milked ...but the rams and billies he left outside in the extensive courtyard.... Sitting down, he milked the ewes and bleating nannies duly in turn, then placed the newborns under their mothers. Without a pause he curdled half the white milk, collected it in braided baskets, and stored it away; the other half he set in bowls, to have it within reach to drink for his supper.

FOOD PROCESSING

Columella, *On Agriculture* 7.8.1–5

Cheese should be made from whole milk that is as fresh as possible, since it quickly takes on a sour flavour if it is left standing or mixed with water. It should generally be curdled using rennet from a lamb or a kid.... When the pail has been filled with milk, it should not be kept for any length of time before heating; and yet the pail ought not to come into contact with an open flame (which is a popular technique with some people) but should be stood near the fire. The moment the liquid has congealed, it should be transferred to wicker filters or baskets or moulds; for it is especially important that the whey be allowed to filter through and separate itself from the solid matter at the first possible moment....

Once the cheese has been removed from the moulds or baskets, to keep it from going bad it is placed on very clean boards in a cool, shady spot, and even still is sprinkled with ground salt to draw out its sour liquid. When hardened, it is compacted by pressing it with great force; it is again lightly sprinkled with absorbent salt, and again compressed with weights. When this procedure has been followed for nine days, the cheese is washed with fresh water and set out in rows on specially designed wickerwork racks, no one touching another, until it is fairly dry. Then, to keep it from becoming too leathery, it is packed tightly onto several shelves in an enclosed space protected from the winds.

In his short note on cheesemaking, Varro (*On Agriculture* 2.11.3–4) includes a list of the principal types of cheeses. For goat cheese in particular, see Vergil, *Georgics* 3.400–403; The milk they extracted at dawn and during the day, they press at night; what they milk in the dark and at the setting sun they carry off in the morning, packed in baskets, when the goatherd goes to town; or they sprinkle it with a little salt and lay it down for the winter.

4.23 Sugar

The principal sweetening agent in antiquity was honey, for descriptions of which see **3.66** and, for its many varieties from the ancient Mediterranean, Pliny, *Natural History* 11.32–42. As the following passage from Pliny reveals, sugar was only vaguely recognised by the Romans.

Pliny, *Natural History* 12.32

Arabia also produces *saccharon*, but the most prized comes from India. It is a kind of honey that has collected in bamboo, white like gum, easily broken by the teeth, in pieces no larger than a hazel nut and used only for medicine.

Note that this is not beet-sugar, but cane-sugar, to which Strabo (*Geography* 15.1.20) also alludes: "Concerning the reeds, Nearchus says that they produce honey, although there are no bees around".

4.24 Roman sauces

The ancients made extensive use of various sauces in their preparation of food, in part to alleviate the blandness of their diet, in part to camouflage the taste of rotting foods that is common in societies without easy access to refrigeration. The most famous Roman version was *garum* or *liquamen*, a fermented, fish-based sauce used not just in main courses but even in desserts and so valued that varieties were shipped around the Mediterranean.[20] Another common base for these sauces was wine in the form of must, which was transformed into *defrutum* (a thick syrup from reduced grape juice), *passum* (raisin wine), and *mulsum* (a mixture of honey and wine, not unlike mead), all of which figure prominently in surviving recipes from the Roman period (Curtis 1991).

Geoponika 20.46.1–2

The Making of Garums

The so-called *liquamen* is made thus. The entrails of fishes are tossed into a vessel and salted; and tiny fishes, especially smelts or tiny red mullets or cackerels or anchovies, or any that seem to be small, are all salted likewise and pickled in the sun, being agitated often. When pickled by the heat, the *garos* is taken from them as follows: a large, dense basket is inserted into the middle of the vessel of the earlier-mentioned fishes, and the *garos* flows into the basket.[21] And thus, the so-called *liquamen* is taken up, being strained through the basket; the remaining residue makes *alix*. [Other methods follow.]

Columella, On Agriculture 12.19–21, 39, 41

Defrutum

Must with the sweetest taste possible will be reduced to a third [of its original volume], and when boiled down, as I have suggested, is called *defrutum*. Once it has cooled off, it is transferred to jars and put down to be aged a year before it is used.

Passum

Mago[22] gives the following recipe for the best *passum*, one that I myself have used. Harvest early-maturing grapes that are fully ripened, discarding the berries that are mouldy or bruised. You want to create a platform of reeds, so take forks or stakes, fix them in the ground at four-foot intervals, and join them together with [a grid of] shoots, then arrange reeds on top of them. Spread out the [bunches of] grapes in the sun, covering them at night to keep the dew off. Then, when they have shrivelled up, pluck off the raisins and throw them into a vat or smaller clay vessel, and add to them enough of the finest must to cover them. On the sixth day, when they have soaked up the must and are saturated with it, gather them into a filter bag, bear down on them in a wine press, and draw off the *passum*.

Mulsum

You can make the finest *mulsum* as follows. Remove from the vat, as soon as you can, the must called *lixivum*, which has trickled out from the very first treading of the grapes[23].... Add ten *librae* of the finest honey to one *urna* of this must, carefully blend the mixture, store it in a flagon, immediately seal it with plaster, and have it placed in a storeroom.... After 31 days you will have to open the flagon, strain the must into another vessel, seal it up again, and store it in the smokehouse.

4.25 Food preservation

In the absence of artificial means of refrigeration, the Greeks and Romans relied principally on salting, smoking, and drying foodstuffs to preserve them for consumption out of season. Salt, then, was a particularly valuable commodity in antiquity, and fortunate was the town situated near natural salt beds like those at the mouth of the Tiber River, just a few kilometres downstream from Rome (see **5.31** for procurement of salt).

In the following passages, Cato first gives detailed instructions for curing ham, Pliny then describes the preservation of seasonal fruits using a variety of methods from around the Mediterranean, Columella discusses exclusion of air, and Varro mentions the importance of cool, dry storage space (cf. **4.13–14** and **4.18**) (Curtis 2016).

Cato, *On Agriculture* 162

You must salt hams in a jar or crock as follows. Once you have bought the hams, cut off their hocks. [Allow] a half-*modius* of powdered Roman salt for each ham. In the bottom of a jar or crock spread a layer of salt, then put in the ham skin-side down, and cover the whole thing with salt. Then put another ham on top and cover it in the same way, making sure that one piece of meat does not touch another. Cover all the hams in this way. Once you have arranged them all, pour more salt on top until no meat is visible, and level it off. When they have been in the salt for five days, remove each one with its own salt; put them back in reverse order and cover and arrange them as before. After 12 days altogether, remove the hams, wipe off all the salt, and hang them out in a breeze for two days. On the third day, use a sponge to wipe them down thoroughly, rub them with oil, and hang them in smoke for two days. On the third day, take them down, rub them with a mixture of oil and vinegar, and hang them on a meat rack. No maggots or worms will touch them.

Pliny, *Natural History* 15.62–65

After the overripe berries are removed with shears, but leaving the hard, hammer-shaped shoot on the branch, the grapes should be suspended in a large, wide-mouthed jar that has just been coated with pitch and then made airtight by sealing its lid with plaster. The same technique works for sorb-apples and pears, all of whose shoots should be smeared with pitch.... Some put them up in jars with a residue of wine, but making sure that the grapes do not come into contact with the liquid; others store apples floating [in wine] in terracotta pans, a method they think imparts the wine's flavour to the fruit. Some prefer all fruit of this kind to be preserved in millet, but the conventional way is to bury it in a two-foot-deep pit on a bed of sand, covered with a clay lid and with soil on top of that.

Some go so far as to coat grapes with potter's clay, dry them in the sun, and then hang them up, washing off the clay when the grapes are to be consumed (they use wine to remove the clay from other kinds of fruits). In the same way they apply a coating of plaster or wax over the choicest apples – but if the fruit has not already ripened, it expands and bursts the crust.... Others devote a separate terracotta jar to each individual apple and pear, seal their lids with pitch, and place them inside a second vessel. And there are even those who nestle the fruit in tufts of wool in boxes that they coat with a mixture of mud and chaff....

The area of the Ligurian coast nearest the Alps first dries its grapes in the sun, then wraps them in rush bundles and stores them in jugs sealed with plaster. The Greeks follow the same procedure, but instead use leaves (of the plane-tree, of the vine itself, or fig leaves) dried for a day in the shade, and they layer grape skins between the fruit. The grapes from Kos and Beirut are preserved in this way, and their sweetness is second to none.

Columella, *On Agriculture* 12.16.5

... they carefully smear with plaster the covers [of storage containers], which have been treated with pitch to prevent the air from entering.

Varro, *On Agriculture* 1.59.2

Importance of cool, dry space: And this reason, too – to make it cooler – they [builders of fruit houses] coat the ceilings, walls, and floors with marble cement.

HONEY AS PRESERVATIVE: FOOD AND ELSEWHERE

Honey, the principal sweetener of antiquity (**3.66**), was used in drink and food, but also as an ingredient in health treatments (**10.49, 104, 112, 115, 118**) and, here, as a preservative for dyes, fruits, and human bodies.

Plutarch, *Alexander* 36. 1–2

When Alexander became master of Susa, he gained 40,000 talents of coined money ... [other items of wealth are mentioned].... Among which they report was discovered 5,000 talents weight of cloth dyed with purple from Hermione, still preserving its freshness and lustre even though it was ten years short of 200 years old. The reason for this, they say, was the dipping of the purple dye in honey and the white dye in white olive oil. For when these substances are used for equal time, they are seen to have brilliancy, pure and gleaming.

FOOD PROCESSING

Columella, *On Agriculture* 12.47.2–4

After describing two other methods using clay and plaster to preserve fruit, Columella turns to the best method: submersion in honey.

[After picking ripe, unblemished quince and wiping off the down and arranging them loosely to prevent bruising, place them] in a new flagon with a very wide mouth. Then, when they have been stowed up to the neck of the vessel, they should be confined with willow-twigs laid across them in such a way that they compress the fruit slightly and do not allow them to be lifted up when they have liquid poured upon them. Next, the vessel should be filled up to the top with the very best and most liquid honey, so that the fruit is submerged. This method not only preserves the fruit itself but also provides a liquor, which has the flavour of honey water and can be given ... to sufferers from fever.... [This method is so sure] that even if there is a small worm in the fruits, they do not deteriorate any further once they have the liquid described above added to them; for such is the nature of honey that it checks any corruption and does not allow it to spread, and this is the reason too why it preserves a dead human body for very many years without decay.

Columella (*On Agriculture* 12.4–16, 43–51) gives exhaustive detail on preservation and includes instructions on making vinegar, brine, and sour milk, pickling herbs, and preserving olives, vegetables, and cheeses as well as fruits.

4.26 Cooking utensils and procedures

We have no comprehensive list of kitchen equipment, cooking utensils, or cutlery from antiquity; most of our evidence for these comes from archaeological artefacts. What follows is a glossary of a few specialised implements and cooking techniques culled from the pages of a Roman cookbook attributed to the first-century AC gourmand Apicius. Though not a complete inventory for the kitchen and *triclinium*, it gives a good impression of the variety of culinary tools in a wealthy Roman household; and from the methods of preparation, for example, we can deduce the existence of a variety of knives and spoons (there being no evidence that a fork was used in antiquity in the kitchen or at table).

Apicius, *On Cooking*

[The Stove]
stationary oven
portable earthenware oven
grill
grid-iron
tiles for baking
hot ashes for cooking
[Utensils for Food Preparation]
mortar and pestle
colander
strainer

bouquet for flavouring
sausage skin for stuffing
mould
various receptacles for liquids
pâté tub
[Methods of Preparation]
chopping, dicing
creaming
marinating
smoking
frying
boiling
steaming
baking
roasting
braising
[Cooking Vessels]
roasting pan
frying pan
saucepan
large cauldron for boiling joints of meat
shallow pan
shallow earthenware cooking dish
shallow bronze cooking dish
small, deep vessel for cooking sauces
deep pan for layered dishes
double-boiler, *bain-marie*
[Serving Utensils]
metal serving dish
circular serving plate
serving dish for mushrooms
small spoon
ladle
flask

Notes

1 Hence the Latin word *pistor*, which comes to mean "miller" and even "baker", actually derives from the "pounding" of *far*, the evolution of a word reflecting the changing occupation (see **4.8**).
2 Compare the prices some 250 years later, in Diocletian's *Edict on Maximum Prices* (**11.136**).
3 See Homer, *Iliad* 5.195.
4 Compare *Natural History* 18.68: "When those varieties of Gallic and Spanish wheat we have already discussed are steeped to make beer, the foam that forms on the

surface is used as leaven, which is why bread from those regions is lighter than others".
5 Line 400.
6 Palladius probably is comparing the treading floor to the nave (or the apse?), the flanking tanks to the aisles. This passage is a difficult one. A good discussion, together with a conjectural reconstruction of the winery, can be found in H. Plommer, *Vitruvius and Later Roman Building Manuals* (Cambridge, 1973); K.D. White, *Farm Equipment of the Roman World* (Cambridge, 1975) 142 gives an abbreviated discussion.
7 The types of presses mentioned in this passage are discussed, with drawings, in White (see here in note 6) 230–232.
8 Cato (*On Agriculture* 18–19) gives full (though confusing) details about the construction of a lever-press (*prelum*), in which pressure is exerted on the bag of grapes set on a platform beneath a horizontal beam pivoted at one end between vertical posts, the other end drawn down by ropes or thongs attached to a windlass turned by handspikes.
9 For an explanation of this bewildering design, see White (see here in note 6) 230–231.
10 *Cinis lixivus*, a surprising additive. In fact, Pliny seems to have conflated two separate passages from Cato: *On Agriculture* 23.2 calls for improving (?) wine by adding *mustum lixivum*, apparently the juice from the first pressing, while 114.1 talks of "conditioning" grapes in advance by covering the vine's roots with ash (*cinis*), which will eventually produce a laxative wine.
11 Compare Pliny, *Natural History* 15.21: "The bitterest olive makes the best oil; ... olives stored on a wooden floor shrivel and lose their quality".
12 Cato (*On Agriculture* 20.1–22.2) gives a detailed description of the design of a *trapetum*: it was basically a stone mortar resembling a Bundt pan, with two rounded millstones attached vertically to a horizontal bar pivoting on the mortar's central column.
13 For the differences in design, see White (see here in note 6) 227–229.
14 Columella here seems to be outlining a method of processing uncrushed olives, which might be a variant of the technique found in Pliny, *Natural History* 15.23:

> It was later discovered that, by washing olives in rapidly boiling water and then immediately putting them under the press while still whole, the lees are pressed out, and then they can be crushed in a *trapetum* and pressed a second time [for the oil].

15 See Pliny (*Natural History* 15.22), who emphasises also the need for cleanliness in the operation: "Oil must be ladled frequently each day, using a cup shaped like a conch shell, into lead kettles (it is spoiled by copper)".
16 Ironically, the Greek word used here for wine (*methu*), in fact comes from the Sanskrit for honey, and has given us our word "mead".
17 Elsewhere (*History* 13.11) Diodorus makes pejorative comments about Celtic beer, a beverage always viewed by the Romans with a mixture of wonder and arrogant scorn, subscribing as they did to the common Mediterranean belief that consumption of wine was an integral part of civilisation:

> The Celts at that time knew nothing of wine from the vineyards or oil produced by olive trees, which we have. As a substitute for wine they used a foul-smelling juice made from barley left to rot in water, and for oil rancid pig-fat with a disgusting smell and flavour.

18 See Pliny, *Natural History* 14.137–148 for a lengthy condemnation of intemperance and some stories of famous drinking bouts.

19 Xenophon (*Memorabilia* 2.1.30) also berates the person who drinks without being thirsty, and who looks everywhere for snow in summertime. Similar disgust at such extravagance is expressed by Seneca (*Natural Questions* 4.13), who describes how snow is compressed to make it last into the summer, and is then brought down from the mountains by pack animals, wrapped in straw that spoils its colour and taste: so even *water* costs money now!

20 Several storage vessels found in the ruins of Pompeii and Herculaneum bear the inscription *liquamen optimum*. The principal source for the recipe is the *Geoponika* but see also Pliny, *Natural History* 31.93–94 and Curtis (1991).

21 Columella, *On Agriculture* 12.38.7 mentions the use of flax rush strainers for making flavoured wine.

22 See **3.3**.

23 See **3.46**.

Bibliography

Diet and food preparation

Amouretti, Marie-Claire, and Jean-Pierre. Brun, eds., *La production du vin et de l'huile en Méditerranée*. Bulletin de Correspondance Hellénique, Suppl. 26. Paris: de Boccard, 1993.

André, Jacques, *L'alimentation et la cuisine à Rome*. 2nd edn. Paris: Belles Lettres, 1981.

Andrews, Alfred C., "Orach as the Spinach of the Classical Period". *Isis* 39 (1948) 169–172.

Andrews, Alfred C., "Oysters as a Food in Greece and Rome". *Classical Journal* 43 (1948) 299–303.

Andrews, Alfred C., "Celery and Parsley as Foods in the Greco-Roman Period". *Classical Philology* 44 (1949) 91–99.

Andrews, Alfred C., "The Carrot as a Food in the Classical Era". *Classical Philology* 44 (1949) 182–196.

Andrews, Alfred C., "Melons and Watermelons in the Classical Era". *Osiris* 12 (1956) 368–375.

Andrews, Alfred C., "The Parsnip as a Food in the Classical Era". *Classical Philology* 53 (1958) 145–152.

Berthiaume, Guy, *Les rôles du mágeiros: Étude sur la boucherie, la cuisine et le sacrifice dans la Grèce ancienne*. Leiden: Brill, 1982.

Blanc, Nicole, and Anne Nersessian, *La Cuisine romaine antique*. Paris: Faton, 1993.

Brothwell, Don, and Patricia Brothwell, *Food in Antiquity: A Survey of the Diet of Early Peoples*. London: Thames & Hudson, 1969.

Bruns, Gerda, *Küchenwesen und Mahlzeiten*. Archaeologia Homerica, II, Q. Göttingen: Vandenhoeck & Ruprecht, 1970.

Cerchiai, Claudia, *L'alimentazione nel mondo antico: I romani nell' età imperiale*. Rome: Istituto Poligrafico dello Stato, 1987.

Curtel, Georges, *La vigne et le vin chez les Romaines*. Paris: Masson, 1903.

Curtis, Robert I., *Garum and Salsamenta*. Leiden: Brill, 1991.

Curtis, Robert I., "Food Processing and Preparation". Pp. 369–392 in John P. Oleson, ed., *The Oxford Handbook of Engineering and Technology in the Classical World*. Oxford: Oxford University Press, 2008.

Curtis, Robert I., "Food Storage Technology". Pp. 587–604 in Georgia L. Irby, ed., *A Companion to Science, Technology, and Medicine in Ancient Greece and Rome, Vol. II.* Chichester: Wiley-Blackwell, 2016.

Dalby, Andrew, *Siren Feasts: A History of Food and Gastronomy in Greece.* London: Routledge, 1996.

Darby, William J., Paul Ghalioungui, and Louis Grivetti. *Food: The Gift of Osiris.* 2 Vols. London: Academic Press, 1977.

Davies, Roy W., "The Roman Military Diet". *Britannia* 2 (1971) 122–142.

Donahue, John F., "Culinary and Medicinal Uses of Wine and Olive Oil". Pp. 605–617 in Georgia L. Irby, ed., *A Companion to Science, Technology, and Medicine in Ancient Greece and Rome, Vol. II.* Chichester: Wiley-Blackwell, 2016.

Donahue, John F., "Nutrition". Pp. 618–632 in Georgia L. Irby, ed., *A Companion to Science, Technology, and Medicine in Ancient Greece and Rome, Vol. II.* Chichester: Wiley-Blackwell, 2016.

Forbes, Robert J., "Food, Alcoholic Beverages, Vinegar", "Food in Classical Antiquity", "Fermented Beverages 500 B.C.–1500 A.D.", "Salts, Preservation Processes, Mummification". Pp. 51–209 in Robert J. Forbes, *Studies in Ancient Technology*, Vol. 3. 2nd edn. Leiden: Brill, 1965.

Forbes, Robert J., "Sugar and its Substitutes in Antiquity", Pp. 80–111 in Robert J. Forbes, *Studies in Ancient Technology*, Vol. 5. 2nd edn. Leiden: Brill, 1966.

Foxhall, Lin, and H.A. Forbes, "*Sitometreia.* The Role of Grain as a Staple Food in Classical Antiquity". *Chiron* 12 (1982) 41–90.

Frankel, Rafael, Shmuel Avitsur, Etan Ayalon, *History and Technology of Olive Oil in the Holy Land.* Arlington: Olèarius, 1994.

Frankel, Rafael, Shmuel Avitsur, Etan Ayalon, "Oil and Wine Production". Pp. 550–569 in Georgia L. Irby, ed., *A Companion to Science, Technology, and Medicine in Ancient Greece and Rome, Vol. II.* Chichester: Wiley-Blackwell, 2016.

Hagenow, Gerd, *Aus dem Weingarten der Antike: Der Wein in Dichtung, Brauchtum und Alltag.* Mainz am Rhein: von Zabern, 1982.

Haussleiter, J., *Der Vegetarismus in der Antike.* Berlin: Topelmann, 1935.

Kritsky, Gene, *The Tears of Re: Beekeeping in Ancient Egypt.* Oxford; New York: Oxford University Press, 2015.

Jasny, Naum, *The Wheats of Classical Antiquity.* Baltimore: Johns Hopkins University Press, 1944.

Matthews, Kenneth D., "*Scutella, Patella, Patera, Patina.* A Study of Roman Dinnerware". *Expedition* 11 (1969) 30–42.

Miller, J. Innes, *The Spice Trade of the Roman Empire.* Oxford: Clarendon Press, 1969.

Moinier, Bernard and Olivier Weller, *Le sel dans l'antiquité, ou les cristaux d'Aphrodite. Realia.* Paris: Les Belles Lettres, 2015.

Murray, Oswyn, and Manuela Tecusan, *In Vino Veritas.* Rome: British School in Rome, 1995.

Nriagu, Jerome O., *Lead and Lead Poisoning in Antiquity.* New York: John Wiley & Sons, 1983.

Paulas, John, "Cooking and Baking Technology". Pp. 570–586 in Georgia L. Irby, ed., *A Companion to Science, Technology, and Medicine in Ancient Greece and Rome, Vol. II.* Chichester: Wiley-Blackwell, 2016.

Petersen, Lauren H., "The Baker, His Tomb, His Wife, and Her Breadbasket: The Monument of Eurysaces in Rome". *The Art Bulletin* 85.2 (2003) 230–257.

Rickman, Geoffrey, *The Corn Supply of Ancient Rome*. Oxford: Clarendon Press, 1980.
Seltman, Charles, *Wine in the Ancient World*. London: Routledge & Kegan Paul, 1957.
Slater, William A., ed., *Dining in a Classical Context*. Ann Arbor: Michigan University Press, 1991.
Solomon, Jon, "Tracta. A Versatile Roman Pastry". *Hermes* 106 (1978) 539–556.
Soyer, Alexis, *The Pantropheon, or, A History of Food and its Preparation in Ancient Times*. London: Simpkin & Marshall, 1853 [repr. London: Paddington Press, 1977].
Sparkes, Brian A., "The Greek Kitchen". *Journal of Hellenic Studies* 82 (1962) 121–137. "Addenda". 85 (1965) 162–163.
Tchernia, André, *Le vin de l'Italie Romaine. Essai d'histoire économique d'après les amphores*. Bibliothèque des Écoles Françaises d'Athènes et de Rome 261. Rome: École Française de Rome, 1986.
Villard, Pierre, "Le mélange et ses problèmes". *Revue des Études Anciennes* 90 (1988) 19–33.
Vivenza, Gloria, "Gli Hedyphagetica di Ennio e il commercio del pesce in età repubblicana". *Economia e Storia* 1 (1981) 5–44.

Presses and hand or animal mills

Amouretti, Marie-Claire, Georges Comet, Claude Ney, and Jean-Louis Paillet, "À propos du pressoir à huile: De l'archéologie industrielle à l'histoire". *Mélanges de l'École Française de Rome* 96 (1984) 379–421.
Callot, Olivier, *Huileries antiques de Syrie du Nord*. Paris: Paul Geuthner, 1984.
Drachmann, Aage G. *Ancient Oil Mills and Presses*. Copenhagen: Levin & Munksgaard, 1932.
Forbes, Robert J., "Crushing". Pp. 138–163 in Robert J. Forbes, *Studies in Ancient Technology*, Vol. 3. 2nd edn. Leiden: Brill, 1965.
Gichon, M., "The Upright Screw-Operated Pillar Press in Israel". *Scripta Classica Israelica* 5 (1979–1980) 206–244. 13 figs.
Jüngst, Emil, and Paul Thielscher, "Catos Keltern und Kollergänge. Ein Beitrag zur Geschichte von Öl und Wein". *Bonner Jahrbücher* 154 (1954) 32–93; 157 (1957) 53–126.
Moritz, Ludwig Alfred, *Grain-Mills and Flour in Classical Antiquity*. Oxford: Clarendon Press, 1958.
Rossiter, John J., "Wine and Oil Processing at Roman Farms in Italy". *Phoenix* 35 (1981) 345–361.

5

MINING AND QUARRYING

A MINING

The use of metals made possible many of the technological advances of the Bronze Age cultures of the Mediterranean world, and intensive exploitation soon exhausted most of the surface deposits of easily smelted oxide and carbonate ores in the region. From Archaic Greek authors we begin to hear of deep shafts and hard-rock mining (cf. **5.13**), and the mines at Laurion in Attica, worked from the Bronze Age onwards, provide evidence for early mining practices. Although prospecting was empirical in character, significant advances were made by both the Greeks and the Romans in identifying the surface indications of deposits and the sequence of geological strata. In the absence of explosives and earth-moving equipment, excavation of open-cast pits [**fig. 5.1**] and mine shafts [**fig. 5.2**] was

Figure 5.1 Pinax with open-pit miners, sixth century BC.
Source: Adapted from Forbes, 1965.7, fig. 33; Berlin 871.

Figure 5.2 The arrangement and operation of mining shafts.

Source: Illustration adapted from Georgius Agricola in *De re metallica*, sixteenth century.

very difficult and dangerous work. Ventilation and drainage [see **fig. 9.1**] were other problems only partly solved by available techniques, and lighting remained primitive. Because of the dangerous and degrading character of the work, miners were often slaves or condemned criminals who had no choice in their employment. Nevertheless, both the Greeks and the Romans identified most of the commercially viable mineral deposits available to them, and they produced enormous quantities of the metals important to their societies: gold, silver, copper, iron, lead, and – to a lesser extent – tin. The result was an economic windfall that gave rise to much legalese regarding ownership, leases, lawsuits concerning corruption and other activities, which often contain material related to the technology being used (see **5.2, 5.10–12**) (Craddock 2016, 2008, Hirt 2010, Healey 1978).

5.1 The technological demands, purposes, and moral consequences of mining

In the late Hellenistic period, it became a philosophical commonplace to denounce the technological trappings of urban civilisation as corrupt, especially those that brought about marked deviation from the "natural" order of things – such as mining, long-distance trade by sea, the use of money, and refined food stuffs. Apparently under the influence of the first-century BC Greek author Poseidonius of Apamaea, as well as Pliny, Seneca, and other Roman authors of the Empire affect philosophical indignation at the violation of the earth by mining, and other aspects of urban life.

Pliny, *Natural History* 33.1–3

We will now speak of metals, the very source of wealth and the prices paid for things, which diligent care seeks out within the earth in various ways. For indeed, in some places the earth is dug for wealth, when the way of life demands gold, silver, electrum, copper, and in other places for luxuries, when gems are wanted, and colours for painting on walls and timbers. Still elsewhere the earth is dug for the sake of rash boldness, when iron is wanted, which in the midst of wars and slaughter is more in demand even than gold. We search through earth's inner parts and live above the excavated hollows, astonished that she sometimes splits open or begins to shake, as if, in truth, this might not possibly be an expression of our sacred parent's indignation.

Entering the bowels of the earth, we look for wealth in the home of the dead, as if the surface we walk upon were not sufficiently kind or fertile. And with all this we search least of all for medicines, for how many mine for the sake of healing? This too, however, she has given us on her surface, along with grain: she is generous and abounding in all things that benefit us. Those things destroy us and drive us to the underworld, which she has hidden away and sunk in the earth, those things that do not occur without being searched for. In consequence, the mind soars aloft into the void in considering what the end result will finally be of emptying her out through all ages, how far greed will penetrate. How innocent, how happy, indeed, even how luxurious life might be if it desired nothing but what was above the earth, in short, nothing but what was already in her grasp.

MINING AND QUARRYING

5.2 Mining forbidden in Italy

The decree Pliny refers to is mysterious, but it may have stemmed from a desire to protect the profits of newly acquired mines in Spain in the second century BC rather than from any philosophical objections. In any case, mines such as the iron mines of Elba remained active in Italy.

Pliny, *Natural History* 3.138

Italy [...] does not yield precedence to any land in the abundance of all kinds of mineral products. Mining, however, is forbidden by an old edict of the Senate ordering that Italy be spared exploitation.

FINDING AND EVALUATING ORES

5.3 Methods of prospecting and origins of the term "metal"

The Romans had an imperfect understanding of the processes of ore formation and deposition, but there was a significant body of practical lore concerning prospecting and the evaluation of ores. Pliny's etymology of the word for metal may possibly be the correct one, since there are no apparent alternative origins for *metall-*, the root of the Greek and Latin words for metal, prospecting, and mining. The term may have arisen from the empirical experience of prospectors.

Pliny, *Natural History* 33.95–98

After these matters let us discuss silver ore, the next madness. It is found only in shafts, and occurs without giving any indication of its presence, since it lacks the shining sparkle gold has. The ore is sometimes reddish, sometimes ashen. It can be smelted only along with black lead or with a vein of lead they call *galena* [lead sulphide], which is very often found adjacent to the veins of silver ore. In this smelting process part of it sinks to the bottom as lead, but the silver floats on top, like oil on water.

Silver is found in nearly all the provinces, but the best of all in Spain. Like gold, it too occurs in sterile soil and even in the mountains, and wherever one vein is found, another is found close by. This, indeed happens for nearly every mineral substance, and it seems to be the reason the Greeks call them "metals" [from the Greek (*alla*) *met'alla*, "one with another"]. It is an extraordinary fact that mines throughout Spain opened up by Hannibal [221–219 BC] still remain in use. They take their names from the individuals who discovered them; among these, the one today called the Baebelo, which provided Hannibal with 300 pounds of silver every day, now dug 1,500 paces [2,200 m] into the mountain. Along this distance water-men are positioned day and night, pumping out the water in shifts measured by lamps and making a stream.

The vein of silver found on the surface is called "the crude". Our predecessors used to stop mining when alum was found: nothing was sought beyond it. Lately, however, a vein of copper found underneath the alum has made men's hopes boundless. The fumes from silver mines are harmful to all animals, but especially to dogs....

5.4 The use of touchstones in assaying ore

Touchstones, usually small tablets of fine-grained, dark schist or jasper, were irreplaceable in antiquity as a means of determining the metal content of ores or alloys in the field (cf. **6.41**). They remain in use even today.

Pliny, *Natural History* 33.126

The stone that they call a touchstone fits in with our mention of gold and silver. According to Theophrastus, it used to be found almost exclusively in the Tmolus River, but now, to be sure, it is found everywhere. Some call it Heraclian stone, others Lydian. It occurs in small pieces, not exceeding four inches in length and two in breadth, and the part which was exposed to the sun is better than the portion which faced the ground. When individuals skilled in the use of these touchstones take a scraping from a vein of ore as a test sample, using the stone like a file, they immediately state the proportion of gold, silver, or copper in it to within a scruple [1.296 g], by a remarkable, unerring calculation.

5.5 Sources of various mineral earths

Copper was the earliest metal to be exploited because it can occur naturally on the surface either in very pure, shiny lumps, or as brightly coloured oxide (cuprite) or carbonate (malachite, azurite) ores. The metallic content of these ores probably was recognised only after they had found application as pigments and medicines. Many other mineral ores or earths are brightly coloured and could be used for the same purposes.

Theophrastus, *On Stones* 49, 51–52

In general, most of these earths, and the most remarkable in character, are found in mines. Some, like ochre [yellow ochre] and ruddle [red ochre], are composed of earth, some, like chrysocalla [massive malachite] and *cyanus* [massive azurite], of a sort of sand, and others like realgar [arsenic disulphide] and orpiment [arsenic disulphide] and substances like them, of powder....

All these are found in silver and gold mines, some also in copper mines, such as orpiment, realgar, chrysocalla, ruddle, ochre, and *cyanus*. *Cyanus is* the rarest, and occurs in the smallest quantities. Some of the rest occur in veins, but ruddle is said to occur in masses....

Sometimes there are mines for ruddle and ochre in the same place, as in Cappadocia, and great amounts are dug out. They say that the danger of suffocation makes things difficult for the miners, since it happens very rapidly or even instantaneously.

5.6 Tin and iron in Britain

Alloying copper with 10 per cent of tin produces bronze, a metal which is both harder than copper and easier to cast. Since there were very few deposits of tin in the Mediterranean world, the tin of Cornwall was famous as early as the Bronze Age and was exported there (cf. **5.20**). Other sources tapped may have been as far away as Afghanistan (Penhallurich 1986).

Caesar, *Gallic War* 5.12

In the midlands of Britain white lead [tin] is found, iron in the maritime regions, but there is little of it. They use imported bronze.

5.7 Inexhaustible iron mines on Elba

Although it was more difficult in antiquity to produce iron from its ores than copper or other metals from theirs, iron ores have the advantage of being much more common at the surface of the earth. This ubiquity was as important to the spread of ironworking as the greater hardness of properly carburised iron. Stories of mines and quarries that replenished themselves after being worked were not uncommon in antiquity but probably reflect only the richness of the deposits (cf. **5.36**). For the smelting of iron from Elba, see **6.21**.

Strabo, *Geography* 5.2.6

Another remarkable thing about Elba is that – with time – the metal-bearing deposits that have been mined are filled up again, just as is said to happen with the flagstone of Rhodes, the marble of Paros, and – according to Cleitarchus – the rock salt of India.

5.8 Sources of lead

Lead, although soft, was another important metal in the ancient economy because it was relatively inexpensive, easy to cast, and resistant to corrosion. Lead was most often produced from *galena* [lead sulphide] as a by-product of silver production (cf. **5.3**). Enormous quantities were produced during the Roman Empire and used in buildings, water-supply systems, and for cooking vessels and containers.

Pliny, *Natural History* 34.164–165

For pipes and sheeting we use black lead, which is dug up with great effort in Spain and all the Gallic provinces but in Britain occurs at the surface of the ground in such abundance that a law forbids the production of more than a certain quantity. These are the names of the different types of black lead: Oviedo lead, Capraria lead, Oleastrum lead. But there is no variation, as long as the dross had been refined away with care. It is a remarkable fact concerning this type of mine alone that they are more productive when reopened after abandonment.... This was recently noted with regard to the Salutariensian mine in Baetica, which used to be rented out at 200,000 *denarii* a year, but after it had fallen into disuse was rented out at 255,000 *denarii*. In the same way, the Antonian mine, in the same province and with an equal rate of rent, has reached a revenue figure of 400,000 *sestertii* [from 200,000 *sestertii*?].

OWNERSHIP AND ADMINISTRATION OF MINES

In the ancient world, the state usually kept mineral rights for itself and either worked mines with slaves under military supervision or leased the workings to civilian contractors. Much of Athens' wealth in the late sixth and early fifth centuries BC was derived from the silver mines at Laurion. A particularly rich strike in the early fifth century was used to build the fleet that defeated the Persian invaders at Salamis (Jones 1982, Mussche 1975).

5.9 Important silver mines at Laurion

Aeschylus, *The Persians* 293–294

ATOSSA: "And what else do they [the Athenians] have? A sufficient store of wealth in their homes?"
CHORUS: "They have a certain fountain of silver, a treasure house beneath the earth".

Xenophon, *Ways and Means* 4.2–3

Now, it is clear to everyone that the mines have been worked for a very long time – at any rate, no one even tries to propose the date at which work began. But even though digging and removal of the silver ore have been going on for such a long time, observe what small portion the dumps are in comparison with the silver-bearing hills still in their natural state. And, so far from contracting in size, they keep finding that the silver-bearing region extends farther and farther.

5.10 Public mining contracts in Athens

In an attempt to avoid corruption, the Athenian state instituted careful procedures for the awarding of mining contracts.

Aristotle, *Constitution of Athens* 47.2

Next, there are the ten comptrollers, elected by lot, one from each tribe. They farm out all the state contracts and rent out the mines and the taxes in concert with the treasurer of the military funds and those elected to the fund for public festivals, in the presence of the council, and they grant it to whomever the council votes for; also the mines sold and the workings let out for three years and the concessions let out for [...] years.

Copies of the leases were set up in the agora for public scrutiny. Some of the inscriptions carved on marble stelae have survived and provide useful information concerning costs, profits, and mining procedures (cf. Crosby 1950).

M. Crosby, *Hesperia* 10 [1941] lines 40–49

[367/6 BC] Mines were leased. In the first (prytany), that of Hippothontis, Dexaicon in Nape at Scopia [The Lookout], where there are boundaries on every side (the property of) Nicias of Kydantidai, (the lessee,) Callias of Sphettos, price 20 *drachmai*; Diacon at Laurion, where the boundaries towards the rising sun are the fields of Exopios, towards the sinking sun is the mountain, (the lessee), Epiteles of Cerameis, price 20 *drachmai*; at Sounion in the (fields) of the sons of Charmylos, where the boundaries at the north are (the lands of) Cleocritus of Aigilia, at the south (the lands of) Leucius of Sounion, (the lessee), Pheidippus of Pithos, price 20 *drachmai*; Poseidoniacon in Nape, (one) of those from the stele, in the (fields) of Alypetos where the boundaries are (the lands) of Callias of Sphettos and Diocles of Pithos, (the lessee), Thrasylochus of Anagyrous, price 1,550 *drachmai*....

5.11 Lawsuits concerning mining activity

Inevitably, there were conflicts in the mining area at Laurion, either between the state and a lessee or between lessees, concerning damages to the workings or diversion of ore. In order to avoid permanent damage to the state-owned shafts or the neglect of the less rich deposits, the procedures for mining were carefully specified.

Demosthenes 37, 7 *Against Pantaenetus* 38

But I think that lawsuits involving mines should be brought against those who are partners in a mine, and who have bored through to the adjacent claim or, in short, those who are working the mines and doing one of those things specified in the law.

Plutarch, *Moralia* 843d

He also brought to trial Diphilus, who had removed from the silver mines the support piers, which hold up the overlying weight and had enriched himself from them, contrary to the laws. And even though the penalty was death, Lycurgus had him convicted.

5.12 Conditions of mine tenure and operation in the district of Vipasca

The Roman state was equally cautious concerning the working of state-owned mineral deposits, and the following two inscriptions provide important details of mining procedures and regulations in second-century AC Portugal, near modern Aljustrel.

Bruns, *Fontes Iuris Romanae*, pp. 293–295, no. 113

If someone has not acted accordingly and has been having ore smelted before having paid the fee as specified above, the occupier's share shall be forfeit, and the procurator of mines shall sell the entire shaft. Whoever has brought evidence that the tenant smelted ore before paying the fee of a half share to the treasury shall receive a fourth part.

Silver mines must be worked according to the scheme that is contained in this law. The fee established by the liberality of the most sacred emperor Hadrian Augustus will be observed, whereby the ownership of the portion that belongs to the treasury will belong to the one who first sets a price for the shaft and deposits 4,000 *sestertii* with the treasury.

If someone has hit a vein of ore in one of a total of five shafts, he shall continue work in the rest without interruption, as has been written above. If he does not do this, another shall have the right of occupation. If anyone after the 25 days allowed for preparation of funding actually has begun regular work but afterwards has left off working for ten days in a row, another shall have the right of occupation. If a shaft sold by the treasury is abandoned for six straight months, another shall have the right of occupation, under the condition that when he extracts the ore the customary half share be set aside for the treasury.

The occupier of shafts shall be permitted to take on whatever partners he wants, as long as each partner contributes a proportionate share of expenses. If anyone does not do so, then the individual who covered the expenses shall have an account of the expenses he covered posted for three days straight in the most public part of the forum and through a herald official notice shall be given to his partners that each should contribute to expenses in proportion to his own share. If anyone has not contributed accordingly or by malice aforethought has done something to avoid contributing or to defraud one or another of his partners, he shall forfeit his share of the shaft and it shall belong to the partner or partners who covered the expenses. Furthermore, those tenants who have incurred expense for a shaft in which there were many partners shall have the right to recover from their partners anything shown to have been incurred in good faith. Tenants may sell to each other in addition those shares of shafts, which they have bought from the treasury and paid for, for as high a price as they can get. If someone wants to sell his share or buy one, he shall give a declaration to the procurator who is in charge of the mines; he is not permitted to buy or sell by other means. Whoever is in debt to the treasury is not allowed to give away his share.

Ore that has been extracted and lies at the shaft must be transported to the smelters by the owners between sunrise and sunset. If someone is convicted of having taken ore from the shafts between sunset and sunrise, he must pay 1,000 *sestertii* to the treasury. If the thief of the ore is a slave, the procurator shall have him beaten and sold under the condition that he be kept always in chains and not reside in any mines or mining territory. The price of the slave shall belong to the owner. If the thief is a free man, the procurator shall confiscate his property and banish him from the mining territory for ever.

All shafts shall be carefully propped and supported, and the tenant of each mine shall provide replacements for rotten timber. It shall be prohibited to touch or injure pillars or supports left for the sake of reinforcement or to do anything with malice aforethought by which these pillars or supports might be less stable [**fig. 5.3**]. If he is a slave, he shall be beaten with whips at the discretion of the procurator and sold by his master on the condition that he not reside in any mine area; if a free man, the procurator shall confiscate his property for the treasury and ban him from the mining area for ever.

Whoever operates copper diggings shall keep away from the ditch that conducts water from the mines and leave a space of not less than 15 feet on either side. It shall be forbidden to interfere with the ditch. The procurator shall allow the driving of an exploratory trench seeking new ore deposits from the drainage ditch, providing that the exploratory trench is not more than four feet wide and deep. It shall not be permitted to look for or to extract ore within 15 feet of either side of the ditch. If someone is convicted of having done otherwise in the exploratory trenches, if a slave, he shall be beaten with whips at the discretion of the procurator and sold by his master on the condition that he not reside in any mining district; if a free man, the procurator shall confiscate his goods and banish him from the mining area for ever.

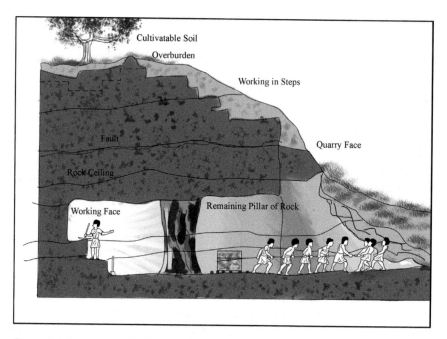

Figure 5.3 Supporting pillar in an underground quarry.
Source: Adapted from Adam, 1994, fig. 21.

Whoever operates silver diggings shall keep away from the ditch that conducts water from the mines and shall leave a space of not less than 60 feet on either side, and he shall keep to those shafts, which he had occupied or received by assignment for his operation, just as they have been prescribed, and he shall not exceed the boundaries, not collect leavings nor drive exploratory trenches beyond the bounds of the assigned shaft....

CIL 2.5181, 58–60 (= *ILS* 6891)

Claims to Diggings, and Permits. Whoever within the mining district of Vipasca [in Portugal] shall claim or occupy diggings or a digging location for the purpose of maintaining his legal claim, according to the law promulgated for the mines, shall within the next two days after claiming or occupying make a declaration with the concessionaire of this tax, or his partner or agent....

5.13 Two early Greek gold mines

As indicated here by Herodotus, the dumps of discarded rock outside a mine shaft could be enormous. These slag heaps and ore dumps remained the property of the state, since it was recognised that both could be processed to yield more metal (see **6.15–16**). Many ancient slag heaps have been re-processed in modern times as well for their metal content. The admonitory story in Pausanias about the small island state of Siphnos, which became rich from a local silver mine in the mid-sixth century BC, indicates the inability of Greek miners to deal with massive flooding.

Herodotus, *Histories* 6.46–47

The revenue [of the Thasians] came to them from the mainland and from mines. From the gold mines of "the Dug Forest" they generally derived 80 talents, from those on the island itself a bit less than this [...] so that the Thasians generally derive from the mainland and the mines 200 talents a year, 300 in the best years.

I myself have seen these mines. The most remarkable by far was that which the Phoenicians discovered [...] on Thasos between the places called Aenyra and Coenyra, opposite Samothrace, a great mountain turned upside down in the search.

Pausanias, *Description of Greece* 10.11.2

A treasury building [at the sanctuary of Apollo at Delphi] was constructed by the Siphnians as well, for the following reason. Their island provided the Siphnians with a gold mine, and the god ordered them to pay a tenth of the proceeds to Delphi. They built the treasury and kept on tithing. When greed, however, led them to omit the offering, the sea flooded the mines and hid them from sight.

5.14 Slave labour in the mines at Laurion

Most individuals who leased a mine, of course, were wealthy enough to leave the actual mining and supervision to others. The difficult and dangerous job of working the mine face naturally was most often given to slaves. A contractor working a mine might hire slaves from their owners for a salary. The author of this passage proposes the creation of a corps of state-owned slaves that would work for public rather than private profit.

Xenophon, *Ways and Means* 4.14–17

For I presume that those of us who care about these things heard long ago that Nicias, son of Niceratus, once owned 1,000 men in the silver mines. He rented them out to Sosias the Thracian on condition that Sosias pay him an obol [one-sixth of a *drachma*] each per day, net, and always furnish replacements to the original complement. Hipponicus, too, had 600 slaves rented out in the same way, which brought him a *mina* [100 *drachmai*] a day, net, and Philemonides 300, bringing half a *mina* and others too as, I believe, each man's means allowed. But why must I speak of things in the past? For now also there are many men in the silver mines let out to rent in the same way. If my proposal were adopted, this alone would be new: just as private individuals have managed a permanent income for themselves by acquiring slaves, so also the city would acquire state-owned slaves until there were three for each Athenian citizen.

Andocides, *Mysteries* 38

For he said that he had a slave at Laurion, and had to go collect his earnings.

5.15 A free worker in a Dacian mine

In more sparsely populated areas of the Roman world free individuals sometimes contracted to work in a mine. The following passage is a contract to work in a mine in the area of modern Romania, written by a scribe for an illiterate miner. The mention of flooding in this passage is a reminder of one of the many hazards encountered in the shafts (cf. **5.20**).

CIL 3, p. 948, no. 10

In the consulship of Macrinus and Celsus [AD 164], 20 May, I, Flavius Secundinus, record at the request of Memmius, son of Asclepius, who claims to be illiterate, that he said he leased and that he did lease his labour in the gold mine to Aurelius Adiutor from today until next 13 November, for 70 *denarii* and subsistence. He will be entitled to receive the salary in instalments. He will be obliged to provide healthy and energetic labour to his above-mentioned employer. If he wants to quit or stop working against his employer's wishes, he will be liable to pay five *sestertii* for each day.... If flooding shall impede operations, he will be obliged to prorate his salary. If at the end of this term the employer shall delay final payment, he will pay the same penalty after three days grace.

5.16 Roman soldiers mining in first-century AC Germany

Where the local population was too scattered or too hostile for employment and slaves were unavailable, ambitious Roman generals sometimes put their soldiers to work in mines.

Tacitus, *Annales* 11.20

In the territory of the Mattiaci, Curtius Rufus, who had opened mines to follow veins of silver, gained the same public distinction [a triumph]. There was little yield, and only for a short time, while the soldiers had the hard work for nothing – digging drainage adits and building beneath the surface works that would have been challenging even in the open air. Exhausted by these tasks, and because many had endured similar things elsewhere in the Empire, the troops composed secret letters in the name of the armies, begging the emperor to assign triumphal honours ahead of time to individuals he was about to entrust with an army.

5.17 The low status of miners and of private wealth derived from mines

Because of the dangerous and degrading character of mining in antiquity and the low social status of the miners, even the private fortunes derived from this activity were felt to be tainted. The source of Crassus' wealth, however, was seen to be even less reputable; he would purchase burning buildings in Rome from their owners at an extortionate discount, then extinguish the fire with his private fire brigade.

Plutarch, *Comparison of Nicias and Crassus* 1.1–2

But first of all, in comparing them, Nicias' wealth was acquired with less prejudice relative to that of Crassus. For although one may not rate the mining business highly, since most of its activities are performed by criminals or barbarians, some of them kept in bonds and dying in constricted and pestilent places, however, in comparison with the public confiscations of Sulla and the letting of contracts during a fire, it appears in a better light. For Crassus exploited these sources of income openly, as one might practice agriculture or money-lending.

MINERS AND MINING TECHNIQUES

5.18 Horrors of gold mining in first-century BC Egypt

The owners of privately-owned slaves had a vested interest in keeping their workers alive and well, but state-owned slaves were often condemned prisoners sent to the mines to be worked to death. The conditions were dreadful.

Diodorus of Sicily, *History* 3.12.1–13.1

Around the frontiers of Egypt and the adjoining territory of both Arabia and Ethiopia there is a region that possesses many large gold mines where much gold is gathered up with great suffering and expense. For while the earth is black by nature, it contains seams and veins of a white stone [quartz] distinguished by its brightness, surpassing in its radiance all the stones that by nature have a bright gleam. Those who oversee the work in the mines extract the gold by means of a multitude of workers, for the kings of Egypt gather up and hand over for gold-mining men who have been convicted of crimes, those taken prisoner in war, along with those who have fallen prey to unjust accusations and have been thrown into prison through spite – sometimes only themselves, sometimes all their relatives too. At one and the same time, the kings exact punishment from those who have been condemned and receive great profit from their labours.

Those who have been handed over, a great number in all, and every one of them fettered with chains, keep busy at their work without ceasing, both by day and all through the night without receiving any rest, carefully guarded against any attempt at flight. For garrisons of foreign soldiers who speak languages different from theirs guard them, so that no one can, through conversation or some friendly communication, corrupt any of those set over him. After burning the hardest of the gold-bearing matrix with a great fire and making it friable, they carry on the process of production by hand. Thousands of the unfortunate creatures crush with a quarrying hammer the rock that has been loosened and is capable of being worked with moderate effort. The workman who assays the ore is in charge of the whole operation and gives instructions to the workers. Of the men chosen for this misfortune, those individuals of outstanding physical strength break up the quartz rock with iron hammers, applying to the work not skill, but force, not cutting tunnels through the rock in a straight line, but wherever there is a vein of the shining rock. These men, then, spending their time in darkness because of the twists and turns in the galleries, carry lamps tied to their foreheads, and after shifting the position of their bodies according to the specific character of the vein, they throw down to the gallery floor the fragments of rock they dig out. And they keep up this work incessantly under the hard supervision and blows of an overseer.

Boys who have not yet reached puberty crawl through the tunnels into the galleries hollowed out in the rock and with great effort collect the ore, which had been thrown down bit by bit, and carry it back to a place outside the mouth of the mine in the open air.

For the rest of this passage see **6.5**.

5.19 A subterranean clay mine on Samos

Some deposits of particularly good diatomaceous earth or pottery clay were mined in cramped underground tunnels by means of the same techniques used to extract mineral ores.

Theophrastus, *On Stones* 63–64

There are more differences between the Melian and the Samian mineral earths. It is not possible for someone to stand upright while digging in the Samian deposits, but he must dig while on his back or side. The vein stretches quite a distance. Its height is around two feet, but its depth is much more. It is surrounded on either side by rock, from which it is extracted. The vein had a stratum running down the middle, which is of better quality than the material around it, and still a second such stratum and likewise a third and a fourth. The last, called "the star", is the best of all. The earth is used mainly, if not exclusively, for treating cloaks.

5.20 Methods of mining silver in Roman Spain

Diodorus' account – probably derived from Posidonius of Apamaea (cf. Strabo 3.2.8–9) – provides a very good summary of the methods, techniques, and results of mining in Spain, the Eldorado of the ancient Mediterranean world (Bird 2004, Domergue 1990, Jones 1980).

Diodorus of Sicily, *History* 5.36–38

But much later, the Iberians learned the specific properties of silver and established important mines. In consequence, by working the best quality and, to be sure, the most abundant silver deposits, they received great income. The manner of mining and refining among the Iberians is as follows. Since the mines are astonishing in their bronze, gold, and silver, those who work the copper mines recover from the ore they dig out a fourth part of pure copper, and some of the untrained individuals who work the silver mines take out a Euboic talent [25.9 kg] in three days, for all the ore is full of compacted, sparkling silver dust. In consequence, one might marvel at the nature of the land and industry of the men who work it. For at first, any untrained individuals whom chance brought worked away at the mines and carried off great wealth, because of the accessibility and abundance of the silver-bearing ore. But later on, when the Romans took control of Iberia, a crowd of Italians filled the mines and bore off great wealth through their lust for profit. After purchasing a multitude of slaves, they turn them over to the overseers of the mine workings, and these, opening shafts up in many places and digging deep into the earth, search for the strata rich in silver and gold. They carry on not only for a great distance, but also to great depth, extending their diggings for many stades and driving on galleries branching and bending in various directions, bringing up from the depths the ore, which provides them with gain.

These mines furnish a great contrast when compared with those of Attica, for the men who work the latter mines and lay out great amounts of money on the workings at times do not receive what they hoped for but lose what they had.... The men who work the mines in Spain, however, pile up from these workings the great wealth they hoped for. For while their first efforts came out well on account of the special excellence of the ground for this type of activity, they are forever discovering still more magnificent veins of both gold and silver. Indeed, all the earth in that vicinity is a web of veins winding in many directions. At a depth they sometimes break in on rivers flowing beneath the surface whose strength they overcome by diverting their welling tributaries off to the side in channels. Since they are driven by the well-founded anticipation of gain, they carry out their enterprises to the end, and – most incredible of all – they draw off the streams of water with the so-called Egyptian screw, which Archimedes the Syracusan invented when he visited Egypt [see **9.32**]. By means of these devices, set up in an unbroken series up to the mouth of the mine, they dry up the mining area and provide a suitable environment for carrying out their work. Since this device is quite ingenious, a prodigious amount of water is discharged with only a small amount of labour, and the whole torrent is easily discharged from the depths into the light of day. One might reasonably marvel at the inventiveness of the craftsman [Archimedes] not only in this, but also in many other even greater inventions celebrated throughout the whole world, each of which we shall discuss carefully in turn when we come to the age of Archimedes.

The individuals, then, who spend their time at the mine workings, produce revenues of incredible magnitude for their masters but themselves wear out their bodies in the diggings beneath the earth both by day and night. Many of them die as a result of the inordinate hardships, for they have no remission or rest from their work, but use up their lives in misfortune, compelled by the blows of their overseers to endure the harsh situation. Some of them live a long time in their distress, holding out through bodily strength and spiritual endurance. Indeed, on account of the magnitude of the suffering, death to them is preferable to life. And although there are many marvels associated with the mines we have spoken of, one might find it not the least remarkable thing that not one has a documented origin, but all were opened up through the Carthaginians' greed when they still ruled Iberia....

Tin also occurs in many parts of Iberia. It is not found on the surface as some have repeated over and over in their histories, but dug out and smelted like silver and gold. There are many tin mines north of Lusitania and on the little islands that lie off Iberia in the Ocean, called the Cassiterides on account of their characteristic product. Much tin is brought over from the island of Britain to Gaul, which lies opposite, and is carried on horses by traders through the heartland of Celtica to the Massalians and to the city called Narbo....

5.21 Methods of mining gold

Pliny's discussion of gold mining indicates once again that, despite the absence of a scientific geology, a great deal of detailed empirical knowledge existed about different types of commercially valuable minerals and ores and about the methods for finding and extracting them. The passages concerned with techniques of hard-rock mining show how miners could create large galleries without the use of explosives. Likewise, the account of diverting streams of water to wash large amounts of gold-bearing gravel into sluices indicates how Roman technology had a major and permanent effect on the topography and ecology of the region (Jones and Bird 1972 and Lewis and Jones 1970).

MINING AND QUARRYING

Pliny, *Natural History* 33.66–78

Gold in our part of the world – passing over Indian gold dug up by ants, or among the Scythians by griffins – is found in three ways: first, in river deposits, as in the Tagus in Spain, the Po in Italy, the Hebro in Thrace, the Pactolus in Asia Minor, and the Ganges in India. No gold is more refined, for it is thoroughly polished by the very flow of the stream and by wear. The other methods are to mine it in excavated shafts or to look for it in the debris of undermined mountains. Both these last procedures must be described.

Men who are looking for gold bring up *segullum;* this is the term given to the earth that signals the presence of gold. It is a deposit of sand, which is washed, and by means of the sediment an estimate is made of the deposit. Sometimes by rare good fortune it is found immediately on the surface of the earth, as recently in Dalmatia during the reign of Nero, when one even yielded 50 pounds each day. When the gold has been found in this way in the upper crust, they call it *talutium* if there is gold-bearing earth beneath it. The mountains of Spain, otherwise dry and unfruitful and in which nothing else is produced, are forced to be bountiful with regards to this commodity.

Gold that is mined in shafts they call "channel" or "trench" gold. It sticks to the white stone [quartz] gravel, ... sparkling in the embrace of the stone. These channels of the veins, which give rise to the term, wander here and there along the sides of the shafts, and the earth is held up with wooden supports....

The third method will have surpassed the accomplishments of the Titans. The mountains are hollowed out by means of galleries driven for long distances by the light of lamps. The lamps also measure the periods of work, since the miners do not see daylight for many months. They call this type of mine *arrugia* [undermined? from Greek *orussein*, "to dig"]. Faults suddenly slip and crush the workers, so that now it seems less audacious to seek pearls and purple dye in the depths of the sea, so much more harmful have we made the earth. In consequence, they leave arches at frequent intervals to hold up the mountains. Beds of flint, which they break up with fire and vinegar, occur in both types of mine. But since this procedure chokes the tunnels with steam and smoke, they more often batter the flint with rams carrying 150 pounds of iron and carry it out on their shoulders by day and by night, passing it through the darkness to the next in line; only the last see daylight. If the flint deposit seems too extensive, the miner follows its edge and goes around it. Nevertheless, the work involved with flint is considered to be relatively easy, for there is an earth composed of a certain type of clay mixed with gravel called *gangadia*, which is nearly impenetrable. They attack it with iron wedges and the rams mentioned above and think nothing is harder – unless the hunger for gold is the most resistant thing of all.

When the work has been completed, they cut down the imposts of the supporting arches, beginning with the furthest from the entrance. A fissure gives the sign, and the only one to observe it is the watchman on the mountain top; with cries and gestures he orders the workmen called out and likewise himself races down. The fractured mountain collapses in a gaping rift with a crash that the human mind cannot conceive and with an equally incredible blast of air. The victorious miners behold the ruin of nature. Still yet, however, there is no gold, and they did not know for certain if there was any when they began to dig; the hope for what they desired was sufficient motivation for such danger and expense.

There is another procedure of equal difficulty and even greater expense: bringing streams along mountain ridges to sluice this rubble, often from 100 miles away. They call the channels *corrugi*, from *corrivatio*, I believe, a "drawing together of streams of water". This too involves 1,000 tasks. The slope has to be such that the water rushes rather than flows, and consequently the water is brought from the highest elevations. Gorges and crevices are spanned by channels carried on support structures. Elsewhere, impassable crags are cut away and forced to provide a place for conduits hollowed out of logs. The man who cuts the stone hangs from ropes; to one watching from far off, the appearance is not so much that of wild beasts, but of birds. For the most part they sight levels and survey the route while suspended, and men make streams flow where there is no room for a man to plant his feet. It is a fault in the sluicing operation if the stream brings in mud as it flows; they call this type of sediment *urium*.[1] For this reason, they guide the water over flint and pebbles and avoid the *urium*. At the head of the slope, on the crest of the mountain, reservoirs are dug out 200 feet on a side and ten feet deep. Five outlets are left in these, nearly three feet square, so that when the gates have been knocked away after the pool has filled, a torrent might burst forth with enough strength to roll rocks away [This procedure is called hushing; **fig. 5.4**].

There is still another task on the plain. Trenches called *agogae* [Greek for "leaders"] are dug for the water to flow through, and have their stepped floors covered with *ulex* [gorse?]. This is a shrub like rosemary, which is rough and holds back the gold. The walls are lined with boards, and the channels are run up steep slopes on props. The earth flowing along in this way glides into the sea, and fractured mountains are washed away. From this cause, Spain has already extended its land mass far out into the sea. The spoil that is dug out of the second type of mine with great effort is washed away in this manner so as not to obstruct the shafts.

The gold sought by sluicing is not refined but is pure from the start. In this way nuggets are found – as also in the shafts – even surpassing ten pounds. The Spanish call them *palagae* or *palacurnae;* the smaller bits they call *baluce*. The gorse is dried and burnt, and its ashes are washed over grassy turf, so the gold might settle out. Certain individuals have related that Asturia, Callaecia, and Lusitania produce 20,000 pounds of gold in this manner every year, and Asturia supplies the most. In no other region of the earth has this bounty lasted for so many centuries....

MINING AND QUARRYING

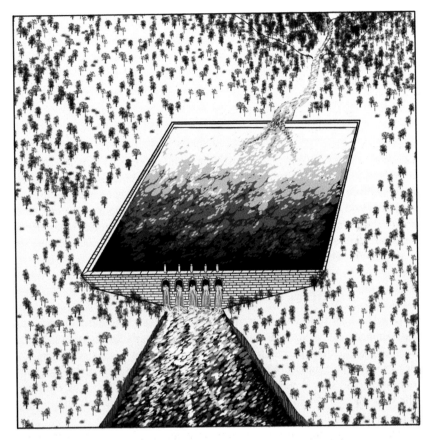

Figure 5.4 Hushing (simplified sketch of outlets, dam, tank, and water).

Source: Adapted from complex description of mining in northwestern Spain in Lewis and Jones, 1970, and Jones and Bird, 1972.

5.22 A roof collapse in a Spanish mine

Even without the use of explosives, there was always danger of rock-falls in the ancient mine shafts, as a result both of improper shoring and of the premature collapse of a support intentionally designed to bring about exposure of an ore face upon its removal (cf. **5.21**).

Statius, *Thebaid* 6.880–885

...just as when the miner of a Spanish hill descends and leaves far behind the light of day and ordinary life. If the overhanging ground shakes and the earth splits open with a sudden crash, he is overwhelmed and lies concealed within as the mountain settles, and his totally broken and crushed corpse does not give back the resentful soul to its native heaven.

MINING AND QUARRYING

5.23 The dangers of fumes in mines

Inadequate ventilation brought many miners a slower and more insidious death through exposure to poisonous fumes, suffocation, or silicosis. The need to constantly replace even the cheapest slaves, as here at a famous mine in Paphlagonia, on the south shore of the Black Sea, ultimately affected the profit of a mine operator.

Strabo, *Geography* 12.3.40

Mt. Sandaracurgium is hollow as the result of mining activity, for the miners have advanced beneath it with great galleries. The revenue farmers used to work them, employing slaves as miners sold in the market because of their crimes. For, in addition to the anguish of the work, they say that the air in the mines is both deadly and hard to bear because of the disagreeable odour of the ore, so the men are quick to die. In addition, they say that often the mine is abandoned because of its unprofitable nature since there are more than 200 workers, but they are continually consumed by sickness and death.

Lucretius, *On the Nature of Things* 6.808–815

Finally, when they follow veins of silver and gold, searching out with iron tools the hidden parts of the earth deep within, what foul smells Scaptensula [a mining town in Thrace] breathes out from below! What harm comes when the gold mines let loose their fumes! What an appearance they give men, and what pallor! Don't you see or hear in how short a time they usually die, and how the source of life fails the men whom the great force of compulsion binds to such work?

5.24 Remedies for bad air in mines

Since the provision of ventilating shafts would have enormously increased the labour involved in opening a shaft, they must have been provided only when the problem became acute and the ore deposits were rich. It is unlikely that the solution mentioned by Pliny could have worked very well without the presence of ventilating shafts.

Theophrastus, *Concerning Fire* 24

For this reason, air that is thick and still also causes suffocation among miners, for it does not give way to exhalation. In consequence they make ventilating shafts, so that the air is thinned by the movement and makes room at the same time that it changes. That thick air is difficult to breathe is obvious and stated without elaboration....

Pliny, *Natural History* 31.49

When well shafts have been sunk deep, fumes of sulphur or alum rush up to meet the diggers and kill them. A test for this hazard is to see if a burning lamp lowered down the shaft is extinguished. If so, air shafts are dug next to the well shaft on both sides to drain off the oppressive fumes. Even in the absence of these pernicious substances, the air itself becomes oppressive as a result of the depth. Continuous ventilation by waving linen cloths remedies this.

5.25 The problem of smoke from fire-setting in mines

In many cases, noxious natural fumes in mine galleries were supplemented by the smoke from fires set to heat the rock and prepare it for splitting. Although Theophrastus' explanation does not make much sense, it is possible that miners in some situations would set a fire to start a draught that would draw fresh air into their gallery through a ventilation shaft or drainage adit.

Theophrastus, *Concerning Fire* 70

Smoke is less annoying when there is a fire in the same place. In consequence, men pestered by smoke in underground galleries do not suffer if they have a fire, for the fire quenches the heat in the smoke. Since the greater is engendered to consume the lesser, when the lesser is quenched, the smoke also is done away with. For this reason, the sun, like fire, assists in the prevention of smoke, for it burns away any of it that happens to come within reach.

5.26 Placer deposits and the methods for working them

Where geological and topographical conditions are suitable, metal deposits are mined naturally by erosion, and the nuggets or flakes of metal collect as placer deposits in the sand or gravel of streambeds. Gold, silver, and tin could be found in quite pure form in such deposits.

Strabo, *Geography* 3.2.9

Poseidonius says he does not disbelieve the story that once, when the forests caught fire, the earth, which incorporated silver and gold, melted and boiled out into the open, because every mountain and every hill is the stuff of money, heaped up by some bounteous good fortune. And he says that someone seeing these places would state that they are the ever-flowing storehouses of nature or the unfailing treasury of an Empire.... But he says that among the Artabrians, who are at the northwest edge of Lusitania [in Spain], the earth effloresces with silver, tin, and white gold (for it is mixed with silver), and that the rivers carry this soil along. The women scrape it together with mattocks and wash it in sieves woven like a basket.

Diodorus of Sicily, *History* 5.27.1–3

In general, silver does not occur in Gaul, but there is much gold, which nature furnishes to the natives without mining or hardship. For in their courses the rivers have sharp bends, and since they dash against the mountains that fringe their banks and erode away great hills, they are full of gold dust. Those who occupy themselves with this business collect it and grind or break up the lumps containing the dust, and washing away the natural earthy elements, they hand it over for smelting in furnaces. Amassing a great quantity of gold in this manner, they use it for ornament – not only the women, but the men as well.

Strabo, *Geography* 11.2.19

It is said that among them [some tribes of Colchis on the Black Sea] mountain torrents swollen by melting snow carry gold down, and the locals catch it with perforated troughs and fleecy hides, and that this is the origin of the myth of the Golden Fleece.... [**fig. 5.5**]

Figure 5.5 Catching gold dust in a cloth.
Source: Illustration adapted from Georgius Agricola in *De re metallica*, sixteenth century.

MINING AND QUARRYING

5.27 Open-cast gold mines in Noricum
Strabo, *Geography* 4.6.12

Polybius says in addition that in his own time among the Taurisci of Noricum, more or less in the hinterland of Aquileia, a gold deposit was found so conveniently formed that someone stripping away the surface soil to a depth of two feet immediately came upon pit-gold, and the excavation never exceeded 15 feet. He says that some of the gold is pure immediately, pieces the size of a bean or lupine, and only an eighth part is lost in the refining process, but some needs more smelting. Nevertheless, it is exceedingly profitable. He says that in the two-month period after the Italiote Greeks began working with the locals, gold suddenly became one-third cheaper throughout all Italy, and after learning this the Taurisci expelled their co-workers and exercised a monopoly. Now, however, all the gold mines are under Roman control. And here too, just as in Iberia, in addition to the pit-gold, the rivers carry along gold dust, not however, in such quantities.

SOURCES OF PETROLEUM, BITUMEN, AND SALT
5.28 Varieties of petroleum

Both Greeks and Romans knew many varieties of petroleum, but they were most familiar with the dry or viscous earthy variety termed bitumen. In its various forms the substance was used as an adhesive, building material, caulking material, fuel, and medicinal substance.

Pliny, *Natural History* 35.178

The nature of bitumen is close to that of sulphur. In some localities it is a viscous substance, in others an earth. The viscous form appears, as we have stated, in the Dead Sea, the earth in Syria around the seaside citadel of Sidon. Both these varieties thicken and coagulate as a dense mass. In fact, there is also a liquid form of bitumen, from Zacynthus, for example, and that imported from Babylon. There, indeed a light coloured variety also occurs. Bitumen from Apollonia also is in liquid form. All of these the Greeks call *pissasphaltos*, from the characteristics of vegetable pitch and bitumen.

5.29 Sources of petroleum in Mesopotamia and Greece

The Greeks and Romans never drilled or mined for petroleum products but made use of very shallow deposits that oozed out to the surface. The richest deposits then, as now, were in the Persian Gulf region, where installations were designed to refine the oil for a variety of uses.

Herodotus, *Histories* 6.119

Ardericca is 210 stades distant from Susa and 40 from the well that supplies three substances. For men draw up for themselves from it asphalt, salt, and oil in the following manner: it is bailed out with a *shaduf*² to which is fastened half a skin in place of a bucket. Making this dip down, the operator draws out the liquid and then pours it into a receptacle. Flowing from this container into another, the liquid is diverted three ways; the asphalt and salt immediately harden. The oil, which the Persians call *rhadinace*, is black and gives off a heavy odour.

Herodotus, *Histories* 4.195

But anything can happen, since I myself have seen pitch brought up from a pool of water in Zacynthus. There are many pools on the island, the largest of them 70 feet across and two fathoms deep. Into it they let down a pole with a myrtle branch tied to its tip and then lift out pitch on the myrtle. It has the odour of asphalt but is, for the rest, better than the pitch of Pieria. They pour it into a basin excavated alongside the pool, and when they have gathered a quantity, they let it flow from the basin into storage jars....

5.30 Collection of bitumen from the Dead Sea

Tacitus, *Histories* 5.6

At a certain time of the year [the Dead Sea] discharges bitumen, which experience has taught the means of collecting, as she teaches all the other crafts. Bitumen is by its own nature a black liquid, but coagulates when sprinkled with vinegar and floats. Those whose job it is catch the bitumen by hand and drag it over the gunwales of a boat; then, without any assistance, it flows in and loads the boat until it is cut away. It cannot be cut with bronze or iron, but it draws away from blood or a cloth stained with a woman's menses. This is what the ancient authors say, but those who know the country relate that masses rich in bitumen are driven and dragged by hand to the shore. Then later, when dried out from the heat of the ground and the sun's force, they are split with axes and wedges, like rocks or wooden beams.

5.31 Sources of natural and artificial salt

Although rock salt dug from mines (cf. **6.11**) seems to have been much less well known in antiquity than sea salt prepared by evaporating seawater, both these techniques constitute a type of mining activity (Currás 2017).

Cato, *On Agriculture* 88.1–2

Earlier, Cato has provided instructions on how to make brine and refine salt using a simple method.

Make white salt thus: once the neck is broken off, fill a clean amphora with clean water, then place it in the sun. Suspend in it a small rush basket with ordinary salt and toss vigorously and top-up from continually. Do this every day regularly until the salt ceases to dissolve for two days ... [until] saturated.... Put that brine in small basins or in shallow pans in the sun. Keep it in the sun continually until it solidifies. Thus refined salt will be produced.

For salt's modern counterpart, pepper, which in antiquity was imported from India, see Pliny *Natural History* 12.26–29.

Pliny, *Natural History* 31.73–85

All salt either is manufactured or occurs naturally; each type is formed in many ways, but there are two causes: condensation and desiccation. It forms in the lake at Tarentum by drying under the summer sun, and the whole pool becomes salt.... All salt from pools occurs as a powder, not in masses. Another type is generated spontaneously from seawater, as foam left on the margin of the shore and on rocks. All this condenses from the spray, and that found on rocks is more pungent.... There are also mountains of natural salt, like Oromenus in India, where it is quarried like stone.... It is also dug out of the earth in Cappadocia, clearly formed by the condensation of moisture. There, indeed, it is quarried like mica [in sheets?]. The very heavy blocks have the local nickname "grains". [...] In Hither Spain too, at Egelesta, salt is quarried in nearly transparent masses.... Every locality in which salt is found is sterile and grows nothing.

There are various types of artificial salt. The type most common and made in the greatest quantity is produced by letting the sea flow into salt pans along with some streams of fresh water. Rain, however, is a particular help, and above all a great amount of sun; otherwise it does not dry out. In Africa, near Utica, they heap up piles of salt like hills, which do not melt with any moisture and can scarcely even be cut with iron tools once they have been hardened by the sun and moon. However, salt is made in Crete without fresh water streams, by letting the sea into salt pans.... It is also manufactured by pouring well water into salt pans.... In Gaul and Germany, they pour salt water on burning wood....

The driest salt is saltiest, the most agreeable, and whitest of all is the salt from Tarentum; for the rest, salt that is especially white is brittle. Rain water makes all salt sweet, but dew makes it more agreeable, and the north wind makes it in quantity. It does not form in a south wind.

B QUARRYING

Like mining, quarrying was largely a technology developed by the Bronze Age and later cultures, once the appearance of metal tools facilitated stoneworking and the evolution of stratified societies fostered the development of stone architecture and luxury objects. Because most quarries were open pits, prospecting, ventilation, and lighting presented fewer difficulties than in mining, and even

MINING AND QUARRYING

drainage was easier to deal with. The development of the compound pulley in the eighth or seventh century BC, along with the bipod crane, made it relatively easy to lift heavy blocks, although land transport with sledges or heavy wagons pulled by oxen remained difficult and expensive [**fig. 5.6**; cf. **fig. 8.1**] (cf. **8.4, 11.35**). Marble and a few other ornamental stones were the only kind traded in large blocks. Long-distance transport had to be carried out in specially reinforced ships, and the most heavily worked quarries were those located near a consuming

Figure 5.6 Transport of a marble block on a sled.
Source: Adapted from Adam, 1994, fig. 31.

city or close to the sea. Work in a quarry was dangerous and unpleasant in antiquity, and in general, few advances in the tools or techniques were made until the invention of the wire stone-saw in the nineteenth century. Sources related to Greek and Roman stone sculpture appear in Chapter 7 of this volume; see also K. Jex-Blake, and E. Sellers, *Pliny the Elder's Chapters* on *the History of Art* (1896. repr. Chicago, 1968), and J.J. Pollitt, *The Art of Ancient Greece: Sources and Documents* (Cambridge, 1990) (Hirt 2010, Fant 2008, 1988).

5.32 A moralistic evaluation of quarrying

If anything, Pliny waxed even more indignant about the Roman trade in coloured marbles than he did about mining (cf. **5.1**). Although the great weight and hardness of stone required heroic measures in the quarries and the construction of special ships, the trade routes during the Empire ran from one end of the Mediterranean to the other. The greatest amount was used for construction, but sculpture workshops consumed much as well.

Pliny, *Natural History* 36.1–3

Nature made mountains for herself as a type of bond for compressing the bowels of the earth and at the same time for holding in check the rushing strength of rivers and breaking the waves of the sea and to restrain with her hardest substance her least quiet parts. We quarry these mountains and drag them away for no other reason than that our pleasure dictates it – mountains which it was once astonishing even to cross. Our ancestors considered it almost a portent that the Alps were climbed by Hannibal and later by the Cimbri: now these very peaks are quarried into 1,000 types of marble. Promontories are laid open to the sea, and nature is made flat. We carry away features, which were meant to serve as barriers for keeping nations apart. Ships are built for the sake of transporting marble, and so here and there over the waves, the wildest portion of nature, are carried mountain peaks.... Each of us who hears the price of these items and sees the massive quantities, which are being dragged around should meditate on how much better life would be without them. Oh, that men should do these things – or rather, endure them – on account of no other purpose or pleasure than to recline surrounded by varicoloured stones…!

FAMOUS QUARRIES
5.33 The marble quarries of Athens and Paros

The quarries of Mt. Pentelicus provided the fine-grained white marble that was used for most Attic sculpture and for the Periclean reconstruction of the Acropolis, famous even in antiquity. To the small city states of pre-Roman Greece, the proximity of a quarry was of the highest importance, since transport was difficult and expensive (cf. **11.35**). Since most quarries were open-pit workings, the quarry workers at Paros mentioned by Pliny must have been following a particularly fine vein of stone.

Strabo, *Geography* 9.1.23

The most famous of the mountains [of Attica] are Hymettus, Brilessus, and Lycabettus, also Parnes and Carydallus. The best quarries for marble, Hymettan and Pentelic, are close to the city.

Pliny, *Natural History* 36.14

All these artists [the Archaic Greek sculptors of Chios], then, used only white marble from the island of Paros, a stone which they used to call *lychnites*, because, as Varro reports, it was quarried in galleries by the light of lamps.[3] Many whiter marbles have since been discovered, lately, in fact, even in the quarries of Luna. A marvel is reported concerning the quarries at Paros; when a single block of stone was split with wedges, [**fig. 5.7**] the stoneworkers found that there was an image of Silenus inside.

Figure 5.7 Splitting a rock with wedges.
Source: After Adam, 1994, fig. 42.

MINING AND QUARRYING

5.34 Roman marble quarries at Carrara

Until discovery of the quarries of fine, bright white marble at Luna, modem Carrara, around 40 BC, all the marble used in Italy had to be imported from the Aegean. The relative proximity of the new source, and the fortunate ease of exploitation and shipping, helped transform Roman architecture in the late Republic and early Empire with a glittering veneer of Greek-like material and style.

Strabo, *Geography* 5.2.5

Of these, note the city and bay of Luna.... The city, to be sure, is not large, but the bay is both very large and very beautiful, containing within it several harbours, all of them deep even close in to shore.... The bay is shut in by high mountains.... The quarries for both white and blue-grey marble are so numerous and so extensive – providing both monolithic slabs and columns – that most of the pre-eminent building projects in Rome and other cities obtain their material from here. And indeed, the stone is easily shipped out, since the quarries lie above and close to the sea, and the Tiber receives the cargo from the sea.

5.35 The Roman trade in coloured marbles

In the first century BC, Roman businessmen and emperors began to exploit sources of coloured marble, granites, and porphyry throughout the eastern Mediterranean and Egypt, responding to a taste for a more colourful, baroque style of architecture (Dodge 1991, 1988).

Strabo, *Geography* 12.8.14

Beyond Synnada is the village of Docimaea [in central Asia Minor], and the quarry for Synnadic marble (for the Romans call it this, but the locals call it "Docimite" or "Docimaean"). Originally the quarry yielded only small lumps, but on account of the present extravagance of the Romans, great columns are extracted in one piece, very similar to alabaster in their variegated colouring. As a result, even though the transport of such heavy burdens to the sea is a problem, nevertheless, columns and slabs of astonishing size and beauty are conveyed to Rome.

Seneca, *Letters* 115.8–9

Smooth and variegated pebbles, picked up on the beach, delight the children, while we enjoy the veins and spots of huge columns, brought here from the sands of Egypt or from the African desert, holding up a portico or some dining hall big enough for an entire people. We admire walls covered by thin marble veneer although we know what they hide beneath.

5.36 Quarries that replenish themselves

Some deposits of highly sought after, brightly coloured marble – such as the source of the so-called "Africano" from Teos in southwestern Asia Minor – were small and difficult to exploit. There was naturally some anxiety that such resources might be exhausted, and – as with mines (cf. **5.7**) – comforting stories arose of quarries that could replenish themselves. In the second passage, a poem written in the sixth century AC, a statue recounts its life history.

Pliny, *Natural History* 36.125

And among the many other marvels of Italy herself is one reported by Papirius Faianus, a man very accomplished in natural science, that marble grows in quarries. The quarrymen also confirm that the wounds in the mountains are filled back in spontaneously. If these reports are true, there is hope that marble will never fall short of the demands of luxury.

Theaetetus Scholasticus, in *Greek Anthology* 16.221

A Median stonecutter with his sharp tools cut me, a dazzling white stone, from a mountain quarry where the rocks grow back again, and he brought me across the sea to make images of men, symbols of steady courage in the face of the Athenians. But when Marathon echoed with the Persian collapse and ships sailed on water stained with blood, Athens, who bears the best children, shaped me into an image of Adrastus, the divinity antagonistic to arrogant men. I counterbalance empty hopes. I am still a Victory for the Athenians, a Nemesis for the Persians.

EXTRACTING AND WORKING STONE

5.37 Christian martyrs in the quarries of Numidia

Like the mines, many of the quarries in the Roman world were owned by the state. The work was just as hard and almost as difficult as that in the mines, so the labour of condemned criminals, military prisoners, or expendable slaves was used in working them (cf. **5.18**, **5.20**). In this case, St. Cyprian recounts the suffering of Christians condemned to work in the quarries because of their beliefs.

St. Cyprian, *Letters* 77.2.4

They also put fetters on your ankles … [The Martyrs] stretched out to sleep on the ground, tired to the bone by their labours…. In the absence of baths, their limbs were filthy, unsightly with dirt and squalor. Little bread was provided, and they lacked clothing to stay warm. Their hair was unevenly cut and unkempt.

5.38 The preparation of marble revetment slabs

Although his comment is moralising and disapproving, Pliny provides important testimony for the use of abrasives to cut thin sheets of marble. The particularly effective sands he mentions were probably a type of corundum, of which emery is a variety. A recently discovered relief shows a water-powered frame saw that was used at one particular site in what is now Turkey [cf. **fig. 2.4**] (Ritti *et al.* 2007).

Pliny, *Natural History* 36.51–53

But whoever first discovered how to cut marble and split luxury into sheets showed harmful ingenuity.[4] This seems to be effected by iron but actually is done by sand, as the saw presses the sand on a very narrow line and brings about the cutting by its very passage back and forth. Sand from Ethiopia is rated most highly.... Later, a no less esteemed sand was found on a certain shoal in the Adriatic Sea, uncovered by low tide and not easy to spot. But now deceitful workmen have dared to cut marble with any sort of sand from any river, a source of waste, which very few notice. For the coarser sand cuts less accurate slices, wears away more of the marble, and by its rough finish increases the work of polishing. Consequently, the revetment slabs are thinner. Again, Theban sand is suitable for polishing, and a compound made from limestone or pumice.

5.39 Other methods of working stone

Both the Greeks and the Romans made use of soft stones that could be shaped into tableware on a lathe. On a larger scale, marble columns could be finished in the same manner (see **8.46** and cf. **fig. 8.7**) (Grewe 2019, Rhodes 1987, Vandeput 1987).

Theophrastus, *On Stones* 41–43

Some stones [...] cannot be carved with iron tools but only with other stones. Generally speaking, there is great variety in the methods of working even the larger stones. For, as has been stated, some can be sawn successfully, others cut, or turned on a lathe.... But there are a number of stones that allow all these methods of working. In fact, there is such a sort of stone [steatite?] in Siphnos, quarried about three stades from the sea, soft enough to be turned on a lathe and cut. But when it has been fired and dipped in oil it becomes extremely dark and hard. They manufacture tableware out of it. All stones of this sort can be worked with iron....

5.40 Some soft stones of commercial value

Pliny, *Natural History* 36.159–162

On Siphnos there is a stone [steatite?], which is hollowed out and turned on a lathe to make containers useful both for cooking food and for serving comestibles. We know that this is also the case with the green stone of Como in Italy. The Siphnian stone, however, is remarkable in that being heated with oil it darkens and becomes hard, although very soft by nature. Such different characteristics can it take on. To be sure, there are outstanding examples of soft stones beyond the Alps. It is said that in Belgic Gaul a white stone is cut with the same saw used for wood, but even more easily, to serve as pantiles or cover tiles or, if so desired, the type of roofing termed "peacock style".[5]

Now these stones can be cut with a saw. However, "mirror stone" [selenite or mica?] – since even this can be termed a stone – by its nature splits much more easily into sheets as thin as might be wanted. Formerly only Hither Spain produced it [...] but now Cyprus as well, and Cappadocia, Sicily, and Africa (where it has recently been discovered). All these sources, however, are inferior to that of Spain. Cappadocia produces the biggest pieces, but they are cloudy.... In Spain it is mined in shafts at a great depth, but it is also found incorporated in rock just below the surface, where it is dragged or cut out. But generally, its formation is such that it can be dug, occurring in separate rough chunks; so far, however, none more than five feet long.... Moreover, the clear variety has the marvellous property of enduring the hot sun or frost, even though it is remarkably soft, and it does not deteriorate, as long as there is no mechanical damage....

5.41 Precautionary weathering of newly quarried building stone

It was considered wise in antiquity to allow quarried blocks to weather for a few seasons and thus expose any possible hidden flaws or stresses. This was particularly true of the soft volcanic tuffs characteristic of the region around Rome.

Vitruvius, *On Architecture* 2.7.5

Since, because of their proximity, necessity compels the use of stone from the quarries of Ruber and Pallene and others very close to the city [Rome], whoever wishes to complete a building without faults must make the following preparations. When construction is necessary, let these varieties of stone be extracted two years ahead of time – in Summer, not in Winter – and remain lying out in the open. The stones that, in the course of this two-year period, are damaged by the effect of the weather are to be thrown into the foundations. The rest, which are tested by nature and undamaged, will be capable of lasting when used in the structure above ground. These procedures must be followed not only for ashlar blocks but also for rubble walling.

5.42 Another application of quarried stone

Not all quarries were set up to produce building stone. Besides the stone vessels and revetments mentioned in **5.38–39**, mill-stones were another common quarry product.

Xenophon, *Anabasis* 1.5.5

The whole region round about [near Pylae, in southeastern Syria along the Euphrates] was barren. The inhabitants quarried mill-stones along the river, shaped them, and took them to Babylon to sell in exchange for grain.

5.43 Quarries, clay-pits, and agriculture

The only honourable (and possibly the only legal) source of wealth for a Roman senator was agriculture. With the explosive growth of Rome in the late Republic came an enormous demand for building stone, partly superseded in the early Empire by the demand for bricks. In supplying this demand from their estates, literal-minded Roman aristocrats could claim that they were simply working their land, not engaging in vulgar commerce.

Varro, *On Agriculture* 1.2.22–23

"Am I, then", I said, "to follow the writings of the Elder and Younger Sasema and understand that the question of how clay-pits ought to be managed is more pertinent to agriculture than mining for silver or other kinds of mines. For doubtless these are carried out in some other field. But since neither stone quarries nor sand-pits are relevant to agriculture, the same goes for clay-pits. This does not mean they should not be worked in land where they are suitable and provide profit …".

Notes

1 "Mountain earth?" from Greek *ouros/oros*, "mountain".
2 Swinging bucket-beam with counterweight; see **9.26**.
3 From the Greek *lychnos*.
4 For the chronology of this discovery, see **8.25**.
5 This term may indicate that the plates overlapped like feathers.

Bibliography

Mining

Ardaillon, Édouard, *Les mines du Laurion dans l'antiquité*. Paris: Thorin et Fils, 1897.
Alishan, Yener K., and Özbal Hadi, "Tin in the Turkish Taurus Mountains, the Bolkardag Mining District". *Antiquity* 61 (1987) 220–226.
Bird, David, "Water Power in Roman Gold Mining". *Mining History: Bulleting of the Peak District Mines Historical Society* 15.4/5 (2004) 58–63.
Burnham, Barry C., and Helen Burnham, *Dolaucothi-Pumsaint: Survey and Excavations at a Roman Gold-Mining Complex 1987–1999*. Oxford: Oxbow, 2004.
Craddock, Paul T., *Early Mining and Metal Production*. Washington: Smithsonian Institution, 1995.
Craddock, Paul T., "Mining and Metallurgy". Pp. 93–120 in John P. Oleson, ed., *The Oxford Handbook of Engineering and Technology in the Classical World*. Oxford: Oxford University Press, 2008.
Craddock, Paul T., "Classical Geology and the Mines of the Greeks and Romans". Pp. 197–216 in Georgia L. Irby, ed., *A Companion to Science, Technology, and Medicine in Ancient Greece and Rome, Vol. I*. Chichester: Wiley-Blackwell, 2016.
Crosby, Margaret, "Greek Inscriptions: A Poletai Record of the Year 367/6 B.C.". *Hesperia* 10 (1941) 15–30.
Crosby, Margaret, "The Leases of the Laureion Mines". *Hesperia* 19 (1950) 189–297.

Currás, Brais X., "The *salinae* of O Areal (Vigo) and Roman salt production in NW Iberia". *Journal of Roman Archaeology* 30 (2017) 325–349.
Davies, Oliver, *Roman Mines in Europe*. Oxford: Clarendon Press, 1935.
Domergue, Claude, *Les mines de la péninsule ibérique dans l'antiquité romaine*. Collection de l'École Française de Rome, 127. Rome: École Française de Rome, 1990.
Edmondson, J.C., *Two Industries in Roman Lusitania: Mining and Garum Production*. Oxford: British Archaeological Reports, 1987.
Elkington, David, "Roman Mining Law". *Mining History: Bulleting of the Peak District Mines Historical Society* 14.6 (2001) 61–65.
Forbes, Robert J., "Ancient Geology", "Ancient Mining and Quarrying", "Ancient Mining Techniques". *Studies in Ancient Technology*, Vol. 7. 2nd edn. Leiden: Brill, 1966.
Healy, John F., *Mining and Metallurgy in the Greek and Roman World*. Ithaca: Cornell University Press, 1978.
Hirt, Alfred Michael, *Imperial Mines and Quarries in the Roman World*. Oxford: Oxford University Press, 2010.
Jones, Geraint D.B., "The Roman Mines at Rio-Tinto". *Journal of Roman Studies* 70 (1980) 146–165.
Jones, John Ellis, "The Laurion Silver Mines: A Review of Recent Researches and Results". *Greece and Rome* 29 (1982) 169–183.
Jones, R.F.J., and D.G. Bird, "Roman Gold-Mining in North-West Spain, II: Working on the Rio Duera". *Journal of Roman Studies* 62 (1970) 59–74.
Lauffer, Siegfried, "Bergmännische Kunst der antiken Welt". Pp. 37–68 in Heinrich Winckelmann, ed., *Der Bergbau in der Kunst*. Essen: Glückauf, 1958.
Lauffer, Siegfried, *Die Bergwerkssklaven von Laureion*. 2nd edn. Forschungen zur antiken Sklaverei, 11. Wiesbaden: Franz Steiner Verlag, 1979.
Lewis, Peter R., and Geraint D.B. Jones, "The Dolaucothi Gold Mines, I: The Surface Evidence". *Antiquaries Journal* 49 (1969) 244–272.
Lewis, Peter R., and Geraint D.B. Jones, "Roman Gold-Mining in North-West Spain". *Journal of Roman Studies* 60 (1970) 169–185.
Mathisen, Ralph W., "Ancient Sources on the Mining and Refining of Gold and Silver during Roman Times". *Journal of the Society of Ancient Numismatics* 7 (1975) 6–11, 15, 28–30; 7 (1976) 43–46, 58–64; 8 (1977) 14–16, 22–24, 34, 39–41; 9 (1978) 10–12, 20, 23.
Mussche, Herman, ed., *Thorikos and the Laurion in Archaic and Classical Times*. Ghent: Comité des Fouilles Belges en Grèce, 1975.
Penhallurick, Roger D., *Tin in Antiquity, its mining and trade throughout the ancient world with particular reference to Cornwall*. London: Institute of Metals, 1986.
Ramin, Jacques, *La technique minière et métallurgique des anciens*. Collection Latomus, 153. Brussels: Latomus, 1977.
Rebrik, Boris M., *Geologie und Bergbau in der Antike*. Leipzig: Deutscher Verlag für Grundstoffindustrie, 1987.
Rosumek, Peter, *Technischer Fortschritt und Rationalisierung im antiken Bergbau*. Bonn: Habelts, 1982.
Rothenberg, Beno, and Antonio Blanco-Freijeiro, *Studies in Ancient Mining and Metallurgy in SW Spain*. London: Institute for Archaeo-Metallurgical Studies, 1981.
Salkield, Maurice Unthank, *Technical History of the Rio Tinto Mines: Some Notes on Exploitation from pre-Phoenician Times to the 1950s*. London: Institution of Mining and Metallurgy, 1987.

Schönbauer, Ernst, *Beiträge zur Geschichte des Bergbaurechtes*. Munich: Beck, 1929.
Shepherd, Robert, *Prehistoric Mining and Allied Industries*. London: Academic Press, 1980.
Shepherd, Robert, *Ancient Mining*. New York: Elsevier Applied Science, 1993.
Weisgerber, Gerd, and Willies, Lynn, "The Use of Fire in Prehistoric and Ancient Mining: Firesetting". *Paléorient* 26.2 (2000) 131–149.

Quarrying and stoneworking

Adam, Sheila, *The Technique of Greek Sculpture in the Archaic and Classical Periods*. London: Thames & Hudson, 1966.
Bedon, Robert, *Les carrières et les carriers de la Gaule romaine*. Paris: Picard, 1984.
Blagg, Thomas F.C., "Tools and Techniques of the Roman Stonemason in Britain". *Britannia* 7 (1976) 151–172.
Bluemel, Carl, *Greek Sculptors at Work*. 2nd edn. London: Phaidon, 1969.
Dodge, Hazel, "Decorative Stones for Architecture in the Roman Empire". *Oxford Journal of Archaeology* 7 (1988) 65–80.
Dodge, Hazel, "Ancient Marble Studies: Recent Research". *Journal of Roman Archaeology* 4 (1991) 28–50.
Dolci, Enrico, *Carrara: Cave antiche, materiali archeologici*. Carrara: Comune di Carrara, 1980.
Dubois, Charles, *Études sur l'administration et l'exploitation des carrières dans le monde romain*. 2 Vols. Paris: Thorin, 1908.
Durkin, Michael K., and Carol J. Lister, "The Rods of Digenis: An Ancient Marble Quarry in Eastern Crete". *Annual of the British School at Athens* 78 (1983) 69–96.
Evely, R. Doniert G., "Some Manufacturing Processes in a Knossian Stone Vase Workshop". *Annual of the British School at Athens* 75 (1980) 127–137.
Fant, J. Clayton, ed., *Roman Marble Quarrying and Trade*. Oxford: BAR, 1988.
Fant, J. Clayton, ed., *Cavum Antrum Phrygiae: The Organization and Operations of the Roman Imperial Marble Quarries in Phrygia*. Oxford: British Archaeological Reports, 1989.
Fant, J. Clayton, ed., "Rome's Marble Yards". *Journal of Roman Archaeology* 14 (2001) 167–198.
Fant, J. Clayton, ed., "Quarrying and Stoneworking". Pp. 121–135 in John P. Oleson, ed., *The Oxford Handbook of Engineering and Technology in the Classical World*. Oxford: Oxford University Press, 2008.
Gorelick, Leonard, and A. John Gwinnett, "Diamonds from India to Rome and Beyond". *American Journal of Archaeology* 92 (1988) 547–552.
Grewe, Klaus, "Recent Developments in Aqueduct Research". *Mouseion Supplement: Festschrift in Honour of Professor John P. Oleson*. Forthcoming (2019).
Heimpel, Wolfgang, Leonard Forelick, and A. John Gwinnett, "Philological and Archaeological Evidence for the Use of Emery in the Bronze Age Near East". *Journal of Cuneiform Studies* 40 (1988) 195–210.
Hill, Peter R., "Stonework and the Archaeologist, including a Stonemason's View of Hadrian's Wall". *Archaeologia Aeliana* 9 (1981) 1–22.
Hirt, Alfred Michael, *Imperial Mines and Quarries in the Roman World*. Oxford: Oxford University Press, 2010.
Klemm, Rosemarie, and Dietrich D. Klemm, *Steine und Steinbrüche im alten Ägypten*. Berlin: Steiner, 1992.

Leotardi, Paola Baccini, *Marmi di cava rinvenuti ad Ostia, e considerazioni sul commercio dei marmi in età romana*. Scavi di Ostia, 10. Rome: Istituto Poligrafico dello Stato, 1979.

Maxfield, Valerie, and Peacock, David, *The Roman Imperial Quarries: Survey and Excavation at Mons Porphyrites 1994–1998*. London: Egyptian Exploration Society, 2001.

Mutz, Alfred, "Die jüdische Steindreherei in herodianischer Zeit. Eine technologische Untersuchung". *Technikgeschichte* 45 (1978) 291–320.

Nylander, Carl, "The Toothed Chisel". *Archeologia Classica* 43 (1991) 1037–1052.

Oustinoff, Elizabeth, "The Manufacture of Cycladic Figurines. A Practical Approach". Pp. 38–43 in *Cycladica. Studies in Memory of N.P. Goulandris*. London: British Museum Publications, 1984.

Rakob, Friedrich, ed., *Die Steinbrüche und die antike Stadt*. Mainz: von Zabern, 1993.

Rhodes, Robin F., "Rope Channels and Stone Quarrying in the Early Corinthia". *American Journal of Archaeology* 91 (1987) 545–551.

Ritti, Tullia, Grewe, Klaus, and Kessener, Paul, "A Relief of a Water-Powered Stone Saw Mill on a Sarcophagus at Hierapolis and its Implications". *Journal of Roman Archaeology* 20 (2007) 138–163.

Schwandner, Ernst-Ludwig, "Der Schnitt im Stein. Beobachtungen zum Gebrauch der Steinsäge in der Antike". Pp. 216–223 in Adolf Hoffmann, Ernst-Ludwig Schwandner, Wolfram Hoepfner, and Gunnar Brands, *Bautechnik der Antike. International Colloquium, 15–17 February 1990*. Mainz: Deutsches Archäologisches Institut, 1991.

Stewart, Andrew F., "Some Early Evidence for the Use of the Running Drill". *Annual of the British School at Athens* 70 (1975) 199–201.

Stocks, Denys A., "Ancient Factory Mass-production Techniques. Indications of Large-scale Stone Bead Manufacture during the Egyptian New Kingdom Period". *Antiquity* 63 (1989) 526–531.

Vandeput, Lutgarde, "Splitting Techniques in Quarries in the Eastern Mediterranean". *Acta Archaeologica Lovanensia* 26–27 (1987–1988) 81–99.

Vollenweider, Marie-Louise, *Die Steinschneidekunst und ihre Künstler in Spätrepublikanischer und Augustäischer Zeit*. Baden-Baden: Bruno Grimm, 1966.

Ward-Perkins, J. Brian, *Marble in Antiquity: Collected Papers of J.B. Ward-Perkins*. Edited by Hazel Dodge, and J. Brian Ward-Perkins. Rome: British School at Rome, 1992.

Weisgerber, Gerd, and Willies, Lynn, "The Use of Fire in Prehistoric and Ancient Mining: Firesetting". *Paléorient* 26.2 (2000) 131–149.

Wycherly, Richard E., *The Stones of Athens*. Princeton: Princeton University Press, 1978.

6

METALLURGY

Once ore had been found and removed from the mines, the metal incorporated in it still needed to be extracted, refined, and prepared in a form appropriate for use, or combined with other metals or given special treatment to improve the final material. Together with the study of the structure and properties of the metals, these procedures form the science of metallurgy.

Although the stages of discovery of metals are obscure, the importance of metals to human development (Raymond 1986) is reflected by their metaphorical use in the literary tradition (cf. **1.1**) and in modern terminology to indicate the stages of human cultural development. The significance of metals and metallurgy is also discernible from the fact that their discovery and development accompanied humankind's evolution into a civilised society and the rise of the first great empires.

The use of metals among humans probably began with the discovery of shiny lumps of the native metal, and then very gradually and by accident, hammering and heating were discovered to improve the basic form. These treatments of the native metal represent the simplest stage of metallurgy. The discovery of the smelting of ores (**6.1**) represents the most important stage, since it led to the search for and refinement of known metals and the recovery of others like lead and antimony that do not occur in native form. Once the basic properties of a few metals were discovered by trial and error, the process of extending these methods to other ores and metals resulted in the improvement and expansion of mankind's knowledge of metals and ores.

Further experimentation with different metals and ores led to the development of alloys and other processes such as casting, but this was done usually with little understanding of the true chemical nature of the ores and metals. The Greeks and Romans understood little about the metals except in terms of their basic external properties. As a result, description of procedures and treatments in our sources is often confused, a problem that is aggravated by the use of the same term to describe several different ores and metals.

Nevertheless, with these advancements, the working of the metal became more complex and the role of the smith increased substantially as knowledge improved and new refining, alloying, and working processes developed. This, in turn, stimulated trade between peoples as raw and worked materials came into greater demand, and probably energised agriculture and other activities to produce more and better items to barter. At the same time the metals began to play important roles in a variety of technologies (**10.1–10.7**). Eventually the metal itself became the standard of exchange, first as a specified weight and then as minted coins (**11.113–11.119**).

Of the three principal roles of metallurgists (smelter, smith, and metalworker) information regarding the smith and his tools is meagre in the literature but much fuller in the archaeological record due to the recovery of tools and furnaces. Metalworking is examined in Chapter 10.

As might be expected, the sources of information for metallurgy are very scattered. The one exception is Pliny the Elder, who has compiled a large amount of information from earlier sources. Unfortunately, some of it is confused or misleading, partly a result of Pliny's methodology, partly of his less than complete research into the science and ores (Craddock 2008, Healey 1978).

WORKING AND ANNEALING NATIVE METALS

The first stage of metallurgy, the working of native metals, required skills different from those used to work bone or stone, because a mass of metal has to be heated occasionally during working to avoid brittleness and cracking. This annealing and hammering also makes metal implements harder. The observation that metals would soften and even melt when heated sufficiently may have been a related discovery.

6.1 Fire and the accidental discovery of metals

The advantages of metal over stone were obvious: colour and gloss, malleability and permanence, the ability of some metals to maintain a sharp edge, fusibility, and especially casting. The characteristics of malleability (when heated) and permanence of shape (when cooled) were similar to those associated with pottery, but metal had the added advantage of reuse and sturdiness. Lucretius attributes the discovery of metal and these characteristics to the action of the forces of nature on native metals, which humans then imitated (Maddin 1988).

Lucretius, *On the Nature of Things* 5.1241–1265

Copper and gold and iron were discovered, and at the same time the value of silver and the potential of lead, when fire and its heat consumed huge forests on a great mountain. The fires may have started from a heavenly lightning bolt or from men waging war upon one another in the forest, men who used fire against the enemy to create fear; or perhaps men wished to extend the rich fields and to turn over the fields for pasturage, or to destroy the wild animals and to grow rich on the plunder. For to hunt with dug pit and fire was practised earlier than to enclose the glade with nets and drive the prey with dogs. No matter what the cause, flaming heat with its terrible din had consumed forests down to the deep roots and scorched the earth with fire, then rivulets of gold and silver, and also copper and lead, ran together and flowed in fiery veins into the hollow places of the earth. When people saw these metals congealed together later, lying on the earth in their splendour with bright colour, they picked them up, captivated by their smooth and gleaming charm, and they saw that they were formed into the same shape of the hollows in which they had left their traces. Then it occurred to them that these metals could be melted with heat and poured into any form and shape, and immediately by hammering, the metal could be drawn out into the beginnings of sharp and elegant edges to make weapons for themselves....

The passage continues, describing uses of the metals and the advantages of bronze and iron. Juvenal (*Satires* 15.165–170) states that metals were first used for agriculture, then for weapons.

6.2 The purity of gold: ease of discovery and recovery

Gold is often regarded as the first metal used by humankind, largely because it occurs in its native form in larger and more widely dispersed quantities than the other metals that occur in native form (silver, copper, and meteoric iron).

Pliny, *Natural History* 33.62

Gold alone, above all other metals, is obtained in the form of nuggets or in detritus. Whereas all other metals found in the mines are brought to perfection by fire, gold is gold right from the start and has its matter perfected at once when it is found thus. For this is the natural way of discovering it.

Strabo, *Geography* 7 Fragment 34

And it is said that the people ploughing the Paeonian soil find nuggets of gold.

See Pliny (**5.21**) for a lengthy description of the recovery of gold without refining by the Romans in Spain. Strabo (*Geography* 3.2.8–11; 4.2.1) describes at length the metallic wealth of Spain and France and the occurrence of native metals (gold, copper, silver, and iron).

[Aristotle], *On Marvellous Things Heard* 25–26 (832a)

In Cyprus they say mice eat iron. And they say the Chalybes, on some island lying to the north of them, gather gold from many of the mice. For on account of this reason it appears they rip apart the mice in the mines.

Pliny (**5.21**) states that ants gather gold in India, griffins in Scythia.

6.3 The ease of working gold, and its special properties

Homer provides an early description of gilding or overlay in this simile as he describes Athena's miraculous transformation of a scruffy Odysseus into a lordly figure. Pliny provides more practical information about the practical properties of gold.

Homer, *Odyssey* 6.232–235

And just like when some skilful man spreads gold over silver, a man whom Hephaestus and Pallas Athena have taught all sorts of skill, and he executes elegant works, thus Athena poured grace down upon his head and shoulders.

Pliny, *Natural History* 33.60–63

Another greater reason for its value is that use wears it away very little, while lines can be drawn with silver, copper, and lead, and they dirty the hands with matter that flakes off. Nor is another material more malleable or able to be divided into more pieces, since an *uncia* can be spread out into 750 or more leaves that are each four digits on a side.... Above all other metals, gold does not suffer from any rust or verdigris, and there is not anything else arising from it that consumes its excellence or diminishes its weight. Moreover, its stability against the effects of salt and vinegar, the vanquishers of material items, surpasses everything and even more than that it is spun and woven in the manner of wool, even without wool (being added). Verrius instructs us that Tarquinius Priscus [616–578 BC] celebrated a triumph in a golden tunic. We ourselves have seen Agrippina, the wife of the emperor Claudius, at a spectacle of a naval battle presented by him, sitting next to the emperor and dressed in a military cloak of pure gold fabric. Now indeed for a long time it has been woven into the cloth called Attalus, the invention of the kings of Asia.

THE EXTRACTION OF METAL FROM ORE

Ores could be washed to extract the metal but smelting, a process that required temperatures attainable only in furnaces, was much more efficient. Higher temperatures allowed the melting of the metal and this led to techniques such as casting.

6.4 Washing silver ore

The silver mines around New Carthage in Spain were very extensive and profitable, employing 40,000 miners and producing as much as 25,000 *drachmai* per day according to Polybius. Washing the ore is the simplest method of removing the unwanted material. Strabo (*Geography* 3.2.10) records similar information about the process.

Polybius, *Histories* 34.9.10–11

... The silver lumps swept down by rivers, Polybius says, are crushed and shaken through sieves into water. Then the sediment is crushed and sifted again, and with the water flowing off it is crushed once more. This process, once the fifth deposit of sediment has been collected and the lead has been spilt out, produces pure silver.

Theophrastus (*On Stones* 58–59) states that the washing process for producing cinnabar was invented about 405 BC by Callias, an Athenian from Laurion (Conophagos 1980).

6.5 Brutal mining conditions in Egypt; cupellation

After describing the brutal nature of work in the mines (see **5.18**), Diodorus continues here with a description of the subsequent process once young boys have removed the broken rock to the entrance of the mine. Pounding the conglomerate in mortars releases the gold from the matrix that contains it. Heating the powder in a cupel is the next step (see **6.6**).

Diodorus of Sicily, *History* 3.13.2–14.4

Then those men over 30 years, taking a defined measure of the quarried stone from these young boys, pound it in stone mortars with iron pestles until they have reduced it down to the size of beans. From these men, the women and older men receive the bean-sized stone and they throw it into one of a series of mills. Then, standing next to the handle in groups of two or three, they grind it until they have reduced their portion to the texture of the finest wheat flour. And since none of them have the means to care for their needy bodies or clothing to cover their shame, there is not a person who can look upon the wretched workers and not feel pity at their excessive sufferings. For neither forgiveness nor relief of any sort exists: not for the sick man, not for the maimed, not for the elderly, not for a woman in her weakness. All are forced by blows to persevere in their labours until, because of their poor treatment, they die in the midst of their sufferings. As a result of the excessive nature of their tortures, the miserable workers believe their future will always be more frightful than the present and therefore look more favourably upon death than life.

In the final stage, the skilled workmen, accepting the finely ground stone, take it off for its final working. For they rub the reduced stone upon a broad wooden board that is slightly inclined and pour water over it. This flow of water dissolves the earthy matter that then runs down the inclined board while the material holding the gold remains on the wood because of its heaviness. After doing this several times they first rub it gently with their hands, then press it lightly with thin sponges to lift out the loose and earthy material until there is only pure gold dust. Finally, other skilled workmen take the recovered material and put it into earthen pots by fixed measure and weight. They mix with it a piece of lead proportionate to the mass, coarse-grained salt, a bit of tin, and add barley bran to it. Then making fitted lids and carefully rubbing them all over with mud, they bake them in a kiln for five days and nights without interruption. Then after allowing them to cool, they find nothing of the other materials in the vessels and recover pure gold with very little waste.

6.6 Refining by cupellation

The cupellation process is one of the oldest and most efficient methods of separating precious metals from base ones. Air was blown across the molten alloys in the cupel (a special porous crucible) and the base metals were oxidised and then absorbed into the walls of the cupel leaving behind the pure gold and silver. Our ancient written evidence for the process is fragmentary.

Pliny, *Natural History* 33.69

The material dug out [of the mines] is crushed, washed, fired, and ground into a dust [**figs. 6.1–2**]. They call the dust from the mortar *scudes*, and the silver that comes out from the kiln *sudor* [sweat]. The dirt which is thrown out of the smelting furnace, in the case of all metals, is called *scoria* [slag]. For gold, this *scoria* is crushed and baked a second time. The crucibles are made of *tasconium*, which is a white, clay-like earth. For no other earth can endure the blast of air, the fire, and the burning material.

Figure 6.1 Milling the ore.

Source: Illustration adapted from Georgius Agricola in *De re metallica*, sixteenth century.

METALLURGY

Figure 6.2 Washing the ore.

Source: Illustration adapted from Georgius Agricola in *De re metallica*, sixteenth century.

Psalms 12: 6

The words of the Lord are pure words; like silver tested by fire, a proof on earth, purified seven times.

6.7 Mercury and the amalgamation process

The amalgamation process, the extraction of gold with mercury, appears to be a Roman discovery. Pliny provides a slightly confused description of the process in which mercury was combined with gold ore and the gangue separated from the amalgam (the solution of gold and silver in the mercury) by forcing the mercury through leather. The gold and silver were then separated from the amalgam by heating. Vitruvius is somewhat clearer.

Pliny, *Natural History* 33.99–100

A mineral called mercury, which is poisonous and always liquid, is found in the veins of silver [in Spain]. It is a poison to everything and ruptures pots by permeating them with destructive decay. Everything except gold floats on it; mercury attracts gold alone to itself. Therefore, it purifies [gold] very well, since it expels all other impurities when repeatedly shaken in clay pots. To separate the mercury itself from the gold, once the foreign matter has been discarded, it is poured out onto leather skins through which it percolates gradually like sweat and leaves behind pure gold. And so, when copper things are gilded, an undercoat of mercury beneath the gold-leaf holds it with the greatest tenacity.... Otherwise, mercury is not found in large quantities.

Vitruvius, *On Architecture* 7.8.1–4

Now I will proceed to explain the treatment of *minium* [here vermilion or sulphide of mercury].... The material, which is called ore, is dug up, then they produce *minium* by treating it. In the veins the ore is like iron, more ruddy in colour, and having reddish dust around it. When it is dug up and beaten with iron tools, it exudes many drops of mercury, which are collected quickly by the miners.

Once this ore has been collected in the workshop it is placed in a furnace to dry because of its large amount of moisture. Next, when the steam stirred up by the heat of the fire condenses on the floor of the furnace, it is discovered to be mercury. After the ore is taken away, the drops which remain because they are too small to be collected are swept together into a container of water where they run together and are combined into one mass. When they are weighed, four *sextarii* of mercury come to 100 *librae*.... Mercury, moreover, is useful in many instances. For neither silver nor brass can be properly gilt without it. And when gold is embroidered in clothing and the garment, worn out with old age, has no decent use, the cloth is put into a clay vessel and burnt over the fire. The ash is thrown into water and mercury is added to it. The mercury then gathers all the particles of gold onto itself and combines with them. Once the water is poured off, the remainder is spread over a cloth and pressed by hand. The mercury, since it is liquid, passes through the texture of the cloth when forced by the pressure, and pure gold is found in the cloth.

Vitruvius has confused vermilion with red lead in this passage, but he provides a good description of the extraction process of mercury. Part of the confusion may rest in the source of the mineral and the workshops, which were originally in the region of Ephesus, on the west coast of Turkey. The use of the amalgamation process is, no doubt, closely connected to the discovery of simple methods of recovering mercury from its ore (**6.9**).

6.8 A layman's understanding of gold refining

Everyone understood the value of gold, and many people had an understanding of the process of refinement. Plato uses gold refining as a simile to demonstrate how monarchy is the best form of government, rule by the few the next best, and rule by the many the least good; all able to be separated just as the most precious metals are separated from earth and stone. The dialogue is between the Younger Socrates (SOC.) and a stranger (STR.).

Plato, *The Statesman* 303d–e

STR.: ... We seem, I think, to be in the same situation as those men who refine gold.

SOC.: How is that?

STR.: Because the workmen first separate the earth and stones and all that sort of material; after that the precious materials, those similar to and mixed with gold such as copper and silver and sometimes adamant [platinum?], they remain and can be removed only by fire. These are separated by troublesome smelting and with testing, leaving what is called unalloyed gold in its purity for us to look upon.

SOC.: Yes, indeed that is said to be the method.

6.9 Recovering mercury

Theophrastus provides the earliest description of mercury prepared from its ore, cinnabar. It is also one of the earliest descriptions of a method to isolate a metal from one of its compounds. Experiments have demonstrated that the process he describes is not merely mechanical, but a true chemical reaction between the ore, the vinegar, and the copper. An amalgam of mercury and copper is produced that would require distillation to obtain pure mercury. Theophrastus does not mention distillation, for knowledge of which we depend on the passage from Pliny; either Theophrastus did not know all the details of the process or the amalgam was used without further refinement.

Theophrastus, *On Stones* 60

Mercury is made when cinnabar mixed with vinegar is pounded in a copper mortar with a copper pestle.

Pliny, *Natural History* 33.123

Hydrargyrum [artificial mercury] is produced in two ways: by grinding *minium*[1] in vinegar with a bronze mortar and pestle or by putting the *minium* in iron, shell-shaped vessels and these in earthenware pans, then covering them with a cooking pot sealed on with clay. Then a fire is set ablaze using pairs of bellows under the pans and the condensation, which becomes the colour of silver and as fluid as water, is collected on the upper cooking pot and is wiped off. This same liquid is easily divided into drops and flows together in slippery fluidity.

6.10 Antimony: extraction and properties

The production of white oxide of antimony from stibnite (the sulphidic ore) was not practised on a large scale for the antimony itself, but for the gold and silver associated with it. Antimony was produced as a by-product that was often confused in antiquity with lead, tin, and arsenic. It was used in medications and as a cosmetic. Pliny has just finished listing some of antimony's medicinal uses and then provides one of the best descriptions of its extraction. Dioscorides (*Pharmacology* 5.84) has a similar description.

Pliny, *Natural History* 33.103–104

Stibnite is burnt in the oven once it has been smeared all over with lumps of ox dung; then quenched with human milk, mixed with rain water, and pounded in mortars. Immediately afterwards the muddy mixture is poured off into a copper vessel, which has been purified with soda [natron]. The dregs are recognised as being very full of lead, and are thrown away once they settle in the mortar. Then the vessel, into which the muddy mixture was poured, is covered with a cloth and left for the night. The next day anything floating is poured out or removed with a sponge. The material that has settled to the bottom is regarded as the best part and, once a cloth has been interposed, it is set under the sun to dry, but not completely. Then it is ground again in a mortar and divided into small tablets. But above all else, a proper measure of heat is necessary so that it does not become lead [metallic antimony].

6.11 Refining lead oxide and lead sulphide

Pliny, *Natural History* 33.106–109

In those same mines [in Spain] there is a substance called scum of silver.[2] There are three types: the best which they call golden, second the silvery, and third the leaden. And for the most part all these colours are found in the ingots themselves. The most approved scum is Attic, the next is the Spanish. The golden scum comes from the vein itself, the silvery from silver, and the leaden from the smelting of lead. The smelting of lead is done at Puteoli, from which it takes its name (*argyritus puteolana*). In addition, each type is made when its own matter has been thoroughly melted and flows down from an upper crucible into a lower one and then is lifted out on small iron spits and twisted around on the spit in the flame itself to make it of moderate weight....

In order to make the scum [lead sulphide] useful, it is melted a second time after the ingots have been broken up into the size of finger rings. Then after heating it with bellows to separate out the charcoal and ash, it is washed with vinegar or wine, which quenches it at the same time. In order to give it radiance, if it is the silvery type, one must break it into pieces the size of beans and boil them in a clay pot filled with water that has wheat and barley added inside new linen cloths; the mixture is boiled until the grains lose their husks. Afterwards they grind it in mortars for six days, washing it three times a day with cold water and, once they have stopped, with hot water and mined salt, adding an *obol* for each *libra* of scum. Then on the final day they store it in a lead vessel. Some people boil it with white bean and pearl barley and dry it in the sun, while others boil it with white bean in wool until it no longer discolours the wool. Then they add mined salt, from time to time changing the water, and dry it on the 40 hottest days of summer. They also boil it in water in a sow's stomach, then take it out and rub it with soda and, as above, grind it in mortars with salt. There are some people who do not boil it, but grind it with salt and wash it by adding water.

Dioscorides (*Pharmacology* 5.87) expands this discussion.

6.12 The earliest reference to poling copper

Pliny records the various forms and blends of copper and bronze and their respective properties and uses. Here he describes poling, a process by which the brittleness of fused copper was lessened and the copper made more malleable through the reduction of copper oxide. The malleable bar copper was produced when poles of unseasoned wood were forced into the molten metal and hydrocarbon gases released from the burning green wood reduced the oxide. This reference to poling hinges on a change in the text from *cribro* (a sieve) to *ligno* (wood for burning), a variant not attested in the manuscript tradition and not accepted by all scholars.

Pliny, *Natural History* 34.95

At Capua copper is smelted with heat not from charcoal but from wood, and then purified with oak firewood after being poured into cold water. In a like manner it is smelted many times, until in the very last stage ten *librae* of Spanish silver-lead [equal amounts of tin and lead] are added to 100 *librae* of copper. In this manner it becomes pliable and takes on a pleasing colour, such as that imparted to other types of copper and bronze by oil and salt.

6.13 Tin production in Britain

Sources of this important metal were not widespread in the Mediterranean world. It was not used in its pure form for anything of much importance, but was essential as an ingredient in alloys with lead (pewter) and copper (bronze). Britain was one of the richest sources of the ore in antiquity and Cornwall continues to be one of the major sources of vein ore. Its significance is clear, as is the impact of commerce in the region arising from merchants travelling to acquire the tin (Muhly 1985).

Diodorus of Sicily, *History* 5.22.2

The inhabitants of Britain living around the promontory called Belerium [Cornwall] are extremely hospitable to strangers and because of their contact with foreign merchants have adopted a civilised way of life. These men work the tin, treating the earth bearing it ingeniously. This earth, which is like rock, has earthy seams in which the workers quarry the stone (*poros*) and they purify it by smelting. Stamping it into the form of knuckle-bones, they convey it to an island lying off Britain that is called Ictis [St. Michael's Mount].

Traders purchased the tin here and transported it to continental Europe. In the following passage, Pliny provides additional information.

6.14 Lead and tin: locations, refining, and properties

Pliny, *Natural History* 34.156–159

The nature of lead, of which there are two types, black [lead] and white [tin], follows. White is the most valuable and was called by the Greeks *cassiteros*. An extraordinary story relates that they hunted for it in the islands of the Atlantic Ocean and transported it in boats made of wicker and stitched with hides. Now, as is well known, it is found in Lusitania and Gallaecia in the uppermost earth, and is sandy and black in colour. It is detected only by its weight. Very small pebbles also occur, especially in torrential streambeds that have dried up. The miners wash this sand and smelt the material that settles in furnaces. It is also found in the goldmines called *alutiae* [washed mines], in which a stream of water washes out black pebbles discoloured with small white spots. These are the same weight as gold and thus remain with the gold in the bowls in which it is collected. Afterwards they are separated in the smelter and the fused metals are melted into white lead. Black lead does not occur in Gallaecia, although neighbouring Cantabria [Biscaya] has an abundance of black lead only; nor does silver come from white lead, but from black lead. Black lead is not able to be soldered to black without the use of white lead, nor can the latter be soldered to black without oil, and not even white to white without black.... There are two sources of black lead: for either it appears in its own vein and bears no other mineral, or it is formed with silver and is smelted with the veins mingled. Of this material that which first becomes liquid in the furnaces is called *stagnum* [alloy of silver and lead]; that liquefied second is called *argentum* [argentiferous lead]; the material left in the furnaces is called *galena* [here, impure lead], which is a third portion of the vein that was originally put into the furnace. After this is again melted down it produces black lead, two-ninths of it being lost.

6.15 Resmelting of slag with improved methods

Furnaces would have originally been simple cavities in the ground so it is not surprising to find that as smelting techniques improved in antiquity it was possible to rework the slag from older periods (see **5.13**). This was especially important once mineral deposits were exhausted in a region.

Strabo, *Geography* 9.1.23

The silver mines in Attica were originally remarkable but now are exhausted. Even so, once the mining operations yielded feeble results, the men working them, by resmelting the old discarded material and the dross, were still able to obtain pure silver since the old furnaces were inefficient.

6.16 Official guidelines for resmelting slag

The following selection is part of a Hadrianic law for mining in Spain, which pertains to the rules for resmelting slag.

CIL 2.5181.46ff. (=*ILS* 6891)

Fee for exploitation of slag heaps and rock dumps. Anyone within the district of the Vipasca mines who wishes to clean, crush, smelt, treat, break, sort, or wash silver or bronze slag, or sand from the slag heaps shovelled out by measure or weight, or who undertakes to do any sort of work in the quarries, shall declare by the third day following what slaves and salaried workers they will send to carry this out and pay the concessionaire [...] each month by the last day of the month: if they do not do this, they shall have to pay double. Whoever shall bring into the mine area diggings of bronze or silver slag from mine dumps at other places shall have to pay one *denarius* per 100 pounds to the concessionaire, his partner, or agent. If something due the concessionaire, his partner, or agent under this section of the law is not paid on the day it comes due, or sufficient security left, he shall have to pay double. It shall be permitted the concessionaire, his partner, or agent to obtain security and the portion of that slag, which is cleaned, crushed, smelted, treated, broken, sorted, or washed and the blocks, which are finished off in the quarries will be made over to him, unless whatever is owed the concessionaire, his partner, or agent is paid. Slaves and freedmen of silver and bronze smelters who are working in the smelting facilities or their masters or patrons are excepted.

The quantity and, above all, the quality of fuel were major concerns in the smelting process since they determined the furnace's temperature in a fundamental manner. Coke, the fuel used today, was unknown in antiquity, and, although lignite was occasionally used (see **2.29**), charcoal was the most common source of fuel for smelting. This fuel was expensive and limited the size of the furnace since large amounts of ore could not be smelted with it. See **2.27** for a lucid and substantially correct description by Theophrastus of the varieties of charcoal (Pliny, *Natural History* 16.23–24 has a less complete account).

EFFECTS OF METALLURGY ON THE ENVIRONMENT

6.17 Clearcutting for fuel

The effects of the refining process on the environment were not necessarily regarded as bad in antiquity. Strabo records one of the advantages: more land became cultivated.

Strabo, *Geography* 14.6.5

In Tamassus [on Cyprus] there are many copper mines in which are found chalcanthite [copper sulphate] and also rust of copper [verdigris], useful for their medicinal properties. Eratosthenes says that in olden times the plains, being overgrown with woods, were covered with trees and not cultivated and that the mines helped somewhat against this, since people would cut the trees for smelting copper and silver....

6.18 Pollution from refining

Xenophon reports one of the disadvantages of refining at Laurion in Attica and Strabo indicates a problem in Spain. No doubt, other areas were adversely affected by the refining process (cf. **6.21**).

Xenophon, *Memorabilia* 3.6.12

"As for the silver mines", Socrates said, "I would think that you have not gone there, and thus cannot say why now less revenue than before comes from them".
"No, since I have not been there", Glaukos said.
"Indeed, for by Zeus", Socrates said, "the region is said to be unwholesome ..."

Strabo, *Geography* 3.2.8 (146)

The Turdetani [of Spain] make their smelting ovens high up [or "with high chimneys"] so that the smoke from the ore may be carried away up in the air; for it is heavy and deadly.

Pliny, *Natural History* 33.122

People in workshops who polish *minium* [cinnabar] tie on their faces loose bladders [as masks], so they do not inhale the deadly dust when breathing and so they can see, nevertheless, over the masks.

Dioscorides (*Pharmacology* 5.94) repeats the advice for those working next to a furnace containing cinnabar.

6.19 An iron-producing society in Asia

Unlike many stories of peoples living off the agricultural produce of the land, the Chalybes in Asia were said to rely totally on iron production to support themselves. The working conditions of these famed iron workers were difficult and the living conditions not much better.

Apollonius of Rhodes, *Argonautica* 2.1001–1007

The Chalybes neither have any concern for ploughing the fields with oxen nor for planting the honey-sweet fruit; nor do they guide the pasturing flocks over the dew-spotted meadow. But they cleave the stubborn iron-bearing earth and exchange their pay for daily sustenance. Not ever does dawn rise up for them without labour, but in the black, murky flame and smoke they endure their ponderous labour.

Strabo offers two "discoverers" of iron and ironworking, both of whom have connections with Crete: the Dactyli inhabiting the area at the base of Mount Ida in Turkey (*Geography* 10.3.22) and the Telchines of Cyprus and Rhodes (14.2.7).

IRON AND STEEL PRODUCTION

Iron and steel production were not very sophisticated in antiquity, largely as a result of the smith's inability to attain very high temperatures or to control lower temperatures precisely in the furnaces. A basic, if somewhat confused, understanding of the process is contained in the literature, based not on theoretical but on empirical knowledge. Thus, the abilities of the individual smith seem crucial to consistent success in the production of good metal. Two basic steps were required: first to smelt the ore and produce a bloom of iron to which carbon was added (modern methods must remove excess carbon) by reheating it in the furnace, and second to quench the white-hot iron in water to harden the metal for use in tools and weapons. In order to make the metal less brittle and tougher, the smith also could temper the iron by gradual and controlled reheating after quenching, a stage apparently invented by the Romans but not thoroughly understood. The final result is a form of steel. Although no single source provides a comprehensive description of the entire process, by assembling a variety of sources it is possible to see the different stages (Craddock 2008, Bakhuizen 1977, Congdon 1971, Vetters 1966).

6.20 The importance of the ore and treatments

Hippocrates in this simile compares superior human health to the smelting of iron.

Hippocrates, *On Diseases* 4.55.29–35

... just as iron is created from stones and earth burnt together. And in the first throwing-in to the fire, the stones and earth adhere to each other with the dross, but at the second and third throwing-in to the fire the dross, now being melted, comes out from the iron. And the outcome that appears before one's eyes is that the iron remains in the fire, having separated from the submissive dross, and becomes firm and dense.

After a discussion of the widespread nature of iron ores Pliny provides details of the special characteristics of ores from different areas and the various treatments during production.

Pliny, *Natural History* 34.143–145

There are numerous varieties of iron. The first difference depends on the type of earth or climate: some lands furnish only iron soft like lead, others a brittle and coppery kind whose use is especially to be avoided for wheels and nails; for these things the former [soft] characteristic is suitable. Another type is good only for short lengths and nails for soldiers' boots; another suffers rust more quickly.... There are also great varieties of smelting furnaces in which a certain kernel of iron is smelted to give hardness to a sharp edge; or by another method to give solidity to anvils or to the heads of hammers. But the main difference is in the water into which the white-hot metal is repeatedly plunged. At different times and locations, the water has proven more useful and made the places renowned for the fame of their iron such as at Bambola and Tarragona in Spain and Como in Italy, although no iron mines exist in those places. But of all the types of iron, first place goes to the Seric [Chinese?] iron; the Seres send it with cloth and skins to us. Second place goes to Parthian iron since no other irons are forged from pure metal; all the rest have a softer alloy welded with them. In places such as Noricum in our part of the world, metal in the vein furnishes this good quality, while in others such as Sulmona it is due to the working, and at others, as we have said, it is due to the water.

6.21 Refining and processing iron on Elba

Populonia (Poplonium) became a major industrial centre in the Hellenistic period once the iron ore of nearby Elba was mined on a large scale. Diodorus provides a vivid picture of the activity on the island and the effect on commerce.

Diodorus of Sicily, *History* 5.13.1–2

There is an island called Aethaleia [Elba] lying off the Etruscan city named Poplonium. The island, about 100 *stadia* distant from the coast, takes its name from the abundance of sooty clouds (*aithalos*) close about it. For it has much iron-rock, which they quarry for melting and casting and for preparing iron. They have a great abundance of the iron ore, and the men working the ore pound the rock and burn the broken stones in cleverly designed furnaces. Then smelting the stones by means of a great fire in these furnaces, they cut the product into moderately sized pieces resembling large sponges in appearance. Merchants, buying and bartering these objects, transport them to Dicaearcheia [Pozzuoli, on the Bay of Naples] and other commercial centres where certain men purchase these cargoes and, having gathered together a large group of metal workers, do further work on the metal to produce all sorts of objects from the iron. Some are fashioned into types of armour, others are cleverly worked into shapes well suited for double-pronged forks, sickles, and other such tools. Since these are then distributed by the merchants to every region, many parts of the world share in the usefulness of these items.

6.22 The production and treatment of a bloom of iron

In a confused account of the production of iron, Pliny describes the production of a bloom of iron (the spongy mass) and its carburisation by hammering. The bloom of metal contains residues of impurities that must be removed by forging, a repeated series of hammering and reheating of the bloom, which produces wrought iron [**fig. 6.3**].

Pliny, *Natural History* 34.146, 149

And it is a remarkable thing, when the vein of ore is melted, the iron flows like water and later breaks up [when solidified] into spongy masses....

Iron heated by fire is ruined unless hardened by blows of the hammer. When it is red-hot it is not suitable for hammering, nor before it begins to turn pale white.

Figure 6.3 Forging the bloom.

Source: Illustration adapted from Georgius Agricola in *De re metallica*, sixteenth century.

6.23 Production of steel

Aristotle records the stage where wrought iron is worked, although he omits the quenching stage. Note that carbon will be added to the bloom by means of the charcoal used to reheat it in the furnace.

Aristotle, *Meteorologica* 4.6 (383a–b)

Wrought iron will indeed soften, and becoming pliable will harden again. They make steel (*stomoma*) in this way: the dross, settled on the bottom, is removed from below; and by repeated processing the iron is purified and the steel created. They do not, however, do it too many times because of the large loss and the reduction in weight of the purified metal. Nevertheless, the better iron is that which has smaller amounts of dross.

Whether the Greek and Roman smelters used fluxes to simplify the removal of the slag is a difficult question. Healey (1978: 185–186) presents the relevant texts with discussion of the problem.

6.24 Quenching and tempering iron

If iron is allowed to cool slowly, it loses the unstable cementite, the compound that gives iron its hardness. Quenching secures the retention of the cementite and thus the hardness of the metal at low temperatures. This process, the plunging of the white-hot metal into a cold bath, was known to the Greeks at an early date, as the following simile from Homer proves. Odysseus and his men have just speared Polyphemus in the eye with a red-hot stake, and Homer compares the noise with that of quenching iron.

Homer, *Odyssey* 9.391–394

Just as when a smith plunges into the cold water a great axe-head or adze to treat it – for this is what gives strength to the iron – and it hisses fiercely, in the same way the Cyclops' eye sizzled around the olive stake.

Although quenching retains the hardness of the iron, it also makes the metal brittle (cf. **11.118**), an unsuitable characteristic for many tools. Therefore, the iron had to be tempered either by hammering, reheating, and quenching a second time, or by annealing (reheating and cooling). The annealing of the iron was a Roman technique that even they did not understand very well. Hippocrates compares how humans can become strong in a simile with this early description of the tempering of steel.

Hippocrates, *On Regimen* 1.13.1–4

Tools of ironworking: with their skills they melt iron by driving the fire with forced air, taking away the existing substance, making it less dense, they strike and hammer it together. And then it becomes strong with the nourishment of some water.

Plutarch offers his opinion of how to make and treat friends, comparing the process to the production of steel. Tempering here involves gradual reheating and cooling to a certain point, then quenching again.

Plutarch, *Moralia* 36 (73c–d)

Just as iron is made dense by cooling and takes on the hardness of steel as a result of first being made soft and relaxed by heat, thus for our friends once they are relaxed and warmed by our approvals, we ought to introduce frankness just like a tempering bath.

6.25 Casting iron

It is believed that iron was not cast in antiquity because the furnaces could not attain temperatures high enough to melt the metal. Some archaeological finds are said to be cast iron, but they are regarded as either mistakenly identified or the result of accidental temperature increases in the furnace. Pausanias and his source may have confused bronze with iron in their discussion of Theodoros who is also associated with bronze casting (Pausanias, *Description of Greece* 8.14.8 and Pliny, *Natural History* 34.83).

Pausanias, *Description of Greece* 3.12.10

This "Skias" [in Sparta] is said to be the work of Theodoros of Samos [sixth century BC], who first discovered how to melt iron and mould statues from it.

6.26 The mythical smithy of Hephaestus and the Cyclopes

After Achilles had lost his own armour, his divine mother, Thetis, implored Hephaestus, the smithy god of the Olympians, to fashion a new set for Achilles. Hephaestus agrees and Homer then provides a glimpse into an early smith's shop.

Homer, *Iliad* 18.468–482

So saying, he left her there and went to his bellows. These he turned towards the fire and ordered them to work. And the bellows, 20 in all, blew on the crucibles, blasting forth strong wind in all degrees of strength to fan the flames, and were at his service as he hurried about here and there; wherever Hephaestus wished them to blow they did, and the work went forward. On the fire he put stubborn bronze, and tin and precious gold and silver. Then he set a great anvil on the anvil-block, and in one hand he grasped his mighty hammer and in the other hand he grasped his tongs.

A description of Achilles' armour follows (**10.7**). Similar descriptions of the smithy of Hephaestus and his Cyclopean helpers occur in Callimachus (*Hymn* 3.46ff.) and Vergil (*Aeneid* 8.424ff.).

6.27 The earliest record of smiths in the Greek world

The earliest written evidence for smiths and metalworking in Greece appears in the Mycenaean Linear B tablets. A surprising number of smiths are recorded, some with slaves. In the lists some smiths (*ka-ke-we*) are allocated specific weights of metals for processing, others are listed but receive no specific allocation.

DMG nos. 253.1–2, 254 (pp. 353–354)

Smiths at A-ke-re-wa having an allocation: Thisbaios: 1.5 kg. bronze; Quhesta-won: 1.5 kg. bronze; [six other names and weights of bronze follow]

And so many smiths without an allocation: Panguosios, Ke-we-to, Wadileus, Petalos.

And so many slaves: (those) of Ke-we-to, Iwakhas, Panguosios, Plouteus.

Smiths of the mistress (at A-ke-re-wa) having an allocation: [four other names with weights follow]

Smiths at A-ka-si-jo having an allocation: Philamenos: 3 kg. bronze; [ten other names with weights follow]; And so much bronze is shared out among them all: 6 kg. So much bronze (in all): 27 kg. And so many smiths without an allocation: [five names follow].

See *DMG* pp. 351ff. for a complete translation of the tablets concerned with smiths. The tablets, most often by the use of ideograms, also mention gold, silver, and lead in addition to bronze.

6.28 Superstition in the workshop

Pollux, *Lexicon* 7.108

Near the furnaces of the copper and bronze smiths it was the custom to hang up or have around some little joke, as an apotropaic charm.

6.29 The earliest surviving reference to a house bellows

The house bellows had a leather, accordion-like bag between two wooden boards. Ausonius compares the death throes of a landed fish to the working of the bellows.

Ausonius, *The Moselle* 267–269

Just as when the blast of air drives on the smithy fires, the woollen valve playing in the beechwood hollow [of the bellows] takes in and confines the winds now by this hole, now by that one.

6.30 A smelting furnace

Although ancient furnaces are rarely described in any detail, Dioscorides supplies us with a reliable account of a working furnace while describing a method of producing *pompholux* (zinc oxide?) (Craddock and Hughes 1985).

Dioscorides, *Pharmacology* 5.75

... *Pompholux* is also made from *cadmia* intentionally by blowing it with bellows in the following manner. In a house with a double roof, build a chimney and close to it, towards the top, a suitable window opening upwards. For the bellows, break through the wall of the room next to the chimney a small opening that leads into the furnace itself. There is also an appropriate door for the workman to enter and exit. Joined to this room is another one for the bellows and the man working the bellows. Then the coals are put into the furnace and are kindled. Finally, the workman, standing nearby, sprinkles on the *cadmia*, which has been beaten into small pieces, from a place above the head of the furnace....

ALLOYS

6.31 Corinthian Bronze

This famous alloy excited the collectors of antiquity. Its ability to be forged in the cold state was attributed to a high tin content, and to maintenance at a given temperature for a period of time, before quenching it in Corinth's famous Peirene spring. Quenching bronze does not make it harder, nor does the water of the spring make much difference as believed in antiquity. Instead it was probably the high tin content and the lengthy exposure to high heat, which made the metal a success[3] (Jacobsen and Weitzman 1988).

Pausanias, *Description of Greece* 2.3.3

... [The water of Peirene] is sweet to drink, and they say that when Corinthian Bronze is red-hot it is tempered by this water, since bronze [the text breaks off here].

In an attempt to explain the characteristics of Corinthian Bronze including its smoothness and unusual deep blue tinge, Plutarch (*Moralia* 395b-d) cites this solution among others, such as being a gold-silver-copper alloy or the effect of the primal element of air. All are incorrect.

6.32 The earliest brass production

Brass, an alloy of copper and zinc, was used for coinage by the Romans from the Augustan period onwards. Its earlier history, however, is not well known. Since zinc occurs only in compounds and was not produced in pure form until the development of distillation in the sixteenth century AC, zinc oxide was often obtained as a by-product when treating ores of lead, silver, iron, and copper; Roman brass was produced from calamine, a compound of zinc oxide. Other zinc ores, contaminated by clay, iron, and calcite oxides, have an earthy appearance that was recognised in antiquity. This "special earth" had the property of giving the colour of gold to copper when smelted together with it. It may be the substance referred to in the following passage, although this has also been identified as a copper-arsenic alloy (Craddock 1978, 1977).

Aristotle, *On Marvellous Things Heard* 62 (835a)

They say the Mossynecian copper is very bright and extraordinarily white. Tin is not mixed with it, but a certain type of earth that is smelted with the copper. They say the man discovering the mixture did not teach anyone; as a result, the copper items made earlier are distinguished in this way, the ones made later are not.

Theophrastus alludes to an alloy of *cadmea* and copper to make brass.

Theophrastus, *On Stones* 49

A very peculiar earth is the one mixed with copper, since in addition to melting and mixing, it also has the remarkable ability to enhance the beauty of copper's colour.

6.33 Production of brass for coinage

Pliny discusses a variety of copper ores with their specific characteristics that make them valuable and popular. He has just ranked some of the ores, named after the owners of the mines. The copper from a mine in Gaul owned by Livia, the wife of Augustus, is ranked third in the list. Brass was used to distinguish among several denominations of Roman coins by means of its distinctive golden colour.

Pliny, *Natural History* 34.4

At present the highest reputation has shifted to Marian copper, which is also called Cordovan copper. Next to Livian copper, this readily absorbs *cadmea* and reproduces the good qualities of gold-copper (*aurichalcum*) in *sestertii* and the two-as pieces, the one-as pieces consisting of Cyprian copper.

6.34 Copper alloys and a method of preventing bronze disease

Pliny, *Natural History* 34.97–99

The following are the proper proportions for making statues and tablets: in the beginning the lumps of ore are melted, then a third portion of scrap copper or bronze – material that has been bought after use – is added to the melted metal. The scrap metal has a special seasoning in it, subdued by rubbing and glittering from habitual use as if softened. Silver-lead is also combined with it in the proportion of twelve and one half *librae* weight per 100 *librae* of melted metal. There is also something called the "mould-making bronze" which is of a very delicate composition since a tenth portion of black lead is added and a twentieth of silver-lead. This is the best way to take on the colour, which is called Grecian. The last type is called "pot-bronze", the name arising from the vessels made of it. It is made of three or four *librae* of silver-lead added to a 100 *librae* of copper. If lead is added to Cyprian copper, the colour of purple, like that on the bordered togas of statues, is produced.

Copper and bronze items attract copper-rust more quickly when rubbed clean than if left alone, unless they are smeared with oil. The best way to protect them, it is said, is to treat them with liquid pine pitch.

Plutarch (*Moralia* 395e-f) cautions against using olive oil on bronze.

6.35 A philosophical explanation for iron rust and its prevention

Pliny has just listed some of the beneficial and detrimental uses of iron: iron ploughs and statues are positive uses, weapons for warfare are negative ones. He has become quite moralising, criticising humans for turning iron to negative purposes; nature is blameless for having created what is perhaps humankind's most useful metal. The second passage discusses rust prevention.

Pliny, *Natural History* 34.141

That same benevolence of nature has thwarted [the strength of iron] by exacting the punishment of rust from iron itself and her same foresight makes nothing more mortal in the world than that which is most dangerous to mortality.

Pliny, *Natural History* 34.150

Iron can be protected from rust with lead acetate, gypsum, and liquid pine pitch. Rust is termed *antipathia* [the natural opposite] to iron by the Greeks. Indeed, they say the same thing can be done by a certain religious rite and that in a city called Zeugma on the Euphrates river there exists an iron chain used by Alexander the Great to make a bridge there. The iron links of the chain, which have been replaced are impaired by rust while the original ones are free from it.

SOLDERS AND WELDING

Besides riveting (cf. **7.3**) and caulking, pieces of metal were also welded and soldered to make larger or more decorative items. Solder involves the use of a metal with a lower melting point than the metals or alloys to be joined, and is classified as hard or soft solder. Hard solders have a higher melting point and are often alloys containing silver or brass; soft solders are often lead alloys. Hard soldering was especially popular for making jewellery. Welding, although not as popular, was widespread and used for objects such as the Roman lead water pipes.

6.36 Types of solder

In the preceding paragraphs Pliny has discussed the natural gold-solder, malachite: how it is obtained, used, and its varieties. Here he describes other solders and concludes with a few observations about fuels for smelting.

Pliny, *Natural History* 33.93–94

Goldsmith have also adopted a gold-solder of their own for soldering gold from which they say all other similar green materials have obtained their names. The material is made from a mixture of Cyprian copper verdigris, the urine of a child below the age of puberty, and soda [sodium carbonate]. The mixture is ground with Cyprian bronze pestles in Cyprian bronze mortars. We Romans call it *santerna* [borax]. This material is used to solder gold, the type called *argentosum* [silvery gold]. If it glitters this is an indication that *santerna* has been added. Contrary to this material, *aerosum* [coppery gold] contracts and becomes dull, and is difficult to solder. For that purpose, a solder is created by adding gold and a seventh part of silver to the materials listed above, and grinding them together....

Such is the solder for gold; for iron, potter's clay; for masses of copper, *cadmea*; for sheets of copper, alum; for lead and marble, resin; but black lead is joined by means of white lead [tin] and white lead to itself with oil; *stagnum* [an alloy of silver and lead] likewise with a solder of copper; silver with *stagnum*. Copper and iron are best smelted using pine wood, but also by Egyptian papyrus; gold with chaff.

6.37 Welding iron

The arrangement of diners at a banquet is the subject of this discussion. Lamprias, one of the banqueters, suggests that rather than assigning places so that like is next to like, it would be better to arrange the diners so that they are with opposites who complement each other. In his last grouping, that of lovers of young boys and girls, he draws a simile with iron. For a more technical description, see **10.4**.

Plutarch, *Moralia* 1.2 (619a)

For heated by the same fire, they will lay hold of each other, just like welded iron – unless, by Zeus, they happen to love the same boy or girl.

DETERMINING THE PURITY OF METALS

Testing for purity was an important concern in antiquity since alloys often looked like pure metals and on occasion were intentionally sold as such. Fire was used to test pure gold on the basis that it would not change colour (impurities caused a change in colour upon cooling). Touchstones functioned by making a streak with the metal to be tested, next to a streak made with gold known to be pure. Comparison of the two allowed the relative impurity of the sample to be detected.

6.38 Copper and gold cups

Aristotle, *On Marvellous Things Heard* 49 (834a)

They say that among the Indians the copper is so bright and pure and without rust, that it is not possible to distinguish it from gold by means of colour. And among the drinking cups of Darius are numerous ones which, if not for their odour, could not otherwise be determined whether they were copper or gold.

6.39 Dishonest practices

Pliny has just finished describing black and white lead (tin) and how they are produced and soldered. Here he describes the possibilities for dishonest practices, an activity made possible since the chemical properties of metals were so poorly understood.

Pliny, *Natural History* 34.160–162

In our time *stagnum* [an alloy of silver and lead] is counterfeited by adding a third part of white copper [brass] into two-thirds part of white lead [tin]. It is also made in another way by mixing equal weights of black and white lead; some people now call the latter *argentarium* [silver mixture]. The same people also give the name "tertiary" to a mixture in which there are two parts of black lead and a third of white lead. Its price is 20 *denarii* per *libra*. Water pipes [of lead] are soldered with this. More dishonest people, having added an equal part of white lead to the tertiary, call it *argentarium*, and any object they wish, they dip in it [for plating]. They set the price of this mixture at 70 *denarii* per *libra*; the price for pure white lead by itself is 80 *denarii* and for black seven *denarii*....

A method invented in the Gallic provinces uses white lead to coat bronze items so that one can scarcely distinguish them from silver; these objects are called *incoctilia* [washed with metal].

6.40 Assaying by fire

Pliny, *Natural History* 33.59

... Gold is the only metal not depleted by fire, and even in blazing fires and on funeral pyres it remains secure. In fact, it improves in excellence the more often it is fired, and fire is a test of gold, since it reddens the gold with a similar colour and causes it to glow the colour of fire itself; they call this *obrussam* [the reddening?].

Pliny, *Natural History* 33.127

There are two different qualities of silver. When scrapings are placed on white-hot iron pans, those retaining their whiteness are approved [as first quality]. Next in quality are those turning red, the black having no value. But fraud has hindered even this test. When the pans are kept in men's urine, the scrapings are stained while being burnt and counterfeit whiteness. There is also some test of polished silver in the breath of a human: if it forms condensation at once and dissipates the vapour from the breath.

Vitruvius, *On Architecture* 7.9.5

Minium is adulterated by the addition of lime. And so, if anyone wishes to test whether a sample is without pollution, it ought to be done as follows. Take an iron plate, place the *minium* on it, and place it on the fire until the plate glows red-hot. When the colour of the *minium* is changed by the glowing heat and is black, remove the plate from the fire. If the *minium*, upon cooling, returns to its original colour, it will be proved to be unadulterated; but if it maintains the black colour, it will demonstrate that it has been adulterated.

6.41 A comparison of assaying methods: fire versus touchstones

Theophrastus, *On Stones* 45–47

Remarkable is the nature of the stone used for testing gold since it seems to have the same ability as fire, which also is used to test gold. On that account some people are in doubt as to the truth of this, but their doubts are not proper for the stone does not test in the same way. Fire tests by changing and altering the colours, while the stone tests by rubbing a mark alongside [another mark]. For the stone seems to have the ability to bring out the essential nature of each sample. They say that a stone much better than the previous type has been discovered. It not only detects refined gold but even gold or silver alloyed with copper, and the amount mixed in a *stater*. There are indications [streaks] for use from the smallest amount: the smallest is [the *krithē*, then the *kollybos*, then] the quarter or half obol; and from these indications the proper proportion is detected. All such stones are found in the river Tmolus. They are smooth in nature, like pebble-counters, flat, but not round, and in size twice as big as the largest pebble-counter. For testing, the top part of the stone that faces the sun differs from the lower surface; the top tests better. This is because the upper surface is drier, for moisture hinders the attraction of the metal. Indeed, even in the heat of the day, testing is worse since the stone exudes moisture causing slippage.

Pliny (**5.4**) repeats sections of Theophrastus, adding that the stones could be found elsewhere, too.

6.42 "Recipes" used in metallurgy: purification, alloys, and testing

Two papyri of the late third or early fourth centuries AC contain over 200 short "recipes" listing the methods and ingredients for working with metals, dyes (**10.50**) and other materials (**10.93**). Rather than dividing the "recipes" into subjects and separating them in this chapter, the following selections are presented together in order to provide a better idea of the content of the papyri. Other selections are also maintained together in their appropriate section.

P.Leid. X (Halleux 1981)

Purification of Tin that is Put into an Alloy of Asem
[artificial electrum]

Taking tin purified from all other things, melt it and let it cool. Having coated it with oil and letting it penetrate, melt it again. Then, having crushed together olive oil and asphalt and salt, coat it and melt it for the third time; and once it has fused, lay it aside after having washed it properly. It will be hard like silver. When you wish to employ it in place of silver plate so that it passes unnoticed and has the hardness of silver, blend together three parts of silver with four parts [of it] and the product will be like silver....

Manufacture of Asem

Twelve *drachmai* of tin, four *drachmai* of mercury, two *drachmai* of Chian earth. Having melted the tin, throw in the crushed earth, then the mercury and stir with an iron, and use....

Manufacture of Asem

Taking small, soft pieces of tin, which have been purified four times, and taking four parts of it and three parts of purified white copper and 1 part of *asem*, smelt them. And once it is cast, clean it a number of times and make what you wish. It will be *asem* of the first quality, so that it will deceive even the artisans....

Preparation of Solder for Gold

Solder for gold is prepared thus: four parts of Cyprian copper, two parts of *asem*, one part of gold. First the copper is melted, then the *asem*, then the gold.

To Determine if Tin Has Been Adulterated

After having melted it, put a sheet of papyrus below it and pour. If the papyrus burns, it contains lead.

Pliny (*Natural History* 34.163) duplicates this test.

The Manufacture of Goldsmith's Solder. How One Makes Goldsmith's Solder for Gold.

Having melted two parts of gold and one part of copper, grind it up. If you wish it to have a better appearance, melt in a bit of silver....

Testing of Gold

If you wish to purify gold [test the purity of gold], remelt or heat, and if it is pure, it will retain the same colour after heating and remain pure like a piece of money. If it appears whiter, it contains silver; if rougher and harder, it contains copper and tin; if black and soft, it contains lead.

Testing of Silver

Heat or melt the silver as with the gold and if it is brilliant white, it is pure and not adulterated. If it appears dark, it contains lead; if it appears hard and orange, it contains copper.

P.Holm. (Halleux 1981)

Manufacture of Silver

Plunge Cyprian copper, which is already worked and laminated for use, into dyer's vinegar and alum and let soak for three days. Then, after mixing in six *drachmai* each of Chian earth, Cappadocian salt, and thin strips of lamellose alum for every *mina* of copper, cast. Cast skilfully and it will be good quality. Add in not more than 20 *drachmai* of good and tested pure silver, which will maintain the whole mixture unspoiled.

Another [for Silver]

Anaxilaus also attributes the following to Democrites. Having crushed very finely common salt together with thin strips of lamellose alum in vinegar and forming pellets, he dried these in the bath chamber for three days. And then after grinding them down he cast them with copper three times and cooled them, quenching them in seawater. Experience will show in the result....

Purification of Tin

The purification of tin, which occurs in the alloy with silver, is as follows: let pure tin cool and, after coating it with oil and bitumen, melt it four times, wash it, and lay it aside cleanly. Add in to the six parts of this, four parts of silver and seven parts of Galatian copper and it will pass inspection as wrought silver.

Notes

1 The extraction and properties of minium, or red lead, are described by Vitruvius (*On Architecture* 7.9.1):

> For the *minium* ore itself, when it is dry, is pounded with iron poles and the wastes are removed by means of repeated washing and heating, so that the

colour is produced. When its strength, which its natural state had contained, has been lost through the removal of the mercury, it becomes soft in nature and weak in power.

2 Litharge, a lead oxide.
3 Giumlia-Mair and Craddock (1993) suggest that Corinthian Bronze was bronze decorated with metal inlays.

Bibliography

Bakhuizen, Simon C., "Greek Steel". *World Archaeology* 9 (1977) 220–234.

Bayley, Justine, "The production of brass in antiquity with particular reference to Roman Britain". Pp. 7–24 in Paul T. Craddock, ed., *2000 Years of Zinc and Brass*. London: British Museum, 1990.

Beagrie, Neil, "The Romano-British pewter industry". *Britannia* 20 (1989) 169–191.

Becker, Marshall J., "Sardinian Stone Moulds: An Indirect Means of Evaluating Bronze Age Metallurgical Technology". Pp. 163–208 in Mirriam S. Balmuth, and Robert J. Rowland, Jr., eds., *Studies in Sardinian Archaeology*. Ann Arbor: University of Michigan Press, 1984.

Benoît, Fernand, "Soufflets de forge antiques". *Revue des Études Anciennes* 50 (1948) 305–308.

Boulakia, Jean D.C, "Lead in the Roman World". *American Journal of Archaeology* 76 (1972) 139–144.

Branigan, Keith, *Aegean Metalwork of the Early and Middle Bronze Age*. Oxford: Clarendon Press, 1974.

Caley, Earle R., *Analysis of Ancient Metals*. Oxford: Pergamon Press, 1964.

Caley, Earle R., *Orichalcum and Related Ancient Alloys*. Numismatic Notes and Monographs, 151. New York: American Numismatic Society, 1964.

Charles, James A., "Early Arsenical Bronzes – A Metallurgical View". *American Journal of Archaeology* 71 (1967) 21–26.

Cleere, Henry F., "Ironmaking in a Roman Furnace". *Britannia* 2 (1971) 203–217.

Cleere, Henry F., "Some Operating Parameters for Roman Ironworks". *Bulletin of the Institute of Archaeology of the University of London* 13 (1976) 233–246.

Coghlan, Herbert H., *Notes on the Prehistoric Metallurgy of Copper and Bronze in the Old World*. 2nd edn. Edited by Thomas K. Penniman, and Beatrice M. Blackwood. Oxford: Oxford University Press, 1975.

Congdon, Lenore O. Keene, "Steel in Antiquity: A Problem in Terminology". Pp. 17–27 in David G. Mitten, and John G. Pedley, J.A. Scott, eds., *Studies Presented to G.M.A. Hanfmann*. Mainz: von Zabern, 1971.

Conophagos, Constantin E., *Le Laurium antique et la technique grecque de la production de l'argent*. Athens: Ekdotike Hellados, 1980.

Craddock, Paul T., "The Composition of the Copper Alloys used by the Greek, Etruscan and Roman Civilizations, 1. The Greeks before the Archaic Period". *Journal of Archaeological Science* 3 (1976) 93–113.

Craddock, Paul T., "The Composition of the Copper Alloys used by the Greek, Etruscan and Roman Civilisations, 2. The Archaic, Classical and Hellenistic Periods". *Journal of Archaeological Science* 4 (1977) 103–123.

Craddock, Paul T., "The Composition of the Copper Alloys used by the Greek, Etruscan and Roman Civilizations, 3. The Origins and Early Use of Brass". *Journal of Archaeological Science* 5 (1978) 1–16.

Craddock, Paul T., "The Metallurgy and Composition of Etruscan Bronze". *Studi Etruschi* 52 (1984) 211–241.

Craddock, Paul T., "Zinc in Classical Antiquity". Pp. 1–6 in Paul T. Craddock, ed., *2000 Years of Zinc and Brass*. London: British Museum, 1990.

Craddock, Paul T., *Early Mining and Metal Production*. Washington: Smithsonian Institution, 1995.

Craddock, Paul T., "Mining and Metallurgy". Pp. 93–120 in John P. Oleson, ed., *The Oxford Handbook of Engineering and Technology in the Classical World*. Oxford: Oxford University Press, 2008.

Craddock, Paul T., and Michael J. Hughes, eds., *Furnaces and Smelting Technology in Antiquity*. London: British Museum, 1985.

Davey, Christopher J., "Some Ancient Near Eastern Pot Bellows". *Levant* 11 (1979) 101–111. 5 figs.

Forbes, Robert J., *Studies in Ancient Technology*, Vols. 8–9. 2nd edn. Leiden: Brill, 1971–1972.

Gale, Noël H., and Zofia A. Stos-Gale, "Cycladic Lead and Silver Metallurgy". *Annual of the British School at Athens* 76 (1981) 169–224.

Giumlia-Mair, Alessandra R., and Paul T. Craddock, *Das schwarze Gold der Alchimisten*. Mainz: von Zabern, 1993.

Halleux, Robert, *Le problème des métaux dans la science*. Paris: Belles Lettres, 1974.

Halleux, Robert, "Les deux métallurgies du plomb argentifère dans l'*Histoire Naturelle* de Pline". *Revue de Philologie de Littérature et d'Histoire Anciennes* 49 (1975) 72–88.

Halleux, Robert, *Les alchimistes grecs*, Vol. I. Paris: Belles Lettres, 1981.

Healy, John F., *Mining and Metallurgy in the Greek and Roman World*. Ithaca: Cornell University Press, 1978.

Jacobson, David M., and Michael P. Weitzman, "What Was Corinthian Bronze?" *American Journal of Archaeology* 96 (1992) 237–247.

Maddin, Robert, ed., *The Beginning of the Use of Metals and Alloys*. Cambridge: MIT Press, 1988.

Modona, A. Neppi, ed., *L'Etruria mineraria. Atti del XII convegno di Studi Etruschi e Italici, 16–20 giugno, 1979*. Firenze: Olschki, 1981.

Muhly, James D., "Sources of Tin and the Beginnings of Bronze Metallurgy". *American Journal of Archaeology* 89 (1985) 275–291.

Muhly, James D., and Theodore A. Wertime, "Evidence for the Sources and Use of Tin During the Bronze Age of the Near East: A Reply to J.E. Dayton". *World Archaeology* 5 (1973) 111–122.

Nriagu, Jerome O., *Lead and Lead Poisoning in Antiquity*. New York: John Wiley and Sons, 1983.

Pleiner, Radomír, *Iron Working in Ancient Greece*. Prague: National Technical Museum, 1969.

Ramage, Andrew, "Gold Refining in the Time of the Lydian Kings at Sardis". Pp. 730–735 in Ekrem Akurgal, ed., *Proceedings, 10th International Congress of Classical Archaeology, 1973*. Ankara: 1973.

Raymond, Robert, *Out of the Fiery Furnace. The Impact of Metals on the History of Mankind*. University Park: Pennsylvania State University Press, 1986.

Roxburgh, Marcus, Stijn Heeren, Hans Huisman, Bertil Van Os, "Early Roman copperalloy brooch production: a compositional analysis of 400 brooches from Germania Inferior". *Journal of Roman Studies* 29 (2016) 411–421.

Ruckdeschel, Wilhelm, and Franz Fischer, "Werkstoffuntersuchungen an Eisenteilen aus dem Fundkomplex 'Römische Militärstation Augsburg-Oberhausen'". *Technikgeschichte* 41 (1974) 187–200.

Schuster, Wilhelm F., *Das alte Metall- und Eisenschmelzen, Technologie und Zusammenhänge*. Düsseldorf: VDI-Verlag, 1969.

Tylecote, Ronald F., *A History of Metallurgy*. London: Metals Society, 1976.

Tylecote, Ronald F., *The Early History of Metallurgy in Europe*. Longman: London, 1987.

Vetters, Hermann, "*Ferrum Noricum*". *Anzeiger der Österreichischen Akademie der Wissenschaften in Wien, Philos.-Hist.* Klasse 103 (1966) 167–185.

Wagner, A., ed., *Silber, Blei und Gold auf Siphnos. Prähistorische und antike Metallproduktion*. Der Anschnitt, Beiheft 3. Bochum: Bergbau Museum, 1985.

Waldbaum, Jane C., *From Bronze to Iron: The Transition from the Bronze Age to the Iron Age in the Eastern Mediterranean*. Studies in Mediterranean Archaeology, 54. Göteborg: Paul Åströms Forlag, 1978.

Wertime, Theodore A., and James D. Muhly, *The Coming of the Age of Iron*. New Haven: Yale University Press, 1980.

7
SCULPTURE

Statues and sculptures are the result of clusters of subsidiary technologies. While in modern parlance the two terms are used interchangeably, Pliny the Elder carefully distinguishes between bronze *statuaria* and marble *sculpturae*, the former related to the Latin verb *statuo*, "to cause something to stand", the latter related to *sculpo*, "to carve, cut, or chisel". The etymology implies that the creation of statues involves an additive process, in which material is accumulated, while sculpting involves a subtractive process, in which material is removed to create the finished piece. The reality was far more complicated than Pliny's dichotomy implies. The complexity of the technological cluster depends on the working material. These materials are typically wood; bone and ivory; stone; clay (terracotta); metal (bronze, gold, silver), or any combination of these. Different materials could be combined, e.g. in the form of surface layers, inlays, decorative or functional attachments, or the assembly of entire components from different materials to one another.

Wood, as an organic material, is most sensitive to environmental influences, such as moisture, mould, or burrowing insects. Sculpting wood is a one-step process, as the artist carves the raw material directly. Subsequent to sculpting, wood requires some surface treatment in the form of protective oil, varnish, or paint. Pausanias calls wood sculptures *xoana* (*xoanon* in the singular), but other authors use the same word for sculptures made of other materials, too (Palagia 2006: 10–11).

Bone and ivory are by-products of hunting and butchering or scavenging. Similar to wood, these are organic materials that are prone to mould and rot. They need to be thoroughly cleaned in order to remove all soft tissue, marrow, and pulp before they are carved. The preparation of modern hunting trophies (deer skulls; boars' tusks) involves boiling in water and the filling of the pulp chamber with wax. The same may have been true in antiquity. A special type of mixed-media sculpture is *chryselephantine* sculpture, made of ivory with gold, usually built on a wooden core. An example was Pheidias' sculpture of Zeus at Olympia.

Clay and stone require prospecting and quarrying. The wide variation among clays and stones (for example, in grain size, mineral content, colour, variegation, brittleness, etc.) requires careful selection and experience on the part of the sculptor. The preferred stone for sculptures was marble. Combinations of stone limbs and heads with bodies of other materials existed as well. These composites are called acroliths.

Clay needs to be mixed with water prior to sculpting and is subsequently fired in a kiln. Firing transforms the soft and malleable clay into hard and durable terracotta (literally "cooked earth" in Italian). The parameters of the firing process in particular (temperature, duration, oxygen content, etc.) are crucial to the quality of the finished product. Countless failed terracotta pieces, so-called wasters, survive in the archaeological record and attest to the difficulty of controlling the firing. Rice (1987: 173: tab. 6.1) records common firing loss rates of up to 50 per cent.

Metal statues are the result of the most complex cluster of subsidiary technologies. The most widespread material used in antiquity for this purpose was bronze, an alloy of copper and tin. Prospecting, mining, smelting, and alloying are technologies employed to procure and manufacture the working material alone. Making a statue requires skills in metalworking (rolling; chasing; riveting; soldering; hammer welding; casting) [**figs. 7.1 a,b**]. Non-cast statues made of bronze sheets hammered into the desired shape around cores of wood and then riveted together are called *sphyrelata* (*sphyrelaton* in the singular). For the casting process, in contrast, the artist needs to create models (usually of wax or clay), cores, and moulds. Moreover, the workshop needs a heat source sufficient to melt significant quantities of bronze (melting point *c.*950 degrees C). Cast-metal statues are prone to manufacturing errors, such as cracks, strains, and distortions due to shrinkage, air inclusions, and differential cooling. Even though casting moulds could be reused a limited number of times, the sheer complexity and expense of making bronze statues may in part explain why Roman reinterpretations of Greek bronze sculptures are frequently executed in marble.

In spite of the complexities, the artists in antiquity had developed good mastery of the processes involved. The number of manufacturing steps from raw material to finished piece, the expense in material and energy, as well as the artistic skill and physical effort required show that statues and sculptures must have been very pricey indeed. Countless surviving examples are evidence that the cost did not detract from the popularity and ubiquity of these pieces both in public and domestic contexts.

Much of the material relevant to this topic appears in Chapter 10, sections A, B, and D (Metalworking, Woodworking, and Ceramics and Glass).

Mattusch (2006) and Palagia (2006) contain detailed descriptions of the procedures involved in making bronze statues and marble sculptures, including the process of copying and transfer from one medium to another. Rice (2015) offers an excellent overview of the technical side of pottery making.

SCULPTURE

(a)

(b)

Figures 7.1 a,b Vase painting of bronze sculptors at foundry.

Source: Drawing of Berlin Foundry Cup, Berlin, Antikensammlung, F 2294, Staatliche Museeen zu Berlin, Germany; Healey, 1978, pl. 73 a, b.

SCULPTURE

7.1 The first sculptors

The desire to represent human and animal forms in three-dimensions goes back thousands of years. In the Aegean region in particular, Bronze-Age sculptures, statuettes, and figures abound. Well known examples include the Cycladic marble figures of the 3rd millennium BC with their almost modernist appearance, and from the 2nd millennium BC the Minoan "snake goddesses" from various sites on Crete, Minoan figures of bull leapers (e.g. a bronze in the British Museum and an ivory in the Heraklion Archaeological Museum) and Minoan and Mycenaean libation vessels (rhyta) in a variety of materials shaped like animal heads (e.g. lion and bull).

Ancient authors attribute the first sculptures to Daedalus, a mythical inventor and craftsman, and father of Icarus. His mythical origins go back to Bronze-Age Crete to the court of King Minos, where he was responsible for devising the wooden cow with which Minos' wife Pasiphae enticed a bull to mate with her and for the labyrinth that held the Minotaur, the offspring of that match, imprisoned. Daedalus' name survives in the term "daedalic", which describes a particular style of archaic Greek statue exemplified, for example, in the "Lady of Auxerre". Pliny is more pragmatic in his focus on the Roman context.

Pausanias, *Description of Greece* 9.3.2

They say that Hera was angry with Zeus for some reason and had retreated to Euboea. As Zeus was unable to change her mind, they say he went to Cithaeron, who was holding power in Plataea, for nobody surpassed Cithaeron in cleverness. So he ordered Zeus to make an image of wood, to wrap it and place it onto an ox cart and to say that he was marrying Plataea, the daughter of Asopus. And Zeus followed Cithaeron's advice. Hera heard the news right away and arrived immediately. When Hera approached the cart and tore the clothes off the image, she was happy at the deceit, having found a wooden image instead of a bride and was reconciled with Zeus. On account of this reconciliation, they celebrate a festival called Daedala because the people of old called the images *daedala*. I think that they called them that before Daedalus, the son of Palamaon was born at Athens and that he got his surname from the *daedala* and not at birth.

Diodorus of Sicily, *History* 4.76.1–3

Daedalus was Athenian by birth, an Erechtheid by name. He was the son of Metion, who was the son of Eupalamus, who was the son of Erechtheus. He surpassed all others in natural talent and zealously pursued the art of building, sculpting, and stone carving. He was the inventor of many tricks of the trade and produced many wondrous works in many regions of the world. In the creation of statues he surpassed all others so much that following generations told stories about how his statues very much resembled living beings. They say that the statues could see and walk and completely retained the condition of the whole body that they seemed to be living beings. As he was the first to render the eyes and to represent the legs in a striding motion, also since he showed the hands stretched out he was righteously marvelled at by other men.

7.2 A very precise wooden sculpture

The following passage provides interesting insight into the sculpting process. Proportionality was important from both an aesthetic and a technical point of view. Polykleitos' canon and his idea of *symmetria*, commensurability of all parts with one another, as well as Vitruvius' modules in architecture demonstrate that the aesthetics recognisable in ancient art and architecture are largely based on mathematical relations. These relations also allowed artists to work independently on individual parts of the same organic form while ensuring that the parts will fit together at the end of the process.

Diodorus of Sicily, *History* 1.98

The most renowned of the ancient sculptors, Telecles and Theodorus, the sons of Rhoecus, spent time among the Egyptians, too. They made a wooden sculpture of Pythian Apollo for the Samians. They say that one half of the sculpture was worked by Telecles in Samos, and the other half was finished by his brother Theodorus at Ephesus. When the two parts were brought together, they fitted so perfectly that the whole work appeared to have been done by one man. This method of working is followed nowhere among the Greeks, but is very much practised among the Egyptians. For among them the proportions of the sculpture are not determined according to how they look to the eye, as among the Greeks, but when they have laid out the stones and, having portioned them, are ready to work them, they take the proportions, from the smallest to the largest parts; for, dividing the structure of the entire body into 21 parts and one-fourth in addition, they render the complete measurements of the figure. Therefore, once the artists have agreed on the size of the statue, they depart from one another and fashion the respective sizes as agreed and with such precision that the accuracy is perplexing. And the wooden sculpture in Samos, in agreement with the artistic ingenuity of the Egyptians, was divided down the middle and cut into two parts from the crown of the head to the groin, the two parts matching each other exactly. And they say that this sculpture is for the most part somewhat like those of Egypt, as it has the arms extended down the sides and the legs in a striding stance.

7.3 The beginnings of cast-bronze statues

According to Pausanias bronze images were first produced using a relatively simple technique (*sphyrelaton*) where separate bronze sheets were hammered around a core and nailed together. The casting of bronze is a much more difficult process. Moulds need to be designed such that air inclusions do not appear in the finished piece. While it is relatively easy to cast, say, a spearhead, the problem increases with increasing complexity of the piece. Moreover, shrinkage in the cooling process makes it difficult to cast pieces true to the intended size. It is no wonder that ancient authors were interested in the inventors of such complex technology.

Pausanias, *Description of Greece* 3.17.6

Pausanias appears to describe an early form of bronze statue, the *sphyrelaton*, which was made by hammering and riveting thin bronze sheets around a wooden core. The technique was replaced by hollow casting.

To the right of the Bronze House [of Athena] the statue of Zeus Most High was erected; the oldest of all the statues as many as were made of bronze. But it was not built in one piece, for each of the parts has been beaten out individually one by one, then each is fitted one to another and nails hold them together so they do not come apart. And they say that the statue was made by Clearchus, a man of Rhegium, whom some say was a student of Dipoenus and Scyllis, while others say he was a student of Daedalus himself.

Pliny the Elder, *Natural History* 34.15

The first bronze image at Rome, I find, was of Ceres paid out of the confiscated property of Spurius Cassius. His own father had him put to death because he wanted to make himself king [in 484 BC]. From statues and images of gods the tradition in many types passes to images of human beings, too. The ancients stained them with bitumen, which makes it is all the more remarkable that later it pleased them to cover the images with gold. I do not know whether this was a Roman invention, but it is certainly not a very old practice in Rome.

Pausanias, *Description of Greece* 8.14.8

The first people to melt bronze and cast an image were the Samians Rhoicus, the son of Phileas, and Theodorus, the son of Telecles.

Pausanias, *Description of Greece* 10.38.6

I have explained earlier in this work that two Samians, Rhoicus, son of Phileas, and Theodorus, son of Telecles, had invented a method for melting bronze with precision. These men were the first to cast bronze. I have not found any surviving work by Thedorus, at least none of bronze. But in the Sanctuary of Artemis at Ephesus, approaching the building that has the paintings, there is a top course above the altar of the so-called Artemis of the First Throne. A number of images are set up on that course, among which that of a woman near the end. It is the work of Rhoicus, and the Ephesians call it the Night.

7.4 Early terracotta sculpture

Some anecdotes ascribe the invention of a particular technology to coincidence.

Pliny the Elder, *Natural History* 35.151

Butades, a potter from Sicyon, first invented at Corinth the modelling of portraits from clay. He did it for his daughter, who had fallen in love with a young man. When that young man was departing to go abroad, she drew the silhouette of his face on the wall by tracing his shadow thrown by a lamp. Her father impressed clay on the outline and made a relief, which he hardened by firing it together with other pieces of pottery. They say that the relief was preserved in the Nymphaeum until Mummius invaded Corinth.

7.5 Other materials

The following passages show the variety of ancient statuary and sculpture. While executions in one material may predate that in another material, these images existed and were visible together, were part of people's daily experience, and formed the background for their daily interactions.

Scholiast on Demosthenes, *Against Androtion* 13

There are three images of Athena in different locations on the Acropolis: one that has existed from the beginning, made of olive wood, which is called Athena *Polias* because the city is hers, a second made of only bronze, which they made after the victory at Marathon, called Athena *Promachos*, and a third made of gold and ivory (*chryselephantine*; see Lapatin 2001).

Pausanias, *Description of Greece* 9.4.1

The Plataeans have a sanctuary of Athena surnamed "the Warlike"; it was built from the spoils that the Athenians gave them from the battle of Marathon. It is a gilded wooden image, but the face, hands, and feet are made of Pentelic marble. It is not much smaller in size than the bronze image on the Acropolis, the one which the Athenians dedicated as first-fruits of the battle at Marathon; Phidias made the Plataean image of Athena, too.

Pausanias, *Description of Greece* 9.10.2

[At Thebes] First at the entrance are stone images of Athena and Hermes, called the Pronaoi. It is said that the Hermes was made by Phidias, the Athena by Scopas. The temple was constructed behind. The image is equal in size to that at Branchidae and does not differ from it in shape. Whoever has seen one of these two images, and learnt who made it does not need much wisdom to realise, when he looks at the other, that it is a work of Canachus. They differ in this: the image at Branchidae is of bronze, while the Ismenian is of cedar wood.

Pausanias, *Description of Arkadia* 8.17.2

After the grave of Aepytus comes a mountain, Cyllene, the highest of the mountains in Arkadia. On top of the mountain is a ruined temple of Hermes Cyllenius. It is clear that the name of the mountain and the god's surname come from Cyllenus, the son of Elatus. As much as I was able to learn, in the old days, these were the trees from which people made images (*xoana*):

ebony, cypress, cedar, oak, yew, and lotus. The image of Hermes Cyllenius is made of none of these, but of juniper. I am guessing its height is about eight feet.

7.6 Steps in making a sculpture

Ancient authors, even the so-called technical writers, were not generally interested in the mechanical details of the crafts they were describing. Passages such as the following are therefore all the more tantalising as they lift the veil on the manufacturing process ever so slightly [cf. figs. **7.2–4**].

Plutarch, *Moralia* 636c

Artists first form their pieces without detail and then add the individual details afterwards, which is why Polycleitus the sculptor said that the work is most difficult when the clay is on the fingernail.

Pausanias, *Description of Greece* 1.40.4

After this, when one enters the precinct of Zeus called the Olympieum [at Megara], there is a temple worth seeing. The image of Zeus, however, is unfinished as work was stopped by the war of the Peloponnesians against the Athenians […]. The face of the sculpture of Zeus is made of ivory and gold, the rest is clay and gypsum

Figure 7.2 Carving a sarcophagus.
Source: Museo Archeologico Lapidario, Urbino, Italy.

SCULPTURE

Figure 7.3 Blacksmithing *putti*.
Source: *Oecus* of the House of the Vettii, Pompeii, Italy.

Figure 7.4 A metal workshop.
Source: After a Fresco of Thetis and Hephaestus in House IX 1,7, Pompeii, Italy.

[…]. Behind the temple lie half-worked pieces of wood, which Theocosmus intended to embellish with ivory and gold in order to finish the image.

SCULPTURE

7.7 Masters of the trade

Praxiteles, an ancient master of the trade, was a generalist creating with equal ability images of marble and of bronze, as Pliny attests. While marble carving and bronze casting require two completely different skill sets, the clay-model-making process that precedes bronze casting is almost identical to sculpting in marble. It is, therefore, not surprising that Praxiteles should have been a master of both techniques.

Pliny, *Natural History* 34.69

Praxiteles, who was more successful and therefore more famous in works of marble, nevertheless also made very beautiful bronze statues.

Pliny, *Natural History* 36.20

I already gave the date for Praxiteles when I wrote about artists working in bronze. But he surpassed even himself in the glory of his work in marble.

7.8 The colour of statues

Ancient statues appear drab today, but in antiquity they were colourful because they were brightly painted or made of materials of differing colours. Most of the colour has eroded over time, and often the colourful details that may have been attached to the surface have broken off or have been stolen if they were made of metal that could be otherwise reused.

Clement of Alexandria, *Protrepticus* 4.48

He used for its execution a mixture of different materials. He had filings of gold, bronze, silver, lead, and tin, too. Not one of the Egyptian stones was missing, either: fragments of sapphire, hematite, emerald, and also topaz. He ground and mixed them all together and coloured the mixture deep blue, which is why the image is rather dark. Having mixed this with the leftover dye from the funeral of Osiris and Apis, he moulded the Serapis.

Plutarch, *Moralia* 674a

We enjoy and marvel at the image of Jocasta, to whose face, they say, the artist mixed silver so that the bronze may take on the look of a person at the point of dying.

7.9 Casting and recycling bronze

Although well-produced wrought iron is harder and cheaper than bronze, the latter has the advantages of greater resistance to corrosion and of a melting point within the reach of ancient technology.

As a result, bronze continued to be used in the Graeco-Roman world for objects in shapes too complicated for production by forging or that would be exposed to the weather – as long as a sharp edge was not a requirement. Particularly large and complicated objects such as statues were cast in pieces and soldered together.

Quintilian, *On the Training of an Orator* 7, Preface 2

For even if all the limbs have been cast, it is not a statue unless they are assembled....

Juvenal, *Satires* 10.56–64

Power, subject to great envy, has overthrown some men, and the long, outstanding list of their offices drags them down. Their statues yield to the tugging rope and fall.... Now the flames hiss, now in the midst of bellows and furnace the head of powerful Sejanus[1] adored by the masses, burns and crackles; then, from the image of the second most powerful man in the whole world they manufacture pitchers, basins, frying pans, and chamber pots.

7.10 Payment for casting models

The various methods of casting metals involved the production by craftsmen or sculptors of open-face moulds in stone or terracotta or of positive models of the final form in clay or wax. Positive models in wax of statues or other complex forms would be encased in clay, then removed by melting to leave a mould into which the metal could be poured; clay models could be used to produce terracotta negatives as moulds for casting directly or for the production of wax positives. The following inscription is part of an account of payment to craftsmen for work on the Erechtheum in Athens, 408/7–407/6 BC.

IG I² 374.248–256

To the clay-modellers sculpting the models of the bronzes for the covering tiles. To Nesos, living at Melite: eight *drachmai*. To another, sculpting a model, the acanthus for the covering tiles, Agathanor living at Alopeke, eight *drachmai*. Total for clay-modellers: 16 *drachmai*.

7.11 An old-testament view on sculpture

Jeremiah, 10:3–4

For the customs of the people are vain: for one cutteth a tree out of the forest, the work of the hands of the workman, with the axe. They deck it with silver and with gold; they fasten it with nails and with hammers, that it move not.

7.12 A large vessel

Bronze tripods and cauldrons are among early pieces produced during the so-called Dark Age. A cauldron such as Herodotus describes is a relatively simple shape, but at the size described, its production is nonetheless remarkable. The body is produced by hammering bronze sheets into the desired shape. The stands and the griffins' heads are more complicated and were usually produced separately by casting and were subsequently attached to the vessel.

Herodotus, *Histories* 4.152

The Samians took the tenth part of their profit, six talents, and made a bronze cauldron in the manner of Argolis. There are griffins' heads arranged all around. They set it up in the Temple of Hera, supporting it with three colossal kneeling bronze figures seven cubits [*c.*3.2 metres] high.

7.13 A flood of statues

Pliny, *Natural History* 34.36–37

What the origin was of representing likenesses will be discussed more conveniently in the context of what the Greeks call *plasticē*, [plastic], for these were earlier than bronze statues. But these have flourished without limit, and if somebody wished to give an account of them it would occupy many volumes. Who could achieve it all? During the aedileship of Marcus Scaurus there were 3,000 statues on the stage of merely a temporary theatre. After conquering Achaea, Mummius filled the city with statues, leaving nothing for his daughter's dowry. For why not mention this along with the excuse? The Luculli, too, brought in many. Mucianus, who was consul three times, states that there are thousands of statues still at Rhodes [different editions give different numbers, 3,000 or 73,000], and it is believed that about as many remain at Athens, Olympia, Delphi. ... Even the works of individual artists are countless in quantity. It is said that Lysippus made 1,500 pieces, all of them of such great quality that each one could have made him famous.

7.14 A wonder of the world

Pliny, *Natural History* 34.41

The most admirable of all statues was the *colossus* of Sol [the Sun-God] at Rhodes, which Chares of Lindus had made – a pupil of the above-mentioned Lysippus. It was 70 cubits [*c.*31 metres] high. The image was toppled by an earthquake 56 years after its construction, but even lying down it is a marvel. Few men are able to embrace its thumb all around, and the fingers are bigger than many a statue. Large holes gape in the broken limbs, and inside one can see large quantities of rocks, by which weight the artist had stabilised the statue. They say it took 12 years to build and cost 300 talents, coming from the abandoned siege machinery left behind by king Demetrius after he had grown tired of a lengthy siege of Rhodes. There are another 100 colossal statues in the same city, smaller than this one, but each one on its own would make a place famous wherever it stands.

7.15 Plaster casts and wax masks

Pliny, *Natural History* 35.153

Lysistratus of Sikyon, the brother of Lysippus, whom I mentioned earlier, was the first who made a model of a person by applying plaster directly to the face, then pouring wax into this plaster mould in order to make corrections on the wax cast. He also began producing true likenesses. Prior to him artists had been striving to make images as handsome as possible.

Polybius, *Histories* 6.53

The image is a mask made to preserve the likeness most closely in terms of both the shape and the outline. When they display these images publicly at sacrifices they adorn them generously, and when a prominent family member dies, they bring them out in the funeral procession, putting them on people who they think most look like the deceased in size and shape.

Pliny, *Natural History* 35.6

Laziness has destroyed the arts, and since there are no portraits of our minds, our bodies are also neglected. In the atria of our ancestors these things were different. Portraits used to be on display, and not statues by foreign artists, either in bronze or marble. Faces moulded in wax were set out on individual chests. These were the portraits that were to accompany family funerals so that whenever somebody passed away, all members of his family who had ever lived were in attendance.

Bibliography

Much relevant bibliography is listed in Chapter 10 (Metalworking, Woodworking, and Ceramics and Glass) and does not appear again here. Only a small body of modern literature focuses on the technological side of ancient statuary and sculpture, but as interest in ancient technology increases, the bibliography is bound to grow rapidly over time. Since the physical properties of the raw materials did not change over the millennia, books on modern metal casting, woodcarving, stonecutting, and pottery are instructive with regards to information about the materials' physical and chemical behaviour as well as potential problems that may occur in the course of production. Safety warnings and advisories in modern work manuals can give researchers an idea of the work conditions to which ancient craftsmen were exposed without protection by modern safety equipment such as steel-toed boots, dust masks, or goggles.

Note

1 A favourite of the emperor Tiberius who fell victim to ambition.

Bibliography

Acton, Peter, *Poiesis: Manufacturing in Classical Athens*. Oxford: Oxford University Press, 2014.

Borg, Barbara E., ed., *A Companion to Roman Art*. Blackwell companions to the ancient world. Chichester: Wiley-Blackwell, 2015.

Burnett Grossman, J., *Funerary Sculpture. The Athenian Agora, 35*. Princeton, NJ: American School of Classical Studies at Athens, 2013.

Daehner, Jens M., and Kenneth Lapatin, eds., *Power and Pathos: Bronze Sculpture of the Hellenistic World*. Los Angeles: J. Paul Getty Museum, 2015.

Friedland, Elise A., Melanie Grunow Sobocinski, and Elaine K. Gazda, eds., *The Oxford Handbook of Roman Sculpture. Oxford Handbooks in Archaeology*. Oxford: Oxford University Press, 2015.

Herz, Norman, and Marc Waelkens, *Classical Marble: Geochemistry, Technology, Trade*. Dordrecht: Kluwer Academic Publishers, 1988.

Hochscheid, Helle, *Networks of Stone: Sculpture and Society in Archaic and Classical Athens*. Cultural interactions: studies in the relationship between the arts, 35. Oxford: Peter Lang, 2015.

Kristensen, Troels M., and Brite Poulsen, eds., *Ateliers and Artisans in Roman Art and Archaeology*. JRA Supplementary series, 92. Portsmouth, RI: Journal of Roman Archaeology, 2012.

Lapatin, Kenneth D.S., *Chryselephantine Statuary in the Ancient Mediterranean World*. Oxford: Oxford University Press, 2001.

Liverani, Paolo, and Ulderico Santamaria, eds., *Diversamente bianco. La policromia della scultura romana*. Rome: Edizioni Quasar, 2014.

Mattusch, Carol C., *Enduring Bronze*. Los Angeles: The J. Paul Getty Museum, 2014.

Palagia, Olga, ed., *Regional Schools in Hellenistic Sculpture*. Oxford: Oxbow Books, 1998.

Palagia, Olga, *Greek Sculpture: Function, Materials, and Techniques in the Archaic and Classical Periods*. Cambridge: Cambridge University Press, 2006.

Rice, Prudence M., *Pottery Analysis: A Sourcebook*. Chicago: The University of Chicago Press, 1987 [2nd edn. 2015].

Russell, Ben, *The Economics of the Roman Stone Trade*. Oxford studies on the Roman economy. Oxford: Oxford University Press, 2013.

Smith, Tyler J., and Dimitris Plantzos, eds., *A Companion to Greek Art* (2 Vols.). Blackwell companions to the ancient world. Chichester: Wiley-Blackwell, 2012.

Spivey, Nigel J., *Greek Sculpture*. Cambridge: Cambridge University Press, 2013.

8

CONSTRUCTION ENGINEERING

Shelter is one of the most important concerns for humans today, and the same concern occupied primitive peoples. Natural shelters such as caves and other protected spaces provided early peoples with their first homes (**8.1–2**), but as humans began to gather in fertile areas, they often had to leave their natural protection behind. If there were no natural shelters in the new areas, they had to provide their own by constructing them.

A variety of factors influenced the types of structures that were created. Technical skill and tools, simple (**8.33, 8.36, 8.56**) to complex (**8.34, 8.51**), were crucial determinants, but the environment and the availability and nature of building materials played a considerable role in the evolution of structures (**8.15–20**). Regions that were cold and rainy (or snowy) had different requirements from hot, dry areas or moist, warm climates (**8.20**). Technology was the key to using more permanent materials: stone, concrete, and kiln-baked brick. Occasionally political conditions could affect the choice of materials; imperial monopolies of quarries and brickyards restricted the availability of some materials on the open market.

Once societies became more permanent and settled, the buildings became more than places merely to eat and sleep. They often retained a symbolic value, which could dictate function, shape, and orientation, requiring planning (**8.16–17**) and protection (**8.44–45, 8.64–65, 8.75–77**). Yet disasters continued to happen (**8.35–36, 8.75–81**). Each of these factors presented problems that technology helped to solve (**8.29**).

This chapter is concerned with two basic elements: the use of available materials and the technical problems involved in using those materials. The types of structures examined are, for the most part, shelters for humans, since other engineering projects are discussed elsewhere: aqueducts and tunnels (Chapter 9), roads and bridges (Chapter 11), harbours and canals (Chapter 11), and fortifications (Chapter 13). Mechanical devices used in the construction process (Chapter 2) and the methods of measurement (Chapter 11) are also examined in other chapters.

Our sources of information are relatively full, but even so there is much archaeological evidence of building technology, which is not explained by the surviving written record. Most of our literary evidence comes from Roman writers, especially Vitruvius and Pliny, but there are also many inscriptions, particularly from fifth- and fourth-century BC Greece, that provide important information regarding the costs of materials and labour (**8.45, 8.46**), and the actual building specifications and techniques used to construct structures (**8.50, 8.54**). Only a small sample can be provided here. (See Martin 1965, Orandos 1966–68, and Adam 1993 for detailed architectural drawings.)

THE ORIGIN OF BUILDINGS AND CONSTRUCTION TECHNIQUES

8.1 The origin of buildings

Shelter was one of humankind's earliest concerns. Vitruvius uses common sense and comparisons with contemporary cultures to trace the evolution of housing among humans.

Vitruvius, *On Architecture* 2.1.2–3, 7

[Once humans had begun to socialise] some people began to make homes from leaves, some to dig caves under hills, some to make places of shelter from mud and small twigs, imitating the nests and building techniques of swallows. Then observing foreign houses and adding new elements to their own ideas, they steadily created better types of huts. In addition, since men were of a nature that inclined towards imitation and learning, every day they showed to each other the achievements of their building, bragging about their discoveries; and thus, with their native talents sharpened by rivalry, the structures improved day by day as a result of better decisions. First, they covered the walls with mud after they had set up forked props connected by twigs. Others, drying mud clods of earth, constructed walls, binding them together with timber and covering them with leaves and reeds to keep off the rain and heat. Then when they had discovered that the roofs were not able to endure the rains during the wintry storms, they made gables and drew off the rain water on sloped roofs covered with clay.... [Vitruvius provides contemporary examples in the Empire].... Then ... they began to build not huts, but houses with foundations and constructed with brick walls or from stone, and roofs of timber and tiles.

8.2 Brick houses and tiles

Pliny, *Natural History* 7.194

At Athens, the brothers Euryalus and Hyperbius first constructed brick kilns and houses; previously caves took the place of houses. Gellius accepts that Toxius, the son of Caelum [the Sky?], was the inventor of clay structures whose model was the nests of swallows. Cecrops named the first town, Cecropia, after himself. It is now the Acropolis at Athens. Some believe that Argos had been founded previously by King Phoroneus, and certain people also add Sicyon, but the Egyptians believe that Diospolis had been founded long before in their own land. Cinyra, son of Agriopa, invented tiles....

8.3 The invention of the stone arch

The chronology of the invention and early use of the arch is plagued with uncertainty. Seneca states that the inventor of the stone arch with voussoirs is Democritus the philosopher (c.460–360 BC). Other evidence reveals that he travelled widely and it has been suggested that he may have observed the arch in Egypt or elsewhere in the East and brought the technique back to Greece.

Seneca, *Letters* 90.32

Posidonius says that Democritus is reported to have invented the arch, in which the curvature of stones, inclined little by little, is held together by the middle stone.

8.4 A Greek describes the pyramids

The remarkable achievements of older civilisations have often amazed later observers. The Greeks and Romans were so impressed with the monumental projects of the Egyptians that they sometimes ignored the earlier spectacular constructions of their own people. This passage from Herodotus provides an enthusiastic account of organisation and manpower in Egypt. Note that Herodotus still understands that the construction of the pyramid was accomplished using human labour.

Herodotus, *Histories* 2.124–125

Cheops, the ruler after Rhampsinitus, brought the country into total misery. For ... he forced all the Egyptians to work for his own advantage, compelling some to drag stone blocks from the quarries in the Arabian mountains to the Nile. Then the blocks were transported across the river by boats and others were ordered to receive and drag them to the Libyan mountains. They worked in groups of 100,000 men for a period of three months each. For ten years the people were oppressed by the construction of the road upon which the stones were dragged; a labour, as it seems to me, not so much less than the construction of the pyramid itself. For the road is five *stadia* long, ten *orguiai* wide, its maximum height is eight *orguiai*, and it is built of polished stone decorated with carved animals. It took ten years, as already mentioned, to construct this road and the underground chambers on the hill where the pyramids stand. He made them as sepulchral chambers for himself and turned the site into an island by leading in a canal from the Nile. It took 20 years to build the pyramid itself. It is square at the base and its height is eight *plethra*, equal to the length of each side.

It is made of polished stone that is most meticulously fitted together, and none of the stone blocks is less than 30 feet long.

The pyramid was made like steps, which some call terraces or courses. Once this first stage was completed, they lifted the remaining blocks of stone by means of devices made from short timbers. They heaved the blocks from the ground up to the first level of steps and when the stone was moved up in such a manner, it was then placed on another device positioned on the first level, and drawn from this one to the second level where there was another device [to move it higher]. For as many levels of steps there may have been as many devices, or there may have been only one device, which was easy to carry and shifted from level to level as they unloaded the stone. Since both methods were related to me, I record them both. The pyramid was completed in the following order: first the highest section, then the next section below it, and finally the lowest parts on the ground.

Pliny (*Natural History* 36.75) offers a less favourable opinion of the Egyptian pyramids, seeing them as a foolish display of wealth. Frontinus (**9.18**) gives a similar opinion not only regarding the pyramids but also about Greek monuments when compared to Roman structures.

8.5 A Greek traveller admires Bronze Age architecture

In an earlier passage, Pausanias describes Mycenae and attributes the construction of its massive walls and Lion Gate to the Cyclopes. In the following passages he describes the walls of Tiryns and one of most impressive engineering feats of the Bronze Age, the tholos tomb. These tombs were built using corbelled vaults rather than true vaults and arches. The Treasury of Atreus at Mycenae is the most famous surviving example of the technique, but many others also exist. The last passage, by attributing the work to mythical beings, indicates a loss of knowledge on the part of later people, an occurrence that we find for other spectacular monuments (Cavanagh and Laxton 1981).

Pausanias, *Description of Greece* 9.36.5, 9.38.2, and 2.25.8

Greeks seem to hold foreign spectacles in greater wonder than their own. For although famous men recount in their writings the Egyptian pyramids in the most exact detail, the Treasury of Minyas and the fortification walls of Tiryns receive not even a brief mention, although they are no less wondrous....

The Treasury of Minyas, a wonder second to none in Greece itself or elsewhere, was constructed in the following manner: it is built of stone and its appearance is rounded since its crown does not rise to too sharp a point. They say the highest stone is the keystone to the whole structure.[1]

The fortification wall of Tiryns, which is the only vestige of the ruins left, is a work of the Cyclopes and is made from unworked stones, each stone being so large that the smallest of them could not be moved even in the slightest degree by a pair of mules. Long ago, smaller stones were fitted in so that each of them is tightly fitted with the large stones.

8.6 A Roman admires an early Roman vault

The Cloaca Maxima, the main sewer of Rome, was originally an open water course, eventually vaulted over in the second century BC. The amazement expressed by Pliny is not merely the enthusiasm of a native; the sewer was regarded as a wonder even by Greeks (Dionysius of Halicarnassus, *Roman Antiquities* 3.67.5). Pliny has just finished describing the strength of the sewer and its ability to handle floods of water, both outgoing floods caused by rain and the incoming floods of an inundated Tiber.

Pliny, *Natural History* 36.106

Above [the vaults] great masses of stone are dragged, but the hollow structures do not cave in. Buildings collapsing either of their own accord or brought down by fires crash down on them, and the earth is shaken by earthquakes, nevertheless they have persevered almost impregnable for the 700 years from Tarquinius Priscus [mistakenly believed to be the original builder].

8.7 The earliest written evidence for construction

Lists of personnel recorded on the Linear B tablets give us our earliest record of the complexity of Bronze Age construction. Evidence for masons and carpenters exists, and some tablets record materials, which may have been used to construct a megaron: chimney, columns, roof beams, nails, cramps, and doorposts (*DMG* 251: pp. 349, 503–504). The first document lists masons who may have been sent to consolidate the defensive works around Pylos. The second provides a list of carpenters.

DMG nos. 41, 47 (pp. 174, 179–80, 422)

Masons who are to build: Pylos two, to Me-te-to^p three, to Sa-ma-ra^p three, Leuktron four....

Men of Aptara: 45 (or more) fief-holders, five carpenters.

CONSTRUCTION MATERIALS: MUD-BRICK TO CONCRETE

8.8 Pisé and mud-brick walls

Pliny, *Natural History* 35.169–173

Are there not walls made from earth in Africa and Spain, which are called *formaceus* [framed walls], because, rather than being built, they are really packed in a form surrounded by two boards, one on either side; and do they not last for ages, undamaged by rains, winds, fires; and are they not sturdier than any quarried stone? Even now Spain looks upon the watchtowers of Hannibal [221–219 BC] and the earthen turrets located on the mountain ridges. Such is the natural substance of earthen sod suitable for the fortifications of our camps and for the embankments against the flood of rivers. At any rate, who does not know that the wicker frames of walls are coated with clay, and built up as if with unbaked bricks?

Bricks must not be made from soil that is sandy, gravelly or, even worse, stony. Instead they should be made from a chalky clay and white or red soil, or even from sand that is definitely masculine [i.e. coarse sand]. The best time for moulding them is the spring, since in midsummer they become full of cracks. They are not approved for use in buildings unless they are two years old. In addition, their pasty clay ought to be softened by soaking before the bricks are moulded.

[Dimensions of three types of bricks are given.].... Smaller bricks are used for private structures, larger for public works in Greece.... The Greeks, except in places where it was possible to make buildings from limestone, preferred brick walls, since they last forever if they are perpendicular. [Examples follow.].... Such structures are not created at Rome, because a 1.5-foot-wide wall cannot support more than one storey, and there is a decree to prevent common walls from becoming thicker; nor does the law of party walls allow it.

Vitruvius (*On Architecture* 2.3.1–3; 2.8.16–17) has similar information on mud-brick composition and characteristics. He does add a few other details.

Vitruvius, *On Architecture* 2.3.4

In addition to these bricks [the three full sizes], there are also half bricks. When these are used in a wall, a course is laid on one face with [full] bricks, on the other face with half bricks, and they are bedded to the [perpendicular] line on each face. The walls are bonded together by the alternate layers, and the middle of the bricks being laid above the [lower] joints makes the wall stable and attractive on both faces.

8.9 On Roman unbaked mud-bricks and baked bricks

Vitruvius has just explained that sun-baked bricks are no longer used in Rome since they are structurally unsuitable to the building requirements (**8.61**). He explains that they are still used outside the city and can be made quite durable by using a few precautions. The passage demonstrates the continued use of traditional building materials even when better materials were developed.

Vitruvius, *On Architecture* 2.8.18–19

When it is necessary to use sun-dried bricks outside Rome, they ought to be used in the following manner so that they are without defects for a long time. At the top of the walls let there be placed a structure of burnt brick under the tiles, and let the structure have a projecting cornice about 1.5 feet in height. Thus, it is possible to avoid the defects, which are usual in these walls. For when the tiles on the roof are broken or cast down by the winds so that water from the rains can leak through, the burnt brick shield will not allow the mud-brick to be damaged and the projecting cornice will throw back the drips outside the vertical face, and in that manner will keep the structures of mud-brick intact. But concerning the burnt brick itself, no one can determine immediately whether it is of the best quality or faulty for use in building because it is only approved if it is firm after it has been laid on a roof and endures bad weather and time. For bricks not made of good clay or not baked long enough show themselves to be defective up there when penetrated by ice and frost. Therefore, the brick that cannot endure the strain on the roofs cannot be strong enough to bear its loads in the wall.

8.10 Wattle and daub construction

Vitruvius, *On Architecture* 2.8.20

I could wish that wattle and daub walls had never been invented. For as great as their advantages are for speed of construction and for increasing space, by the same degree they are a public disaster. They are like torches ready to be burnt. It seems better, therefore, to spend on walls of burnt brick and pay more, than to be in danger by economising with walls of wattle and daub. And in such walls, cracks are also made in the plaster coating because of the arrangement of the uprights and cross pieces. For when they are daubed, they take up moisture and swell, then contract as they dry and break the solidity of the plaster by their shrinkage. But since speed or lack of resources or partitions on an unsupported space forces some people to use this method, it should be done as follows: let the foundation be laid down so that the wall is not in contact with the rough stone and pavement; for when walls are sunk in these, they become rotten with age, then settle and lean forward, breaking the surface of the plaster covering.

8.11 Types of Greek masonry

Vitruvius and Pliny often provide similar information, although Pliny may sometimes be confusing or omit part of a discussion, as the following passages demonstrate (Cooper 2008).

Pliny, *Natural History* 36.171–172

The Greeks construct walls from hard stone or *silex*[2] blocks of the same size, as though using bricks. When they build in this way, they call the type of construction isodomic ["equal courses"]. But when the courses are built in varying thickness, they are called pseudo-isodomic. A third style is *emplecton* ["interwoven"] where only the faces are dressed and the rest of the material is laid at random. It is imperative that the joints be made to alternate so that the middle of a block covers the seam of the previously laid blocks, even in the core of the wall if possible, and if not, then at least on the faces. They call the walls with interiors packed with rubble, *diatonic*. *Reticulatum* ["network"];

CONSTRUCTION ENGINEERING

[**fig. 8.2**] construction, which is very common in Rome, is subject to cracks. Masonry must be done according to rule and level, and should be perpendicular when tested by the plumb.

Vitruvius (*On Architecture* 2.8.5–7) has a similar description of Greek practice, adding that these styles use mortar and are not merely ashlar blocks laid without a binding agent, and that the desire for rapid construction has resulted in the erection of less stable walls by the Roman builders.

Vitruvius, *On Architecture* 2.8.7

The Greeks, however, do not build in such a manner. Instead they lay their blocks level with every second block lengthwise into the thickness of the wall,[3] and they do not fill in the middle of the wall. Rather they construct a solid, continuous mass for the thickness of the walls throughout, from facing to facing. In addition, they regularly insert single blocks that run through the entire thickness of the wall. These stones, which show at each end, are called *diatoni* and by means of their bonding power they contribute very much to the stability of the walls.

8.12 The use of plaster of Paris and the astonishing strength of mortared rubble construction

The lack of understanding of the chemical properties of minerals resulted in confusion on the part of ancient builders and writers. In the following passage the term "gypsum" is used to describe both the natural mineral and plaster of Paris, which is produced by heating the mineral. At the end Theophrastus may, in fact, be referring to quicklime, which generates more heat than dehydrated gypsum. If so, the following passage represents a combination of descriptions by the writer who gives other indications that he did not have first-hand knowledge of the process.

Theophrastus, *On Stones* 65–66

The varieties of gypsum are of a peculiar nature in that they are more stone-like than earthy. In fact, the stone is similar to oriental alabaster, although it occurs in small lumps and therefore cannot be cut into large pieces. Its stickiness and heat are remarkable when it is moistened. It is used for building, being poured around the stone itself or anything else of the sort that one wishes to bond. Workmen, after chopping it up and pouring water on it, stir it with sticks, for it cannot be done with the hand because of the heat. It is moistened immediately before use, since if it is done a little beforehand the material hardens and cannot be divided. Remarkable too is its strength; for when stones [of buildings] are broken and come apart, the gypsum in no way loosens. Indeed, often parts of a structure have collapsed and been swept away while a suspended section stays in place, held together by the bonding [force of the gypsum].

Dioscorides, *Pharmacology* 5.115

Quicklime. *Calx viva* is made in the following manner. Taking the shells of the sea creatures named "*bucina*", cover them in the fire or put them into a well-heated oven and let them remain there all night. The next day, if they have become very white take them out, but if not heat them again until they become white. Afterwards, having dipped them in cold water, cast them into a new pot and, having stopped it up carefully with cloth, leave them alone there for one night. In the morning take the product out after the full treatment. This material is also made from flints or pebble stones that are burnt, and from common marble, which is preferred before all other materials. But all *calx* has a fiery, biting, and burning incrusting character....

See **8.58** for the construction of a lime kiln.

8.13 Ingredients and specifications for Roman concrete

Opus caementicium, Roman concrete, is regarded as a principal factor in the transformation and development of Roman construction. The stages of its creation and evolution are lost, but by the first century of the Empire, the best composition and treatments had been formulated (Lechtman and Linn 1987, Lamprecht 1994, Brandon *et al.* 2014).

Pliny, *Natural History* 36.175–177

There are three types of sand: quarry sand, to which a quarter part of lime must be added, and river sand and sea sand, to which a third part of lime must be added. If a third part of crushed potsherds also is added, the substance will be better.... The principal cause of collapsed buildings at Rome is that the rough stones are laid without [proper] mortar because of the pilfering of lime. The slurry [a lime paste] also becomes better the older it is; in the laws concerned with the old buildings there is even a regulation preventing a contractor from using a slurry less than three years old. As a result, no cracks disfigure the plaster of those buildings. Of the wall coatings, stucco never has enough brilliance unless three coatings of sand mortar and two of marble stucco are applied. In places where swamps or briny air might damage buildings, they can be made more fit with an undercoating of [crushed] potsherds. In Greece, furthermore, sand mortar for plaster is kneaded with wooden poles in a trough before it is applied. The test of marble stucco is in the kneading: until it no longer clings to the trowel; in whitewashing, on the other hand, the slaked lime should cling like glue. When slaking, the lime should be in lumps.

Vitruvius, *On Architecture* 2.4.1–5.1

In cement structures it is necessary first to enquire concerning the sand, that it is suitable to mix into mortar and that it does not have earth mixed in with it. The following are the types of quarried sands: black, grey, red, and carbuncular. Of these, the one which makes a crackling noise when rubbed in the hand or struck is best; while the one that is earthy will not be rough enough. Likewise, if it is covered up in a white cloth, then shaken up or pounded, and it does not soil the cloth and the earth does not settle into it, then it is suitable. But if there are no sandpits from which it can be dug, then it must be sifted out from riverbeds or from gravel or even from the seashore. But these have the following defects when used in buildings: the wall dries with difficulty and this type of wall does not allow continuous loading – it requires interruptions in the work – and it cannot carry vaults. But even more, when seashore sand is used in walls and stucco is applied onto them, a salty residue leaches out and destroys the surface. But quarried sand dries quickly in the buildings, the plaster coating is permanent, and it can carry vaults. Here, however, I am speaking of sand that is recently taken from the sandpits. For if it is taken out and lies too long, weathered by the sun and moon and hoar frost, it breaks down and becomes earthy. As a result, when it is thrown into the masonry it is not able to bind the rubble, but the rubble sinks and falls down because the walls cannot support the loads. But freshly quarried sand, although it exhibits such great excellence in buildings, is not so useful in plaster, because with its richness the lime mixed with the straw cannot dry without cracking on account of the strength of the sand. River sand, on the other hand, although useless in *signinum* [waterproofing work] because of its fineness, attains a solidity in plaster when worked by polishing tools. After considering the account of the sources of sand, one must be careful that, in regards to lime, it is burnt from white rock, whether [hard] stone or [softer] silex. The lime from close-grained, harder stone will be most useful in structural forms, while that made from porous stone will be best in plaster. Once it has been slaked, then let the mortar be mixed three parts quarried sand to one of lime; or if river or marine sand is thrown in, two parts sand to one of lime. These will be the proper proportions for the composition of the mixture. Furthermore, if anyone adds a third part of crushed and sifted burnt brick into the river or marine sand, he will make the composition of the material better to use.

8.14 Pozzolana: the special ingredient of Roman concrete

The characteristics of pozzolana, the most important ingredient of Roman concrete, are described by several authors. This sandy volcanic ash was named after the area near the Bay of Naples in which it was first found and quarried. Unfortunately, the inability of the ancient builders to analyse the material properly meant that a simple difference of colour might exclude a deposit from use. The hydraulic properties of pozzolana made it especially attractive for a culture based around the shores of the Mediterranean.

Pliny, *Natural History* 35.166

But other creations belong to the earth itself. For who could marvel enough that on the hills of Puteoli [Pozzuoli] exists a dust – so-named because it is the most insignificant part of the earth – that, as soon as it comes into contact with the waves of the sea and is submerged, becomes a single stone mass, impregnable to the waves and every day stronger, especially if mixed with stones quarried at Cumae.

Vitruvius, *On Architecture* 2.6.1

There is a kind of powder, which by its nature produces wonderful results. It is found in the neighbourhood of Baiae [near Puteoli] and in the lands of the municipalities around Mount Vesuvius. This material, when mixed with lime and rubble, not only furnishes strength to other buildings, but also, when piers are built in the sea, they set under water. Furthermore, this seems to occur for the reason that under these mountains there are both hot soils and many springs, which would not exist unless deep down there were great fires burning with sulphur, alum, or pitch. Therefore, the fire and vapour of the flame within, spreading and burning through the fissures, make that earth light; and the tufa created there rises up and is without moisture. Thus, when three substances formed in a similar manner by the strength of fire are brought together into one mixture, they suddenly cohere into a single mass because the water is absorbed. And quickly having been hardened by the moisture, they become solid, and neither the waves nor the force of water can dissolve them.

GEOGRAPHY, TOPOGRAPHY, AND CLIMATE

8.15 Site selection and its influence on materials

Vitruvius has just explained the fundamental principles of architecture in terms of its theoretical subdivisions: order (ground plan, elevation, and perspective), arrangement, proportion, symmetry, decoration, and distribution [economy]. Each of these has some impact upon building technology, but the economy has an immediate influence on the technology employed to build a structure since the architect is expected to adapt his expertise to the conditions imposed upon him by location.

Vitruvius, *On Architecture* 1.2.8

Distribution is the proper management of materials and of the site, and the thrifty and wise balancing of expenses on the works. This will be accomplished, if in the first place the architect does not demand those things, which cannot be found or obtained without great expense. For not everywhere is there a supply of quarry sand, hewn stone, fir, spruce, or marble. Different things are produced in different areas and their transport is difficult and expensive. Where no quarry sand exists, washed river or sea sand must be used; and the lack of fir or spruce can be rectified by the use of cypress, poplar, elm, or pine. Other remaining problems must be resolved by similar methods.

CONSTRUCTION ENGINEERING

CITY PLANNING

Early growth of urban culture usually resulted in unplanned, even chaotic, environments that are readily seen in the remnants of many urban sites. With the growth of stronger centralised power, towns and cities began to be arranged differently and attempts were made to regularise both their boundaries and internal zones. The Etruscan and Roman use of the *pomerium* established a sacred boundary of the urban area that was gradually increased at Rome as the city grew. Varro (cf. Livy, *History of Rome* 1.44) incorporates military significance to the *pomerium* by using the terms *fossa* and *murus*. Tacitus relates the basic method and initial route of the *pomerium* at Rome, finishing with active changes from his own time.

8.16 The *pomerium*

Varro, *On the Latin Language* 5.143

Many people founded towns in Latium using the Etruscan ritual; that is, with yoked cattle, a bull with a cow on the inside, they drive a furrow around in a circle with a plough [to create an enclosure], which they do on an auspicious day for religious reason, so that they might be fortified by a *fossa* [ditch] and a *murus* [wall]. The place from which they had dug out the earth, they called the *fossa* and the earth heaped inwards they called the *murus*. Inside them, was the *orbis* [circle], the beginning of the *urbs* [city].... Stone markers of the *pomerium* stand both around Aricia and around Rome.

Tacitus, *Annales* 12.24

... I think it worthwhile to understand the origin [of the city-foundation] and the *pomerium*, which Romulus set down. A furrow for marking out the [boundary of the] town was started from the Forum Boarium – where we see the bronze statue of a bull, since that is the kind of animal yoked to the plough – in order to encompass the great altar of Hercules; then, stones were set at regular intervals along the foot of the Palatine Hill to the altar of Consus, then to the old *curiae* [places of assemblies], and then to the shrine of the Lares, then to the Roman forum, although the Roman forum and the Capitol are thought to be added to the city not by Romulus but by Titus Tatius. Afterwards the *pomerium* was enlarged along with Rome's good fortune. And the boundaries, which Claudius in his time set out, are easily recognised and are registered in the public records.

8.17 Planning the city: Hippodamus of Miletus

Hippodamian grid plans regulated building arrangements through the use of straight streets that crossed at right angles and allotted areas to specific functions. Although the Greeks attribute this concept to Hippodamus, much earlier cities exist with such planning. Nevertheless "Hippodamian" is still used to describe the orthogonal grid plan of ancient cities. The sense that everything belongs to its own proper space, however, is inherent in both Greek and Roman plans (for Roman Forts see **13.12**), and is reflected in Alexander's attempt to lay out Alexandria in the last passage (cf. Arrian, *Anabasis* 3.1.5–2.2).

Aristotle, *Politics* 2.21–2, 29–34.1267b

The Milesian son of Euryphon, Hippodamus, who devised the apportionment of [the elements of] cities and laid out the Piraeus [in streets], ... was the first of those not engaged in politics who attempted to say something about the best form of constitution. He was for arranging a city with a population of 10,000, divided into three classes. He made one class of craftsmen, one of farmers, and the third of armed defenders. And he divided the space [of the city] into three parts: sacred, common, and private. From which, sacred land was to supply the customary offerings to the gods, the common land from which the defenders were to live, and private land to be owned by the farmers.

Plutarch's description of the initial attempt to plan Alexandria provides later evidence for 'arranging' of cities according to a rational plan; the measuring out the site and drawing of a plan with chalk seems realistic, while the use of grain on the ground seems a bit fanciful. The colony at Carthage, which uses boundary-stakes, probably reflects the more normal method of survey.

Plutarch, *Alexander* 26.2–4

They say that after having prevailed over Egypt he wished to leave behind a great and populous Greek city having united it under his own name, but had not yet begun to measure off and enclose some site [selected] on the advice of his architects ... [lines from the *Odyssey* in a dream compel him to seek out the island of Pharos, which he visits the next morning and finds an ideal location for his city] ... having said that Homer was admirable in other ways and was a very wise architect, he ordered the plan of the city to be drawn so as to conform with the site. As there was no white earth [chalk] at hand, taking some barley-groats they marked out on the dark flat earth a rounded hollow, of which straight sections took up the inner circumference from the edge, just as in the design of a *chlamys* [military cloak], the lines beginning from the edges and uniformly narrowing from the larger [outer] part. While the king was taking delight in the design, suddenly birds from the river and the lagoon, infinite in number and of every sort and size, swooping down upon the place like clouds left not a bit of the barley-groats, so that even Alexander was disturbed at the omen. [He is reassured that this is a good omen of Alexandria nourishing all peoples.]

Plutarch, *Gaius Gracchus* 11.1

In Libya, moreover, in regards to the colony of Carthage, the one that Gaius named Junonia – 'Heraea' in Greek – they say that many adverse signs from the gods appeared. [Several bad omens are listed]; … and a sudden, powerful wind scattered the sacrificial victims laid upon the altars and dispersed them beyond the boundary-stakes marking out the future colony's design; and wolves set upon the boundary-stakes themselves and, tearing up and carrying them off, they went far away.

8.18 The change from wooden to stone temples

Pausanias describes one of the earliest monumental temples built in Classical Greece. Although writing over 700 years after the temple was initially built, Pausanias sees one column of wood among the stone ones. This probably reflects the transition from the wood of the original structure to the more durable stone: a change dependent upon technological and economic conditions. The gradual replacement of wood by stone columns in the temple is verified by archaeological evidence on the site (Cooper 2008).

Pausanias, *Description of Greece* 5.16.1

It remains after this for me to describe the temple of Hera and the numerous distinguished objects in it.… The temple's style is Doric, and columns stand all around it. In the back room (*opisthodomos*) one of the two columns is made of oak [the rest are stone].

8.19 Transport from the quarry: technique and cost

Diodorus, describing the construction of a marvellous temple at Engyon (modern Gangi) in Sicily, reveals some of the difficulties experienced if materials are not available locally. But compare **5.41** for a discussion of problems arising from local stone. Cost and transport are the two major problems. He makes it clear, however, that with sufficient funds almost any obstacle could be overcome.

Diodorus of Sicily, *History* 4.80.5–6

For not having suitable stone on their own territory, they imported it from their neighbours at Agyrium, although the cities were 100 *stadia* apart and the road, by which it was necessary to transport the stones, was rough and very difficult to travel upon. For this reason, after constructing four-wheeled wagons, they transported the stone by using 100 pairs of yoked oxen. In fact, because of the abundance of sacred possessions they were so well-off that they overlooked the expense.

A variety of inscriptions also gives us information about transport expenses (**8.52**, **11.35**), and Vitruvius (**11.34**) provides an account of the technology involved in moving the material from the quarry to the site.

CONSTRUCTION ENGINEERING

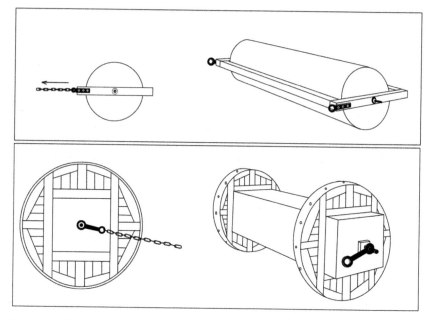

Figure 8.1 Clever ways to move heavy marble pieces.
Source: Adapted from Adam, 1994, fig. 33.

8.20 Accommodating building design to climate and location

Climatic conditions affected not only people (**1.15**) but also the types of structures built and used by them. Buildings were adapted to counter the adverse weather conditions and sometimes this resulted in the development or introduction of different building technologies.

Vitruvius, *On Architecture* 6.1.2, 12

In the north, buildings should be vaulted (*testudinatus*) and completely enclosed, and not exposed but with an orientation towards the warm sun. On the other hand, in southern areas oppressed by the heat where the sun beats down, buildings should be constructed that are more open and oriented to the north or the northeast. Thus, we can modify by art what nature by herself would harm.... If it is a fact that different regions should be matched to various classes according to climate, with the result that the very nature of the peoples born in them are also dissimilar in mental and physical character, then we also should not hesitate to construct buildings appropriately adapted to the characteristics of nations and races, since we have from nature herself a skilled and ready guide.

8.21 Types of floorings

Pliny, *Natural History* 36.184–189

Paved floors skilfully embellished with a type of painting originated among the Greeks and lasted until mosaic work superseded them.... I think the first paved floors made were those which we now call *barbarica* [foreign or rough] and indoor floors. In Italy they were packed down using rods; at any rate, this can be deduced from the name itself. At Rome a diamond patterned floor first was laid in the Temple of Capitoline Jupiter after the beginning of the Third Punic War [149 BC], but fine mosaic pavements were common and very popular before the Cimbrian War [113–101 BC]....

The Greeks, who roof their houses in the same manner, invented open-air flooring, an easy method in a warm area but more uncertain wherever the rains freeze. It is essential that two pairs of joists be laid across each other and that their ends be fastened lest they warp. A third part of crushed potsherds should be added to fresh broken stone, then the broken stone, in which two-fifths part of lime have been mixed, must be tightly packed to a thickness of one foot. Then a core six digits thick is applied and large square stones not less than two digits thick placed on it. A slope of 1.5 *unciae* in ten feet is to be maintained and it is to be polished carefully with a grindstone. To cover the floor with oak planks is considered unprofitable since they warp, but on the other hand it is believed beneficial to layer straw or fern below to lessen the negative effect of the lime on the wood. It is also essential to lay down round pebbles as a foundation. Tiled floors with a herringbone pattern are made in a similar manner.

Still there is one type of floor that must not be neglected, the *graecanicum* [Greek style]. Once the earth has been rammed down, a layer of rubble or crushed potsherds is laid on it. Next, once some charcoal has been closely trodden on it, a substance mixed from coarse sand, lime, and ashes is spread on top to a thickness of 0.5 foot. It is completed according to the ruler and level; and although it has an earthy appearance, if it is polished with a grindstone, it possesses the look of a black pavement. Mosaic floors had come into use as early as Sulla [82–79 BC], for one made of small cubes, which he had laid in the shrine of Fortuna at Praeneste is still preserved today. Subsequently driven from the ground, such pavements were promoted to vaulted ceilings, and made of glass. This is a recent innovation. Certainly Agrippa, in the baths, which he built at Rome [33 BC], painted the brickwork of the hot rooms with wax-based pigments, and the rest with stucco. He undoubtedly would have put glass mosaics on the vaults if they had already been invented....

Vitruvius (*On Architecture* 7.1.1–7) provides additional information on flooring methods.

8.22 Walls and wall-facings

Vitruvius, *On Architecture* 2.8.1–2, 4

There are the following types of wall-facings: *reticulatum* [network], which everyone uses now, and the old style called *incertum* [irregular work]. *Reticulatum* is the more pleasing of these, but is prepared in such a way that it makes cracks, because it has beddings and joints spread out in every direction. The irregular stones of the *incertum*, however, lying one course above another and imbricated, produce a wall that is not pretty but is stronger than the *reticulatum*. Furthermore, both types have to be built with very small stones so that the walls will last longer once they have been thoroughly saturated with a mortar of lime and sand. For since the stones used are soft and porous in nature, they dry up the moisture by drawing it out of the mortar. When the amounts of lime and sand are

Figure 8.2 Opus reticulatum.
Source: Adapted from Sear, 1989, fig. 40.

abundant and in excess, however, the wall has more moisture and does not become frail quickly; it is held together by these elements. But as soon as the watery strength is sucked out of the mortar through the looseness of the rubble, and the lime separates from the sand and is dissolved, at the same time the rubble cannot adhere with them and the wall will in time become ruinous.... But if anyone wishes to avoid falling into this disaster, he should leave the middle hollow behind the orthostates [the facing slabs], and on the inside build walls two feet thick out of red squared stone or burnt brick or lava in courses, and bond their facings with these by means of iron clamps and lead. The work is constructed, therefore, not as a heap but in an organised fashion and can last forever without defect, because its beddings and joints, all settling together and bonded at the seams, do not push the work outward nor allow the orthostates, which have been bonded together to slip.

8.23 Slaking lime for stucco

In the next two passages Vitruvius sets out the necessary treatment of lime for its use on the walls and vaults of buildings. The careful treatment he records produced a very stable surface upon which magnificent Roman frescoes were applied. Great numbers of Roman frescoes have been recovered from Pompeii and Herculaneum, many in remarkable states of preservation, at least partly a result of their stable foundations.

Vitruvius, *On Architecture* 7.2.1–2

Next it is necessary to explain white stucco work. This material will be suitable if the best lumps of lime are slaked well before they are needed. Then, if any lump has not been burnt in the kiln long enough, it will be forced by the long period of slaking to throw off its heat and will be thoroughly burnt to an even quality. For when it is taken fresh rather than completely slaked, it blisters upon application since it has little uncooked bits that are hidden in it. These little bits, once they finish their slaking on the wall, break up and destroy the polish of the stucco. But when proper attention is given to the slaking and more care paid to the preparation for the work, a hoe is applied to the slaked lime in the trough in the same manner that wood is chopped. If the hoe comes into contact with uncooked bits of lime, it is not yet tempered; and when the iron is withdrawn dry and clean, this indicates the lime is weak and thirsty; but when it is rich and properly slaked, it sticks around the tool like glue and proves by this means that it is tempered.

8.24 Stucco work on vaults

Vitruvius has just described the construction of wooden centring for the vault (**8.31**). Here he describes the process of completing the vault's construction and decoration using plaster and stucco.

Vitruvius, *On Architecture* 7.3.3–6

Once the vaults are in place and interwoven with the reeds, the lower surface is to be plastered, then sand mortar applied, and finally afterwards polished with chalk or marble powder. Once the vaults have been polished, the cornices have to be set beneath them and they, obviously, ought to be as slight and delicate as possible since, when large, they slip down under their own weight and cannot support themselves. Gypsum, which sets too soon and does not allow the work to dry uniformly, should not be used in the stucco cornice, but powdered marble should be evenly applied. In addition, the arrangements of the ancient builders must be avoided in these vaults, since the projecting surfaces of those cornices jut out and are dangerous on account of their weight....

Once the cornices are finished, the walls are to be plastered as roughly as possible, and later when the rough plaster is nearly dry, use sand layers to bring it into form so that the lengths conform to ruler and line, the heights to the plumb, and the angles to the square; ... a second and third coat are to be applied as the previous layer dries. As a result, the better the foundation of sand mortar layers, the stronger and more durable the solidity of the plaster. Once not less than three layers of sand mortar, in addition to the rough plaster, have been applied, then it is necessary to make up the material for the layers of powdered marble. This mortar is to be tempered so that when mixed it does not cling to the trowel, but the iron comes out free and clean from the trough. After a thick layer has been spread and is drying, apply a second thinner coat. When that has been worked over and well-rubbed, lay on an even finer coat. Thus, the walls will be solid with these three coats of sand mortar and the same number of marble, and will not possibly be subject to cracks or any other defect....

Vitruvius continues, describing how pigment is added to the last coat while still damp; he concludes by cautioning against fewer coats, which results in less stable surfaces. Cato, *On Agriculture* 128 offers instructions for a simple exterior plaster.

8.25 Marble veneers

The limited supply of marble in Rome until the last century of the Republic promoted the popularity of veneers that provided the appearance of solid marble at a fraction of the cost. Pliny suggests that marble veneers may have been invented on the west coast of Asia Minor about 350 BC, but did not reach Rome until almost three centuries later when the technique was used to decorate private residences. Pliny is very critical in general of the use of marble in private residences. See **5.38** for the technique of cutting thin sheets of marble.

Pliny, *Natural History* 36.47–50

The cutting of marble into thin slabs may have been invented in Caria. The oldest example, as far as I can find, is at Halicarnassus in the house of Mausolus [died 353 BC], where the brick walls were covered with Proconnesian marble.... Cornelius Nepos relates that Mamurra [*c.*60–50 BC] was the first man at Rome to cover whole walls of his house, which was on the Caelian Hill, with marble veneer.... The stage of the theatre of Marcus Scaurus [58 BC], I think, was the first to have marble walls, but I cannot say easily whether they were veneered or solid polished blocks, as at present are the walls of the temple of Jupiter Tonans on the Capitoline Hill. For I have not yet discovered traces of marble veneer in Italy at that time.

ROOFING PROBLEMS AND SOLUTIONS

Trussed roofs were not common in the Greek period, and roofing large structures with the post and lintel design caused substantial problems (Hodge 1960). Large beams were difficult to obtain (note the size of beams for the Arsenal in **8.57**), and they needed heavier and more frequent supports. The size of stone beams was limited to a maximum axial spacing of less than 9 m, and generally less than 7 m, because of stone's poor tensile qualities. As a result, the interior of the structure was either cluttered with more supports or left unroofed. In the Roman period, the use of vaults and domes provided greater uninterrupted interiors.

8.26 Massive wooden beams

This passage demonstrates that huge wooden beams were available, even if infrequently, for transverse and ridge beams. Pliny, however, does not state that they were to be used as roof beams.

Pliny, *Natural History* 16.200–201

The largest tree ever seen at Rome down to the present day is believed to be the one that Tiberius Caesar had exhibited as a wonder on the deck of the ship used in his mock sea-battle ... and was finally used in the amphitheatre of the Emperor Nero. It was a beam of larch, 120 feet long and of a uniform two-foot thickness.... Within our memory there was also an equally wonderful tree left in the porticoes of the Saepta by Marcus Agrippa, which remained from [the construction of] the Diribitorium; it was 20 feet shorter than the other and 1.5 feet thick.

Pliny then discusses large timbers used for the masts of ships. Dio Cassius (*Roman History* 55.8.3–4) describes the Diribitorium as the largest roofed structure ever built by his time (*c.* AD 200).

8.27 A temple too large to roof

Strabo errs regarding the size of the temple of Apollo at Didyma since the temples of Hera at Samos, Artemis at Ephesus, and the Olympieum at Agrigento in Sicily are all larger. The very impressive remains of Apollo's temple give no indication of roofing.

Strabo, *Geography* 14.1.5

The Milesians constructed the largest temple of all, but it remains without a roof because of its size.

8.28 A roof imitating a tent

The idea of a tent roof seems to be an old one and is probably the underlying concept for the use of awnings over theatres and amphitheatres.

Plutarch, *Life of Pericles* 13.5

The Odeum of Pericles, in its interior arrangement, had many seats and columns. Its roof was made sloping on all sides and descending from a single peak, an arrangement which, they say, was an imitation of the tent of the King of Persia. Pericles had charge of this also.

8.29 A large roofed structure with usable interior space

Unlike most Greek religious rites that were celebrated outside, the cult at Eleusis required interior space for the use of the initiates. Initially only an enclosure wall separated the rites of the mystery cult from the outside world, then a roof was added. Vitruvius is aware that emphasis on interior space is not normal in Greek architecture; the lack of exterior columns dramatises this oddity.

Vitruvius, *On Architecture* 7, Preface 16

At Eleusis Ictinus roofed over the temple (*cella*) of Ceres and Proserpina, a chamber of immense size, so that there was more scope for sacrifices. He used the Doric order without exterior columns.

8.30 A wooden roof constructed without nails

Pliny, *Natural History* 36.100

Also, at Cyzicus is a large building called the *bouleuterium* [council house] with rafters that have no iron nails and are arranged in such a manner that the beams can be removed and put back without the aid of scaffolding (cf. the Pons Sublicius **11.24**).

8.31 The importance of carpentry for making vaults

Besides the obvious information on the use of stucco, this passage is also important as one of the few that mention the need for skilful carpentry and additional supports when pouring concrete vaults and domes [**fig. 8.3**]. (Lechtman 1987).

CONSTRUCTION ENGINEERING

Vitruvius, *On Architecture* 7.3.1–2

So, when vaulting is required, it is necessary to proceed as follows. Parallel laths [thin slats of wood] are to be positioned not more than two feet apart, preferably of cypress wood rather than fir that is quickly spoiled by age and decay. These laths, once they have been arranged into a curved shape, are to be secured to the floor above or to the roof by fastening them to wooden ties with many iron nails. The ties are to be fashioned from a material unaffected by decay, age, or moisture. A material such as boxwood, juniper, olive, oak, cypress, and similar types except for common oak, which creates cracks by warping in the works in which it is used. To bend the laths into the required shape once they are in place, use cords made of Spanish broom to bind them to Greek reeds that have been pounded flat. Mortar mixed from lime and sand is to be spread on the top of the vault immediately so that if anything drips from the floors or roofs above, they will be held up....

Alternative materials and directions follow.

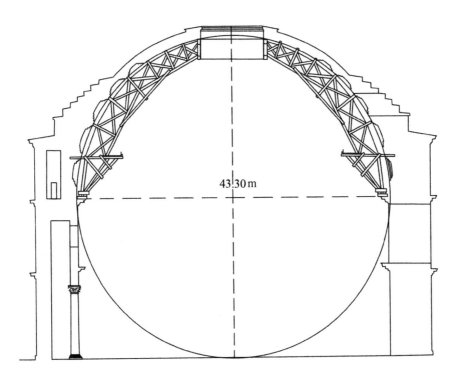

Figure 8.3 Reconstructed centering of the Pantheon.
Source: After Adam, 1994, fig. 443.

8.32 The dome of Hagia Sophia

The greatest dome of antiquity is still intact on the Pantheon in Rome. We do have some contemporary descriptions of its dome (Dio Cassius, *Roman History* 53.27.2; Ammianus Marcellinus, *History* 16.10.14), but almost no written evidence survives regarding its construction. Fortunately, this is not the case with the second great dome of antiquity, that of Hagia Sophia (mid-sixth century AC) in Istanbul. Although this structure lies beyond the chronological scope of our book, its building tradition originates in the Roman period in structures like the Pantheon. A short passage regarding its rebuilding (between AD 558 and 562) concludes this section on roofing. For other building collapses see **8.78–81**.

Agathias, *Histories* 5.9.2–5

The Emperor Justinian tried to restore the many public buildings damaged by the disaster [an earthquake in AD 557]. Some structures were unsound while others had already collapsed. He was especially concerned about the Great Church of God; this had previously been burnt by the people [AD 532] and he had rebuilt it right from the foundations in a conspicuous and wonderful form, and further enhanced it with a much greater size, with majestic proportions, and with the lavish adornment of coloured marbles. He covered it with baked brick and mortar, binding it together with iron in many places, but using very little wood so that it would no longer easily be fired. The architect and builder of all this was Anthemius, whom I have already mentioned. This time, however, as the result of an earthquake, the church had lost the central part of its roof – the part that towers over the rest [i.e. the dome] – and the emperor repaired it again, more securely, and raised it to a greater height. Since Anthemius had been dead for a long time, Isidore the Younger and the other engineers examined among themselves the earlier design and, by observing what was still standing, they determined the faults of the section that had collapsed. They left the arches on the east and west sides as they were originally, but in regard to the north and south ones they extended inwards the convex part of the structure and made them slightly broader, so that they fitted more closely with the other arches and observed the harmony of equal sides [forming a square]. In this manner they were able to cover over the irregularity of the void and to take away a bit of that space that produced an oblong plan. Thus upon these arches they placed again the circle or hemisphere or whatever else they call it [the dome], which rises up in the middle [of the structure]. Henceforth the dome became naturally more regular and more beautifully curved, corresponding to the [proper geometrical] figure. But it was narrower and steeper so that it did not inspire the onlooker as fully as the old one did, although it was much more firmly set.

The rebuilt dome may have been more secure, but it was not as pleasing.[4]

CONSTRUCTION ENGINEERING

SIMPLE TOOLS OF THE TRADE

8.33 Early uses of the chalk line

Homer's simile compares two equal lines of military forces joining battle. In a gloss on *stathme* (chalk string) the taut, thin string, smeared with red or black pigment, is snapped to lay a straight-edge on the wood in preparation for sawing.

Homer, *Iliad* 15.410–13

But just as the chalk string makes straight a ship's timber in the hands of an experienced carpenter, who indeed knows well of all manner of craft through the counsels of Athena, so evenly was strained their war and battle.

Odysseus begins to build his raft as he prepares to leave Ogygia, Kalypso's island.

Homer, *Odyssey* 5.243–45

[After Kalypso provides Odysseus with a bronze axe and adze she leaves him among the trees.] ... then he began cutting trees, and his work went quickly. He felled twenty trees in total and shaped them with the axe. Then he skilfully shaved them and made them straight according to the string.

8.34 Surveying and levelling

The *groma* (*Ferramentum*)

Frontinus, *The Science of Land Measurement*, Campbell pp. 12, 19–29 (T 15.3 = L 26.9)

The Roman land surveyors, the *agrimensores/gromatici*, not only survey large areas of land for colonies, camps, distributions, and allotments but were continually involved in decisions regarding urban and rural property rights. In addition to the *dioptra* and *chorobates* [figs. 8.4–5], the simple *groma* produced excellent results as a result of the training and high standards of precision maintained by the profession (cf. **11.57, 13.12**) (Grewe 2019, Campbell 2000, Dilke 1971).

[Terrain is not just cultivated land but varied with mountains, rivers, swamps, broken ground, etc.].... Nevertheless it is necessary for every portion of land, even the smallest, to be subject to the mastery of the land surveyor and to be encompassed according to its own individual demands by the application of right angles. And so, most particularly, we must anticipate by what use of the *ferramentum* [*groma*, here] we might overcome whatever might be encountered. Next, we must apply attentiveness to the measuring, preferably so that the line of the measuring of the determination might match up with the results from the lengths of the sides [a textual problem]. First, we must use the *ferramentum* and aim straight, once it has been balanced in regards to all movements, and then sight with the eye from every projection (*corniculi*) through the strings or threads, once they have been stretched taut by the weights and thus have been standardised among themselves, until one sees only the closest thread with the view of the other thread now totally obscured. Then we must fix the markers and recover them – after the *ferramentum*, in the meantime, has been relocated from the last marker [and retuned] to the same movement with which it was being controlled – and we must continue the straight line that was begun to a turn or to the end. Moreover, at all turns let the plumb-line indicate a place of intersection [of two lines].

THE *CHOROBATES*

Vitruvius, *On Architecture* 8.5.1

Now I will explain how the supplies of water to country residences and to walled towns ought to be accomplished. The first step is to take the level, which is done with the *dioptrae*, water levels, or the *chorobates*. It is done more accurately with the *chorobates* than with the *dioptrae* and levels, which are deceptive. The *chorobates* is a straight board about 20 feet long. At its furthest ends it has legs, made exactly alike, that are fastened at right angles to the ends of the long board. Crosspieces, joined by tenons, connect the long board and the legs. These crosspieces have lines accurately drawn to the perpendicular upon them and plumb

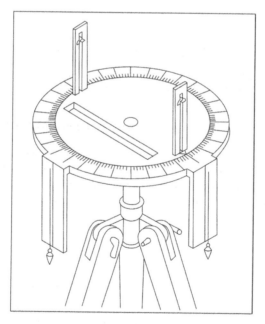

Figure 8.4 Dioptra.

Source: After Adam, 1994, fig. 1.

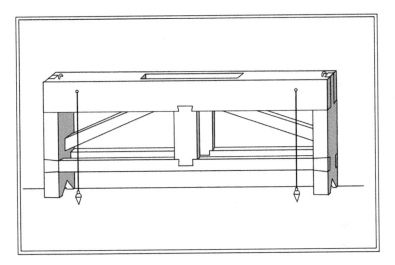

Figure 8.5 Chorobates.
Source: Adapted from Adam, 1994, fig. 16.

lines hang from the long board over each of the drawn lines. When the board is in position, those plumb lines that strike the marked lines both equally and at the same time show the level position of the *chorobates* (Grewe 2019).

8.35 Failure to use proper tools

In his discussion of the reasoning process, Lucretius compares faulty thought processes based upon mistaken senses with a house that has been poorly constructed because of faulty building tools.

Lucretius, *On the Nature of Things* 4.513–519

Finally, as in a building, if the original straight-edge is deformed, or if a faulty square departs from straight lines, or if a level is a trifle off in any part, the entire structure will necessarily be made imperfectly and be leaning over, warped, sloping, leaning forward, leaning backward, and everything askew. The result is that some parts seem about to collapse soon, and some do collapse, all betrayed by the initial faulty decisions. Thus, your understanding of matters must be warped and faulty whenever it is produced from false senses.

8.36 Failure to use a plumb bob

Cicero is supervising the work being done on the villa of his absent brother. As might be expected, Cicero has his own ideas and is critical of the workmanship of the hired help.

Cicero, *Letters to his Brother Quintus* 3.1.1–2

The pavements appear to be laid correctly. Some vaults that I did not approve of, I have ordered to be changed.... Diphilus, the architect, has erected columns that are neither perpendicular nor opposite each other. He will have to pull them down, of course. Eventually he will learn the use of the plumb-line and measuring tape.

THE ARCHITECT

Our knowledge about architects and their profession in antiquity is surprisingly meagre considering the large number of structures erected and the variety of treatises that were produced. Vitruvius' writings, the only major surviving study, provide us with a comprehensive statement about the ideal training of an architect. Otherwise, to develop our understanding of the ancient architect we are reduced to assembling small bits of information recorded by historians, poets, and encyclopaedic writers (Coulton 1977, Holloway 1969, MacDonald 1977).

8.37 On the training of architects

The architect is equipped with knowledge supplied by many disciplines and various types of learning that are used in the other arts. His own service arises from both practice and theory. Practice is the continuous and regular habit of use, which is completed by the hands in whatever material is necessary according to the purpose of a design. But theory is the ability to demonstrate and explain things constructed in accordance with skill and planning.

Vitruvius, *On Architecture* 1.2–10

Thus, architects who have aimed at obtaining manual skill without scholarship cannot attain a position of authority corresponding to their labours, while those who trusted only in theories and scholarship clearly pursue shadows rather than reality. But the men who have a thorough knowledge of both, like men equipped with full armour, acquire sooner and with authority what they want.

The two following elements are found in all things, but especially in architecture: that which is signified, and that which gives it significance. The thing signified is the subject proposed about which we speak; and that which gives significance is a demonstration explained through the principles of knowledge. On account of which, it seems that the man who professes that he is an architect ought to be trained in both areas. He must, therefore, be both naturally talented and willing to learn. For neither natural talent without instruction nor instruction without natural talent can produce the perfect artist. He should be a man of letters, be skilful with the sketching pencil, be learned in geometry, know much history, have listened carefully to the philosophers, and understand music. He also should know the opinions of those skilled in the law, be familiar with astronomy and the theories of the heavens, and not be ignorant of medicine.

The following are the reasons why this is so. An architect ought to be a man of letters so that he can produce a more lasting record in his treatises. Second, he should have knowledge of drawing so that he can more easily, with coloured examples, draw the effect of the work he wishes. Geometry also furnishes many aids to architecture; it first teaches us the use of the ruler and compass, from which the laying out of buildings on their sites is made especially easy through the correct application of the squares, the levels, and straight alignments. Again, through optics, lighting is brought into buildings from appropriate areas of the sky. By arithmetic not only are the costs of buildings calculated and measurements indicated, but difficult problems of symmetry are also resolved by geometrical theories and methods.

An architect should have a wide knowledge of history because in their structures architects often design many decorative elements that they should be able to explain to those enquiring why they were made and what they signify.... [The use of Caryatids and Persian statues in the place of columns is explained.]

As for philosophy, it makes the architect high-minded but not arrogant; rather he is courteous, fair-minded and trustworthy without avarice, which is very important since no work can be accomplished without honesty and incorruptibility. He should not be greedy nor have a mind occupied with the procurement of gifts, but he should preserve his dignity with seriousness and by having a good reputation. For these are the precepts of philosophy. In addition, philosophy explains the nature of things, which in Greek is called *physiologia*, which must be learnt by careful study because it has many and diverse natural problems. [Difficulties with water supply follow.].... Also, anyone who reads the works of Ctesibius or Archimedes and the others who have composed treatises of the same sort will not be able to understand them unless he has been trained in these matters by philosophers.

An architect must also know music so that he understands the theory of harmony and mathematics, and besides this, so that he can properly make adjustments to the ballistae, catapults, and scorpions....

Theatres also contain copper vessels, called *echeia* in Greek, that are placed in niches under the seats according to a mathematical principle. By this means a variety of noises are combined to form musical symphonies or harmonies through the vessels, which are apportioned in the circular cavea according to fourths, fifths, and the octave. The result is such that when the voice of an actor comes into contact with them, the power is increased and it reaches the ears of the audience more clearly and sweetly....

The architect must also know the art of medicine on account of the different regions of the earth, called *climata* in Greek, and because of the changes in weather, healthy and pestilent locations, and the uses of water. For no dwelling can be suitable without considering these conditions. The architect should be familiar with laws, especially those necessary for structures with party walls, in regards to the limit of water dripping from eaves, drains, and lighting. And the supply of water and other things of a similar manner ought to be known by architects, so that, before they begin buildings, they are careful not to leave disputes about the completed structure for the owners. It also should be known so that when writing the contracts, the architect is able to protect both the employer and contractor by means of his knowledge. For if the contract is written skilfully, each party can be released by the other without injury. From astronomy, moreover, the architect finds the east, west, south, and north as well as the nature of the heavens: the equinox, the solstice, and the courses of the heavenly bodies. If one has no understanding of these things, he will not be able to comprehend the theory of sundials.

8.38 Shortages of architects and craftsmen; and mismanagement

Pliny the Younger, *Letters* 10.39.1–6; 10.40

[Pliny addresses Trajan.] The greater part of the theatre at Nicaea, my Lord, has been built and, although still unfinished, has already consumed more than ten million *sestertii*, as I have been told, for the accounts of the work have not been examined. I fear that it may be money spent in vain. For the structure is sinking and gapes open with huge cracks, either because the soil is damp and soft or the stone itself is poor and crumbly. It is certainly worth considering whether it should be finished, abandoned, or even demolished. For the foundations and substructure, by which the theatre was to be held up, do not appear so solid to me, although they are expensive.... These same citizens of Nicaea have also begun to rebuild their gymnasium ... but an architect – to be sure, a rival of the one who started the structure – contends that the walls, even though 22 feet thick, cannot support the loads placed on them because the core, which is packed with rubble, is not overlaid with brick.... I am forced to request from you, not only on account of the theatre but also because of the bath at Claudiopolis [unfinished and already very expensive], that you send an architect to examine whether it is more useful, considering what has already been spent, to complete the works according to the initial plans by whatever means we can, or whether it is better to set them right by changing something, or whether it is best to move them, so that we do not lose more money by trying to protect what has been spent already.[5]

[Trajan addresses Pliny.] You decide what to do. You cannot lack architects. Every province has men who are both skilled and talented. And do not believe that they can be sent more quickly from Rome, since they usually come to us from Greece.

8.39 The competitive spirit of architects

As in the previous passage, hints of rivalry among architects appear throughout the Greek and Roman periods. The next two passages demonstrate that this rivalry could lead to new technological developments while simultaneously producing other negative results. In the first, Diodorus, having detailed the reputation and inventiveness of Daedalus, then explains why he went into exile.

Diodorus of Sicily, *History* 4.76.4–6

Although Daedalus was admired on account of his ingenious constructions, he had to flee from his native land [Attica] once he was found guilty of murder for the following reason. Talos, the son of Daedalus' sister, was being educated in the home of Daedalus while he was a young boy. But being more talented than his teacher he first invented the potter's wheel; then later, coming by chance upon the jawbone of a serpent and using it to saw through a small piece of wood, he copied the jaggedness of the teeth to create a saw made of iron. With this he cut the wood in his work and was regarded as having discovered a tool most useful for the building craft. He also invented the carpenter's tool for drawing a circle and certain other cleverly devised tools for which discoveries his reputation increased greatly. But Daedalus, becoming jealous of the boy and thinking that the youth's fame would rise far above that of the teacher, murdered the boy.

Rulers of complex empires are generally thought so occupied with affairs of state that they have no time to spend on building projects other than to endorse the plans of others. This is probably not entirely true as demonstrated by Hadrian (117–138), whose active involvement in planning structures is related by several sources. Hadrian's personal contribution is not accepted by all scholars, but others do acknowledge his involvement and see an innovative mind at work. The following passage illustrates two characteristics of Hadrian: first his willingness to challenge tradition, second his anger when criticised for mistakes. The pumpkins (literally, "squash", *kolokuntai*) may refer to types of segmented domes such as those later built in Hadrian's Villa at Tivoli. The second criticism regarding the size of the cult statue had already been levelled against Pheidias (Strabo, *Geography* 8.3.30), who had purposefully constructed his overpowering Zeus.

Dio Cassius, *Roman History* 69.4.1–5

Hadrian first exiled and then put to death the architect Apollodorus who had built several monuments of Trajan in Rome: the forum, the odeum, and the gymnasium. The reason given was that Apollodorus had committed some sort of offence, but the true reason was that once when Trajan was consulting with him about something to do with his buildings, Apollodorus said to Hadrian, who had interrupted with some comment, "Go away and draw your pumpkins [*kolokuntas*]. You know nothing about these matters". For it happened that Hadrian at that time was priding himself about some such drawing. When he became emperor, therefore, he remembered the insult and would not endure Apollodorus's outspokenness. Hadrian sent him the plans of the Temple of Venus and Roma to show that it was possible to create a great work without Apollodorus's help, and asked if the structure was satisfactory. Apollodorus replied, ... in regards to the cult statues, that they had been made too large for the height of the cella. "For should the goddesses wish to get up and leave", he said, "they will not be able to do so". After he wrote all this so bluntly, Hadrian was both annoyed and deeply pained because he had committed a blunder, which could not be corrected. And he held back neither his anger nor his pain, but killed Apollodorus.

Aulus Gellius, *Attic Nights* 19.10.1–4

Presentations before a client by competing architects, as described in the following passage, were probably a common practice for large or complex projects.

... Several builders (*fabri aedium*) were standing next to him [the client Cornelius Fronto]. He had summoned them to construct new baths and they were now showing him various plans of baths, drawn on delicate parchment sheets. Once he had selected out of these one layout and plan for the work, he asked how much it would cost to complete the whole project. After the architect (*architectus*) had replied "About 300,000 *sestertii*", one of Fronto's friends added, "And another 50,000, more or less".

THE GREEK TEMPLE

Vitruvius provides much theoretical description of how temples are constructed: the proportions and relationships of the various elements in the structure. Some of his information can be related to technology since it was employed to resolve certain problems that hindered architectural development. A few examples are provided in the following selections, and more can be found in Book 3 of Vitruvius. See Aulus Gellius (**8.39**) for plans of a more complex structure (Cooper 2008).

8.40 The modular design of the Doric temple

Vitruvius has just explained some aesthetic problems of the Doric order, which he concludes are unjustified if his system is used. He now describes how tetrastyle and hexastyle Doric temples would be built. This explanation demonstrates that ancient architects and engineers did not require elaborate plans, drawings, and models to construct buildings with traditional and simple forms (Coulton 1977).

Vitruvius, *On Architecture* 4.3.3–6

Let the front of a Doric temple, in the place where the columns are set, be divided into 27 parts if it will be tetrastyle, 42 parts if hexastyle. One part out of these will be the module, which is called *embater* in Greek; and once this module is fixed, all parts of the work are produced by calculations [based upon it]. The thickness of the columns will be two modules, the height with the capital 14. The height of the capital will be one module, its width two and one-sixth modules. Let the height of the capital be divided into three parts, of which one will make the abacus with the cymatium; the second the echinus with annulets; the third the necking.... The height of the architrave, with the taenia and guttae, is to be one module; the taenia a seventh of a module; the length of the guttae, under the taenia and corresponding to the width of the triglyphs, should hang down one-sixth of a module including their regula. The breadth of the architrave at the soffit should likewise correspond to the necking at the top of the column. Above the architrave the triglyphs with their metopes should be placed: the triglyphs 1.5 modules high, and 1 module wide in front. They should be arranged so that they are placed over the centres on the corner columns and those in the middle, and 3 in the middle intercolumniations on the front and back with 2 in the remaining intercolumniations. Thus, with the middle spaces extended, there will be an entrance without impediment for those approaching the statues of the gods. The width of the triglyph is to be divided into 6 parts, of which 5 parts in the middle and 2 half parts on the right and left edges are to be marked by the ruler. Let one section in the middle form a *femur* [the flat space between the grooves], which is called *meros* in Greek. Channels are to be cut alongside it, to fit the point of a square, and in succession other femurs are to be made on the right and left of the channels. Half channels are to be placed on the far edges. Once the triglyphs have been arranged thus, the metopes, which are between the triglyphs, are to be as high as they are wide. In addition, on the far corners let there be half metopes inserted, which are half a module wide.... The capitals of the triglyphs should be made in a sixth part of a module. Above the capitals of the triglyphs the cornice should be placed with a projection of two-thirds of a module with a Doric cymatium at the bottom and another at the top. Moreover, the cornice with its cymatia will be half a module in height. On the bottom part of the cornice, perpendicular to the triglyphs and the middle of the metopes, the straight lines of the *viae* and the rows of the *guttae* should be divided so that there are six *guttae* in the length and three in the breadth. The remaining spaces, because the metopes are wider than the triglyphs, are to be left plain or thunderbolts are to be carved; and at the very edge of the cornice, a line, which is called the *scotia*, should be incised. All the rest, like the tympana, the simas, and the cornices, are to be finished as described above for the Ionic buildings.

8.41 Origin of the Corinthian capital in nature

In the first chapter, the inspiration of Nature was believed to be one of the most important stimuli for the development of human civilisation (**1.12–14**). In this passage Vitruvius demonstrates that humankind continued to imitate nature in order to solve problems. He has just finished attributing the proportions of the Doric order to the proportions of a man's body, the Ionic order to the characteristics of a woman.

Vitruvius, *On Architecture* 4.1.9–10

The first discovery of the Corinthian capital is said to have happened in the following manner. A young girl, a Corinthian citizen, just of marriageable age was attacked by a disease and died. After her burial, her nurse collected and put into a basket some drinking cups, which the girl had taken pleasure in while alive. She then carried them to the grave monument and placed them on top of it, covering the basket with a roof tile so that it might last longer even though exposed to the weather. By chance the basket was placed above the root of an acanthus plant. Although the root of the acanthus was pressed down by the weight on its middle, in the springtime it put forth leaves and stalks. The stalks grew up along the sides of the basket and since they were pressed down by the corners of the tile as a result of the weight's force, they were forced to form the curves of volutes at the outer edges. Callimachus, who on account of the elegance and delicacy of his marble art had been named *katatechnos* by the Athenians, was passing by the monument and noticed the basket and the young leaves growing round it. Delighted with the style and the novelty of the form, he made columns after that example for the Corinthians and established their proportions. Then he assigned the rules for the Corinthian order in finished works.

8.42 Optical refinements

The size, style, and setting of a temple created small but detectable visual problems for the ancient architect. Vitruvius describes some of the problems and their solutions that the Greeks had devised centuries earlier: the refinements mentioned here had already been incorporated into the Parthenon in Athens (447–432 BC). Vitruvius has just finished describing the different proportions of columns in terms of a diameter to height ratio. In the following sections he advocates fine-tuning the proportions of the temple by adjusting column diameters, contracting the space between columns, employing *entasis*, and giving the stylobate an upward curve (see Hasselberger 1999 and Goodyear 1912).

Vitruvius, *On Architecture* 3.3.11–13, 3.4.5

The thickness of the shafts must be increased in proportion to the increase of the space between the columns. In the araeostyle temples, for example, if a ninth or tenth part[6] is given to the thickness, the column will appear thin and scanty because through the width of the intercolumniations the air will consume and diminish, at least in appearance, the thickness of the shafts. On the other hand, in pyconostyle temples if the thickness is an eighth part, because of the crowding and narrowness of the intercolumniations, it will create a swollen and unpleasing appearance. Therefore, it is necessary to follow the symmetry appropriate for each style of work. The corner columns also ought to be made thicker than the other columns by a fiftieth part of their diameter, because they are cut into by the air and seem more slender to the viewers. Therefore, what escapes the eye must be compensated by calculated adjustments. [Specific examples are given according to different heights.] ... Because of the variation in height, which the eye has to climb, these adjustments are added to the thicknesses of columns. For the eye pursues pleasant visions, and unless we satisfy its desire by the adjustment and strengthening of the modules so that what is lost is compensated for by adjustment, an unpolished and dissatisfying appearance will be presented to the viewers. With regards to the enlargement made at the middle of columns, which is called *entasis* among the Greeks, an illustrated calculation of it will be added at the end of the book to show how a delicate and appropriate form can be executed.... The stylobate must be levelled in such a way that it increases towards the middle by the *scamilli impares* [unequal edges of the stylobate risers]; for if it is laid level, it will seem to the eye to be concave.

8.43 Foundations

Vitruvius, *On Architecture* 3.4.1–2

Let the foundations of those works be excavated from the solid earth and down to solid ground, if it can be found,[7] as much as seems appropriate from the size of the work, and let the whole foundation be built in a structure as solid as possible. Above ground, let the walls be erected under columns that are 1.5 times thicker than the columns will be, so that the lower sections are more stable than the upper sections; they are called *stereobates* for they take the load. The projections of the column bases are not to extend beyond its foundation.... But if solid ground is not found and the site is marshy or loose earth right to the bottom, then it must be dug out and cleared, and constructed with piles of alder, olive, or charred oak that have been driven closely together (like those piles used for bridges) by machines.

Diogenes Laertius, *Aristippus* 2.103

A Samian, Theodorus the son of Rhoikos, recommended placing a layer of charcoal beneath the foundations of the temple in Ephesus. For since the site was very wet, he said that the charcoal, which is free of wood-fibre, would create a solid layer that water could not penetrate.

See the next passage for Pliny's version of this story. Herodotus (*Histories* 5.16) describes a village built on piles in the middle of a lake.

8.44 Problems concerning the Temple of Diana at Ephesus

Size, beauty, and wealth were the hallmarks of this ancient wonder of the world, but Pliny, who elaborates on the information provided by Diogenes Laertius above, regards technological solutions to problems as one of the reasons for the structure's fame. Vitruvius (**11.34**) records the clever manner in which Chersiphron brought the drums and blocks from the quarry to the temple site (cf. **2.33–2.35** and **2.38** for the use of simple machines to move the heavy weights involved in monumental construction). The present passage also gives an indication of the pressures that could be exerted upon an architect.

Pliny, *Natural History* 36.95–97

A real wonder of Greek grandeur still stands, the Temple of Diana at Ephesus, which all of Asia took 120 years to build. They constructed it on marshy ground so that it would not be affected by earthquakes or threatened by subsidence. On the other hand, to avoid setting the foundations of so great a mass in shifting and unstable ground, they were underlaid with packed charcoal and then with fleeces of wool. The length for the entire temple is 425 feet and its width 225 feet. There are 127 columns each 60 feet high and given by different kings. Scopas carved one of the 36 columns that were decorated with relief. Chersiphron was the architect in charge of the work. That he was able to raise the epistyle blocks of such a massive structure is a remarkable accomplishment. He achieved that with wicker baskets filled with sand that he heaped up into a gently graded ramp stretching to the top of the column capitals. Then by gradually emptying the bottom baskets, the blocks gently settled into place. He encountered the most difficulty with the lintel, which he was placing over the doors; for this was the largest block and it did not settle in its bed. The artist in his anguish considered committing suicide as his last solution. They say that, in this state of mind and worn out, the goddess for whom the temple was being constructed appeared before him in his sleep that very night. She was urging him to continue living since she herself had placed the stone. In the light of the following day this was the case; the block appeared to have corrected itself by means of its own weight.

8.45 Building the second temple in Jerusalem to withstand earthquake

Conflict between rivals had delayed the work on the House of God until Darius, the present Persian king, finds the original decree of Cyrus, which Darius quotes to re-initiate the construction. The three layers of stone and the layer of timbers have been may have helped to absorb and redirect seismic waves; an early example of the base isolation concept (Stathis 1995 and cf. **8.44**).

Ezra 6:3–4

In the first year of King Cyrus, Cyrus the king issued a decree: "Concerning the House of God at Jerusalem: let the temple, the place where sacrifices are offered, be rebuilt and let its foundations be retained, its height being 60 cubits [about 27 m] and its width 60 cubits; with three layers of huge stones and one layer of timbers".

8.46 Immense columns easily turned

Most scholars agree that the following passage describes not a structure on Lemnos, but the Temple of Hera on Samos, since neither is there a suitable structure on Lemnos nor were the architects Rhoecus and Theodorus natives of that island. On the other hand, the forest of columns of the Samnian temple provides an appropriate alternative. The columns may have been suspended horizontally from pivots at either end, to allow shaping while turning about their long axes, as on a lathe.

Pliny, *Natural History* 36.90

The Lemnian labyrinth, which was similar to the Cretan one, was more remarkable only because of its 150 columns, the drums of which were so well-balanced when suspended in the workshop that they could be turned by a child on their axes. The architects who constructed it were all natives: Smilis, Rhoecus, and Theodorus.

8.47 Raising an obelisk in Rome

Roman emperors brought almost two dozen obelisks from Egypt to Rome and other places during the first four centuries of the Empire. Being monoliths of exceptional size was obviously one of their attractions. In the passage Ammianus describes the transportation and erection in the Circus Maximus in AD 357 of Tuthmosis III's obelisk, the largest (32.50 m; 455 tons) of all the ones taken by the Romans. In the late fourth century AC Theodosius removed the twin of this obelisk to Constantinople's hippodrome, where it remains [fig. 8.6].

Ammianus Marcellinus, *History* 14.4.13–15

But Constantine ... having ripped away the huge mass [the obelisk] from its foundations ... patiently let it lie for a long time while appropriate equipments

Figure 8.6 Raising the obelisk of Theodosius.
Source: Relief *in situ*, Istanbul, Turkey.

were being prepared for its transportation. Once conveyed down the channel of the Nile and unloaded at Alexandria, a ship of extraordinary size, until now, was manufactured, which was to be propelled by 300 oarsmen ... finally having been loaded onto the ship and brought over the sea and up the flow of the Tiber, ... it was brought to the village of Alexander at the third milestone outside the city [Rome]. From there, having been put on the *chamulci* [machines: perhaps cradles or slings] and gently drawn through the Ostian Gate and by the Public Pool, it was carried into the Circus Maximus. After this only the raising remained, which was hoped only barely if not impossible to accomplish. But it was done thus: to tall beams, which were heaped high and spaced-out vertically – so that you might perceive a grove of cranes – enormous, long ropes were fastened in the form of a manifold weave concealing the sky with its excessive density. To these ropes was attached that mountain itself, fashioned with its primordial written characters, and gradually extended up through the lofty void, hanging for some time, and with many thousands of men turning *metas* [capstans here] just like millstones, it was finally setup in the middle of the hollow [of the circus].

8.48 A secret door to treasures

Pausanias, *Description of Greece* 9.37.5–7

They say that these two boys, once they had grown up, were quite skilful in constructing sanctuaries for the gods and palaces for men. For they built both a temple to Apollo and a treasury for Hyrieus. They made one of the stones in the latter structure in such a way that they could remove it from the outside. They continually kept taking things from the items stored there.

The two brothers met with an untimely end. One, caught in a snare inside the treasury, was decapitated by his sibling to spare the trapped brother from torture and himself from exposure. Subsequently the surviving brother was swallowed up by the earth.

THE TECHNOLOGIES INVOLVED IN MAJOR BUILDING PROJECTS

8.49 The use of diverse technologies

Plutarch provides an abbreviated account of the many technologies employed in a building project like the Periclean building programme on the Acropolis in Athens.

Plutarch, *Life of Pericles* 12.5–7

Pericles proposed to the people great building projects and elaborate artistic designs for works that would take so much time that the people at home would have a pretext to gain aid from and to share in the public funds not less than the sailors and border guards and soldiers. The materials used were stone, bronze, ivory, gold, ebony, and cypress wood; the craftsmen who elaborated and worked upon these materials were builders, moulders, bronze-workers, stonecutters, dyers, gold and ivory workers, painters, embroiderers, and engravers; in addition there were suppliers and transporters of the materials, like the merchants, sailors, and navigators by sea and, by land, the wagon-makers, the breeders of draught animals, drivers, rope-makers, linen-weavers, road-builders, and miners. And each craft, just as a general has his own army, kept its own private throng of workers organised like an army, as an instrument and body of service. To sum up, the advantages were apportioned to every age and type of person through the distribution of wealth in this manner.

Cato (**3.5**, **3.7**) provides a similar description and list of expenses for the construction of a farmstead.

8.50 Details of the tools used to work marble blocks

A long inscription (early second century BC) provides some of our most detailed information regarding the dressing and placement of large floor blocks (here for the Temple of Zeus Basileus in Lebadeia, Boeotia). In the latter part of the inscription several masonry tools are mentioned for different types of work that provides some sense of how the Greeks achieved such fine jointing between the stone blocks of their structures. The text uses the rare term *anathyrosei*, the technical term describing the very exact dressing of a block's outer edges that was carefully set to another block, while the recessed, more roughly worked interior of the face did not contact other block. (Bundgaard (1946) for commentary and fuller translation; Dworakowska (1979) for tool identification and use; for the comparable stages of working a column drum see **fig. 8.7**).

Building contract from Lebadeia, *IG* 7.3073

(89) .... For the temple of Zeus Basileus. For the peristyle outside the cella: the dressing and setting of the floor [of hard Lebadaean stone] for the long side ... (97) Number: 13 [blocks]. Measurements of these will be the length, breadth and thickness following those in the long sides that are finished, and against which these will be set.

Having received the stones beside the temple where they have been approved as sound and having the [proper] measurements for finishing to the stipulated size, the contractor, using a close-toothed, sharpened chisel, will first work [the under-faces of] the bases of all the stones [until] true, without distortion, unfractured, and tested with minium, [so that] (104) everything rests upon the supporting stones and the *hypeuthynteria* (the levelling course), not less than two feet from the coming joint.

Using a coarse-toothed chisel he will dress the middle contact-faces [so that] everything [will adhere] to a continuous straight-edge no shorter than the length of the stone being worked upon. The [straight-edge's] width is not to be less than six *dactyloi*, its height not less than ½ *pous*.

(110) And he is to undercut from the under-face of all the floor blocks that part resting upon the *hypeuthynteria*, the determination of width and depth from the side-facing, working the trimmed surfaces as is written above concerning the under-sides and making no larger margin than a small *dactylos* in the undercut along the *hypeuthynteria*.

(115) He will also finish all the side-faces of the floor blocks, tested with minium, true, without distortion, unfractured, vertical, at right angles [to one another] according to the [mason's] square; using a smooth, sharp broad-bladed chisel and testing with minium [he will make] solid [joins] over at least nine *dactyloi* around three sides [of the face]. (120) Using coarse-toothed chisels, the middle contact face [will be worked], and he is to finely abrade [*anathyrosei*] all the [edges of the] side-faces with a true [rubbing] stone, testing with minium and rubbing true all the [rubbing] stones as often as we may order by verifying according to the true [rubbing] stone standard gauge in the sacred enclosure.

(125) He shall also finish off the side-faces of the floor blocks, which are in position and finished, against which he will set [the 13 new blocks]. Having stretched out the cord on the upper side [... text missing ...] in a straight [line]

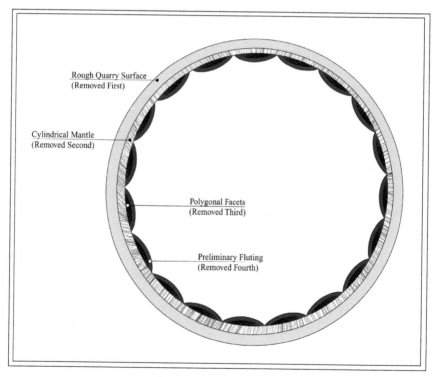

Figure 8.7 Stages in shaping a column drum.
Source: Adapted from Dinsmoor, 1975, fig. 65.

from the left and in the vestibule from the long side, (130) and having marked off the lines in the presence of the architect, he is to remove the surplus [stone] using a pitching tool, making the width the one given [by the building], making all [the stones] true, sharp-angled, and he is to test the edge of all those 13 placed floor blocks with minium against (135) a long straight-edge of not less than 20 *podes*. Thickness of the straight-edge: six *dactyloi*; height: ½ *pous*; and making the minium test with a smooth, sharp pitching tool, making the stones true, without flaw, tested with minium over a depth of not less than nine *dactyloi*, after first having incised guides along the joints on each of the stones vertically according to the [mason's] square and according to the marked-off line along which he is to finely finish the smoothing. [Further instructions concerning the working of the stone and the placement of the blocks follow.]

8.51 Large constructions require large machines: hoists and cranes

Vitruvius provides some of our best descriptions of the hoists and cranes/used by the Romans and Greeks to move and to lift the massive blocks used in monumental construction. Only one description is provided below, but Vitruvius continues to describe changes to the machinery for more massive constructions [**figs. 8.8–9**]. Passage **11.81** also indicates their utility for moving the cargoes of ships. See also Hero, *Mechanica* 3.2–5 and Drachmann (1963: 97–102) (Coulton 1974).

Vitruvius, *On Architecture* 10.2.10

First, we will commence with those [machines] that must be made for the completion of temples and public buildings. They are made thusly. Two timbers are procured according to the weight of the loads. They, fastened together at the top by a *fibula* [fastener, bolt, brace] and stretched apart at the bottom, are set up with their elevation maintained by ropes arranged all round and attached to their tops. At the topmost point is fastened a *troclea* [pulley block], which some call a *rechamus* [block]. Two small discs [sheaves] are enclosed in the block, which turn on small axles. A drawing rope is carried over the sheave at the top, then let down and passed round a sheave of a block below. But then it is brought back to the bottom sheave of the upper block, and so it goes down to the lower block and is bound fast in a hole of that block. The other end of the rope is brought back between the lowest sections of the machine.

Socket-pieces are affixed to the back faces of the squared timbers at the place where they are spread apart, into which the ends of the windlass are inserted so that the axles may turn easily. Near the ends of the windlass are two holes, arranged in such a way that handspikes can be fitted into them. But to the bottom block are fastened iron pincers, whose prongs are fitted into the stones, which have been bored. When the end of the rope is fastened to the windlass, which is moved by turning the handspikes, the rope is drawn out by winding itself around the windlass, and thus raises the load to the [proper] height and settings for the construction.

CONSTRUCTION ENGINEERING

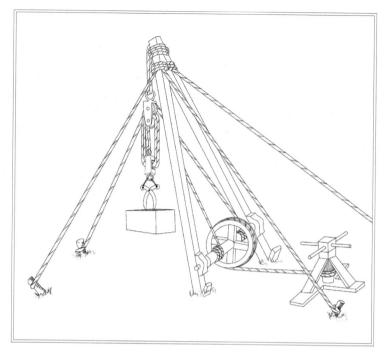

Figure 8.8 A-frame hoist with capstan.

Source: Adapted from Lancaster, 1999, fig. 11 and Adam, 1994, fig. 89.

Figure 8.9 Tomb of the Haterii.

Source: Lateran Collection, Museo Gregoriano Profano, Vatican Museums.

This is the explanation of the mechanism, which revolves about with three sheaves, that is called a *trispast* [a three-pulleyed hoist]. Indeed, when two sheaves in the lower block and three in the upper are turning, that is called a *pentaspast* [a five-pulleyed hoist]. But if ever machines have to be made for greater loads, it will be necessary to resort to timber both longer and thicker and, for that reason, with larger fasteners at the top, and to arrange properly for the turnings of the windlass at the bottom. Once these things have been arranged, let forestays be placed in front, lying slack; let the backstays [carried] over the shoulder-pieces of the machine be arranged at a distance, and, if there is nothing where they can be fastened, let sloped-back piles be dug down [driven] deeply, and let them be compacted by ramming down all round, and let the ropes be made fast to them. [Descriptions of larger machines follow].

INSCRIPTIONS DOCUMENTING MONUMENTAL BUILDINGS

A substantial number of inscriptions have survived, which provide details regarding expenses and step-by-step instructions for monuments that were either about to be constructed or were already under construction. The information obtained from them reveals not only the control of projects by administrative boards but also details about construction technology and the variety of trades that were brought together in a project.

8.52 Building accounts of the Parthenon

The inscriptions that describe the financial records and payments to workers can be quite detailed. Several inscriptions survive concerning the grand structures erected on the Acropolis in Athens. The Parthenon, the most famous structure of them all, was begun in 447/446 BC and completed by 433/432 BC. Accounts from each year were inscribed and posted. The best-preserved records are from 434/433 BC, near the end of the construction period (Burford 1963, 1969, and Randall 1953).

IG 1³ 449.1–41

[Accounts of] the commissioners for whom Anticles was secretary; in the fourteenth council [since the beginning of construction], for which Metagenes was the first to function as secretary; in the archonship of Crates at Athens.

Receipts of this year	1,470 *drachmai*
Surplus from the previous year	74 *stateres*
Electrum *stateres* of Lampsacus	27 1/6 *stateres*
Electrum *stateres* of Cyzicus	25,000 *drachmai*
From the treasurers who manage the assets of the Goddess ...	1,372 *drachmai*
Price of the gold sold: a weight of 98 *drachmai*	1,309 *drachmai*
Price of the ivory sold: a weight of 3 talents	60 *drachmai*

Expenditures:	202+ *drachmai*
Purchases	1 *obolos*
Wages	1,926 *drachmai*
To the workers [quarrying] stone at Pentele	2 *oboloi*
and loading it onto wagons	16,392 *drachmai*
To the sculptors of the pedimental sculpture	a stipend
To those hired by the month	2 *oboloi*
Surplus from this year:	
Electrum *stateres* of Lampsacus	74 *stateres*
Electrum *stateres* of Cyzicus	27 1/6 *stateres*

8.53 Expenses for the Erechtheum (409/408 BC)

The following excerpt lists the funds distributed for two stages of the work: first the setting into place of the frieze blocks, work that involves hoisting, setting, clamping, and the final working to obtain a tight join; second, the dressing of the top surface so that it is ready for the next course of stone.

IG 1³ 475.1–20

[For work on the south wall:] for setting in place blocks 8 feet long, 2 feet high, at five *drachmai* per 4-foot length, to Simon living in Agryle, for 3 blocks: 30 *drachmai*. For setting in place blocks 6 feet long, 2 feet high, 1 foot thick, to Simon in Agryle, for 5 blocks: 37 *drachmai* 3 *oboloi*. For setting in place a block 2 feet high, 1 foot thick, 2 feet long, to Simon in Agryle, for 1 block: 2 *drachmai* 3 *oboloi*. For setting in place backing-stones for these, two blocks of Pentelic marble, from the stoa, 3.75 feet long, 2 feet high, 0.75 foot thick, at 2 *drachmai* 4 *oboloi* per block, to Simon living in Agryle, for 2 blocks: 5 *drachmai* 2 *oboloi*. For setting in place other backing-stones between the wooden beams, of Aeginetan stone, from the stoa, 4 feet long, 2 feet wide, 1.5 feet thick, at 2 *drachmai* 5 *oboloi* per block, to Simon living in Agryle, for 8 blocks: 22 *drachmai* 4 *oboloi*. For dressing the top surface of these blocks, for 14 four-foot lengths, at 3 *drachmai* 3 *oboloi* per four-foot length, to Phalakros living in Ko [...], 49 *drachmai*.

8.54 Securing blocks

A better description of the process of joining blocks by cutting holes for metal dowels and clamps, which were then secured by pouring molten lead, is provided by an inscription from Eleusis. Other references to bonding and clamping are in **8.22** and **8.59**.

IG 2/3² 1666.38ff. (fragmentary)

... The metopes are to be made ... and lifted up and fitted into place ... and bound [with clamps] and secured with dowels and then lead is to be poured all around and the top is to be dressed [to receive the next course]...

CONSTRUCTION ENGINEERING

8.55 A part of the workforce

Elsewhere in the inscription of the Erechtheum accounts we have a record of the varied workforce (carpenters, roofers, painters, sculptors, gilders, etc.). The following passage provides one example: the work of carpenters (cf. **7.10**).

IG 1³ 475.54–71

Wages: to the sawyers working by the day, to two men, at 1 *drachma* per day each, for 12 days, to Rhadios living in Kollytos and his helpers: 24 *drachmai*. For sawing 14 eight-foot lengths of wooden beams, for 84 cuts at 2 *oboloi* per cut, to Rhadios living in Kollytos: 28 *drachmai*. To a sawyer for sawing a wooden beam 24 feet long, for 5 cuts at 1 *drachma* per cut, to Rhadios living in Kollytos: 5 *drachmai*. For sawing a wooden beam for the conduit, to Rhadios living in Kollytos: 1 *drachma* 2 *oboloi*. To carpenters working by the day, for making the conduit and placing it under, for nine days, to Kroisos, the son of Philokles: 9 *drachmai*. For making and hewing smooth straight-edges and working by the day, for 10 days, to Gerys: 10 *drachmai*. For adjusting a straight-edge and working by the day, for 2 days, to Mikion: 2 *drachmai*.

8.56 Simple technology: a generally overlooked labourer is honoured

The everyday work done by animals of burden is usually taken for granted both in antiquity and today, but some owners and workers certainly had favoured animals that became known for their particular characteristics. Most famously, although more than a mere beast of burden, Alexander the Great's horse Bucephalus was celebrated on a heroic level by Alexander, who named a city after him. In the next passage Cato has been castigating people who mistreat or discard their slaves and animals, then turns to a series of good men who treat their "workers" well when they become old.

Plutarch, *Cato* 5.3

While building the Hekatompedon [Parthenon], those mules deemed most steadfast in their work, the Athenians released free and able to ranging at large in order to pasture. One of the mules, they say, having gone back down to the works of its own accord, trotted alongside the yoked teams hauling the wagons onto the Acropolis, sometimes even leading the way, as if urging and encouraging them on. As a result, the Athenians voted the mule was to be maintained at the public expense as long as it lived.

8.57 The Arsenal (*Skeuotheke*) of Philon (347/346 BC)

Philon was active in the second half of the fourth century BC and during that time was responsible for two great projects: the construction of the arsenal in the Piraeus and the addition of a magnificent porch to the Telesterion at Eleusis. Inscriptions survive that provide detailed instructions for the construction of both structures. Philon himself also wrote a treatise on temple proportions and a commentary on his arsenal, but both have been lost. The following passage, although lacking the theoretical analysis of his commentary, does offer us a complete set of instructions for building large ancient structure (Jeppesen 1958).

IG 2² 1668

The gods. The specifications by Euthydomus, son of Demetrius of Melite, and Philon, son of Exekestides of Eleusis for the stone arsenal for the storage of sailing equipment. To build the arsenal for storing the sailing equipment in Zea, starting from the propylaion out from the agora and running behind the ship sheds, which have a common roof, to be 4 *plethra* in length, 55 feet in width inclusive of the walls.

Cutting a trench into the ground that is 3 feet at its deepest, and having completely cleared the rest of the area, the builder shall pave over the stone foundation and erect the structure level and upright according to the stonemason's rule. He shall also lay the pavement for the piers [or columns?], leaving 15 feet space from each wall, inclusive of the thickness of the pier. The number of piers in each row will be 35, leaving a passage for the people through the middle of the arsenal. The distance between the piers will be 20 feet. He shall make the foundation 4 feet thick, setting the stone blocks, like a mat, [alternately] crosswise and by length.

He shall build the walls of the arsenal and the piers from stone from the Piraeus, having established a *euthynteria* [a levelling course] for the walls with blocks 3 feet wide, 1.5 feet thick, and 4 feet long; but at the corners 4 feet and 3 palms. And on the *euthynteria* he shall set orthostates around the middle of the *euthynteria*, 4 feet long, 2.5 feet 1 *dactyl* thick, and 3 feet high; but those at the corners will take their length from the measure of the triglyphs, leaving doors in the short walls of the arsenal, two doors on each, 9 feet wide. And he shall construct a *metopon* [middle wall?] on each end between the doors, 2 feet wide and extending 10 feet into the interior. The builder shall also make returns in the wall up to the first piers, against which each door will open. On top of the orthostates he shall build the walls with squared blocks, 4 feet long, 2.5 feet wide, and 1.5 feet thick; but those at the corners will take their length from the measure of the triglyphs. He shall make the height of the wall from the *euthynteria* 27 feet including the triglyph under the *geison*, the doors 15.5 feet high.

He shall set up lintels of Pentelic marble, 12 feet long, with a width equal to the walls and two courses high, once he has set doorposts of Pentelic or Hymettian marble and thresholds of Hymettian.

He shall also set in place a *geison* over the lintels, 1.5 feet high. He shall make windows round about in all the walls opposite each intercolumniation and with three in the short walls at each end, 3 feet high and 2 feet wide. And he shall insert fitted bronze gratings into each window. He shall set *geisa* around all the walls and he shall make pediments and place *geisa* on the pediments.

He shall set up columns [or piers?] on the stylobate, which is horizontally level to the *euthynteria* and composed of blocks 1.5 feet thick, 3 feet and 1 palm wide, and 4 feet long. The thickness of the lower part of the columns will be 2 feet 3 palms, the length including the capital will be 30 feet, each with seven drums 4 feet long, except the first of 5 feet. He shall set capitals on the columns of Pentelic marble and on the columns he shall set and fasten wooden epistyle blocks, 2.5 feet wide and at most 9 palms high, 18 in number on each row of columns. And he shall place tie-beams on the piers over the passageway, with a width and height equal to the epistyle blocks.

He shall put in place rafters, 7 palms wide and 5 palms 2 *dactyloi* high without the sloping surface, having set bases underneath for the tie-beams, 3 feet long, 1.5 feet wide. And he shall fit the rafters to the tie-beams with dowels. He shall place roof timbers, 10 *dactyloi* thick, 3 palms 3 *dactyloi* wide, separated from each other by 5 palms. He shall place planks on top, 0.5 feet wide and 2 *dactyloi* thick, separated from each other by 4 *dactyloi*, and placing sheathing on them, 1 *dactyloi* thick and 6 *dactyloi* wide, fastening them with iron nails and plastering them, he shall roof the structure with Corinthian tiles fitted to each other. And he shall place over the doors on the dividing walls on the inside a stone roof of Hymettian marble.

And he shall set doors fitted to the doorways on the arsenal, making them of bronze on the outside. He shall pave the floor with stones on the inside that all fit each other and make it true and level on top. He shall partition each intercolumniation with two stone orthostates, 3 feet high, and in the space between set up barred latticed gates. And he shall make roofs in the middle, upon which the tackle will rest, on the inside of the piers on each side up to the wall, arranged on every single pier and extending to the wall on both sides with support beams, 5 palms wide, 1 foot high, overlapping on the wall 3 palms, and beside the piers he shall set stone doorposts. And on the support beams he shall set roof timbers, seven on each room, filling the space up to the piers, 3 palms wide and 0.5-foot thick, and he shall floor every room with flat planks, putting them together and tightly joining them, 3 feet wide and 2 *dactyloi* thick.

And he shall make shelves, upon which the [trireme] braces and equipment will rest, along each wall, two in height and returning along the side walls, and he shall make a return at the columns in each room. He shall make the height from the roof 4 feet, the upper shelf separated from the other by 5 feet. He shall set uprights from the lower ceiling to the upper ceiling, 0.5-foot wide, 6 *dactyloi* thick, support beams of the same thickness holding apart the uprights. He shall place down a continuous floor beam, one on each side, 6 *dactyloi* on all sides [square], and on these he shall put tightly glued boards (*pinakes*), 4 feet long, 3 feet wide, 2 *dactyloi* thick, and he shall nail down the closely fitted boards evenly onto the floor beams. He shall also make wooden ladders to go up to the shelves and he shall make chests for the sails and for the white tarpaulins, 134 in number, made according to the model. And he shall set one in front of every column, and one in the room directly opposite. He shall make those resting against the wall of the room open from the front of the enclosure, those against the columns to open on both ends of the enclosure, so that if someone goes through he can see all the equipment there is in the arsenal. And so that there may be a coolness in the arsenal when he builds the walls of the arsenal, he shall leave spaces in the joints of the wall blocks, as the architect may bid. Those hired shall complete all these things according to the specifications and the measurements and the model, as the architect specifies, and they shall deliver over each of the tasks at the times arranged by contract.

8.58 Instructions for building a lime kiln on the farm

Cato's simple, practical directions for building a lime kiln probably indicates that the structure was not uncommon; its handy location, like compost pits, indicates fairly vigorous use. The Romans burnt lime principally to create quicklime, a key ingredient to make plaster, mortar, and cement (**8.13**) for a variety of uses from pot repairs, to wall decoration to constructing massive structures. Whether the lime was used as a mineral fertiliser is not clear from the texts, which emphasise quicklime's use in construction (Dix 1982).

Cato, *On Agriculture* 38

Make a lime kiln ten feet wide, 20 feet high, continuously reducing it to three feet wide at the top. If you cook [calcine] through one stoking-door, make a large hollow inside so that it is big enough to contain the ash and does not have to be emptied out. Design the kiln wisely and construct the foundation of the furnace pounded together for the entire lower section of the kiln. If you cook with two stoking doors, there is no need for the hollow since when the ash needs to be thrown out, one stoking-door can be used for discarding while the fire is in the other. Be careful not to let the fire lapse, but it is always burning. So, take heed that neither at night nor at any other time the burning suffers interruption. Load good stones, as white and as alike in colour as possible, into the kiln.

When you make the kiln, make the upper constriction come downward steeply. Once you have excavated sufficiently, create a place for the kiln so that it is as deep and as little exposed to the wind as possible. If you have only a kiln at minimal depth where you build, construct the top from brick or from concrete, and plaster over the top with clay on the outside. When you set in the fire, if flames emerge anywhere except the round opening at the top, plaster over with mud. Be on guard that the wind does not gain access at the stoking-door and be especially careful of the south wind there.

The sign that the lime is cooked is when the upper stones have been cooked; likewise, the lowest stones, once cooked, will collapse, and the fire will send out less smoke.

If you are unable to sell firewood and sticks and have no stone from which to cook lime, cook up charcoal from your wood.

8.59 A building contract at Puteoli

Few Roman building contracts survive; the best example is an inscription from Puteoli dated to the end of the second century BC. The local magistrates drafted the very detailed document for the construction of a wall in front of the Temple of Serapis. Potential contractors are required to supply sureties (both people and property), respect guidelines, dimensions and quality standards, and finish the work to the satisfaction of the magistrates as well as a council. Since payment is to be made in two instalments, half once sureties have been given during the contract and half upon completion, this seems a fixed-price contract. It is written in three columns (Anderson 1997: 72–75).

CIL 1.698 = *ILS* 5317

In the ninetieth year since the founding of the colony, when the duovirs were N. Fufidius, son of N., and M. Pullius, in the consulship of P. Rutilius and Cn. Mallius [105 BC].

(Column 1)

The Second Contract of Public Works. The contract for the construction of a wall in the building lot which is in front of the Temple of Serapis across the road. He who undertakes the work must provide bondsmen and register their estates as securities according to the decision of the duovirs.

In the building lot across the road, let the contractor open a gap [for a gateway] in the middle of the wall, which is near the street; let him make it 6 feet wide and 7 feet high. From that wall he shall project towards the sea two *antae* that are 2 feet long and 1 foot and 0.25 thick. Above the gap let him place a lintel of hard oak, 8 feet long, 1 foot and 0.25 wide, and 0.75 foot high. Above that and the antae let him project out from the wall on either side for 4 feet, hard oak topping-beams (*mutuli*), 2.67 feet thick and 1 foot high. Above let him attach the decorated sima with iron. Above the topping-beams let him place two fir cross-beams, 0.5-foot thick each way.

(Column 2)

And let him fasten them with iron. Let him cover them over with rafters cut from fir, 0.33-foot thick each way; let him place them not more than 0.75 foot apart. And let him put in place fir panel-work that he shall make from timber 1 foot wide. And let him place the fir wood facing, 0.75-foot wide and 0.5-inch thick, and a *cymatium*, and let him fasten them with flat iron [clamps]. And he shall roof the doorway with six rows of roof tiles on each side. Let him fasten all the lower rows of roof tiles to the facing with iron and let him put on a coping. Let the same person make two latticed doors with winter oak doorposts, and he shall erect, close, and pitch them in the manner that was done for the Temple of Honour. As for the wall that is in the furthest part of the sanctuary, the same person shall make that wall 10 feet high with the coping. The same person shall wall up the doorway in the wall, which is now the entrance into the building lot, and the windows that are in the wall near the lot. And let him add a continuous coping on the wall that is now near the street; and all the walls and copings that have not been coated, let him make them properly coated with [a mortar of] lime and sand, smoothed, and whitewashed with moist lime. The substance that he will use in the structure, let him make from clay mixed with one-quarter part of slaked lime. And let him lay rough tiles that are not larger than the dry rough tiles weighing 15 pounds, nor make the corner tiles higher than 4.5 inches.

(Column 3)

And let him give back a clean site according to the requirements of the work. And likewise, in regards to the shrines, altars, and statues that are in the building site, let him lift, move, arrange, and set up all the ones singled out, to a place that will be designated by the decision of the duovirs. Let him carry out this whole project by the decision of the duovirs and the former duovirs who usually sit in council at Puteoli, so long as not fewer than 20 are present when the matter is deliberated. Whatever 20 members on oath approve, let it be approved; what the same men do not approve, let it not be approved. The day [for the beginning] of work: the first of next November. The day for payment: a half part will be given when the estates [for securities] have been satisfactorily registered; the other half part will be paid when the work is completed and approved. Gaius Blossius, son of Quintus, [undertakes the contract for] 1,500 *sestertii*, and pledges [himself as surety]....

DOMUS AND *INSULAE*: DOMESTIC CONSTRUCTION IN ROME

Technology helped to alleviate the housing crisis in Rome as the population increased in the last centuries of the Republic and in the Imperial period. Land was expensive and scarce, but by building less expensive apartment buildings, often with shops on the ground floor and living space higher up, additional residences were made available for the masses. Some of these structures may have been over six storeys high; vertical construction of this magnitude could not be safe or very economical without new technological innovations [**fig. 8.10**].

8.60 Building an atrium house: technological advantages and disadvantages of different styles

Vitruvius is most interested in describing the plans and proportions of houses (*domus*), but he also provides some insight into the reasons that particular styles were used and the technological limits imposed on the construction.

Vitruvius, *On Architecture* 6.3.1–2

The inner courts of houses are of five different styles, named according to form: Tuscan, Corinthian, Tetrastyle, Displuviate, and Vaulted. The Tuscan ones are those in which the beams across the breadth of the atrium have cross-beams on them, and valleys running down from the angles of the walls to the angles of the beams; thus, the runoff of the rainwater is directed by the beams into the roof-opening [*compluvium*] in the middle. In the Corinthian style the beams and roof-openings are arranged in the same manner, but the beams are run in from the walls and are supported on columns all around. The Tetrastyle courts are those that, by supporting the beams at the angles with columns, benefit in usefulness and stability; because neither are they forced to support a great span nor are they loaded by the cross-beams. But Displuviate courts are those in which the gutter beams, supporting the opening, drain the rainwater outwards. They are most suitable in winter residences since the roof-openings, being raised up, are not a hindrance to the lighting of the dining rooms. But there are great troubles maintaining them because the pipes intended to hold in check the rainwater flowing down around the walls do not take in the water flowing out of the channels quickly enough, and consequently backing up they overflow and ruin the woodwork and walls of this style of house. Vaulted courts are used when the span is not great, and they provide spacious residences in the storeys above.

8.61 Grand housing for the masses

Vitruvius describes the positive changes in living arrangements made possible by the construction of apartment buildings, referred to as *insulae*, "islands" built of concrete and baked-brick [**fig. 8.11**].

Vitruvius, *On Architecture* 2.8.16–17

Since even very powerful kings have not disdained walls constructed of sun-baked bricks, even though their incomes and plunder would have frequently permitted walls made not only from quarried and squared stones, but even from marble, for this reason I do not think that buildings made with brick walls should be rejected, as long as they are properly roofed. But I will explain why that type of construction should not be employed by the Roman people in the city [of Rome], and I will not omit the theories and reasons for it. The public laws do not allow a thickness of more than 1.5 feet for walls constructed in a public place. Moreover, the other walls are built of the same thickness so that the space does not become too cramped. But sun-dried brick that is 1.5 feet wide, unless two or three bricks thick, cannot support more than one storey. Yet because of the majesty of this city and the immense crowding of its citizens it is necessary to furnish countless dwellings. Consequently, since a horizontal space cannot receive such a great multitude to live in the city, the circumstances themselves have forced us to find relief by constructing high buildings. And so, by means of stone pillars, burnt brick walls, and partition walls of rubble, tall structures have

Figure 8.10 Reconstruction of *Insula Dianae* at Ostia.
Source: Adapted from a model at the Museo de la Civiltà Romana, Rome, Italy.

Figure 8.11 Wall painting with bricklayers.
Source: Adapted from fresco in tomb of Trebius Justus on the Via Latina, Rome, Italy.

been erected that have been provided with wooden planked floors one after another, producing views from the upper storeys of the utmost advantage. As a result of the residences multiplied by the various storeys high in the air, the Roman people have available excellent housing.

8.62 A famous apartment block

The Insula Felicles, a huge apartment block near the Flaminian Circus, must have been an impressive structure. Not only is it identified by name in the Regionary Catalogues (Region 8), which is very rare in itself, but Tertullian also compares it to the immense, towering homes of gods.

Tertullian, *Against the Valentinians* 7

First of all, the Roman poet Ennius simply spoke about the "great dining rooms of heaven", so-named from their proud location or because he had read in Homer that Jupiter dined there. But as for the heretics, it is marvellous how great the storeys upon storeys, and how many heights upon heights, they have suspended, raised up, and spread out into dwelling places for their gods. Even the Ennian dining room has been arranged in the shape of small rooms by Our Creator, with room after room erected on top of each other, and assigned to each god by as many staircases as there were heresies. The world has become a place to rent. You might think that the Insula Felicles was such a great arrangement of storeys.

CIL 4.138 = *ILS* 6035

Insula Arriana Polliana of Gnaeus Alleius Nigidius Maius: shops with attached stalls in front, equestrian [expensive] lodgings on the upper level, and a house are for rent from the next Kalends of July [the first of July]. Let the renter contact Primus, slave of Gnaeus Alleius Nigidius Maius.

8.63 The less attractive side of high-rise living

The next two passages prove that these types of lodgings in Rome were not always regarded with the admiration voiced by Vitruvius. The insulae were neither entirely safe nor stable. Even though various attempts were made to improve their structural integrity and to fireproof them (**8.75–77**), the insulae were still susceptible to collapse and fire at the end of the first century AC. Technology solved some problems, but introduced others at the same time.

Cicero, *Letters to Atticus* 14.9

I summoned Chrysippus because two of my shops (*tabernae*) have fallen down and the rest are cracking with the result that not only the tenants, but even the mice have run away. Others say this is a disaster, but I think it's not even an inconvenience.... But, nevertheless, on the advice and suggestion of Vestorius, a plan for rebuilding has been started that will make this loss profitable.

Juvenal, *Satires* 3.190–202

Who at cool Praeneste, or at Volsinii located among the wooded hills, or in simple Gabii, or on the sloping summit of Tibur, who of these people fear or have feared in the past the collapse of their house? We, however, live in a city that, for the most part, is balanced on slender reeds; for thus does the landlord support the tottering structures and, once he covers over the gaps of old cracks, he bids [his tenants] to sleep securely in a ruin ready to fall. One ought to live in a place where there are no fires, no nightly terrors. Already Ucalegon is calling for water, already he is shifting about his pitiful belongings, and although the third storey of your building is now full of smoke, you don't know it; for if the alarm is given from the bottom floors, the last to burn will be the person who is sheltered from the rain by the roof tiles alone, up where the gentle doves lay eggs....

PREPARATIONS AND ACTIONS TO FIGHT FIRES

The private fire brigade created by Crassus (Plutarch, *Crassus* 2) for enriching himself was soon taken over by Augustus (Strabo, *Geography* 5.3.7) who established a city-brigade, the *vigiles*, to control fires in Rome. Regulations also demanded that light, relatively simple, fire-fighting equipment be available in the buildings themselves in an attempt to limit damage as quickly as possible. Ignoring the regulations or poor response time resulted in the need for heavier equipment and drastic measures to stop the fire from spreading (cf. **8.75–77**).

8.64 Equipment for fighting fires

Paul, in *Digest* 1.15.3–4

[Paul states that] it should be understood that the Prefect of the *Vigiles* must be up the entire night, making the rounds with his men, properly booted, and provided with grappling hooks and pick-axes, and they are to take care to warn each lodger ... to have a ready supply of water ready in the upper room.

Ulpian, in *Digest* 33.7.12.16–18

[In this discussion defining *instrumentum* [equipment] "belonging" to a house] ... Pegasus says "the *instrumentum* of a house is what is intended to keep out the weather and fight fires ..." Most authorities as well as Pegasus state that in addition vinegar, which is prepared for extinguishing fires, and likewise, patchwork cloths, siphons [pumps], and poles and ladders, wicker mats, sponges, fire buckets, and twig brooms are included [in the *instrumentum* of the house].

CONSTRUCTION ENGINEERING

8.65 Stopping the spread of fire in crowded cities

When fires burnt out of control, buildings were knocked or pulled down to block the fire or to create firebreaks. Catiline's threat to destroy the Republic and save himself is framed in a metaphor of such a fire recorded by Sallust and Cicero; Tacitus and Suetonius cite the practice to fight the Great Fire of AD 64, although Suetonius's negative account implies destruction for Nero's gain rather than sound fire-fighting practice.

Sallust, *Catiline* 31.9

Then that man [Catiline], in his burning rage, shouts out "Since indeed I am surrounded on all sides by enemies and driven headlong, I will extinguish my own conflagration by pulling down everything in ruin".

Cicero, *On Behalf of Murena* 51

[About to be charged, Catiline threatens Cato that] should a raging fire be roused up against his own station and properties, he would extinguish it not with water but by pulling everything down in ruins.

Tacitus, *Annales* 15.40

At last on the sixth day, an end was made to the fire at the base of the Esquiline Hill. Buildings had been torn down over a vast area so that this fire of unremitting ferocity would encounter only open ground just as if a bare sky.

Suetonius, *Nero* 38.1

[Suetonius blames the fire on Nero's greed] ... and having used war machines [catapults?] to bring down certain granaries in the area around the Golden House, whose properties he very greatly desired, they were set on fire.

HEAT AND SHADE: THE COMFORTS OF CIVILISED LIVING

The comfort of indoor living and recreation is partially the result of controlled temperature and lighting. The following passages reveal a combination of these two functions as well as two dissimilar attitudes towards the convenience: moralising disgust and open admiration of the luxury.

8.66 Hypocaust systems for heating

Although hypocaust systems are most often associated with bath buildings, they were used in a variety of structures. Seneca provides the earliest reference to a hypocaust system in a dwelling.

Seneca, *On Providence* 4.9

The man whom the glazed windows have always liberated from the breeze, whose feet are kept warm between warm applications that are frequently changed, whose dining rooms are regulated by heat underneath and circulating through the walls, this man the light breeze will not touch without danger.

8.67 Warmth in the country house

In the Laurentine villa of the Younger Pliny, planning, passive solar heating, and hypocaust systems all played a part in maintaining a comfortable living environment. The following excerpts are only samples of some of the heated rooms at the villa. Pliny (*Letters* 5.6.24–25) describes similar features in his other villa at Tifernum.

Pliny the Younger, *Letters* 2.17.4–5, 7–9, 11, 23

A portico [next to the atrium] in the form of the letter D includes a small but pleasing space. This provides an excellent retreat against storms since it is protected by glazed windows and even more by overhanging eaves.... The angle enclosed by a bedroom and dining room retains and augments the purest sunshine. This serves as a room for winter and also as the gymnasium of my household.... On the corner is another bedroom built in a curved shape, which follows the course of the sun by means of all its windows.... Then comes a series of sleeping rooms, which you enter through a passage that has an elevated floor and pipes in the walls to receive hot air and circulate it at a regulated temperature here and there to the rooms.... Attached to the bath is the oiling room, the *hypocauston* [for dry heat], and the *propnigeion* [for damp heat] for the bath, then two small rooms more elegant than they are expensive that lead to a wonderful heated pool from which swimmers can view the sea.... Attached to a different bedroom is a second small heating system [*hypocauston*], which by means of a narrow opening either releases or holds back the heat below as the situation demands in regard to the room. Beyond are an antechamber and another bedroom stretched out into the sun to catch its force immediately when it rises; and it retains the sun, although at an oblique angle, until after midday.

8.68 How a hypocaust works

The following passage of Statius (along with Section 23 of the preceding passage by Pliny) explains how the hypocausts (hot air rooms) actually work. Generally, they are in a lower floor with the furnace, but Pliny implies that the furnace is below the hypocaust at his villa [**fig. 8.12**].

Statius, *Silvae* 1.5.45–46, 57–59

Everywhere there is much light where the sun shines through the roof with all its rays, and yet is inferior and is burnt by another heat.... Why now should I speak

CONSTRUCTION ENGINEERING

Figure 8.12 Hypocaust system.

Source: Adapted from Adam, 1994, figs. 624 and 634.

about the floors laid on the ground that will eventually hear thudding balls, where the languid heat wanders about the structure and the hypocausts breathe out a delicate warmth?

8.69 Large windows in the baths

Seneca, *Letters* 90.25

We know that certain things have been developed only within our own memory, such as the use of paned windows, which permit the transmission of clear light by means of transparent tile, and such as the vaults of baths and the piping placed in the walls through which the heat is circulated to heat evenly both the lowest and highest areas of the room.

For another discussion of windows in baths, see **9.40**. Vitruvius, *On Architecture* 5.10.1–5 provides more information regarding the construction and arrangement of baths. Varro, *On Agriculture* 3.7.3 mentions "Punic windows" that allow light but block snakes and predators from the *peristeron* (a dove-cote); construction not presently understood.

8.70 Awnings to shade theatres and the forum

The difficulty of protecting large numbers of people in theatres and open spaces from the heat of the sun was resolved by the use of immense awnings.

Pliny, *Natural History* 19.23–24

Stretched linen produced shade in the theatres, a device which Q. Catulus first invented when he dedicated the Capitol [after 83 BC]. Then Lentulus Spinther is said to have first supplied fine linen awnings in the theatre at the games of Apollo. Soon after this, the dictator Caesar covered the entire Forum Romanum and the Sacred Way from his house, and the slope right up to the Capitol, which seemed a more remarkable thing than even a gladiatorial event. Then even without any games, Marcellus, the son of Augustus' sister Octavia, during his aedileship in the eleventh consulship of his uncle [23 BC], from the first of August shaded the Forum with awnings so that the litigants could meet in more healthy conditions.... And lately awnings the colour of the sky and decorated with stars have even been stretched with ropes over the amphitheatres of the Emperor Nero. They are red in the inner courts of houses and they shelter the moss from the sun; for other things, white has remained the consistent favourite.

REMARKABLE STRUCTURES

8.71 A unique amphitheatre

Pliny has just finished describing the amphitheatre of Scaurus (58 BC), a structure so lavish and expensive that no one could match it. Instead, Curio builds a unique structure (50 BC) to win his place of fame among the builders of Rome. Note that the arrangement described by Pliny cannot function as stated: the two theatres could not have rotated to form an amphitheatre without the addition of another section or device.

Pliny, *Natural History* 36.117

Curio had to use his wits to devise something [wonderful].... He constructed two very large wooden theatres next to one another, propped up on the swinging point of separate pivots. Before midday a performance of plays was given in both of them while the theatres faced in opposite directions so that the actors would not drown out each other. Then suddenly they were swung around – and after the first days, it was done even with some of their audience still sitting – and with the corners coming together against each other, Curio made an amphitheatre and put on gladiatorial contests, spinning the Roman people themselves into greater danger [than the gladiators].

Pliny is amazed that the Roman people would trust themselves to such a contraption, which soon failed to rotate.

8.72 Metal structural elements

The appearance of iron beams in the lintel of the Propylaia at Athens and in the under surface of the architraves on the Temple of Zeus Olympius at Agrigento are rare examples of attempts to use metal's tensile strength in ancient structures. The structural property of metal was never exploited to its full potential in antiquity although its use for clamps and dowels was common (**8.22**, **8.54**, **8.59**). Instead, metal was often employed in a decorative manner when lavish display was desired (Cooper 2008, Lancaster 2008).

Pliny, *Natural History* 34.13

The ancients even made lintels and doors in temples out of bronze. I have found also that Gnaeus Octavius [169 BC], who won a naval triumph over the Persian king, made a double portico near the Flaminian Circus, which is called the Corinthian portico after the bronze capitals of its columns. It was also resolved that the shrine of Vesta should be roofed with an outer shell of Syracusan bronze. And the capitals of the columns put up by Marcus Agrippa [27 BC] in the Pantheon are Syracusan bronze also. Moreover, it has even been used in a similar manner for private opulence: among the charges the quaestor Spurius Carvilius brought against Camillus was that he had the doors of his house covered with bronze.

Historia Augusta, Caracalla 9.4–5

The following passage provides a very unusual and obscure use of metal in the Baths of Caracalla. The *cella solearis* is probably the hot room (*caldarium*) of the bath, which in this instance seems to have had an elaborate metal grid associated with it – probably tall frames for window panes designed to allow solar heating of the space and tanning. The uncertainty expressed by the writer reflects a time when knowledge has been lost, a period of lesser ability (cf. **8.5**).

Caracalla left public works at Rome including the extraordinary baths named after himself. The architects [of the present day] state that its *cella solearis* [lit. sun room] cannot be reconstructed in the manner by which it was built originally. For it is said that the entire vault rested on a bronze or copper lattice enclosure placed underneath it, but so great is the size that those men learned in mechanics deny that it could be done (De Laine 1987 and 1997).

8.73 Street lighting in Antioch

The street lighting of Antioch was justly famous in antiquity and is mentioned by several authors in passing. Ammianus describes how it thwarted the nocturnal activities of the short-lived ruler, Gallus Caesar (AD 351–354).

Ammianus Marcellinus, *History* 14.1.9

... And taking with him a few men armed with concealed weapons, Gallus used to wander about through the tavernas and street corners in the evening, demanding in the Greek language, in which he was very much an expert, what everyone thought about the Caesar. He used to do this confidently in a city where the brightness of the light through the whole night was comparable to the splendour of the day. At last, being recognised frequently, and believing that he would be conspicuous if he continued to come out in the same manner, he did not come out in public except in the light of day and then only to conduct the affairs that he thought appeared serious.

8.74 Palace security in Rome

Fear of assassination was relatively common by rulers in both the Greek and Roman worlds. Sometimes heirs like Dionysious II were isolated and passed their time making toys or furniture (Plutarch, *Dion* 9), but guards, security routines and simple restrictions regarding potential weapons (**10.111**) are more often cited in the sources. Here the Emperor Domitian is credited with some palace upgrades to safeguard his back once his assassination has been predicted.

Suetonius, *Domitian* 14.4

But as the period of critical danger approached, greatly agitated he adorned the walls of the porticoes, in which he had forbidden others to walk about, with polished crystal [phengite] stone, by the sheen of which everything that happened behind his back could be perceived beforehand through their reflections.

BUILDING CODES: FIRE AND FLOOD PROTECTION

The great size of buildings and the cramped and haphazard condition of Rome created a situation in which buildings often burnt and collapsed. Plutarch (*Crassus* 2.4) relates that Crassus, one of the wealthiest men of the late Republic, made part of his fortune buying up burning buildings for a small fee, then dowsing them using his own fire brigade and having his own architects and builders repair them. The emperors expanded these actions, both through both physical constructions and through regulation, for more benevolent reasons. Similarly, actions against earthquake and flooding were continually being updated and new regulations produced. These all had consequences for the building technology when imposed upon builders, and although disasters continued to occur, perhaps they were a bit less devastating (cf. **8.64–65**) (Adrete 2007).

8.75 Attempts to overcome nature

Aurelius Victor, *[Epitome] On the Caesars* 13.12–13

At that time, more destructively by far than under Nerva, the Tiber flooded with incredible destruction of the nearby structures; in addition, severe earthquakes, terrible plague, famines, and fires occurred throughout many provinces. Trajan provided relief for all these disasters through carefully considered solutions, decreeing that the height of buildings not surpass 60 feet on account of the ease by which they collapse and their devastating costliness whenever such disasters should come to pass.

Pliny, *Letters* 8.17.1–2, 6

In this letter to Caecilius Macrinus, Pliny vividly describes the devastation of a flood, noting that the foresight of Trajan has eased the problem in one area, but nature has unleashed more than his preparations can handle. His final sentences are pertinent to the modern world and its technological attempts to prepare for the worst – often unsuccessfully.

Is the weather possibly as savage and stormy where you are? Here we have gales and floods one after another. The Tiber has overflowed its riverbed and poured in depth over its low-lying banks, even though being drained by the channel that our most far-seeing emperor made, it overwhelms the valleys and inundates the fields and wherever level ground exists water is seen instead of land ... [descriptions of devastation and death follow and then a request about Caecilius' situation].... But it differs very little whether you should endure the actual disaster or dread [its potential arrival], other than that there is a limit to suffering the disaster, while apprehension has none. For how much you might suffer you would know once it has happened, but how much you would dread in anticipation can continue on without end.

8.76 Fires and fire prevention

Strabo (*Geography* 5.3.7) states that Augustus had limited the height of buildings in Rome to 70 feet, and Vitruvius (**8.61**) implies regulations under Augustus, but they did not prevent the great conflagration under Nero. Nero himself tried to correct the problem and may have stimulated the use of vaults in construction by not permitting the use of wooden beams. Suetonius (*Nero* 16.1 and 38) summarises the information of Tacitus, who also offers a more positive interpretation of the fire and the attempts to quell it.

Tacitus, *Annales* 15.43

[After the fire in AD 64] the rest of the city, which had survived the [construction of the Golden] Palace, was rebuilt not as after the Gallic fire [390 BC] with no order and scattered about, but with lines of streets measured out, wide roadways, restricted heights for buildings, open areas, and with porticoes added to the front of insulae for protection ... the buildings themselves, in a specified part of each building, were to be made stable without wooden beams and composed of solid Gabine or Alban stone, the stones that are impervious to fire ... [the water supply is to be regulated].... And everyone was to have equipment accessible for firefighting. Nor were there to be common walls, but each building was to be enclosed by its own walls. These regulations, which were accepted for their utility, also brought an elegance to the new city. [The measures had both supporters and detractors.]

8.77 Fireproofing of wood

The dangers of fire in the *insulae* of the city were still present in the second century AC, even after further regulations by Trajan (Aurelius Victor, *Epitome* 13.13). A block of houses on fire has just been described and the rhetorician Antonius Julianus (mid-second century AC) relates a method of fireproofing wood.

Aulus Gellius, *Attic Nights* 15.1.4–6

"If you had read the nineteenth book of the Annals of Quintus Claudius, that best and most faithful writer", he said, "surely Archelaus, the prefect of King Mithridates, would have explained to you by what remedies and ingenuity you might prevent fire so that not any of your wooden buildings would burn, even if attacked and penetrated by flames". I earnestly enquired what this marvel of [Quintus Claudius] Quadrigarius was. He replied, "In short, in that book I discovered it written that when Lucius Sulla was attacking the Piraeus in Attica and Archelaus, the prefect of King Mithridates, was making sallies out of the town against him, Sulla was unable to burn a wooden tower constructed for the sake of defence, even though it had been completely surrounded on all sides with fire, because it had been smeared with alum by Archelaus".

Fires continued to destroy these multiple dwellings (Justinian, *Digest* 9.2.27.8) even though laws were devised to make fire-fighting equipment available in the structures (*Digest* 33.7.12.18–19).

STRUCTURAL DISASTERS

Although our sources record some deaths caused by the collapse of houses and other small structures, more widespread are descriptions of disasters involving huge public structures. The accounts themselves are interesting for the actions taken by the people in control, but they also demonstrate that even with established technologies, the carelessness of the builder or architect could wreak havoc. Materials, structural flaws and mismanagement (**8.38**) are all documented repeatedly in the sources (see Oleson, 2011).

8.78 The importance of materials

Reasons for architectural failures appear in many sources, usually in a straightforward manner by authors like Vitruvius and Pliny (cf. **8.8 to 8.14** for proper construction). Caligula is credited with crafting an interesting architectural metaphor about faulty mortar to criticise the "poor quality" of Seneca's style. The metaphor's meaning is revealed by Pliny and Vitruvius provides proper mixtures (**8.13**). Vitruvius also describes structural problems (**8.22**) arising from different building techniques and materials.

Suetonius, *Caius Caligula* 53.2

Yet Caligula, mocking the loose and smooth style of writing [i.e. rhetoric] so much, said that Seneca, then at his greatest popularity, "wrote nothing but detached essays", and that "it was sand without lime".

8.79 The collapse of a private house

Cicero, when relating how the art of remembering was discovered, states that Simonides of Ceos (c.556–467 BC) was summoned from the house of the Thessalian Scopas by the Dioscuri, who were grateful to him for honouring them in his poetry. Scopas has just reneged on the agreement to pay Simonides for composing a song. Quintilian (*Institutes of Oratory* 11.2.11–16) expands upon the story.

Cicero, *On the Orator* 2.86.353

A short time afterwards, they say, a message arrived for Simonides that he should go outside; two young men were standing at the door, who were urgently summoning him. Simonides stood up and went out, but saw no one. In the short interval, that chamber where Scopas was dining collapsed and he and his relatives were crushed to death in the catastrophe.[8]

8.80 Unreliable structures: a breeding ground for panic

As Pliny the Elder (**8.71**) demonstrated, not much confidence was placed in some of the large theatres. Panic was, no doubt, not too far distant at any public event.

Suetonius, *Augustus* 43.5

Once at a public show [in the Theatre of Marcellus] in honour of his grandsons when the people were overcome by fear of collapse, Augustus was unable to calm and reassure them in any manner. So, he moved from his own seat and sat down in that part of the theatre that was most suspect.

8.81 A deadly spectacle

In the following passage, Tiberius has been recalled from Capri because of a disaster at an amphitheatre on the mainland. There was good reason to fear some of these structures.

Suetonius, *Tiberius* 40

Tiberius was immediately recalled by the people with unremitting entreaty because of a disaster when the amphitheatre at Fidenae collapsed during a gladiatorial show and more than 20,000 people perished.

CONSTRUCTION ENGINEERING

CONSTRUCTION PROJECTS AND URBAN GREATNESS

Many passages describe and praise the greatness of a region or city in terms of the grand structures and projects that have been completed for the needs of the population. Few of these passages have much to do with technology per se, but they do provide a testament to the impact of construction technology upon the people of antiquity. Often magnificent projects are envisioned and initiated by powerful individuals such as Pericles, Demetrios Poliorcetes, and Augustus, but sometimes the visions are only realised much later if at all.

8.82 Grand engineering projects

Alexander the Great and Julius Caesar, two of most powerful leaders of the ancient world, both died before great projects associated with them were begun or not completed if begun. Alexander's grand plans for the future were revealed after his death to the Macedonians; the extravagance and complexity are so outrageous that they are often regarded as later fiction but Alexander did control the resources to carry them out. Many of the plans of Julius Caesar were attempted and sometimes carried-out by later rulers and governments after his death (e.g. **9.12** and **11.14**). Such passages provide an indication of one of the forces behind massive technological activity.

Plutarch, *Alexander* 18.4.4–6

[Alexander, in preparation for an expedition to the western Mediterranean, is recorded as listing massive military forces and supplies and wished] ... to construct a coastal road in Libya as far as the Pillars of Herakles [Gibraltar] and, in accordance with such a great expedition, to construct harbours and dockyards at advantageous locations, to build six expensive temples ... [other costly projects follow].... Once the plans were made known, the Macedonians ... seeing that the enterprises were excessive and formidable decided not to complete any of the ones mentioned.

Vitruvius (*On Architecture* 2. Preface 2–3) records the proposal for a massive statue of Alexander shaped from Mount Athos; Alexander is favourably disposed but ultimately refuses on practical grounds.

Suetonius, *Julius Caesar* 44.1 and 3

For concerning Rome's embellishment and fitting-out, [Caesar undertook many great activities such as] ... to drain dry the Pomptine Marshes, to drain out Lake Fucinus, to build a paved road from the [Adriatic] sea across the Apennine ridge to the Tiber, to excavate [a canal] through the Isthmus [of Corinth].

Plutarch, *Julius Caesar* 58.4–5

... [Caesar] set to work to dig [a canal] through the Corinthian Isthmus, having put Anienus in charge of this project, and planned for the Tiber to run straight from the Rome in a deep channel, bending at Circaeum to empty into the sea at Tarracina, as a safeguard and convenience for the back and forth commerce with Rome. In addition, he also contrived to drain the marshes around Pomentium and Setia to reclaim the productive plain for many thousands of men, to build breakwaters along the sand spits where the sea is nearest to Rome, to clear out all the hidden obstructions endangering anchorage at the shore of Ostia, and to construct harbours and trustworthy havens for great maritime ventures. And all these things were in preparation.

8.83 Criticisms and objections to ambitious engineering plans

After a particularly bad flood of the Tiber in AD 15, which resulted in much loss of life and property, Tiberius arranged an investigation with the intent to prevent future flooding. In his description of a discussion by the Senate, Tacitus reveals an ambitious plan consisting of at least three large engineering projects to reduce the flow of the Tiber (for Tiber floods see Aldrete 2007). As in the modern world, various objections about the potential disruptions to other areas resulted in the rejection of the plan. For other criticisms of projects see Suetonius, *Nero* 31 and Tacitus, *Annales* 15.42.

Tacitus, *Annales* 1.79

The next question raised in the Senate by Arruntius and Ateius was whether, in order to control the floodings of the Tiber, the rivers and lakes, by which the floods gain strength, should be diverted. Delegations from municipalities and colonies were given a hearing. The people of Florentia begged that the Clanis River not be diverted from its natural channel and redirected into the River Arnus, which would bring ruin upon them. Then the people of Interamna set out similar arguments, arguing that Italy's most fertile plains would be devastated if the River Nar, divided into streams as was planned, inundated the area and became a great lake. Nor did the people of Reate remain silent, but made objections to damming Lake Velinus where it discharges into the Nar since it would burst over the neighbouring lands. They argued that nature had taken care of the interests of mortals exceedingly well, giving rivers their own mouths, their channels, as well as their source and domain. And that consideration must be given to the religions of the allies, who had consecrated sacred rites, groves, and altars to their hereditary rivers. Indeed, they argued that Tiber himself, deprived of his neighbouring rivers, would not want to flow with dampened glory. Whether the pleas from the colonies were effective, or whether it was the difficulty of the work, or just the superstition, the result was that it was conceded to the motion of Piso, who had advocated no change.

8.84 Physical structures and historical fame

The following three passages have been selected to conclude this chapter: one from the fifth-century BC Greek historian, Thucydides, a second by Pliny in the first century AC, and the third by the second-century AC traveller, Pausanias.

Thucydides, *The Peloponnesian War* 1.10.2

Thucydides is considering whether the traditions passed down to his time regarding the Trojan War can possibly be correct. How could the ruler of a small place like Mycenae be the commander of an expedition so large and glorious as that described by Homer? He reflects upon two examples from his own day: Sparta and Athens.

For if the city of the Lacedaemonians [Sparta] should become deserted, and only the temples and foundations of buildings remain, I think that in a much later time the people would have serious doubts about its power compared to its reputation. And yet they occupy two-fifths of the Peloponnese and lead the whole of it, as well as many allies elsewhere. For all that, it is neither a regularly formed city nor decorated with costly temples and buildings, but is inhabited like a village in the old manner of Hellas; it would appear more inferior than it is. But if Athens were to suffer the same fate, the city would be thought twice as powerful as it is from the appearance of the site.

8.85 The wonder of Rome seen through her buildings

Clearly Pliny is a proud citizen, but Rome really had become the pre-eminent city of the Mediterranean world and its position was proclaimed through the great building programmes undertaken by the magistrates, generals, and emperors of the previous 800 years. Similar praises of Rome's splendour occur in a variety of other authors, both Greek and Roman.

Pliny, *Natural History* 36.101

But indeed, it is appropriate to move on to the wonders of our own city and to examine the virtues taught through 800 years, and to demonstrate that here too we have conquered the world; a thing which will seem to have occurred almost as often as the wonders being described. For if the whole physical mass of our city was collected and piled into one heap, there would be no other magnitude towering up so high: it would seem as if some other entire world was being described in a single location.

Pliny continues by describing a few of the great constructions of the city and some of their costs: circuses, basilicas, fora, temples, theatres, and sewers.

8.86 One definition of a city in the Roman world

By the time of the Roman Empire, the physical appearance of a city was certainly important. Hundreds of inscriptions survive that provide evidence of how quickly some cities were built up with structures donated by their wealthy inhabitants. Others, however, did not fare as well.

Pausanias, *Description of Greece* 10.4.1

It is 20 *stadia* from Chaeronea to Panopeus, a city [*polis*] of the Phocians, if one can call a place a city where the inhabitants have no government buildings, no gymnasium, no theatre, no market-place, no water falling down into a fountain, but live in hollow shelters like huts in the mountains, right on the cliff.

Notes

1 Corbelled vaults do not use a keystone.
2 Marble or hard limestone.
3 That is, using header and stretcher construction.
4 For additional passages on Hagia Sophia see C. Mango, *The Art of the Byzantine Empire:312–1453* (Englewood Cliffs, NJ, 1972) 72–102.
5 Cf. Pliny the Younger, *Letters* 10.40.3.
6 Eight and a half is the recommended ratio elsewhere.
7 Ruined structures often were levelled and used as fill or slightly raised, artificial foundations.
8 The bodies were so disfigured that identification was possible only by Simonides remembering the seating arrangement.

Bibliography

General surveys

Clark, Michael L., "The Architects of Greece and Rome". *Architectural History* 6 (1963) 9–22.
Fitchen, John, *Building Construction before Mechanization*. Cambridge: MIT Press, 1986.
Hill, Donald, *A History of Engineering in Classical and Medieval Times*. London: Croom Helm, 1984.
Hoffmann, Adolph, Ernst-Ludwig Schwandner, Wolfram Hoepfner, and Gunnar Brands, eds., *Bautechnik der Antike. International Colloquium, 15–17 February 1990*. Mainz: Deutsches Archäologisches Institut, 1991.
Kostof, Spiro, "The Practice of Architecture in the Ancient World: Egypt and Greece". Pp. 3–27 in Spiro Kostof, ed., *The Architect: Chapters in the History of the Profession*. New York: Oxford University Press, 1977.
Müller, Werner, *Architekten in der Welt der Antike*. Leipzig: Koehler & Amelung, 1989.

Bronze Age construction

Arnold, Dieter, *Building in Egypt. Pharaonic Stone Masonry*. Oxford: Oxford University Press, 1991.
Badaway, Alexander, *Ancient Egyptian Architectural Design: A Study of the Harmonic System*. Berkeley: University of California Press, 1965.
Badaway, Alexander, "The Periodic System of Building a Pyramid". *Journal of Egyptian Archaeology* 63 (1977) 52–58.
Besenval, Roland, *Technologie de la voûte dans l'orient ancien*. 2 Vols. Paris: ADPF, 1984.
Cavanagh, William G., and Robert R. Laxton, "The Structural Mechanics of the Mycenaean Tholos Tomb". *Annual of the British School at Athens* 76 (1981) 109–140.
Clarke, Somers, and Reginald Engelbach, *Ancient Egyptian Masonry, The Building Craft*. London: Oxford University Press, 1930.

Graham, James W., *The Palaces of Crete*. 2nd edn. Princeton: Princeton University Press, 1987.
Hult, Gunnel, *Bronze Age Ashlar Masonry in the Eastern Mediterranean: Cyprus, Ugarit, and Neighboring Regions*. Studies in Mediterranean Archaeology, 66. Göteborg: Paul Åströms Förlag, 1983.
Rosalie, David, *The Pyramid Builders of Ancient Egypt. A Modern Investigation of Pharaoh's Workforce*. London: Routledge & Kegan Paul, 1986.
Shaw, Joseph W., *Minoan Architecture: Materials and Techniques*. Annuario della Scuola Archeologica di Atene, 49. Rome: Instituto Poligrafico dello Stato, 1973.
Spencer, A. Jeffrey, *Brick Architecture in Ancient Egypt*. Warminster: Aris & Phillips, 1979.

Greek construction

Andronikos, Manolis, "Some Reflections on the Macedonian Tombs". *Annual of the British School at Athens* 82 (1987) 1–16.
Bundgaard, Jens A., "The Building Contract from Lebadeia. Observations on the Inscription IG1 VII 3073". *Classica et mediaevalia; revue danoise de philologie et d'histoire* 8 (1946) 1.43.
Bundgaard, Jens A., *Mnesikles: A Greek Architect at Work*. Copenhagen: Gyldendal, 1957.
Bundgaard, Jens A., *The Greek Temple Builders at Epidauros*. Liverpool: Liverpool University Press, 1969.
Burford, Alison M., "The Builders of the Parthenon". *Greece and Rome*, Suppl. to 10 (1963) 23–35.
Campbell, Ian, "*Scamilli Impares*: A Problem in Vitruvius". *Papers of the British School at Rome* 48 (1980) 17–22.
Cooper, Fredrick A. "Greek Engineering and Construction". Pp. 225–255 in John P. Oleson, ed., *The Oxford Handbook of Engineering and Technology in the Classical World*. Oxford: Oxford University Press, 2008.
Coulton, John J., *Ancient Greek Architects at Work. Problems of Structure and Design*. Ithaca: Cornell University Press, 1977.
Coulton, John J. (1974), "Lifting in Early Greek Architecture", *The Journal of Hellenic Studies*, 94 (1974) 1–19.
Doxiadis, Constantinos A., *Architectural Space in Ancient Greece*. Cambridge: MIT Press, 1972.
Drachmann, Aage G., *The Mechanical Technology of Greek and Roman Antiquity: A study of the literary Sources*. Copenhagen, Munksgaard; Madison, University of Wisconsin Press; London, Hafner, 1963.
Dworakowska, Angelina, "Notes on the Tools Mentioned in the Building Contract from Lebadeia: Kolapter, Xois, Leistrion". *Archeologia* 29 (1979) 16–23.
Heisel, Joachim P., *Antike Bauzeichnungen*. Darmstadt: Wissenschaftliche Buchgesellschaft, 1993.
Hellmann, Marie-Christine, "Le vocabulaire architectural grec: Bilan de plus de cent ans de recherches". *Revue des Études Grecques* 102 (1989) 549–560.
Hodge, A. Trevor, *Woodwork of Greek Roofs*. Cambridge: Cambridge University Press, 1960.
Holloway, R. Ross, "Architect and Engineer in Archaic Greece". *Harvard Studies in Classical Philology* 73 (1969) 281–290.

Jeppesen, Kristian, *Paradeigmata. Three Mid-Fourth Century Main Works of Hellenic Architecture Reconsidered.* Aarhus: Aarhus University Press, 1958.

Lawrence, Arnold W., *Greek Architecture.* 4th edn. Rev. by R.A. Tomlinson. Harmondsworth: Penguin Books, 1983.

Martin, Roland, *Manuel d'architecture grecque*, I: *Matériaux et techniques.* Paris: Picard, 1965.

Müller-Wiener, Wolfgang, *Griechisches Bauwesen in der Antike.* Munich: Beck, 1988.

Orlandos, A., *Les matériaux de construction et la technique architecturale des anciens grecs.* 2 Vols. Paris: De Boccard, 1966–1968.

Randall, Richard H., Jr., "The Erechtheum Workmen". *American Journal of Archaeology* 57 (1953) 199–210.

Stiros, Stathis C., "Archaeological evidence of antiseismic constructions in antiquity". *Annali di Geofisica*, 38, 5–6 (1995) 725–736.

Tomlinson, Richard A., "Architectural Context of the Macedonian Vaulted Tombs". *Annual of the British School at Athens* 82 (1987) 305–312.

Roman construction

Adam, Jean-Pierre, *Roman Building Materials and Techniques.* London: Batsford, 1993.

Aldrete, Gregory S., *Floods of the Tiber in Ancient Rome.* Baltimore: Johns Hopkins University Press, 2007.

Anderson, James C. Jr., *Roman Architecture and Society.* Baltimore and London: Johns Hopkins University Press, 1997.

Anon., *Le Projet de Vitruve. Objet, destinataires et réception du* De Architectura. *Actes du colloque international organisé par l'École Française de Rome.* Collection de l'École Française de Rome, no. 192. Paris: de Boccard, 1994.

Blake, Marion E., *Ancient Roman Construction in Italy from the Prehistoric Period to Augustus.* Washington: Carnegie Institution of Washington, 1947.

Blake, Marion E., *Ancient Roman Construction in Italy from Tiberius through the Flavians.* Washington: Carnegie Institution of Washington, 1959.

Blake, Marion E., *Roman Construction in Italy from Nerva through the Antonines.* Edited by Doris T. Bishop. Memoirs of the American Philosophical Society, 96. Philadelphia: American Philosophical Society, 1973.

Brandon, Christopher J., Robert L. Hohlfelder, Michael D. Jackson, and John P. Oleson, (ed. John P. Oleson), *Building for Eternity: The History and Technology of Roman Concrete Engineering in the Sea.* Oxford: Oxbow Books, 2014.

De Laine, Janet, "The 'Cella Solearis' of the Baths of Caracalla: A Reappraisal". *Papers of the British School at Rome* 55 (1987) 147–156.

De Laine, Janet, "Structural Experimentation: The Lintel, Corbel and Tie in Western Roman Architecture". *World Archaeology* 21:3 (1990) 407–424.

De Laine, Janet, *The Baths of Caracalla: A Study in the Design, Construction, and Economics of Large-Scale Building Projects in Imperial Rome*, Portsmouth, RI: Journal of Roman Archaeology Suppl. Series no. 25, 1997.

Dilke, Oswald A.W., *The Roman Land Surveyors. An Introduction to the Agrimensores.* Newton Abbot, 1971.

Dix, Brian, "The Manufacture of Lime and Its Uses in the Western Roman Provinces". *Oxford Journal of Archaeology* 1 (1982) 331–345.

Evans, Edith, "Military Architects and Building Design in Roman Britain". *Britannia* 25 (1994) 143–164.

Grewe, Klaus, "Urban Infrastructure in the Roman World". Pp. 768–783 in Georgia L. Irby, ed., *A Companion to Science, Technology, and Medicine in Ancient Greece and Rome, Vol. II*. Chichester: Wiley-Blackwell, 2016.

Grewe, Klaus, "Recent Developments in Aqueduct Research". *Mouseion Supplement: Festschrift in Honour of Professor John P. Oleson*. Forthcoming (2019).

Haselberger, Lothar, "Old Issues, New Research, Latest Discoveries: Curvature and Other Classical Refinements". Pp. 1–68 in Lothar Haselberger, ed., *Appearance and Essence. Refinements of Classical Architecture: Curvature*. Philadelphia: University of Pennsylvania Press, 1999.

Knell, Heiner, *Vitruvs Architekturtheorie. Versuch einer Interpretation*. Darmstadt: Wissenschaftliche Buchgesellschaft, 1985.

Lamprecht, Heinz-Otto, *Opus Caementicium. Bautechnik der Römer*. 3rd edn. Düsseldorf: Beton-Verlag, 1994.

Lancaster, Lorenzo, "Roman Engineering and Construction". Pp. 256–284 in John P. Oleson, ed., *The Oxford Handbook of Engineering and Technology in the Classical World*. Oxford: Oxford University Press, 2008.

Lechtman, Heather N., and Linn Hobbs, "Roman Concrete and the Roman Architectural Revolution". Pp. 81–128 in W. David Kingery, ed., *Ceramics and Civilization*, Vol. 3: *High-Technology Ceramics—Past, Present, and Future*. Westerville, OH: American Ceramic Society, 1987.

Lugli, Giuseppi, *La tecnica edilizia romana, con particolare riguardo a Roma e Lazio*. 2 Vols. Rome: Giovanni Bardi, 1957.

MacDonald, William L., "Roman Architects". Pp. 28–48 in Spiro Kostof, ed., *The Architect: Chapters in the History of the Profession*. New York: Oxford University Press, 1977.

MacDonald, William L., *The Architecture of the Roman Empire*, I: *An Introductory Study*. Rev. edn. New Haven: Yale University Press, 1983.

Meiggs, Russell, *Trees and Timber in the Ancient Mediterranean World*. Oxford: Oxford University Press, 1982.

O'Connor, Colin, *Roman Bridges*. Cambridge: Cambridge University Press, 1993.

Oleson, John P., "Technical Aspects of Etruscan Rock-Cut Tomb Architecture". *Römische Mitteilungen* 85 (1978) 283–314.

Oleson, John P., "*Harena sine calce*: Building Disasters, Incompetent Architects, and Construction Fraud in Ancient Rome". *Commentationes Humanarum Litterarum* 128 (2011) 9–27.

Oleson, John P., and Graham Branton, "The Harbour of Caesarea Palestinae: A Case Study of Technology Transfer in the Roman Empire". *Mitteilungen, Leichtweiss-Institut für Wasserbau* 117 (1992) 387–421.

Rodriguez Almeida, Emilio, *Forma Urbis Marmorea. Aggiornamento generale 1980*. 2 Vols. Rome: Edizioni Quasar, 1981.

Storz, Sebastian, *Tonröhren im antiken Gewölbebau*. Mainz: von Zabern, 1993.

Ward-Perkins, John B. *Roman Imperial Architecture*. 2nd edn. Harmondsworth: Penguin, 1981.

Wilson, Roger J.A., "Terracotta Vaulting Tubes (*tubi fittili*): On Their Origin and Distribution". *Journal of Roman Archaeology* 5 (1992) 97–129.

9

HYDRAULIC ENGINEERING

Thales of Miletus (*flor. c.*580 BC), standing nearly at the beginning of Greek science and philosophy, perceptively chose water as the primary constituent of matter (see **9.1**). Water is ubiquitous and essential to life, and it can easily be observed to condense and to rarefy in passing into its solid and gaseous states. But even more important, ensuring its presence or absence was a major practical problem in urban design and agricultural development, and this struggle must have appealed to Thales' strong bias towards applied science. He seems to have grappled eagerly with practical problems, and our sources attribute to him skills in physical science, political and economic theory, engineering, geography, astronomy, and mathematics. Thales' apotheosis of water was an exaggerated response, but all human dealings with the substance in antiquity, however prosaic, were based on its primary characteristics of significant weight and subtle, liquid state. In its customary liquid form water flows downhill and collects at the lowest level it can reach. It was important, therefore, in obtaining water for human use, either to intercept the flow above the level where it was needed and guide it to the point of use in a gravity-flow channel, or to lift it up directly to the point of use or to a conduit from the most convenient pure source.

Since water is essential for the maintenance of life and health in large urban centres, we can expect rich evidence for methods of obtaining it in markedly urbanised Greek and Roman cultures and in the intensive agricultural exploitation on which they were based. The tunnel of Eupalinos at Samos (**9.13**) and the magnificent aqueducts of ancient Rome (**9.13–25**) represent sophisticated realisations of the response to the need for a water supply. Although relatively simple in principle, gravity-flow conduits require careful surveying and solid construction in order to be successful. As the Latin texts make clear, the continuing jobs of upkeep and administration were equally complex and crucial. The Hellenistic Greeks and the Romans also had at their disposal a variety of mechanical water-lifting devices designed to lift water to a point where it could conveniently be used, or to remove it from where it was not wanted. Animal power and water-power could be used to propel several of these devices. Without the assistance of efficient mechanical pumps, the phenomenal agricultural yield of the Nile valley would have decreased dramatically, the grain freighters on which Rome

HYDRAULIC ENGINEERING

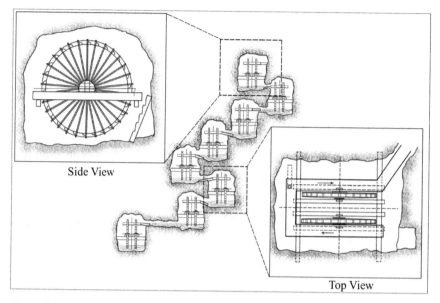

Figure 9.1 Waterwheels in Rio Tinto Mine, Spain.
Source: Adapted from Palmer, 1926, figs. 69 and 74.

depended for food would have sunk, the rich mines [**fig. 9.1**] in Spain and Romania could not have been worked, and fire would have constituted an even greater hazard to Roman cities than it already did. It is indicative of the depth and sophistication of this technology of guiding and lifting water that the principles of the Roman aqueducts and the very designs of the mechanical pumping devices are still in use today, virtually unchanged (Mithen 2012, Wilson 2008, Tölle-Kastenbein 1990).

9.1 A philosophical assessment of water

Aristotle, *Metaphysics* 1.3.3–5 (983b)

The earliest philosophers for the most part thought of the first principles of all things in the form of a material substance: that which all extant things partake of, the primary substance from which they are born and back to which they finally dissolve, the essence of which persists, although altered in its compounds. They call this the element and first principle of that which exists.... Thales says that water is the first principle (on account of which he declared that the Earth floats on water). He may have derived such an assumption from the observation that everything depends on sustenance that is moist, and that heat itself is generated by moisture and lives upon it, ... and because the seeds of everything have a moist nature....

WATER SOURCES AND THEIR QUALITY
9.2 Finding water and judging its quality

The methods of water-divining cited by Vitruvius sound chancy, but they undoubtedly are based on experience. Pliny records the same procedures (*Natural History* 31.4349). Since water usually was drunk without any chemical treatment in antiquity, the natural flavour and salubrity, affected by the character of the subsoil and the type of source, were important considerations. If the quality was doubtful, salt was added (see **9.5**).

Vitruvius, *On Architecture* 8.1.1–2, 4–6

Water is an absolute necessity for life, for pleasure, and for daily use. It will be more convenient if the streams are out in the open and flowing, but if they do not flow out above ground, then sources must be sought and collected underground. The following technique must be tried: one lies on the stomach before sunrise in the places where it is to be sought and, with the chin resting on and supported by the ground, examines the area. In this manner, when the chin is not moved, the glance will not wander higher than it should, but with precise definition it will describe a level line throughout the area. Then excavation is to be carried out in those places where mists curl upwards and rise into the air, for this sign cannot appear above dry ground.

Those who seek water must also take note of the local geology, for there are certain places where it springs up. In clay the amount is poor and meagre and near the surface: it will not have the best taste. The supply also is poor in loose gravel but is found at a greater depth. It will be muddy and harsh. In black earth, however, meagre drops and beads of moisture are found, which, accumulating after winter storms, collect in hard, impermeable spots. These have the best taste. In gravel, to be sure, moderate and unpredictable sources are to be found that also have a notable sweetness. In coarse gravel, sand, and red sand, the supply is more certain and has a good taste. The water from red rock is both copious and good, unless it slips through faults and leaks away. But at the feet of mountains and in flinty rocks supplies are richer and more abundant, colder, and more healthful. Water from springs on level ground is salty, unpleasant, tepid, and bitter, unless it flows underground from mountains and breaks out in the middle of fields and there, protected by the shade of trees, provides the sweetness of mountain springs....

Wherever these indications show up, trial must be made in the following manner. Let a spot three feet square be dug to a depth of no less than five feet, and around sunset let a bronze, or lead bowl, or basin be placed in it. Whichever of these is used, it must be spread with oil on the inside and placed upside down and the top of the pit covered with reeds or branches and earth heaped up above. On the following day it is to be opened, and if drops and beads of moisture are found within the pot, the place will provide water.

Likewise, if a pot of unbaked clay is placed in this pot and covered in the same manner, if the place contains water, the pot will be wet when opened and close to collapsing with the damp. And if a woollen fleece is placed in the pit and water is squeezed from it the next day, it indicates that the place has a water supply. In addition, if a lamp is filled with oil, trimmed, lit, placed in the pit, and covered up, and on the next day has not burnt out but has some oil and wick left and is itself found moist, it signifies that the place has water. For all heat draws moisture to itself. Likewise, if a fire is kindled in this place and the earth gives off a cloud of steam when thoroughly heated and scorched, the place will supply water.

When these procedures have been tried and the indications found, which have been described above, then a well must be sunk in that spot. If a spring is found, more wells must be dug around the first and all connected to one spot by means of tunnels....

The following passage describes an instrument, now called a hydrometer, that is used to determine a liquid's density to ascertain its alcohol or sugar content. Consisting of a buoyant tube, the instrument floats vertically in the liquid whose density is to be determined. The higher the density of the fluid, the higher the buoyancy of the tube. The Greek word ῥοπή *rhopē* is generally translated as "weight", but the instrument clearly measures the liquid's density, a concept not common in antiquity. We translate, therefore, *rhopē* as "heaviness", a suitable compromise between weight and density.

Synesius, *Letter* 15

(To Hypatia) I am in the altogether bad situation that I need a hydroscope. Order that one is made from bronze and assembled.

The tube is cylindrical and has the shape and size of a flute. Along a straight line, it has notches by which we can estimate the heaviness (*rhopē*) of the water. A close-fitting cone caps it on one end so that both the tube and the cone have a common base. This is the plumb (*baryllion*). So, when you immerse the tube in water, it will stand upright and allow you to count the notches, which indicate the heaviness.

9.3 A water diviner at Rome, AD 507/511

Vitruvius' comments are intended for the most part for application in rural landscapes, where water supplies were required for private, domestic use. There were, however, also professional water diviners in the Roman world who could be hired to find water sources suitable for use in urban or suburban areas. The central part of this letter, addressed by Theodoric to Apronianus, Count of the Private Domains, bristles with echoes of Classical texts on water-divining: Vitruvius, *On Architecture* 8.1; Pliny, *Natural History* 31.43–49; and Palladius, *On Agriculture* 9.8.1, 4, 6. It is interesting that the city centre was still well-supplied with water, but that the suburbs suffered from its lack.

Cassiodorus, *Letters* 3.53.1 and 6

Through Your Honour's report we have learnt that a water diviner (*aquilegus*) has come to Rome from Africa – where that art is always cultivated with great zeal on account of the region's aridity – who in drought-stricken localities can provide water supplies that spring from the earth [or can be raised from low places], so that his service renders habitable places dried up by a severe insufficiency of water.

... For although the city of Rome has a bountiful supply of running water and is blessed with springs and the extremely copious flow of the aqueducts, nevertheless one might find many localities in the vicinity of the city that seem to have need of this man's skill. He is deservedly detained who is found to be indispensable even for a portion of the job. An engineer (*mechanicus*), however, must surely be assigned to this diviner to raise the water that he finds and by his own skill bring up what cannot move upwards by its own inherent nature.

9.4 Some water sources

Polybius describes a man-made seepage gallery, a so-called *qanat*, in Media. These structures originated in the ancient Near East, most likely in the area of modern Iran. *Qanats* tap aquifers and make the water accessible at convenient points. The Romans adopted the technique and subsequently brought it all the way to Spain, at the opposite end of the Mediterranean region. The account also demonstrates the strategic importance of water access, or denial thereof, on military campaign.

According to the second passage, by Pliny, the best water comes from wells, as long as these are in constant use and are sheltered from the sun.

Polybius, *Histories* 10.28

Arsaces expected that Antiochus would come as far as this place (Media), but that he would not dare to cross the adjoining desert with such a big army, especially due to the lack of water. For in this region there is no surface water, although there are numerous subterranean channels linked with vertical shafts throughout the desert unknown to unfamiliar persons. A true account of these channels has been handed down by the inhabitants according to which, while the Persians held power over Asia, they granted for five generations the enjoyment of the land to inhabitants who had brought in spring water from areas previously not irrigated. Since numerous and abundant streams flowed down from Mount Taurus, these people endured all kinds of expense and toil to construct these underground channels coming from far away. They did so in such a way, that even now the people who use the water do not know where the channels start and where they catch their water. When, however, Arsaces saw that Antiochus was making ready to cross the desert, he planned at once to collapse and destroy the shafts. But King Antiochus, when this was reported to him, sent Nicomedes ahead with 1,000 horses. He found that Arsaces had withdrawn with his army, but came upon some of his cavalry in the act of plugging the shafts that connected to the underground channels. They attacked them and, having routed and put them to flight returned to Antiochus. The king crossed the desert and arrived at the city called Hekatompylos, which is situated in the centre of Parthia and derives its name from the fact that it is located at the crossroads of the routes leading to all the surrounding regions.

Pliny, *Natural History* 31.38–39

From which type of source, then, is the most commendable water obtained? From wells, of course, as I see them arranged in towns, but [specifically] from those, which through frequent dipping have the advantage of circulation and clarity produced by filtration through the earth. This is enough for wholesomeness. For coolness, both shade is necessary and that the well be open to the sky. Above all, one observation – and the same is relevant to continual flow: the spring should issue from the bottom of the well, not its sides. It can also be arranged artificially that the water be cool to the touch, by squirting it up high or letting it fall from an elevated spot to strike against and absorb the air.

9.5 Instructions on digging wells and building cisterns

Under the best of circumstances the excavation of a well required great effort. In the neighbourhood of Rome, and across much of the volcanic landscape of Central Italy, there was the added complication of poisonous fumes trapped in the strata of the pyroclastic rock. Even the technologically simple procedure of catching the run-off from natural precipitation involved the careful preparation of cisterns designed both to hold the water without any leakage and to keep it as fresh as possible.

Vitruvius, *On Architecture* 8.6.12–15

But if we are to create sources from which water is drawn, it is necessary to dig wells. One must proceed systematically in digging wells, and with intelligence and experience one must consider the way the natural world works, since the earth has many and varied substances in it. For it is, like everything else, made up of four elements: first, it is itself earthy; and second, of the liquid element, it has springs of water; third, heat, by which sulphur, alum, and tar are generated; and fourth, it has great currents of air. These currents, when heavy, pass through the intervening porous earth to a well excavation and overcome the men digging there, since by nature of the gas they stop up the breath of life in their nostrils. And so those who do not quickly escape die there.

Precautions, then, must be taken against this by the following methods. A lighted lamp is to be lowered, and if it continues burning one can enter the shaft without danger. But if the flame is snuffed out by the power of the gas, then ventilation shafts are to be dug next to the well on either side. In this way the gas vapours will be dissipated through the shafts as through nostrils. When these have been finished and water is reached, then the shaft is enclosed with a wall, but without blocking up the veins.

But if the ground is hard or the water too deep, then run-off from roofs or higher ground must be collected in cement cisterns. These are the procedures for cement work. The cleanest and sharpest sand is prepared, flint aggregate is broken into pieces of no more than a pound, and the hottest possible hydrated lime is mixed in a trough – five parts sand to two of lime. A trench as deep as the level of the desired walls is tamped down with iron-clad wooden beams. When the walls have been poured, the earth in the middle is removed to the bottom of the walls. After this space has been levelled, the floor is poured to the agreed-upon thickness. These cisterns however, will constitute a much more healthful water system if they are built in a series of two or three so that the water can be changed by slow movement. For when the silt has a place to settle, the water becomes clearer and keeps its taste free of extraneous smells. If not, it will be necessary to add salt to clarify it.

Pliny more or less repeats Vitruvius' instructions, but with the additional detail of a strainer to help keep the water pure.

Pliny, *Natural History* 36.173

Cisterns should be constructed with five parts of pure, coarse sand, two parts of the most caustic unslaked lime possible, with flint aggregate weighing no more than a pound. Both floors and walls alike are to be compacted with iron rods. It is more useful if cisterns are built in pairs, so that the pollutants might settle out in the first and the pure water pass through a strainer into the second.

9.6 How to tap a submarine fresh-water spring

Even today the small rocky island of Arwad (ancient Arados), 4 km off the coast of Syria and about 1 km in circumference, must depend on rainwater or imported water to support its population. In antiquity, when the strategic island was even more thickly settled than at present, the inhabitants found a way to extract fresh water from some nearby submarine springs. The turbidity characteristic of brackish water was probably first noted by divers collecting sponges or shellfish. Pliny, *Natural History* 5.128 repeats the same story.

Strabo, *Geography* 16.2.13

Aradus lies off a stretch of coast without harbours and open to the waves, ... 20 *stadia* off the mainland. It is a sea-girt rock, about seven *stadia* around, thickly settled. It has been so heavily populated even up to the present, that the inhabitants live in houses of many storeys.... They obtain some of their water from cisterns filled by the rain, some from the mainland. In time of war they get water from the straits a little landward of the city, where there is a fast-flowing spring. A wide-mouthed vessel of lead contracting to a narrow base with a moderate hole is inverted and let down over this from the watering boat. A leather pipe (or rather one should say a bellows) is fastened around the base to receive the water forced up from the spring through the vessel. The first water forced up is sea water, but they wait for the discharge of pure, drinkable water, catch as much as is needed in containers kept at the ready, and convey it to the city.

IRRIGATION AND DRAINAGE
9.7 The Nilometer at Elephantine

In general, agriculture in the Graeco-Roman world depended on the techniques of drought farming rather than on irrigation. Both the Greeks and the Romans, however, were familiar with the use of the annual inundation of the Nile for irrigation in Egypt, and some of the principles involved may have spread to Greece as early as the late Bronze Age. The Nilometer at Elephantine in Upper Egypt was a device that measured the height of the flood and allowed calculation of the proper moment for opening the inlets to the diked fields. A similar account appears in Diodorus of Sicily, *History* 1.36.7–12.

Strabo, *Geography* 17.1.48

The Nilometer is a well, built of ashlar masonry on the bank of the Nile, in which are indicated the highest, lowest, and average inundations of the Nile. For the water in the well rises and sinks along with the river. In consequence, there are signs on the wall of the well shaft to designate the height of the complete and of the other inundations. Those who watch these marks signal to the rest so they might know. For they know far in advance from these marks and the dates what the inundation will be and make predictions. This is helpful not only to the farmers with respect to the division of the water, embankments, canals, and other such things, but also to the authorities with respect to the public revenues.

9.8 The irrigation system of Mesopotamia

In Egypt the inundation came just before the proper season for planting, and the object was to fill as many of the fields as possible with the silt-bearing water. In the Tigris and Euphrates river valleys the flood season came as the crops matured, and farmers diked their fields and built drainage canals to keep the flood under control. From the early Bronze Age on, only a complex system of dikes and canals allowed cultivation of the rich alluvial soil. Alexander the Great's campaigns in this area revealed a tradition of hydraulic technology already three millennia old.

Strabo, *Geography* 16.1.9–10

Several rivers flow through the country, the largest of them being the Euphrates and the Tigris .... The Persians, therefore, in a rational desire to prevent voyages up the rivers, from fear of outside invasions, to this end had prepared artificial barrages. But as he went up the river Alexander dismantled as many of these as he could, especially those by Opis. He also paid attention to the canals, for the Euphrates floods in early summer, beginning to rise with the spring, when the snows of Armenia melt. Necessarily it forms stagnant pools and inundates the ploughed fields, unless someone distributes the excess flow and surface water by means of trenches and canals, just as it is done for the Nile flood in Egypt. From this situation, then, originated the canals. They require a great deal of work, for the soil is deep, soft, and yielding. In consequence, it is easily eroded by the streams and leaves the plains naked, while the silt easily fills the canals and blocks up their mouths....

Perhaps it is not possible to prevent such flooding absolutely, but it is the role of good rulers to provide every possible assistance. This is how they help: on the one hand they forestall the bulk of the flooding by means of dikes, and on the other prevent the deposition of sediments caused by the silt by dredging the canals and keeping their mouths unclogged. Now the dredging is easy, but the diking requires the work of many hands. For since the soil is yielding and soft, it does not hold up against the silt driven against it but caves in and carries the silt along with it, making the breach difficult to close off. And there is also need for speed in quickly closing the canals and preventing the water from draining out of them completely. For when they dry up during the summer, they dry up the river as well, and the river, when lowered, cannot supply the sluices at the moment they are needed, most of all in the summer, when the land is hot as blazes and burnt up. It's all the same whether the crops are overwhelmed by the abundance of water or perish of thirst in a waterless field.

9.9 A solution for the dry climate of Greece

Plato makes clear reference in this passage to the principles of run-off irrigation: blocking the flow of rain-fed streams either to form pools for irrigation or to replenish the ground-water in a small valley by both holding back the topsoil and allowing water to soak in. Techniques of this sort allowed agriculture to flourish in antiquity even in the deserts of North Africa and the Near East, but they were never applied extensively in Greece itself. Plato is describing here the situation in an "ideal" state. Property holdings may have been too small on average to make it worthwhile, or the limestone bedrock characteristic of the Greek landscape may have been too porous to allow extensive storage of ground-water. Where there was sufficient water for irrigation, the system of earthen water channels described by Aristotle could be used.

Plato, *Laws* 6.761a–b

They shall attend to the rain water so that it might not damage the land but rather be of benefit to it as it flows down from the high ground to the hollow valleys among the mountains. They shall cut off their outflow with dikes and channels, so that they might receive and absorb the rainfall, forming streams and springs in all fields and regions below them, providing even the driest places with an abundant supply of pure water.

Aristotle, *On the Parts of Animals* 3.5 (668a)

[The system of blood vessels in the human body] is just like the water channels that are prepared in gardens. From one source or spring they branch out into many channels and then into others, throughout the garden....

9.10 Land reclamation at Thisbe in Boeotia, Greece

Although Greece in general was badly off with respect to water resources, particularly poorly supplied with perennial streams and rivers, there were some local exceptions. At Thisbe, for example, and at nearby Orchomenos, surface water had to be controlled and diverted with dikes and canals, as in the river valleys of Mesopotamia and Egypt. Greek legend even suggests that this hydraulic technology was transferred to Greece by the Egyptian Danaus, who settled at Argos in the Bronze Age, became king of Argos, and solved the local problem of water supply (see Apollodorus, *Mythological Library* 2.1.4). Danaus' 12 daughters killed their husbands and were put to work in the underworld drawing water in leaky buckets, possibly an allusion to the typically Egyptian task of irrigating with a *shaduf* (Horace, *Odes* 3.11; Lucian, *Dialogues of the Sea Gods* 6); the *shaduf* is described in **9.26**.

Pausanias, *Description of Greece* 9.32.3

On account of the quantity of water, nothing would prevent the plains between the mountains from becoming a marshy lake if they had not made themselves a strong dike down the middle. Every year they divert the water alternately to one or the other side of the dike and farm the land on the opposite side.

9.11 A remarkable Bronze Age drainage system

In the Mycenaean period the fertile land under the Copaic Lake northwest of Thebes was reclaimed by a complex hydraulic engineering system employing drainage works that included chambers, tunnels, and dikes. Strife and neglect at the end of the Bronze Age resulted in the disablement of the drainage system, and although several failed attempts to repair it in antiquity and later were made, not until the late nineteenth/early twentieth century was the Copaic Basin again reclaimed. Strabo's first passage provides a memory of the Bronze Age achievement, Pausanias a memory of its disablement at the end of the Bronze Age, while the second passage of Strabo provides evidence of an attempt to clear the system.

Strabo, *Geography* 9.2.40

They say the place, which Lake Copais covers today, used to be dry ground and was cultivated in all kinds of ways by the Orchomenians, who lived nearby.

Pausanias, *Description of Greece* 9.38.6–8

Seven *stadia* from Orchomenos is a small temple of Hercules and a small statue. Here are the sources of the Melas River, and the Melas empties into the marshy Cephisian [Copaic] Lake. The lake generally covers the most part of the Orchomenian area, but in the winter season, when the wet, south-west wind has swept forth, the water engulfs even more of the land. The Thebans assert that the Cephisus River was diverted into the Orchomenian plain by Hercules, and that for a time it discharged into the sea from under the mountain, until Hercules blocked up the chasm through the mountain … [Pausanias dismisses Hercules's action using Homer's lines.].... Nor is it likely that the Orchomenians would not have found the chasm, and, having broken up the work of Herakles, restored the ancient channel to the Cephisus outlet, since they were wealthy up to the to the Trojan war.

Strabo, *Geography* 9.2.18

[Strabo records a chasm opening up and producing a subterranean channel which drained Lake Copais].... And then when once more the passage ways were filled up, Crates, the mining engineer, a man from Chalcis, ceased clearing away the blockages because of the Boeotian's internal strife, although, as he himself says in a letter to Alexander [the Great], many other places had already been restored, among which some people include the ancient site of Orchomenos while others include Eleusis and the Athens on the Triton River in Boeotia.

9.12 Hydraulic-engineering projects of Claudius

The Emperor Claudius was a bold and imaginative patron of civil-engineering projects. The Claudian harbour and aqueduct are still impressive today even in their ruined state, but his draining of the Fucine Lake elicited the most comment by contemporaries. Like the plain near Thisbe mentioned above (**9.10**), the large, flat-bottomed valley southeast of Rome occupied by the Fucine Lake was a closed catchment area surrounded by mountains. In order to reclaim the rich inundated land for agricultural use, the emperor undertook to cut a permanent drainage tunnel through the mountains. The project was lengthy, difficult, and expensive, and even the celebrations at the discharge end of the drain on opening day almost ended in disaster.

Suetonius, *Claudius* 20.1–2, 32

He brought to completion public works great in scale and utility rather than in number. The following are particularly notable: an aqueduct begun by Caligula, also a drainage tunnel for the Fucine Lake, and the harbour at Ostia. These last two, to be sure, even though he knew that the tunnel had been refused by Augustus to the Marsi, who urgently requested it, and that the port had often been considered by the Deified Julius but given up on account of the difficulty. The Aqua Claudia brought into the city on stone substructures cool and copious water sources, of which one is called Caeruleus and the others Curtius and Albudignus, along with the stream of the Anio Novus, and he distributed them into a very large number of beautifully decorated basins. He attacked the Fucine Lake no less in the hope of profit as of glory, since there were certain individuals who promised that they would drain it at their own expense if the fields uncovered were granted them. After 11 years, partly by digging a canal, partly by tunnelling through the mountain, he finished the project with difficulty, even though 30,000 workers were employed continuously, without a break.... He even gave a banquet above the outlet of the Fucine Lake and was nearly overwhelmed when the water was let out with a rush and overflowed.

Pliny, *Natural History* 36.124–125

I, indeed, consider the tunnel dug through a mountain to drain the Fucine Lake among the most memorable accomplishments of the same Claudius, although maliciously neglected by his successor. The expense was indescribable and the workers numberless over so many years because, where the mountain was of earth, the spoil from the channel had to be cleared out through vertical shafts by means of winches, and elsewhere the rock had to be cut away. All of this great effort, which can be neither comprehended nor described except by those who have seen it, took place in darkness.

I pass over the Ostia port project, also the roadway driven through mountains, the Lucrine Lake separated from the Tyrrhenian Sea by a mole, and so many bridges built at so much expense.

PIPELINES AND AQUEDUCTS
9.13 Tunnels and other water systems

The Greeks were by no means the first Mediterranean people to construct aqueducts as an aid to urban water supply, and as a result of the hydrology, topography, and political structure of Greece their projects usually were small in scale. One exception was the water-supply system devised for the rich island city-state of Samos in the sixth century BC, which included a tunnel 1,036 m long and about 1.8 m high by 1.8 m wide. The successful execution of the tunnel, which was cut from both ends at once, was made possible by the Greek skill in geometry and surveying (Kienast 1995).

Herodotus, *Histories* 3.60

I have extended my discussion of the Samians somewhat because they have completed the three greatest projects among all the Greeks. The first of them is a tunnel open at either end, carried right through the base of a mountain 150 fathoms high. The length of the channel is seven *stades*, the height and width both eight feet. And along its whole length another channel has been cut, 20 cubits deep and three feet wide, through which the water brought from a great spring is conducted in pipes and comes to the city. The tunnel's architect was Eupalinos of Megara, son of Naustrophus. [The second work was the harbour breakwater, and the third the temple of Hera.]

The so-called inscription of Nonius Datus (*CIL* 8.2728) gives an interesting glimpse into the details of a botched tunnelling project through a mountain in Roman North Africa. According to the text, two teams who were tunnelling towards each other from opposite directions with the goal of meeting in the middle both veered off to the right so that the two lots missed each other underground. The project was at the point of being abandoned altogether when Nonius Datus was recalled, and he managed to fix the problem by establishing a cross-link between the two lots (Cuomo 2011, Grewe 2008).

Inscription of Nonius Datus, *CIL* 8.2728

...I came to Saldae and met with Clemens, the procurator. He took me to the mountain where they were lamenting the failed project. They thought it would have to be abandoned because the tunnel, as it had been dug, was already longer than the mountain was wide. It became clear that the excavation had strayed from the straight line so that the upstream portion had veered southward, to the right, and the downstream portion had likewise veered to the right, but northward. And, therefore, the two excavation lots were going astray as they had not followed a straight line. The straight line had been marked with stakes on the mountain surface, running east-west. [...] When I had assigned the work so that everybody knew what their respective portion in the excavation was, I set up a competition between soldiers of the navy and Gallic mercenaries, and in this way, they met in the middle of the mountain. I, therefore, who had first surveyed the elevations and had determined the route of the aqueduct, arranged the construction according to the plan, which I had given to the procurator, Petronius Celer. After completion of the project, the procurator Varius Clemens dedicated the work by admitting the water.

The relatively modest quantities of water provided by public water systems in Archaic and Classical Greece generally were designed to feed public fountains. Although utilitarian in purpose, the draw basins could be surrounded by impressive architectural monuments or sculpture. This type of visible public benefaction was a favourite of the Greek tyrants of the sixth century BC. The contrast between the single famous fountain of Megara and the hundreds of elaborate draw basins in imperial Rome (see **9.17**) is striking (Glaser 1983).

Pausanias, *Description of Greece* 1.40.1

There is in the city [Megara] a fountain, which Theagenes built for them... While he was tyrant, Theagenes built the fountain, which is worth seeing on account of its size and ornament and multitude of pillars.

9.14 The design and execution of a municipal aqueduct

The Romans were the intelligent and creative heirs of the long Greek tradition of finding, guiding, storing, and pumping water. The nature of the Roman political system, however, their skill in building with arches and concrete, and the great size of the Roman urban centres fostered continued developments in hydraulic technology. By the early third century BC Rome already had a system of gravity-flow conduits (aqueducts) carrying significant quantities of water from the nearby mountains, the predecessor of the vast system developed in the late Republic and the Empire. None of the principles involved in the Roman gravity-flow or pressurised water systems is new, but the scale and pervasive applications of the systems go far beyond any Greek model. As this passage makes clear, the intent of the system was delivery of water to public fountains and baths, but private individuals could purchase the right to pipe water directly into their houses. Some of the terms used in paragraphs 5 and 6 are not entirely clear, but Vitruvius obviously is concerned with the high water pressure at the base of a pipeline crossing a valley (an inverted siphon) (Lewis 1999, Hodge 2002, Grewe 2019).

Vitruvius, *On Architecture* 8.6.1–11

Water can be conducted in three ways: by flow in masonry channels, lead pipes, and terracotta pipes. Here are their specifications. If in channels, the construction must be as solid as possible, and the stream bed must have a uniform slope of no less than six inches in every 100 feet. The channel is to be vaulted over so that the sun does not touch the water at all. When it reaches the city walls, a reservoir is to be built, and adjoining the reservoir a triple tank for receiving water. Three pipes of equal bore are to be installed in the reservoir, leading to the receiving tanks, which are connected in such a manner that when the two outside tanks overflow, they pour into the middle tank.

Pipes run from the middle tank to all the basins and fountains, from the second to the baths, that they might provide an annual public income, and from the third to private homes. In this way water for public use will not be lacking, for private parties will not be able to draw it off, since each has its own separate supply from the source. I have set up these divisions so that those who draw water off to their homes for private use might by their rents help the maintenance of the aqueducts by contractors.

But if there are hills along the course between the city and the water source, the following procedure is used. An underground channel is to be dug with the uniform slope described above. If the bedrock is tuff or hard stone, the channel is to be cut directly in it, but if it is earth or sand, a vaulted channel with floor and walls is to be built in the tunnel and the water carried through it in this manner. Vertical shafts are to be cut from the surface every 120 feet.

If, however, the water is to be brought in lead pipes, first a reservoir is to be built at the source, then the gauge of the pipe is to be decided in accordance with the amount of water, and these pipes put in place from the first reservoir to one inside the city walls. Pipes are to be cast in lengths of no less than ten feet. If they are number-100 pipes, each is to weigh 1,200 pounds, if number-80, 960 pounds, if number-50, 600 pounds, if number-40, 480 pounds, if number-30, 360 pounds, if number-20, 240 pounds, if number-15, 180 pounds, if number-10, 120 pounds, if number-8, 100 pounds, and if number-5, 60 pounds. The pipe gauges receive their names from the breadth of the sheets in digits before they are rolled up. For when a pipe is made from a sheet which is 50 digits wide, it is called a number-50, and the rest in the same way.

This aqueduct, then, which is to be made of lead pipes, will have the following arrangement. If the source is on the level of the city without any higher intervening hills capable of interrupting it, but with low spots, it is necessary to build it up to an even level as with the flow in channels. And if the way around these depressions is not long, a detour is made, but if they are unbroken, the water course will be directed along the sunken area. When it comes to the bottom, it is carried on a low substructure to give it as long a level course as possible: this, then will be the *venter* ["belly"], which the Greeks call *koilia*. Then when it comes up against the hill, the long stretch of the *venter* prevents a sudden burst of pressure: the water is forced up to the height of the hilltop.

But if a *venter* has not been built in the low areas nor a level substructure, but an elbow, the water will burst out and split the pipe joints. In addition, *colluviaria* [a word of uncertain meaning; possibly "vents" or "drains"] are to be built along the *venter* to allow air pressure to dissipate. Those who conduct water through lead pipes in this manner will be able to effect it very nicely by these schemes, because the descents, detours, *ventres*, and ascents can be carried out in such a way when the sources and the city are on the same level.

Likewise, it is useful to set up reservoirs every 24,000 feet, so that if a fault should turn up somewhere neither the whole load nor the entire system may be damaged, and it may be easier to find where it has occurred. But these reservoirs cannot be placed on a descent, or on the level *venter*, or on the ascent, or anywhere in the valleys, but on an uninterrupted level course.

But if we wish to incur less expense, we must proceed in the following manner. Terracotta pipes with walls no less than two digits thick are to be made in such a way that they are flanged at one end, so that one pipe can slide into and join tightly with another. Their joints, furthermore, are to be smeared with unslaked lime worked up with oil. At the bends at either end of the level portion of the *venter* a block of red stone is to be placed right at the elbow joint and bored out, so that the last pipe along the descent can be jointed to the stone along with the first of the level *venter*. In the same way, also at the uphill slope the last pipe of the level *venter* is to be jointed to a hollow block of red stone, and the first of the rising pipeline jointed in the same manner.

Once the level course of the pipes has been balanced in this way at the descent and the rise, it will not be displaced. For in channelling water, air pressure so powerful is customarily generated that it can even break the stone blocks apart, unless the water is let in slowly at first and in small amounts at the source and controlled at the elbows and turns by tie bands or the weight of sand ballast. Everything else is to be set up as for lead pipes. Furthermore, when water is first sent in from the source, ashes should be put in first so that the joints might be plugged with ash wherever they have not been sufficiently caulked.

Aqueducts employing terracotta pipes have these advantages. First, that if some defect occurs, anyone can fix it. Second, that the water from terracotta pipes is much more healthful than that from lead pipes. Lead seems to make water harmful for this reason, that it generates lead carbonate, and this substance is said to be harmful to the human body. So if what is generated by it is harmful, it cannot be doubted that it is itself not healthful.

Lead workers can provide us with an example, since their complexions are affected by a deep pallor. For when a blast of air is used in casting lead, the fumes from it infiltrate the parts of the body and, subsequently burning them up, it deprives the limbs of the virtue of their blood. And so it seems that water should in no way be carried in lead pipes if we wish to keep it healthful. Our everyday dining can show that the flavour of water from terracotta pipes is better, for everybody, even when they have tables piled high with silver dishes, nevertheless uses pottery to preserve the taste.

9.15 Specifications for pipes and pressurised systems

Like Vitruvius, Pliny is concerned with the contamination of drinking water by lead pipes, but considers it the most suitable conduit for a reliable pressurised system. The insistence that such a system should frequently rise to the original level lest the flowing water lose momentum has no scientific basis, but – in combination with elevated reservoirs – seems a reasonable precaution to keep the occasional failure of a joint or pipe from immediately affecting the whole system.

Pliny, *Natural History* 31.57–58

For the rest, the most suitable way of conducting water from a spring is by means of terracotta pipes with walls two digits thick and box joints arranged so that the upper fits into the lower and is smoothed with unslaked lime and oil. The fall of the water should be at least one-quarter inch every 100 feet. If it should come by tunnel there must be access shafts every 240 feet. Water that must form a jet should come in a lead pipe. It rises up as high as its source. Should it come by a rather long route, the conduit must go up and down frequently, so that the momentum is not lost. It is proper for pipes to come in ten-foot lengths, and if they are five-digit pipes, to weigh 60 pounds, if eight-digit, 100 pounds, if ten-digit, 120 pounds and so on in this progression. A ten-digit pipe is so-called because the breadth of its sheet before it is rolled up is ten digits, a five-digit pipe is made from one half as wide. There must be a five-digit pipe at every bend along a hilly course, where the pressure must be contained, likewise reservoirs as conditions demand.

9.16 The infrastructure of the metropolis of Rome

Strabo, a native of the Greek East, makes perceptive comments on the differences between Greek and Roman imperial urban centres. The Greeks, in their search for security and local self-sufficiency, neglected the refinements essential to the infrastructure of a Roman city, made possible by the peace and the resources of the Empire. Strabo quite rightly links aqueducts with sewers: since the flow of an aqueduct cannot be turned off (except at the source, often 20 or 30 km away), drains had to be provided to carry off the constant overflow from fountains, baths, and homes, at the same time incidentally washing away the city's sewage. The assertion that most houses possessed a piped-in water supply may not be too much of an exaggeration, to judge from Frontinus' complaints about illegal diversion of the public water supply (**9.21**) (Koloski-Ostrow 2001, Evans 1993).

Strabo, *Geography* 5.3.8

These benefits, then, the nature of the place supplies the city, but through foresight the Romans have added others. For while the Greeks seem to have been particularly successful in setting up their cities, because they endeavoured to make them beautiful, strategically located, with harbours and productive soil, the

Romans gave particular attention to areas the Greeks neglected: paved roads, aqueducts, and sewers capable of washing the filth of the city out into the Tiber. Furthermore, they have also paved the roads in the countryside, adding both cuts through hills and viaducts across valleys so that wagons can take on a ship-load. And the sewers, vaulted over with ashlar masonry, in some places have room left for wagons loaded with hay to pass. So great is the amount of water carried in by the aqueducts that rivers seem to flow through the city and the sewers, and nearly every house has cisterns, pipes, and bounteous fountains. Marcus Agrippa expended the greatest care on these, although he beautified the city with many other structures as well. The old Romans took little interest in the beauty of Rome, involved, so to speak, with other greater and more necessary undertakings, but the later Romans, and particularly those of our present-day, have not fallen behind even in this regards, and they have filled the city with many beautiful structures.

9.17 Rome's water-supply system: a wonder of the world

The statistics quoted by Pliny emphasise the pervasive character of the distribution network within the city: it served large numbers of ornate but functional public fountains, hundreds of public baths, and included even the high elevations within the city. Pliny also makes clear that an efficient drainage system is necessary when water is constantly flowing into the city centre on aqueducts (de Kleijn 2001, Aicher 1995).

Pliny, *Natural History* 36.121–123

But we must speak of marvels that a true evaluation will find unsurpassed. Quintus Marcus Rex, when ordered by the Senate to rebuild the channels of the Appia, Anio, and Tepula aqueducts, brought to Rome within the term of his praetorship a new aqueduct named after himself, driving underground channels through the mountains. Agrippa, too, while aedile, after adding the Virgo and repairing and putting in order the other aqueducts, constructed 700 basins, along with 500 fountains and 130 reservoirs, many of them magnificently decorated, and added 300 bronze and marble statues to these works, and 400 marble columns: all this in the space of a year. In the report of his aedileship, he himself adds that he celebrated games for 59 days and that admission to all 170 baths was made free: these are now infinitely more numerous at Rome.

But the latest project, begun by Gaius Caesar and finished by Claudius, has surpassed the previous aqueducts in expense, inasmuch as the Curtian and Caerulean springs, and the Anio Novus, were brought in from the fortieth milestone at a height sufficient to serve all the city's hills. A total of 350,000,000 *sestertii* was paid out for this project.

Now if someone shall carefully appraise the abundance of water in public buildings, baths, pools, channels, houses, gardens, and suburban villas, the distance the water travels, the arches that have been built up, the mountains tunnelled, and the level courses across valleys, he will acknowledge that nothing more marvellous has ever existed in the whole world.

Pliny, *Natural History* 36.104–6

At that time [first century BC] elderly men marvelled ... in particular at the drains, the most remarkable engineering feat of all.... Seven streams flow through the city and meet in a single channel, and rushing down like mountain torrents, they are forced to snatch up and carry off everything. When they are swollen besides by the volume of rainwater, they shake the bottom and sides of their channels. Sometimes the backwash of the flooding Tiber enters the drains, and the opposing streams of water struggle within them, but the unyielding strength of the construction nevertheless holds up. Great blocks of stone are dragged along the streets above, yet the tunnels do not collapse....

9.18 An evaluation of the water-supply system of Rome

Frontinus, a general and statesman of the second half of the first century AC, served for several years during the reigns of Nerva (AD 96–98) and Trajan (AD 98–117) as the official in charge of Rome's water supply. As an efficient, reform-minded bureaucrat, he kept careful records of his enquiry into abuses within the system and prepared an account of his term that fortunately has survived. He claims to have written the account to serve as a handbook for subsequent administrators. The document is an invaluable source of information on procedures and problems within the water department in Rome, and the alternately testy and satisfied comments by the author clearly reveal his character and the political dimension of the work. His pithy comparison of the water system with earlier architectural marvels is justly famous as an indication of a fundamental Roman attitude (Blackman and Hodge 2001).

Frontinus, *On the Aqueducts of Rome* 1.16

With such numerous and indispensable structures carrying so many waters, compare, if you please, the idle pyramids, or else the indolent but famous works of the Greeks.

9.19 Frontinus' administrative initiatives: the courses of Rome's aqueducts

By the late first century AC, nine aqueducts served the city, each with its own special characteristics of height, volume, clarity, temperature, and taste. Frontinus addressed directly the problems specific to each system.

Frontinus, *On the Aqueducts of Rome* 1.17–19

It has seemed to me not incongruous to describe as well the length of each aqueduct's channel, according to the type of structure. For since the greatest part of this office consists in their protection, the director ought to know which requires the greater expenditure. Indeed, merely looking over details did not satisfy our sense of duty, but we took care also to have plans of the aqueducts made from which one can see where the valleys are, and how big, where rivers are crossed, where conduits placed on mountainsides require a greater and constant care in protecting and reinforcing their channels. From it springs the useful result that we can have the structures directly in view, so to speak, and deliberate as if standing beside them.

All the aqueducts reach the city at different elevations. As a result, some discharge at higher places, while others cannot reach any sort of elevation, for on account of frequent conflagrations the hills have gradually grown higher with fill. There are five whose head can reach every point in the city, but of these some are driven by more pressure, others by less. The highest is the Anio Novus, next the Claudia, the Julia has third place, the Tepula fourth, then the Marcia, which at its source is equal to the level of the Claudia. But the ancients brought water channels to the city at a lower elevation, either because the art of levelling had not yet been brought to refinement or because they intentionally placed them underground to make them harder for the enemy to cut, since they were still frequently waging war on the Italians. Now, however, in certain places, where a channel has collapsed through age, as a short-cut the subterranean circuit of valleys is left aside, and the channel is carried across on substructures or arches. The Anio Vetus has the sixth place in elevation, but it would likewise be capable of supplying even the higher points of the city if it were lifted upon substructures or arches where the situation of valleys or depressed areas requires it. The Virgo comes after it in elevation, then the Appia. These could not be raised to any great height since they were brought in from the vicinity of the city. The Alsietina, which serves the region across the Tiber and particularly low-lying areas, is the lowest of all.

Six of these streams are received in covered cisterns inside the seventh milestone of the Via Latina, where they throw their sediment as if catching their breath after the run. Their volume is measured as well, by means of calibrated scales set up in the same place.

9.20 Specifications for pipes and adjutages at Rome

As Frontinus himself makes clear, there was dispute concerning the precise dimensions of the various gauges of pipe used in the water system at Rome. The ambiguities, resulting both from the empirical character of Roman science and from deficiencies in mathematics, were exploited by unscrupulous personnel in the water department to divert water for sale to private individuals. The complicated science of hydromechanics was still in its infancy, but Frontinus obviously understood in this case the important role played by velocity, and ultimately by pressure, in determining the discharge of a conduit.

Frontinus, *On the Aqueducts of Rome* 1.23–25, 29, 31, 33–36

Since I have gone through the builders of each aqueduct and their dates, besides the point of origin and length of the channels and their respective elevations, it does not seem to me inappropriate to add as well some details and to point out how great is the abundance of water, which is sufficient not only for public and private uses and applications but truly even for pleasure; also to point out by how many reservoirs it is distributed and to which regions – how much outside the city and how much within – and of this last how much is delivered to basins, how much to ornamental fountains, how much to public buildings, how much in the emperor's name, and how much for private uses. But I think it is reasonable, before we bring up the terms *quinaria, centenaria,* and other adjutages by which water is metered, to indicate as well what their origin is, what their capacities, and what each term signifies; and, once I have set down the rules by which their proportions and basis are computed, to show how I found discrepancies, and what method of correction I followed.

The adjutages for water have been set up according to standards of either digits or inches.... Now the digit is agreed to be the sixteenth part of a foot, the inch the twelfth part. But just as there is a difference between the inch and the digit, so also there is no single standard for the digit itself. One is termed square, another round. The square is larger than the round by three-fourteenths of itself, the round is smaller than the square by three-elevenths of itself, clearly because the corners are omitted.

Later on, an adjutage termed *quinaria* came into use in the city, based neither on the inch nor on either of the digits and excluding the earlier standards. Some think it was introduced by Agrippa, others by workers in lead influenced by the architect Vitruvius. Those who make Agrippa the inventor say it took its name from the fact that five of the former adjutages, or perforations, so to speak, through which water used to be distributed when there was little of it, were gathered up in a single pipe. Those who favour Vitruvius and the lead workers depend on the fact that a flat lead sheet five digits wide makes a pipe of this dimension when rolled up. But this is inexact, since when it is rolled up the inner surface contracts and the exterior stretches. It is most probable that the *quinaria* is so-called from its diameter of five-fourths of a digit, a standard which continues in the following adjutages as far as the size 20 pipe, the diameter of each increasing by the addition of a single quarter of a digit: so in the size 6 pipe, which is six-quarters of a digit in diameter, and the size 7 pipe, which is seven, and successively with a like increment as far as the size 20 pipe....

For larger pipes the standard is based on the number of square digits contained in the cross-section – i.e. the opening – of each adjutage, from which sum the pipes get their names. For those which have a cross-section – i.e. a round opening – of 25 square digits are called number-25 pipes. Likewise, number-30 pipes and so on with equal increments of five square digits up to the number-120 pipe...

The gauging of the pipes from the number-5 up to the number-120 is consistent for all the adjutages just as we have shown.... But the water-men (*aquarii*), while they agree with the public calculation of most, have made alterations in four adjutages: the number-12, number-20, number-100, and number-120....

And so, since they diminish the number-20 pipe by which they always distribute water and enlarge the number-100 and 120 pipes by which they always receive it, they divert in the case of the number-100 pipes 27 *quinariae* and in the case of the number-120 pipes 86 *quinariae*....

The total number of adjutages is 25....

Let us remember that every water course, in cases where it comes from a higher place and pours into a reservoir within a short distance, not only corresponds to its adjutage but even surpasses it, while in cases where it comes from a less elevated spot, that is with less pressure, and has a longer course, by the slowness of its flow it falls short of the proper measure. So for this reason it needs either restriction or assistance at delivery, respectively.

But the position of the pipe also has an effect. Placed at a right angle and level, it keeps the proper measure. Turned towards the flow of water and sloping downward, it draws away more than the proper measure. If it is oriented oblique to the passing water and sloping downstream, i.e. less suitable for drawing water, it takes it in slowly and in small volume. The *calix*, however, is a bronze adjutage pipe that is set into the channel or reservoir and pipes connected to it. Its length ought to be no less than 12 digits, its cross-section such a capacity as has been specified. The rationale seems to be that bronze is hard and less malleable and cannot easily be expanded or contracted.

9.21 The fraudulent diversion of public water in Rome

It is clear from Frontinus' description of easily detected abuses that he took over a government department accustomed for some time to corruption at all levels. His comment that the tearing up of illegal pipes brought in a considerable quantity of scrap lead shows that he took decisive action, probably leaving hundreds of shopkeepers and home owners scrambling to make new arrangements. Plutarch, *Themistocles* 31, demonstrates that theft from the public water supply was not only a Roman problem.

Frontinus, *On the Aqueducts of Rome* 2.75–76, 112–115

The next irregularity is that one measure is adopted at the beginning of a water course, another, significantly smaller, at the reservoirs, then the smallest at the point of distribution. This is the result of fraud by the water-men, whom we have caught drawing off water from the public aqueducts for private use. But in addition, many landowners whose fields the aqueduct passes by tap the water channels, and in consequence it comes about that they choke off the flow of the public aqueducts for the sake of private parties, even for irrigating their gardens.

Nothing more or better can be said about crimes of this sort than what Caelius Rufus stated in his oration "On the Water Supply". I wish that we did not have the violations to prove that the water supply is still now wrongly appropriated with equal boldness: we have uncovered irrigated fields, workshops, even attic apartments, and, finally, houses of ill repute, all fitted out with a constant supply of water under pressure....

.... In many reservoirs I have found certain adjutage pipes larger than had been authorised, and of these some that had not even been stamped. For whenever a stamped adjutage pipe exceeds its legitimate measure, it exposes the dishonesty of the official who stamped it. When, however, it is not even stamped, clearly everyone is found at fault, particularly the one accepting it, but also the overseer. In certain reservoirs, although adjutage pipes of legitimate size had been stamped, pipes of large diameter were attached immediately to them. From this it resulted that the water was not restricted by the regulation length of pipe [about nine inches], but was squeezed through a short narrow section and easily filled the wider pipe directly following. Therefore this additional precaution should be taken, that whenever an adjutage pipe is stamped, the adjoining pipe also should be stamped for the distance, which we stated was indicated in the Senate's decree [50 feet stated in section 105] For not until then will the overseer lack any excuse, when he knows that only stamped pipes may be installed.

In installing adjutage pipes, it is necessary as well to take care that they are set up in a line and that no adjutage is placed lower than another, or higher: the lower draws more, the upper less, because the stream of water is taken up by the lower. The pipes of some reservoirs lacked adjutages altogether. These pipes are termed "uncontrolled" and are enlarged or contracted at the pleasure of the water-men.

There is still another intolerable type of fraud practised by the water-men: when water rights are transferred to a new owner, they put a new hole in the tank but leave the old one for drawing water for sale. I believe, then, that this practice should be among the first corrected by the water commissioner. For it pertains not only to the protection of the water supply itself, but also to the care of the reservoir, which is weakened when tapped frequently and without good reason.

Likewise, that form of obtaining money, which the water-men call "puncturing" is to be eliminated. There are extensive areas scattered throughout the city through which pipes pass, hidden beneath the pavement. I have learnt that these, tapped here and there by men called "puncturers", provide water through branch pipes to all the businesses along their course and bring it about that only a small measure of water reaches the points of public access. How much water has been stolen in this way I estimate from the fact that a not insignificant amount of lead has been brought in from tearing up this sort of branch pipe.

9.22 An account of the water totals entering Rome

Frontinus' account of the total amount of water supplied by the aqueducts is frustrating, because he uses the *quinaria* as his unit of measure (see **9.20**). This was not a direct measure of volume, but of capacity, the amount of water that would be discharged under pressure by a pipe 1¼ digits in diameter. Naturally this amount will vary according to circumstances, and estimates of its value range from 75,000 to 340,000 litres every 24 hours. The total of 14,018 *quinariae* in consequence could range from *c.* 1,000,000 m³ to 4,750,000 m³/day. Modem calculations of the total supplied by Rome's water system run between 680,000 m³ and 900,000 m³/day. It is obvious, however, that Frontinus feels his reforms had brought about significant gains.

Frontinus, *On the Aqueducts of Rome* 2.77–78, 87–88

There remains the task of setting out in order the discharge of the aqueducts (which on taking office we found totalled up and, so to speak, in a lump sum – in fact, even attributed to the wrong aqueducts) individually by name and by regions of the city. I know that the summary of this information can seem not only dry but even baffling, however I will set it down as briefly as possible so that the regulations, as it were, of this office might not lack anything. Those who find it sufficient to know the totals may pass over the details.

The distribution, then, of the 14,018 *quinariae* has been calculated so that the 771 *quinariae* that are diverted from some aqueducts to help supply others and appear twice in the summary of distribution figure only once in the summation. Of this total 4,063 *quinariae* are delivered outside the city: 1,718 *quinariae* in the name of Caesar, 2,345 to private parties. The 9,955 *quinariae* remaining are distributed within the city to 247 reservoirs. From these are delivered in the name of Caesar, 1,707½ *quinariae;* for public uses, 4,401 *quinariae:* of this last sum, to ... military camps, 279 *quinariae*, to 75 public buildings, 2,301 *quinariae*, to 39 ornamental fountains, 386 *quinariae*, to 591 public basins, 1,335 *quinariae*....

This is the summary calculation of the water supply of Rome up to the Emperor Trajan and the manner in which it was distributed. Now by the foresight of this most careful ruler whatever was fraudulently diverted by the watermen or lost by carelessness has increased the supply, as if by the discovery of new water sources, and the flow has been nearly doubled....

The queen and mistress of the world day by day feels the results of this concern on the part of her most dutiful ruler Trajan, and her health will feel it all the more as the number of reservoirs, public works, ornamental fountains, and basins increases. No less profit accrues to private parties from the increase in his grants: even those who drew an illegal supply in fear now enjoy an official grant in security. Not even the overflow water is lost: the appearance of the city is changed and neat, the air cleaner, and the causes of the oppressive atmosphere for which the city air was so infamous among the ancients have been removed....

9.23 Variations in the quality and applications of aqueduct water

Because of variations in the quality of the water, most aqueduct streams were kept separate so that flow could be directed to appropriate applications. Irrigation and sewer flushing were lowest on the scale.

Frontinus, *On the Aqueducts of Rome* 2.92

It was decided therefore to keep all the aqueducts separate, and besides that each of them be regulated in such a way that first of all the Marcia might serve only for drinking, and then the rest – each according to its own particular quality – should be allotted suitable applications, so that the Anio Vetus, for many reasons (for the lower down a stream is tapped, the less healthful it is), might be applied to the irrigation of gardens and for the more base tasks of the city proper.

9.24 Administration of the water supply at Rome

Like all efficient bureaucrats, Frontinus issued regulations for the operation of his department. The stipulations are clear and efficient, but it is not known whether his successors followed his recommendations.

Frontinus, *On the Aqueducts of Rome* 2.103, 105, 107, 109

Now I will append what the water commissioner must observe and the laws and decrees of the Senate that pertain to setting up his procedures. As for the right of private parties to draw water, it is to be watched that no one draw it without the authorisation of Caesar – that is, that no one draw public water that has not been granted and that no one draw more than has been granted. For in this way we will bring it about that the full measure, which we have stated has been regained, can be assigned to new fountains and to new imperial grants. In both cases, however, great care must be exercised against the various types of fraud. Careful frequent rounds should be made of the aqueducts outside the city to inspect the grants. The same must be done for reservoirs and public fountains, so that the water might flow day and night without a break....

Whoever wishes to draw water for private use must obtain a grant and bring a permit from the emperor to the water commissioner. The commissioner then must promptly execute the grant.... The contents of the permit must also be made known to the overseers, lest they try to cover up their negligence or fraud by pleading ignorance. The superintendent must summon the levellers and arrange to have an adjutage pipe of the granted gauge stamped, and he should diligently direct his attention to the scale of gauges, which we discussed above, and take notice of their positioning, that it not be left to the discretion of the levellers to approve now an adjutage of larger bore, now of smaller, according to the influence of the parties involved. Nor should the option be left them of joining any sort of lead pipe immediately to the adjutage pipe, but for 50 feet one of the same gauge as that stamped on the adjutage pipe....

The right to a water grant does not pass to an heir, a buyer, or any new proprietor of the estate....

When water rights become vacant, an announcement is made and entered in the records, which are consulted so that grants to applicants might be made from vacant water rights. It used to be the custom to turn off such a water supply immediately, so that in the meantime it might be sold either to the owners of the estate or even to other parties. It seemed more humane to our ruler to allow a period of 30 days grace and avoid cutting off an estate suddenly....

9.25 Regulations for maintaining Rome's aqueducts

The detailed recommendations for maintenance of the water system are a strong reminder of the complicated and delicate character of the system. The high density of water and the destructive potential of the moving liquid carried by the aqueducts made continuous surveillance and maintenance essential. Even a small leak could rapidly cause serious damage to a conduit or the substructure supporting it.

Some stretches were more susceptible to damage than others and some seasons more propitious for repairs. Repairs were complicated by the need to shut off the water at the source of the channel or to divert it around the repair and by the paramount consideration of causing as little disruption as possible to the overall public water supply. Regulations concerning the easements on private land were intended to avoid the possibility of damage to the channels and allow easy access for maintenance and repair.

Frontinus, *On the Aqueducts of Rome* 2.116–125, 127

The topic of aqueduct maintenance remains. But before I begin to speak about it, I must explain a few things about the companies of slaves that have been set up for this purpose. There are two companies: one public and one belonging to Caesar. The public one is older, as we have said, left by Agrippa to Augustus, and by him made public. It contains about 240 men. The total of Caesar's company is 460; Claudius set it up when he brought his aqueduct into the city.

Each company, then, is subdivided into several different types of workers: overseers, reservoir-men, inspectors, pavers, plasterers, and other craftsmen. Some of these ought to be outside the city at tasks that are not great undertakings but nevertheless seem to require an immediate remedy. The men in the city, at their stations by the reservoirs and fountains, will apply themselves to various jobs, in particular to sudden emergencies, so that from several regions an abundant water supply can be diverted to the neighbourhood where its immediate assistance is needed. Both of these companies, amply large, used to be drawn away to private projects by favouritism or the negligence of their supervisors. I have undertaken in the following way to recall them to some discipline and to their public duties: I write down the day before what is to be done next, and a record is made of what has been done each day.

The stipend for the public company is paid by the state treasury, an expense, which is compensated for by income from rents on water rights. These are levied on places or buildings in the vicinity of aqueducts, reservoirs, ornamental fountains, or basins, an income of nearly 250,000 *sestertii* ... Caesar's company receives its stipend from the imperial purse, and the same fund pays for all the lead and all the expenses relating to conduits, reservoirs, and basins.

Since we have related what seems to concern the slave companies, we will turn – as promised – to the maintenance of the aqueducts: a matter worthy of more earnest care, since it is the special expression of the greatness of the Roman Empire. The numerous and extensive works are subject to continuous decay, and they should receive care before major repairs are needed. Very often, however, they should be maintained with prudent moderation because one cannot always trust those who seek either to construct works or to extend them. The water commissioner, therefore, ought to be equipped not only with knowledgeable experts but also with personal experience, and he should not make use only of the architects of his own office but also call upon both the trust and the perception of many others, that he might determine what must be executed immediately, what should be put off, and, further, what should be done by contractors, what by craftsmen from the bureau.

The need for repairs stems from these causes: there is damage from the careless behaviour of land owners, from age, from the force of storms, and as a result of faulty workmanship, which has appeared more often in recent years.

Those sections of aqueducts that are supported on arches or are applied to the sides of hills and those that pass over a river on arches are generally most affected by age or the force of storms. In consequence, these structures must be set in order with anxious haste. The underground channels, not exposed to frost or heat, are less subject to damage. Defects, however, are of two kinds: either they can be remedied without turning off the flow or they cannot be made good without diverting the flow, such as repairs that have to be carried out in the channel itself.

This last type of problem arises from two causes: either the water channel is constricted by the precipitation of a deposit that sometimes forms a hard crust, or else the lining of the channel is damaged, allowing leaks by which the sides of the channel and the substructures necessarily are injured. Sometimes the very piers constructed of tufa also collapse beneath so great a weight. Problems with the water channels should not be rectified during the summer, lest their use be interrupted at the time it is most needed, but rather in the spring or autumn, and with the greatest possible speed – everything, of course, being made ready ahead, so that the flow might be interrupted for as few days as possible. Everyone is aware that this is to be done for single aqueducts at a time, so that the city might not lack water through simultaneous diversion of several.

Repairs that ought to be carried out without interrupting the flow of water consist primarily of masonry work, which should be done well and in the right season. The suitable season for masonry work runs from the first of April to the first of November, but it would be best to leave off during the hottest part of the summer because moderate weather is needed if the masonry is to cure and harden into a single mass. A too intense sun detracts from the material no less than frost does, and no type of construction requires more diligent care than that meant to contain water ...[1].

No one, I think, will doubt that the stretches of the aqueducts closest to the city, i.e. those which from the seventh milestone have been built of ashlar masonry, should be watched over with particular care, both since they are large in scale and because each carries several water channels. If it were necessary to interrupt them, the city would lose the greater part of its water supply. There are solutions, however, even for this difficulty: a temporary construction is built up to the level of the channel being taken out of use, and a conduit of lead troughs across the gap of the broken channel makes it continuous again. Furthermore, since nearly all the channels were laid out across private land and the provision for future expenses seemed difficult, a resolution of the Senate was passed which I append:

... be it resolved that when the channels, conduits, arches are repaired that Augustus Caesar promised the Senate he would repair at his expense, earth, clay, stones, potsherds, sand, wood, and other materials of which there is need for this repair may be taken, removed and brought from the estates of private parties, whence each of them most expeditiously can be taken, removed, and brought without injury to the private parties, and the value estimated by an honest man is to be paid, and as often as it is necessary, passage be granted and lanes left open through private land for removing all these construction materials for the purpose of repairing the aqueducts, without injury to private parties....

... be it resolved that a space of 15 feet be left clear on either side of fountains, arches, and walls, that five feet be left on either side of underground channels and conduits within the city and within buildings adjacent to the city, and that henceforth it is not permitted to build in this area a tomb or any building, or to plant a tree, and if there are already trees within this space, they shall be cut down, unless connected with a villa or enclosed by buildings.

Ovid, *Metamorphoses* 4.121–24

And as he was girt, he plunged his sword into his side and without delay, in death, drew it from the searing wound. As he lay stretched out on the ground, the blood gushed high as when a damaged lead pipe splits open, and from the small, hissing crack shoot long streams of water and in jets burst through the air.

WATER-LIFTING DEVICES
9.26 The mechanics of a *shaduf*

The diversion, confinement, and guidance of water were all important procedures in both pre-Classical and Graeco-Roman land reclamation, irrigation, and urban water supply. The technology of lifting water, however, did not become mechanised, and thus capable of increased output, until the Hellenistic period. Even then, the difficulty of the task restricted the output and thus the application, of pumping devices. One device already in use from the fourth millennium BC was the *shaduf* (the modern Arabic name), a hinged pole with a bucket suspended from one end and a counterweight from the other. Although a low-output device, the *shaduf* is simple and inexpensive, so it continued in use alongside more sophisticated pumps in the Graeco-Roman period, and has remained in active use up to the present [**fig. 9.2**]. For more illustrations of this device and the other water-lifting devices mentioned below, see Oleson 1984.

[Aristotle], *Mechanical Problems* 28.857a–b

Why do they construct the *shaduf* at wells the way they do? For they add to the wooden beam a lead weight, the bucket itself having weight whether empty or full. Is it that the work is divided into two moments (for it is necessary to dip it and to haul it up again) and it happens to be easy to send the empty bucket down, but hard to haul it up full? There is an advantage, then, in letting it down a little more slowly, in proportion to the great lightening of the load as one draws it up. The lead weight or stone attached to the end of the pole accomplishes this. For the individual lowering the bucket must overcome a greater weight than if he were to let down the empty bucket alone, but when it is full, the lead pulls it up,

Figure 9.2 Egyptian wall painting with *shaduf*.
Source: Adapted from Tomb of Ipuy, c.1295–1213 bc; White, 1984, fig. 163.

or whatever weight has been attached. So overall, both actions are easier for him than in the previous circumstances.

9.27 Militant oxen working a Persian irrigation device

Plutarch, quoting information from the Hellenistic historian Ctesias of Cnidus, mentions another water-lifting device probably developed in the Bronze Age: the *cerd* still in use throughout Asia, consists of a large leather bag with a self-dumping mechanism, raised and lowered in a well by draught animals walking up and down a sloping ramp. It is this oscillating motion that facilitated calculation of the work accomplished by the punctilious oxen in the royal gardens of the Persian king.

Plutarch, *Moralia* 21.974c

Some animal natures have an understanding of numbers and the ability to count – for example the cattle around Susa. For in that region they water the royal garden with buckets raised by wheels. The number of bucketfuls is defined: each cow raises 100 every day, and it is not possible either to trick them or to force them to raise any more. Just to make sure, their keepers often try to add to the number, but the cows halt and will not do any more once they have produced the allotted number, so carefully do they calculate and remember the sum. Ctesias the Cnidian relates this in his history.

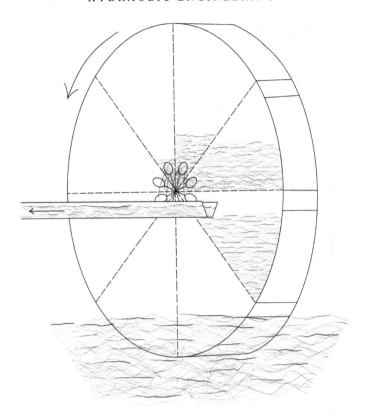

Figure 9.3 Wheel with compartmented body (axle not shown).
Source: Adapted from Landels, 2000, fig. 15.

9.28 Several wheel-like devices for raising water

The best single surviving description of the water-lifting devices developed in the Hellenistic period at the Museum in Alexandria is Vitruvius, *On Architecture* 10.4–7. In this passage he describes the wheel with compartmented body, wheel with compartmented rim, bucket chain, water screw, and force pump. These devices, too, have remained in use up to the present. Although more complicated and expensive to construct than the *shaduf* or *cerd*, the compartmented wheels could raise water higher, more efficiently, and in greater quantity. They could be driven by draught animals through an angle gear (see **9.30**), by men treading the rim, or even by water-power. Such wheels and bucket chains were put to work lifting water for irrigation, drinking, baths, and industrial uses, and in draining mines, dry-docks, and excavation areas.

Vitruvius, *On Architecture* 10.4.1–4, 5.1

Now I will explain the devices that have been invented for raising water, how the various designs are contrived. First, I will speak about the *tympanum* ("drum"; **fig. 9.3**). This device, to be sure, does not lift water to a great height,

HYDRAULIC ENGINEERING

but it discharges a great amount quickly. The axle is turned on a lathe or made by means of compasses and its ends capped with iron sheeting. Around the middle it has a drum made of planks joined together, and it is mounted on beams, which have iron bearings to carry the axle ends. In the interior of this drum are set eight radial partitions running from the axle all the way to the circumference, dividing the interior into equal sections. Around its outer surface are fixed planks with six-inch openings for receiving the water. Likewise, close to the axle there are small round holes in one side corresponding to each compartment. When this device has been pitched ship-fashion, it is set in motion by men treading it. It scoops up the water through the openings around the circumference and discharges it through the circular openings near the axle into a wooden trough connected with a conduit. In this manner a quantity of water is provided to gardens for irrigation or to salt pans for dilution.

When, however, the water has to be raised higher, the same principle will be put to use in this manner. A wheel will be built around the axle, of a large enough diameter so that it can reach the height that is required. Rectangular compartments will be fixed around the circumference of the wheel and made tight with pitch and wax [**fig. 9.4**]. Thus, when the wheel is turned by men treading it, the containers will be carried up full to the top of the wheel and on their downward turn will pour out into a reservoir what they have themselves raised [**fig. 9.5**].

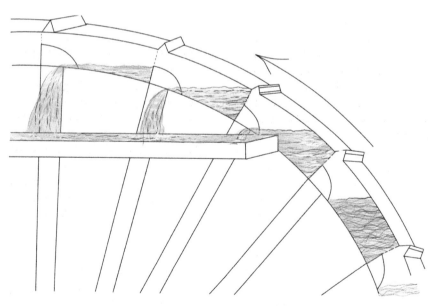

Figure 9.4 Wheel with compartmented rim (detail of compartments).
Source: Adapted from Landels, 2000, fig. 16.

HYDRAULIC ENGINEERING

Figure 9.5 Pair of compartmented wheels (axles and frame not shown).
Source: Adapted from Landels, 2000, fig. 17.

But if a supply is required at still greater heights, a double iron chain will be set up, wound around the axle of the same sort of wheel and allowed to hang down to the lowest level, with bronze buckets the capacity of a *congius* suspended from it [**fig. 9.6**]. Thus, the turning of the wheel, by winding the chain over the axle, will carry the buckets to the top, and as they are borne over the wheel they will necessarily turn over and pour out into a reservoir what they have raised.

Wheels of the same design as has been described above (in **4.3**) can also be set up in rivers. Around the circumference are fixed paddles, which, as they are struck by the force of the river, move along and cause the wheel to turn. And in this manner drawing up the water in compartments and carrying it to the top without the use of labourers for treading, the wheels are turned by the force of the river itself and provide what is needed.

9.29 A Hellenistic water-powered bucket chain

Among the works of Philo of Byzantium, a mechanician and scientist of the late third century BC, is a book of pneumatic devices, including water-lifting machinery. Some of the devices may go back to the mid-third-century BC mechanician Ctesibius of Alexandria, whose works are lost, but even Philo's book survives complete only in an Arabic translation. Vitruvius may also have drawn upon

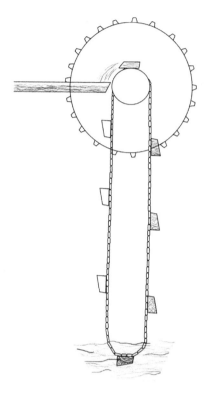

Figure 9.6 Tread-wheel driven bucket chain.

Source: Adapted from Landels, 2000, fig. 18.

the original works of Ctesibius for his bucket chain (see **9.27**), but Philo's description is more detailed. It also has the added feature of a waterwheel drive, although in the form presented it probably is incapable of realisation.

Philo of Byzantium, *Pneumatics* 65[2]

Construction of another elegant device [**fig. 9.7**]. Let us build yet another device that might be used for many other functions. With it water can be lifted from rivers or other places in order to deliver it to elevated places to water gardens and farms. This water can also be lifted to flow into fortresses and elevated hidden places. The river that is to be used for irrigation with this device must have a strong current flowing downhill, copious enough in relation to the water, which this device lifts.

Let us construct a rectangular building similar to a tower. Its proportions are such that it is not weakened by its height, and it is removed a certain distance from the river so that the mass of the river's water might not enter the space from which the water is drawn. Let the front and back part be spaced so that this

HYDRAULIC ENGINEERING

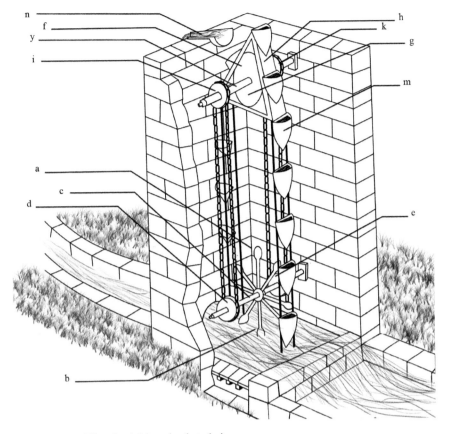

Figure 9.7 Paddle-wheel driven bucket chain.
Source: Adapted from Carra de Vaux, 1903, 210.

construction is restricted to the place where the water is drawn. A wooden floor is placed on these foundations, resting on masonry, and the water is directed over it. A trench is cut from the river up to this building. This trench is one and a half fathoms deep below the level line, that is below the water level in the trench. The sides of the trench are solidly built, and its bottom made of lime and plaster, carefully worked, up to where it empties into the basin. This basin has two walls six cubits long, and its width is such that the device, which discharges the water for irrigation, is fixed within it.

This device is set up on a very solid cross-shaft, an axle, which carries pulley wheels two cubits in diameter. Each end of the axle is clad and fitted into a square bearing member presenting a socket in which it can turn easily. The entire device is solidly fitted, because the movement is strong. Another solid axle is placed in the upper part of the tower, similar to that which we have described in its lower part. The irrigation wheel is in the middle of this axle. Its diameter is four cubits. Let there be a triangular device made of copper: its sides have such length as results when they are tangent to the radius of the waterwheel, and its width is one cubit. At each end of the axle are identical pulley wheels, like the pulley wheels, which we described in the lower part. The waterwheel, on which is the triangular device, is placed in the middle, and the pulley wheels are fixed to the axle.

The lower waterwheel is labelled *a b*, its axle *c*, the pulleys *d e*. The upper wheel is labelled *f g*, the triangular device *h*, the [upper] axle *y*, and the [upper] pulley wheels *i k*.

It is also necessary to prepare an iron structure similar to a vertebral column, fastened to the pulley wheels, whose length is such that, being placed on the triangular structure, it comes to within one cubit of the basin floor. The length of each of its segments is also one cubit. It is hinged by means of iron nails.... Then prepare some rectangular copper or wooden buckets, pinned to this device, and jointed on their inner sides. They are labelled *m*. Let the iron structure be placed around the triangular device, as we have said. The pinned buckets are labelled *m*. If the axle is made to turn by force, the triangular device turns, and the buckets rise up, full of water. The portion of the construction that holds the buckets necessarily falls on the corners of the triangle, so that when the wheel turns and the buckets have filled themselves, they empty. They empty above point *n*. Beneath the spot where the buckets empty, place a container to receive the water and make it flow towards the channel placed on masonry pillars, as we have described.

It is left to explain how the axle moves without anyone approaching it and lifts the water by means of the buckets. The device must dip in the water we have mentioned, coming from the trench. Ducts are contrived, which empty on the containers of the waterwheel. Let them be thick and strong. These ducts are so arranged that when the containers are full, the lower axle moves with great force. When this lower axle moves with force and vibrates, the upper one moves as well because of the chains on which the buckets are mounted. Four containers fill on each section of the waterwheel device, and each has a capacity of two *kouz*. The discharge depends on the abundance or scarcity of the water.

It should be noted that, of the water wheels, the largest are the water wheels with the triangular device. If the apparatus has enough force to lift 20 buckets, it should have a height of 60 cubits, and the raising of the buckets will be easy. Enough water must be left in the basin for the buckets to be immersed and fill up. The water that is in excess of this amount ought to be drained off by another trench going downhill. This apparatus is built as we have said. Here is the figure.

9.30 A gear-driven bucket chain in Roman Egypt

Animals could be used to drive water wheels and bucket chains through an angle-gear drive: the animals walked in circles yoked to a turning, vertical axle carrying a large horizontal wheel with upright, peg teeth around its circumference (a crown gear). These teeth meshed with those of a second crown gear oriented at a right angle to the first, carried on a turning horizontal axle that also carried the waterwheel or chain of containers. As the animals walked in a circle, the horizontal wheel drove the vertical wheel that turned the axle of the water-lifting element. This gear train, probably invented in the second century BC, also appears in the water mill described by Vitruvius (see **2.13**). The device is still used for water-lifting today, under the Arabic name of *saqiya*, essentially unchanged from the arrangement observed by Sulpicius Severus around AD 405.

Sulpicius Severus, *Dialogues* 1.13

So, when I entered the edge of the desert – I had with me as a guide one of the monks who knew the area well – at about 12 miles distance from the Nile we came to a certain old hermit living at the foot of a mountain. There was a well at that spot, a very rare thing in those parts. He possessed an ox whose only task was to raise water by driving a machine fitted with wheels (*machina rotalis*), for the well was reputed to be about 1,000 feet deep or more. There was a garden there, well-supplied with numerous green vegetables – this, indeed, quite against the nature of the desert, where everything is dried out, burnt up by the heat of the sun, and bears no seed or a feeble root. But in the case of that saintly old man the labour he shared with his animal together with his own hard labour produced this result: for the copious watering gave such fertility to the sandy soil that we saw the vegetables of that garden flourishing and bearing fruit marvellously.

9.31 The design and construction of a water screw

Ancient sources tell us explicitly that the water screw was invented by the famous third-century BC scientist Archimedes (see **9.32**, **5.20**; cf. Dalley and Oleson 2003). This exquisitely simple and efficient pumping device may be an offshoot of the wheel with compartmented rim, modified by Archimedes in the light of his research on spirals. The whole device turns as a unit, motivated by a worker treading the barrel (see **9.33**): as the spirals turn, they lift discrete quantities of water and eventually dump them from the upper end of the tube. The water screw is still in use as a pump today, executed in modern materials and modified for propulsion by a motor or by a hand crank at the upper end of the axle.

Vitruvius, *On Architecture* 10.6.1–4

There is besides a screw design, which lifts a great quantity of water but does not raise it as high as the wheel. The contrivance is prepared in the following fashion [**fig. 9.8**]. A beam is taken, as thick in inches as it is long in feet. This is rounded off like a cylinder. The circumference of either end will be divided up by a compass into fourths and eighths, and thus into eight parts. And let these lines be placed in such a fashion that when the beam is oriented horizontally the radii at either end correspond precisely. The length of the beam is to be marked off into segments equal to one-eighth of the circumference, and again, after laying the beam flat, lines are to be drawn from one end to the other, corresponding exactly. There will in consequence be squares of equal size described on the exterior. Where the longitudinal lines are drawn in this fashion, they will intersect the cross-lines, and these intersections will be marked as points.

When these lines have been marked perfectly, a thin strip cut from willow or osier is coated with liquid pitch and affixed to the first point of intersection. It is then stretched out at an angle across the succeeding intersections of the longitudinal and circumferential lines and continues on in the same manner, passing across each point in order and in twisting around is fixed to each, until – passing along from the first to the eighth part – it arrives at and is fixed to the same longitudinal line to which its beginning was attached. In this manner it advances as far in length up to the eighth point as it does obliquely past the eight points. Strips of wood fixed obliquely in the same manner over the whole surface, at each intersection of the longitudinal and circumferential lines form channels wound around the eight divisions of the thickness in a perfect and natural imitation of a snail shell.

Other strips are fixed upon this track in the same manner, one upon the other, smeared with liquid pitch, and are stacked up until the height of the whole stack equals one-eighth of the length. Planks are fixed around the circumference of the spirals to cover them. Then these planks are smeared with pitch and bound with iron bands, so that they may not be dislodged by the effect of the water. The ends of the shaft are capped with iron. On the right and left of the screw, beams are placed with crosspieces at each end fixed to both. In this manner the screws can be turned by men treading them.

The mounting of the screw will be at such an angle that it corresponds to the construction of a Pythagorean right-angled triangle: that is, so that the length is divided into five units and the head raised three of the same, and in consequence the distance from the foot of the perpendicular to the lower end of the screw will be four of these units. The arrangement by which this should be done is shown in a diagram at the end of the book in its proper place. In the same place I have recorded as clearly as I could the water-lifting devices made of wood – that they might be better known: how they are constructed and how driven so as to provide countless benefits by their rotation.

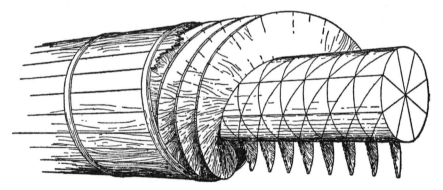

Figure 9.8 Design for a Vitruvian water screw.
Source: After Morgan, 1914, 295.

9.32 The original application of the water screw

Since Archimedes invented this versatile machine during a visit to Egypt (see **5.20**), it seems likely that it was intended originally for the specific agricultural application described in this passage, based on information in the lost Hellenistic historian Agatharchides of Cnidus. The characteristic features of the water screw – low lift, constant force whatever the angle of lift, high volume, simplicity, and low susceptibility to clogging – were ideal in the Egyptian situation, which involved rural irrigation with the muddy waters of the Nile.

Diodorus of Sicily, *History* 1.34.2

Since the Nile Delta is formed by river alluvium and is well-watered, it produces fruit of all sorts in great quantity; for the river in its annual inundation always deposits new mud, and the inhabitants easily irrigate the whole region by means of a certain device which Archimedes the Syracusan invented, called the "screw" on account of its design.

9.33 The method of working a water screw

There is no good evidence that the crank was used to power the water screw in antiquity and considerable evidence for the use of treading. Philo of Alexandria here examines the paradox involved in this type of propulsion.

Philo of Alexandria, *On the Confusion of Tongues* 38

Compare the screw, the water-lifting device. There are some treads around the middle on which the husbandman steps whenever he wants to irrigate his fields, but naturally he keeps slipping off. To keep from continually falling, he grasps something sturdy nearby with his hands and clings to it, suspending his whole body from it. In this way he uses his hands as feet and his feet as hands, for he supports himself with his hands, which are generally used for working, and he works with his feet, which customarily serve as supports.

Both compartmented wheels and water screws were used for mine drainage in the Roman world, and extensive remains of both types of devices have been found in ancient mine shafts. The ancient historian and philosopher Posidonius, quoted by Diodorus (*History* 5.37.3–4; **5.20**), describes an extensive drainage operation in progress in a Spanish mine of the early first century BC.

9.34 The Ctesibian force pump

Technically the most complex of the ancient mechanical water-lifting devices, the force pump was invented by Ctesibius of Alexandria in the mid-third century BC. Unlike the suction pump, which depends on the effect of atmospheric pressure to lift the water, the force pump sits in the liquid it is intended to lift, and pushes it up a discharge tube by the movement of pistons working in a pair of cylinders. One-way flap valves keep the water moving through the system in the proper direction. The earliest description we have, in the Arabic translation of Philo of Byzantium's *Pneumatica*, may be derived from the original Ctesibian design. Although the description certainly may have been altered during transmission, this hypothesis is supported by certain primitive features of the pump's design. The most striking feature is the total independence of the two halves: each cylinder has a separate pump handle and discharge tube, and no attempt has been made to assure a constant flow by operating the two pistons in a reciprocal manner. Since the only advantage in having two pistons lies in the possibility it provides for reciprocal motion, it is possible that Ctesibius was grasping for this solution or that he had reached it but the design was altered during manuscript transmission. In addition, the pump has the look of a theoretical demonstration device: the pistons are not set in a well or reservoir but within kettles, which must be filled artificially, and no advantage is taken of the jetting form of delivery unique to the force pump. Practical experience later altered both of these details. Unfortunately, the design of the valves is not described in detail, but it was probably the simple flap valve borrowed, for example, from the air bellows. For another description of the force pump, in Vitruvius, and an illustration, see **2.47** (Stein 2014).

Philo of Byzantium, *Pneumatics* appendix 1, chapter 2[3]

Another device for making water rise by an elegant procedure [**fig. 9.9**]. Take two copper pots, each three spans in diameter and two cubits high. Let *a b* be these pots. Within each we set a solid and vertical pump body *c d*, at the base of which we open the intake valve *e*. We also fit to it a piston, *o f*. We give to the pump body a protuberance at *g*, in which opens the discharge valve *w*. Then we

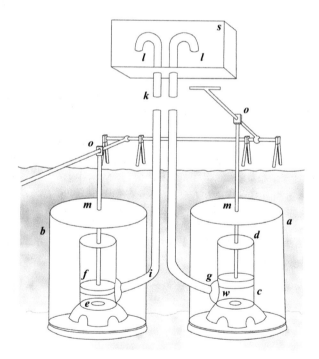

Figure 9.9 Philo of Byzantium's force pump.
Source: Adapted from Carra de Vaux, 1903, 217.

take two pipes, which we mount on the protuberance, below the discharge valve; each has a height of ten cubits. They are labelled *i k*. At the top of the piston on the outside, at point *o*, we place a rod, the lever of which is worked, and we attach to this lever two joints, as we did for the well. At the opening of the two pots we place a cover *m*.

Of necessity it must happen that, when the piston is raised, the water is sucked from the pot into the pump body, since the intake valve is raised by the air. At that moment the water is drawn and enters the pump body. When, on the contrary, the lever is depressed, the intake valve closes, the discharge valve opens, and the water mounts in the pipes, the terminations of which are at points *l*. It is emptied from here into a reservoir, which receives it in *s*. There must always be water in the two pots. This is what we wished to explain. See the figure.

9.35 A Roman fire-extinguisher of the first century AC

The force pump was the only ancient water-lifting device with a jetting delivery (in fact, its name, *siphon*, probably is meant to express the squirting sound involved), and in consequence it found some special applications. The most spectacular of these was extinguishing fires, although Isidore of Seville also mentions its use to clean high ceilings. From a safe distance, pumps with swivelling nozzles, probably mounted on carts carrying a reservoir (like their nineteenth-century equivalents) could direct a stream of water at the flames. The device must have been standard equipment, since the corps of firemen in Rome included numerous personnel called *siphonatores* (see *CIL* 6.1057–58, 2994, 31075).

Hero, *Pneumatics* 1.28

The force pumps used in conflagrations are made as follows [**fig. 9.10**]. Take two vessels of bronze *a b c d*, *e f g h*, having the inner surface smoothed off to fit a piston (like the cylinders of water-organs), *k l*, *m n* being the pistons fitted to them. Let the cylinders communicate with each other by means of the tube *q o d f* and be provided with valves, *p* and *r*, such as have been explained above, within the tube *q o d f* and opening upwards from the cylinders. In the bases of the cylinders open up the circular apertures *s* and *t*, covered with the polished discs *u w* and *x y*, through which insert the spindles *z* and *z* soldered to or in some way connected with the bases of the cylinders and provided with shoulders at the extremities that the discs may not be forced off the spindles. To the centre of the pistons fasten the vertical rods *A* and *B* and attach to these the beam *C D*, working at its centre about the stationary pin *G*, and about the pins *E* and *F* at the upper ends of rods *A* and *B*. Let another vertical tube *H J* communicate with the tube *q o d f* and branch at *J* into two arms, which are provided with pipes fitted inside one another, through which it forces up the water, such as were explained in the description of the machine for producing a water-jet by means of the compressed air [in the same work, 1.10]. Now, if the cylinders, provided with these additions, be plunged into a vessel containing water *L M N O* and the beam *C D* be made to work at its extremities, which move up and down about the pin *G*, as they descend, the pistons will drive out the water through the tube

HYDRAULIC ENGINEERING

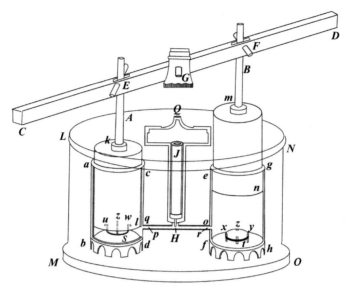

Figure 9.10 Force pump with swivelling nozzle.
Source: Adapted from Schmidt, 1899, fig. 29.

H J and the moveable nozzle **Q**. For when the piston **m n** ascends, it opens the aperture **t** as the disc **x y** rises and shuts the valve **r**. But when it descends, it shuts **t** and opens **r**, through which the water is driven and forced upwards. The action of the other piston, **k l**, is the same. Now the small pipe **Q**, which waves backwards and forwards ejects the water to the required height but not in the required direction, unless the whole device be turned around – which in emergencies is a tedious and difficult process. In order, therefore, that the water may be ejected easily to the spot required, let the tube **H J** consist of two tubes, fitting closely together lengthwise, of which one must be attached to the tube **a b c d** and the other to the part from which the arms branch off at **J**, and thus, if the upper tube is turned around, the inclination of the mouthpiece **Q**, the stream of water can be directed to any spot we please. The upper joint of the double tube must be secured to the lower to prevent its being forced from the machine by the violence of the water. This may be effected by clamps in the shape of the letter [upper case] gamma soldered to the upper tube and sliding on a ring, which encircles the lower.

9.36 An effective Roman bilge pump

Although one literary passage has survived recording the use of the water screw as a bilge pump (from Moschion, see Athenaeus, *Philosophers at Dinner* 5.207, 208), the archaeological remains all reveal force pumps in this position. This seems to be the device envisioned in the following description of the fate of an elderly sailor who was low man in the crew of a grain freighter forced to put out from Sardinia during the winter. While they were trying to ride out a storm in the lee of an island, the anchors were lost, so the crew quickly abandoned ship in the lighter, leaving the old man busy at the bilge pump within the hull. His faithless shipmates were lost immediately, but since the old man fortunately had converted to Christianity only a short time before the voyage, divine assistance helped him run the ship single-handed. All wooden-hulled ships leak, and the combined effect of small leaks along a large hull can be serious. The increase in size of the ships carrying grain and other bulk goods in the Hellenistic Greek and Roman imperial cultures was in large part made possible only by the increased discharge capacity of new mechanical pumping devices.

Paulinus of Nola, *Letters* 49.1, 2, 3, 12

They left behind one of the complement of sailors, an old man assigned to bilge pumping, either forgetting him in their fear or despising his life as worthless. Meanwhile the ship was carried out to sea without sailors or anchors....

When the old man, who was unaware that he had been deserted, noticed the vessel pitch and roll, he emerged from the bowels of the ship to see everything deserted, just sea and sky on every side....

He had scarcely touched the rigging to work it when immediately the top sail spread out on its yard filled with wind and the ship gathered headway. Water slipping through seams in the hull threatened to sink it, but after one or two draughts of the small bilge pump the water was gone, the ship dried out, and he had nothing to do....

I ask you, how were the position of bilge-pumper – very lowly even among sailors – his poor appearance and the skin cloak of the Sardinians prejudicial to this old man?

9.37 A water-lifting installation in Roman Egypt

Various devices could be used in combination to raise water, according to the special circumstances involved. At the Roman legionary camp of Babylon, the water screws probably were used to bring water up the river banks to the camp, while the wheels were used to draw water from the river itself, since they could be driven by its current.

Strabo, *Geography* 17.1.30

But now Babylon [near present-day Cairo] is a camp for one of the three legions guarding Egypt. There is a ridge running from the camp down to the Nile along which wheels and screws bring water up from the river. One hundred and fifty prisoners are kept busy at this work.

9.38 Irrigation devices for kitchen gardens

Pliny makes it clear that irrigation (by means of little conduits or by water-driven compartmented wheels?) from a passing stream is preferable to irrigation from a well. The supply of water is greater, and less effort is required to obtain it. If a well must be used, however, several alternatives are possible for obtaining the water.

Pliny, *Natural History* 19.60

There is no doubt that the gardens should adjoin the farmhouse, and above all that they should be kept irrigated by a passing stream, if there happens to be one. But if not, they should be irrigated from a well by means of a pulley or force pumps or the bailing action of a *shaduf*.

ROMAN BATHS

9.39 The design and construction of a Roman bath

As we have seen above (**9.14**), the first priority of Roman water systems, according to Vitruvius, was the provision of public fountains. The second priority, however, was the public bath, an institution that had no real counterpart in Greece before the Hellenistic period. At Rome the daily bath had become an important hygienic and social ritual by the late second century BC. During the Republican period, both the public and private baths were dark, cramped, utilitarian structures with no architectural pretensions. During the Empire, however, immense, carefully planned structures with extraordinarily rich decoration were provided for public use in most Roman cities, both in the East and the West. Furnaces heated the water. Hot air beneath floors suspended on pillars and in hollow tiles running up the walls provided radiant heating to some parts of the bath [see **fig. 8.12**]. Custom dictated bathing first in cold, then warm, and finally the hot water, then back to warm and cold baths again. In this passage Vitruvius emphasises the procedures of construction in the various parts of a bath building. His insistence that the hot rooms of a bath divided into men's and women's sections should be adjacent so that they can share a furnace is virtually the only explicit statement of an interest in energy efficiency extant in Graeco-Roman literature (DeLaine 1997, Weber 1996, Nielsen 1990; for the Greek bath see Lucore and Trümper 2013).

Vitruvius, *On Architecture* 5.10

First of all, a site as warm as possible must be chosen, that is, turned away from the north and east. Further, hot and warm bath areas are to receive their light from the direction of the winter sunset – or if the configuration of the site does not allow it, in any case from the south – because the favourite time for bathing is fixed between noon and evening. And one likewise must see to it that women's and men's hot baths are adjoining and have the same orientation: for in this way it will be brought about that there is a common heating system for both of them and their fittings. Three bronze tanks are to be installed over the furnace, one for the hot bath, another for the warm bath, a third for the cold bath, and so arranged that the amount of hot water, which flows from the warm tank into the hot will be replaced by the same amount flowing from the cold tank into the warm. The vaulted ducts are to be heated from a common furnace.

The hanging floors of the hot rooms are to be made in this way: first, the ground is to be paved with tiles 18 inches on a side, sloping towards the furnace in such a way that when a ball is thrown in it cannot stop inside but rolls back to the furnace door by itself. In this way the heat will more easily spread out beneath the floor. On this surface, piers of bricks eight inches square are to be built in such a pattern that tiles two feet square can be placed above them. These piers are to be two feet high, put together with clay kneaded with hair, and the two-foot tiles are to be placed on them to carry the pavement.

The vaults will be more serviceable if made of concrete, but if made of wood, tiling is to be applied to their lower surface. This is to be done in the following manner. Iron rods or bows are to be made up and hung with iron hooks from the timbers as close together as possible. These rods or bows are to be so arranged that tiles without flanges can rest upon and be supported by pairs of them, and in this way all the vaults are to be finished off resting on iron. The upper joints of these vaulted ceilings are to be smeared with clay kneaded with hair, the lower surface, however, which faces the pavement, first is to be plastered with a mixture of lime and potsherds, then smoothed off with stucco or fine plaster. Such vaulted ceilings in hot rooms will wear better if made double, for in this way water vapour will not be able to rot the wood of the timber work but will be dissipated between the two vaults.

Let the dimensions of the baths suit the size of the crowd. They should be planned in the following manner. Let the breadth be two-thirds of the length, not counting the room with the basin and tank. The basins should be placed below the light source so that those standing around it might not darken it with their shadows. The rooms containing the basins ought to have enough space that when first comers have taken their places around the basins, those waiting their turn might be able to stand in order. The width of the tank between the back wall and the front edge should be no less than six feet, of which the lower step and the seat occupy two.

The *laconicum*, or sweat room, should be adjacent to the warm room. The dome should spring at a height equal to the width of the room. A window is to be left in the centre of the dome and a bronze disc hung from it by chains. By raising and lowering the disc, adjustment is made to the sweating: it too should be circular, so that the force of the heat and the steam might be diffused equally from the centre over the rounded curve of the vault.

9.40 Old and new style Roman baths

Seneca, a philosophical moralist, does his best to bewail the enervating luxury of the public bath of his day (c. AD 50), and contrasts with it the "healthy" simplicity of the bath in a villa built by Scipio Africanus (236–184/3 BC). The emphasis on the immense windows in the modern version is significant. The problem of lighting large interior spaces in antiquity was a very difficult one, so natural light was used wherever possible. Furthermore, the careful orientation of the hot rooms towards the south-west (see **9.41**) meant that sunlight could be used to heat the rooms with large windows in a clever early application of passive solar energy. See also **8.69**.

Seneca, *Letters* 86.8–12

In this bath of Scipio's there are the narrowest slits, hardly windows, cut into the stone walls to admit light without damaging the fortifications. But these days people say baths are fit only for moths (*balnea blattaria*) if they have not somehow been laid out so as to receive the sun all day long through the broadest of windows, unless one can bathe and get a tan simultaneously, or unless from the bath one has a view over fields and sea. Thus the baths that were thronged and admired when they were dedicated are now avoided and shoved into the category of the old fashioned as soon as luxury has prepared something new, by which she outdoes herself. But formerly the baths were both few in number and not equipped with any refinement. For why should something be elaborately equipped, which cost a *quadrans* [a fourth part of an *as*, a small amount] for admission and was invented for use, not pleasure? The water was not poured over bathers then, nor did it always run fresh, as if from a hot spring, and they didn't think it mattered how clear the water was they left their filth behind in.

But, good God! how pleasant it is to enter those dark baths, covered with a common sort of roof, baths whose heating you know your Cato as aedile adjusted with his own hand, or Fabius Maximus or someone of the Cornelii. For even the noblest aediles used to carry out this duty, entering these baths, which accommodated the populace, and enforcing a standard of cleanliness and a serviceable and healthy degree of heat – not the temperature, which is a recent innovation, like an inferno, so much so that a slave convicted of some crime might well be "bathed alive". For it seems to me that there is now no difference between "the bath is burning" and "the bath is warm".

Of what rustic behaviour some now accuse Scipio because he did not let the daylight into his hot room through wide, glazed windows, because he did not broil himself in the full sun and wait around until he could stew himself in the bath. Unfortunate man! He didn't know how to live. He didn't wash with filtered water, but often with water that was cloudy, and when it rained rather hard, nearly muddy.... Indeed, if you knew the truth, he didn't bathe every day. For we are told by those who have recounted the customs of old Rome, that they used to wash only their arms and legs daily, since these attracted dirt in the course of their labour; as for the rest, they washed all over once a week.

9.41 An opulent Roman bath

In contrast to the censorious attitude of Seneca, the Greek Satirist Lucian describes the luxuries of a bath building with obvious approval. His emphasis on the carefully arranged sequence of spacious rooms, the colourful marble revetments, and the flood of light accurately reflect the priorities of the design. The mention of rooms for strolling and conversation reveals the social character of this institution very much in tune with the society it served. Other descriptions of Roman baths can be found in Statius, *Silvae* 1.5, Martial, *Epigrams* 6.42, Sidonius Apollinaris, *Letters* 2.2.4–9.

Lucian, *Hippias* or *The Bath* 4–8

The site was not level, but quite sloping and steep, and when he undertook it, excessively low on one side. But he made the one side level with the other, not only building a firm platform for the whole work and confirming the security of the superstructure by putting in foundations, but also reinforcing the whole with buttresses, quite steep and for safety's sake close together. The structure is proportionate to the size of the site, conforms very well to the sensible degree of furnishing, and holds to the principles of lighting.

The doors are high up, approached by broad steps with a longer tread than rise, for the convenience of those entering. A common hall of good size receives you as you enter, with sufficient space for servants and attendants to wait. On the left are the lounging rooms, these also very fit for a bath, elegant retreats filled with light. Next to them is a room beyond the needs of the bath but essential for the reception of the more prosperous. Next, dressing rooms sufficient for those removing their clothes on either side of a very high and very well-lit hall containing three cold-water swimming pools. It is adorned with Laconian marble, and in it are two white marble statues of ancient workmanship: Hygeia and Asclepius.

As you leave this hall, another receives you, slightly warmed, not jarring you suddenly with its heat. It is oblong, rounded at either end. After it on the right is a particularly bright hall offering gentle rub-downs with oil for those who come in from the exercise ground. It has doors at either end adorned with Phrygian marble. Next after this is another hall, the most beautiful of all, very suitable for standing around or sitting and for lingering without danger, or strolling around with particular profit. It too shines with Phrygian marble up to the roof peak. Next in succession the hot passageway receives you, adorned with Numidian marble. The hall after it is very beautiful, flooded with light and warmed with a colour like that of purple hangings. It offers three hot basins.

After bathing, you do not have to go back again through these same rooms but directly to the cold room through a slightly warmed chamber. All of these rooms enjoy great illumination and contain full daylight. In addition, the height of the rooms is suitable, and their width proportionate to their length, and everywhere great grace and beauty appear.... This may be in particular the effect of the light, the brightness, and the windows. For Hippias, as a truly wise man, caused the cold room to be built facing north, although not completely lacking a southern exposure, and those requiring much heat he laid out to south, east, and west. Why should I describe to you in addition the exercise grounds, and the arrangement common to both cloakrooms, which have direct passages to the bath rather than roundabout ones, for the sake of convenience and to avoid harm?

Let no one suppose that I selected an insignificant accomplishment, proposing to adorn it with my rhetoric. For I think it is a sign of no small intelligence to conceive of new patterns of beauty for common things; such is the accomplishment the marvellous Hippias provided for us. It has all the virtues of a bath: utility, convenience, good illumination, proportion, harmony with the site, provision for safe enjoyment; and furthermore, it is adorned with the other marks of careful planning: two lavatories, numerous exits, and two devices for telling time, one a water clock with a chime like a bellowing bull, the other a sundial.

9.42 The tax base for support of public baths at Rome

Luxurious public baths were expensive both to build and to maintain. Admission was either free or nominal in cost, and endowments left by the emperors or by private individuals for their upkeep were not always sufficient. The massive quantity of wood required to fuel the furnaces is a factor often overlooked by the modern investigator, but it constituted a major expense and a significant logistical problem for the original patrons and administrators. The provision of olive oil for night-time illumination constituted a remarkable luxury in the baths described here.

Historia Augusta, Severus Alexander 24.5–6

He [Severus Alexander] imposed a very lucrative tax on the makers of trousers, linen weavers, glass workers, furriers, locksmiths, silversmiths, goldsmiths, and other trades and directed the proceeds to be used to maintain for the use of the people both the baths that he had himself built and those built in the past. He also allotted forests to supply the baths, and he added oil for illuminating them, since previously they did not open before dawn and closed before sunset.

9.43 Regulations for the management of public baths

The regulations for the management of public baths at the Roman mining town of Vipasca in Spain have survived in an inscription from the reign of Hadrian (117–138). Private individuals would bid for the right to operate the service and recouped their rent by charging admission at a rate fixed by the provincial governor. All the lessee's responsibilities are carefully spelled out. For the portion of the inscription relevant to mining techniques, see **5.12**.

CIL 2.5181, 19–31

On managing the baths. The lessee of the baths or his partner, as agreed in the lease running until 30 June next, will heat the bath every day entirely at his own expense and must open them up from dawn through the seventh hour for women and from the eighth hour until two hours after sunset for men, at the pleasure of the procurator who presides over the mines. He must provide a proper flow of water to the bath up to the very brim of the hot tubs and lip of the basins, for both the women and the men. The lessee shall charge males half an *as* each, and women one *as* each. Freedmen and slaves in the service of the procurator or who receive a salary from him are exempt, as are minors and soldiers. The lessee or his partner or agent at the termination of the lease must return intact the fittings of the bath and everything that was consigned to him, except what has been worn out by age. Every 30 days he must properly wash, polish, and smear with fresh grease the bronze fixtures he uses. If unavoidable problems or damages impede the proper use of the baths, the lessee must pro-rate the payments for that period. If he does anything besides this for the sake of operating said baths, he will be owed no reduction. The lessee is not permitted to sell wood, except for branch trimmings that are not suitable for fuel. If he does anything against this stipulation, he will pay the treasury 100 *sestertii* for each load. If this bath is not kept open properly, the procurator of the mines will be permitted to fine the lessee up to 200 *sestertii* for every occasion it is not properly kept open. The lessee shall at all times have wood stored up sufficient for ... days...

9.44 A Roman hot-water heater

Private individuals could provide themselves with hot water economically with a heater of this type, essentially a small furnace containing a spiralling tube for the water.

Seneca, *Speculations about Nature* 3.24.2–3

It is customary to construct containers shaped like serpents and cylindrical milestones and many other forms, in which we mount thin-walled bronze pipes in descending spirals so that in passing over the same fire a number of times the water might flow through a length of pipe sufficient to make it hot. In this way water flows in cold and pours out hot. Empedocles thinks that the same happens beneath the earth, and the people of Baiae believe he is right, since their baths are heated without fire. From a spot seething with heat, hot air flows into them and, passing through conduits there, heats the walls and the basins of the bath. All the cold water, then, is changed to hot in its passage, and it does not acquire a flavour from the steam conduit, for it flows by in a closed channel.

ROMAN TOILETS

9.45 Roman urinals

Urine, because it contains ammonia, was an important ingredient in the tanning and fulling processes. Human urine was easily collected (as opposed to animal urine) in ceramic jugs located in the streets. The smell must have been unpleasant to say the least, as the following passage by Martial attests. Vespasian famously quipped, however, that the taxes levied from these public urinals have no smell (Suetonius, *Vespasian* 23).

Martial, *Epigrams* 6.93

Thais smells worse than a miserly fuller's old pot, recently cracked in the middle of the street, worse than a rutting he-goat, worse than a lion's maw, worse than a dog's hide dragged from across the Tiber, worse than a chick rotting in an aborted egg, worse than a jar of vile, spoilt *garum*. In order to deceive and change this stench into a different odour, whenever she takes off her clothes and goes into the bath, she is green with depilatory cream or lies concealed under a layer of chalk and vinegar or covered with three or four layers of thick bean meal. Even though she thinks she's safe by 1,000 tricks, when she has done it all, Thais still smells like Thais.

Macrobius, *Saturnalia* 3.16.15

[Wealthy young men] play dice with devotion, are smeared with unguents, surrounded by courtesans. When the tenth hour has arrived, they order a slave be called, and they send him to the Comitium to find out what happened in the Forum, who argued in favour, who against, how many tribes voted for, how many against. Then they go to the Comitium lest they have to plead for themselves. On their way, there is not a pot in the alley that they don't fill, indeed, since their bladder is full of wine.

9.46 Communal latrines

Roman public toilets were places of communal interaction. They consisted of an open space with wooden or stone seats arranged around the walls. Stall partitions did not exist. Socialising, conversation, and even scrounging a dinner invitation from the seat neighbour was common, as Martial demonstrates in the following passage. Continuously running water under the seats – frequently the grey water from adjacent bath buildings – carried off all waste immediately. In front of the seats, a second runnel of water allowed the patrons to wash their hands. The remains of Roman latrines, large and small, public and private, survive from all over the Empire. Most notable are examples at Pompeii and Ostia (Jansen 2011, Koloski-Ostrow 2015).

Martial, *Epigrams* 11.77

Vacerra spends hours in all the latrines, sitting all day long. He does not want to shit, he wants to dine.

9.47 Roman personal hygiene

Instead of toilet paper, one could use a sponge attached to the end of a stick, the so-called *xylospongium*, to clean oneself. Here Seneca discusses death and the will of not only great men, but men of low status, who choose their own time and way to die, taking to hand whatever tool is available if it provides this freedom of choice. He describes the suicide of a man, given a private moment to relieve himself, before he was supposed to enter the arena to fight wild animals. (see also **10.110**).

Seneca, *Letters* 70.20

Recently, in a school for beast-hunting gladiators, a German, while getting ready for the morning's spectacle, withdrew in order to relieve himself – he was allowed to do nothing else in private without a guard. There he crammed, all the way into his gullet, that stick, the one located there with its obscene clinging sponge to clean [the anus], and with his throat blocked he squeezed away his breath. This was an insult to Death. Yes, indeed, neither decorous nor decent. What is more foolish than to die squeamishly?

Notes

1 See **8.12–14** for a discussion of the character and use of Roman mortared masonry and concrete.
2 Translated from Carra de Vaux 1903: 185–188.
3 Carra de Vaux 1903: 192–194.

BIBLIOGRAPHY

The past two decades have seen an immense surge of publications on ancient hydraulic technology. The following bibliography is a subjective selection of only some of the most prominent works. It has been updated from the first edition mainly by adding edited works, to which numerous eminent scholars have contributed. Any one of the contributions contains individual voluminous bibliographies.

Bibliography

General studies

Bioul, Anne-Catherine, "A propos des fontaines antiques". *Acta Classica* 53 (1984) 274–279.
Biswas, Asit K., *History of Hydrology*. Amsterdam: North-Holland Publishing, 1970.
Bonneau, Danielle, *Le Régime administratif de l'eau du Nil dans l'Égypte grecque, romaine et byzantine*. Leiden: Brill, 1993.
Bonnin, Jacques, *L'Eau dans l'antiquité. L'hydraulique avant notre ère*. Paris: Eyrolles, 1984.
de Haan, N. and G.C.M. Jansen, eds., *Cura Aquarum in Campania*. Leuven: Peeters, 1996.
Fahlbusch, Henning, *Vergleich antiker griechischer und römischer Wasserversorgungsanlagen*. Braunschweig: Leichtweiss Institut, 1982.
Forbes, Robert J., "Water Supply". Pp. 149–194 in Robert J. Forbes, *Studies in Ancient Technology*, Vol. 1. 2nd edn. Leiden: Brill, 1964.
Grewe, Klaus. "Tunnels and Canals". Pp. 319–336 in John P. Oleson, ed., *Handbook of Engineering and Technology in the Classical World*. Oxford: Oxford University Press, 2008.
Jansen, Gemma C.M., ed., *Cura Aquarum in Sicilia*. Leuven: Peeters, 2000.
Mithen, Stephen. *Thirst: Water & Power in the Ancient World*. London: Weidenfeld & Nicolson, 2012.
Ohlig, Christoph, and Tsvika Tsuk, eds., *Cura Aquarum in Israel II*. Clausthal-Zellerfeld: Papierflieger, 2014.
Ohlig, Christoph, Yehuda Peleg, and Tsvika Tsuk, eds., *Cura Aquarum in Israel*. Siegburg: DWhG, 2002.
Oleson, John P., *Greek and Roman Mechanical Water-Lifting Devices: The History of a Technology*. Toronto: Toronto University Press, 1984.
Rouse, Hunter, and Simon Ince, *History of Hydraulics*. Iowa City: Iowa Institute of Hydraulic Research, 1957.
Smith, Norman A.F., *A History of Dams*. London: Peter Davies, 1971.
Smith, Norman A.F., *Man and Water: A History of Hydro-Technology*. London: Peter Davies, 1976.

Tölle-Kastenbein, Renate, *Antike Wasserkultur*. Munich: Beck, 1990.
Weber, Marga, *Antike Badekultur*. Munich: Beck, 1996.
Wikander, Örjan, ed., *Handbook of Ancient Water Technology*. Leiden: Brill, 2000.
Wilson, Andrew I. "Hydraulic Engineering and Water Supply". Pp. 285–318 in John P. Oleson, ed., *Handbook of Engineering and Technology in the Classical World*. Oxford: Oxford University Press, 2008.
Wiplinger, Gilbert, ed., *Cura Aquarum in Ephesus* (2 Vols.). Leuven: Peeters, 2006.
Yegül, Fikret, *Baths and Bathing in Classical Antiquity*. Cambridge: MIT Press, 1992.

Greek

Carra de Vaux, Bernard, "Le livre des appareils pneumatiques et des machines hydrauliques de Philon de Byzance d'après les versions arabes d'Oxford et de Constantinople". *Académie des Inscriptions et des Belles Lettres: Notice et extraits des mss. de la Bibliothèque nationale, Paris* 38 (1903) 27–235.
Cook, John M., "Bath-Tubs in Ancient Greece". *Greece and Rome* 6 (1959) 31–41.
Crouch, Dora P. *Water Management in Ancient Greek Cities*. Oxford: Oxford University Press, 1993.
Ginouvès, René, *Balaneutikē. Recherches sur le bain dans l'antiquité grecque*. Paris: de Boccard, 1962.
Glaser, F., *Antike Brunnenbauten (KPHNAI) in Griechenland*. Vienna: Österreichische Akademie der Wissenschaft, 1983.
Kienast, Hermann J., *Die Wasserleitung des Eupalinos auf Samos*. Bonn: Habelt, 1995.
Lang, Mabel, *Waterworks in the Athenian Agora*. Princeton: American School of Classical Studies at Athens, 1968.
Lucore, Sandra, and Monika Trümper, eds., *Greek Baths and Bathing Culture: New Discoveries and Approaches*. BABESCH Supplement 23. Leuven: Petters, 2013.
Murray, William M., "The Ancient Dams of the Mytikas Valley". *American Journal of Archaeology* 88 (1984) 195–203.
Tölle-Kastenbein, Renate, *Das archaische Wasserleitungsnetz von Athen*. Mainz: von Zabern, 1994.

Roman

Aicher, Peter J., *Guide to the Aqueducts of Ancient Rome*. Wauconda, WI: Bolchazy-Carducci, 1995.
Blackman, Deane R., "The Volume of Water Delivered by the Four Great Aqueducts of Rome". *Papers of the British School at Rome* 46 (1978) 52–72.
Blackman, Deane R., and A. Trevor Hodge. *Frontinus' Legacy*. Ann Arbor: University of Michigan Press, 2001.
Brodner, Erika, *Die römischen Thermen und das antike Badewesen. Eine kulturhistorische Betrachtung*. Darmstadt: Wissenschaftliche Buchgesellschaft, 1983.
Callebat, Louis, "Le vocabulaire de l'hydraulique dans le livre VIII du 'De architectura' de Vitruve". *Revue de Philologie* 47 (1973) 313–329.
Calvet, Yves, and Bernard Geyer, *Barrages antiques de Syrie*. Paris: de Boccard, 1992.
Cuomo, Serafina, "A Roman Engineer's Tales". *Journal of Roman Studies* 101 (2011), 143–165.

de Haan, Nathalie and Gemma C.M. Jansen, eds., *Cura aquarum in Campania: Proceedings of the Ninth International Congress on the History of Water Management and Hydraulic Engineering in the Mediterranean Region.* BABESCH Supplement 4. Leuven: Peeters, 1996.

de Kleijn, Gerda. *The Water Supply of Ancient Rome.* Amsterdam: Gieben, 2001.

DeLaine, Janet, *The Baths of Caracalla: A study in the Design, Construction, and Economics of Large-Scale Building Projects in Imperial Rome.* Portsmouth, RI: Journal of Roman Archaeology Suppl. Series no. 25, 1997.

Eschebach, Hans, "Die Gebrauchswasserversorgung des antiken Pompeji". *Antike Welt* 10:2 (1979) 3–24.

Evans, Harry, *Water Distribution in Ancient Rome.* Ann Arbor: University of Michigan, 1993.

Fabre, Guilhem, Jean-Luc Fiches, Philippe Leveau, and Jean-Louis Paillet, *Pont du Gard: Water and the Roman Town.* Paris: CNRS, 1992.

Fagan, Garrett G. *Bathing in Public in the Roman World.* Ann Arbor: University of Michigan Press, 1999.

Fassitelli, Enzo Fabio, and Luca Fassitelli, *Roma: Tubi e valvole. Tubi e valvole nel mondo.* 13th edn. Milan: Petrolieri d'Italia. n.d. [*c.*1990].

Garbrecht, Günther, E. Werner Eck, Gerhard Kühne, Henning Fahlbusch, and Bernd Gockel, *Wasserversorgung im antiken Rom.* Munich, Vienna: R. Oldenbourg, 1982.

Grewe, Klaus, *Planung und Trassierung römischer Wasserleitungen.* Wiesbaden: Chmielorz, 1985.

Grewe, Klaus, *Licht am Ende des Tunnels.* Mainz: von Zabern, 1998.

Grewe, Klaus, "Tunnels and Canals". Pp. 319–336 in John P. Oleson, ed., *The Oxford Handbook of Engineering and Technology in the Classical World.* Oxford: Oxford University Press, 2008.

Grewe, Klaus, "Urban Infrastructure in the Roman World". Pp. 768–783 in Georgia L. Irby, ed., *A Companion to Science, Technology, and Medicine in Ancient Greece and Rome, Vol. II.* Chichester: Wiley-Blackwell, 2016.

Grewe, Klaus, "Recent Developments in Aqueduct Research". *Mouseion Supplement: Festschrift in Honour of Professor John P. Oleson.* Forthcoming (2019).

Hauck, George, *The Aqueduct of Nemausus.* Jefferson, NC: McFarland, 1988.

Heinz, Werner, *Römische Thermen: Badewesen and Badeluxus im römischen Reich.* Munich: Hirmer, 1983.

Hodge, A. Trevor, "A Plain Man's Guide to Roman Plumbing". *Classical Views/Echos du Monde Classique* 26 (1983) 311–328.

Hodge, A. Trevor, *Roman Aqueducts and Water Supply*, 2nd edn. London: Duckworth, 2002.

Jansen, Gemma C.M., ed., *Cura Aquarum in Sicilia: Proceedings of the Tenth International Congress on the History of Water Management and Hydraulic Engineering in the Mediterranean Region, Syracuse May 1998.* BABESCH Supplement 6. Leuven: Peeters, 2000.

Jansen, Gemma C.M., Ann Olga Koloski-Ostrow, and Eric M. Moormann, eds., *Roman Toilets.* BABESCH Supplement 19. Leuven: Peeters, 2011.

Kessener, H. Paul M. *Roman Water Distribution and Inverted Siphons.* Doctoral dissertation, University of Nijmegen, 2017.

Koloski-Ostrow, Ann Olga, ed., *Water Use and Hydraulics in the Roman City.* Dubuque: Kendall/Hunt, 2001.

Koloski-Ostrow, Ann Olga, ed., *The Archaeology of Sanitation in Roman Italy.* Chapel Hill: The University of North Carolina Press, 2015.

Kreiner, Ralf, and Wolfram Letzner, eds., *SPA. Sanitas Per Aquam. Tagungsband des Internationalen Frontinus-Symposiums zur Technik und Kulturgeschichte der antiken Thermen. Aachen, 18.-22. März 2009*. BABESCH Supplement 21. Leuven: Peeters, 2012.

Lewis, Michael J.T. "Vitruvius and Greek Aqueducts". *Papers of the British School at Rome* 67 (1999) 145–172.

Manderscheid, Hubertus, *Bibliographie zum römischen Badewesen*. Munich: Wasmuth, 1988.

Manderscheid, Hubertus, *Dulcissima Aequora – Wasserbewirtschaftung und Hydrotechnik der Terme Suburbane in Pompeii*. BABESCH Supplement 13. Leuven: Peeters, 2009.

Minonzio, Franco, ed., *Problemi di macchinismo in ambito romano*. Como: Commune di Como, 2004.

Neudecker, Richard. *Die Pracht der Latrine: Zum Wandel öffentlicher Bedürfnisanstalten in der kaiserzeitlichen Stadt*. Munich: Pfeil, 1995.

Neuerburg, Norman, *L'architettura delle fontane e dei ninfei nell'Italia antica*. Naples: Macchiaroli, 1965.

Nielsen, Inge, *Thermae et Balnea*, 2 vols. Aarhus: Aarhus University Press, 1990.

Prager, Frank D., "Vitruvius and the Elevated Aqueducts". *History of Technology* 3 (1978) 105–121.

Rogers, Dylan Kelby, *Water Culture in Roman Society. Brill Research Perspectives in Ancient History*. Leiden: Brill, 2018.

Shaw, Brent D., "Lamasba: An Ancient Irrigation Community". *Antiquités Africaines* 18 (1982) 61–103.

Smith, Norman A.F., "Attitudes to Roman Engineering and the Question of the Inverted Siphon". *History of Technology* 1 (1976) 45–71.

Stein, Richard, *The Roman Water Pump: Unique Evidence for Roman Mastery of Mechanical Engineering*. (Monographies Instrumentum 48). Montagnac: Monique Mergoil, 2014.

Van Deman, Esther B., *The Building of the Roman Aqueducts*. Washington: Carnegie Institution, 1934.

Wiplinger, Gilbert, ed., *Cura Aquarum in Ephesus*, 2 Vols. BABESCH Supplement 12. Leuven: Peeters, 2006.

Wiplinger, Gilbert, ed., *Historische Wasserleitungen, Gestern – Heute – Morgen*. BABESCH Supplement 24. Leuven: Peeters, 2013.

Wiplinger, Gilbert, ed., *De Aquaeductu Atque Aqua Urbium Lyciae Pamphyliae Pisidiae – The Legacy of Sextus Julius Frontinus: Tagungsband des Internationalen Frontinus-Symposiums Antalya, 31. Oktober – 9. November 2014*. BABESCH Supplement 27. Leuven: Peeters, 2016.

Wiplinger, Gilbert, and Wolfram Letzner, eds., *Wasserwesen zur Zeit des Frontinus – Bauwerke – Technik – Kultur: Tagungsband des Internationalen Frontinus- Symposiums Trier, 25.-29. Mai 2016*. BABESCH Supplement 32. Leuven: Peeters, 2017.

Other cultures

Dalley, Stephanie and John P. Oleson, "Sennacherib, Archimedes, and the Water Screw: The Context of Invention in the Ancient World". *Technology and Culture* 44.1 (2003) 1–26.

Evenari, Michael, Leslie Shanan, and Naphtali Tadmor, *The Negev. The Challenge of a Desert*. 2d edn. Cambridge, MA: Harvard University Press, 1982.

Goblot, Henri, *Les Qanats: Une technique d'aquisition de l'eau*. Paris: Mouton Éditeur, 1979.
Lamon, Robert S., *The Megiddo Water System*. Chicago: University of Chicago Press, 1935.
Oleson, John P., "Strategies for Water Supply in *Arabia Petraea* during the Nabatean through Early Islamic Periods: Local Adaptations of the Regional 'Technological Shelf'". Pp. 17–39 in Jonas Berking, ed., *Water Management in Ancient Civilizations.* Berlin: Edition Topoi, 2018.

10

HOUSEHOLD CRAFTS, HEALTH AND WELL-BEING, AND WORKSHOP PRODUCTION

Protecting the body from the elements with coverings was one of the earliest human needs that required technological innovation. In the Greek and Roman periods, textiles and leathers were by far the most common. The most important textiles were produced from two animal fibres (wool and silk) and two plant fibres (cotton and linen); other sources played minor roles (hemp, asbestos, reed, hair from various animals). Similarly, the majority of leather was supplied by cattle, sheep, and pigs, although hides from other animals continued to be used.

The production of raw materials is mostly covered in the chapter on agriculture and animal husbandry (Chapter 3). Once harvested, the raw materials underwent a variety of treatments to make the finished fabric suitable for use. Textile production began with cleaning and softening, then mordanting and dyeing in order to ready the fibres for spinning and weaving into cloth. In leather production, the tanning of hides and skins required the removal of residual flesh and fat and of the epidermis from the hide in order for the middle corium to receive the tanning agent, which preserved the corium layer and made it waterproof. The leather treatment was completed by rolling, applying grease to help with pliability and water-resistance, dyeing, and other processes to improve the appearance of the leather. Finally, the leather could be cut and processed into clothing, shoes, weaponry, and a multitude of other products.

The humble character of ceramics changed in the Bronze Age when evidence for large-scale production indicates the presence of professionals. After the fast wheel appeared in Greece in the ninth century BC, production speed increased once more, and more precise shapes appeared. Advances in the procedures for preparing clays, decorating with slips and glazes, and firing ceramics in kilns during the seventh and sixth centuries BC laid the groundwork for the magnificent black- and red-figure pottery of the sixth and fifth centuries AC. Mould-made vessels with relief were popular from the Hellenistic period into the Late Roman Empire. Vitreous lead glazes were known from the first century BC but were never widely used. In every period, however, the most common ceramics were heavy, undecorated wares used for cooking, eating, or storing food.

The technique of making glass was discovered probably as a by-product of the high temperatures involved in smelting copper in the Early Bronze Age, but although glassmaking is much later than ceramic production, the two technologies are similar, involving the alteration of common natural substances – in the case of glass, pure quartz sand and potash (potassium carbonate) or soda (sodium carbonate) by employing the heat of a kiln. Early glass products were granular or opaque, but in the Iron Age the techniques for producing brightly-coloured objects by dipping, hand-moulding, or winding threads of glass were well understood. In the Hellenistic period, perhaps influenced by ceramic techniques, glass bowls were created by pressing discs of translucent, coloured glass into moulds. The invention of glass-blowing in the first century BC, which stimulated the production of elaborate shapes and transparent glass, made glass relatively inexpensive and common throughout the Roman Empire. During the Imperial period vessels were also decorated by cutting with an abrasive lap wheel.

Applied or practical chemistry is the chemistry of use, rather than the chemistry of theory, which has no immediate application. Applied chemistry was characterised by misinformation and guesswork, largely because of a lack of understanding of the qualities of materials that were utilised. In ancient Greece and Rome applied chemistry was based on observation of changes that, most often, were introduced by heat. Although of great impact on many aspects of technology (metallurgy, perfumes, glass, ceramics, painting, and food production), the superficial observations and empirical methods used in antiquity to acquire knowledge resulted in much uncertainty and misinformation. Nevertheless, progress was made in the preparation of pitch, soaps, inks, dyes and pigments, cleaning agents, cosmetics and perfumes.

Greek and Roman medical knowledge and practice initiated, in many ways, the transformation of Western understanding of diseases and injuries and the treatment of patients. Notions and methods of the Greeks and Romans were commonly employed, for good or ill, throughout the Middle Ages and into the Early Modern Period. The technological aspect of ancient medicine resulted in the creation of medical instruments made of various materials (iron, bronze, steel, ivory, wood, bone, stone, gold) and in a staggering variety of sizes and shapes for specific procedures. The sources provide evidence for simple personal hygiene (strigil), methods to improve one's speaking or singing, simple skin treatments (pimples), prosthetics (feet, hand, wigs, teeth), drugs (for pain, sleep, antidotes), and the tools for simple (nose fractures) and complex (dislocations and fractures to trepanation) surgical procedures. In several instances, specific instruments (a traction table, called Hippocrates' Bench; the Spoon of Diocles) were created to deal with injuries. All were intended to improve lifestyle and enhance enjoyment of life as much as possible.

The chapter also presents texts dealing with dissection and vivisection, procedures developed from other technologies but applied to medical purposes, and concludes with a section on the methods of preservation (mummification, immersion in honey, use of herbs and spices) and the treatment of the body after death (trophies, entombment, cremation, and deification).

In the Greek and Roman world, specialised labour and the need for vast quantities of raw materials and products resulted in large numbers of people organised into groups for manufacturing. The demand for larger volumes and products of superior quality led to further specialisation of labour and to piecework, where individuals could more quickly produce separate components that would later be assembled by other workers, a process viable only in urban or palatial settings.

The texts provide only general evidence for large-scale, organised production, but documentation of individuals' sources of wealth indicates its existence. A few texts also imply a factory-like manufacturing system or the presence of large workforces assembled to create products such as bricks, pots, and buildings. Some archaeological evidence, like the bakery reliefs on the Tomb of Eurysaces, and remains of similar facilities at Pompeii, provides additional clues, but evidence for such production in other contexts (fulling or tanning) is meagre. Large numbers of workers are documented also in the chapters on food production, quarrying and mining, construction, and textile production.

A METALWORKING

Early in the Bronze Age metalworkers discovered that alloying copper with tin produced a metal that not only was harder than copper, but melted at a lower temperature and gave off fewer gas bubbles when cast. Although the cutting edges of bronze tools or weapons might be hardened with hammering, through the Roman period casting in moulds remained the main technique for shaping anything bronze other than sheeting. The production of hollow-cast bronze statues was a particularly complex process. Bronze sheeting was shaped and thinned by hammering or by turning on a lathe to form vessels in a wide variety of shapes. Mould-cast handles, spouts, and rims could then be attached by rivets or solder, repoussé decoration could be hammered up from behind, and further decoration applied in the form of chasing, inlay, or patinas. Wires were produced by drawing thin pieces of metal through a swage-block, or by twisting thin strips. Gold and silver were worked in much the same manner, but brass had to be cast. Although lead did not take well to shaping by hammering, sheets could be poured in moulds and formed into vessels, pipes, or tanks by rolling, bending, and soldering or riveting. Metal objects could be plated with tin by dipping or by splashing the liquid metal on to the surface. Gold leaf could be applied by using mercury to help the amalgamation (see **6.7**).

Because iron could not be cast in antiquity, it had to be shaped by hammering, yielding the characteristic image of the smith working at his forge. Since it was difficult to produce large sheets in this manner, metal cooking utensils and vessels in antiquity continued to be made of bronze. Complex iron objects could be produced, however, by piecing together separate units with rivets or by hammer welding.

Since no handbooks of metalworking survive from antiquity, the passages collected here – with the exception of those from Pliny the Elder – are mainly snippets of information from literary or historical texts (Mattusch 2008).

HOUSEHOLD CRAFTS, HEALTH/WELL-BEING, WORKSHOP PRODUCTION

10.1 Early images of the smith

Homer envisaged his "smiths" as part of a world in which bronze was the primary metal, but their tools – even when said to be made of bronze – and procedures often reflect the knowledge of iron-working. Hephaestus, crippled by a deformed foot, has the overdeveloped upper body and underdeveloped legs fostered by the need to stand working at an anvil, a physical characteristic typical of blacksmiths in the Early Modern Period. The anonymous smith in the second passage carries the typical tools of iron forging: anvil (obviously much smaller than the modern variety), hammer, and tongs. The term *chalkeus* used here continued to be applied in later Greek to bronze smith, blacksmith, or metalworker in general (Mattusch 2008).

Homer, *Iliad* 18.369–379, 410–411

Hephaestus rose from his anvil block, a monstrous bulk, limping, but his slender legs moved with agility.[1]

Homer, *Odyssey* 3.432–434

... and the smith came, carrying in his hands the bronze tools, implements of his craft: the anvil, hammer, and well-made tongs, with which he worked the gold.

10.2 The blacksmith as an object of wonder

Although this passage refers to an event that occurred in the second half of the sixth century BC, the activities of a blacksmith at his forge were still an object of significant interest, compared in this case with the discovery of some enormous bones thought to be those of Orestes, son of Agamemnon [fig. 10.1].

Figure 10.1 Vase painting with blacksmith.

Source: Adapted from sixth century BC. Black-figure Painting on Vase from Orvieto, Italy; Singer *et al.*, 1956.2, fig. 29; Museum of Fine Arts, Boston, Accession number 01.8035.

Herodotus, *Histories* 1.68

Lichas, ... through wit and good luck, found [the tomb of Orestes] at Tegea. At that time dealings with Tegea were unrestricted, and entering a smithy, he watched iron being forged and marvelled at what he saw being done. The smith, noticing that he was astonished, stopped his work and said, "Spartan, although you are so astonished with ironworking, if you had seen what I did, you would really have something to wonder at. For, when I wanted to construct a well in this courtyard, in my digging I came upon a coffin seven cubits [3.5 m] long".

10.3 An early reference to the bow drill

The bow drill, in which an upright drilling stick held in a stone socket at its upper end is turned by a strap or string wrapped once around it, probably originated in the Mesolithic period. It continued to be used through the Roman period as a means of drilling a wide variety of hard and soft materials and large or small objects. For especially large or delicate objects, one individual might direct the drill, another pull the pair of straps or the bow. Frequently, only one direction of the oscillating rotation was used for the cutting, and pressure on the drill was released during the return rotation. Odysseus and his men are here using a large, heated stake to put out the Cyclops's single eye, and the sizzling of the blood elicits a second simile concerned with tempering steel (see **6.24**).

Homer, *Odyssey* 9.383–390

Taking the olive-wood stake with its sharpened point, they thrust it into his eye, while I, leaning hard on the other end, spun it around, just as a man bores a ship's timber with a drill – those below take hold of the strap on either side and make it spin, and it runs round and round without stopping – just so we took the fiery stake and spun it around in his eye, and the blood flowed out around it as it turned.

10.4 An early example of iron welding at Delphi

Since cast iron was unknown to the Graeco-Roman world, iron objects had to be forged in one piece or assembled with rivets or welding. Hammer welding is preferable in most situations, but it is a tricky process, and large or complicated objects were a tour de force of the blacksmith's art.

Pausanias, *Description of Greece* 10.16.1

Of the offerings, which the kings of Lydia sent, there was nothing left except the iron stand for the cauldron of Alyattes [ruled *c.*619–560 BC]. This was the work of Glaucus of Chios, the man who discovered how to weld iron. Each of the stand's plates is attached to another, not by pins or rivets, but the weld alone holds it, and it is this, which bonds the iron. The form of the stand is very much like a tower, tapering upwards from a broad base. Each side of the stand is not entirely enclosed, but there are iron cross-braces, just like the rungs of a ladder. The upright iron plates flare outward at the top and provide a seat for the cauldron.

HOUSEHOLD CRAFTS, HEALTH/WELL-BEING, WORKSHOP PRODUCTION

10.5 An ancient "Swiss army knife"

The "Delphic knife" Aristotle mentions seems to have been a type of knife combining several functions, perhaps several kinds of blades, or a blade and spoon. His prejudice against it is based on philosophical principles rather than considerations of practicality or expense. **Fig. 10.2** for common types of knives.

Aristotle, *Politics* 1.1.5 (1252b)

Nature makes nothing in a stingy fashion, like cutlers make the Delphic knife, but one thing for one purpose. For each tool is finished to the highest degree of perfection if it serves one task rather than many.

10.6 Divine instruction for fine gold work

Not unexpectedly, Homer credits expensive, fine metalwork to divine agency. Archaeological remains testify to the delicateness of such production in Bronze Age Greece and elsewhere (La Niece 1983).

Homer, *Odyssey* 6. 232–35

And as when a man overlays silver with gold, a cunning workman whom Hephaestus and Pallas Athena have taught all manner of craft, and full of grace is the

Figure 10.2 Roman relief with knife seller.
Source: Adapted from a second-century AC relief, Museo de la Civiltà Romana, Rome, Italy.

work he produces, even so the goddess shed grace upon his [Odysseus] head and shoulders.

10.7 Decorative relief work on the shield of Achilles

Thetis has just asked Hephaestus to make a new set of armour for her son Achilles to replace that lost with his companion Patroclus. The shield probably is a creation of Homer's imagination, but the motifs and techniques most likely reflect works of art he had seen, and they can be paralleled in surviving artefacts from the Bronze Age through the Roman period. Iron is not mentioned; the body of the shield probably was made of hides with a surface of bronze sheeting. The figured scenes were worked in relief on this surface and coloured with gilding and inlays of silver, tin, glass paste, and probably copper. For the first part of the metalworking process, see **6.26** (Alexander 1979).

Homer, *Iliad* 18.483–565

First, he made a large, strong shield, decorating it skilfully in every part. Around it he forged a shining rim, triple, glittering, and from it a silver shield strap. The shield itself had five layers, and on it he fashioned many entrancing images with skilful art.

On it he represented the earth, the heavens, and the sea, the tireless sun and the full moon, and all the constellations that crown the sky … two cities of mortal men, … two glittering armies. He placed black fallow-land on it, and rich, wide plough-land, ploughed three times, and many ploughmen there were driving their yokes of oxen back and forth, and the earth grew black behind them like ploughed soil, although it was of gold. He rendered this marvel outstandingly.

On it he made as well a beautiful vineyard in gold, heavy with grape clusters; the grapes along it were black, and from one end to the other supported on silver props. Around it he drove a trench of blue glass and a fence of tin …

10.8 A handy form of gold or silver ingot

Herodotus, *Histories* 3.96

The king [Darius, 521–486 BC] stores this tribute in the following manner. He has it melted down and poured off into large wine jars. After having filled the vessels, he breaks them away, and, whenever he needs funds, he hacks off as much as is required on each occasion.

10.9 Roman luxury in silverware and fine metalwork

Pliny nearly always describes luxurious objects or habits with a censorious tone, and this passage, listing some of the finer "brand names" of luxury tableware, is no exception. The names are derived from the names of the shop owners or craftsmen who produced them, although most undoubtedly depended heavily on anonymous workshop assistants. Some of these vessels could be of enormous size. Pliny indicates that designs were painted on the surface of silver vessels before engraving. Compare Martial, Epigrams 4.39 (Vickers 1994, Strong 1966).

Pliny, *Natural History* 33.139, 145

The fickleness of human taste brings about remarkable changes in silver vessels, since no one workshop finds favour for long. First, we demand Fumian silver, then Clodian, then Gratian – for we even take over shop names for use at our tables – then vessels decorated in low-relief and with a rough surface where the silver has been engraved along the painted lines of the design. Now we even put trays in our sideboards to carry the food, and other vessels we decorate with openwork, so that the file might have wasted as much silver as possible. The orator Calvus [82–*c*.47 BC] complained that cooking pots were made of silver, but we have conceived the taste for adorning carriages with silver, and in our own day Nero's wife Poppaea gave orders for her favourite team of mules to be shod with gold....

In fact, slightly earlier than these events [83/2 BC] silver trays were made that weighed 100 pounds, and it is established that at that time there were more than 150 of them in Rome.... During the reign of Claudius, his slave Drusillanus, called Rotundus, the steward of Nearer Spain, owned a tray weighing 500 pounds, for the manufacture of which a workshop had been constructed first, and, as part of the set, 8 side dishes of 250 pounds – I ask you, how many of his fellow slaves would carry them in, and for what banqueters?

10.10 Silver polish and protective finishes

Types of silver vessels mentioned above naturally had to be polished in order to remain attractive; this was generally accomplished by rubbing with cloth, sometimes assisted by the addition of a polishing compound. As a result of this treatment, the relief decoration on otherwise well-preserved Roman silver plate often is blurred or abraded. With bronze objects, however, this treatment could spoil a stable, natural patina and foster corrosion. In the absence of shellac, substances such as wax, *amurca* (the watery juice from olive pulp after the oil has been removed) and mineral oil were used to protect silver or bronze vessels or statuary and iron implements from oxidation.

Pliny, *Natural History* 35.199

Another clay-like substance is called "silversmith's earth", since it puts the shine back in silverware.

Pliny, *Natural History* 34.99

Objects made of bronze are more susceptible to corrosion when scoured clean than when left alone, unless they are thoroughly rubbed with oil. It is said that the best way to protect them is with liquid pine tar. It has long been the custom to use that metal to ensure the longevity of commemorative monuments, by inscribing official enactments on bronze tablets.

Cato, *On Agriculture* 98

You may also smear *amurca* on all types of bronze objects, but first scour them well. After you have anointed something, polish it when you want to use it. It will be shinier, and corrosion will not be a problem.

Pliny, *Natural History* 35.182

Another use of *bitumen* [mineral oil] is its application as a coating on bronze vessels, which increases their resistance to fire. We have also pointed out that it was the custom to use it to give a patina to bronze and to coat statues. Blacksmiths also favour it in their workshops as a coating for iron nail heads, and for many other applications.

B WOODWORKING

The technology of woodworking is even older than that of metalworking, and much of the information about types of wood, their applications, and the methods of harvesting trees found in Theophrastus and Pliny probably originated in the Bronze Age or earlier. Even the arrival of iron tools had little effect on the techniques of working wood other than speeding up the procedures and allowing somewhat greater precision – particularly in sawing. The carpentry tools found in Bronze Age and Greek or Roman contexts are virtually identical in function, and for the most part the shapes changed only in response to the way iron had to be worked. One tool apparently new to the Roman period was the plane – probably invented in the first century AC – which required a very hard, sharp blade to function properly. We are fortunate that long and detailed discussions of types of wood, their functions, and methods for harvesting and working them have survived in Book 5 of Theophrastus, *Enquiry into Plants*, which forms the basis for much of Book 16 of Pliny's *Natural History*. Pliny's discussion is favoured here for much of this material, since – although derivative – it is focused more on technology than on botanical theory. For the discussion of woods used for charcoal or firesticks, see **2.22** and **2.27** (Ulrich 2007, 2008, Liversidge 1976).

10.11 The deforestation of Greece

The arid climate of Greece, its thin soil and rugged topography, and the intensive herding of goats meant that forests could not regenerate once they were cut. By the end of the Bronze Age most of the Greek landscape was stony scrub land. From that time onward, timber of any size had to be imported from northern Greece or the Black Sea region. Memories of the original landscape, however, lingered (Meiggs 1982).

Plato, *Critias* 111 b–c

At that time [the Late Bronze Age] the land was untouched and its mountains were high rolling hills of plough-land, and it had plains covered with rich soil, now called "fields of stone", and much timber in the mountains – of which even now traces are visible. For there are some mountains that now provide sustenance only for bees; but it is not such a long time since trees were cut there as rafters for very large buildings, and the roofs are still sound. And there were many tall cultivated trees besides, and they produced extraordinary pasturage for flocks.

Theophrastus, *Enquiry into Plants* 5.2.1

Some distinguish among regions and say that the best wood imported into Greece for the use of carpenters is that from Macedonia, for it is free of knots, straight-grained, and full of sap. The second best is that from the Black Sea, the third that from Rhyndacus [in northwestern Asia Minor], the fourth that from the land of the Aenianes [northern Thessaly]; the worst is that of Parnassus and of Euboea, for it is full of knots, rough, and quick to rot.

10.12 The tallest known trees

Several sections of Pliny's *Natural History* provide a handbook of information – much of it drawn directly from Theophrastus, *Enquiry into Plants* Book Five – about trees, timber, and the applications of different varieties of wood. Italy was not as overpopulated or as arid as Greece, and in the fourth century BC Theophrastus could refer to Latium as forested with large trees (5.8.3). Although the rapid spread of agriculture and urbanisation in the peninsula during the Republic had left little old-growth forest, very large trees were still harvested from time to time, particularly from the mountains (cf. **8.26**).

Pliny, *Natural History* 16.195, 200

... The larch and silver fir are the tallest and straightest of all trees. Fir is preferred for ships' masts, because of its light weight....

It is thought that the largest tree seen in Rome down to the present time is the one which Tiberius Caesar set up as a marvel.... This was a larch-wood beam 120 feet long, and 2 feet thick along its entire length, giving an idea of the almost incredible original height to these who estimated the length to its crest.

10.13 The fibrous nature of wood; working techniques

Pliny, *Natural History* 16.184–192

There are fibres and veins in the flesh of some trees. It is easy to tell the difference, for the veins are broader and of lighter colour than the fibres. Veins are present in wood that is easily split. For this reason, it happens that, when you put your ear at one end of a beam of any length whatever, the sound even of a graver tapping on the other end can be heard, since it follows the longitudinal channels. By this method one can detect whether the wood is twisted and broken up by knots....

A piece of wood floats more or less horizontally, each part sinking deeper the closer it was to the root. Certain types of wood have fibre without veins and are composed only of thin filaments. These split the easiest. Others, like the olive and vine, which lack fibre, are more prone to breaking than splitting....

The best time to cut down trees that are to be peeled, like smoothed trunks for temples and other applications of round pillars, is when they bud. At other times the bark cannot be removed, rot sets in underneath, and the wood becomes dark. The best time for cutting beams or timbers from which an axe strips the bark, is from the winter solstice to the arrival of the west wind in spring.... It is of great importance to take account of the moon, and people refuse to cut trees except from the twentieth to the thirtieth of the month. Everyone, however, is in agreement that the best moment for felling timber is at the moon's conjunction with the sun.... For Tiberius Caesar certainly prescribed such conditions for the cutting of larches in Raetia for restoring the gangway of the installation for sham naval battles after it had burnt down.... Some people, not without profit, cut into the pith around the circumference of trees and leave them standing so that all the sap might drain out....

Theophrastus, *Enquiry into Plants* 5.5.1, 6

Some types of wood are easy to work, others difficult. Varieties that are soft are easy to work, above all lime wood; those that are hard, full of knots, and having a compact and twisted grain are difficult. The most difficult woods to work are those of the holm oak, and the knotty portions of the fir and silver fir. The softer portion of any given wood is always better than the harder, for it is fleshier, and as a result carpenters can mark out the planks right away. Poor iron tools can cut hard wood better than soft, for in soft wood they lose their edge – as I said with regard to lime wood – but hard woods even sharpen them.

Types of lumber are "split", "trimmed", and "round". It is called split when, in cutting it up, they saw it down the middle, trimmed when they trim off the outside; clearly, round indicates it has not been worked at all. Of these, the split is resistant to further splitting, because the core has been exposed, dries out, and dies, but the trimmed and round are susceptible to splitting – particularly the round, since the core is included in it. Indeed, there is no type of wood that will not split.

10.14 Various types of bark and wood and their uses

Although cork is a particularly spectacular example, there were many other types of bark that the Greeks and Romans put to use for a variety of rural applications. The Latin word for bark – *liber* – is the same as the word for book, suggesting that it was the earliest Roman writing material as well (see Pliny, *Natural History* 13.69).

Pliny, *Natural History* 16.34–42

The cork tree is very small, with a few poor acorns, and only its bark is useful. This is very thick and will grow back after being removed; flattened out, the sheets can be as large as ten feet square. It is used principally on ships, anchor ropes and fishermen's nets, for stoppers on jars, and for women's winter shoes as well. For this reason, the Greeks quite rightly call it the "bark tree".... Where the holm oak does not grow, they use cork tree wood in its place, especially in workshops that produce carts.

The bark of the beech, lime, fir, and pitch pine is also in great use among country folk. They make vessels of it and baskets, and a certain type of wider container for bringing in the newly harvested grain or grape clusters, also hut eaves....

The most suitable shingles are made from oak, the next quality from the other acorn-bearing trees and the beech. Shingles are most easily produced from all trees, which yield resin, but these last the shortest time – except for those of pine. Cornelius Nepos reports that Rome was roofed with shingles as late as the war with Pyrrhus [281 BC]....

The habitat of the fir tree, sought out for boats, is the same – high in the mountains, as if it had fled the sea.... It is an excellent wood for beams and the numerous applications in daily life.

10.15 More types of wood and their applications

Pliny, *Natural History* 16.206–232

The next hardest wood [after ebony, box, nettle tree, and Valonia oak] is cornel, although it cannot be brought to a smooth finish on account of its weak grain. But this wood is hardly useful for anything other than the spokes of wheels or if something has to be wedged in wood or fixed with tree-nails as hard as iron ones.... As for the rest, most of these different types of wood, but especially the oak, are so hard that they cannot be drilled before being soaked, and even in this condition a nail driven into them cannot be pulled out. In contrast, cedar does not hold a nail.

Elm keeps its rigidity most stoutly. For this reason, it is the most suitable wood for the hinges and frames of doors, since it warps the least. Only it must be inverted, so that the crest is by the lower hinge, the root above....

Certain woods, however, last longer in some applications than in others: elm is stable in the open air, Valonia oak when buried in the earth, and oak when submerged in water. When oak is used above ground it causes cracks in the structure by warping. Larch and black alder are especially good in damp conditions. Valonia oak is rotted by sea water.... The alder, in contrast, never rots when driven into the ground in swamps, and carries any amount of weight.

In our region some types of wood split of their own accord. For this reason, architects prescribe that they should be cured in a coating of dung to keep contact with the air from doing harm. Fir and larch are capable of carrying weight even when placed across a gap; oak and olive bend and give way to the load.... Pine and cypress hold out the best against rot and worms. Walnut flexes easily. Beams are also made of it. When it breaks, it gives notice ahead of time by a creaking noise.... Pine, pitch pine, and alder are hollowed out as pipes for carrying water, since they will last for very many years when buried in the ground. These same pipes, however, age rapidly if they are not quickly covered up, and they are markedly stronger if there is moisture outside as well.

Fir is the strongest wood for uprights, also very suitable for roof eaves and for whatever sort of inlaid work one might want.... It is notable for the fibrous shavings stripped off by the rapid strokes of a plane, always curling themselves up into spirals like a vine. Moreover, of all woods it is the one most easily bonded with glue, so much so that it splits where it is solid rather than at the join.

Indeed, gluing is also of great importance for those items that are covered by thin sheets of veneer or some other type. Wood with a grain of thread-like veins is approved for veneering, ... since in every kind of wood a grain with gaps and curls spurns the glue. Certain types of wood cannot be glued either to themselves or to some other woods. Oak is an example.... All the woods we have spoken of as pliant can be bent for any sort of project, ... while those that are somewhat moist can be drilled or cut. Indeed, dry wood gives way even beyond the point where you drill or saw, while green wood – with the exception of oak and box – resists more stubbornly and clogs saws whose teeth are set without skill in an even line. For this reason, saw teeth are bent alternately in opposite directions, to throw the sawdust.

Ash is the wood most easily worked for any purpose whatever. It is better than hazel for spears, lighter than the cornel cherry, more pliant than the service-tree. In fact, the Gallic ash even has the reliance and light weight essential for chariot construction. Elm would rival it if its weight were not a drawback. Beech also is easy to work, but easily broken, and soft. Cut in thin sheets for veneer, it is flexible and is the only material serviceable for book boxes and desks. The holm oak also can be cut into very thin sheets and has a not disagreeable colour, but it is particularly reliable in applications in which it is subject to friction: for example, as axles carrying wheels. Ash is chosen for this application too, for its pliancy, likewise holm oak, for its strength, and elm, for both reasons. Wood is also used in small pieces for the tools of various trades, and a prominent tradition holds that the most serviceable hafts for tools that bore are made of wild olive, box, holm oak, elm, and ash. These same woods are suitable for mallets, and pine and holm oak for larger ones.... Cato recommends that levers be made of holly, laurel, or elm, Hyginus recommends that handles for agricultural tools be made of hornbeam, holm oak, or turkey oak.

Citrus, turpentine tree, the different types of maple, box, palm, holly, holm oak, elder root, and poplar are the principal varieties of wood cut into sheets and used as a veneer over another kind of wood.... This was the first application of trees to the demands of luxury, that one wood be cloaked in another, and the outer surface of a cheaper wood be made from a more expensive one.

10.16 Some applications of reeds, rushes, and the willow

Reed-beds and coppiced willows were important resources, particularly for farmers, although Pliny also lists military and urban applications. Many of the applications he mentions can still be seen today virtually unchanged.

Pliny, *Natural History* 16.156–178

Among the plants that prefer a cold climate, it may be appropriate to mention the aquatic shrubs. Reeds will hold the principal spot among these, indispensable for the activities of war and peace, and likewise pleasing for our diversions. The northern peoples roof their homes with reed thatch, and such roofs last for ages. In the rest of the world they furnish very light ceilings for rooms. Reeds are also useful as pens....

No reed is more suited for arrow shafts than that which grows in the Reno at Bologna. It has the most pith, proper weight for flying as well as a balance steady even against gusts....

Indeed, there is another type of reed completely hollow, which they call the flute reed. It is very useful for making flutes since it has no pith or pulpy filling.

Bird-catching reeds from Palermo have the best reputation; for fishing rods, those from Arbasa in Africa.

In Italy the main application of the reed is as a vine prop.

The willow has many different kinds of applications. They send out poles of great length useful for vine-trellises and at the same time strips of bark for withes. Some grow supply shoots useful for tying, others very thin shoots for weaving baskets of notably fine texture, and yet others stronger shoots for weaving heavy baskets and a great many agricultural utensils. Once the bark has been stripped away with light working, the whiter variety is made into vessels more capacious than can be made with leather. It is likewise very suitable for weaving luxurious easy chairs.

From rushes are made thatch and mats. With the outer covering removed, they serve as candles and funerary torches. In certain regions they are stronger and more rigid: rivermen on the Po and even African fisherman at sea hoist their sails over rush boats.

10.17 Luxury furniture of citrus wood

An item of particular luxury at Rome and throughout the Mediterranean world in the late Republic and early Empire was a table, the top of which consisted of a slab of wood from a burr or a section of the trunk of a citron tree (Greek *thyon*; modern *Callitris quadrivalvis* Vent.). Particular emphasis was placed on the colour, size of the slab, and the complexity of the grain seen in its polished surface. The high value of the wood led to the use of piecing, veneering, and even forgery (Mols 1999, Richter 1966).

Pliny, *Natural History* 13.91–101

Adjacent to Mt. Atlas is Mauretania, which produces very many citrus trees and the mania for citrus wood tables.... There exists today a table that belonged to Marcus Cicero, for which he paid 500,000 *sestertii*, despite his small means and – what is more surprising – the early date, and there is a record of one belonging to Gallus Asinius purchased for a million.... Up to now, the dimensions of the largest have been a diameter of 4½ feet and a thickness of 3 inches on one made of two semicircular pieces of wood for King Ptolemy of Mauretania. The skilful concealment of the join makes it a greater marvel than if nature had produced it. Again, a diameter of 3 feet 11¾ inches and a thickness of 11¼ inches throughout on a solid slab table carrying the name of Namius, a freedman of the emperor. In this section mention should be made of a table belonging to the Emperor Tiberius more than 4 feet 2¼ inches in diameter and 1½ inches thick, covered with a citrus wood veneer....

The special quality of citrus wood tables is a wavy or hooked pattern in the grain.... The highest value of all, however, is placed on the colour of the wood; that resembling mead pleases the most.... Next comes size: at present whole trunk sections are in fashion, and several trunks joined together to make a single table.

The natives bury fresh wood in the ground and smear it with wax, but woodworkers cover it with heaps of wheat for a week at a time, with a week's interval in between. It is remarkable how much weight is drawn off by this method.... The best way to maintain and polish such tables is by rubbing with a dry hand, especially after a bath. They are not harmed by spilt wine, having been created as wine tables.

Already Theophrastus ... held this tree in high esteem, relating accounts of beams of this wood in old temples, and the virtual imperishable nature of the material in buildings, unaffected by all causes of decay....

10.18 Some Greek woodworking tools

Most of the tools used for working wood were already known in the Bronze Age, and with the use of iron they took on the shapes that, for the most part, are still in use today. The plane, however, which requires a very sharp and durable blade, may have been invented and certainly only became common in the first century AC. The brace and bit (essentially a form of crank) was an invention of the medieval period. The oscillating bow drill and hand-held gimlet were used to bore small diameter holes. Shark skin or abrasive powders were used instead of sandpaper. The author of these two epitaphs for woodworkers wrote during the early Hellenistic period (Ulrich 2007, 2008, Goodman 1964).

HOUSEHOLD CRAFTS, HEALTH/WELL-BEING, WORKSHOP PRODUCTION

Leonidas of Tarentum, *Greek Anthology* 6.204

Theris of the skilful hand, when ceasing to practise his craft, dedicated to Athena his straight cubit rod, his rigid saw with its curved handle, his axe and sharp-edged plane, and his rotating drill.

Leonidas of Tarentum, *Greek Anthology* 6.205

These are the tools of the carpenter Leotichus: the toothed files, swift devourers of wood,[2] chalk lines and ochre, and the two-headed hammers beside them, the rulers stained with ochre, the bow drills and rasp, and this heavy axe with its handle, lord of the craft; also, the easily turned augurs and swift gimlets, and these four borers for setting dowels, and the adze for final smoothing. These, then, the man dedicated to Athena who gives grace to the work, when he ceased to practise his craft.

10.19 Epitaph for a proud wood cutter

The individual commemorated by this epitaph found at Athens may have cut wood either for fuel or for lumber.

IG 1² 1084

This is the beautiful monument of Manes Orymaios, who was the best of the Phrygians in all Athens. And by Zeus I never saw a better wood cutter than myself. I died in the war.

10.20 Various terms for craftsmen who work with wood

The Latin word *carpentum* mentioned in Isidore – from which the English word "carpenter" descended – in fact was borrowed by the Romans from the Celtic word for a special type of cart. Perhaps because of the plentiful supplies of high-quality wood in Gaul, the Gallic woodworkers were particularly skilled in the production in wood of sophisticated vehicles and machinery such as force pumps and water mills.

Isidore, *Etymologies* 19.19.1–2

Concerning woodworkers. A craftsman in wood generally is called *lignarius* [from *lignum*, wood]. Carpenter is a special name, for he makes only the cart called *carpentum*, just as a *navicularius* is so-called because he is a builder and craftsman only of ships (*naves*), ... and likewise the builder we call *tignarius* [from *tigna*, timber] because he erects buildings with wood.... All wood is called "material" (*materia*) because something is made from it; whether you use it for a threshold or a statue, it constitutes the "material".

Gaius, in *Digest* 50.16.235

We term builders (*fabri tignarii*) not simply those who trim timbers (*tigna*), but all who build.

10.21 Skilful Gallic coopers

The Celtic peoples made common use of barrels as containers earlier than the Greek or Roman cultures did, perhaps because of the ready supply of wood in northern Europe. Because pitching is not necessary to keep a properly constructed barrel watertight, Strabo may be assuming the use of pitch simply because it was necessary to seal the ceramic containers familiar to him in the Mediterranean world.

Strabo, *Geography* 5.1.12

The country has marvellous pitch distilleries. The jars indicate the abundance of wine, for the wooden ones [i.e. barrels] are larger than houses. The abundance of the pitch is a great help towards their proper pitching.

10.22 The production and uses of resin and wood pitch

Resin and wood pitch, like bark or charcoal, were by-products of timber. They were important materials, used in various forms as a sealant, adhesive, food additive, medicine, preservative, and fuel. For a description by Theophrastus (*Enquiry into Plants* 9.3.1) of the actual method of distilling the pitch, see **10.88**.

Pliny, *Natural History* 16.38, 52–60

Six varieties of trees in Europe produce pitch.... On the sunny side of a pitch tree an opening is made, not a puncture but a cavity where the bark has been removed to leave an aperture two feet long at most and at least 18 inches above the ground.... Later all the fluid from the whole tree trickles down into the sore.... When the flow stops, an opening is made in another part of the tree in the same way, and then another. Afterwards the entire tree is cut down and its pith burnt....

In Europe, pitch is distilled from the pitch pine by heating it. It is used for protecting ships' tackle and many other applications. The wood is chopped up and heated in ovens surrounded on all sides on the exterior by fire. The first exudation flows off in a pipe like water; in Syria this is called "cedar oil", and it has such strength that in Egypt they make use of it for embalming human corpses.

The next distillation is thicker and now yields pitch. This, in turn, is collected in bronze cauldrons and thickened with vinegar, used as a coagulant, and takes the name "Bruttium pitch". It is useful simply for [lining] storage jars and other pots, differing from the other type of pitch in its viscosity, also in its reddish colour and because it is greasier than all the rest. It is made from pitch-resin cooked with hot stones in barrels made of strong oak, or, if these are not available, by piling up a heap of wood, as in making charcoal.

This is ground up like flour, but is somewhat darker in colour and added to wine. If the same resin is cooked up gently with water and clarified, it takes on a red colour, becomes viscous, and is called "distilled pitch". But in general, the inferior resin and the bark are set aside for this. Another formula produces "intoxication resin": the untreated flower of resin is picked, along with many thin, short chips of wood, and crushed to small fragments in a sieve. It is then steeped in boiling water until it is melted.

The grease derived from this becomes a special resin and one hard to find except in a few places in the foothills of the Italian Alps. It is useful for medical purposes....

We must also mention that the Greeks call pitch scraped off sea-going ships and mixed with wax "live pitch".... It is much more efficacious for all the applications for which the various types of pitch and resin are useful....

Birch trees are tapped on the side facing the sun, not with an incision, but with a wound resulting from removal of the bark at the most 2 feet long, to be at least a cubit above the ground. Nor is the trunk itself spared, as in other situations, since the chips are useful. To be sure, the chips closest to the surface are most valued, those deeper in add a bitter taste. Afterwards, all the sap from the whole tree runs together into the wound; the same for the pitch pine. When it ceases to flow, the tree is tapped in the same way at another spot, and then another. Afterwards the whole tree is cut down and its main portion burnt.... The best pitch is obtained everywhere from sunny locations with a northern exposure.... Certain trees produce the product abundantly the year after being tapped, others on the second year, still others on the third, for the wound fills up with resin, not with bark or a scab....

C TEXTILES AND LEATHER

One of the first concerns of humankind must have been for coverings to protect the body from the elements, a practice so old that later attributions to specific inventors for the most part are imaginary (**10.23**). Although a variety of materials have been used for clothing, during the classical period garments were most often made from textiles and leathers. The most important textiles were made from two animal fibres (wool and silk) and two plant fibres (cotton and linen); other sources played minor roles (hemp, asbestos, reed, hair from various animals). The hides of a variety of animals were used as clothing from an even earlier date, but in the classical world, cattle, sheep, and pigs had become the main sources of leather products.

Although the raw materials were generally not treated by technological means until they had been harvested, and the production of these raw materials might be considered part of agriculture and husbandry, a few examples of early treatment can be cited (**10.32** and **10.41**). Once harvested, the crude materials underwent a variety of treatments to make the fabric suitable for use: from cleaning and softening (**10.31–37**) to mordanting and dyeing (**10.39–50**). At this point the textile was ready for spinning and weaving into cloth (**10.34** and **10.51–61**) (Harlizius-Klück 2016, Harlaw and Nosch 2014, Gleba and Pásztókai-Szeőke, eds., 2013, Wild 2008).

The tanning of hides and skins (**10.64–66**) required an initial treatment to remove the epidermis and flesh layers of the hide, leaving the middle corium layer in a manner that opened the structure to receive the tanning agent. The application of tanning agents preserved the corium layer and made it waterproof. The final stage of treatment involved finishing the leather by rolling, applying grease to help with pliability and water-resistance, dyeing, and other treatments to improve the appearance of the leather. It was then ready to be cut and used for clothing, shoes, weaponry, and a multitude of other products[3] (Driel-Murray 2008, Waterer 1976, Forbes 1966).

TEXTILES, DYES, AND DYEING

10.23 The inventors of textiles

Pliny, *Natural History* 7.196

… Woven fabrics were invented by the Egyptians, the dyeing of woollens by the Lydians at Sardis, using the spindle in wool production by Closter [the Spinner], son of Arachne [the Spider], linen and nets by Arachne, fulling by Nicias of Megara, and shoemaking by Tychius of Boeotia.…

10.24 Textiles become the domain of the woman

Lucretius, *On the Nature of Things* 5.1350–1360

Plaited cloth appeared before woven garments, which in turn came after iron since the loom is built with iron. Nor is it possible to make the spindles, the treadles, the shuttles, and the noisy leash-rods so smooth in another way. And nature forced men to work the wool earlier than the female sex – for the entire male sex is far better in skill and much more clever – until the hardened farmers turned it into a reproach, so that the men willingly yielded it to female hands and at the same time they themselves undertook the hard labour and toughened their bodies and hands in the hard work.

10.25 Pallas Minerva: the patron goddess and instructor

Ovid, *Fasti* 3.815–824

Now you lads and young girls, pray to Pallas; for those who win the favour of Pallas will be learned. After the favour of Pallas has been obtained, let the girls learn how to soften [card] the wool and unload the full distaffs. She also teaches how to run the shuttle through the upright warp and to make denser the loose work with the comb. Revere this goddess, you who take out the stains from damaged clothes; revere her, you who ready the bronze cauldron for the fleeces. Nor will anyone who is in disfavour with Pallas make shoes well, not even if he is more skilled than Tychius....

10.26 Varieties of wool in the Roman world

Pliny, *Natural History* 8.190–193

The most-praised wool is Apulian and the one that is called "the wool of the Greek flock" in Italy, elsewhere Italian wool. Milesian sheep maintain the third place. The Apulian fleeces are short in regards to their wool, and not notable except as woollen cloaks.... No white fleece is esteemed more than that from the region of the Po....

Sheep are not shorn everywhere, since in some places the custom of plucking still exists. There are many types of colours: ... Spain has special sheep with black fleeces, Pollentia near the Alps has white, Asia has red fleeces that they call Erythrean, Baetica the same, Canosa tawny, and Tarentum fleece its own dark colour....

Istrian and Liburnian is closer to hair than wool, unsuitable for clothing with a soft nap; and the same with the one which Salacia in Lusitania commends with its chequered pattern. There are similar ones around Piscinae in the province of Narbonne and also in Egypt, from which clothing worn by use is darned and made durable for a long time again. And the coarse hair of a shaggy fleece has a very ancient popularity in carpets.... Wools that are self-felted make clothing and, if vinegar is added, even resist iron, and indeed even fire, the most recent way of cleaning them.... Frieze cloaks began within the memory of my father, and cloaks with hair on both sides within my own.... Fenestella writes that togas made of smoothed cloth and Phrixian wool began in the latest years of divine Augustus' reign. The togas made from closely woven poppy-cloth have an older origin; they were already noted by the poet Lucilius in the case of Torquatus [mid-fourth century BC].

10.27 Ritual cleanliness necessitates the use of linen

Occasionally cult practice and religious beliefs influence the clothing and appearance of devotees. Such is the case in the following passage where a philosopher discusses the superiority of linen over wool, his own dress having come under attack.

Philostratus, *Life of Apollonius* 8.7.5

How is linen better than wool? Indeed, the latter is clipped from the meekest animal and one valued by the gods, who don't deem themselves above being shepherds; and, by Zeus, either the gods or stories once considered it worthy to have a golden form. On the other hand, linen is sown everywhere and anywhere, nor is there any talk of gold concerning it. But nevertheless, since it is not plucked from a living creature, the Indians believe it pure, as do the Egyptians, and because of this property Pythagoras and I used it when lecturing, praying, and sacrificing. And it is a pure material under which to sleep, since, to those people living as I do, dreams bring truer revelations of themselves.

10.28 Cotton

Several authors comment upon the "cotton trees" of India and Egypt. The material, however, was rare in the Mediterranean area, and accounts of the fibre are sometimes confusing.

Pliny, *Natural History* 12.38–39

In the same island [Bahrain] on a higher plateau, there are wool-bearing trees, but in a manner different from the Chinese trees [silk production], since on these the fruitless leaves, if not a bit smaller, would look like vine leaves. They bear gourds about the size of quinces which burst at maturity and disclose balls of down from which they make [fustian] clothing with costly linen. They call the tree *gossypinus*.... Juba relates that the trunk has a down about it and that its cloth is superior to the linen of India.

Pliny, *Natural History* 19.14–15

Pliny offers this slightly different version of cotton in a later passage.

The upper part of Egypt, towards Arabia, grows a shrub which some people call *gossypios* [cotton], more often called *xylos* [wood]. And from this, the name *xylina* is given to cloth made from it. It is small and bears a fruit similar to a bearded nut, of which the down from the inner silken fibre is spun. Nor are there any other threads brighter, softer, or silkier. Clothing made from it is very pleasing to the priests of Egypt.

The difficulties of obtaining strong cotton warp thread resulted in fabric made from two materials: linen warp with cotton weft-threads. This "fustian" fabric, named after the suburb of Fostat in Cairo, may be one of the causes for the confusion in terms applied to linen and cotton in antiquity. Theophrastus (*Enquiry into Plants* 4.7.7) provides similar information, adding that both expensive and cheap fabrics were made from the "wool". See, too, brief references to cotton in Herodotus, *Histories* 3.106 (the earliest mention); Pollux, *Lexicon* 7.76; and Strabo, *Geography* 15.1.21 (694). Philostratus (*Life of Apollonius* 2.20; 3.15) also records that *byssus* (cotton) was imported into Egypt from India for sacred rites. Cotton and even asbestos (**10.31**) have been suggested for the identity of the material described by Philostratus; conflation of the two is quite possible.

10.29 Silk

The silk known to the Greeks was produced by a local silkworm grown on several of the islands, the most famous being Cos. Using the account of Aristotle and additional muddled comments of Pliny

(*Natural History* 11.76–78) about the Coan silk industry, scholars have suggested that the Coan material *is tusseh silk*, from two moths: the *Pachypasa otus* and the *Saturnia pyri*.

Aristotle, *History of Animals* 5.19 (551b)

From a certain large larva, which has horns of a sort and is different from the others, first there arises a caterpillar after the larva metamorphoses, then a cocoon, then from this a *nympha* [or moth?]. It goes through all these forms in six months. Some of the women unwind the cocoons of these animals, reeling the thread off, and then they weave [fabric from it]. Pamphile, the daughter of Plates, is said to have been the first to weave on Cos.

Lucan, *Pharsalia* 10.141–143

Although other types of "silk" had been used for centuries, the superior silk from China only became available in large quantities during the Roman Empire. Cleopatra's extravagant meeting with Caesar is being described.

Her white breasts were visible through the Sidonian fabric, made tight by the comb of the Seres [Chinese]; the needle of an Egyptian worker had loosened the fabric and relaxed the threads by stretching out the cloth.

Procopius, *History of the Wars* 8.17

Although trade brought Chinese silk into the Mediterranean world, it was not until the reign of Justinian that sericulture was begun in Byzantium, eliminating the Persian middlemen.

At about this time [AD 522] certain monks came from India.... They had spent a long time in the region above the many nations of India in a land called Serinda, and in this place had precisely learnt by what means it was possible to produce raw silk in the land of the Romans.... The monks said that certain worms were the producers of the raw silk, that nature was their teacher and also forced them to work continually, and that although it was impossible to transport the living worms here to Byzantium, their offspring were both transportable and little trouble. The offspring of these worms, they said, were a countless number of eggs from each one. Men bury these eggs in dung a long time after they were produced, and after heating them a sufficient time in it, they produce living creatures. [Justinian sends the monks back.] ... Once they were in Serinda again, they brought the eggs [of the silk worm] back to Byzantium, and bringing about their transformation into worms in the manner described, they fed them on the leaves of the mulberry. And from that act, subsequent raw silk production was established in the land of the Romans.

10.30 Hemp

Herodotus, *Histories* 4.74

The Scythians have hemp (*kannabis*) growing in their land, which very much resembles flax except the hemp is thicker and taller by far. This grows both of itself and by being sown; and the Thracians even make outer garments from it, which are very similar to linen, so that no one, other than someone very skilled in hemp, could distinguish whether it was linen or hemp. Whoever has not yet seen hemp will think the outer garment is linen.

Herodotus then describes the practice among the Scythians of inhaling the smoke from incinerated hemp seeds to induce intoxication.

Pliny, *Natural History* 19.173–174

Hemp is very useful for ropes. It is sown when the west wind comes in the spring; and the thicker it grows together, the thinner its stalks. Its seeds, when they have ripened, are stripped off after the autumn equinox and dried by the sun or the wind or with smoke. The hemp itself is gathered after the vintage and cleaned by peeling the bark off in the lamplight. The best type from Alabanda [in Caria] is especially useful for hunting nets. Three grades are made there: the one near the bark or the pith is considered inferior, while the most highly praised is that from the middle between the bark and pith, which is called *mesa* [middle]. The second best type comes from Mylasa [in Caria]. As far as height is concerned, Rosean hemp in the Sabine territory grows as high as trees.

10.31 Other fibres

Pliny describes the treatment and uses of esparto grass and asbestos.

Pliny, *Natural History* 19.27–30

In Cartagena [in Spain] esparto is used by the rural peoples for bedding, for fires and torches, for shoes and the clothing of herdsmen.... For other uses it is laboriously torn out from the ground by people wearing leggings and woven gauntlets on their hands and legs and using bone and oak tools. It is done now almost into the winter, although most easily from mid-May into June, the time when it ripens. After gathering it is heaped into bound bundles for 2 days, then on the third day untied and spread in the sun to dry. Next the esparto is bundled again and put under cover. Afterwards it is softened by soaking, preferably in sea water, but even in fresh water if the other is lacking. After drying in the sun once more, it is moistened again. If it is urgently required, it is soaked in a tub of warm water and dried standing up, thus securing a saving of labour. Finally it is pounded to make it useful, and it is unsurpassed especially in water and the sea – they prefer ropes of hemp on dry land – since once submerged, esparto is actually nourished.... It has a natural quality for repair, and however old a piece may be, it can be joined again with a new piece. Nevertheless, someone who wishes to comprehend this marvellous fibre must appreciate how much it is used in every land for the rigging of ships, for machines used in building, and for the other necessities of life.

The long, filament-like fibres of asbestos led to the belief that it was a plant. Its most famous characteristic, resistance to fire, is described in several sources (Dioscorides *Pharmacology* 5.138).

Pliny, *Natural History* 19.19–20

Now a linen has even been found that is not consumed by fire. They call it "live" linen, and I have seen table napkins made from it glowing on the hearths at banquets, gleaming more brilliantly once the dirt had been burnt off than ever possible using water. Funerary tunics for kings made from this fabric separate the ashes of the body from the rest of the ashes. It grows in the deserts and scorched lands of India, where no rain falls, among deadly snakes, and it becomes accustomed to live with the burning heat. It is rarely found and difficult to weave because of its short length, and its colour is red usually, although it glows white in fire. When found, its value equals that of very fine pearls. It is called *asbestinon* [inextinguishable] by the Greeks from the evidence of its own character. Anaxilaus records that a tree wrapped in this linen can be felled so that nobody hears since the blows are deadened. Thus, first place in the entire world goes to this linen.

10.32 Mycenaean production

Wool, linen, and leather are all cited in the Linear B tablets, although their processing receives only brief mention.

DMG nos. 211, 217, 219, 224, 226–227, 317 (pp. 316, 319, 321–322, 490)

16 cloaks of *ko-u-ra* type, 26 2/3 measures of wool; 1 cloth of *to-na-no* type, 3 measures of wool; 4 edged cloths, 26 measures of wool.
24 cloths with coloured *o-nu-ke*, 372 with white *o-nu-ke*, 14 dyed cloths, 42 of the colour of *pa-ra-ku*, 1 grey one. So many cloths (in all?): 149.
Linen clothes from D.: 1 cloak, 1 tunic.
21 purple double cloaks.
4 woollen cloths to the young (or new) fuller Didumoi.
20 woollen cloaks which are to be well boiled.

For Augeiateus: 4 skins for saddle-bags. For ... (?): 2 skins. For Augeiateus: ... skins (as) straps. For Augeiateus: 4 skins (as) bindings of pack-saddles; 4 panniers of basketry. Mestianor: 1 skin (as) fastenings of a hamper; 3 red skins for.... (For) Wordieia: 2 (skins as) sandals; 10 pig's skins; 1 hide as wrappers for 3 pairs (of sandals?); 4 hides (as) laces of sandals; 2 deer skins (as).... (For) Amphehia: 1 pig's skin with fringes underneath; 1 deer skin with pig's skin underneath. For Myrteus: 1 sheep's (skin?).... For Myrteus: 1 goat's skin for sandals.

The crude materials used to make textiles all needed cleaning and preparation to accept dyes. Wool was the most impure material, containing dirt and grease amounting to 50 per cent of the weight in some cases; cotton was the purest, having as little as 4 per cent impurities. Wool was cleaned by washing and beating out the dirt with sticks, a process that also made the wool easier to card since the fibres were loosened. It was then combed or carded; combing was also used for silk and flax. Dyeing was the next usual step before the wool was spun and woven; cotton and linen were dyed before weaving. Most of this was done in the home, but eventually professional fullers prepared much more uniform woollens for the market.

10.33 Jacketing the sheep to produce fine fleece

Special sheep were carefully cared for by the owners. Stalls were paved with stone to drain urine, the sheep had a special diet, and double the number of shepherds watched them. Varro records all of this as well as the practice of jacketing the sheep (Frayn 1984, Moeller 1976).

Varro, *On Agriculture* 2.2.18

Jacketed sheep are those which, on account of the excellent quality of their wool (such as those at Tarentum and Attica), are covered over with jackets (*pelles*) so that the wool is not stained, since staining makes it more difficult to dye, wash or bleach the wool properly.

Varro later states that after shearing, the jackets are greased inside and placed back on the sheep.

10.34 Processing wool

Normally the sheep were sheared, but occasionally the wool was plucked from them (**10.26** and Varro, *On Agriculture* 2.11.9). Once this was done, the wool had to be cleaned. Aristophanes supplies a clear account of this in his simile describing politics in Greece. The simile begins when a magistrate asks Lysistrata how the women could settle the complex political problems of the Peloponnesian War.

Aristophanes, *Lysistrata* 567–586

LYS: Just as with a skein of wool when it is tangled, we lift it up thus and draw it carefully this way and that from the spindle. That's how we will resolve this war, if we're allowed to, disentangling through our ambassadors this way and that.

MAG: Are you out of your minds? Do you think you can stop such serious matters with your skeins of wool and spindles?

LYS: Ah, if only you had it in you to manage politics in the way that we handle this wool of ours.

MAG: How then? Come on, explain it.

LYS: Just like a fleece in a bath, once you have washed out the sheep-dung from the polis [city] it's necessary to beat out on a table the wretched bits and to pick out the burrs; then the conspirators and the closed associations aiming at power for themselves, card them out and pluck off the heads. Next card into a basket everyone remaining, mixing together the common goodwill; the metics [resident aliens], any friendly foreigners, and anyone in debt to the Treasury, mix these in too. And, by Zeus, the *poleis*, as many as are colonies of this land [Athens], each one independent. They lie scattered about for us to gather up, just like flocks of unspun wool [ready for the distaff]. From all these, after taking up the flock, gather them together and combine hither into one and then make a large ball of spun yarn, and from this weave a warm cloak for the people.

10.35 The raw materials used by the fuller

Two inscriptions, one of which is quoted here, provide evidence of the large amounts of water used by fullers when processing wool. Both record the construction of a canal about 8,400 feet long with a cross-section of 41 feet square at Syrian Antioch during the reign of Vespasian (AD 69–78). It was financed by the inhabitants along the route.

D. Feissel, *Syria* 62 (1985) 79–83

... Construction of a fullers' canal and barriers that diverted the very same watercourse, approved by Marcus Ulpius Traianus the legate, ... was carried out by the metropolis of Antioch city block by city block in the year 122 [AD 73–74]. From the Orontes River to the opening at the base of Mount Amanus it is 14 *stadia* long and 41 feet square. Equal assignment of the construction was made among the city blocks calculated according to the ratio of people and the blocks' length, width, and depth, by which each city block was made responsible.

Seneca, *Speculations about Nature* 1.3.2

You will see the same thing, if ever you were to observe a fuller: since, when he has filled his mouth with water and has lightly squirted the clothing spread out on little stretchers, he seems to produce a variety of colours in that sprinkled air, the sorts of colours that usually glitter in the sun.

Pliny, *Natural History* 28.91; 28.66

The presence of ammonia made urine an important detergent agent for cleaning material before dyeing.

They say urine is very useful to fullers.... And smears of black ink are removed by means of this. Men's urine is good for gout, as proved by fullers, who say that on that account they are never afflicted by that ailment.

Vespasian even went so far as to tax the collection of urine (Suetonius, *Vespasian* 23). Several minerals were also used as detergents. After describing location and some properties, Pliny lists four types of sulphur, of which two have applications for fulling.

Pliny, *Natural History* 35.175, 196–198

They call the second type lump-sulphur (*glaeba*), a familiar substance in the shops of fullers. There is only one use for the third type, for smoking wool from beneath, since it gives whiteness and softness to the wool; this kind is called *egula*....

There is another use for Cimolian earth in cloth. The kind called *sarda* [calcium montmorillonite], which is brought from Sardinia, is used only for white cloths and is useless for multicoloured ones. It is the cheapest of all the types of Cimolian earths. More valuable are Umbrian [kaolinite] and the one they call *saxum* [bentonite or quicklime].... Umbrian is not used except for furbishing cloths.... The process is as follows: first the cloth is cleaned with *sarda*, then fumigated with sulphur, and then the cloth, which is colour-fast, is scoured with Cimolian earth. The sulphur reveals a counterfeited colour by blackening and spreading, while natural and valuable colours are only darkened by the sulphur and are softened and enlivened with a sort of brightness by the Cimolian earth. The *saxum* is more useful for white clothing after the sulphur, but is bad for coloured garments. In Greece they use Tymphaean gypsum in place of Cimolian earth.

10.36 Olive oil to treat new linen cloth

Homer is describing the excellence of the Phaeacian women in regards to their tasks at weaving. He interjects a line that may refer either to a wash given to linen to make it glossy or to the fact that the linen is so closely woven the oil will not soak through.

Homer, *Odyssey* 7.107

... and from the closely woven linen the soft olive oil drips down.

10.37 Preparation of flax for weaving

The process of separating linen fibres from flax stalks has remained the same for millennia.

Pliny, *Natural History* 19.16–18

Among us the maturity of flax is recognised by two indications: the swelling seed or the appearance of a yellowish colour. Then it is pulled up and bound into hand-sized little bundles and then hung in the sun to dry with the roots turned upward for a day, then for five more days with the heads of the bundles turned in towards each other so that the seed may fall into the middle.... After the wheat harvest, the actual stalks of the flax are soaked in water warmed by the sun, although they have to be sunk under a weight since they are very light. The loosening outer coat is an indication they are soaked, and again they are turned upside down as before to be dried in the sun. Once they are dried up, they are pounded on a stone with a tow-hammer. The part nearest the skin is called oakum, an inferior linen more nearly fit for lamp wicks. It also is combed, however, with iron claws until the entire outer skin is peeled off, and the peeled skin is used for [heating] ovens and furnaces. For the pith, there is a variety of grades according to whiteness and softness. To spin flax is even respectable for men; there is an art to combing out and separating it: 15 *librae* should be carded out of 50 *librae* of bundles. At that point it is polished in the thread, and once out of the water is repeatedly beaten on a stone, and then when made into a fabric again pounded with clubs, since it is always made better by such rigorous treatment.

Dioscorides, *Pharmacology* 5.134

In Egypt is found ... *lapis morochthus* [Soapstone or French Chalk], which the linen makers use for the whitening of the cloth, since it is soft....

10.38 A monopoly on linen weaving in an Egyptian workshop

A late third-century BC papyrus provides a variety of instructions to the oeconomus regarding his duties for industries under his power. It is clear that a certain production standard was expected and that strict control over the industry was exercised in Ptolemaic Egypt.

P.Tebt. 703.87–117

Also travel to the weaving-sheds in which the linen is woven and do your utmost to see that the greatest number of looms are at work, and that the weavers are completing the embroidered materials assigned to the nome. If any are in arrears for the pieces ordained, they shall pay the value as established by the decree for each kind. Especially take care that the linen is the best quality and that the threads are made according to specification. Travel also to the washing sheds in which the flax is washed and the ... [several words missing] ... and make a list, and send it so that there is castor oil and natron for washing at hand.... [A fragmentary section speaks of accounts and surplus amounts that are to be credited to subsequent months.].... Let all of the looms that are not working be transported to the metropolis of the nome and once they are assembled together in the storehouses, let them be sealed.

In antiquity dyes could be so precious that the recipes for specific ones were often maintained as family secrets. The natural dyes, most of them organic, were difficult to reproduce in terms of exact colour, partly a result of the equipment used and partly a result of the impurity of the dyes themselves. A wide variety of materials was used to produce the dyes: animals, insects, plants and occasionally minerals; sometimes the ancient writers confused these sources.

10.39 Mordanting: pre-dyeing treatment in Egypt

To make the colour of the dye fast, the dyer either treated the material before dyeing or used mordant dyes that gave fast colours. True mordants were usually a type of natural alum, while materials (such as urine), which were not true mordants but detergents with an alkaline nature, dissolved acid dyes into proper dyeing solutions (Robinson 1969).

Pliny, *Natural History* 35.150

In Egypt they also colour fabric by an extraordinarily noteworthy process, since after they have thoroughly rubbed white cloth, they smear into it not colours, but compounds that absorb colour. When they have done this, the cloth doesn't show it, but once it has been plunged into a vat of boiling dye, the cloth is drawn out in a minute, dyed. It is a remarkable thing that in the cloth a variety of colours is produced by this process, although only one colour is in the vat. These colours change according to the quality of the components used; nor is it possible to wash them out afterwards. Thus, the vat, which would without doubt mix together the colours if dyed materials were thrown into it, produces several colours out of one and dyes the material while being boiled. In addition, the fabrics, as they are heated, become stronger for their uses than if they were not heated.

10.40 Uses of alum

Pliny, *Natural History* 35.183–185

In Cyprus there is a white and a darker alum, and although the difference in colour is small, there is a great difference in use since the white and liquid type is most useful for dyeing wools with a bright colour, while the dark type is best for dark and sombre colours. Gold is also cleaned with the dark type.... Liquid alum also has the essence of an astringent, a hardener, and a corrosive.... It checks odour from the armpits and perspiration.

10.41 Soapwort as a dyer's mordant

Pliny, *Natural History* **24.96; 19.48**

Radicula [soapwort], which I have stated is called *struthion* by the Greeks, also prepares the wools for the dyers.... *Radicula* has a juice used only for washing wools, contributing in a remarkable degree to their whiteness and softness.

10.42 Remarkable properties of the water at Hierapolis

The unusual character of the mineral water at Hierapolis (in modern Turkey) was utilised in antiquity for a variety of purposes including spas and other health-related functions. Strabo reveals that it also had properties valuable for dyeing. The alkaline qualities of the water functioned as a detergent and permitted the dye to adhere properly to the fibres.

Strabo, *Geography* **13.4.14**

The water at Hierapolis is also wonderfully suited to the dyeing of wools, so that the wool dyed with madder-root is comparable to the ones dyed with the kermes and with sea-purple.

10.43 Mordanting before shearing the sheep

No sources describe explicitly the mordanting of wool still on the sheep, but perhaps the animals mentioned here were made to wade through a solution to pre-treat the fleece.

Pliny, *Natural History* **8.197**

Already we have seen fleeces of living animals that are dyed purple, scarlet, crimson, ... as if luxury forced them to be born thus.

10.44 The utility of plants and animals for dyes

Pliny has already related the marvellous utility of plants in regard to food, perfume, and ornament. He now proceeds to describe some of the dyes obtained from plants. Pliny makes a common mistake, stating that the *coccum* is a berry when in fact it is a type of insect, the kermes. He also notes the dangers facing the divers searching for the murex. In *Natural History* 21.45–46 he continues his criticism of luxurious scents and the most prominent dyes: the reds and purples (Forbes 1964).

Pliny, *Natural History* **22.2–4**

Indeed, I have noticed that some foreign peoples use certain plants on their bodies both to make themselves more attractive and to maintain traditional rites. Certainly, among the barbarian peoples, the women stain their faces with one plant or another, and even the men among the Dacians and the Sarmatians tattoo their own bodies. In Gaul a plant similar to the plantain grows that is called *glastum* [woad], with which the wives and daughters-in-law of the Britons stain their whole body. Then, dyed a colour resembling that of the Ethiopians, they march unclad in certain religious processions. We also know that garments are dyed with a marvellous [vegetable] dye and – although I pass over the other berries of Galatia, Africa, and Lusitania – the *coccum* is reserved for [the scarlet dye of] the military cloaks of our generals. In addition, Transalpine Gaul by means of their plant dyes produces Tyrian purple, oyster purple, and all the other colours. [To obtain these colours] nobody needs to seek the murex oyster in the depths, searching a sea bottom untouched even by anchors in order to snatch up the murex while offering himself up as bait to the sea creatures.... Although they complain that the vegetable dye washes out with use, except for this complaint, luxury could have been furnished more brilliantly, and certainly more safely.

10.45 Expensive dyes: production pollutes the air

Strabo describes the island of Tyre, which was connected to the mainland by a mole while under siege by Alexander the Great. Devastated first by Alexander, then by an earthquake, the city recovered as a result of its citizens' maritime activities, especially the production of one of the most luxurious of all ancient dyes: Tyrian purple.

Strabo, *Geography* 16.2.23

... But it overcame such disasters and restored itself both by its people's seamanship ... and by its purple dye [production]. For Tyrian purple has clearly proved itself by far the most beautiful of all dyes. The shellfish come from nearby, and the other things used in dyeing are easily acquired. Although the great number of dye-works makes the city unpleasant to live in, nevertheless it makes the city rich since its people are so enterprising.

10.46 The murex and production of its dye

Pliny provides a long, detailed description of the sources of purple and the preparation of the dye. He begins with some comments regarding the transitory nature of the product and its outrageous expense, then turns to the shellfish.

Pliny, *Natural History* 9.125–141

Purples [shellfish that produce purple] live for 7 years at most.... They come together in the spring time, and by their rubbing against each other they discharge a sticky fluid of a certain waxiness. The murex also does this in a similar manner, but it has that flower of purple, sought out for dyeing garments, in the middle of its throat: here is the white vein with a bit of liquid from which that valuable dye, glimmering with the colour of a dark rose, is drained; but the rest of the body yields nothing. They try to capture them while alive because they vomit up the juice when they die. And from the larger purples they also steal away the liquid by pulling off the shell, while the smaller ones they crush alive with their shells, since only in that case do they spew it out. The best of Asia comes from Tyre, of Africa from Meninx and from the Gaetulian coast of the Ocean, and of Europe from Laconia. [Pliny describes some uses and other sources for the dye.].... Purples are captured in little, tightly woven baskets similar to fish baskets that are thrown into the deep water. Inside is bait, molluscs that snap closed, as we see mussels do. When these are half-dead they are returned to the sea to gape greedily as they revive; the purples seek them out and attack them with extended tongues. But when the molluscs are pricked by their sting they close themselves up and squeeze together on the shellfish biting them. Thus the purples are lifted up, hanging on their own greed.... After the vein, of which we have already spoken, has been removed, it is necessary to add salt to it, a *sextarius* for every hundred *librae*. Three days is proper for steeping, since the fresher the salt the stronger it is. Preferably in a lead container, a mixture of 50 *librae* of dye to every amphora measure of water should be heated by a level and moderate vapour brought by a pipe from a furnace far removed. In this way the flesh that adheres to the veins is gradually deposited. Then in about 10 days, once the cauldron has been strained, a cleanly washed fleece is immersed as a test, and the liquid is heated until a suitable temperature is achieved.... The wool soaks for 5 hours and is immersed again after being carded, until it soaks up all the juice.... But to produce Tyrian purple, the wool is first saturated with sea-purple in a light green cauldron and then transformed with *bucinum*. Its highest praise rests in its colour likened to congealed blood, blackish at first glance but glistening when held up.

The history of the dye in Rome and its expense are then described. Pliny also describes (*Natural History* 13.136) a marine plant called *phycos* in Greek: "The one growing on the rocks around the island of Crete is also used for a purple dye. The most highly praised comes from the north side of the island". The great cost of purple dye (over 1,000 *denarii* per *libra*) meant that imitations were produced in substantial numbers: Pliny again (*Natural History* 35.44–45) records one in use during the Roman Empire.

10.47 Indigo and a test for purity

Pliny, *Natural History* 35.46

Indigo is of the second greatest importance and comes from India. A slime adheres to the scum of reeds, and when separated it is black, but upon dilution it returns to a marvellous mix of purple and blue. Another form exists floating in the pots of the purple dye-works; the scum of purple. Those who adulterate it stain pigeon droppings with real indigo or colour Selinusian earth or *anularia* [a white earth?] with woad. It is tested with a burning coal, since it gives off a flame of bright purple if genuine and, while it is smoking, an odour of the sea. For this reason, some people think it is collected from coastal crags. The cost of indigo is 20 *denarii* per *libra*.

10.48 Walnut dye for wool and hair

Pliny, Natural History 15.87

Wools are dyed with the shell of walnuts, and the hair of the head is reddened by the young nuts when first forming. This was discovered when hands were stained extracting them.

10.49 Preservation of dyes

Plutarch describes some of the riches obtained by Alexander after he captured Susa, the capital of the Persian King Darius. In addition to coin and luxurious furniture, dyes were also discovered.

Plutarch, *Alexander* 36.1–2

There, they say, was also found purple dye from Hermione, 5,000 talents weight, which, although stored for 190 years, still preserved its colour fresh and even looking new. They say the reason for this is that the dye was made with honey for the purple, with clear olive oil for the white dye; for these substances after equal amounts of time possess a brilliance that is pure and gleaming.

10.50 Recipes for dyes and mordants

Several papyri in Leiden preserve more than 110 recipes for making dyes and mordants, extracting metals and counterfeiting metals and precious stones. Many recipes refer to a variety of the processes used in the dyeing process: washing and cleaning, mordanting, and colouring, as well as combinations for preparation of the dyes: both hot and cold. As with other recipes from these papyri (**10.93**), they have been kept together to provide a better understanding of the varied contents of these very long and very important documents (Halleux 1981 and Caley 1926 and 1927).

P.Leid. X (Halleux 1981)

Dilution of Alkanet. Alkanet is diluted with pine-cones, the inside of the Persian nuts [walnuts], purpura, beet juice, lees of wine, camel urine, and with the inside of citrons.

Fixation of Alkanet. Finely crush navelwort and alum that have been mixed in equal parts and throw in the alkanet.

Styptic Agents. Melantheria [iron and copper sulphate], scorched vitriol, alum, chalcitis, cinnabar, lime, bark of pomegranate, pods of the acacia tree, urine with aloe. These serve in dyeing.

Manufacture of Purple. Having chopped up Phrygian stone into small pieces, put it to boil, and after soaking the wool leave it until it cools. Then throwing into the vessel a *mina* of seaweed, put it to boil and throw in the wool; after allowing it to cool, wash it in sea water. This is the first washing, and the Phrygian stone, which has previously been chopped up, is cooked until it becomes purple.

Dyeing with Purple. Wet lime with water and let it stand for one night; having decanted, place the wool into the liquid for one day. Take it out and dry it. And after wetting alkanet with vinegar, put it to boil and throw in the wool; it will come out dyed purple. Boiling in water and natron releases the purple colour. Then dry the wool and dye it further in the following manner: boil seaweed in water, and when it has been exhausted, throw in a bit of copperas [green vitriol] by the judgement of your eye to develop the purple, and then dip the wool and it will be dyed. If you add too much copperas, it will become darker.

10.51 The Three Fates spin the thread of life

The most famous spinners were the Three Fates who jointly decided the course of a person's life: Lachesis who assigned the lot, Clotho who spun the thread, and Atropos who severed it. The pervasive nature of their role is emphasised by their work at this common household task.

Catullus, *Poems* 64.310–319

Their hands solemnly plucked at their eternal labour. The left hand held the distaff wrapped about with soft wool; next the right hand, carefully drawing out the fine threads with upturned fingers fashioned them, then twisting, the right hand twirled the spindle weighted with a smooth whorl on the downward thumb; and then with their teeth continually plucking they made the product even and smooth and bitten bits of wool clung to their dried-up lips, bits which previously had been sticking out from the smooth yarn. Before their feet the soft fleeces of the shining wool were protected in small wicker baskets.

10.52 Spinning while doing other tasks

Women were expected to conduct themselves in an industrious manner and to do as many things as possible at the same time, at least in the real world of the small family. If Herodotus is to be believed, such industry was rare and highly valued in the Persian world. He relates a story about a Paeonian woman from the area near the Strymon River in Thrace whom Darius saw when they were both in Sardis. The situation has been artificially contrived by the woman's brothers who wished to be established as rulers by convincing Darius of the value of such women.

Herodotus, *Histories* 5.12

The brothers dressed their sister as elegantly as they could and sent her with a jar on her head for water, leading a horse by a bridle around her arm, and spinning flax [all at the same time].... She, coming to the river, watered the horse, then, once he was watered and the jar filled with water, she came back on the same road, bearing the water on her head and leading the horse on her arm and twirling her spindle.

10.53 Home spinning and weaving

Home spinning and weaving were never entirely replaced by production in large shops, but Columella is probably being idealistic here. Earlier (*On Agriculture* 12 Preface 9–10), he criticised both the luxury and the lack of industry of women who no longer plied the trade at home once they were able to purchase expensive clothing made by others.

Columella, *On Agriculture* 12.3.6

On rainy days or when it is cold or frosty, when a woman cannot go out into the open to do fieldwork, wool should be readied and carded so that she can turn back to the working of wool; thus, more easily she will be able to carry out proper wool-working and to demand the same from others. It harms nothing if garments are made at home for herself, for the stewards, and for slaves held in esteem, so that the accounts of the estate's owner are less burdened.

The production of such home workshops was sometimes on a very large scale (**10.133**). Strabo (*Geography* 5.1.7) provides evidence of a similar industry for clothing produced at Padua for export to Rome, while Suetonius (*Augustus* 73) reveals that even an emperor might wear clothing spun by the women of the house.

Pollux, *Lexicon* 10.124–125

Pollux provides a list of the necessary pieces of equipment that should be in the women's quarters to facilitate weaving. The *onos* and the *epinetron* were ceramic covers fitted and shaped to provide a stable base on the upper leg (from the knee to the thigh) for twisting the rove before hand spinning the thread.

The most essential equipment for use on a daily basis has already been discussed. In a few words one can pull together the rest of those items that pertain to the women's quarters: woven baskets and baskets with narrow bottoms and the smaller types of both, the *onos* upon which they spin and the *epinetron*, and the spindle and the circular whorls, the skeins of yarn, the weaver's shuttle and the comb of the loom; and the upright loom as well as the side beams of the loom; and the weaver's rod [to attach the alternate threads of the warp] and the beam along with the vertical beams of the loom [between which the web hangs down] and the long beams of the loom [between which the web is stretched]; and the stone weights [for the warp threads] and the loom weights, and the flat blade [to strike the woof threads home].

10.54 The strange looms of Egypt, and other wonders

Herodotus marvelled not only at the monuments but also at the customs of Egypt and its inhabitants; he regarded everything in Egypt as the reverse of what it was elsewhere. He is describing some of the wonders which he has experienced, noting the peculiar climate and the unique nature of the Nile. He continues with some of the unique customs of the inhabitants themselves, including the use of gravity to tighten the woof on a vertical double-beamed loom.

Herodotus, *Histories* 2.35

Among these people, the women buy in the market place and conduct trade while the men stay at home and weave. And while other people push the woof upwards, the Egyptians push it downwards. The men carry burdens on their heads, the women on their shoulders. Women urinate standing upright, men squatting.

10.55 An old-fashioned warp-weighted loom

One of the earliest types of looms was a vertical warp-weighted loom that required the weaver to stand, since the weaving is done at the top as the cloth is wound to maintain a constant working height. During the Roman Empire the vertical double-beamed loom, which allowed the weavers to sit as they worked, was introduced in the West (Wild 1987, Hoffman 1964, Carroll 1983).

Servius, *On the Aeneid* 7.14

… Our ancestors used to weave standing, just like today when we see the linen weavers.

A false etymology is suggested in the next passage, but it provides evidence for the vertical warp-weighted loom.

Festus, *On the Significance of Words* (Excerpts of Paulus 276 M.; 381 Th.)

The manly garments named *rectae* ["upright tunics"] are those which are woven upwards by standing weavers.

10.56 A weaving contest

The arrogance of Arachne, a mortal woman from Asia Minor, resulted in her eventual transformation into a spider, in which form she continued to weave her web. Ovid describes the operation of a vertical loom [**fig. 10.3**] in the contest. Earlier he had related some of the other stages of the textile industry such as the use of the distaff, spindle, and needle.

Ovid, *Metamorphoses* 6.53–60

Without delay they set up the twin looms in different places and stretch them with the fine warp. The web is bound to the beam, reed separates the thread [of the warp], the woof is threaded through the middle [of them] with sharp shuttles, which their fingers help through, and once led between the threads [of the warp], the notched teeth pound [it into place] with the hammering sley. Both hasten along, and with their mantle girded about their breasts they ply their skilled hands, their eagerness making the labour light.

Elaborate and colourful scenes are then woven by each of the contestants.

HOUSEHOLD CRAFTS, HEALTH/WELL-BEING, WORKSHOP PRODUCTION

Figure 10.3 Vase painting with vertical loom.

Source: Adapted from Attic Vase in Metropolitan Museum of Art, New York.

10.57 Criticism of the a simple account of the evolution of weaving

Seneca the Younger in this series of letters provides advice on how to become a more devoted Stoic. Pertinent here, they also provide valuable insights into Roman life, including the description of technologies, such as this passage on weaving, which also contains, as in other letters, the mild criticism of Posidonius' account.

Seneca, *Letters* 90.20

Behold Posidonius ... when he wishes to describe [the process of weaving] the way in which first some of the threads are twisted and some are drawn from the soft, dense mass. Then how the upright warp stretches the lengthwise threads with hanging weights; next how the woof, its thread introduced, softens the hardness of the web holding fast on either side, and is forced by the batten to come close together and to unite with the warp. He even stated that the early skill of the weaver's shop was discovered by wise men, having forgotten that the subtler style was created afterwards ... [lines from **10.56** are quoted here]. What indeed, if it had happened that he had experienced these weavings of our time, clothing produced with which nothing is concealed, clothing about which I don't say there is no protection for the body, but no protection even for modesty!

10.58 The origin of embroidery

Pliny, *Natural History* 8.196

The Phrygians invented the technique of making garments with a needle [embroidery], which as a result are called Phrygian. King Attalus of Asia also invented threading gold into garments, whereby came the name Attalic garments. Weaving different coloured patterns was especially practised at Babylon, which gave its name to that type. But Alexandria established the one woven with many threads, which is called *polymita* [damasked], while division into diamonds was introduced by Gaul.

10.59 The variety of cloth and leather workers in Rome

Plautus makes fun of the expenses of the rich. He describes a wealthy mansion, virtually placed under siege by the multitude of workers demanding payment for services rendered.

Plautus, *The Pot of Gold* 505–522

Now to whatever place you come you see more wagons at the mansions than in the country when you go to the country estates. But even this is beautiful compared to when your bills and debts come looking for payment. The cleaner, the gold embroiderer, the jeweller, the woollen worker all linger about. Also standing around are the shopkeepers who make borders for women's garments, tunic sellers, dyers of flame colour, violets, and browns; the makers of sleeved garments lounge about as do the *murobatharii* [makers of women's clothing?] and retailers of linens, shoemakers; the squatting cobblers, slipper- and sandal-makers linger on, and the makers of mallow garments; makers of breast bands and at the same time the makers of little girdles hang on. Then when you may think they are all taken care of, others push their way in, hundreds seeking their money: weavers, the makers of ornamental hems, and makers of little boxes; they stand in your halls carrying their little bag for offerings. They are brought in and money is paid out. You may think you have finally paid them all off, when in come the saffron dyers or some other pest: always there is a wretch seeking something.

10.60 The winter costume of a Greek farmer

Hesiod, *Works and Days* 536–546

Then protect your skin, as I instruct you, by clothing it in both a soft outer cloak and well-fit tunic, and take care to weave a thick woof on well-spaced warp. Clothe yourself in this so that your hair keeps its place and doesn't stand upright and bristle all over your body. Upon your feet bind boots made of the hide of a slaughtered ox; well-fitted and lined with felt on the inside. And when the cold season comes stitch together with ox gut the skins of kids in their first year to throw over your back to escape the rain. On your head above, wear a well-made hat of felt to protect your ears from the wet.

10.61 Fine linen nets for hunting

Linen gradually began to replace wool as the preferred cloth in Italy during the Empire. As a result, more areas were turned over to the cultivation of flax, and it was used for more items (see **3.68-69**, **3.73**, **10.27**).

Pliny, *Natural History* 19.10–12

Flax from Zoela [in Spain] comes into Italy and is especially useful for hunting nets.... Flax from Cumae in Campania also has acquired its own reputation for nets used to capture fish and birds, and the same material is used for hunting nets.... The nets from Cumae will cut the bristles on a boar and even prevail against the keen edge of iron. And we have already seen netting of such fineness that it passes through a man's ring with its closure cords, so slight that one man can carry enough to encircle a wood. Nor is that the most wondrous thing, rather every string of these nets consists of 150 threads.... This amazes people who don't know that in a breastplate of an Egyptian king called Amasis ... each string consisted of 365 threads.

LEATHER

10.62 An unusual source of leather

Hides and pelts from a variety of animals were used for clothing and other leather equipment. The sources are obvious in most cases, but some were unusual. The Scythians were a brutal and bloodthirsty people whose habits appalled the Greeks. Herodotus has just related that a Scythian warrior drinks the blood of his first victim and brings the decapitated head to the Scythian king to prove his right to share in the division of the booty.

Herodotus, *Histories* 4.64

He scalps the head in the following manner: cutting around it in a circle about the ears and taking hold, he shakes out the head. Then after scraping off the flesh with the rib of an ox he kneads it until soft with his hands, and once it is made supple, he keeps it as a hand-towel that he hangs from the bridle of the horse he himself rides. He takes delight and pride in this since the man who has the most skin hand-towels is judged the best man. Many of the Scythians also make from the skins outer cloaks to wear, stitching them together like the skin coats of peasants. Still others, flaying their dead enemies' right hands – nails and all – make coverings for their quivers, since human skin is stout and bright, almost the brightest white of all skins. Many others flay the entire bodies of men and, stretching the skins on wooden frames, carry them about on horseback.

10.63 Curing

Curing is the process used to arrest the decay of the hide, accomplished by drying or salting, wet or dry. During this process the hair and flesh are removed, and the skin is rendered malleable. Urine and dung were often used to help remove the hair, and the skin was beaten to increase malleability and prepare it to receive the tanning agent. Here Pausanias describes a disadvantage for the Ozolian Locrians who once wore uncured hides: a conclusion reached by associating their name with *ozein* (to smell); most interpretations of the name involved odours, usually foul (cf. Strabo, *Geography* 9.4.8 and Plutarch, *Moralia* 2.294f.).

Pausanias, *Description of Greece* 10.38.3

And another story is told that the first people here were aboriginals, who, not yet knowing how to weave clothes, used to make themselves protection against the cold from the uncured hides of wild animals, turning the shaggy side of the hides to the outside for the sake of a becoming appearance. As a result, their own skins were surely as foul-smelling as the hides they wore.

Pliny, *Natural History* 23.140

The leaves of mulberry, soaked in urine, remove hair from hides.

10.64 The stench of tanning

The combination of dead animal bodies and chemicals made the tanner's workplace one of offensive smells. There are many references in the literature to the stench associated with tanners and their work.

Artemidorus, *Interpretation of Dreams* 1.51; 2.20

The tannery is an irritant to everyone. Since the tanner has to handle animal corpses, he has to live far out of town, and the vile odour points him out even when hiding.... The vultures are companions to the potters and the tanners since they live far from towns and the latter handle dead bodies.

10.65 Tannins

Many of the numerous vegetable products used for tannin were also employed to dye textiles and leathers (**10.44** and **10.50**). Minerals used as tannin and dyes are mentioned relatively rarely, but alum and salt was the combination most widely used. The process was not well understood and, it is likely that selection of tannin was based upon practical experience. Dioscorides (*Pharmacology* 1.106–120), noting their astringent properties, provides the most complete list of vegetable tannins and dyes; a few follow.

Dioscorides, *Pharmacology* 1.106–111

Dyer's Oak. Each part of the oak has an astringent power, but the most binding layer is that which lies between the bark and the stock.... Kermes Oak ... soaked in water until it becomes tender and rotten, and being applied for a whole night, dyes the hair black, after being made clean with Cimolian earth.... Oak Galls ... also make the hair black after first being macerated in vinegar or water.... Tanning Sumac ... is the fruit called *Rhus coriaria*, which is so-named because tanners use it for the thickening of their hides; ... the decoction dyes the hair black.

Pliny (*Natural History* 35.190) comments on how significant alum is for other functions in life and in finishing hides and wools.

10.66 A method of tawing

Homer describes the struggle between the Greeks and the Trojans over the body of Patroclus. He compares the battle to a primitive method of tawing, during which a hide soaked in fat is pulled to and fro by men standing in a circle in order to stretch the skin and to make the fat penetrate into the pores.

Homer, *Iliad* 17.389–393

As when a man gives to his people for stretching a great bull's oxhide, all dripping with fat; once they have taken it, they stand round in a circle and stretch it. And straightway the moisture goes out and the fat sinks in with the many tuggings, and the hide is fully stretched to the utmost.

10.67 *Amurca:* a pest control and leather treatment

Pliny, *Natural History* 15.33–34

Lees of olive oil (*amurca*) are rubbed on the floors used for threshing grain to keep ants and cracks away. Indeed even the clay of the walls and the plasters and pavements of granaries, and even wardrobes are sprinkled with lees of olive oil as protection against woodworms and the damage of animals.... Reins and all leather items and shoes and axles are greased with boiled lees, and bronze implements to give them a more elegant and pleasing colour as well as protection against bronze disease....

10.68 Country clothing

Homer and Hesiod (**10.60**) both provide descriptions of what may have been typical costumes for people living in the country, no doubt much of it made in a rough and ready manner by the wearer or the immediate family. Homer provides two descriptions:

Homer, *Odyssey* 14.23–24

[Eumaeus, the swineherd, then Laertes, the father of Odysseus] himself was constructing boots about his feet, cutting the well-coloured oxhide....

Homer, *Odyssey* 24.226–231

Odysseus found his father alone in the orderly orchard, digging around a tree. He was dressed in a filthy tunic, patched and pitiful. About his shins he had fastened patched oxhide leggings to avoid scratches, on his hands he had gloves because of the thorns, and a goatskin cap above on his head....

Compare the elaborate list of clothing that seemed appropriate to a guest of Odysseus' stature (Homer, *Odyssey* 24.276–278).

10.69 Poor leather shoes that stretch

In antiquity, retailers and producers were not above taking advantage of careless or ignorant consumers. Aristophanes reveals a few tricks that the unscrupulous tradesman could use on unsuspecting buyers of leather goods.

Aristophanes, *Knights* 315–321

SAUSAGE MAKER: If you aren't a master at stitching the soles of footwear, then I'm not one at sausage-making. You who sell the hides of oxen that are in a sorry state, cunningly cutting them underneath so that they look thick and stout to the rustics; and before the day is out, the shoes are more than two palms wide.

NICIAS: By Zeus that's exactly what he did to me, providing my friends and fellow citizens with a great laugh: for before I made it to Pergasae, I was swimming in my shoes.

10.70 A leather raincoat

Martial, *Epigrams* 14.130

Although you may take to the road when the sky is everywhere clear, never lack a leather overcoat in case of sudden showers.

10.71 Making rope from hides

Cato, *On Agriculture* 135.3–5

... Lucius Tunnius from Casinum and Gaius Mennius, son of Lucius Mennius from Venafrum make the best rope for presses. Eight good native hides, ones that have been freshly dressed and have a minimum of salt ought to be employed for a rope. They should first be dressed and smeared with fat, then dried. At the start, the rope ought to be 72 feet long; it should have 3 splices and 9 leather thongs, which are 2 digits wide, on each splice. When twisted it will be 49 feet long. After fastening, 3 more feet will be lost, the remaining length will be 46 feet. When stretched it will add 5 feet, the length will be 51 feet. When stretched, rope for presses should be 55 feet long for the largest presses, 51 feet for smaller ones. The proper length of leather thongs on a cart is 60 feet, cords 45 feet; leather reins on a cart 36 feet, for a plough 26 feet; leather traces 27.5 feet, leather yoke straps on a plough 19 feet, lines 15 feet; on a plough, leather yoke straps 12 feet, and line 8 feet.

10.72 Some atypical leather beds and cushions

The passage from Plutarch seems to be our only evidence for ancient waterbeds. The youthful emperor Elagabalus was noted for his perverse sense of humour. Among the more harmless of his pranks were the trick pillows mentioned in the *Historia Augusta*. Leather cushions filled with straw could also be used for flotation (see **11.70**).

Plutarch, *Alexander* 35.14

For the Babylonian soil is very fiery ... and in the hot season the inhabitants sleep on leather bags filled with water.

Historia Augusta, Elagabalus 25.2–3

For some of his less exalted companions he used to set out inflated leather bags instead of couches, and while they were dining, he would deflate them in such a way that often diners would suddenly find themselves under the table.

D CERAMICS AND GLASS

CERAMIC PRODUCTION

No handbooks or lengthy technical discussions of pottery manufacture have survived in Greek or Latin literature, probably because of the humble character of the craft and the fact that its products, unlike those of metalworking or quarrying, were not central to the interests of the elite. The production of pottery vessels was a common domestic activity in the Neolithic period, but by the Late Bronze Age professionals had for the most part taken over. The fast wheel, which appeared in Greece only in the ninth century BC, sped up production and allowed more precise shapes. Advances in the procedures for preparing clays and firing ceramics during the seventh and sixth centuries BC laid the groundwork for the magnificent Attic pottery of the sixth and fifth centuries BC that was decorated with figured scenes in black-figure and red-figure technique. Vessels with relief decoration produced by moulds – probably in imitation of silverware decorated with repoussé figured scenes – and covered with glossy black or red slip became popular as table ware in the Hellenistic period and remained in favour through the Late Empire. Vitreous lead glazes were known from the first century BC but were never widely used. At every period the most common ceramics were heavy, undecorated wares used for cooking, eating, or storing food stuffs (Jackson and Greene 2008, Noble 1988).

10.73 The free-turning potter's wheel

Potter's wheels that turned on their own were a relatively new development at the time Homer's poem was composed in the eighth century BC. Previously, pots had been centred and smoothed on discs or mats that did not turn freely on a pivot.

Homer, *Iliad* 18.599–601

These, then, were running very lightly on skilful feet, as when a potter sits down and tries his wheel, suited to his hands, to see if it will run.

Plutarch, *Moralia* 20.588f *(On the Sign of Socrates)*

There is no reason to be surprised at this [the quick response of the soul to the mind], when we see that large merchant ships are brought about by small steering oars and that potter's wheels spin smoothly at the light touch of a fingertip.

10.74 The perils of kiln firing

In this poem the numerous disasters that can befall pots in a kiln [**fig. 10.4**] are named metaphorically, as gods. The conditions implied are gas explosions from inadequately wedged clay, over-fired and under-fired vessels, the collapse of a stack of vessels, and a blast of heat from the peephole. The dialogue gives the impression that the potters were hardworking craftsmen living on the margins of poverty.

Figure 10.4 Pinax showing a potter at the kiln.
Source: Adapted from a *pinax*, Louvre, Paris, France.

[Herodotus], *Life of Homer* 32

On the next day some potters firing a kiln-full of delicate pottery saw Homer leaving and called to him, having learnt that he was a wise man. They asked him to sing to them, promising to give him some pottery and anything else they had. And Homer sang to them the following song, entitled "Potter's Kiln": "If you pay me, potters, I will sing. Come here Athena, and stretch out your hand over the kiln. May the *kotylai* [two-handled cups] and all the *kanastra* [low, one-handled cups] become nice and black, and be baked through and the price put on them please the buyer and many be sold in the market place, many in the streets and bring a great profit. And as for me, may I sing to them. But if you potters turn to shamelessness and try to cheat, I then call down upon your kiln the destroyers: *Grinder*, together with *Crasher*, *Unquenchable*, *Shatterer*, and *Lord of the Unbaked*, who gives the craft many problems. Ravage stoking hole and chamber, and let the whole furnace be churned up while the potters wail loudly. Like the chewing of a horse's jaws, may the kiln grind to powder all the pots within it…. And if anyone peeks in, may his whole face be scorched, so that all might learn to act fittingly".

10.75 Poor instruction by poor potters

Like most crafts in antiquity, the skills of a potter were passed on by the master to apprentices or children. In this dialogue, Socrates, who speaks first, echoes a typical elite attitude by assuming that a successful potter will abandon his craft for a life of leisure.

Plato, *Republic* 4.421d–e

"See, then, whether the following factors corrupt the other craftsmen so as to make them inferior".

"What are they?"

"Wealth", I said, "and poverty".

"But how?"

"In this way. Do you think it likely that a potter who became rich would still wish to engage in his craft?"

"Not at all", he said.

"But he would become more lazy and careless than before?"

"Much more so".

"Therefore, he becomes a worse potter?"

"And very much so", he said.

"And further, if poverty keeps him from furnishing tools or some other thing necessary for his craft, he will produce worse products, and his sons or whomever else he instructs will turn out to be worse workmen".

"How not?"

"From both conditions, then, poverty and wealth, the products of the crafts become worse and the craftsmen too".

"Apparently".

10.76 Types of pottery and centres of production

Pliny's comments give an excellent idea of the variety and ubiquity of ceramic products in the Roman world, and of the long-distance reputation and trade in fine table wares. Most of the table wares he mentions were mould-made, with raised decoration, and covered with a glossy red slip (Kanowski 1989, Higgins 1976).

Pliny, *Natural History* 35.159–161, 163

Nor are we sated by the presence everywhere of pottery products, with jars devised to hold wine, pipes for water, flue ducts for baths, tiles for roofs, fired bricks for walls and foundations, and items turned on wheels, because of which King Numa set up a seventh guild for potters. Indeed, many even prefer to be buried in pottery tubs after death like Marcus Varro.... The greater part of the human race uses pottery vessels. Among table wares, the Samian is praised even now. Arretium in Italy also holds a high rank, and – for cups alone – Surrentum, Hasta, Pollentia, and in Spain, Saguntum, in Asia Minor, Pergamum. In Asia

Minor, Tralles has its special products, and Mutina in Italy. Since even nations become famous in this way, these products too are carried this way and that across land and sea from workshops renowned for the potter's wheel. At a temple in Erythrae even today are displayed two amphorae dedicated on account of their thin fabric, resulting from a contest between a potter and his apprentice as to who could throw earthenware with the thinner wall. The pottery of Cos is particularly praised for this characteristic, the pottery of Hadria for durability.... But, by Hercules, Vitellius during his reign had a dish produced for 1,000,000 *sestertii* for which a kiln had to be built out in the fields....

10.77 The prominence of Athens in ceramic production

Although ceramic production was not in fact "invented" at Athens, the fame of the ceramics produced there in the sixth and fifth centuries BC was still known to Athenaeus 600 years later, as demonstrated by his quotation of the poet Critias.

Athenaeus, *Philosophers at Dinner* 1.28c

"And she who set up her noble trophy at Marathon [Athens] invented the potter's wheel and the offspring of clay and oven, highly renowned pottery, that useful steward". Attic pottery is very well regarded.

10.78 *Pithoi* (large jars) used for temporary housing

From the Bronze Age through the Late Roman period, large pottery jars (*pithoi*) were produced to serve in groups as storage containers for bulk foodstuffs. They could be several metres high and hold thousands of litres. Allusion is made here to a squatter who took up residence in a large disused *pithos* lying on its side.

Aristophanes, *Knights* 792–793

And how you love him, you who watched him without pity for 8 years as he took up residence in *pithoi* and nests and little towers.

10.79 A pottery workshop in Roman Egypt

This contract for the lease of a pottery, preserved as a papyrus document from Oxyrhynchus dated to AD 243, is an indication of the sort of everyday information concerning ceramic production that has not survived in the literary sources quoted above. The pottery will produce for the most part amphorae, large two-handled transport jars for wine, to be made watertight by lining them with pitch on the interior. Translation adapted from *Oxyrhynchus Papyri* 50 (1983) 236–237.

P.Oxy. 3595

To Aurelia Leontarous and Aurelia Plousia and however she is styled through Aurelius (...) dorus their guardian, from Aurelius Paesis, son of Hephaestas and

Thaisous, who lives in the village of Senepta, a potter of wine jars. Of my own free will I undertake to lease for two years from the current month Thoth of the present seventh year your potting establishment for making wine jars, in the large farmstead of your estate around Senepta, together with its storerooms, kiln, potter's wheel and other equipment, on condition that annually I make for you, fire, re-fire, and coat with pitch so-called Oxyrhynchite *four-choes* jars to a total of 15,000, 150 double *keramia*, and 150 *two-choes* jars, while you provide the friable earth [clay?], sandy earth [temper?], and black earth, sufficient fuel for the kiln, water for the cistern, and – for coating with pitch – 26 talents of pitch in weight by the measure of Aline for the 10,000 jars. I provide for myself sufficient potters, assistants, and stokers and receive payment for the single *keramia* only, 32 *drachmai* per 100 and as a special payment for the 10,000 jars, two *keramia* of wine and two *keramia* of vinegar.... I shall manufacture during the winter and hand over the aforementioned jars on the drying floors of the said pottery, well-fired and coated with pitch from bottom to rim, not leaking and excluding any that have been repaired or are blemished ... and at the end of the contract period I shall hand over the said pottery free of ash and potsherds.

GLASS PRODUCTION

The technique of making glass was discovered much later than that of ceramics, probably as a by-product of the high temperatures involved in smelting copper in the Early Bronze Age. In many ways, however, the technologies are similar, involving the alteration of common natural substances – in the case of glass, pure quartz sand and potash (potassium carbonate) or soda (sodium carbonate) by the heat of a kiln. The earliest products were granular or opaque, but by the early Iron Age the techniques for producing brightly-coloured objects of a homogeneous texture by dipping, hand-moulding, or winding threads of glass were well understood. In the Hellenistic period, perhaps under the influence of ceramic techniques, the production of bowls by means of pressing discs of translucent, coloured glass into moulds became popular. Glass vessels did not become inexpensive and common, however, until the invention of glass-blowing in the first century BC. This technique also allowed the production of elaborate shapes and transparent glass. During the Imperial period vessels were also decorated by cutting with an abrasive lap wheel. (Stern 2008, Price 1976, Forbes 1966)

10.80 The process and raw materials of glassmaking

The materials and procedures for making glass are relatively straightforward, and Pliny and Strabo give simplified but generally accurate accounts of the invention of the technology in the Bronze Age Near East and its spread to Europe in the Roman period. Phoenician workshops are among the earliest. Because of the cost of transport and relative fragility of the product, workshops moved to locations where proper raw materials could be found near major markets. Josephus, however, notes that in the first century AC the special sand from a beach near Ptolemais was being shipped out to glassworks by boat. The role of soda is that of a flux which fosters the melting of the individual particles of sand (Henderson 1985).

Pliny, *Natural History* 36.190–194

In that portion of the province of Syria that is called Phoenicia and borders on Judaea, among the foothills of Mt. Carmel, there is a swamp called Candebia. The Belus river is believed to flow from it and in five miles reaches the sea by the colony of Ptolemais.... It is muddy and deep, and the sand is not visible except at low water. These deposits are rolled by the waves and gleam once the impurities have been rubbed away.

Then also, it is believed, they are made astringent by the salty bite of the sea and ready for use. The beach is no more than half a mile long, but for many centuries this sand alone was used for the production of glass. There is a story that soda [natural sodium carbonate] merchants were driven here in their ship. While they were scattered along the shore preparing a meal, not finding any stray stones to set their pots on, they instead put beneath them lumps of soda from the ship. When these were heated and mixed with the beach sand, streams of an unknown translucent liquid flowed out, and this was the origin of glass.

Soon, since man's skill is ingenious, he was not content to mix just soda, but magnetite, since it is believed to attract to itself the melted glass just as it does iron. In the same way, shining stones came to be added to the melt in many places, then shells and pit sand. There are authors who state that the glass in India is incomparable. The heating is done with light, dry wood, and copper and soda (preferably Egyptian) are added to the melt. Like bronze, glass is melted in a series of furnaces, and dull blackish lumps are formed. Molten glass everywhere is so sharp that before any pain is felt it cuts to the bone whatever part of the body it strikes. The masses are melted again in the workshops and coloured, then some of the glass is shaped by blowing, some ground on a lathe, some engraved like silver. Sidon was once renowned for these workshops, since indeed she even invented glass mirrors.

This was the old procedure for making glass. But in addition, nowadays white sand carried down by the Volturnus River in Italy is taken from a six-mile stretch of beach between Cumae and Liternum, where it is the finest grained, and reduced to powder in a mortar or mill. Then it is mixed with three parts of soda by weight or measure, melted, and poured into other furnaces. There it becomes a lump, which is called soda-sand, and is remelted and becomes a lump of pure and shining glass. To be sure, these days even in the Gauls and Spain sand is compounded in the same manner.

Strabo, *Geography* 16.2.25

Next is the great city of Ptolemais [Akko], which formerly was called Ake.... Between Ake and Tyre there is a stretch of sandy beach yielding sand for glass.

It is said, however, that it is not smelted there but is carried off to Sidon to be worked. Some say that only the Sidonians have sand suitable for glassmaking but others that all sand everywhere can be fused. I was told by glass-workers at Alexandria that there is also a certain vitreous earth in Egypt without which it is not possible to bring to completion varicoloured and costly designs, as if some glass-workers need one mixture, others another. It is said that in Rome also many techniques have been found both for producing colours and for facilitating manufacture, in the case of transparent glass ware, for example. For at Rome it is possible to purchase a bowl or cup for a copper coin.

Josephus, *Jewish War* 2.189–190

About two *stadia* distant from the city [Ptolemais] flows the stream Belus. The Tomb of Memnon stands on its bank, and nearby is a very remarkable spot 100 cubits [across] – a circular basin that produces sand suitable for making glass. Many boats are put in here for the sand, and whenever they empty it out, the hollow is filled up again by the wind....

10.81 Production and use of soda (sodium carbonate)

Pliny, *Natural History* 31.106, 109, 110

I must not put off a description of soda, which is not too different from salt.... In Egypt it is produced more abundantly by artificial methods, but is of a lower quality, dark and stony. It is produced in nearly the same manner as salt, except that for salt one lets the sea into salt pans, but for soda they divert the Nile into soda pits. When the Nile is rising, these are dry, but when it is falling, they are wet with a solution of soda for a period of 40 days.... If in addition it has rained, they let in less water from the river and gather the soda as soon as it has begun to thicken, to keep it from percolating back into the soda pits.... It lasts a long time stored in heaps.... The best soda is the most powdery, since in this way the froth is best. However, the lower quality is good for some applications, as in dyeing purple cloth and dyeing in general. The soda is of particular use in glass manufacturing....

10.82 The many uses and colours of glass

The ability to produce vivid and lasting colours in glass through the addition of mineral and metal substances developed very early, and by the Hellenistic period the palette was rich. Clear glass, in contrast, was difficult to produce because the sand had to be completely free of iron, which gives glass a greenish-yellow tinge. Pliny notes the danger of temperature shock, but is confused about cullet, fragments of broken glass that were an important ingredient in new glass. He may mean that a broken glass vessel could not be repaired by welding the fractures. The water-filled sphere probably served as a type of focusing lens (cf. **2.4**)

Pliny, *Natural History* 36.198–199

Artificial obsidian glass for table ware is produced by a type of dyeing process, also a type of completely red, opaque glass called "blood-red ware". White glass also is produced, and murrine ware, and glass that counterfeits sapphires and lapis lazuli and all the other coloured stones, for there is now no other substance more yielding or adaptable, even for painted decoration. Clear glass, however, as close as possible in appearance to rock crystal, is held in the highest esteem. In fact, it has replaced the use of gold and silver vessels for drinking. It cannot, however, stand heat, unless a cold liquid is put in first.

Nevertheless, glass balls filled with water and placed in the sun become hot enough to set clothing on fire. Heating only causes broken pieces of glass to adhere; they cannot be completely melted again except into distinct drops, as when the little round counters are made, which are sometimes called "eyeballs", some of them with various multicoloured patterns....

Pliny, *Natural History* 37.29

Glassware is now able to mimic rock crystal to a marvellous degree, but, surprisingly, the increase in its own value has not diminished the value of the stone.

10.83 The first steps in the manufacture of cane glass

A type of patchwork glass was manufactured in the Hellenistic and Roman world by breaking multicoloured rods of glass into uniform sections, which were then melted together into a disc in a furnace and shaped on a mould. It is possible that this short (and possibly incomplete) poem gives an account of the first stage of the process, the production of a coloured rod. Glass rods were also used to produce coloured cubes for floor, wall, and vault mosaics in homes and bath buildings (see **8.21**) (Grose 1984).

Mesomedes, *Greek Anthology* 16.323

The workman dug and gathered up material for the glass and put the iron-hard lump in the fire. The glass, heated through by the all-devouring flames, ran out like wax. Men marvelled to see the long coil trickling from the fire and the workman trembling in fear it should fall and break. He put the lump on the tip of the double forceps.

10.84 A poetic description of glass-blowing

This poem, preserved on a papyrus of the third or fourth century AC, is the only detailed description surviving of the technique of blowing glass, invented in the Near East in the mid-first century BC. The technique is essentially the same as that in use today, and shapes could be blown freehand into moulds. Translation adapted from R.A. Coles, *Oxyrhynchus Papyri* 50 (1983) 58 (Stern 2007, Grose 1977).

P.Oxy. 3536

[There are several gaps in the text.] ... fabricating for human use.... First he heated the very point of the iron, then snatched from nearby a lump of bright glass and placed it skilfully within the hollow furnace. And the crystal, as it tasted the heat of the fire, was softened by the strokes of Hephaestus like.... He blew in from his mouth a quick breath ... like a man essaying the most delightful art of the flute. The glass received the rush of breath and became swollen out front of itself like a sphere. It would accept a second rush of the divine blast, for swinging it often like a herdsman swings his crook, he would breathe into....

10.85 The advantages of cheap glassware

Although the major advantages of glass-blowing are that it permits production of thinner vessel walls and a much wider variety of shapes, the technique also sped up production enormously and made most glass relatively cheap. Martial suggests that it also could stand temperature changes better than an expensive imported (cast?) crystal. The term "fearless" sounds like a brand name.

Martial, *Epigrams* 12.74

While you are waiting for the Egyptian fleet to bring you your crystal glassware, accept these cups purchased near the Flaminian Circus. Who is the more fearless: the cups, or those who send them? But there is a two-fold advantage in cheap glassware, Flaccus: no thief thinks twice about these relief-ware cups, and hot water doesn't break them. What about the fact that the guest can drink without worrying the steward, and that shaking hands don't fear a slip? This too is something: you will drink a toast in these cups, Flaccus, if you've got to break your glass afterwards.

Martial, *Epigrams* 14.94

We are plebeian relief-decorated cups of "fearless" glass, and hot water does not crack our crystal.

10.86 A story about "flexible" or "unbreakable" glass

The invention of blown glass, which was shaped by bending and stretching, seemed such a marvel that Romans of the first century AC wondered why the major remaining drawback of the material – brittleness – could not be overcome as well. Gossips hostile to the Emperor Tiberius (who reigned 14–37) concocted the following story to show that such a technological advance had in fact been made but suppressed; it is repeated in one form or another by two other authors (Dio Cassius, *History* 57.21.6, and Petronius, *Satyricon* 50–51; cf. **14.23**).

Pliny, *Natural History* 36.195

It is said that during the reign of Tiberius a formula for making glass flexible was worked out and that the craftsman's workshop was destroyed to keep the metals bronze, silver, and gold from falling in value. This story has for a long time been more frequently told than documented. But what does it matter, since during Nero's reign a technique of the glass-maker's art was developed to produce so-called "stoneware" (*petroti*), a pair of small cups selling for 6,000 *sestertii*.

10.87 The legal responsibilities of glass-workers

Glass blanks could also be decorated or shaped by grinding wheels. In the third century AC a taste developed, particularly in Roman Germany, for glass cups with a raised lacework of surface decoration produced by laboriously grinding the pattern into thick glass blanks with lap-wheels [**fig. 10.5**]. As the following legal text shows, even the blanks were valuable (Doppelfeld 1961).

Ulpian, in *Digest* 9.2.27.29

If you handed over the blank for a *diatretum* cup to be made up and someone broke it through inexperience, he is liable for the extent of the loss. But if he did not break it through inexperience, but it already had flaws which made it susceptible to cracking, he can be acquitted. Consequently, most craftsmen, when given material of this sort, are accustomed to make an undertaking not to work it at their own risk.

E APPLIED CHEMISTRY

Practical or applied chemistry is the chemistry of use, rather than the chemistry of theory, which has no immediate application. Although theoretical advances

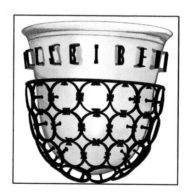

Figure 10.5 Diatretum cup.

Source: Milan Cage Cup, fourth century AC, Civiche Raccolte Archeologiche e Numismatiche, Civico Museo Archeologico, Milan, Italy).

were made in chemistry during the seventh to fourth centuries BC by Greek philosophers, applied chemistry was filled with misinformation and guesses, largely because of a lack of understanding of the materials that were utilised. In antiquity applied chemistry was based upon observation of changes, changes that were most often introduced by some element of heat. Although it had a great impact upon many aspects of technology (metallurgy, perfumes, glass, ceramics, painting, and food production), the superficial observations and empirical methods used in antiquity to gain knowledge resulted in much uncertainty and misinformation regarding chemistry in general. Not until the sixteenth century AC was applied chemistry put on a firm scientific footing.

This section includes a wide variety of products that utilise chemistry in their creation; other examples can be found in metallurgy (**6.9**, **6.11**, **6.42**), construction (**8.12**), and textiles (**10.35**, **10.39**, **10.46**) (Faure 1987).

10.88 Methods of preparing pitch

Pitch was used for a variety of important tasks in antiquity: for pitching wine containers, waterproofing roofs and walls, careening ships, coating ships' tackle, as a material to produce a type of pigment for painters, as pitch-plaster to remove body hair, to coat metal objects to protect them from corrosion, and for medicinal purposes. As a result, its procurement was essential.

Theophrastus, *Enquiry into Plants* 9.3.1

They produce pitch by fire in the following way. They prepare a level piece of ground like a threshing floor, with a sump at the centre, and tamp it smooth and hard. Then they split and stack the wood in the same way charcoal-burners do, except the latter do not have a sump. The billets are placed on end, next to each other and stacked up as high as needed for the amount. They say it is finished when the heap is 180 cubits around and 50 or at most 60 high – or 100 in circumference and 100 high if the wood is rich in pitch. Having built the heap in this way and covered it over with wood, they conceal it with a covering of earth so that the glow of the fire cannot be seen at any point, for if this happens, the pitch is lost. They kindle the heap through the access passage, then fill that up too with wood and earth. Wherever they see smoke seeping out, they climb up ladders and keep piling earth on, so that the gleam of the fire might not be seen. A drain made right through the heap for the pitch allows it to flow out into a pit about 15 cubits away. The stream of pitch flowing out is cold to the touch. The heap burns for two days and two nights ... and all this time they keep watch.

Pliny (**10.22**) duplicates some of the information and adds details regarding further treatment with vinegar and heat to produce tar and resin.

10.89 Ancient fire retardant and fire-extinguishers

Aeneas Tacticus, *Treatise on the Defence of a Besieged City* 34

If the enemy attempts to set anything on fire with their powerful incendiary equipment it is necessary to extinguish it with vinegar. For not easily can it then be re-ignited. Preferably smear [the objects] with birdlime, since fire does not attack this.

Theophrastus (*On Fire* 61) and Pliny (*Natural History* 33.94) note the fire-resistant properties of birdlime, vinegar, mistletoe, and eggs. For other precautions against fire, see **2.25** and **8.64–65**.

10.90 Preparation of beeswax

Wax was employed in numerous processes in the classical world: casting statues and wax portraits, writing, painting, medicine, the protection of walls and weapons, and of course ear plugs. Its source was the honeycomb, but suitable preparation of the wax was necessary to produce a pure product without debris.

Pliny, *Natural History* 21.83–84

Wax is made once the honeycomb is pressed, but first it is purified with water and dried in the shade for three days. On the fourth day, having been liquefied in a new clay pot on the fire with the water covering the combs, it is strained in a plaited basket. The wax is boiled again in the same pot with the same water and then poured into other cold water that is in pots smeared all around with honey. [Types of wax are listed.] ... Punic wax is made in the following manner: yellow wax is exposed to the open air several times, then boiled in water taken from the open sea to which soda has been added. Then they collect by spoonfuls the "flower", that is, the whitest bits that they then pour into a pot containing a bit of cold water, and they boil it down again by itself in the sea water. Next they cool the vase itself with water. Once this has been done three times, they dry it by sunlight and moonlight on a rush mat in the open; the latter makes it whiter, while the sun dries it. They cover it with thin linen fabric so the sun will not liquefy the wax. The whitest wax is made if it is re-boiled again after exposure in the sun.

10.91 The earliest mention of soap

In the classical world, soap was not in common use for cleansing the body. Oil, with or without mechanical agents such as sand, bran, juice, pumice, or ash, was used instead. The terminology is problematic, but the passage of Theocritus may be regarded as possibly the earliest reference to soap. Praxinoa, a testy mistress, addresses her slave.

Theocritus, *Idylls* 15.29–32

Move yourself: bring some water quickly. I have to have water first, but she brings the soap (*smema*)! Give it here anyway! Not too much, you thief! Pour on the water. You wretch, why are you spilling it on my dress? Stop, enough! I've washed as much as the gods allow.

10.92 Composition of Gallic soap

The exact meaning of Pliny's "soap", used as a pomade, is uncertain; it may be simply an unguent or ointment. The ingredients, however, are suitable for use as a cleansing agent.

Pliny, *Natural History* 28.191

Soap (*sapo*), an invention of the Gallic peoples for dyeing their hair red, is also beneficial [for treating sores]. It is made from suet and ash – the best from beech ash and goat suet – in two kinds, thick and liquid. Both types are used among the Germans, more by the men than the women.

10.93 Cleaning agents

A variety of "recipes" for cleaning different objects has been preserved in two papyri of the late third or early fourth century AC. Other recipes from these papyri are described in metallurgy (**6.42**) and textiles (**10.50**) (Caley 1926, 1927).

P.Leid. X (Halleux 1981)

Cleaning Copper Objects. After boiling some beets, carefully clean the copper and silver objects with the juice. The beets are boiled in water.

P.Holm. (Halleux 1981)

As with *P.Leid. X*, this is another very long papyrus with more than 150 recipes, many concerned with colouring, dyeing, and cleaning gems and imitation metals.

Another. This mixture also polishes [cleans] papyrus sheets, which have been written upon, so that they seem like they had never been written upon. Taking *aphronitron*,[4] dissolve it in water. Then put into it, once the nitron solution has formed, one part of any sort of raw earth and one part of Cimolian earth, and cow's milk so that the whole mixture becomes glutinous. Then mix in juice of mastic and coat [the papyrus or pearl] using a feather and let it dry. Finally peel it off and you will find the whiteness. If the pearls are a deep yellow, coat them again; if it is for a papyrus, coat only the letters.

Pliny, *Natural History* 28.66 recommends urine as an ink remover.

HOUSEHOLD CRAFTS, HEALTH/WELL-BEING, WORKSHOP PRODUCTION

10.94 The manufacture of artificial gems

Numerous recipes for the production of artificial gems are also recorded in the Stockholm Papyrus. A few follow.

P.Holm. (Halleux 1981)

The Dyeing of Carnelian: Dissolve alkanet in oil, then throw in the blood of a pigeon, fine Sinopian earth, and a bit of vinegar so the blood does not coagulate. Immerse the transparent stone [selenite?] in it and, sealing it, place it in the dew for 10 days. If you wish to make it with a white border, once the stone has been worked, wrap horse hair around it, bind it all up, and throw it into the dye solution....

The Manufacture of Pearls: Taking an easily crushable stone, such as mica, pulverise it. And taking cow's milk and tragacanth, soften it for 10 days. When it becomes soft to the touch, pour it into another vessel until it has the thickness of glue. Melt Etruscan wax and add to this the white of an egg and mercury: two parts of mercury, three parts of stone, and one part for each of the other ingredients. Once mixed, knead the wax and stone with the mercury and the tragacanth and the chicken's egg, then soften and blend it with all the liquid. Finally make, according to a model, the stone [pearl] that you would manufacture. It quickly petrifies. Make deep and spherical forms and, while damp, pierce them, and let them solidify. Then polish it well with a tooth [ivory?] and, if handled as one ought to, it will be superior to a natural pearl.

The Dyeing of Chalcedony: The treatment of crystal so that it appears like chalcedony. Taking a smoky crystal, make small stones that are well-finished. Take and heat them gradually in a darkened shelter up to the moment when they seem to have the fire within themselves; heat them in the slag of a goldsmith. Take and dip them into cedar oil with untreated sulphur and leave them to soak in the dye until morning.

Manufacture of Pearl: Etch a piece of crystal in the urine of a virgin boy and a tablet of alum. Then put it in mercury with the milk of a woman.

Etching of Stones: A corrosive for all stone. Alum and natron in equal amounts are boiled in an equal amount of water. Then the small stones are etched. Heat them earlier for a brief time near the fire, then dip them into the corrosive. While the solution is boiling, do this for a certain period, up to three times altogether, dipping and soaking them again up to three times, but no more lest the small stones fracture.

10.95 Vinegar's role in antiquity's most expensive meal

The acidic nature of vinegar was often exaggerated, as it is here (see also **5.21**), but its diverse uses were significant (e.g. **6.9**, **10.22**, **10.26**, **10.50**, **10.65**, **10.79**, **10.88**, **10.89**, **10.94**). In this story Cleopatra has wagered that she could spend an unbelievable fortune on a single banquet. A large, but ordinary first course has been served and then this second course.

Pliny, *Natural History* 9.120–121

Cleopatra ordered the second course to be brought in. According to instructions, the servants placed before her only a single bowl of vinegar, which was the harsh and powerful kind able to dissolve pearls into solution. She was wearing on her ears that truly unique and wonderful work of nature [the two largest pearls in antiquity]. And while Antony was waiting to see what she would do, she removed one and immersed it. Then once it had been dissolved she drank it down.

The umpire saves the other pearl by declaring Cleopatra winner of the wager. Since no such vinegar exists, it has been suggested that Cleopatra may have just swallowed the pearl and recovered it later.

10.96 The four essential pigments

Many of the pigments used by the Greeks and Romans were native materials that required little treatment other than the removal of impurities and grinding. A few did require chemical treatments. Pliny here objects that more colours do not make better paintings: earlier and superior artists made do with a restricted palette.

Pliny, *Natural History* 35.50

The most famous painters – Apelles, Aetion, Melanthius, and Nicomachus – made those immortal works with only four colours: whites from Melinum, ochres from Attica, reds from Pontic Sinopis, blacks made from *atramentum*. Even so each of their paintings was worth the wealth of entire towns.

10.97 White pigment production

Theophrastus has just finished discussing natural and artificial ochre and cyanus and their manufacture. He then describes the manufacture of white lead, which was used as the white pigment in antiquity.

Theophrastus, *On Stones* 56

Lead about the size of a brick is put in jars above vinegar. When it acquires a thick coat, which it does usually in ten days, the jars are opened, and a type of mould is scraped from the lead; and the lead is placed in the same way again and again until it is all consumed. The part that has been scraped off is pounded in a mortar and continually filtered off. The white lead is the material finally left at the bottom.

10.98 Verdigris

Theophrastus continues to describe other pigment production.

Theophrastus, *On Stones* 57

Verdigris is manufactured in the same manner. Red copper is placed above wine-lees [grape residues], and the material that collects on it is scraped off; for the verdigris appears there as it forms.

Dioscorides (*Pharmacology* 5.79) and Pliny (*Natural History* 34.110–113) elaborate upon this method.

10.99 Sources for bootblack and ink

The best black pigment was produced from carbon. Although the fine soots from glassmaking and torches were favoured (Dioscorides, *Pharmacology* 5.160–161), carbon was retrieved from numerous sources.

Pliny, *Natural History* 35.41–43

Black pigment also will be put among the manufactured colours, although it also originates from the earth in two ways. For either it emanates from the earth in the manner of saltwater brine, or a sulphuric coloured earth is itself approved for this material. Painters have been discovered who violated graves and excavated the charred remains to obtain the pigment; all inconvenient and recent methods. For it can be made from soot in a variety of ways, from burnt resin or pitch, on account of which smokehouses (*officinae*) have even been constructed that do not allow the smoke to escape. The most highly praised type is made in the same way from the pitch pine tree. It is adulterated with the soot of furnaces and baths, which is used for writing books. Some people boil down the dried lees of wine, and state that, if the lees are from a good wine, the black pigment will be like Indian ink. The very famous painters Polygnotus and Micon made black pigment from the skins of grapes, calling it *tryginon* [made from lees]. Apelles devised and produced from burnt ivory the type called *elephantinum*. The composition of a black pigment imported from India is still unknown to me. One kind is even made with the dyes from the black dust, which sticks to bronze pots. Another is produced from the burnt logs of pitch pine after the charcoal is pounded in a mortar.... But the treatment of all black pigment is completed by exposure to the sun and the kind used for writing books is mixed with gum, that for painting walls with glue. The black pigment that has been dissolved in vinegar washes out with difficulty.

Pliny, *Natural History* 34.123–125

The Greeks have made a connection between copper and shoemakers' black by their name for it; for they call it *chalcanthon* [copperflower]. Nor is the nature of anything else so remarkable. It occurs in the wells or pools of Spain that have the type of water that holds the substance in solution. That water is mixed with an equal amount of fresh water, then boiled and poured into wooden tanks. Stretched by rocks and hanging from fixed beams over these tanks are ropes to which the slime adheres in glassy droplets similar in appearance to grape clusters. Once it is taken off, it is dried for 30 days. Its colour is an extremely remarkable and lustrous blue, and it is thought to be glass. When dissolved, it makes a black pigment for dyeing leather. It is also made in a variety of ways: by digging trenches in that type of earth, from the sides of which the solution trickles and forms icicles in the winter cold. They are called *stalagmias*, and there is no other purer kind.... It is also made in hollows of the rocks when the rainwater and slime run together and freeze, and it is made in the manner that salt is, when the very hot sunshine evaporates the fresh water mixed with it.

PERFUMES, OILS, AND COSMETICS

Perfumes, oils, and other cosmetics have a long history and an importance that perhaps even exceeds that of today. In the classical world, the practice of using these cosmetics may have come from the East and Egypt where they had long been known, at least some of the more famous ones. Elaborate and expensive materials and procedures were used to produce a wide variety of scents and other cosmetics. Needless to say, moralists did not always appreciate the expense or result, but the wide array of products recorded by Pliny the Elder and Theophrastus provide adequate proof of their importance. Unfortunately, the methods of preparation are not fully documented (Forbes 1965).

10.100 Places of origin and changes of fashion

The names of perfumes often were derived from place of origin or from the plant producing the scent. As with most things, their popularity was subject to change with the times.

Pliny, *Natural History* 13.4–5

... The most-praised perfume from ancient times was made on the island of Delos, in later times at Mendes [in Egypt]. Nor did this happen as a result only of the mixing and combination, since the same juices prevailed or lost favour in various ways and places. For a long time, the iris perfume of Corinth was very much favoured, later that of Cyzicus, and similarly the attar of roses from Phaselis, although Naples, Capua, and Palestrina took away this fame. Saffron oil from Soli in Cilicia was greatly praised for a long time, then that of Rhodes....

HOUSEHOLD CRAFTS, HEALTH/WELL-BEING, WORKSHOP PRODUCTION

10.101 The importance and introduction of scent at Rome

Pliny was one of the most vocal critics of the extravagant use of perfume and its fleeting impact.

Pliny, *Natural History* 13.20, 24

This stuff is the most useless of all the luxurious goods; for even pearls and gems pass to an heir, and clothing lasts for some time, while unguents give out on the spot and die in their hour of use ... and they cost more than 400 *denarii* per *libra!* [Pliny records some of the extravagant uses.] ... I could not easily state when they first made their way to the Romans. It is certain that when King Antiochus and Asia were conquered in the five-hundred and sixty-fifth year of the city [189 BC], the censors Publius Licinius Crassus and Lucius Julius Caesar ordained by proclamation that no foreign unguents were to be sold....

Pliny (*Natural History* 13.1–3) attributes the invention of perfumes to the Persians. This was widely believed in antiquity, but references in Homer indicate a much earlier date of origin.

10.102 The manufacture of oil-based perfumes

Theophrastus, *Concerning Odours* 14–19

The composition and preparation of perfume is aimed entirely at the preservation of the odours. As a result, perfumes are set into oil, since it lasts a very long time and at the same time is most fit for use. And yet by nature it is very poorly fit for receiving an odour because of its viscosity and greasiness; this is especially so for the most greasy ones such as almond oil, while sesame oil and olive oil are the least receptive. [A variety of oils is described.] ... They use spices in all perfumes, some to draw up the oil to thicken it and some to infuse their odour. For they thicken somewhat all the oils so that they take the scent better, just as for wool for dyeing. The thickening is done with the less strong spices, then later they throw in the one from which they wish to obtain the odour, since the last one added always prevails, even if in smaller portion. Thus if a *mina* of myrrh is thrown into a *kotyle* of oil and later two *drachmai* of cinnamon are added, the two *drachmai* of cinnamon prevail.... The most receptive oil, which is the [Egyptian] *balanos*, is also the longest lasting ... since the oil that is most receptive combines and becomes one [with the spice] ....

10.103 Heat extraction of scents

Heat extraction was the preferred method for producing perfumes, although some perfumes, such as that made from the iris, did use cold extraction.

Theophrastus, *Concerning Odours* 21–22

Nearly all spices and fragrant scents, except flowers, are dry, hot, astringent, and pungent. Some also have a certain bitterness, as previously stated, such as iris, myrrh, frankincense, and perfumes in general. But the most common of their strengths are astringency and heat production, which they truly produce. Exposed to fire, they all become astringent, but some take on their special odours when cold and not exposed to fire. And it seems that just as with vegetable dyes where some are dyed hot and some are dyed cold, so it is with odours. But in all cases the cooking to produce both the astringent quality and the special odour is done in vessels standing in water and not touching the fire itself. This is so, because the heating must be gentle, and the waste would be great if contact with the flames happened; and even more, it would smell of burning.

Theophrastus then describes the benefits of using heat. Pliny adds a briefer recipe for making perfumes.

Pliny, *Natural History* 13.7

The recipe for making perfumes consists of two things: the juice and the solid part. The former usually consists of types of oils, the latter of types of scents. They call the oils *stymmata* [thickeners] and the scents *hedysmata* [sweeteners]. Meanwhile a third element, neglected by many, is colour; for the sake of which cinnabar and alkanet should be added. A sprinkle of salt preserves the nature of the oil ... resin or gum is added to perpetuate the scent in the solid, since it evaporates very quickly and disappears if these are not added.

10.104 A few recipes for perfumes

The great number of scents recorded in ancient texts indicates the importance of perfumes and unguents, but unfortunately most of the recipes seem to have been closely guarded secrets. Rarely do we have more than general descriptions of a scent and its ingredients.

Theophrastus, *Concerning Odours* 25

They add in the suitable spices for each of the perfumes. Into *kypros*, after mixing in sweet-scented wine, they add cardamom and *aspalathos* [a spiny shrub, yielding fragrant oil]. Into rose, they add ginger-grass, *aspalathos*, and sweet flag; likewise mixed.

Pliny, *Natural History* 13.8

Metopium ... is an oil pressed out from bitter almonds in Egypt, to which are added the oil of unripe grapes and olives, cardamom, rush, sweet flag, honey, wine, myrrh, balsam seed, *galbanum*, and terebinth resin.

10.105 Protection for perfumes

Precautions were taken to protect perfumes from the light and heat of the sun, their two worst enemies.

Theophrastus, *Concerning Odours* 40

A hot season, place, or the sun destroy perfumes if they are so located. On account of this the perfumers seek out upper rooms, not facing the sun but as shaded as possible. For the sun and a hot place diminish the odour and in general cause the perfumes to lose their character more than is done by the cold. The cold and frost, even if they make them less fragrant by their contracting action, nevertheless do not rob away their strength completely. For the worst destruction, as with wines and other juices, is for its natural heat to be altered.

Pliny duplicates the information of Theophrastus, but also adds one caution about testing perfume.

Pliny, *Natural History* 13.19

When being tested the perfumes are taken on the inverted hand, lest the heat of the fleshy part spoil it.

10.106 Cosmetics: a useless expense

The condemnation of expensive perfumes (cf. **10.101** and Martial 10.116) was merely one of a multitude of criticisms levelled at the cosmetic industry in antiquity. Attempts to make oneself appear more youthful and attractive by using cosmetics were also often attacked. For hair dye see **10.48, 10.65, 10.92**.

Lucian, *Greek Anthology* 11.408

You dye your hair, but never will you dye your old age nor smooth out the wrinkles of your cheeks. So, don't plaster over your entire face with white lead so that you have a mask and not a face; for it is no help. Are you crazy? Never will orchil [rouge] and white lead paint turn Hecuba into Helen.

Juvenal, *Satires* 6.464ff. and 9.12ff. also attacks cosmetics.

F HEALTH AND WELL-BEING

Greek and Roman medical knowledge and practice initiated, in many ways, the transformation of Western understanding and treatment of patients. The technological aspect of ancient medicine resulted in the creation of medical instruments fashioned from materials such as iron, bronze, steel, ivory, wood, bone, stone, and gold in a staggering variety of sizes and shapes (**10.116–117**) attuned to specific procedures and patients. The sources provide evidence for simple hygiene (**10.108–111**), for simple aesthetic treatments (**10.112**), for methods to improve oneself for speaking or singing (**10.113**), for prosthetics (**10.114**) such as feet, hand, wigs, teeth, and braces, for drugs (**10.115**) to help with pain and sleep or to counter poisons, and for the tools and methods to deal with dislocations and breakage, from a simple broken nose to serious and complex dislocation and breakage and for brain operations that employed drills for trepanation(**10.117**). In several instances specific instruments were created to deal with specific injuries and ailments. Most famous are the traction table, Hippocrates' Bench (**10.117**), and an instrument to remove arrows, the Spoon of Diocles(**10.118**), but adaptations of other instruments, represented by their sheer number of variations, indicate continual creative inspiration at work. All were intended to improve one's health and to enhance one's lifestyle so as to enjoy life as much as possible.

Medical technology also borrowed from other technologies in order to advance understanding of the human body. For example, flaying, dissection, and vivisection (**10.119**) likely arose from the technologies associated with the skinning, butchering, and tanning of skins of animals, but were applied to humans for medical purposes to enhance knowledge and treatment of human.

The chapter concludes with treatment of the body: first, some of the procedures such as mummification, immersion in honey, and the use of herbs and spices employed to preserve the physical remains of some bodies (**10.120–122**) and, last, the ultimate destination of the bodies whether cremated, entombed or deified (**10.123–124**).

10.107 Greek beginnings: Aesculapius and his sons

After recognising that even uncivilised peoples have various tools for the aiding of wounds and diseases, Celsus celebrates the special contributions of the Greeks to the art of medicine, beginning with Aesculapius and his sons and moving through other great practitioners up to Hippocrates of Cos, the first practitioner to remove medicine from philosophical study. He concludes with Herophilus and Erasistratus as his last examples of men who made major advances in treatment. For 'need' as the mother of medicine see **1.25**.

Celsus, *On Medicine 1 Proem* 1–4

Just as agriculture promises nourishment to healthy bodies, the Healing Art promises health to the sick. Indeed, this Art is lacking nowhere at all, for even the least experienced peoples are acquainted with herbs, and other things near at hand for helping with wounds and diseases. But, nevertheless, it has been refined among the Greeks much more than among the other races – and not, indeed, among them from their earliest beginnings, but only for a few ages before ours. Aesculapius, above all, is celebrated as its most ancient practitioner, and since he improved this science, up to that point primitive and homespun, with somewhat more refinement, he was numbered among the gods. Next, his two sons,

Podalirius and Machaon, who followed the commander Agamemnon to the Trojan War, supplied considerable aid to their fellow-soldiers. Homer lays out, however, that they did not render any type of assistance against the pestilence or the various types of diseases, but to a large extent only that they usually healed by the knife and by medicaments. From which evidence it appears that they attempted only these aspects only of the Healing Art, and that they were the oldest practitioners.

BASIC HYGIENE, APPEARANCE, AND IMPROVEMENT

10.108 Oil and strigil

Soaps (**10.91–92**) were not much used by the Greeks and Romans. Instead, they preferred oil which Plutarch (*Alexander* 57.5) records as a gift from the gods for refreshment after toil. The strigil, a curved metal tool with handle and concave blade, was used to remove the dust and oil from the body, which was then lightly rubbed with oil again. One could conduct the process alone, but the sources imply it was easier to have help.

Historia Augusta, Hadrian 17.5–7

Often, he bathed with the common people, out of which this bathing-joke became well known. On a particular occasion when he had seen a certain veteran, known to him from military service, scraping his back and the rest of his body on the wall, he inquired why he gave himself up to be scoured by the marble. When Hadrian learnt that this was done because he did not have a slave, Hadrian provided him with both slaves and the expense of keeping them. But on another day when many old men were scouring themselves on the wall in order to stir up the princeps' generosity, he ordered them to be called out and to rub one another down in turn.

The physical appearance of Augustus displays a well-known problem (see next passage) with the strigil.

Suetonius, *Augustus* 80

[In addition, it is said that] many parts of his body were marred with hard, scabby eruptions and with certain thick patches of skin arising from the itching in his body and the constant and vigorous use of the strigil [**fig. 10.6**].

The discussion is focused on the role of the bath and cleanliness, and how to deal with the rough scraping of the strigil.

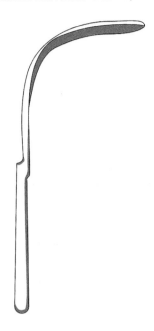

Figure 10.6 Strigil, 21 cm.
Source: Milne.

Hippocrates, *On Regimen in Acute Diseases* 18

It is better not to use friction, but if friction is used, it is necessary to employ a hot unguent [*smegma:* paste of olive oil and an alkali] in greater abundance than is the norm, and not a little is to be poured out, and water quickly to be poured out afterward ... others should pour water and rub him ... and it is necessary to use sponges rather than scraper [strigil], and the body should be anointed when not too dry.

10.109 The humble sponge

Sponges, readily available from the Mediterranean, were used in diverse ways (see Index): to fight or prevent fires, as a medicinal compress, to avoid heat stroke, to plug ears, for sound absorption, to erase, and for cleaning of households (Homer *Odyssey* 1.111, 20.151, 22.439 and 453) and of oneself (Voultsidadou 2007).

Homer, *Iliad* 18.414

[Hephaestus awaits Thetis] and with a sponge he wiped his face and his two hands.

10.110 The toilet sponge

Archaeological evidence exists for fig trees planted near latrines to provide shade, privacy and, based upon contemporary rural practice, leaves to use as toilet paper. Probably grasses and other vegetal and animal substances were used since they seem less traumatic than documented stones and pottery fragments. The hand was available as well. A significant step towards comfort was made with the introduction of the sponge attached to a stick that would be left in a bucket of water, perhaps with vinegar. Martial makes a simple reference to the instrument and the reference in the papyrus – indicates common use. See **9.47** for a much more dramatic and, definitely, non-hygienic use (Koloski-Ostrow 2015 and Jansen *et al.* 2011).

Martial, *Epigrams* 12.48.5–8

[To the host of a lavish dinner who is attempting to win Martial's favour.]

Nevertheless, your dinner is a sumptuous one, I confess, most sumptuous, but tomorrow it will be as nothing; on the contrary today, indeed straightaway, it is nothing other than whatever the wretched sponge of the accursed stick might know, or any dog or a piece of pottery stuck to or set on the road.

P.Mich. VIII 471. 29–30

He gave less consideration to me than to a sponge-stick, but [turned to] his own business and his own affairs.

10.111 Barbers: tools and methods

By severely wounding herself Porcia proves her loyalty to her husband, Brutus, who is conspiring against Julius Caesar.

Plutarch, *Brutus* 13

Having taken up the little knife with which the barbers trim away fingernails, and having driven out all the attendants from her bedroom, she made a deep cut in her thigh that produced a large flow of blood. And soon she was consumed by intense pain accompanied by shivering and high fever from the wound.

The paranoia of the Elder Dionysius was so great that family members were stripped and re-clothed before being allowed into his presence. Given the activity of Porcia in the previous passage, his concern with barber tools is not surprising.

Plutarch, *Dion* 9.3

The Elder Dionysius was so distrustful ... that he did not even allow the hair on his head to be trimmed with the barber's larger knife [scissors?; **fig. 10.7**], but one of his workmen was accustomed to singe the hair of his head with a live ember.

Figure 10.7 Shears/scissors, 10 cm.
Source: Pompeii.

10.112 Pimple treatment

Concerns about appearance, whether baldness, skin eruptions or scarring, were as bothersome in antiquity as they are today. We have a few texts about removal of hair using plasters (**10.88**) and chemical deodorants (**10.40**), but much more evidence about skin care, especially for females.

Celsus, *On Medicine* 6.5

It is almost folly to treat pimples and lenticular spots and freckles, but care for their looks cannot be torn away from women. But of these blemishes, which I have mentioned above, the pimples and lenticular spots are commonly known, although there is also a more rare type which the Greeks call *semion*, since this lenticular spot is more red and irregular. Freckles, which are nothing more than roughness and are ignored by most people in fact; they are nothing more than a roughened and fibrous discolouration. The others appear only on the face, while the lenticular spots sometimes arise on other parts of the body.... Pimples are most easily removed by an application of resin to which not less than the same amount of easily split alum and a bit of honey has been added. Galbanum and soda remove a lenticular spot when they are in equal measure and have been pounded in vinegar until they reach the consistency of honey. The skin is to be smeared with this concoction, and after a considerable number of hours, it is to be washed off the next morning and the spot lightly anointed with oil. A resin, to which a third part of rock-salt and a little honey has been added, removes freckles.

For all the above and also for coloured scars, that compound is useful, which is said to have been invented by Trypho the father. In it are equal parts of the dregs of bennut oil, bluish Cimolian chalk, bitter almonds, barley and vetch meal, along with white soapwort and mellilot seeds. All which are ground together with very bitter honey, and are smeared on at night and washed off in the morning.

10.113 Improvements to body and mind
Vocal abilities

Public speaking, whether for political or legal reasons or for entertainment, was a necessity for success. Two famous individuals demonstrate different techniques for improvement: one for oratory, one for singing.

Plutarch, *Demosthenes* 11

Demetrius of Phaleron, saying that he heard this from Demosthenes himself in his old age, records that Demosthenes overcame his lack of clear speech and lisping and that he endued himself with articulate speech by taking pebbles into his mouth and reciting speeches at the same time. [other techniques follow].

Suetonius, *Nero* 20

Nor did Nero neglect any of those aids which the masters of that profession persistently practice, for the objective of preserving and strengthening the voice. But flat on his back he would bear a lead slab on his chest, be cleansed by vomiting and enemas, and abstained from fruits and foods detrimental to the voice.

Intellectual ability

Plutarch suggests that the senses are an intellectual distraction and records an extreme, even if false, example of how to focus oneself better by restricting visual enticements.

Plutarch, *Concerning Curiosity* 12

The story is false that Democritus voluntarily put out his eyes by fixing them on a burning mirror and receiving the reflected heat from it, but it records that he did this so that his eyes, always summoning confusion outside, might not yield up his intellect to it, but might permit his intellect to stay safe inside and to pass the time on mental contemplations, just like blocking-up windows looking onto the street. And this is truer than all else, that those men employing their intellectual capacity the most, engage their sense-perception the least.

10.114 Prosthetics and other devices

The importance of movement prompts several references to prosthetic feet, but hands are also mentioned several times as are other body parts such as false teeth as well as less mobile accoutrements like wigs. The artificial replacements for injured and missing elements were usually of wood and bronze or copper, but bone, ivory and, sometimes, more precious metals were employed. The general consideration of comfort vs. discomfort accounts for some of the aids that were created to make one's life and work better.

Pindar, when tracing lineage of Dionysios of Syracuse to Pelops, provides the earliest reference to Pelops' ivory shoulder that replaced the one eaten by the grief-stricken Demeter when his body was reassembled.

Pindar, *Olympian Odes* 1.26–27

[Poseidon fell in love with Pelops] when Clotho took him out of the pollution-free cauldron, his radiant shoulder gloriously distinguished with ivory.

Teeth and wigs

Archaeological evidence for tooth removal and false teeth is relatively common from the Etruscan and Roman periods. Partial dentures were made from human or animal teeth fastened together with gold bands.

Here, Cicero discusses limits upon funerary burial and other practices, but notes an exception.

Cicero, *On the Laws* 2.24 (60)

[In addition, there was a law:] "let no gold be added" [into the grave with the dead], but note how another law makes this humane exception: "But for anyone whose teeth were fastened with gold, he will be buried or cremated with the gold, but let him be without offence".

Martial, *Epigrams* 9.37

Although you yourself are at home, Galla, you are decked-out as if you are in the middle of the Subura. Your hair was made far away and at night you lay aside your teeth in the same manner as your silks. You lie stored away in 100 little [cosmetic] boxes – your face doesn't even sleep with you. Nevertheless, you give me a sign with that eyebrow, which was brought out for you in the morning.

A woman's beautiful hair has been dyed (see **10.106**) and treated with heat and various procedures to such an extent that it has been so damaged that a wig is necessary. One famous supply resource is mentioned here.

Ovid, *Amores* 1.14.25–50

How patiently your locks offer themselves to fire and steel,
so that its woven curl might become a tortured ball.
I often cried, "That is a crime, a crime to scorch those locks …"
No mistress of magic herbs has wounded you,
no Thessalian hag has soaked you in malign water.
Nor has the force of sickness harmed you – banish that foreboding thought!
Nor has an envious tongue thinned your dense hair.
You yourself provided the poison to mix on your head.
Now Germania will send the hair from its captives to you
and you will be secure with that booty from a conquered people.
But, O how often you will blush when someone praises your hair,
and you will say: "Now I am being judged by the purchase price,
I don't know whether that man praises the Sygambrian [Germanic] woman instead of me.
Yet I remember when that fame was mine".

Feet

An incredible story is related about Hegesistratus, a Greek diviner from Elis who had helped the Persians, then been captured. He faced a dire future at the hands of his Greek captors.

Herodotus, *Histories* 9.37.1–4

The Spartans, having taken Hegesistratus, kept him chained by his foot, having sentenced him to death since they had suffered so many grievous injuries from him. Being in such a bad situation, one in which he was running the risk of death and, even before his death, enduring many dreadful pains, he performed a deed greater than mere words can describe. For, since he was bound fast in an iron-bound foot-stocks, he got possession of an iron tool, somehow smuggled in, and immediately contrived the most courageous action of all those we have known. For having given full weight to how the rest of him might get free of his foot, he cut off his own foot at the instep. Having done this, since he was being guard by sentries he tunnelled through the wall of the structure and escaped towards Tegea. [The Spartans, finding the half-foot are amazed and pursue; he hides during the day and on the third night arrives at Tegea, where he is sheltered.] … Once he was healed and had procured a foot of wood for himself, he became an outright enemy of the Lacedaemonians.

In his diatribe, Lucian uses this example to comment upon the Collector who has fine books but does not understand them.

Lucian, *The Ignorant Book Collector* 6

Recently there was a rich man in Asia, whose misfortune resulted in both feet being amputated, rotting away as a result of the freezing cold, I believe, when one time it happened that he had to make a journey through the snow. To continue, he had suffered a pitiful blow, and attending to his bad luck he had a pair of wooden feet made that once he bound them on, he was able to walk with his house slaves, by leaning upon them. But that man kept doing a silly thing, for he continued to buy the most stylish, most beautiful boots and he paid the utmost attention to them so as that his wooden feet would be embellished with the most beautiful footwear.

Hands

Pliny describes one of the earliest uses of a prosthetic hand that has been made to replace the one lost by M. Sergius during the Second Punic War. The iron prosthetic allowed him to hold a shield and continue to fight in several subsequent battles.

Pliny, *Natural History* 7.29

No one, in my mind at least, can justly rank any man above M. Sergius, although his great-grandson, Catiline, taints the esteem of his name. In his second campaign Sergius lost his right hand. In two campaigns he was wounded 23 times, with the result that he had next to no use in either hand or either foot. Only his spirit was intact. Yet, even though a crippled soldier he fought in many later campaigns. He was twice captured by Hannibal (for he battled not just any ordinary foe) twice he escaped his chains, although kept in chains or shackles every day for 20 months. With his left hand alone, he engaged in combat four times, and on a single day two horses were disembowelled while he was riding them. He fashioned a right hand of iron for himself and, with it bound fast [to his arm], he joined battle and delivered Cremona from its siege, safe-guarded Placentia, and captured 12 enemy camps in Gaul, all of which are confirmed from his speech during his praetorship when the attempt was being made by his colleagues to disqualify him from the sacred rites as being a cripple.

A finger brace

Simple supports, such as walking sticks, that help with weakness are often noted in passing, probably implying widespread use; fingerstalls, perhaps, as well.

Suetonius, *Augustus* 80

Occasionally he endured the feeling that the index finger of his right hand was so feeble, when it was numbed and shrunken by cold, that it scarcely functioned for his writing, even with the reinforcement of a horn fingerstall.

An ingenious defence against dogs

Among the laws attributed to Solon is one that employs a special harness to address the problem of unprovoked dog attacks.

Plutarch, *Solon* 24

Solon also enacted a law concerning injuries from four-footed beasts, in which it was decreed that a dog having bitten someone and subsequently having been fettered with a dog-collar [attached to a pole] three cubits long, must be turned over to the injured person – a clever device for personal safety.

10.115 Drugs: sleeping potions, anaesthetics, poisons, and antidotes

Surgery without anaesthetic seems to have been relatively commonplace (Hippocrates, *The Physician* 5 [IX 210–212]), but wine and several concoctions were used to ease pain and to help patients recover (Cilliers 2012).

Typical battlefield treatment without anaesthetic

An arrow has struck Alexander during the campaign against the Malli in the Punjab. This extraction method was probably not uncommon.

Plutarch, *Alexander* 63.3–6

[During the melee] … one man standing a little farther away shot his arrow from his bow, so well strung and powerful that the missile pierced Alexander's chest armour and was stuck into the ribs about his breast. [Two guards defend Alexander, who is wounded again and again, until the Macedonians swarm in to take him to the safety of his tent.] And straightaway with difficulty and very laboriously, having sawn off the wooden arrow shaft and thus the chest piece was only barely set free. They then came to the cutting out of the arrow-head, which was embedded in one of his ribs. It is reported the head was three *dactyloi* wide and four long. On this account, as it was being taken out Alexander convulsed [from pain] to the point of passing out, very close to death, but nevertheless he recovered.

Benefits of drugs

Many concoctions, some quite dangerous, were used to ease pain and to help patients recover. Galen provides a general statement regarding the benefits of medications, while Celsus and Dioscorides provide a plethora of recipes of which a few are provided (cf. **10.35** for the health benefits of urine).

Galen, *On the Composition of Drugs According to Places* 12.966 14K (Von Staden: 418 T249)

Herophilus observed that drugs are just like the hands of the gods.

Sleeping potions

Celsus, *On Medicine* 2.32.1

Poppy, lettuce, most particularly the summer ones whose tender stalk is just then full of milk, the mulberry, and the leek are used to attain deep sleep.

Celsus, *On Medicine* 3.18.12

[For those having difficulty sleeping, but who need it to get well.] If, nonetheless, the patients are unable to sleep, some [doctors] endeavour to induce sleep by providing water in a drink in which poppy or hyoscyamus have been boiled down, while others put apples of mandrake under the pillow, and others apply either cardamom balsam or sycamine tears to the forehead.

Celsus, *On Medicine* 5.25.1

There are also many pills, and are made for various purposes. They call those that alleviate pain through sleep anodynes. To use them, unless great necessity makes it urgent, is a risky undertaking for they are made from powerful medicaments, which are inimical to the stomach. Still, for all that, one pill has the power to aid digestion. It consists of poppy-tears and galbanum, 4 grams each, and of myrrh, castor, and pepper, 8 grams each, from which it is enough to take a pill the size of a vetch. Another mixture, worse for the stomach but stronger for sleeping, is made from these things: 1 gram of mandrake, 16 grams each of celery-seed and likewise of hyoscyamus seed, which are ground up after soaking in wine. One of the size set out above is sufficient to take.

Plutarch, *Dion* 6

When the Elder Dionysius was sick and seemed near to death, Dion attempted to speak with him about his children by Aristomache, but the doctors who wished to ingratiate themselves with the one about to inherit the realm, did not give him the opportunity. But, as Timaeus records, when Dionysius demanded a sleeping potion, they gave him one that deprived him of any perception of his senses, bringing on sleep with death next.

Dioscorides mentions 20 soporifics and poisons, but only mandrake is recommended as a consistent anaesthetic for surgery and cautery, and it was often used as a soporific. Hemlock, the executioner's drug used upon Socrates, was also utilised in several forms: a juice to induce sleep, a salve or plaster for various other ailments, and the leaves for relieving swelling, pain or flux. The seed contains the neurotoxin, which in light doses has a sedative effect, but causes death in large doses. Plato (*Phaedo* 117e-118a) describes the progress of the drug's effects on Socrates: paralysis, chill, lack of verbal response, death.

Dioscorides, *Pharmacology* 4.75.2–7

Concerning Mandrake: To extract a juice, the green bark of the root is mashed and then put under a press. It will have to be baked in the sun and, after the congealing, stored away in a ceramic container.... Others boil the roots in wine down until a third remains and put it up in storage once it has been strained, using a small *kyathos* [less than 50 ml] for those who cannot sleep and for those suffering great pain and is one for those whom they want to make insensible [to pain] when cutting them [operating on] or cauterising them ... [other uses are provided: for removal of bile, as an eye treatment, painkiller and stool-softener, for menstruation, as an abortifacient, and even for softening ivory to make statues; but too much causes death].... [About 150 ml of another variety of mandrake root mixed with sweet wine] is to be given to those about to be operated upon or cauterised, as said above, for they are not conscious of the pain because of their semi-comatose state.... The doctors also use another type, *morion*, whenever they are preparing to make incisions or to cauterise.

An indication of widespread knowledge of anaesthesia?

St. Hilary does not seem to have had any special interest in medicines, suggesting his description of anaesthesia – perhaps through the use of opiates – was common knowledge in his day (mid-fourth century AC). He has been discussing the nature of the human body and its ability to sense cold, heat, pleasure, and pain.

St. Hilary of Poitiers, *On the Trinity* 10.14 (329)

When grave necessity demands that part of the body has to be cut away, the active spirit is lulled to sleep by a medicated drink, and the mind, which had been engrossed with the maiming incisions, is brought about into a death-like forgetfulness of its own senses. And then the limbs, ignorant of the pain, are cut off; the sensation of the flesh is dead and escapes every thrust of the deep incision since the sensation of the soul inside is asleep.

Antidotes

Of several recipes employed as antidotes to poison, this is the most famous.

Celsus, *On Medicine* 5.23

But the most famous antidote is Mithridates', which, by taking it daily, that king is said to have to have rendered his body safe against dangers of poisons. In it are these ingredients: 1.66 grams of costus, 20 grams of sweet flag, 8 grams each of hypericum, tree-gum, sagapenum, acacia juice, Illyrian iris, and cardamon, 12 grams of anise, 16 grams each of Gallic nard, gentian herbal root and dried rose-leaves, 17 grams each of poppy-tears and rock-celery [parsley], 20.66 grams each of wild cinnamon, hartwort, darnel, and long pepper, 21 grams of storax, 24 grams each of castor oil, frankincense, cytinus hypocistis juice, myrrh and gum of opopanax, 24 grams of malabathron leaves, 24.66 grams each of flower of round rush, terebinth resin, galbanum, and Cretan parsnip seeds, 25 grams each of nard and opobalsam, 25 grams of pennycress, 28 grams of Pontic rhubarb root, 29 grams each of saffron, ginger, and cinnamon. Once these ingredients have been ground up, they are mixed with honey. A piece the size of a Greek nut [walnut] is sufficient against poison and is given with wine.

10.116 Materials, sizes, and variations

Archaeological and literary evidence reveal that a wide variety of materials were employed to manufacture the physicians' equipment: iron, steel, bronze, tin, lead, gold, silver, copper, bone, ivory and horn, wood and stone. The following passages indicate some of the general considerations regarding tools created for specific procedures.

Hippocrates, *The Physician* 2.14–16

Use bronze for nothing other than instruments, for it seems to me to be vulgar ornamentation to use such materials for vessels.

Hippocrates, *The Physician* 2.19–21

All instruments must be well-fitted for their use in size, weight and fineness.

Specific size is stated for many different tools. The next passages deal with raspatories, bleeding cups, and cauteries, but sizes for trepans and catheters are mentioned in later passages and the texts are filled with references to different sizes and shapes for scalpels, chisels, forceps, and probes among others. For a small sample of variety see **figs. 10.9** and **10.10.1–11**.

Galen, *On the Method of Curing* Vol. 10 p. 445

Simple fractures opened right up to the diploë require the use of narrow raspatories, as they are called now. There should be numerous types of different sizes so that never is there a lack of the most effective tool for the work. Once the injured bone has been exposed according to customary practice, first the broader raspatory is to be used, then after it a narrower one, and thus the others, one after another in sequence to the narrowest. And it is necessary to use the last, narrowest one in the diploë itself.

Hippocrates, *The Physician* 7 (IX 212)

I note two methods for using cups [**figs. 10.8–9**]. For when the discharge is of a deeper origin in the flesh, the cup should be small in the mouth and rounded in the belly, and the section for the hand neither elongated nor heavy. Such a cup functions so as to draw in a direct line, and draw up towards the surface the deeply seated discharges [*ichors*]. But when the condition is more dispersed on the body, the cups, otherwise resembling the others, ought to have a large mouth. For thus it is able to draw the findings from a more extensive area into that distressed section [to be evacuated].... Being heavy they sink into the upper [afflicted] areas ... and by this means, the broad-mouthed cups draw together from further away much more [discharge] from the surrounding flesh.

The variety of cauteries, usually iron but occasionally copper alloy, gold, silver or bronze, is indicated by the abundance of descriptive terms applied to them: fine, small, big, thick, and sword-shaped or wedge-shaped. Two other unusual shapes are provided below.

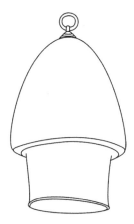

Figure 10.8 Bleeding cup, 15 cm.
Source: Naples Museum.

IG 2² 1534, fr. A, r. 61

Kallimachos Thymaitades dedicated this hollow cauterising rod.

Paul of Aegina, *Medical Compendium in Seven Books* 6.48.1

Stretching up the skin lying on the spleen with hooks, and going right through with a long cautery turned in the fire we will cauterise it, so that with one application two eschars are created. And we will do this three times so that all six eschars are made. But Marcellus by using the so-called trident or trident-shaped cautery, made the [six] eschars with one application.

Figure 10.9 Relief of a box of scalpels with flanking bleeding cups, 43×33 cm.
Source: Acropolis at Athens.

KNIVES/SCALPELS

Figure 10.10a Scalpel, 8.7 cm.
Source: Museum at Le Puy-en-Velay.

Figure 10.10b Scalpel, 14 cm.
Source: Louvre.

PROBES

Figure 10.10c Probe, 18 cm.
Source: Milne.

Figure 10.10d Spatula probe, 18 cm.
Source: Naples.

Figure 10.10e Spatula probe, 15.5 cm.
Source: Mainz.

Figure 10.10f Ligula, 18.4 cm.
Source: Milne.

Figure 10.10g Bifurcated probe, 15 cm.
Source: Herculaneum.

FORCEPS

Figure 10.10h Vulsellum, 5.8 cm.
Source: St. Germain.

Figure 10.10i Uvula forceps, 20.2 cm.
Source: Basel.

Figure 10.10j Bone forceps, 21 cm.
Source: Naples.

BONE LEVER

Figure 10.10k Bone lever, 15.5 cm.
Source: Naples.

10.117 Specific instruments and devices for procedures

Catheters: size matters

Celsus, *On Medicine* 7.26

[Obstruction or collapse of the passage for urinating requires treatment.] Accordingly, larger and smaller bronze tubes [**fig. 10.11**] are made so they are suitable for every body type. The surgeon should have three sizes for males, two for females: of the masculine ones, the largest 15 fingerbreadths long, the medium 12, the smallest nine; of the feminine ones, the bigger nine, the smaller six. It is necessary that they are a bit curved, but more so those for men, and they must be completely smooth and neither too fat nor too slender. [Procedure for insertion follows with instructions for drainage and removal of the catheter once the urine has been drained.]

Speculum

This fearsome looking device, a vaginal speculum [**fig. 10.12**], has a long history and has been used, more or less, in this form from the Roman period until relatively recently; rectal specula are also mentioned in earlier Greek texts. The basic form consists of a *priapiscus* with two to four valves that are opened or closed by a handle attached to a screw mechanism.

Aëtius of Amida, *Medical Books* 16.89.1–18

Let the examiner, sitting on the right, conduct the examination with a speculum that is age-appropriate. It is necessary, before the examination, to measure with a probe the depth of the vagina in order that the *priapiscus* [stem] is not too long so as to compress the uterus. [If too long, pads are used to support the speculum.] Then it is necessary to insert the *priapiscus* of the speculum with the screw upward, and with the speculum being controlled by the examiner, the screw is turned by the assistant so that the vagina is dilated by the spreading flattened arms [valves] of the *priapiscus*.

Figure 10.11 Bladder sound (same shape and size as catheter, but solid not hollow), 15 cm.

Source: Mainz.

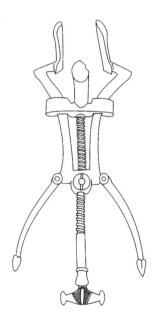

Figure 10.12 Vaginal speculum, 23 cm.

Source: Naples.

Drills for trepanation

Plutarch's text recording the caustic comment of Cato the Elder is evidence of a successful procedure of great antiquity; trepanned skulls have been recovered that date from as early as the Neolithic period (Brothwell 1981, Missos 2007, Potts 2015).

Plutarch, *Cato* 9

Once, when three Romans were selected to go to Bithynia as ambassadors, of whom one was gouty, the second had a hollow cut all round in his head from trepanning, and the third was thought to be a dolt, Cato, mocking them, said that the embassy being sent off by the Romans had neither feet, nor head, nor heart.

Celsus provides a long description of the tools and procedure for trepanation (cf. Hippocrates, *On Wounds in the Head* 1.21 and Galen, *On the Method of Curing* Vol. 10 pp. 447–448.8–18 K). Besides the particular drill described in his passage, bow drills (**10.3**) or thong drills [**fig. 10.13**] were sometimes used in the procedure as well as for wounds where a point or weapon is lodged in the bone (Paul of Aegina, *Medical Compendium in Seven Books* 6.90.5).

Celsus, *On Medicine* 8.3

Bone is excised in two methods; if what is damaged is very small, with the *modiolus*, which the Greeks call a *choinikion*; if a larger area, then with trepans. I will set forth both methods.

The *modiolus* is a hollow cylindrical iron device, which is serrated on its lower edges; a pin, surrounded by an inner disc, runs through its centre. There are two types of trepans: one is similar to the one that craftsmen use, the other has a longer central pin that begins from a sharp point, then suddenly becomes broader, and again at the other end becomes even smaller than that above.

If the disease is in a limited area, which the *modiolus* can encompass, that device is more suitable. And, if there is bone decay below, the central pin is lowered into the opening; if black bone instead, a small cavity is made with the angle of a chisel to receive the pin, so that with the pin being fixed, the *modiolus* when being spun is unable to slip. Then it is whirled around like a trepan by means of a strap. [Even pressure is needed to penetrate or to prevent jamming; light lubrication using rose oil or milk is also suggested.] At the moment when a pathway has been impressed by the *modiolus*, the central pin is drawn out, and the *modiolus* operated by itself. Then, when the health of the lower bone has been recognised by the bone dust, the *modiolus* is removed.

But if the disease is too broad to be encompassed by the *modiolus*, the operation must be conducted with a trepan. A hole is made with it at the exact boundary of the diseased and sound bone, then not so far away a second and a third until the whole area to be excised is encompassed by these holes; and at those points the bone dust also indicates how deep the trepan is to go. Next the excising chisel is forced through from one hole to the next by striking the bone with a small mallet, which cuts out each intervening bone section, and so a ring is made like the smaller circle engraved by the *modiolus*. [The excising chisel should be used to remove all diseased bone, which is tested using a probe; deep, diseased bone should use the trepan since a wider opening is required] ... and lest it become excessively hot, it should be dipped frequently in cold water.

[Care is necessary during boring not to go too deeply.] ... On that account then, the strap must be drawn more gently and the left hand must be held up and moved away more often, and the depth of the hole must be inspected so that we may ascertain whenever some bone is being broken through, and not endanger [the patient] lest the cerebral membrane be injured by the sharp point [which could cause death].

Once the holes have been made, the intervening sections are to be excised with the same technique, but much more carefully so the corner of the chisel does not harm the cerebral membrane, until a suitable access has been created to insert a shield, which the Greeks call a *meningophylax*, for the membrane. It is a thin bronze plate, its edges slightly turned up, smooth on the exterior side; this, having been so inserted that the exterior side is closer to the brain, is gradually brought under that section being broken by the chisel. And, if it takes a blow from the corner of the chisel, it prevents deeper penetration. And as a result of

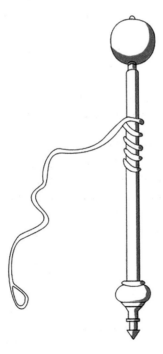

Figure 10.13 Thong drill.
Source: After Vidius.

this safeguard, the surgeon continues striking the chisel with the mallet more boldly and more safely until the bone, having been cut on all sides, is lifted by the same plate and is able to be removed without any injury to the brain. [The remaining bone edges must be filed smooth and dust removed so that the skin will grow to cover the hole properly.]

Dislocations and breaks

The Bench of Hippocrates utilised tension and leverage to aid in setting bones and dislocations, a forerunner of modern traction devices. Alas, as with other technological inventions intended to benefit humans, the traction-concept was later used in the rack for torture. Other authors (Oribasius 49.26 and Paulus Aeginieta 6.118) also describe the bench, yet problems regarding its form are still debated. (See Hippocrates, *On Fractures* 13 for a simpler solution for less serious injuries.)

Hippocrates, *On Joints* 72

It already has been mentioned earlier that it is worthwhile, for anyone practising medicine in a crowded city, to acquire a wooden, quadrilateral bench, which will suffice if it is six cubits long or a bit more, has a width of about two cubits, and a depth of a hand-span [c.0.5 cubit]. Then down its length it must have a groove cut from end to end so that the mechanical device is not higher than suitable. It must also have short, sturdy supports securely fitted-in and a winch at each end. Next, on half of the wooden bench (although nothing prevents it being through the entire bench) it will suffice that five or six long grooves are cut into it, leaving a space of four *dactyloi* [fingerbreadths] between each of the grooves, which are three *dactyloi* in breadth and the same depth.

At the centre of the wooden bench it is necessary to have a deeper mortise-cutting, about three *dactyloi* square. And into this mortise whenever there seems to be a need, a wooden fitting is to be inserted that has been made properly [in its lower section] for the mortise, but well-rounded in the upper section [above the mortise]. It is to be inserted, whenever it seems beneficial, between the perineum and the head of the thigh bone.

This wooden fitting, once set, prevents the body from yielding to the traction being exerted towards the feet. In fact, sometimes this wooden fitting itself functions as a substitute for the counter-extension upwards. And sometimes, when the leg is being stretched in both directions, this same wooden fitting, adapted to be effective to this or that side, can be suitable to lever out the head of the thigh bone towards the outside section.

For this reason the grooves are cut so that a wooden lever, having been inserted into whichever groove may be appropriate, can be levered-up either to the sides of the joint-heads or upon the heads themselves, exerting pressure simultaneously with the extension of the traction, whether it is advantageous to be levered to the outside or inside part, and whether it is advantageous for the lever to be rounded or broad. For one is suitable for some joints, the other for other joints.

This leverage along with the extension of traction is serviceable for the setting of all joints concerning the legs. Thus, as concerns the present subject, it is suitable that the lever is to be rounded, but for an outwardly dislocated joint it is suitable for it to be flat. It seems to me that no joint, no matter the sort, lacks being able to be put in place from these mechanical devices and forces.

Setting a broken nose

A common problem in antiquity that used simpler treatment and tools.

Celsus, On Medicine 8.5

[A description of variations of broken noses precedes the treatment.] Whatever happens in regards to the cartilage, it must be raised up gently either with a probe placed under it or being compressed with two fingers, one on either side; then linen, bundled lengthwise into a roll lengthwise and covered with supple fine leather sewn around, is to be forced within; or in the same manner another put together from a dry roll of scraped lint; or a large quill smeared with gum or joiner's glue, and wrapped round with supple fine leather, which will not allow the cartilage to settle down.... [If broken across both nostrils equal-sized rolls in each nostril; if bent to one side, that side requires a thicker roll.]....

And on the outside a supple strap is to be place around the head; its middle smeared with a mixture of the finest flour and the soot of incense. And it must be carried over behind the ears and stuck to the forehead by its two ends. For that mixture adheres to the skin just like glue once it has hardened, and skilfully encases the nose. But if that, which has been inserted, is painful, as happens most particularly when the interior cartilage is fractured, the nostrils having been raised are to be held in place by the strap alone.

Then after 14 days the strap itself is to be taken off. It is loosened with warm water, and then the nose is to be fomented daily the same way. [Other treatments follow.]

10.118 Wounds

The following passages deal with different aspects of wounds: removal of the cause of the wound and closure of the wound.

The Dioclean *cyathiscus* (spoon)

After discussing smaller weapons and the option of pulling out or pushing through, Celsus turns to removal of larger projectiles, describing a special tool created for this purpose.

Celsus, On Medicine 7.5

But if a widespread missile has been lodged into the body, it is not prudent to extract it through a second incision, lest we also add a large wound to the first large wound itself. For this reason, it must be extracted by the type of metal instrument that the Greeks call the Dioclean *cyathiscus* because it was invented by Diocles, whom I have already set out among the greatest physicians of antiquity.

[The instrument is composed of] two iron or even copper blades [*lammina*]; hooks are turned backward and downward from one blade-end on each of its sides; from the other blade-end, which is enlarged on its sides, the edges are smooth and turned up in that section, whereby it is curved, and on the upper part it is perforated there.

This second blade, facing away, is sent down close by the weapon, and then when it has come to the lowest point of the weapon, is turned a bit so that it catches the weapon in the instrument's own perforation. Once the point is in the perforation, the fingers are used to place the hooks of the other blade under it in that very place, and then at the same time both that metal instrument and the weapon are drawn out.

The next two passages discuss methods to close wounds using "medications".

Celsus, *On Medicine* 5.2

[The following materials] agglutinate a wound: myrrh, frankincense, gums – particularly acanthus gum; fleawort, gum-tragacanth, cardamon, bulbs, linseed, nasturtium-cress; white of egg, glue, isinglass [fish-glue]; white vine, snails pounded with their shells, heated honey, a sponge either squeezed out of cold water or out of wine or out of vinegar; virgin wool squeezed out of the same things; if the cut is slight, even cobwebs.

Celsus, *On Medicine* 5.19

And out of the plasters none perform greater use than those immediately applied to bleeding wounds, which the Greeks call *enhaema* [styptic medicines]. For these, curb inflammation unless great strength rouses it, and even so they diminish its onset. In addition, they agglutinate wounds which submit to it, and induce a scar in them. And as the plasters consist of medicaments which are not fatty, on that account they are named *alipe* [without fat].

Of these, the best is the one called *barbarum*. It has 12 grams of shaved verdigris, 80 grams of litharge of silver, 4 grams each of alum, dried pitch, and dried pine-resin, to which is added a half *sextarius* each [250 cc.] of oil and vinegar. [Many other plaster recipes follow.]

10.119 Dissection and vivisection

The skinning and butchering of animals for food and clothing probably led to examination of the bodies to improve efficiency of effort, eventually culminating in a search for signs from deities and knowledge of how the body functioned through dissection and vivisection (Nutton 2013, pp. 130–141 for Alexandria).

In his work, Aristotle indicates common knowledge of dissection being practised on both animals and humans for a variety of reasons, but all are intended to obtain knowledge.

Aristotle, *On the Parts of Animals* 3.4 (667b)

… when animals die from disease, and from such problems previously mentioned, a diseased affliction of the heart presents itself for those animals once they are cut open.

Aristotle, *On the Parts of Animals* 4.2 (677a)

[Followers of Anaxagoras wrongly attribute diseases to the gallbladder.] For almost all of the people suffering these types of diseases have no gallbladder at all, as would become apparent were they to be cut open.

Vivisection

Improvement of tools, the techniques to use them and an understanding of anatomy resulted in public demonstrations of the flaying of live animals to demonstrate a scholar's expertise and, for a brief period, the vivisection of humans for the sake of knowledge. Only a few years after Aristotle, Herophilus and Erasistratus, two notable physicians at Alexandria in the court of the Ptolemies, not only dissected the dead but also conducted vivisection on living humans to improve the understanding of the human body. The historical procedure has a mythological precedent in the punishment of Marsyas by Apollo after his defeat in their contest.

The Flaying of Marsyas

The "lesson" is recalled throughout antiquity via imagery on pots and statuary, as well as in several texts. Here Apollodorus provides the basic myth, while Ovid is more graphic and medically-minded.

[Apollodorus], *Library* 1.4.2

Apollo also slew Marsyas, the son of Olympus. For Marsyas, having found the pipes which Athena had cast off since they made her face unsightly, entered into a musical contest with Apollo. Having agreed that the victor should carry out his wish on the loser, the trial took place. Apollo contending for the victory, turned his lyre upside down and ordered Marsyas to do the same thing. When Marsyas could not do it, Apollo was declared the victor and utterly destroyed Marsyas, hanging him from a tall pine tree and stripping off his skin.

Ovid, *Metamorphoses* 6.383–391

[Various characters are telling stories] Another storyteller remembered the satyr, Marsyas defeated with his reed pipe of Tritonian Minerva, upon whom Apollo,

Latona's son, inflicted his punishment. Marsyas cried "Why do you strip me out of myself?" "Aahhh! I repent", he screamed. "Aahhh! A pipe is not worth so much!" While he is screaming, the living skin is flayed from the surface of his body, he is nothing other than a single wound. Blood flows everywhere, the exposed sinews are visible, and the trembling veins beat without any skin [to conceal them]: you can count the pulsing internal organs, and the fibres of the lungs glimmering in his chest.

Historical vivisection

In the third century BC, the Library of Alexandria was a focal point of learning for intellectuals. For millennia Egypt had practised mummification, which utilised elements of dissection, and now with the patronage of the Ptolemies, Greek medical scientists used this favourable convergence to dissect both animals and humans to learn inside out the secrets of anatomy.

Debate continues whether Herophilus and Erasistratus vivisected human subjects, but the conduct of humanity throughout history favours acceptance of the texts as do the gruesome charges against them, which led to prohibition of dissection in the West until the Renaissance. Whatever the case, the research of the two physicians reached a level of anatomical knowledge not improved upon for 1,500 years.

Celsus, *On Medicine* I Proem 23–26

[Pain and diseases internally require knowledge of the internal elements of human anatomy.] Thus, it is necessary to cut open the bodies of dead people and to examine thoroughly their internal organs and intestines. They hold that Herophilus and Erasistratus accomplished this best by far, when they laid open living men, criminals received out of prison from the kings. And while the criminals were still breathing, the physicians observed the internal organs that nature previously had concealed, their position, colour, shape, size, arrangement, hardness, softness, smoothness, contact points, protuberances and depressions of each, and whether any part is inserted into or is received into another. [Justification of the necessity for live-examination is given.] Nor is it, as most people imagine, cruel by the sufferings of criminals, and only a few of them, to seek out remedies for innocent people of all future ages.

Celsus later offers his own opinion about dissection and vivisection, ending with a comment about the traditional examination of wounds as the method to improve knowledge.

Celsus, *On Medicine* 1 Proem 74–75

But to cut open the bodies of living men while still alive is both cruel and unnecessary; cutting open the bodies of the dead is a necessity for the learner since the position and arrangement, which a corpse exhibits better than does a living and wounded man, must be learnt. But for the rest, which only can be learnt from the living, the practice itself through the very treatment of the wounded will impart the knowledge, a bit more slowly but much more mildly.

Criticism of human vivisection continued. Almost five centuries later, Herophilius' practices are regarded as vile and, perhaps, even useless.

Tertullian, *On the Soul* 10.4

Then there is Herophilus, that surgeon or 'butcher', who cut apart 600 men to search for [the secrets of] nature, who was vexed with, to the point of hating, human beings to get knowledge of them. I am in doubt whether he explored all the internal parts of man properly, since death itself changes those things that had been alive, especially when the death is not a simple and natural one, but the death itself arising from making mistakes in the midst of the very artifice of dissection.

10.120 Preservation: trophies and symbols

Many of the methods to preserve foodstuffs and other items (**4.18, 4.25, 10.49, 10.67**) were also employed to preserve human remains, but flayed human skin will seem a bit bizarre to a twenty-first century audience (Forbes 1965).

Flayed human skin

Herodotus records the existence of the skin of Marsyas' flayed body; seemingly a reminder of the dangers of arrogance (cf. **10.62**).

Herodotus, *Histories* 7.26.3

And in the agora [of Celaenae in Phrygia] the skin of Marsyas the Silenus is suspended. The tradition of the Phyrgians maintains that it was flayed and hung there by Apollo.

Before the Battle of Leuctra (371 BC), the Theban leader Pelopidas dreamt that he was commanded to sacrifice a red-haired virgin to ensure his success. His diviners and generals provide a list of human sacrifices that have led to success. One, Pherecydes, a late-sixth century BC philosopher, had his skin preserved, possibly as a type of relic of knowledge.

Plutarch, *Pelopidas* 21.2

Later Pherecydes, the wise man, was sacrificed by the Spartans and his flayed skin was guarded by their kings according to some oracle.

The Roman Emperor, Valerian, was captured in AD 260 by Shapur I. Lactantius, who records terrible deaths for Christian persecutors like Valerian, provides one account of his captivity in which

Valerian was used as a living mounting-stool when Shapur boarded his carriage or mounted his horse. Lactantius' conclusion depicts Valerian's end as a perpetual warning to the enemies of both Shapur and Christianity.

Lactantius, *On the Deaths of the Persecutors* 5

After Valerian finished such a shameful life in that vile fashion [as a stool], his skin was stripped away and deprived of its internal organs, and his hide was dyed with vermilion in order to be put in the temple of the barbarian gods. This was to be a remembrance of a most illustrious triumph that might be displayed forever to our ambassadors so that when they viewed the flayed skin of a captured emperor among the Persian gods they would not have excessive confidence in the strength of Rome.

10.121 Treatment of the dead body: the role of fragrance and spices

Herbs and spices for preservation/longetivity

Achilles, ready to return to battle, is worried about the putrefaction of Patroclus's body in his absence. Thetis uses nectar and ambrosia, the divine sustenances that provide energy to the immortals (Hesiod, *Theogony* 794–799) to preserve his body. Such myths may be part of the tradition that led to the use of aromatics in funerary rituals for preservation and to overcome the smell of corruption.

Homer, *Iliad* 19.23–39

Now indeed I will arm myself, but in the meantime wretchedly I fear lest flies might enter the wounds inflicted by bronze weapons on the corpse of the valiant son of Menoetius, and that they might breed worms and treat his corpse unbeseemingly, for his life is slain out of him and all his flesh will rot. Then the goddess, silver-footed Thetis, answered him: "Child, do not let these things worry your heart. From him I will endeavour to fend off the savage tribes, the flies that feed upon men slain in war. For even if he were to lie dead for the full course of a year, always his flesh will be unspoilt, or even better than it is...." Thus, having spoken, she filled Achilles with unshakeable courage, and for Patroclus she infused ambrosia and ruddy nectar through his nostrils, that his flesh might be uncorrupted.

Odysseus and his men are concealed in incredibly rank seal skins waiting to ambush Proteus, the Old Man of the Sea. Success occurs only by the goddess masking the odour of rotting flesh – a simple forerunner of the ostentatious expenditure of aromatics later (Pliny, *Natural History* 12.41 (18) and Plutarch, *Sulla* 38.2).

Homer, *Odyssey* 4.440–446

One after another Eidothea lay us down in ambush, and threw a seal skin upon each man. Then the ambush would have become most desperate, for the overpowering deathly stench of the sea-reared seals weakened us terribly – for who would lay down beside a sea-beast? – but she herself rescued us and contrived a great refreshment, for beneath the nose of each man she dabbed ambrosia whose very sweet fragrance destroyed the stench of the sea-beast.

10.122 Treatments of the corpse

Some foreign practices adopted/adapted by Greeks and Romans

Cicero, *Tusculan Disputations* 1.45 (108)

The Egyptians embalm their dead, and keep them safe at home; the Persians having smeared them all round with wax bury them in order to preserve their bodies as long as possible. [Other foreign practices, such as leaving the bodies for beasts or dogs, follow.]

Mummification

The Greeks and Romans were familiar with Egyptian mummification procedures and practised it upon some famous individuals. The process removed the brain and almost all the inner organs before the flesh was desiccated. Diodorus of Sicily (*History* 1.91–93) discusses the accompanying rituals (Aufderheide 2003 and Papageorgopoulou 2009).

Herodotus, *Histories* 2.86–88

[For embalming] there are men solely occupied with this occupation and who have this particular skill. Whenever a dead body is brought to them, these men show wooden models of corpses, examples rendered with paint, to the men who brought the body. [Three grades of diminishing quality are described for the selection process.]....

[In the most perfect, most expensive process] they first draw out the brain through the nostrils with an iron hook, extracting the majority of it thus and flushing out the remnants with drugs. Next, making a lengthwise cut along the flank with a keen Ethiopian stone knife, they take out the entire contents of the abdomen via the cut, cleansing and rinsing it with palm wine and in turn they percolate it with bruised incense. Then, after filling the body cavity with pure ground myrrh and casia and all the other aromatic spices, except frankincense, they sew up the cut. Having done these things, they embalm the body, completely covered with natron, for seventy days; it is not possible to embalm for more days. When the seventy days have passed, having washed the body, they wrap the entire corpse with cut swathes made of fine linen, smearing the under-sides with acacia-gum, which the Egyptians mostly use instead of glue. At that point the concerned party, having received back the body, makes a wooden, hollow mould like a man, in which they enclose the corpse, and having shut it up thus they keep it safe in a coffin-chamber, standing it upright against a wall....

[In the second, less expensive process] they prepare as follows. The embalmers fill their syringes [**fig. 10.14**] with cedar oil with which they fill the body cavity of the dead man, neither cutting open the body nor removing the organs, but injecting the oil through the anus and holding the fluid back from running out. They then embalm the body for the prescribed days, and on the last day they let out from the body cavity the cedar oil which they introduced earlier. It has such great power so as to bring out with itself at the same time the now-liquefied bowels and internal organs. But the natron also melts away the flesh, and, in truth, only the skin and bones of the body remain. Then the embalmers give back the dead body in this condition, with no further consideration.

The third process for embalming is this one, which prepares the bodies of those with less resources. The embalmers, having cleansed the body cavity with a purgative-solution, embalm the body for the seventy days and then the embalmers give it back to be taken away.

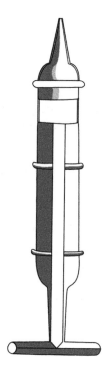

Figure 10.14 Syringe.
Source: Milne.

Embalment of a notable Greek and Roman

Several passages indicate clearly that Alexander the Great was embalmed, although it seems this was not the initial plan for his body. Nero's wife, after an unexpected death, was also embalmed with great extravagance.

Alexander's body, seven days after his death, is still reportedly without corruption.

Q. Curtius Rufus, *History of Alexander* 10.10.13

I provide what is the tradition rather than what is believed, ... that those who had entered saw his body corrupted with no decay, with not even the least bit of bluish colour. The vitality too which comes from a breathing person had not yet left his face. And thus, the Egyptians and Chaldeans were ordered to care for his body according to their custom. At first, they didn't dare to put their hands upon him, as if he was still breathing. Then having prayed that it might be morally and legally proper for mortals to handle a god, they cleaned out and purified the body, the golden sarcophagus was filled with aromatic herbs and spices....

Alexander's body waited almost two years before it began its long journey, eventually ending up at Memphis before its final transference to Alexandria years later. Beside the natural preservation and work mentioned in the previous passage, special arrangements were made for his transportation.

Diodorus of Sicily, *History* 18.26.3

First, they fitted together a coffin well suited for the body, made of hammered gold, and about the body they filled the space around the body with strong, aromatic herbs and spices to render the body sweet smelling and imperishable. Upon this casket, a gold cover had been placed, one fitting closely and clasping the upper rim.

The mummification was obviously a success and the remains were viewed for centuries by later powerful leaders.

Suetonius, *Augustus* 18

About the same time [30 BC] Octavian, once the sarcophagus and body of Alexander the Great had been brought out of its inner sanctuary, subjected them to his intense gaze, then paid homage to Alexander by crowning the body with a golden diadem and with scattered flowers. When asked whether he would like to view the Mausoleum of the Ptolemies he replied that he had wished to see a king, not dead bodies.

Dio Cassius, *Roman History* 51.165

And after these things, Augustus saw the body of Alexander, and coming close, touched it so that part of the nose, as they relate, broke into pieces.

Tacitus records that the pregnant Poppaea died from an unpremeditated kick by Nero.

Tacitus, *Annales* 16.6

Her body was not consumed with fire, as was Roman tradition but, in the custom of foreign kings, after being packed full with aromatic herbs and spices, was embalmed. It then was taken into the Tomb of the Julii [the Mausoleum of Augustus].

Other methods to preserve the body

Use of honey by itself does not seem unusual if the casual references in Plutarch, Columella (**4.25**), and Varro (*Mennipean Satires* Frag. 81/82) are indicative of general knowledge, but wax and, perhaps, even plasters were used to prevent decay.

After King Agesilaus (444–360 BC) died in Libya, the voyage to Sparta required preservation of his body. One can compare Lord Nelson, whose body was placed in a cask of brandy mixed with camphor and myrrh to transport it back to England after Trafalgar.

Plutarch, *Agesilaus* 40.3

The Spartan custom was to bury the other dead soldiers and leave their bodies in foreign lands, but to carry the bodies of their kings back home. Since there was no honey at hand, the Spartans who were there with Agesilaus poured melted wax over his dead body and carried it back to Lacedaemon.

10.123 Repositories of the dead
Tombs: simple, magnificent, and mystifying

Pausanias provides a simple evolution of impressive tomb. The first is in Arcadia in Greece, the second in modern Turkey and the last in Jerusalem, and probably refers to Helena of Adiabene and the Tomb of the Kings.

Pausanias, *Description of Greece* 8.16.3–5

I was particularly in haste to see the tomb of Aepytus because Homer, in his lines on the Arcandians, makes note of it: "the monument of Aepytus". It is a mound of earth, not very large, surrounded by a circular base of stone. Alas, since Homer had not seen a more noteworthy monument, quite reasonably it was a wonder to him.... But I know many praiseworthy tombs of which I will mention two: the one at Halicarnassus and the one in the land of the Hebrews.

The one at Halicarnassus made for Mausolus, King of the Halicarnassians, is such an immense size and so admired for all its construction and decoration that the Romans greatly admiring it call the remarkable tombs on display in their own lands mausolea.

For the Hebrews, in the city of Solomon [Jerusalem] there is the tomb of Helen, a native woman, which the Roman Emperor razed to the ground. It had been cunningly contrived in the tomb so that the door, which as for every tomb is stone, does not open before the year brings on the same day and the same hour. At that time the door is opened by mechanism alone, and after not much time, also closes by itself. This happens only at this time, but at another time trying to open it you could not open it, but by using force before the proper time you would break it.

The sarcophagus

Corpses could be deposited in simply-dug graves, but the wealthy had other options, including elaborate tombs in which bodies in finely decorated sarcophagi were deposited. Here, Pliny provides the etiological origin for the term sarcophagus: flesh-eater.

Pliny, *Natural History* 36.27 (17)

At Assos in the Troad they quarry a stone called "sarcophagus" that has laminated veining. The flesh of the deceased people entombed in that stone are consumed within forty days, leaving the teeth [and bone]. The writer Mucianus also records that mirrors, strigils, clothing, and shoes buried with the dead become stony. In Lycia and in the East, there are stones of the same type, which also corrode the flesh when fastened to living bodies.

10.124 Inhumation vs. cremation: from pragmatic to extravagant

According to Pliny cremation began to replace inhumation among Romans during the late Republic for practical reasons.

Pliny, *Natural History* 7.55 (54) (187)

To consume the body with fire is not an old tradition among the Romans, since they were accustomed to be interred in the earth. But after they learnt that bodies, which had been buried in the distant foreign wars, were being rooted out, the custom of cremation was instituted. Many families, however, still practised the ancient rites. In the Cornelian family, for example, nobody was given to be cremated before Sulla the Dictator, who wished it since he feared a punishment similar to that of Caius Marius' cadaver, which had been dug up and cast out [by Sulla].

To become a Roman god: pyre cremation and enactment

Many Roman emperors transitioned from mortal to immortal; the road to the ultimate destination of godhood is documented for the magnificent funeral of Septimius Severus (died AD211).

Herodian, *History of the Empire after Marcus* 4.2

It is custom for the Romans to make gods of the emperors who have died with sons or other successors, calling this great honour "apotheosis" [deification]. They begin the mourning with a combination of festive holiday and religious ceremony that is observed throughout the whole city.

First, they bury the body of the dead emperor amidst an incredibly costly funeral in the customary practice of humans. Then, having fashioned an image of wax, which is identical in all respects to the dead man reposing on a large, ivory couch lifted up high, they set it out in the entrance of the palace, draping gold-embroidered coverings from it. This image lies pallid in the manner of a sick man.

[For seven days subdued upper class women and men sit on either side of the wax image, with daily visits from physicians who announce that the emperor's condition is worsening. When announced that 'he' is dead, the image is taken to the old Forum for mourning.]

After this, having lifted it up, they carry the couch out of the city into the so-called Field of Mars, where, in the widest section of the field, a rectangular structure with equal sides had been constructed in the form of a house, utilising no other material, but is solely a construction of very large wooden beams.

Its entire interior is filled with firewood and its outside is decorated with gold-embroidered hangings, ivory images, and coloured paintings. Upon it rests a second structure, comparable in shape and decoration, but smaller, and it has open windows and little gateways. And there is a third and a fourth level, always smaller than the one resting below. At the climax, the last, smallest structure completes the edifice.

One might compare its shape with the construction of lighthouses, which lying at the harbours along the coast guide to safety the ships being led at night by their fire. Most people call them *pharoi*...

Bringing up the couch into the structure they set it down in the second level, then they carry up and pour out in heaps every kind of aromatic and incense that the earth produces, and all those fruits, herbs, and juices dashed together for their sweet fragrance. [Once the structure is filled with herbs and spices various rituals are conducted outside around the structure.]....

When these rites have been completed, the successor to the Empire, taking up a torch, puts it to the structure, and the rest of the people set fire to it on all sides. Everything is easily and quickly kindled by the fire; all the masses of firewood and the heaps of fragrant stuffs.

Out from the uppermost and smallest structure, as if from a battlement, an eagle flies forth, soaring with the flames into the sky. The Romans believe it bears the soul of the emperor from the earth into heaven, the realm of the gods. And from that point he is worshipped with the rest of the gods.

G LARGE-SCALE, ORGANISED PRODUCTION

In antiquity the development of specialised labour and the need for vast quantities of raw materials and products meant that it was sometimes very profitable for large numbers of people to be organised into groups for manufacturing. The beginnings of large-scale production are evident in the division of labour (**1.26–1.27**) and the development of workshops where apprentices would learn their trade (**10.132**). The demand for larger volumes and products of superior quality led to specialisation of labour and eventually to piecework, where individuals could more quickly produce separate components that would later be assembled by other workers (**10.127, 10.129**). This is a viable alternative only in urban or palatial settings (**10.125–10.127**).

Unfortunately, our literary sources are less than plentiful regarding this aspect of technology. The paucity of literary evidence and the absence of the mention of industrial production in some lists of occupations and classes of workers (Aristotle, *Politics* 4.3.11–15 [1291a-b]) indicate that such arrangements were not common. Occasionally, however, the sources of people's wealth are recorded, proving the existence of such large-scale operations. In addition, we have a few scattered passages that actually describe some of these establishments in general terms (**10.128, 10.133, 10.136**). Otherwise we have only a few brief notices and short inscriptions that imply a factory-like manufacturing system or the existence of large workforces assembled to create products such as bricks, pots, and buildings (**10.140**).

Archaeological evidence such as the sepulchral monument of Eurysaces in Rome with its sculptural reliefs detailing the workings of a large bakery operation in the early Empire and the remains of such operations at Pompeii provide vivid proof of their existence. This evidence, however, is not sufficient to reveal how widespread the operations were and must be used with care especially when considering undertakings such as tanning and fulling, for which we have virtually no literary evidence.

Some occupations that did use substantial numbers of workers in large-scale production techniques are examined elsewhere: construction (**8.49**), quarrying and mining (**5.18, 5.23**), food production (**2.19**), and textiles (**10.3, 10.38**) (Wilson and Flohr 2016, Wilson 2008).

10.125 Large-scale production in the Mycenaean palaces

Although the documents from the Mycenaean world are not completely understood, the appearance of groups of people listed as workers or producers of specific products makes it probable that they had been assigned to produce the goods either for palace members or for trade. The appearance of large centralised palaces (cf. **8.5**, **8.7**) meant that an increased demand for goods, such as food and clothing, would be placed upon the production system; specialisation of labour was one way to meet the demand of this urban population centre. The menial tasks and the phrase "under instruction" in the following passages probably imply lists of slaves. In the following, italicised letters indicate unknown numbers or words.

DMG nos. 1, 4, 21, 26, 28 (pp. 158–159, 164, 166–167)

7 grain-grinding women, 10 girls, 6 boys.

21 spinning-women, 25 girls, four boys; 1 to-.

Ri-jo-nian women: x *da-*, x *to-* ...; 3 young women under instruction, x older girls, ... 3 boys under instruction, x older boys ...

Measurers of grain: 24 Ka-pa-ra-de women, 10+ boys, 8 Ko-ro-ki-ja women, x boys, 21 Cnidian women, x boys.

At Metapa: ... women barley-reapers. 6 women reapers, their father a slave and their mother among the Kytherans; 13 women reapers ...; 3 women reapers, their father a slave and their mother a slave of Diwia; 1 woman reaper, her mother a slave and her father a smith; 3 women reapers, their mother a slave and their father a smith.

Other occupations mentioned in the tablets include carders, flax-workers, bath-attendants, headband-makers, musicians, and sweepers.

10.126 A type of assembly-line production?

The Linear B tablets provide lists of some of the components for chariots, indicating that some of the elements were provided in large numbers by specialised groups. The extent of this type of piecework production is unknown, but the wheels and axles at least seem to have required special attention.

DMG nos. 252, 265, 273, 278, 282 (pp. 349–350, 366–368, 371–372)

Thus the woodcutters contribute to the chariot workshop: 50 saplings, 50 axles. And the Fields of Lousos (contribute) so many: 100 axles, and so many saplings: 100.

3 horse-(chariots without wheels) inlaid with ivory, (fully) assembled, equipped with bridles with cheek-straps (decorated with) ivory (and) horn bits. The workshop (of) Kolkhidas.

80 horse-(chariots) not inlaid, not (fully) assembled. The workshop of Alexinthos.

... 1 (?) pair of wheels of elm-wood, with tyres; 1 pair of bronze wheels; 3+ pairs of bronze-bound wheels; 41 and a half (?) pairs of wheels of *ki-da-pa* wood, with tyres; 40+ pairs of wheels of willow-wood, with studs.

22 and a half pairs of new wheels of willow-wood, with studs, of better quality. The workshop (of) Kolkhidas.

DIVISION OF LABOUR AND SPECIALISATION

The gathering of people to form urban units permitted, and indeed encouraged, specialisation of labour (**1.26**). There are many ancient allusions to this division of labour, but few really set out the idea of production line assembly and its advantages. Instead, most tend to recognise its benefits in general terms. A few passages do reveal the next logical step: the specialised manufacture of components of the whole (Acton 2014, Hopper 1979).

10.127 Quality through assembly-line production

Xenophon has just described the marvellous varieties of foods and their excellent preparation in the kitchen of the Persian king. He now explains how this is possible by comparing the preparation process to other trades that use specialised labour. Note that the emphasis is upon quality, not greater volume, and upon the urban environment.

Xenophon, *Cyropedia* 8.2.5

Nevertheless, that this is so should not be regarded as a marvel. For just as the other crafts are especially perfected in the large cities, in the same manner the food at the king's residence is brought to perfection by much diversification. For in the small towns, the same worker makes chairs, doors, ploughs, tables, and often this same man builds houses, and he is happy indeed when he gets enough work to support himself. Obviously it is impossible for a man crafting many things to do them all well. In the large cities, however, because the many have need of each craft, even one trade alone is enough to support a man, and oftentimes, not even an entire trade: for example, when one man makes shoes for men, another shoes for women. There are even places where a man makes a living solely by stitching shoes, another by cutting out the leather, another by shaping the uppers, while another does none of these things but only puts these parts together. So by necessity, the man who applies himself to a very narrow field of work has to do it in the best possible way.

10.128 A large (?) shoemaker's shop

Aeschines is attacking the moral character of Timarchus, one of his political opponents. According to the speech, Timarchus was a spendthrift who consumed fortunes, including the inheritance left by his father, which included land, debts owed to him, and several slaves working in shops for him. The number of shoemakers in this shop is not great, but may in fact represent a sizeable establishment for ancient Athens.

Aeschines, *Against Timarchus* 1.97

... And besides land and houses there were nine or ten house slaves, craftsmen of leatherwork [shoemakers], each of whom paid a fee to Timarchus of 2 *oboloi* per day, and the foreman of the shop who paid 3 *oboloi*. In addition to these men there was a woman [slave?] skilled in making fine flax fabrics and producing them for the market, and a man skilled in embroidery.

10.129 Piecework speeds up production

A large number of inscriptions about the silver industry record various specialised occupations: engravers, gilders, polishers, embossers, and casters. But not until Saint Augustine do we have a passage that explicitly states their relationships. Even then it is difficult to postulate anything more than a small shop with a few workers. In this passage Augustine compares the lesser pagan gods and the obscurity of craftsmen who complete parts of a work to well-known gods and master craftsmen: the message is that obscurity is better than reputation since, once known, one can become the object of ridicule (cf. Harris 1980, Hodge 1971).

Saint Augustine, *The City of God* 7.4

[The anonymous multitude of lesser gods are comparable to] workmen in the silversmiths' quarter, where one vessel passes through the hands of many craftsmen to bring it to perfection, although it could have been perfected by one master craftsman. But it was believed that the only way to take advantage of a large number of workmen was to have individuals learn different aspects of the craft quickly and easily, so that they all would not have to become accomplished in the entire craft slowly and with difficulty.

10.130 Specialised production of chandelier parts

This brief passage clearly demonstrates specialisation of the Roman craftsman in some manufacturing processes. Pliny indicates that entire shops produced individual parts, something that was unusual in antiquity.

Pliny, *Natural History* 34.11

The island of Aegina especially endeavoured to produce only the upper part of chandeliers, just as Taranto produced only the stems. And thus, the praise of the workshops for these items is shared. Nor is there any shame attached to buying them for a price equivalent to the pay of a military tribune.

Pliny concludes the section with an anecdote of the high price paid by a woman for one of these chandeliers and a humpback fuller, for whom she eventually conceived a lusty passion.

10.131 Shops that dominate the market

The ascendancy in the market of specific styles of silver named after a producer may indicate factory-style production. It is difficult, although not impossible, for a small shop to dominate the market.

Pliny, *Natural History* 33.139

The fickleness of human nature changes in miraculous ways the fashions for dishes made from silver. No type of workshop remains long in favour; now Furnian, then Clodian, now Gratian [are in demand in the market].

10.132 An early start at apprenticeship

A mythical story regarding Daedalus was related above (**8.39**). Here Plato provides a theoretical base to the childhood training of a craftsman. Of course, even if true, such stories do not necessarily imply large-scale production, but they do provide one of the possible avenues to specialisation and the means by which workers could gather and form larger establishments if they perceived a beneficial situation. In this passage an Athenian gives his definition of education to the Cretan, Clinias. See also the description of Lucian's apprenticeship as a stonecutter (**14.13**).

Plato, *Laws* 1.643b–c

I say that any man who will be good at anything must study that special activity right from childhood, both while playing and in earnest, using what is appropriate in each of the activities. In regard to the man who will be a good farmer or builder, the latter must play at constructing toy buildings, the former, in turn, at cultivating; those rearing these children must make available to each child little implements modelled upon real ones. In addition, they should have early education in those areas that ought to be taught, such as a carpenter being taught in play to measure and calculate or the soldier taught to ride or something else similar. By means of these childhood games we ought to try to turn the pleasures and desires of the children in the direction of the final occupation where they will end up.

10.133 Large-scale production in the home

Much craft activity took place in the home on a small scale, but this could expand to large-scale operations if desired. Although Xenophon may be describing a fictitious situation in the difficult times after the Peloponnesian War, no reason exists to doubt that such arrangements developed in some homes especially in food and textile production, the two occupations most commonly associated with the home. Xenophon, in fact, does mention the manufacturing successes of several men, perhaps real examples of home production. The extract begins just after Socrates has suggested to Aristarchus that he should set his 14 female relatives in the house to work in order to relieve the financial problems caused by the war and confiscations. Aristarchus is surprised at the suggestion: since his relatives are freeborn and have only a liberal education, they are not slaves trained as artisans; Socrates gradually shows him the error of his thinking.

Xenophon, *Memorabilia* 2.7.5–6,12,14

"What are artisans?" Socrates said, "Those who know how to make something useful?"

"Are barley-groats useful?"

"Very much".

"What about bread?"

"No less so".

"What of men's and women's cloaks, little tunics, capes, and sleeveless tunics?" Socrates said.

"Indeed yes, all these are useful". Aristarchus replied.

"Then", said Socrates, "don't the women in your household know how to make these things?"

"I think they can make them all".

"Really? Don't you know that by producing one of these things, barley-groats, Nausicydes sustains not only himself and his family, but many pigs and cattle in addition, and has so much extra that he often assumes public duties for the city; that by producing bread Cyrebus feeds his entire family and lives lavishly; Demeas of Collytus by making capes; Menon by making cloaks; and most of the Megarians keep well nourished by making sleeveless tunics?"

"Yes, by Zeus", Aristarchus replied, "for these men have purchased barbarians that they can force to make what is useful".

"But I have freeborn people and relatives".

[A monologue by Socrates stressing useful knowledge and productive work follows. Aristarchus is convinced to put the women to work.] ...

As a result of this, capital was furnished and wool purchased. The women ate lunch while working, and once finished they ate their dinner.... [A happy workplace is the outcome except for Aristarchus who seems idle. Socrates compares him to a dog protecting sheep.]

... [Socrates is speaking] "And therefore you [Aristarchus] tell these women that like the dog you are their guardian and manager, and on account of you they live, working safely and pleasantly, being harmed by no one".

10.134 Papyrus: an example of large-scale manufacture

Pliny is discussing the various grades of paper and how they came to be named after Romans. He has already listed the three superior grades (Augustus, Livia, and hieratic for religious treatises), and now turns to the paper of fourth quality.

Pliny, *Natural History* 13.75

The next grade had been given the name of "amphitheatre paper" from its place of production [in the amphitheatre at Alexandria]. The shrewd workshop of Fannius at Rome took on this grade, and out of common-quality paper produced paper of the first quality, making it finer by careful processing; and at the same time gave it its name [Fannian paper]. The paper that was not treated thus remained in its own grade as amphitheatre paper.

10.135 Woollens: another example

When a specific region corners the market by supplying a single product to the consumer, large-scale production may be indicated, even though the methods of manufacture are not explained. Here Strabo describes the wealth of Liguria (Moeller 1976).

Strabo, *Geography* 5.1.12

As for the wool, the regions around Mutina and the River Scultenna produce the soft grade, the best by far of all the wools; Liguria and the area of the Symbri produce the coarse grade from which most of the households of the Italians are clothed; the regions around Patavium produce the middle grade from which expensive carpets and shaggy coverings and everything of this sort that is woolly on either both sides or one is made.

10.136 Shields and furniture

The production of arms seems to have been one of the areas in which large-scale production paid off. This would be true especially in times of war when the need for weapons was great and rapid

Figure 10.15 Vase painting showing slave hauling furniture.

Source: Red-figure *pelike*, fifth century BC, Gela, Sicily; Ashmolean Museum, Oxford, UK, AN1890.

production was an advantage, although we have some references to production in less hectic times. Here Demosthenes also lists another industry in which his family was involved: furniture making. This trade involves a variety of materials and ways to work them (cf. Juvenal, *Satires* 11.120ff.) that might be done more economically and efficiently by a team of workers assembled in one place. Demosthenes is listing the estate left as his inheritance, which has been swindled from him by his guardians [**fig. 10.15**].

Demosthenes 27, *Against Aphobus* 1.9–10

My father, men of the jury, left two workshops for crafts (*ergasteria*), neither one small. One produced swords and employed 32 or 33 slaves that were worth up to 5 or 6 *minai* each and none worth less than 3 *minai*, from which he received an untaxed income of 30 *minai* per year. The other produced furniture and employed 20 slaves, given to him as a security for 40 *minai*, and which gave him an untaxed income of 12 *minai* per year. Apart from these things he left ivory and iron, which were used in manufacturing; and wood for the couches, worth about 80 *minai*; and oak-gall for dye and copper bought for 70 *minai*.

Lysias' father Cephalus had been a wealthy shield-maker, first at Syracuse and then in the Peiraeus. When he died the business passed to Lysias and his brother, and seems to have thrived during the turbulent times of the late fifth century BC. Eventually "The Thirty" seized the lucrative business, putting Lysias to flight and executing his brother. Lysias has just finished describing the interment and death of his brother as he pleads his case. The shields and slaves that he mentions belong to the arms-factory.

Lysias, *Against Eratosthenes* 12.19

They took 700 shields of ours, and they took all that silver and gold, the copper and the decorations and the implements, and the women's clothing, which they did not expect to get; and of the 120 slaves, they took the best of them [the skilled ones?] and turned the rest over to the state....

10.137 Military production

Assemblages of workers with assorted vocations are recorded for monumental building projects in numerous inscriptions and some literary passages (**8.49–59**). These groups, however, rarely represent a coherent body of workers, but are haphazard or temporary arrangements. A major exception occurs in standing armies, especially those on the move during the Roman Empire.

Vegetius, *Military Science* 2.11

Moreover, the legion had carpenters, masons, cartwrights, smiths, painters, and the rest of the artisans for constructing winter quarters, for preparing machines, wooden towers, and the other weapons by which the enemies' cities are attacked and our own defended. And they also produce new or repair battered weapons, vehicles, and all the other types of machines of war. They have shield-factories, cuirass-factories, and bow-factories, in which bows, arrows, helmets, and all other types of weapons are manufactured. For this was a special concern, that anything which seemed essential for the army should not be lacking in the camp, to the extent that they even had sappers who, in the custom of the Bessi [of Thrace], once they had dug a tunnel under the earth and penetrated beneath the foundations of the wall, unexpectedly emerged to capture the enemies' cities. The particular officer in charge of all these men was the prefect of the workmen.

10.138 *Minium* production

The extraction of *minium*, a red mineral pigment, is described earlier in this passage (**6.7**). Here Vitruvius notes its refinement in the workshops (*officinae*) that have been relocated from Ephesus to Rome. Pliny (*Natural History* 33.118) adds a few additional details, including the fact that this expensive product, virtually a government monopoly, had its price fixed by law.

Vitruvius, *On Architecture* 7.9.4

Moreover, the workshops that were in the mines at Ephesus have been moved to Rome, because that type of vein was discovered later in regions of Spain. The ores from these mines are transported to Rome and treated there by public contractors (*publicani*). The workshops are between the temples of Flora and Quirinus.

10.139 Pigments

Although Vitruvius (*On Architecture* 7.9–14) explicitly demonstrates shop production only for *minium* and blue (*caeruleum*) pigments, presumably on a large scale, other pigments were probably manufactured in similar shops owned by people such as Vestorius, a banker in Puteoli and friend of Cicero.

Vitruvius, *On Architecture* 7.11.1

The proportions of blue were first discovered at Alexandria and afterwards Vestorius also established a place for its manufacture at Puteoli. [The process of obtaining blue follows.]

10.140 Bricks and tiles

Information contained in thousands of brick and tile stamps provides some idea of the large-scale nature of this industry in the Roman Empire. The stamps vary, but can include a considerable amount of information: the names of the owner where the clay is obtained or where the kilns are located, the manager of the kiln, the maker of the brick, the date, other names and phrases. From the stamps it has been estimated that many yards employed over 20 workers and some had several hundred. By the third century AC brick production had almost become an imperial monopoly as various brickyards were confiscated or left to the imperial family. Note the involvement of women in the industry (Tapio 1975) [**fig. 10.16**].

Figure 10.16 Firing of bricks in a kiln.
Source: Adapted from Adam, 1994, fig. 140.

CIL 15.415–419 = *ILS* 8661 a–e (time of Commodus)

a   From the claylands [territorial districts] of the Propetiani, Metilius Proculus was the wholesaler [?], Zos. An. was the manager.
b   From the estate of Hortensius Paulinus, the wholesaler was Metilus Proculus.
c   From the estate of Hortensius Paulinus, from the claylands of the Propetiani, the wholesaler was Aurelia Antonia.
d   From the estate of Hortensius Paulinus, an illustrious man (*clarissimus vir*), Egnatius Clemens was manager, Valerius Catullus was wholesaler.
e   From the estates of the heirs of illustrious men of an illustrious woman, Passenia Petronia, Valerius Catullus was wholesaler.

CIL 15.184 = *ILS* 8662 (AD 203)

Terracotta work from the estate of Gaius Fulvius Plautianus propraetor, an illustrious man, consul for the second time, the stamp of Fulvius Primitivus.

CIL 15.761 = *ILS* 8663

Terracotta work from the estate of our mistress, leased by Publicia Quintina.

CIL 15.650a, 651a = ILS 8664a–b

a. One-and-a-half-foot long roof tiles from the claylands of Julia Procula.
b. Two-foot long roof tiles from the claylands of Julia Procula.

CIL 15.213

Terracotta work from the estates of our lord emperor [Septimius Severus] from the claylands of the Favorianae.

Similar stamps and inscriptions on clay lamps, lead pipes, and marble blocks and columns have been used to suggest large-scale production in the Roman Empire for these industries. This is certainly true for some operations, but small workshops also stamped their products and were not part of a large-scale industry.

10.141 Security in a frankincense factory at Alexandria

Factory production usually involves workers who do not share fully in the profits. Such a situation could create an atmosphere in which pilfering became a great temptation, especially when the product was small and valuable. As a result, much as today, security measures were put into place. Pliny has just finished describing the honest practices of the owners at the collecting stage.

Pliny, *Natural History* 12.59

But, by Hercules, at Alexandria where the frankincense is processed, no amount of care is sufficient to guard the workshops. The workers' loin-cloths are sealed tight, masks or tightly meshed nets put on their heads, and they leave nude; so much less trust is held among those processing the fruit than among those collecting it in the forest.

Notes

1 For the rest of this passage, see **2.48**.
2 It has been suggested that this phrase is an allusion to either a plane or a saw, but neither emendation is a strong one. See Andrew S.F. Gow, and Denys L. Page, *The Greek Anthology. Hellenistic Epigrams*, Vol. 2. Cambridge, 1965, 316–317.
3 The translation of words concerning colours and the names of some plants and animals is very difficult. Items that were similar in appearance were often not differentiated in antiquity or were labelled erratically (**10.28**).
4 A type of sodium carbonate.

Bibliography

Metalworking

Alexander, Shirely M., "Ancient and Medieval Gilding on Metal – The Technical Literature". Pp. 65–71 in Denise Schmandt-Besserat, ed., *Early Technologies*. Malibu: Undena, 1979.

Bol, Peter C., *Antike Bronzetechnik. Kunst und Handwerk antiker Erzbildner.* Munich: Beck, 1985.

Born, Hermann, and Aliki Moustaka, "Eine geometrische Bronzestatuette im originalen Gussmantel aus Olympia". *Athenische Mitteilungen* 97 (1982) 1723.

Branigan, Keith, *Copper and Bronze Working in Early Bronze Age Crete.* Lund: Åström, 1968.

Brown, David, "Bronze and Pewter". Pp. 25–41 in Donald E. Strong, and David Brown, eds., *Roman Crafts.* New York: New York University Press, 1976.

Carroll, Diane L., "Wire-Drawing in Antiquity". *American Journal of Archaeology* 76 (1972) 321–323.

Carroll, Diane L., "A Classification for Granulation in Ancient Metalwork". *American Journal of Archaeology* 78 (1974) 33–39.

Carroll, Diane L., "Antique Metal-Joining Formulas in the *Mappae Clavicula*". *Proceedings of the American Philosophical Society* 125 (1981) 91–103.

Cave, John, "A Note on Roman Metal Turning". *History of Technology* 2 (1977) 78–94.

Charles, James A., "The First Sheffield Plate". *Antiquity* 42 (1968) 278–285.

Formigh, Edilberto, "Modi di fabbricazione di filo metallico nell'oreficeria etrusca". *Studi Etruschi* 47 (1979) 281–292.

Gehrig, Ulrich, "Frühe griechische Bronzegusstechniken". *Archäologischer Anzeiger* (1979) 547–558.

Hauser, Kurt, and Alfred Mutz, "Wie spannten die römischen *vascularii* (Dreher) ihre Werkstücke". *Technikgeschichte* 40 (1973) 251–269.

Haynes, Denys, *The Technique of Greek Bronze Statuary.* Mainz: von Zabern, 1992.

Higgins, Reynold A., *Greek and Roman Jewellery.* 2nd edn. London: Methuen, 1980.

Hunt, Leslie B., "The Oldest Metallurgical Handbook: Recipes of a 4th-century Goldsmith". *Gold Bulletin* 9.1 (1976) 24–31.

Jex-Blake, Katherine, and Eugénie Sellers, *Pliny the Elder's Chapters on the History of Art.* Rev. edn. of 1896 edition. Raymond Schoder, ed., Chicago: Argonaut, 1968.

Klumbach, Hans, ed., *Spätrömische Gardehelme.* Munich: Beck, 1973.

La Niece, Susan, "Niello, an Historical and Technical Survey". *Antiquaries Journal* 63(1983) 279–297.

Lang, Janet, and Michael J. Hughes, "Soldering Roman Silver Plate". *Oxford Journal of Archaeology* 3 (1984) 77–107.

Mattusch, Carol C., *Classical Bronzes: The Art and Craft of Greek and Roman Statuary.* Ithaca: Cornell University Press, 1996.

Mattusch, Carol C., "Metalworking and Tools". Pp. 418–438 in John P. Oleson, ed., *The Oxford Handbook of Engineering and Technology in the Classical World.* Oxford: Oxford University Press, 2008.

Mutz, Alfred, *Die Kunst des Metalldrehens bei den Römern. Interpretationen antiker Arbeitsverfahren auf Grund von Werkspuren.* Stuttgart: Birkhäuser, 1972.

Pollitt, Jerome J., *The Art of Ancient Greece: Sources and Documents.* Cambridge: Cambridge University Press, 1990.

Riz, Anna E., *Bronzegefässe in der römisch-pompejanischen Wandmalerei.* Mainz: von Zabern, 1990.

Schwandner, Ernst-Ludwig, Gerhard Zimmer, and Ulrich Zwicker, "Zum Problem der Öfen griechischer Bronzegiesser". *Archäologischer Anzeiger* (1983) 57–80.

Strong, Donald E., *Greek and Roman Gold and Silver Plate.* Ithaca, NY: Cornell University Press, 1966.

Vickers, Michael, *Artful crafts. Ancient Greek silverware and pottery*. Oxford: Oxford University Press, 1994.
Vittori, Ottavio, "Pliny the Elder on Gilding". *Endeavour* 3 (1979) 128–131.
Weisgerber, Gerd, and Christoph Roden, "Römische Schmiedeszenen und ihre Gebläse". *Der Anschnitt* 37 (1985) 2–21.
Weisgerber, Gerd, and Christoph Roden, "Griechische Metallhandwerker". *Der Anschnitt* 38 (1986) 2–26.
Wells, H. Bartlett, "The Position of the Large Bronze Saws of Minoan Crete in the History of Tool Making". *Expedition* 16.4 (1974) 2–8.
Wolters, Jochem, *Die Granulation: Geschichte und Technik einer alten Goldschmiedekunst*. Munich: Callwey, 1983.
Zimmer, Gerhard, *Griechische Bronzegußwerkstätten. Zur Technologieentwicklung eines antiken Kunsthandwerkes*. Mainz: von Zabern, 1990.

Woodworking

Béal, Jean-Claude, *L'arbre et la forêt, le bois dans l'antiquité. Paris:* de Boccard, 1995.
Chapman, Hugh, "A Roman Mitre and Try Square from Canterbury". *Antiquaries Journal* 59 (1979) 403–407.
Gaitzsch, Wolfgang, and Hartmut Matthäus, "Schreinerwerkzeuge aus dem Kastell Altstadt bei Miltenberg". *Antike Welt* 12.3 (1981) 21–30.
Gaitzsch, Wolfgang, and Hartmut Matthaus, *"Runcinae* – Römische Hobel". *Bonner Jahrbücher* 181 (1981) 205–247.
Goodman, William L., *The History of Woodworking Tools*. London: Bell, 1964.
Liversidge, Joan, "Woodwork". Pp. 155–165 in Donald E. Strong, and David Brown, eds., Roman Crafts. New York: New York University Press, 1976.
Makkonen, Olli, *Ancient Forestry: An Historical Study*. Part I. *Facts and Information on Trees*. Acta Forestalia Fennica, 82. Helsinki: Suomen Metsatieteellinen Seura, 1968.
Makkonen, Olli, *Ancient Forestry, An Historical Study*. Part 2: *The Procurement and Trade of Forest Products*. Acta Forestalia Fennica, 95. Helsinki: Suomen Metsatieteellinen *Seura*, 1969.
Martin, Roland, "Note sur la charpenterie grecque d'après *IG*, H.2, 1668, 11. 45–59". *Revue des Études Grecques* 80 (1967) 314–324.
Meiggs, Russell, *Trees and Timber in the Ancient Mediterranean World*. Oxford: Clarendon *Press*, 1982.
Mols, Stephan T., *Wooden Furniture in Herculaneum*. Amsterdam: Gieben, 1999.
Mulliez, Dominique, "Notes sur le transport du bois". *Bulletin de Correspondance Hellénique* 106 (1982) 107–118.
Richter, Gisela M.A., *The Furniture of the Greeks, Etruscans, and Romans*. London: Phaidon Press, 1966.
Rieth, Adolf, "Werkzeuge der Holzbearbeitung: Sägen aus vier Jahrtausenden". *Saalburg Jahrbuch* 17 (1958) 47–60.
Rival, Michel, *La charpenterie navale romaine: matériaux, méthodes, moyens*. Paris: CNRS, 1991.
Ulrich, Roger B., *Roman Woodworking*. New Haven: Yale University Press, 2007.
Ulrich, Roger B., "Woodworking". Pp. 439–464 in John P. Oleson, ed., *The Oxford Handbook of Engineering and Technology in the Classical World*. Oxford: Oxford University Press, 2008.

Weeks, Jane. "Roman Carpentry Joints: Adoption and Adaptation". Pp. 157–168 in Sean McGrail, ed., *Woodworking Techniques before AD 1500*. Oxford: British Archaeological Reports, 1982.

Textiles

Barber, Elizabeth J.W., "New Kingdom Egyptian Textiles. Embroidery vs. Weaving". *American Journal of Archaeology* 86 (1982) 442–445.

Barber, Elizabeth J.W., *Prehistoric Textiles. The Development of Cloth in the Neolithic and Bronze Ages, with Special Reference to the Aegean*. Princeton: Princeton University Press, 1991.

Barber, Elizabeth J.W., *Women's Work: The First 20,000 Years. Women, Cloth, and Society in Early Times*. New York: Norton, 1994.

Bieber, Margaret, "Charakter and Unterschiede der griechischen und römischen Kleidung". *Archäologischer Anzeiger* 88 (1973) 425–447.

Brøns, Cecilie and Marie-Louise Nosch, *Textiles and Cult in the Ancient Mediterranean. Ancient textiles series 31*. Oxford: Oxbow Books, 2017.

Burnham, Dorothy K., "Coptic Knitting: An Ancient Technique". *Textile History* 3 (1972) 116–124.

Carroll, Diane Lee, "Warping the Greek Loom: A Second Method". *American Journal of Archaeology* 87 (1983) 96–98.

Carroll, Diane Lee, Dating the Foot-Powered Loom: The Coptic Evidence". *American Journal of Archaeology* 89 (1985) 168–173.

Crowfoot, Grace M., "Textiles, Basketry, and Mats". Pp. 258–281 in Charles Singer, Eric J. Holmyard, and Alfred. R. Hall, eds., *History of Technology*, Vol. 1: *From Early Times to Fall of Ancient Empires c.500 BC*. Oxford: Clarendon Press, 1954.

Crowfoot, Grace M., "Of the Warp-weighted Loom". *Annual of the British School at Athens* 37 (1936–1937) 36–47.

Day, Florence E., "Aristotle: *Ta bombukia*". Pp. 207–218 in *Studi Orientalistici in onore di Giorgio Levi della Vida*, Vol. 1. Rome: Istituto per l'Oriente, 1956.

Droß-Krüpe, Kerstin, ed., *Textile Trade and Distribution in Antiquity. Textilhandel und -distribution in der Antike. Philippika 73*. Wiesbaden: Harassowitz Verlag, 2014.

Dunand, Françoise, "L'artisanat du textile dans l'Égypte Lagide". *Ktéma* 4 (1979) 47–69.

Forbes, Robert J., "The Fibres and Fabrics of Antiquity"; "Washing, Bleaching, Fulling and Felting"; "Dyes and Dyeing"; "Spinning, Sewing, Basketry and Weaving". Robert J. Forbes, *Studies in the History of Technology*, Vol. 4. 2nd edn. Leiden: Brill, 1964.

Forbes, William T.M., "The Silkworm of Aristotle". *Classical Philology* 25 (1930) 22–26.

Frayn, Joan M., *Sheep-rearing and the wool trade in Italy during the Roman period*. Liverpool: Cairns, 1984.

Gaitzsch, Wolfgang, "Antike Seilerei". *Antike Welt* 16 (1985) 41–50.

Griffith, Francis L., and Grace M. Crowfoot, "On the Early Use of Cotton in the Nile Valley". *Journal of Egyptian Archaeology* 20 (1934) 5–12.

Gleba, Margarita and Judit Pásztókai-Szeőke, eds., *Making Textiles in Pre-Roman and Roman Times. Peoples, Places, Identities. Ancient Textiles Series Vol. 13*. Oxford; Oakville: Oxbow Books, 2013.

Halleux, Robert, *Les alchimistes grecs*, Vol. 1. Paris: Belles Lettres, 1981.

Harlizius-Klück, Ellen, "Textile Technology". Pp. 747–767 in Georgia L. Irby, ed., *A Companion to Science, Technology, and Medicine in Ancient Greece and Rome, Vol. II*. Chichester: Wiley-Blackwell, 2016.

Harlow, Mary and Marie-Louise Nosch, eds., *Greek and Roman Textiles and Dress: An Interdisciplinary Anthology*. Ancient Textiles Series, 19. Oxford; Philadelphia: Oxbow Books, 2014.

Hoffmann, Marta, *The Warp-Weighted Loom. Studies in the History and Technology of an Ancient Implement*. Studia Norvegica, 14. Oslo: Universitetsforlag, 1964.

Jones, A. Hugo M., "The Cloth Industry under the Roman Empire". *Economic History Review* 13 (1960) 183–192.

Jongman, William M., *Economy and Society of Pompeii*. Amsterdam: Gieben, 1988.

Killen, John T., "The Wool Industry of Crete in the Late Bronze Age". *Annual of the British School at Athens* 59 (1964) 1–15.

Moeller, Walter O., *The Wool Trade of Ancient Pompeii*. Leiden: Brill, 1976.

Pietrogrande, A.L., *Le fulloniche*. Scavi di Ostia, 8. Rome: Istituto Poligrafico dello Stato, 1976.

Riederer, Josef, "Römische Nähnadeln". *Technikgeschichte* 41 (1974) 153–172.

Robinson, Stuart, *A History of Dyed Textiles*. Cambridge, MA: M.I.T. Press, 1969.

Wild, John P., *Textile Manufacture in the Northern Roman Provinces*. Cambridge: Cambridge University Press, 1970.

Wild, John P., "Textiles". Pp. 167–177 in Strong, Donald E., and David Brown, eds., *Roman Crafts*. New York: New York University Press, 1976.

Wild, John P., "The Roman Horizontal Loom". *American Journal of Archaeology* 91 (1987) 459–471.

Wild, John P., "Textile Production". Pp. 465–482 in John P. Oleson, ed., *The Oxford Handbook of Engineering and Technology in the Classical World*. Oxford: Oxford University Press, 2008.

Winter, John G., and Herbert C. Youtie, "Cotton in Graeco-Roman Egypt". *American Journal of Philology* 65 (1944) 249–258.

Wipszycka, Ewa, *L'industrie textile dans l'Égypte romaine*. Warsaw: Polskiej Akademii Nauk, 1965.

Leather

Busch, Anna Lisa, "Die römerzeitlichen Schuh- und Lederfunde der Kastelle Saalburg, Zugmantel und Kleiner Feldberg". *Saalburg Jahrbuch* 22 (1965) 158–210.

Driel-Murray, C. van, "The Production and Supply of Military Leatherwork in the First and Second Centuries AD. A Review of the Archaeological Evidence". Pp. 43–81 in Mike C. Bishop, ed., *Production and Distribution of Roman Military Equipment*. Oxford: BAR, 1985.

Driel-Murray, C. van, "Tanning and Leather". Pp. 483–495 in John P. Oleson, ed., *The Oxford Handbook of Engineering and Technology in the Classical World*. Oxford: Oxford University Press, 2008.

Forbes, Robert J., "Leather in Antiquity". Pp. 1–79 in Robert J. Forbes, *Studies in Ancient Technology*, Vol. 5. 2nd edn. Leiden: Brill, 1966.

Goette, Hans Rupprecht, *"Mulleus – Embas – Calceus*. Ikonographische Studien zu römischem Schuhwerk". *Jahrbuch des Deutschen Archäologischen Instituts* 103 (1988) 401–464.

Lau, O., "Schuster und Schusterhandwerk in der griechisch-römischen Literatur und Kunst". Diss. Bonn, 1967.
Morrow, Katherine D., *Greek Footwear and the Dating of Sculpture*. Madison: University of Wisconsin Press, 1986.
Reed, R., *Ancient Skins, Parchments and Leathers*. London: Seminar Press, 1972.
Salonen, Armas, *Die Fussbekleidung der alten Mesopotamier nach sumerisch-akkadischen Quellen*. Annales Academiae Scientiarum Fennicae, ser. B., 157. Helsinki: Academia Scientiarum Fennica, 1969.
Waterer, John W., "Leatherwork" Pp. 179–193 in Strong, Donald E., and David Brown, eds., *Roman Crafts*. New York: New York University Press, 1976.

Ceramics

Arafat, Karim, and Catherine Morgan, "Pots and Potters in Athens and Corinth: A Review". *Oxford Journal of Archaeology* 8 (1989) 311–346.
Bailey, Donald M., "Pottery Lamps". Pp. 93–103 in Donald E. Strong, and David Brown, eds., *Roman Crafts*. New York: New York University Press, 1976.
Brodribb, Gerald, *Roman Brick and Tile*. Gloucester: Sutton, 1987.
Cuomo di Caprio, Ninina, "Pottery Kilns on Pinakes from Corinth". Pp. 72–82 in Herman A.G. Brijder, ed., *Ancient Greek and Related Pottery*. Amsterdam: Allard Pierson Museum, 1984.
Cuomo di Caprio, N., *La ceramica in archeologia. Antiche tecniche di lavorazione e moderni metodi d'indagine*. Rome: "L'Erma" di Bretschneider, 1985.
Evely, Don, "The Potter's Wheel in Minoan Crete". *Annual, British School of Archaeology in Athens* 83 (1988) 83–126.
Farnsworth, Marie, "Draw-Pieces as Aids to Correct Firing". *American Journal of Archaeology* 64 (1960) 72–75.
Farnsworth, Marie, and Ivor Simmons, "Coloring Agents for Greek Vases". *American Journal of Archaeology* 67 (1963) 389–396.
Georgiou, Hara, "Minoan Coarse Wares and Minoan Technology". Pp. 75–92 in Olga Krzyszkowska, and Lucia Nixon, eds., *Minoan Society*. Bristol: Bristol Classical Press, 1983.
Gericke, Helga, *Gefässedarstellungen auf griechischen Vasen*. Berlin: Bruno Hessling, 1970.
Higgins, Reynold A., "Terracottas". Pp. 105–109 in Donald E. Strong, and David Brown, eds., *Roman Crafts*. New York: New York University Press, 1976.
Jackson, Mark, and Kevin Greene, "Ceramic Production". Pp. 496–519.in John P. Oleson, ed., *The Oxford Handbook of Engineering and Technology in the Classical World*. Oxford: Oxford University Press, 2008.
Johnston, Alan W., *Trademarks on Greek Vases*. Warminster: Aris & Phillips, 1979.
Kanowski, Maxwell G., *The Containers of Classical Greece: A Handbook of Shapes*. London: Queensland University Press, 1984.
Mayes, Philip, "The Firing of a Pottery Kiln of a Romano-British Type at Boston, Lincs". *Archaeometry* 4 (1961) 4–18.
Noble, Joseph V., *The Techniques of Painted Attic Pottery*. 2nd edn. London: Thames & Hudson, 1988.
Noll, Walter, "Techniken antiker Töpfer und Vasenmaler". *Antike Welt* 8:2 (1977) 21–36.
Noll, Walter, *Alte Keramiken und ihre Pigmente. Studien zu Material und Technologie*. Stuttgart: 1991.

Peacock, David P.S., *Pottery in the Roman World: An Ethnoarchaeological Approach.* London: Longman, 1982.
Rice, Prudence, *Pottery Analysis*, 2nd edn. Chicago: Chicago University Press, 2015.
Scheibler, Ingeborg, *Griechische Töpferkunst. Herstellung, Handel und Gebrauch der antiken Tongefässe.* Munich: Beck, 1983.
Seiterle, Gérard, "Die Zeichentechnik in der rotfigurigen Vasenmalerei". *Antike Welt* 7:2 (1976) 3–10.
Vickers, Michael J., and David Gill, *Artful Crafts: Ancient Greek Silverware and Pottery.* Oxford: Clarendon Press, 1994.
Winter, Adam, *Die antike Glanztonkeramik. Praktische Versuche.* Mainz: Philipp von Zabern, 1978.

Glass

Abrami, Michael, "Eine römische Lampe mit Darstellung des Glasblasens". *Bonner Jahrbücher* 159 (1959) 149–151.
Baldoni, Daniela, "Una lucerna romana con raffigurazione di officina vetraria. Alcune considerazioni sulla lavorazione del vetro soffiato nell'antichità". *Journal of Glass Studies* 29 (1987) 22–29.
Barag, Dan, "Mesopotamian Glass Vessels of the Second Millennium BC. Notes on the Origin of the Core Technique". *Journal of Glass Studies* 4 (1962) 8–27.
Bezborodov, M.A., *Chemie und Technologie der antiken und mittelalterlichen Gläser.* Mainz: von Zabern, 1975.
Butcher, Sarnia A., "Enamelling". Pp. 43–51 in Donald E. Strong, and David Brown, eds., *Roman Crafts.* New York: New York University Press, 1976.
Doppelfeld, Otto, "Das Kölner Diatretglas und die anderen Netz-Diatrete". *Gymnasium* 68 (1961) 410–424.
Forbes, Robert J., "Glass". Pp. 112–236 in Robert J. Forbes, *Studies in Ancient Technology*, Vol. 5. 2nd edn. Leiden: Brill, 1966.
Grose, David F., "Early Blown Glass". *Journal of Glass Studies* 19 (1977) 9–29.
Grose, David F., "Glass Forming Methods in Classical Antiquity: Some Considerations". *Journal of Glass Studies* 26 (1984) 25–34.
Haevernick, Thea E., and P. Hahn-Weinheimer, "Untersuchungen römischer Fenstergläser". *Saalburg Jahrbuch* 14 (1955) 66–73.
Harden, Donald B., "Ancient glass, I: Pre-Roman". *Archaeological Journal* 125 (1968) 46–72.
Harden, Donald B., "Ancient glass, II: Roman". *Archaeological Journal* 126 (1969) 44–77.
Harden, Donald B., "New Light on the History and Technique of the Portland and Auldjo Cameo Vessels". *Journal of Glass Studies* 25 (1983) 45–54.
Henderson, Julian, "The Raw Materials of Early Glass Production". *Oxford Journal of Archaeology* 4 (1985) 267–291.
Kisa, Anton, *Das Glas im Altertum.* 3 Vols. Leipzig: Hiersemann, 1908. Labino, Dominick, "The Egyptian Sand-Core Technique: A New Interpretation". *Journal of Glass Studies* 8 (1966) 124–127.
Lierke, Rosemarie, *"Aliud torno teritur.* Rippenschalen und die Spuren einer unbekannten Glastechnologie. Heisses Glas auf der Töpferscheibe". *Antike Welt* 24 (1993) 218–234.

Noble, Joseph V., "The Technique of Egyptian Faïence". *American Journal of Archaeology* 73 (1969) 435–439.

Oliver, Andrew, Jr., "Millefiori Glass in Classical Antiquity". *Journal of Glass Studies* 10 (1968) 48–70.

Oppenheim, A. Leo, ed., *Glass and Glassmaking in Ancient Mesopotamia*. London: Associated University Presses, 1988.

Pilliger, Renate, *Studien zu römischen Zwischengoldgläsern*, 1: *Geschichte der Technik und das Problem der Authentizität*. Vienna: Österreichische Akademie der Wissenschaften, 1984.

Price, Jennifer, "Glass". Pp. 111–125 in Donald E. Strong, and David Brown, eds., *Roman Crafts*. New York: New York University Press, 1976.

Stern, E. Marianne, "Ancient Glass in a Philological Context". *Mnemosyne* 60 (2007) 341–406.

Stern, E. Marianne, "Glass Production". Pp. 520–550 in John P. Oleson, ed., *The Oxford Handbook of Engineering and Technology in the Classical World*. Oxford: Oxford University Press, 2008.

Welzel, Josef, "Zur Schleiftechnik der römischen Diatretgläser". *Gymnasium* 86 (1979) 463–473.

Applied chemistry

André, Jacques, "La résine et la poix dans l'antiquité: Technique et terminologie". *L'Antiquité Classique* 33 (1964) 86–97.

Caley, Earle R., "The Leyden Papyrus X". *Journal of Chemical Education* 3.10 (1926) 1149–1166.

Caley, Earle R., "The Stockholm Papyrus". *Journal of Chemical Education* 4.8 (1927) 979–1002.

Faure, Paul, *Parfums et aromates de l'Antiquite*. Paris: Fayard, 1987.

Forbes, Robert J., "Chemical, Culinary, and Cosmetic Arts". Pp. 238–298 in Charles Singer, Eric J. Holmyard, and Alfred. R. Hall, eds., *A History of Technology*, Vol. I: *From Early Times to Fall of Ancient Empires* c.*500 BC*. Oxford: Clarendon Press, 1954.

Forbes, Robert J., "Cosmetics and Perfumes in Antiquity," "Salts, Preservations Processes, Mummification," "Paints, Pigments, Inks and Varnishes". Pp. 1–50, 164–209, 210–264 in Robert J. Forbes, *Studies in Ancient Technology*, Vol. 3. 2nd edn. Leiden: Brill, 1965.

Forbes, Robert J., "Dyes and Dyeing". Pp. 99–150 in Robert J. Forbes, *Studies in Ancient Technology*, Vol. 4. 2nd edn. Leiden: Brill, 1964.

Graeve, Volkmar von, and Frank Preusser, "Zur Technik griechischer Malerei auf Marmor" *Jahrbuch des Deutschen Archäologischen Instituts* 96 (1981) 120–156.

Gwei-Djen, Leo, J. Needham, and D. Needham, "The Coming of Ardent Water". *Ambix* 19 (1972) 69–112.

Halleux, Robert, *Les alchimistes grecs*, Vol. 1 Paris: Belles Lettres, 1981.

Jones-Lewis, M. "Pharmacy". Pp. 402–417 in Georgia L. Irby, ed., *A Companion to Science, Technology, and Medicine in Ancient Greece and Rome, Vol. I*. Chichester: Wiley-Blackwell, 2016.

Levey, Martin, *Chemistry and Chemical Technology in Ancient Mesopotamia*. Amsterdam: Elsevier, 1959.

Partington, James Riddick, *Origins and Development of Applied Chemistry*. London: Longmans & Green, 1935.

Schmauderer, Eberhard, "Seife und seifenähnliche Produkte im klassischen Altertum". *Technikgeschichte* 35 (1968) 205–222.
Singer, Charles, *The Earliest Chemical Industry*. London: Folio Society, 1948.
Stulz, Heinke, *Die Farbe Purpur im frühen Griechentum*. Stuttgart: Teubner, 1990.
Taylor, Frank S., *A History of Industrial Chemistry*. London: Heinemann, 1957.
Wylock, Michel, "La fabrication des parfums à l'époque mycénienne d'après les tablettes de Pylos". *Studi micenei ed egeo-anatolici* 11 (1970) 116–133.

Health and well-being

Aufderheide, Arthur C., *The Scientific Study of Mummies*. Cambridge: Cambridge University Press 2003.
Bliquez, Lawrence J., *The Tool of Asclepius: Surgical Instruments in Greek and Roman Times: Studies in Ancient Medicine Volume: 43*. Leiden and Boston: Brill Academic Press, 2014.
Brothwell, Don R., *Digging up Bones: The Excavation, Treatment and Study of Human Skeletal Remains*. Ithaca: Cornell Universtiy Press, 3rd edn. 1981.
Bouras-Vallianatos, Petros and Sophia Xenophontos, eds., *Greek Medical Literature and its Readers: From Hippocrates to Islam and Byzantium. Publications of the Centre for Hellenic Studies. King's College London*. London; New York: Routledge, 2018.
Caldwell, L. "Gynecology". Pp. 360–370 in Georgia L. Irby, ed., *A Companion to Science, Technology, and Medicine in Ancient Greece and Rome, Vol. I*. Chichester: Wiley-Blackwell, 2016.
Cilliers, Louise, "Anaesthesia and Analgesia in Ancient Greece and Rome (*c.*400 BCE–300 CE)". Pp. 31–44 in Helen Askitopoulou, ed., *History of Anaesthesia VII*. Herakleion: Crete University Press, 2012.
Craik, Elizabeth, ed., *Hippocrates. Places in Man. Greek text and translation with Introduction and Commentary*. Oxford: Clarendon Press, 1998.
Dioscorides Pedanius of Anazarbus, *De Materia Medica: Being an Herbal with Many Other Medicinal Materials Written in Greek in the First Century of the Common Era*. Trans. by Tess A. Osbaldeston and Robert P.A. Wood; Johannesburg: Ibidis, 2000.
Forbes, Robert J., "Salts, Preservations Processes, Mummification," "Paints, Pigments, Inks and Varnishes". Pp. 164–209 in Robert J. Forbes, *Studies in Ancient Technology*, Vol. 3. 2nd edn. Leiden: Brill, 1965.
Harris, William V., ed., *Popular Medicine in Graeco-Roman Antiquity: Explorations. Columbia studies in the classical tradition, 42*. Leiden; Boston: Brill, 2016.
Israelowich, Ido, *Patients and Healers in the High Roman Empire*. Baltimore: Johns Hopkins University Press, 2015.
Jansen, Gemma C.M., Ann O. Koloski-Ostrow, and Eric M. Moormann, eds., *Roman Toilets: Their Archaeology and Cultural History*. Babesch, Annual Papers on Mediterranean Archaeology, Supplement 19. Louvain, 2011.
Jackson, Ralph, "Roman Doctors and Their Instruments: Recent Research into Ancient Practices". *Journal of Roman Archaeology* 3 (1990) 5–27.
Jones-Lewis, M. "Pharmacy". Pp. 402–417 in Georgia L. Irby, ed., *A Companion to Science, Technology, and Medicine in Ancient Greece and Rome, Vol. I*. Chichester: Wiley-Blackwell, 2016.
Koloski-Ostrow, Ann Olga, *The Archaeology of Sanitation in Roman Italy: Toilets, Sewers, and Water Systems*. Chapel Hill: University of North Carolina Press, 2015.

Laes, Christian, *Disabilities and the Disabled in the Roman World: A Social and Cultural History*. Cambridge: Cambridge University Press, 2018.

Le Blay, F. "Surgery". Pp. 371–385 in Georgia L. Irby, ed., *A Companion to Science, Technology, and Medicine in Ancient Greece and Rome, Vol. I*. Chichester: Wiley-Blackwell, 2016.

Longrigg, James, *Greek Medicine from the Heroic to the Hellenistic Age. A Source Book*. London: Duckworth, 1998.

Michaelides, Demetrios, ed., *Medicine and Healing in the Ancient Mediterranean World. Including the proceedings of the international conference with the same title, organised in the framework of the Research Project INTERREG IIIA: Greece/Cyprus 2000/2006, Joint Educational and Research Programmes in the History and Archaeology of Medicine, Palaeopathology, and Palaeoradiation, and the 1st International CAPP Symposium 'New Approaches to Archaeological Human Remains in Cyprus'*. Oxford; Philadelphia: Oxbow, 2014.

Milne, John Stewart, *Surgical Instruments in Greek and Roman Times*. London: The Clarendon Press: 1907. Reprint New York: Augustus K. Kelley, 1970.

Missios, Symeon M.D., "Hippocrates, Galen, and the uses of trepanation in the ancient classical world". *Neurosurg Focus* 23 (1): E11, 2007: 1–9.

Mitchell, Piers D., ed., *Sanitation, Latrines and Intestinal Parasites in Past Populations*. Farnham; Burlington VT: Ashgate, 2015.

Nutton, Vivian, *Ancient Medicine*. London: Routledge, 2nd edn, 2013.

Papageorgopoulou, Christina, Nikolaos I. Xirotiris, Peter X. Iten, Markus Baum, Martin Schmid, and Frank Rü, "Indications of Embalming in Roman Greece by Physical, Chemical and Histological Analysis". *Journal of Archaeological Science* 2009 36, 35–42.

Potts, D.T., "An Archaeological Meditation on Trepanation". Pp. 463–492 in Brooke Holmes and Klaus-Dietrich Fischer, eds., *The Frontiers of Ancient Science: Essays in Honor of Heinrich von Staden*. Berlin, Munich, Boston: Walter de Gruyter, 2015.

Samama, Evelyne, *La médecine de guerre en Grèce ancienne. De diversis artibus, 98*. Turnhout: Brepols Publishers, 2017.

Steger, Florian, *Asklepios: Medizin und Kult*. Stuttgart: Franz Steiner Verlag, 2016.

Van Tilburg, Cornelis, *Streets and Streams: Health Conditions and City Planning in the Graeco-Roman World*. Leiden: Primavera Pers, 2015.

Von Staden, H., *Herophilus. The Art of Medicine in Early Alexandria*. Cambridge: University Press, 1989.

Voultsiadou, Eleni, "Sponges: an historical survey of their knowledge in Greek antiquity". *Journal of the Marine Biology Association of the United Kingdom* 87 (2007) 1757–1763.

Large-scale production

Acton, Peter, *Poiesis: Manufacturing in Classical Athens*. Oxford; New York: Oxford University Press, 2014.

Grose, David, "The Formation of the Roman Glass Industry". *Archaeology* 36:4 (1983) 38–45.

Harris, William V., "Roman Terracotta Lamps. The Organization of an Industry". *Journal of Roman Studies* 70 (1980) 126–145.

Tapio, Helen, *Organization of the Roman Brick and Tile Industry in the First and Second Centuries AD*. Annales Academiae Scientiarum Fennicae, 5. Helsinki: Suomalainen Tiedeakatemia, 1975.

Hodge, A. Trevor, "A Roman Factory". *Scientific American* 263.5 (1991) 58–64.

Hopper, Robert J., *Trade and Industry in Classical Greece*. London: Thames & Hudson, 1979.

Jones, A. Hugo M., "The Cloth Industry under the Roman Empire". *Economic History Review* 13 (1960) 183–192.

Jongman, William M., *Economy and Society of Pompeii*. Amsterdam: Gieben, 1988.

Moeller, Walter O., *The Wool Trade of Ancient Pompeii*. Leiden: Brill, 1976.

Prachner, Gottfried, *Die Sklaven und Freigelassenen im arretinischen Sigillatagewerbe*. Wiesbaden: Franz Steiner, 1980.

Wilson, Andrew, "Large-Scale Manufacturing, Standardization, and Trade". Pp. 393–417 in John P. Oleson, ed., *The Oxford Handbook of Engineering and Technology in the Classical World*. Oxford: Oxford University Press, 2008.

Wilson, Andrew, and Miko Flohr, eds., *Urban Craftsmen and Traders in the Roman World*. Oxford: Oxford University Press, 2016.

11

TRANSPORT AND TRADE

A LAND TRANSPORT

The early development of transport by water meant that for communities located in coastal or riverine areas, no pressing need to develop important land routes existed. Of course, local roadways and paths were important and wheeled vehicles used them (**11.2**), but inter-community pathways were expensive if they were more than tracks. Only civilisations such as those of the Persians and the Romans, with their strong systems of central administration, built and maintained very extensive networks of roads and bridges (**11.1** and **11.5**).

Other areas such as Greece were heavily reliant upon water transport because of geographical considerations and had little use for major roadways, especially since transport by land was more expensive than by water (**11.45, 11.87**). Little needed to be done, however, until the regular use of wheeled vehicles became widespread and more substantial roads and bridges were required. Even with increased urbanisation, with the exception of Sacred Roads and special roads built for unusual tasks such as heavy transport from quarries (**8.4, 11.35**), the roadways of classical Greece were not very impressive. The lack of a strong centralised government in Greece probably contributed to this situation, but even during the Roman Empire, Greece had few substantial roads if the comments of Pausanias (**11.3**) are taken at face value.

Besides military uses (**11.11**), developed road systems were needed by centralised powers for communication. Indeed, the development of postal systems (**11.26–28**) to link the outlying areas of an Empire with the seat of government was partially responsible for improved roadways [**fig. 11.1**] since speed of communication rested upon a service using the fastest means available: horsemen on reliable roads. The great distances involved meant that such postal systems required rest and supply stations at regular intervals, which eventually were expanded as merchants and other travellers began to follow the routes.

The increase in travel also created a need for reliable maps for those who were making journeys. Numerous writers record the considerable attention given to make more accurate charts, strip-maps and globes (**11.51–59**) (French 2016, Alcock *et al.* 2012, Quilici 2008, Chevllier 1976, Forbes 1965).

TRANSPORT AND TRADE

ROADS

11.1 The royal road of Persia

The Greeks, who lacked anything comparable, were impressed with the great road from Susa to Sardis that was maintained by the Persian king. Messages sent by horse could travel the distance (2,600 km) in nine days (**11.26**). Although not a paved roadway, the road was suitable for wheeled vehicles, but they occasionally did get stuck in the mud (Xenophon, *Anabasis* 1.5.7).

Herodotus, *Histories* 5.52–53

The nature of the road is as follows. All along it are royal rest stops and excellent lodgings, and the entire road runs through inhabited and safe country.... In all there are 111 stages with as many rest stops on the road going up from Sardis to Susa. If the Royal Road has been measured properly in *parasangs* and the *parasang* is equivalent to 30 *stadia*, as it certainly is, then from Sardis to the royal palace called Memnon there are 13,500 *stadia* since there are 450 *parasangs*. And for people travelling at a rate of 150 *stadia* each day, just 90 days will be consumed.

11.2 Early roads for chariots in Greece

Although Homer does not directly discuss the roads of his day, his mention of cross-country travel in chariots implies that more than mere paths were available. The same may be true during the Mycenaean period since we have evidence of chariots in the Linear B Tablets (**10.126**). Here, Telemachus, the son of Odysseus, is able to make a trip from Pylos to Sparta, a distance of more than 75 km, in two days.

Homer, *Odyssey* 3.475–486

"My sons, bring forth the beautifully maned horses for Telemachus, and yoke them to the car so that he can take to the road".

Thus, spoke Nestor, King of Pylos, and they clearly heard and obeyed, quickly yoking the swift horses to the car. Then the mistress of the stores put in bread, wine, and dainties, such as those eaten by beloved kings of Zeus. Telemachus then stepped up into the beautiful chariot. Peisistratus, a son of Nestor and leader of men, mounted into the chariot beside him and took the reins in his hands. He whipped the horses onward and they willingly sped to the plain, leaving the lofty citadel of Pylos behind. All day long the horses shook the yoke about their necks.

11.3 The poor roads of Greece

Even in the Roman period, the road system of Greece was not developed to the extent of other areas, perhaps a result of Greece's topography and geography or its reduced importance in the Roman Empire. Pausanias often expresses surprise that roads are fit for wheeled traffic, although Livy (*History of Rome* 7.37) records that as early as 190 BC good roads existed in Macedonia.

TRANSPORT AND TRADE

Pausanias, *Description of Greece* 1.44.6; 8.54.5; 10.32.8

Originally, as they say, the road from Megara to Corinth was made for unencumbered and agile men to travel over, but the Emperor Hadrian made it wider, and it was even suitable for chariots to pass each other when they met...

The road to Argos from Tegea is very well-suited for the carriage, and is really a superb highway...

The road across Parnassus is not entirely mountainous, but is even suitable for carriages.

11.4 Location, location, location

As paths, tracks and roadways developed, they were intended to make work and life easier. With increased centralisation of power and resources, surveying of best routes began to be conducted usually with this in mind. Nevertheless, difficult and arduous roadways continued to be built when special considerations came into play, which ignored safety and convenience.

Strabo, *Geography* 14.2.23

But Mylasa is sited in an exceedingly prosperous plain with a mountain that possesses a very beautiful quarry of white stone running down from its summit. This quarry is of not of little benefit, but has abundant fine stone at hand for structures and especially for constructing temples and other public works; accordingly, this city is as beautifully adorned with stoas and temples as much as any other. But one can only marvel in disbelief at the man who threw down the city's foundation at the base of the steep and commanding crag. Indeed, one of its commanders in amazement at the result, is reported to have said, "If the man who founded this city was not afraid, was he not even ashamed?"

11.5 Greek and Roman attitudes towards roads

One of the reasons for the disparity between the roads of the Greeks and Romans was the fundamental difference in their attitudes towards utilitarian projects. The Greeks did not ignore them, but they simply did not place the importance upon them that the Romans did (**9.16**) (Raepsait 2016, 2008, Crouwel 2012,1993).

Strabo, *Geography* 5.3.8

The Romans have especially provided for those things that the Greeks took little account of, such as the laying of roads and the building of aqueducts and sewers capable of washing out the refuse of the city into the Tiber. And they have laid their roads through the countryside, by employing cuts into hills and embankments in valleys, so that their wagons are able to take ships' cargo.

Compare Strabo's description (*Geography* 7.7.4ff.) of the Via Egnatia, a major road built by the Romans through northern Greece to connect the eastern part of the Empire with Rome.

11.6 Early Roman road contracts

Livy records the first paving of the roads within the city of Rome in 174 BC.

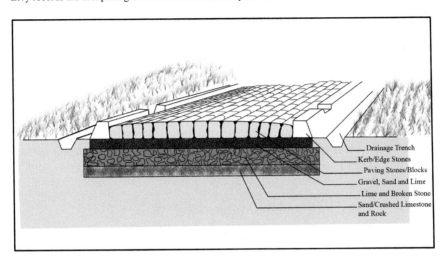

Figure 11.1 Roman road construction.
Source: Adapted from Forbes, 1965.2, fig. 37.

Livy, *History of Rome* 41.27.5

First of all, the censors contracted for paving the roads in the city [Rome] with flint (*silex*), and for laying the foundation and shoulders of roads outside the city with gravel, and for making bridges in many places….

The maintenance of the roads at Rome was an important concern for her citizens and we find a variety of attempts to urge different wealthy citizens to maintain them (Suetonius, *Augustus* 30). This inscription from Rome (44 BC) reflects the most common method by which the roads were maintained; a Hellenistic inscription (*SEG* 13.521 = *OGIS* 483) relates similar regulations for Pergamum's citizens.

CIL 1.593.20–55 = *ILS* 6085

In regards to the roads that are, or will be, in the city of Rome or within 1 mile of the city of Rome, where it is continuously inhabited, it shall be the duty of all people before whose building any road will run to maintain that road to the satisfaction of the *aedile* to whom that part of the city will be assigned by this law. And that same *aedile* shall take care that all people required by this law to maintain a road in front of their building shall maintain that road to his satisfaction, and that no water shall stand in that place, by which that road would be of less convenient use to the public….

In the case of anyone who is required by this law to maintain a public road in front of his building, who shall not maintain it to the satisfaction of the *aedile* concerned, then it shall be the duty of the *aedile* at whose discretion the road ought to be maintained to contract for the maintenance of that road....

Nothing in this law is meant to prevent the *aediles* – and the *quattuorviri* for cleaning the streets within the city, and the *duoviri* for cleaning the roads outside but within 1 mile of the city of Rome – from taking care to clean the public roads and having the power in that matter in all respects as they are or shall be required by the laws, plebiscites, or decrees of the senate.

Anyone before whose building a footpath will be located shall be required to keep that footpath properly paved for the whole length of the building with whole stones tightly packed to the satisfaction of the *aedile* whose authority shall be in that part of the roads by this law.

11.7 Fraudulent road-building exposed

As with anything involving money, vigilance was required on the part of citizens and magistrates to make sure that the conditions of road contracts were met by the builders. Failure to do so sometimes resulted in shoddy and virtually useless roads.

Tacitus, *Annales* 3.31

Corbulo, after declaring [*c.*AD 25] that many roads throughout Italy were broken apart and impassable as a result of the deception of the contractors and the carelessness of the magistrates, willingly undertook their prosecution. The result was not so much a benefit to the public as a catastrophe to the many people whose property and reputation he ravaged with his condemnations and auctions.

Dio Cassius (*Roman History* 59.15.3–5) states that the emperor later restored the property to those who had lost it.

11.8 Popularity obtained by building roads

At the beginning of the last quarter of the second century BC, Gaius Gracchus had become one of the favourites of the people of Rome by means of a series of popular reforms and programmes. His interest in the road system was unusual.

Plutarch, *Gaius Gracchus* 7.1–2

He was especially anxious about road-building, paying attention to utility as well as to that which was beneficial to grace and beauty. For the roads were carried straight through the country without wavering, and were paved with quarried stone, and made solid with masses of tightly packed sand. Hollows were filled up and bridges thrown across whatever wintry streams or ravines cut the roads.

And both sides were an equal and parallel height with the result that the road for its entire course had a level and beautiful appearance. Besides these things, he measured the whole road mile by mile – the mile is a bit less than eight *stadia* – and set up stone columns as distance indicators. He also placed other stones on either side of the road at lesser intervals so that it would be easier for those people who had horses to mount them from the stones without requiring a groom to help.

11.9 Inscriptions commemorating road construction and repair

Inscriptions cut either on scarps or on 2-m-high stone cylinders, which served as milestones in antiquity, functioned as permanent memorials to the builders and restorers of roads. They could also include information regarding financing, special conveniences, and distances from Rome, provincial capitals, or regional capitals. The first inscription below is from a milestone of a Republican Consul in Lucania; the rest are imperial. The underlined names in the passages below give the form of the emperor's name most familiar to the modern reader.

CIL 1.638 = *ILS* 23

[Publius Popillius, consul, son of Gaius; 132 BC] I made the road from Regium to Capua and I placed on that road all the bridges, milestones, and sign-posts (*tabellarii*). From here to Nuceria there are 51 miles; 84 to Capua; 74 to Muranum; 123 to Consentia; 180 to Valentia; 231 to the Strait at the Statue; 237 to Regium. Total from Capua to Regium 321 miles.

CIL 3.8267 = *ILS* 5863

The Emperor Caesar Nerva Trajan Augustus Germanicus [AD 100] ... built this road by cutting through mountains and eliminating the curves.

IGR 1.1142

The Emperor Caesar Trajan Hadrian Augustus [AD 137] ... constructed the New Road of Hadrian from Berenice to Antinoopolis [in Egypt] through the safe, flat country and equipped it at regular distances with abundant cisterns, places to rest, and garrisons....

CIL 3.199 = *ILS* 5864

The Emperor Caesar M. Aurelius Antoninus Augustus Armeniacus and the Emperor Caesar L. Aurelius Verus Augustus Armeniacus [AD 160s] rebuilt the road destroyed by the force of the water, by cutting through the mountain under the direction of Julius Verus, legate with rank of propraetor of the Province of Syria and their friend, at the expense of the Abileni.

CIL 10.6854 = *ILS* 5822

The fourth milestone of the repaved road [the Via Appia].¹ The Emperor Caesar M. Aurelius Antoninus [Caracalla; AD 216] ... repaired the road with new hard stone (*silex*) by which it became more firm for travelling, for a distance of 21 miles using his own funds. Before this the road had been paved unprofitably with [soft?] white stone and had become dilapidated. Seventy-one miles from Rome.

CIL 8.22371 = *ILS* 5869

The Emperor Caesar M. Antonius Gordianus Pius Felix [AD 239] ... repaired the road along with the bridges that had fallen into ruin from old age and heavy rains.

11.10 The Appian Way 900 years after construction

Other passages demonstrate the need for road maintenance (**11.6**, **11.9**, and *Digest* 43.11.1), but if maintained, Roman roads could last almost indefinitely. Procopius attributes too much to the original builder in this passage and makes a mistake regarding the source of materials (quarries have been found along the road), but the praise is well-placed, as surviving sections of the road prove.

Procopius of Caesarea, *History of the Wars* 5.14.6–11

For the unencumbered man, the Appian Way is a five-day trip; for it runs from Rome to Capua. The width of this road is such that two wagons meeting can pass each other, and of all the sights, it is particularly worth seeing. For all the stone, being both millstone and hard by nature, Appius [censor in 312 BC] cut in another place far away then brought it hither; for it is found nowhere in this territory. First, he worked the stones smooth and flat, and cut them into polygonal shapes. Then he bound them to each other without putting in mortar or anything else. They were so firmly bound together and close to each other that they gave the appearance, not that they were fitted to each other, but that they grew together. And after the passing of so long a time, and after being traversed by many wagons and all the animals every day, neither have the joints separated in any way nor have any of the stones been worn out or made thinner; no, they have not even lost any of their fine finish. Such, then, is the Appian Way.

11.11 Road-builders on the march

The importance of good roads for armies on campaign was so great that major forces often carried not only the necessary personnel, but also the materials to provide level roads that were made as straight as possible. The "road-builders" for the Persian army were so important that men rejected from one of the branches of the army were converted into road-builders. They marched ahead of the army in order to prepare the road, especially for wheeled vehicles, before delays could develop.

Xenophon, *Cyropaedia* 6.2.36

You commanders of the road-builders have from me a list of the men rejected for service in the spearmen, the archers, and the slingers. While on the march, those men rejected from the spearmen must carry the wood-cutting axe, those from the archers the mattock, and those from the slingers the shovel. With these tools they are to proceed in squadrons in front of the wagons, so that if there is need for road-building you can start work immediately and so that if I need something I know where they are when the time comes.

Exaggerated belief regarding the strength of vinegar and its particularly cold nature (see **10.95**) may have prompted its use or reputed use to break apart rocks, but any cold liquid will help crack heated rock. Hannibal and his army have been caught in the deep snow on a mountainous cliff during their march over the Alps in 218 BC. They have just made camp.

Livy, *History of Rome* 21.37.2–3

Then the soldiers were ordered to make the cliff passable, their only possible road out. Since it was essential to cut through the rock, after immense trees growing about had been cut down and lopped off, they built a huge pile of logs and, when the strength of the wind was suitable for making it start to burn, set it on fire. Then they made the heated rocks crumble by dowsing them with vinegar. In such a manner, once it had been made dry by the fire, they threw open the cliff for passage with iron tools and eased the steepness of the slope with gentle cutbacks so that not only the baggage animals, but even the elephants could be led down.

Polybius (*Histories* 3.55.7) gives a simpler version of the story. For examples of armies using road-builders to provide direct access through difficult terrain see Thucydides, *Peloponnesian War* 2.98.1 and Arrian, *Anabasis* 1.26.1.

11.12 A poem in praise of Domitian's Road

Few poets have composed poems to commemorate roads, but the Via Domitiana is a notable exception. It was built in AD 95 to replace a poor road along the coast from Rome to Naples. Since the route of the well-built Via Appia turned inland at Sinuessa, this new road represented a considerable shortcut for those travelling straight to Naples. Statius first describes previous conditions for travellers and then relates the benefits of the new road and how it was constructed; the verses are a rare literary account of the construction of a Roman road [**fig. 11.1**].

Statius, *Silvae* 4.3.20–55

The Emperor Domitian, vexed at the slow journeys of his people and at plains that lengthen the entire trip, banishes the long windings and solidifies the troublesome sand with a new infusion, rejoicing to bring the home of the Euboean Sibyl and the Gauran valleys and sweltering Baiae [places close to Naples] nearer to the seven hills [of Rome].

Here once a dejected traveller, carried on a single axle, swayed back and forth on the swinging pole of the carriage while the wicked earth sucked in the wheels, and in the middle of plains the Latin people shuddered at the evils of a sea voyage. Nor were chariots quick on the road, but the silent ruts slowed their impeded journey while the sluggish quadruped, complaining of the excessive weight, crept along under its high yoke. But now a journey that used to wear away a solid day is completed in scarcely two hours....

Here the first labour was to begin ditches and to cut out the borders of the road and to excavate the ground with a very deep void; next to fill in the emptied ditches with other materials and to prepare the interior for the high ridge [the pitch of the road] so that neither the soils give way nor the treacherous foundation provide an unstable bed for the rocks that have been pressured [by traffic]; finally to bind the road with [notched?] stone blocks forced in on both sides and with closely packed wedge-shaped stones. Oh, how many gangs are at work at the same time! These men cut down forests and clear the mountains, another group smoothes down boulders and fashions beams with iron tools; others bind the rocks and weave together the work with baked sand and lowly tufa and this group dries up the thirsty pools and draws the smaller streams far away by their labour.

See Galen, *On Methods of Healing* 9.8 (Kühn 10, pp. 632–633) for a general account of Trajan's improvements of the roads within Italy.

11.13 A highway tunnel at Puteoli

In order to speed traffic around the Bay of Naples and to avoid a sea voyage or a long-shore route around the promontory of Pausilypum, Octavian, the future Emperor Augustus, commanded that tunnels be dug under the spurs of the Apennines. One was dug at Cumae and the other between Naples and Puteoli. They are about 3 m wide, 3–22 m high, and about 1,000 m (Cumae) and 700 m (Puteoli) long. Strabo and Seneca offer conflicting accounts, but the archaeological evidence suggests that Strabo has confused his tunnels and is actually describing the tunnel at Cumae rather than the one at Puteoli.

Seneca, *Letters* 57.1–2

When I had to return to Naples from Baiae I easily convinced myself that there was a storm so that I didn't have to undergo another ship voyage; and yet the entire road was so full of mud that I may be regarded as having made a voyage nevertheless. On that day I had to endure the full fate of athletes: after the anointing, the sand sprinkle followed upon us in the tunnel at Naples. Nothing is longer than that prison, nothing more dim than those torches, which allowed us to see not through the darkness, but to see the darkness itself. Indeed, even if the place had light, the dust – an oppressive and annoying thing even in the open – would consume it; how much worse in a place where it rolls back on itself and, when it is enclosed without any breeze, blows back against those people by whom it was stirred up! Thus, we endured two inconveniences simultaneously that were polar opposites: on the same road and on the same day we struggled with both mud and dust.

Strabo, *Geography* 5.4.7

Here there is also a tunnel, the mountain between Puteoli and Naples having been tunnelled just like the one to Cumae, and a road many *stadia* long has been opened that is able to accommodate teams that meet. And at many places, windows have been cut out so that the light of day is brought down along deep shafts through the mountain.

11.14 The *diolkos* at the Isthmus of Corinth

Although numerous proposals and some attempts were made to cut a canal through the Isthmus of Corinth, it was not until 1893 that the feat was accomplished. Instead, the ancient Greeks and Romans portaged their goods across the Isthmus when they wished to avoid the dangerous circumnavigation of the Peloponnese. A track for the portage (*diolkos*) was developed at least as early as the sixth century BC, perhaps by Periander. In the first passage, the Spartans and their allies seem poised to invade Attica by land and sea, during the Peloponnesian War. The allies are to assemble on the Corinthian side of the Isthmus (Werner 1997 for its history).

Thucydides, *The Peloponnesian War* 3.15.1

... And the Lacedaemonians themselves arrived first, and on the Isthmus prepared hauling-machines to carry the ships overland from Corinth to the sea near Athens in order to attack simultaneously by sea and land.

Strabo, *Geography* 8.2.1

The Isthmus at the *diolkos*, where they haul ships overland from one sea to the other, as I have said, is 40 *stadia* wide.

Pliny briefly describes the perils of circumnavigation of the Peloponnese then recounts the attempts to construct a canal through the Isthmus.

Pliny, *Natural History* 4.10

Because the journey around the Peloponnese is long and hazardous for ships with a mass that prohibits them from being transported across the Isthmus on wagons, King Demetrius, the dictator Caesar, and the emperors Gaius and Nero attempted to dig through the narrow part, undertaking to build a shipping canal.

For examples of successful canal projects, see **11.109–112**.

TRANSPORT AND TRADE

11.15 Special roads and maintenance

Information regarding utilitarian roads used for heavy transport is sparse (**8.4**) and occasionally only the road's absence is noted. Epigraphic evidence provides some data regarding Sacred Roads used for religious processions. These routes often benefited from bridges (**11.25**), paving, and special maintenance. Here the Amphictyones, the board of the Delphic oracle, have recorded their provisions for the upkeep of Sacred Roads (380–379 BC).

CIG 1688

... and the Amphictyones shall repair the bridges, each his own part, and they shall take care that they are not damaged. And the ambassadors of the Amphictyones shall maintain the roads, whatever is needed for them, and they shall punish [those people who damage the roads]....

11.16 Road ruts

The ruts (Latin *orbitae*; Greek *ogmoi*) that are often seen in paved roads are usually not the result of neglect and long use [**fig. 11.2**]. Instead they were cut into the paving stones to provide a means of steadying and directing the wheels of a vehicle and, where they existed, were regarded as the true road. Abundant archaeological evidence demonstrates a relatively consistent depth and gauge for the ruts of a region, implying a type of railway. Unfortunately the literature supplies only indirect evidence for these purposefully cut guidelines; often moral rules are compared to the ruts and tracks that keep one on the road.

Figure 11.2 Typical street in Pompeii with wheel ruts.
Source: Adapted from Adam, 1994, figs. 649, 650, 652.

Quintilian, *On the Training of an Orator* 2.13.16

But guidelines are also helpful for a young man if they show him the direct track, and not the only one (*orbita*). For whoever believes it a dastardly crime to deviate from the main track must be willing to take the slower path much like tightrope walkers.

Juvenal, *Satires* 14.33–37

Occasionally a few young men reject the bad examples set for them; ... the majority, however, don't jump the track but are guided by the footsteps of their fathers and are dragged along the established ruts (*orbitae*) of ancient vice.

Many poets (cf. Vergil, *Georgics* 3.292–293) use the *topos* of leaving the main or well-travelled road to make their own way.

Nicander, *Theriaca* 367–371

But when the Dog Star dries up the water ... then the snake, with tongue hissing, frequents the parched ruts (*ogmoi*) in the roadways.

11.17 Travel by canal to avoid land travel

Canals were not common in the mountainous terrain of Italy, although when available, they were often used to provide a break in tedious journeys by land. They also offered the added incentive of sleep to the travellers while moving along the canal at night (note Strabo's comment, **11.112**). Horace describes a less than pleasant outing on the canal that ran from Forum Appii to Tarentum through the Pomptine marshes. He has already travelled the 70-odd kilometres from Rome to Forum Appii and looks forward to a relaxing evening journey by mule-pulled barge.

Horace, *Satires* 1.5.9–23

Already night was beginning to draw the shadows over the earth and to scatter the stars in the sky. Then slaves fire loud insults at the boatmen, the boatmen at the slaves: "Pull in here!" "You're packing hundreds in!" "Wait, that's enough!" While the fare is collected and the mule harnessed, a whole hour is lost. Wretched gnats and the swamp's frogs turn aside sleep; the boatman, sloshed on a lot of wine that has turned, sings about his girlfriend left behind and a fellow traveller earnestly joins in. Finally, the weary traveller falls asleep and the lazy boatman puts his mule to pasture and ties the reins to a stone, then flat on his back begins to snore. It's already daylight when we realise the barge is not moving, not, at least, until one excitable traveller jumps out and beats the mule and boatman about the head and back with a willow club. It was almost ten when we finally disembarked....

TRANSPORT AND TRADE

BRIDGES

The movement of military forces across a large landmass entailed dealing with rivers and other bodies of water. Our sources are most abundant regarding these military bridging projects, but commercial, social, and religious aspects of society also required that communication overland be made easier by the use of bridges. Philostratus (**1.12**) offers a natural model for the origin of bridges.

11.18 Xerxes' pontoon bridge over the Hellespont

The Persians were adept at bridging water courses. Darius had spanned the Bosporus and then the Danube in order to attack the Scythians (Herodotus, *Histories* 4.87ff.); his successor Xerxes, after one failed attempt, finally bridged the Hellespont and led his forces against the Greeks. His first set of bridges (one with papyrus cables, a second with flaxen cables) had been destroyed by a storm, whereupon Xerxes had the Hellespont whipped, fetters thrown into it, and the overseers of the bridges beheaded before new crossings were ordered to be built (Hammond and Roseman 1996).

Herodotus, *Histories* 7.36

They made the bridges in the following manner. They lashed together pentecontrers and triremes – 360 to support the bridge on the Euxine Sea side and 314 for the bridge on the other side – at an angle to the Black Sea and with the current of the Hellespont in order to maintain an even tension on the cables. Once they had lashed them together, they lowered very large anchors at both ends, since east winds blew upon the end closest to the Black Sea and west and south winds blew upon the end towards the west and the Aegean. And they left a narrow opening for passage through the pentecontrers and triremes so that anyone so wishing could sail on small craft both into and out of the Black Sea. Having done these things, they stretched the cables by twisting them taut with wooden windlasses. Nor did they any longer keep the two types of cables separate, but combined two flaxen and four papyri cables for each bridge. The thickness and fine quality was the same for them all, but the flax was heavier in its proportion: a *pexus* of it weighed one talent. Once the strait was bridged, they sawed wooden logs, making them equal to the width of the floating bridge. Next, they set them in order on top of the taut cables and, having set them one after the other there, they fastened them. After these things were done, they layered on brushwood, then, having set the brushwood in order, they layered on earth and tramped it down. Finally, they erected paling on both sides to prevent the draught animals and horses from looking down at the water and becoming afraid.

Xerxes also had a canal cut for his ships through Akte (Athos), the eastern finger of the Chalcidice (Herodotus, **11.110**).

11.19 Caesar's bridge on piles over the Rhine

Although the Romans also used pontoon bridges (**11.22**), their sense of occasion sometimes compelled them to create more permanent structures. Caesar himself records the method he used to span the Rhine for his invasion.

Caesar, *Gallic War* 4.17–18

Caesar had decided to cross the Rhine for the reasons given earlier; but to cross by boats he deemed not safe enough and ruled it worthy neither for himself nor the Roman people. And so, even though he was confronted by the greatest difficulty for making a bridge because of the river's width, swiftness, and depth, nevertheless he decided that he had to make the effort or else not lead his army across. He used the following method for the bridge. At intervals of 2 feet, he joined pairs of timbers that were 1.5 feet thick, sharpened a bit at their bases, and measured for the depth of the river. Having lowered these into the river with machines, he fixed and rammed them down using pile drivers, not quite perpendicular in the manner of piles, but leaning forward and sloping so that they inclined with the natural flow of the river. In addition, he planted two piles opposite these at an interval of 40 feet downstream, fastened together in the same manner but turned into the force and flow of the river. These two rows were kept firmly apart by inserting into their tops, beams two feet thick, which were the same length as the distance between the piles, and that were supported with pairs of braces at the outer side of each pile. As a result of this combination of holding apart and clamping together, so great was the stability of the work and its character that the greater the force of the water rushing against it, the more tightly its parts held fastened together. These beams were interconnected by timbers laid at right angles, and then these were floored over with long poles and wickerwork. In addition, piles were driven at an angle into the water on the downstream side, which were thrust out underneath like a buttress and joined with the entire structure to take the force of the river. Similarly others were placed a little bit above the bridge so that if tree trunks or vessels were sent by the barbarians to knock down the structure, the force of those objects might be diminished by these defences and prevent the bridge from suffering harm.

Ten days after the timber began to be collected the bridge was completed and the army was led across.

Herodotus (*Histories* 5.16) describes the bridge and huts of a lake-town built on piles.

11.20 A small premade portable pontoon bridge

Nicias, a wealthy Athenian benefactor in late-fifth century BC Athens, provided much support for religious ceremony, festivals and dramatic contests. Here Plutarch implies a special approach arranged to avoid an unseemly arrival among the eager crowd.

Plutarch, *Nicias* 3.5

When Nicias conducted the festal embassy [to Delos], he disembarked onto the island of Rheneia with his chorus, the sacrificial victims, and the other equipment. Then using a bridge of boats, which had been made to measure at Athens and was splendidly adorned with gildings, and providing them with dyed materials and with garlands and tapestries, he bridged the narrow strait between Rheneia and Delos during the night. At daybreak, leading his sacred procession in honour of the god, and his chorus adorned in lavish splendour and singing, he disembarked onto the shore via the bridge.

11.21 Trajan's bridge over the Danube

Trajan's architect Apollodorus built an even more impressive stone bridge across the Danube. At first regarded as a major advantage for the Romans, it was soon partially dismantled in an attempt to maintain a strong defensive position on the right bank of the river.

Dio Cassius, *Roman History* 68.13.1–6

Trajan constructed a stone bridge over the Danube.... It has 20 piers of squared stone 150 feet high (excluding the foundations) and 60 feet wide. These stand 170 feet apart from each other and are connected together by arches. How is one not to marvel at the cost expended on them? And the method by which each of them was set into such a deep river, in the eddying water and on the muddy bottom? For it was not possible to divert the flow anywhere. I have already spoken about the width of the river – not that it runs at the same width for its whole length since it spreads two and sometimes three times as wide – but the narrowest and most suitable width for bridging in the region is at this point. Yet by as much as the river falls from a great flood at this point and is first restricted in a narrow strait and then flows out becoming wider again, by that much it becomes more violent and deep; and that has to be considered when judging the difficulty of constructing the bridge.... [At present] only the piers are standing ... since Hadrian removed the superstructure, fearing that if the guards at the bridge were overcome by the barbarians there would be an easy crossing into Moesia.

11.22 Pre-fabricated bridges: military methods for crossing rivers

After explaining the best ways for an army to ford a river, including dividing it into shallow channels (cf. Herodotus, *Histories* 1.189), Vegetius describes several methods to cross navigable rivers.

Vegetius, *Military Science* 3.7

But navigable rivers are crossed either by piles fixed into the bottom of the river with planks positioned on top of the piles or, in a sudden emergency, they are crossed by fastening together empty caskets and covering them with boards. [Horses swim across led by their riders.] ... But a more convenient way has been discovered. The army now carries in its train on carriages, *monoxyli*, which are small boats a bit wider than usual and hollowed out of single logs. These are very light on account of the type and quality of wood used. In addition to the boats, the army carries sufficient planks and iron nails that are ready for immediate use. Thus, without delay a bridge can be constructed that exhibits the solidity of a stone bridge for some time even though it is lashed together by means of cables, which have to be available for that reason.

A more detailed method of bridging rivers by using boats is offered by Dio Cassius.

Dio Cassius, *Roman History* 71.3.1

The channels of rivers are very easily bridged by the Romans since this is always practised by the soldiers on the Danube, the Rhine, and the Euphrates just like any other military exercise. The manner of construction, which is not familiar to everyone, is as follows. The boats, by which the river is to be bridged, are flat-bottomed and are anchored a little bit upstream, above the spot where the bridge will be built. When the signal is given, they release one boat to drift in the current near the bank they are holding. When it has floated into a position opposite to the place to be bridged, they throw into the stream a wicker basket filled with stones, attaching a cord so it serves as an anchor. The boat, secured in this way, stays near the bank and a base is paved immediately up to the landing with the boards and materials for the bridge, which the boat carries in abundance. Then they release another boat at a little distance from the first, and another from that one, until they have driven the bridge to the opposite bank.

11.23 A contract to repair bridges in Egypt

The maintenance and repair of existing wooden bridges was a concern for any community along a waterway. Catullus (*Poems* 17) describes a decrepit wooden bridge in Italy, but in Egypt the number of bridges was substantial and the work immense. This fragmentary papyrus refers to work in the Fayum about 245 BC.

P.Petr. 3.43 (2)

In the second year of the rule of Ptolemy, son of Ptolemy, and Arsinoe [other names follow] ... at Crocodilopolis in the Arsinoite nome. A contract was given out from the royal treasury by public auction through Hermaphilus, the oeconomus, in the presence of Theodorus, the engineer (...) [for the following works]: to dismantle the two bridges at Ker (...) and to revet the bank with brushwood against the underlying parts and to make the opening at the top 8 *pecheis* wide and to revet with brush along the (...) for a distance of 35 *schoenia* from the bend; to dismantle the two bridges at (...) and again to revet the bank with brushwood against the underlying parts and to make the opening at the top 14 *pecheis* wide; to dismantle the bridge at Hiera Nesus and again to revet the bank with brushwood against the underlying parts and to make the opening at the top 8 *pecheis* wide, and to revet with brush the 5 *schoenia* inundated by the water.

The papyrus lists more than 10 other bridges with similar instructions and concludes with penalties for failure to complete the contract according to its terms.

11.24 The first bridge at Rome: the Pons Sublicius

According to tradition Ancus Marcius, the fourth king of Rome, constructed the first Roman bridge across the Tiber: the Pons Sublicius. The all-wood bridge was damaged many times in antiquity, but was always repaired without iron or bronze. It is most famous as the bridge defended by Publius Horatius Cocles against the Etruscan army until it could be destroyed.

Dionysius of Halicarnassus, *Roman Antiquities* 3.45.2

Marcius is also said to have constructed the wooden bridge over the Tiber, which custom required be bound without bronze and iron, being held together by the wooden timbers themselves.

Many ancient authors refer to this unusual structure (Plutarch, *Numa* 9; Livy, *History of Rome* 1.33; Pliny, *Natural History* 36.100).

11.25 Inscriptions about bridges

Custom, and perhaps religious restrictions, determined the materials used for the Pons Sublicius. In the first inscription that follows, dating from 421/420 BC, religious function limited the size of this bridge near Eleusis, restricting it to pedestrian use, and may be the reason for using the specified building materials.

*SIG*³ 86

When Prepis, son of Eupheros, was secretary. The boule and people decreed – the tribe Aigeis was in prytany, Prepis was secretary, Patrocles presided – Theaios moved that the Rheitos on the side towards the town be bridged with stones, which had been consecrated to the old temple and which remained unused from the wall, so that the priestesses can carry the sacred statues very safely. They shall make the width five feet so that wagons cannot pass through but pedestrians can walk to the rites. They shall cover over the Rheitos according to the plans of Demomeles, the architect. If they are not.... [The stone is broken here.]

The next inscription commemorates Trajan's completion of the great bridge over the Tagus River near Alacantra in Spain (AD 106), providing an abbreviated but important insight into the concept of municipal cooperation for constructing major structures that served more than one community: 11 *municipia* (towns with inhabitants possessing Latin rights) contributed to this magnificent bridge, which still stands 45 m above the river.

CIL 2.759 and 760 = *ILS* 287 and 287a

To the Emperor Caesar Nerva Trajan Augustus Germanicus Dacicus, Son of the Deified Nerva.... The peoples of the *municipia* of the province of Lusitania who have completed the construction of the bridge once the contributions were collected are the Igaeditani, Lancienses Oppidani, Talori, Interannienses, Colarni, Lancienses Transcudani, Aravi, Meidubigenses, Arabrigenses, Banienses, Paesures.

HIGHWAY SERVICES

One of the principal non-military reasons for the construction of great roadway systems was the post, in reality, a transport system. Unlike today, the system was reserved for royal or imperial messages in antiquity, even though the cities along the route often had to bear the expense of maintenance.

11.26 The Persian pony express

The formulation of the system was attributed to Cyrus (550–530 BC), who established posting stations about 25 km apart on the road from Susa to Sardis (**11.1**). This provided the fastest long-distance communication in antiquity, a feat that was obviously admired by the Greeks.

Xenophon, *Cyropaedia* 8.6.17–18

In regards to the magnitude of his Empire, we have also discovered another device of Cyrus by which he learnt more quickly the state of affairs at any distance. For after examining how long a journey a horse, which was ridden hard, could finish in a day, he established posting stations (*stathmoi*) at just such distances and equipped them with horses and the men to look after them. And at each of the places he stationed the proper men to receive and pass on the dispatches, and to take charge of the exhausted horses and men, and to furnish fresh ones. [Sometimes riders travelled by day and night.] ... This is undeniably the fastest travel by land possible for humans....

The improvement of the postal system is attributed to Darius I (521–486 BC), and several passages reflect the efficiency of the system under his successor Xerxes, including one from the Old Testament (Esther 8.10). More vivid, however, are the messages sent back to Susa by Xerxes (486–465 BC) as he invaded Greece. He used the postal system to relay news to Susa first of his sack of Athens, and then of his own disaster at Salamis.

Herodotus, *Histories* 8.98

There is no mortal man who can accomplish a journey faster than these Persian messengers. The idea was invented by the Persians. For it is reported that as many days as there are for the entire trip, so many are the horses and men posted, a horse and a man for each day's journey. Not snow, not rain, not heat, not night hinder these men from covering the stage assigned to them as quickly as possible. The first rider passes the dispatch to the second, the second to the third, and so on along the line....

The remains of a third-century BC day book from an intermediate postal station in Egypt record the day and hour of the delivery of rolls to the receiver. The Hellenistic system in Egypt was obviously a very closely regulated service (*P.Hib.* 110; translated in the Loeb series, *Select Papyri* 2.397).

11.27 The Roman postal system

Augustus improves on the earlier systems by using better roads and transporting not only the messages, but the messenger for the entire journey in case clarification was needed. The use of vehicles meant that people other than the messenger could be transported by the service, opening the way for abuse of the system.

Suetonius, *Augustus* 49.3

And in order to enable what was happening in each province to be announced and known more quickly and clearly, at first Augustus stationed young men at moderate intervals along the military roads, then vehicles. The latter arrangement seemed more advantageous, since the same men who bring the messages from a place can also be questioned if needed.

11.28 Abuse of the Roman postal system

Use of the postal system was closely guarded, partly since the communities along the route had to finance it. A variety of attempts were made by emperors to curb abuse but the continued use by unauthorised persons demonstrates how poor their control was. The Emperor Claudius seems to have attempted several solutions, all of which failed.

CIL 3.7251 = *ILS* 214

Tiberius Claudius Caesar Augustus Germanicus ... proclaims: Although I have often tried to lighten the burdens of furnishing transport for both colonies and municipalities, not only of Italy but also of the provinces and likewise the cities of each of the provinces, and although I think that I have found enough remedies, nevertheless these have not sufficiently counteracted the wickedness of men.... [The stone breaks off here.]

Pliny reveals his caution when he bends the rules and provides his wife with a ticket to return home for a family emergency.

TRANSPORT AND TRADE

Pliny the Younger, *Letters* 10.120

Up to this time, Lord [Trajan], I have never given a travel pass (*diploma*) to anyone for their convenience, nor have I sent any matter except for your official business. A certain necessity, however, breaks my long-standing practice, since I thought it would be cruel to deny the use of the service to my wife once she learnt of her grandfather's death and wished to rush to her aunt.... I have written this lest I seem insufficiently grateful to you, ... but because of your confidence in me, I have not hesitated to approve the matter as if I had consulted you, since if I had consulted you I would have acted too late. [Trajan accepts the explanation.]

11.29 Requisitioned transport: attempts to control abuse

Requisitioned transport of local vehicles, animals and manpower, which were major resource for imperial officials throughout the Empire, was constantly abused by unsanctioned agents as implied in the following inscription. But it also documents some of the resources commandeered by imperial officials for their duties of office, in this case not speedy mounts or oxen for quarry-work, but hard-working teams of men and other animals used for transportation of materials, here, in a rugged area of Pisidia (Mitchell 1976, Leone 1988).

S. Mitchell, *Journal of Roman Studies* 66 (1976) 107–108

Sextus Sotidius Strabo Libuscidianus, *legatus pro praetore* of Tiberius Caesar Augustus says:

It is the most unjust thing of all for me to tighten up by my edict that matter which the Augusti – one the greatest of gods, the other the greatest of emperors – have guarded against most diligently lest anyone should make use of transport vehicles without payment. But since the insolence of certain people calls for immediate punishment, I have set up in the individual towns and villages a formula for those services which I judge must be provided, that will be heeded, or, if neglected, will be punished not only with my own power but with the majesty of the best emperor from whom I received the mandates ... [about the matter]....

The people of Sagalassus must provide a service of ten wagons and just as many mules for the necessary uses of people passing through, and are to receive, from those who use the service, ten asses per *schoenum* [a local measure of distance] for each wagon and four asses per *schoenum* for each mule, but if they prefer donkeys, they should give two in place of one mule for the same price.... [The option to sub-contract to others in nearby villages is included]. In addition, they will be obliged to provide the means of transport as far as Cormasa and Conana.

The right to use this service, however, will not be for everyone, but for the procurator of the best emperor and his son, granted the use of up to ten wagons, or three mules in place of a single wagon or two donkeys in place of a single mule on the same occasion, for which at the same time they are to pay the price fixed by me ... [senators, equites and those on military service are provided with reduced resources but the same payment rate and the same ratio of exchange: 1 wagon and its mule = 3 mules = 6 donkeys]....

I want nothing to be provided to those who transport grain or anything else of that sort either for their own profit or for their own use, nor is anyone to be provided with a baggage animal for their own [use] or [the use] of their freedmen or of their slaves. [The inscription finishes with billeting arrangements.]

11.30 Roadside services

Long journeys on regular routes, whether for the postal system or for private travel, required facilities for rest and supplies (cf. **11.1**). These services gradually improved as travel increased, but the prudent traveller carried supplies in case of emergency. In the first passage – Service Exits on the Road to Hades – Aristophanes provides some amusing insights into the thoughts of travellers making long journeys. Here Dionysus is on his way to Hades, duplicating the journey made earlier by Hercules, whom he visits. Dionysus questions Hercules regarding the facilities along the route, just as any traveller would question another traveller familiar with a trip.

Aristophanes, *The Frogs* 108–115

But the reason I came, dressed up like you, was so that you might advise me about the hosts with whom you had dealings when you went down to get Cerberus, in case I need them. And tell me too about the havens, the bakeries, the brothels, the resting-places on the road, the branches off the road, the fountains, the roads, the cities, the lodgings, the landladies, where you found the fewest bugs.

Travel in desert areas required consistent and reliable water stops. Strabo describes man-made watering holes along the desert road through Mesopotamia from Syria to Seleucia and Babylon.

Strabo, *Geography* 16.1.27

The road from the crossing of the Euphrates to Scenae is 25 days long. There are camel drivers who maintain rest places that are sometimes well-provided with reservoirs, usually cisterns, sometimes with water that has been brought in.

The wear and tear upon wheeled vehicles during long journeys and campaigns necessitated carrying along tools and materials to repair damaged vehicles, since travellers usually had to rely upon themselves. The Persian Cyrus makes that clear as he prepares for his army's expedition.

Xenophon, *Cyropaedia* 6.2.33–34

We must also have spare wood for both the chariots and the wagons. For by necessity many parts are worn out by continual use. We must also have the most

TRANSPORT AND TRADE

essential tools for all these matters; for craftsmen will not be found everywhere and there are few men who cannot make something work for a day.

VEHICLES AND HORSES

Many varieties of vehicles were developed and used in the Greek and Roman worlds. Some were quite luxurious, others very utilitarian (Crouwel 2012, 1993).

11.31 Names for wheeled vehicles

Etymological definitions in antiquity are notoriously inept, but they often provide us with a better understanding of a term. Such is the case for the names of vehicles defined by Varro.

Varro, *On the Latin Language* 5.140 (31)

A *vehiculum* [wagon], in which beans or something else is transported, is so named because it is *vietur* [woven together] with osiers or because *vehitur* [carrying is done] by it. A shorter wagon is labelled by some, as it were, an *arcera* [covered wagon].... Because the wagon is made from boards like an *arca* [strong box], it is called an *arcera*. A *plaustrum* [heavy cart] is so named from the fact that, unlike those mentioned above, it is totally open, not just partially, and the things carried in it such as stones, wooden beams, and building materials, *perlucent* [shine forth].

Isidore of Seville, *Etymologies* 20.12 provides a similar account of the names of vehicles.

11.32 An early wagon with an upper section

Travel by chariot has already been described (**11.2**) by Homer. Here he describes a vehicle with a superstructure that may be either an awning or, less likely, a cargo box of some sort. Alcinous is speaking to his daughter, Nausicaa.

Homer, *Odyssey* 6.69–70

Go along. The slaves will prepare your stately and well-wheeled wagon, furnished with its upper part (*hyperteria*).

11.33 Luxurious travel by litter

In his attack upon Verres, Cicero discusses the duties of a provincial governor, which include travelling around the province, a very wearying task at the best of times. Verres is accused of travelling only after winter has ended and never on horseback, perhaps a picture closer to the truth for many governors than the hardy Romans would like to admit.

Cicero, *Against Verres* 2.5.11.27

For, as was the custom for the kings of Bithynia, he was borne about in a litter carried by eight men, in which there was a transparent Maltese cushion stuffed with roses. He himself wore one garland on his head, a second around his neck, and put to his nostrils a small bag made from the finest linen with minute mesh holes and full of rose petals. In such a manner, once his journey was completed when he had come to a certain town, he was carried directly up to his bedchamber in the same litter.

11.34 Special frames for heavy transport

Some of the problems and unique features of the Temple of Diana at Ephesus have already been described (**8.43–44**). The difficulty of transporting the immense column drums and architraves for the temple was also worthy of note [see **fig. 8.1**] (Snodgrass 1983, Burford 1960).

Vitruvius, *On Architecture* 10.2.11–12

Moreover, it is not even a digression to describe an ingenious device of Chersiphron. For when he wished to bring down the shafts of the columns from the quarries to the Temple of Diana at Ephesus, he tried the following method since he had no confidence in his two-wheeled wagons; he was afraid that their wheels would bog down on account of the huge size of the loads and the field-like softness of the roads. So he framed and enclosed the columns using four timbers that were 4 *unciae* square, two of them being placed transversely that were as long as the column, and then on the ends of the columns he fastened iron pivots with lead, like dowels, and fixed sockets in the wood to encompass the pivots. Next, he bound the ends with wood slats so that the pivots were enclosed in the sockets and turned freely. Thus, when the yoked oxen pulled, the columns rolled by turning on their pivots and sockets without hindrance. Then after they had transported all the shafts in this manner and it was time to transport the architraves, Metagenes, the son of Chersiphron, transferred the method of moving the shafts to the bringing down of the architraves. For he made wheels about 12 feet in diameter and enclosed the ends of the architraves in the middle of the wheels. In the same way he fixed the pivots and sockets in the ends. Thus, when the 4-*unciae* frames were drawn by the oxen, the pivots that were enclosed in the sockets turned the wheels, while the architraves being enclosed like axles in the wheels reached the building without delay in the same manner as the shafts.

11.35 A "little cart" and a sled used for heavy transport

Inscriptions also provide some detailed information regarding the vehicles used to transport the heavy loads from quarries to construction sites. The "Little Cart" (*hamaxis*) mentioned at the beginning of the text may be a nickname for the large vehicle implied by the rest of the inscription, which refers to construction at Eleusis in 327/6 BC.

Note: missing words of the inscription are indicated by three periods (...); *dr.=drachma/ai*, *ob.=obolos/oi*.

IG 2² 1673.11–43

... For preparing the "Little Cart" (...) for the transport of stone we took four measures of rope from the shipyards and (...) apart from the wood which we took; of these ropes we cut up two into cords for suspension (...) payment to the man cutting and making up the cords for suspension: to Simias from the Cerameicus living at (...).

From (...) the rope-maker, five each of 45 *pecheis*, and two, one of 35 *pecheis* and one of 32 *pecheis*; and (...) each 35 *pecheis*, worth 65 *dr.*; 10 metal hoops from Callicrates, the iron-merchant, worth 5 *dr.* 5 *ob.* (...) six beams to use for bars from Calliphanes, the wood-merchant, worth 12 *dr.*; ten iron rods (...) into the city to Simias, the muleteer, 7 *dr.* 3 *ob.*; rope for the winding-mechanism from Theocles living in the Peiraeus (...) worth 19 *dr.*; of rope, 3 talents for the cords for suspension from Callianaxides living in the Peiraeus, worth 90+ [*dr.*] (...)....

(...) axles broken in the year, 17; for undoing the iron rings and for fastening others of the same size and the (...) paid 2 *dr.* each for as many axles as were undone, and 5 *dr.* each for as many new ones as were fastened, [for all] the axles, paid to Mnesilochus from Coile: the value of all the axles worked upon in the year, 119 *dr.*; nails, which are needed, 68; iron rings, 4; for breaking up the (...) the same man paid for the ships' anchor beams [to be used] for the transport of stone, 150 *dr.*; from Aristonus Cholleidai each *stater* (...)....

(...) rush cords for the "Little Cart", 5; for putting together ropes rent apart (...) for re-twisting a broken cord, to Syrus, the rope-maker, 3 *dr.* 3 *ob.*; two nails for the round beams from Pitthidus (...) for unfastening an axle-block and fastening on a sound one, paid (...) binding on six new rings to the axles from Mnesilochus from Collytus.... [The inscription continues with directions for the transport of column drums from the quarry to Eleusis.]

Although the *Mechanics* written by Hero have survived intact only in Arabic, a few fragments have been preserved in the Greek compilation of Pappus (a Diocletianic writer). One fragment records the use of a sled to transport heavy items.

Hero, *Mechanics* 3.1 (Pappus, *Mathematical Collection* p. 1130)

In the following we shall record, from the third book of Hero, machines constructed for ease and utility, by which also great loads are moved. He states that the things moved along the ground are dragged on tortoises. The tortoise is a framework fastened together from squared timbers that have turned-up ends. The loads are placed upon these and either compound pulleys or the ends of ropes are fastened to the ends of the timbers. These are then pulled either by hand or go to winches, which are turned to slide the tortoise along the ground on the rollers or boards placed underneath it. For if the load is small, one has to use rollers, but if it is larger, boards since they slide easily and the revolving rollers are dangerous once the heavier load builds up speed. Some people use neither rollers nor boards, but after placing solid wheels on the tortoise they haul the loads.

The Theodosian Code records decrees restricting the size of large vehicles (8.5.17) and the carrying capacities of others (8.5.30).

11.36 Pomp and display on an unimaginable scale

Festive occasions with processions of costumed characters and participants, along with their various iconographical accoutrements and different animals, implements and vessels played an important role in the religious year of the Greeks and Romans. Great triumphal processions of victorious Roman general and their armies attesting to the power of Rome and the support of the gods are some of the best known examples of extravagant display, but the account of the magnificent procession of Ptolemy Philadelphus with around 100,000 participants and an overwhelming abundance of incredibly costly regalia and paraphernalia almost defies belief. A few extracts from the long account are included here to stimulate one's thoughts about the technologies employed to make the ostentatious display possible: metalwork, weaving, dyeing, pipe making, mechanical innovation, bracing and support features, and so on (Rice 1983, Fortmeyer 1988, and Coleman 1996).

Athenaeus, *Philosophers at Dinner* 5.197e–202f (excerpt of Callixenus of Rhodes)

[The excerpt begins with the description of a very large and sumptuous tent filled with costly statues, paintings, hangings, plants and flowers].... Of the Dionysiac procession, the Sileni, keeping back the masses, advanced first, some dressed in purple cloaks, some in scarlet ones. Following them were Satyrs, 20 in each division of the stadium, who were bearing lamps interwoven with gold made of ivy-wood. After them the Nikes with golden wings carried six-cubit-high incense burners adorned with branches of ivy-wood interwoven with gold. The Nikes, clothed in chitons embroidered with animal figures, also had a great deal of golden ornament about themselves. And after them a six-cubit-high altar followed, a double altar, densely covered with ivy leaf foilage interwoven with gold, having a golden, vine-leaf crown, girded with with an ivory-white band ... [more participants in expensive gilded and dyed costumes follow as well as a famous priest and the *technitai* [artists] of Dionysus]....

And next in order to them were carried the Delphic tripods, the prizes for the chorus-leaders of the contestants; the one for the boys' contest nine cubits in height, and the other for the leader of the men, 12 cubits in height.

After them was a four-wheeled wagon, 14 cubits long and eight cubits wide, drawn by 180 men; upon it was a 10-cubit-high statue of Dionysus pouring libations of wine from a golden cup, wearing a purple chiton that reached to his feet; and he was clad in a purple, gold-embroidered himation ... [massive gold implements and expensive spices are located around the statue and the wagon is followed by an elaborate Dionysian procession].... After them another four-wheeled, eight-cubit-wide wagon was drawn by 60 men, upon which was a statue of Nysa, eight cubits high in a sitting posture and clothed in a yellow-coloured chiton embroidered with gold and wearing a Laconian himation. And this statue rose up by mechanism, without any one applying a hand, and it poured libations of milk from a golden *phiale* [a special bowl], and then it sat down again....

Next another four-wheeled wagon, 20 cubits long by 16 in width, was drawn by 300 men. On it a wine-press, 24 cubits in length and 15 in breadth and full of grapes, had been constructed; and 60 Satyrs were treading on the grapes, singing to the flute in praise of the wine-press; and Silenus presided over them while the sweet new wine was streaming over the whole road. Next a four-wheeled wagon, 25 cubits long and 14 in width, was borne along; it was drawn by 600 men. And upon it was a wineskin stitched together from leopard skins, holding 3,000 measures [*amphoras*] of wine. And this too streamed forth, being let loose little by little, down the whole road. And Satyrs and Sileni, all 120 being garlanded followed it, some carrying large vessels of wine, others drinking bowls, and others the famous Thericlean wine cups, all made of gold. [much extravagant display continues, including a cave on a wagon with two flowing fountains, and a great variety of exotic animals and other wagons with statues and valuables].

... And on other four-wheeled wagons a golden thyrsus 90 cubits long was carried, and a silver spear 60 cubits long, and on another a golden phallus, 120 cubits long, engraved and wrapped with garlands interwoven with gold, having on the end a golden star, whose circumference was six cubits.

So indeed of all the many cunningly wrought things which we have reported in this procession, we have emphasised only those in which there was gold and silver. But there were many other items well worth report [he continues with exotic animals, wagons bearing stautues of kings and gods, 300 musicians with gold harps, thousands of expensively ornamented bulls, more large statues of gods and rulers, gold and ivory thrones and more gold etc.]....

Also paraded were a golden breastplate, 12 cubits in size, and another silver one of 18 cubits, having on it two, ten-cubit-long, golden thunderbolts, and a garland of oak-leaves set with precious stones; and 20 golden shields, and 64 gold panoplies, and two golden greaves three cubits long, and 12 golden *lekanae*, and an almost endless number of *phialae*, and 30 wine-serving vessels, and 10 large anointing containers, and 12 *hydriae*, and 50 large dishes for barley loaves, and various tables, and five stands for gold plate vessels, and a solid gold horn 30 cubits long. And all these items of gold plate were in addition to those carried in the procession of Dionysus. Then there were 400 wagons of silver plate, and 20 wagons of gold plate, and 800 of aromatic spices and herbs....

11.37 The design of a hodometer on a carriage

Although theoretically a workable device, the careful description of Vitruvius' hodometer has been labelled an armchair creation rather than a description of a functioning mechanism. The sizes implied for both the teeth and the drums suggest a large, impractical instrument.

Vitruvius, *On Architecture* 10.9.1–4

The deliberation of my book now turns to a useful device of the greatest ingenuity passed down by our ancestors, by which we are able to know how many miles we have travelled on a journey, whether sitting in a carriage on the road or sailing by sea. This will be possible as follows. Let the wheels, which are on the carriage, be 4 feet in diameter, so that when a wheel has a mark made upon it and begins to move forward from that mark to make a revolution on the surface of the road, by arriving at that mark from which it began the revolution, the measured distance of 12.5 feet will have been completed.

Once these have been prepared accordingly, let a drum having a single tooth extending outside the face of its circumference be securely attached to the inner side of the hub of the wheel. In addition, above this let a box be firmly fastened to the body of the carriage, a box having a revolving drum set on edge and attached to an axle. On the outside surface of the drum there are 400 teeth equally spaced to mesh with the tooth on the lower drum. Moreover, let another tooth be fixed to the side of the upper drum so that it projects beyond the [other] teeth.

In addition, above, let there be placed a horizontal drum, toothed in the same way and mounted in a second box, with teeth meshing with the tooth, which shall have been fixed on the side of the second drum. And on that drum let holes be made, as many as the number of miles – more or less affects nothing – a carriage can make during a day's journey. And in all these holes let round stones be placed, and in the cover of the drum or the box let there be made one hole having a channel, by which the stones that have been placed in the drum, when they come to that spot, can fall one by one into the body of the carriage and a bronze receptacle placed underneath.

Thus when the wheel, going forward, carries with it the lowest drum, and at every revolution its tooth, by striking the teeth of the upper drum, forces the upper drum to move with the result that when the lower drum will have turned 400 times, the upper drum will be turned once and the tooth, which is fixed to its side, will move forward one tooth of the horizontal drum. Since, therefore, with 400 revolutions of the lower drum the upper will be turned one time, the progress made will be a distance of 5,000 feet, that is, of one mile. Hence, each of the stones that fall will announce by their ringing that one mile has been travelled. The number of stones collected from below, in their total, will indicate the number of miles for the day's journey.

11.38 A practical hodometer and its uses

Writing more than a century after Vitruvius, Hero describes a more feasible hodometer that may have actually been successfully used.

Hero of Alexandria, *Dioptra* 34

I believe that my dioptrical treatise should also explain how to measure the distances on land by the so-called hodometer, since this device records distances for us through the turning of the wheels as one rides in a carriage rather than forcing us to measure laboriously and slowly using a chain or cord. Since our predecessors have explained some ways by which this is accomplished, we can judge the device both by their description and by our own.

First there shall be a casing, like a small box, which will contain all the fittings about to be described. In the base of the little box attach a bronze disc, *ABCD*, with eight pegs attached. An opening shall be cut in the bottom of the little box for these pegs so that a pin – attached to the hub of one of the wheels of the carriage – at every revolution of the hub projects into the hole in the bottom of the little box and pushes forward one of the eight pegs in such a way that the next one takes up the same position as the previous peg; and so on indefinitely. Consequently, when the wheel has made eight turns, the disc with the pegs will have made a single revolution.... [Hero describes the rest of the gearing, adding some cautions.]

In order to avoid opening the little box to examine the teeth of each disc whenever we want to know the length of the road, I shall explain how the length of the road can be determined by using pointers turning round on the surface of the little box. For although the previously mentioned toothed discs are set in such a manner that they themselves do not touch the sides of the little box, their axles do project to the outer part of the walls. These projections should be square so that they can carry pointers by means of square holes. Then when the disc is turned with the axle, the pointer will also be turned....

11.39 Criticism of luxurious vehicles

Extras were not always regarded favourably, especially for owners who were little liked or under attack. The Emperor Commodus was despised by many writers who tried to criticise everything he did. Here, among the luxurious items owned by Commodus that were eventually sold to raise money, the writer lists an array of vehicles with different amenities.

Historia Augusta, Pertinax 8.6–7

And there were carriages (*vehicula*) made with the most recent advances in the art, with entwined and carved wheels and with carefully devised seats that through an opportune turning would at one moment avoid the sun, at the next face the breeze; and others for measuring the road and showing the hours and even others appropriate for his vices.

11.40 Ancient shock absorbers

The magnificent funeral carriage created to transport the body of Alexander the Great from Babylon back to Macedonia was an extraordinary work of art and may have been incorporated into the funerary monument at Alexandria where his body was ultimately laid to rest. In addition to the expensive materials and workmanship, the carriage had an unusual shock-absorbing device, which is poorly understood. Diodorus has just finished describing the decoration. Road-workers and mechanics accompanied the clumsy carriage.

Diodorus of Sicily, *History* 18.27.3–4

The platform of the chariot under the vaulted chamber [which held the body of Alexander in a gold sarcophagus] had two axles, which turned four Persian wheels. The hubs and spokes of the wheels were gilded while the part that ran along the ground was iron.... Halfway along the length of the axles, a pole was ingeniously fitted in the middle of the vaulted chamber, so that by means of this device the chamber was able to remain unshaken by the jolts from the uneven places.

11.41 How to mount a horse

Riders often had the aid of either a groom or a mounting stone (**11.8**) to mount their horse. Xenophon describes the proper method to mount a horse without any aid.

Xenophon, *Art of Horsemanship* 7.1–2

First, then, the rider should hold at the ready the rope of the horse's halter, either attached to the chin-strap or to the ring on the kerb chain, in the left hand, and loose enough that he doesn't jerk the horse whether he plans to mount by holding the mane near the ears or by springing up with the help of his spear. He should hold the reins beside the withers along with the mane in his right hand, so that when mounting he doesn't jerk the mouth of the horse with the bit in any way. When the rider lifts up to mount, he should raise his body with the left hand and, stretching out the right hand, raise himself at the same time. Mounting in this way will not present a clumsy appearance with bent limbs, even from behind. He should not put his knee on the back of the horse, but throw the leg completely over the right side. When he has brought the foot round, then let him set his buttocks down on the horse.

11.42 A first-century BC "horseshoe"

In antiquity the metal horseshoe was not nailed onto the hoof as in modern practice. Instead a leather slipper with a metal sole was slipped on and off as needed. Luxury models using valuable metals are recorded by Suetonius (*Nero* 30) and Pliny (*Natural History* 33.140). Here Catullus uses a simile to express his annoyance with a fellow townsman who has treated his young wife in an inappropriate manner.

Catullus, *Poems* 17.23–26

Now I want to fling him head-first from your bridge, to see if he is able suddenly to wake up his obtuse lethargy and leave his indolent mind in the heavy muck, as a mule leaves her iron slipper in the sticky quagmire.

11.43 Snowshoes, sleds, and crampons

Strabo describes the winter conditions of the peoples living in the southern Caucasus. Although not living on the summits, they were able to pass over the mountains after inventing a number of aids to help them. (**13.8** for early pitons).

Strabo, *Geography* 11.5.6

The summits are impassable in the winter, but the people climb them in the summer, binding on broad shoes made of untanned hide, like drums, and with sharp spikes because of the snow and ice. They come down by sitting on skins with their loads and sliding down, as is the custom in Atropatian Media and on Mount Masius in Armenia, although there they also fasten spiked wooden discs to the soles of their shoes.

11.44 Distance and speed records by land

The distances covered by the postal services (**11.26**) were never bettered in antiquity on a regular basis, but several speedy journeys by wheeled vehicles were worthy of note. Cicero reports that news of the death of his client's father was brought by a freedman to the enemy of his client, from Rome to Ameria. In the second passage, Pliny describes the longest recorded one-day trip by carriage.

Cicero, *On Behalf of Sextus Roscius Amerinus* 7.19

And although the father had been killed after the first hour of the night, the messenger came here to Ameria by first light. During ten nocturnal hours he flew over 56 miles by light two-wheeled vehicles....

Pliny, *Natural History* 7.84

Tiberius Nero traversed the longest journey by carriage that was completed in a day and a night when hurrying to his sick brother Drusus in Germany. It was 182 miles.

11.45 The cost of land transport

The much greater expense of land transport versus transport by water no doubt influenced technological development to some extent; land transport was avoided for goods whenever possible. This example accounts for the expense of land transport of an everyday item; for the expense of transporting luxury goods, see **11.129**.

Cato, *On Agriculture* 22.3

A mill can be purchased in Suessa for 400 *sestertii* and 50 *librae* of oil. The expense for assembly is 60 *sestertii*; for transport using oxen and six men with drivers for six days the cost is 72 *sestertii*.... At Pompeii a completed one is purchased for 384 *sestertii*; transport is 280 *sestertii*.

TRANSPORT AND TRADE

URBAN STREETS AND TRAFFIC

11.46 Smyrna

Strabo describes Smyrna as one of the most beautiful cities of Asia Minor, partly a result of its regular and paved streets, which, nevertheless, do have one defect.

Strabo, *Geography* 14.1.37

The regular division of Smyrna by its streets is remarkable; the roads are as straight as possible and paved with stone.... One mistake, and not a small one, made by the engineers was that when paving the roads they did not supply them with underground drainage. Instead filth covers their surface, especially in a rain storm when the refuse discharges into the streets.

11.47 The inventor of the urban grid plan

The grid plan for streets is attributed to Hippodamus of Miletus, although archaeological evidence proves the system was used well before the traditional dates of his activity (cf. **8.16–17** for city planning).

Aristotle, *Politics* 2.5.1 (1267b)

Hippodamus, the son of Euryphon, a Milesian, is the man who invented the dividing up of cities and cut up [into a grid plan] the Peiraeus.

Aristotle (*Politics* 7.10.4 [1330b]) repeats the statement, adding that straight streets are most convenient in times of peace, but unsuitable for defensive measures.

11.48 Lack of street lighting

Street lighting (**8.73**) was not at all common in antiquity: only the greatest cities of the Empire utilised it. Instead, travel at night even within a city or town could be somewhat hazardous especially for a person in a strange area or someone returning from a night on the town (cf. **2.24** for an early solution).

Petronius, *Satyricon* 79

Neither was there any torch as an aid to show the road to us as we wandered, nor did the silence of the midnight hour offer any hope of meeting up with [someone carrying] a light. In addition, our inebriation and lack of knowledge of the place would have confused us even in daylight. And so, after nearly an hour of dragging our bleeding feet over all the sharp stones and the fragments of jutting, big-bellied pots, finally we were saved by the cleverness of Giton. For the cautious lad, since he feared getting lost even in the clear light of day, had marked all the posts and columns with chalk, the lines of which vanquished the darkest night and by their extraordinary whiteness showed the way for us who were lost.

11.49 A law restricting wheeled traffic in cities

The improvement of roads and the rapid growth of urban areas during the late Republic and Empire created a nightmare in terms of the traffic converging on these focal points. Several traffic laws were instituted in an effort to alleviate some of the difficulties. This passage forms part of a long law promulgated by Julius Caesar to regulate urban matters.

CIL 1.593.56–67 = ILS 6085

On those roads that are or shall be within the city of Rome among those places where habitation shall be continuous, no one, after the first day of next January shall be permitted in the daytime – after sunrise or before the tenth hour of the day – to lead or drive any freight wagon except when it is necessary to bring in or transport material for the sake of building the sacred temples of the immortal gods, or for the sake of building public works, or where, in carrying out a public contract for demolition, it shall be necessary for the good of the public to carry material out of the city and out of those places, and in situations for which specified persons shall be allowed for specified causes to drive or lead freight wagons by this law.

On those days when the Vestal Virgins, the *Rex sacrorum*, and the flamens shall be required to ride in wagons in the city for the sake of the public sacrifices of the Roman people, and when wagons shall be necessary for the sake of a triumph on the day someone will have the triumph, or where wagons shall be required for games publicly celebrated at Rome or within one mile of the city of Rome, or for the procession at the circus games ... for the sake of those causes and on those days nothing in this law is intended to prevent wagons from being led or driven in the daytime in the city.

Nothing in this law is intended to prevent wagons that will have been brought into the city at night, if returning empty or carrying away refuse, from being drawn by oxen or draught animals in the city of Rome or within one mile of the city of Rome after the sun has risen in the first ten hours of the day....

Juvenal (*Satires* 3.232ff.) complains about both the night-time and daytime traffic of Rome.

11.50 Other efforts to curb urban traffic

Suetonius, *Claudius* 25.2

By an edict, Claudius announced that travellers could not pass through the towns of Italy except on foot, in a sedan, or in a litter.

Historia Augusta, Marcus Antoninus 23.8

Marcus forbade anyone to ride a horse or drive a vehicle in Italian cities.

MAPPING

As exploration, conquest, and commerce expanded the limits of the Mediterranean world, more accurate maps and charts were demanded by travellers. The development and improvement of devices like the hodometer (**11.37–38**) resulted in continual improvements upon earlier efforts. (Especially chapters by O.A.W. Dilke in Harley and Woodward, eds., 1987, French 2016, Irby 2016, Talbert 2008).

11.51 The earliest recorded map-maker

No doubt earlier maps existed, but Anaximander (611–546 BC) is credited with the first map of the world.

Diogenes Laertius, *Anaximander* 2.1.2

Anaximander was the first man to draw the circumference of land and sea, and he also made a sphere [globe].

11.52 Problems with maps

Even by the fifth century BC criticism of existing maps was becoming common. Travellers such as Herodotus were beginning to collect much more information and realised how inaccurate contemporary maps were.

Herodotus, *Histories* 4.36

But I have to smile when I see all those people drawing up maps of the earth and not one explaining it sensibly. They draw the River of Ocean streaming all around the earth, itself being round as if made with a compass, and making Asia and Europe of equal size.

Herodotus then gives a lengthy description of the world as he knows it.

11.53 The spherical shape of the earth

Many shapes were proposed for the earth throughout antiquity, an important aspect of map-making, since the conclusions reached had a major impact upon the appearance of maps that were produced. Strabo offers his reasons for believing the earth was a sphere, a belief already accepted by some scholars for centuries.

Strabo, *Geography* 1.1.20

It is only necessary to sum up briefly the indications of whether something comes from the sense of observation or intuitive knowledge, if indeed it does.

Such an example is that the earth is sphere-shaped: the suggestion comes rather indirectly from [the law that] bodies tend towards the centre and each body inclines towards its own centre of gravity, a suggestion that comes straight from the phenomena observed at sea and in the heavens; for both our sense of observation and intuitive knowledge can bear witness to this. For obviously the curvature of the sea interferes with the sailors so that they are unable to see in a straight line the distant lights at an elevation equal to the eye. But if raised higher than the eye, they are visible, even if at a greater distance from the eyes; and likewise, if the eyes themselves are raised to a height, they see what was previously hidden.... And for the sailing towards land, the landfall area becomes continually more visible, and land that seemed low at first continually rises.

Aristotle, *On the Heavens* 2.13.293b-294a provides two theories regarding the shape of the earth: either spherical or flat and shaped like a drum; see Pliny, *Natural History* 2.161–62 for other theories.

11.54 The shape and size of the inhabited world
Strabo, *Geography* 2.5.9

... The total breadth of the inhabited world would be less than 30,000 *stadia* from south to north. But its length is calculated at about 70,000, this is from the setting to the rising sun – from the headlands of Iberia to the headlands of India – measured out by land and sea journeys.... Thus, the length is more than twice the breadth and its shape is described as like that of a *chlamys* [cloak].

11.55 Making an accurate map or globe
Ptolemy, *Geography* 1.20.1–22.6

... Making a map on a sphere gives the likeness of the form of the earth ... but it is neither convenient to make it sufficiently large to make room for the many things that have to be properly placed, nor is it possible to fix one's gaze on the whole model all at once.... A map on a flat surface is entirely free of these problems, but it requires a certain modification to correspond to the spherical form to make the related distances on the unfolded form equivalent to the real ones....

Precautions to be taken for a map drawn on a flat surface

On account of these problems, it would be well to take care that the lines representing the meridians are straight, and to draw those representing parallels as arcs of circles about one and the same centre. For a centre point set at the North Pole, the straight meridian lines will have to be drawn so that, above all, the appearance of the sphere is almost retained in respect to both nature and form, with the meridian lines again meeting the parallels without bending and still converging at that common pole. Since it is not possible to preserve the proportions of the sphere through all the parallels, it is sufficient to take care to maintain the ones through Thule and the equator so that the boundaries around our latitudes may be accurately proportioned. And the parallel through Rhodes, on which the most comparisons of longitudinal distance have been made, can be divided according to its proportion to the meridian – that is, in accordance with a ratio of almost 4:3 of equal parts of the circumference – as Marinus has done, so that the better-known length of the inhabited world would be in proper proportion to its breadth. The intricate manner of doing this I shall make clear after I have demonstrated how a map should be delineated on a sphere.

How the inhabited Earth ought to be delineated on a sphere

The decision of the map-maker regarding the number of items to be set down will determine the size of the sphere; a function also of his ability and desire, since the larger it is, the more the detail represented and the more clear the product. But however big it is to be, once we determine its poles we carefully attach to it through the poles a semicircle that is raised just a bit from the surface so that it does not rub the sphere when it revolves. The semicircle should be narrow so that it does not cover much area. One of its edges should extend exactly between the points marking the poles so that we may draw the meridians with it. After dividing it into 180 parts, we will indicate the numbers, making our start from the middle where it intersects the equator. And likewise, after drawing the equator and dividing one of its semicircles into 180 equal parts, we will furnish them with numbers, making our start from that boundary through which we will draw the westernmost meridian. Now we will make our map on the basis of the tables of degrees of both longitude and latitude according to each one of the places being indicated, and on the basis of the divisions of the semicircles: the equator and the movable meridian. This is done by moving the latter semicircle about the sphere to the degree of longitude indicated, that is, to the division of the equator corresponding to the number, and measuring the distance from the equator according to latitude by means of those divisions on the meridians, and then placing marks corresponding to the indicated number of degrees in the same manner as for a star map on a solid sphere. Similarly, it will be possible to draw meridians at intervals of as many degrees of longitude as we choose, by using the divided edge of the semicircle as a ruler. It will also be possible to draw parallels at intervals as great as we choose, by setting the marker next to the number on the edge of the meridian that indicates the proper distance, and turning it with the semicircle as far as the meridians defining the boundaries of the known world.

11.56 Making globes and charts

Strabo tries to simplify the procedure set out by Ptolemy.

Strabo, *Geography* 2.5.10

In order most closely to imitate the truth by means of man-made models one must make the earth a sphere.... But since a large sphere is required so that the very much smaller section of it, a fraction of the whole, can be adequate to receive clearly the appropriate representations of the inhabited world and to present a proper appearance to onlookers, it is better to build one of sufficient size if that can be done. It should be no less than 10 feet in diameter. But if a globe of so great a size, or not much smaller, cannot be constructed, one should sketch down the drawing on a level surface not less than 7 feet long. For it makes little difference if, in place of circles – the parallels and meridians – we draw straight lines, by which we indicate the zones of the earth (*climata*), the winds and the other distinctions, and the qualities of the parts of the earth in relation both to each other and to the heavens. We can draw horizontal parallel lines for the parallel circles and vertical straight lines for the vertical circles since our intelligence can easily convert the model and magnitude seen by our eye on the level surface to the globular and spherical models.

11.57 Verbal mapping and surveying

Often the most important information from a map was an accurate record of distances between locations, much like the function of many milestones (**11.9**). The famous Peutinger Table, a thirteenth-century copy of a third- or fourth-century AC original, provides such information in the visual form of a strip map 6.82 m long (originally *c*.7.44 m) and 0.34 m wide. Pliny conveys a similar image of part of India through words as does the small sample from the later, and much longer, *Itinerarium Antonini*.

THE ANCIENT EQUIVALENT OF A STRIP MAP

Pliny, *Natural History* 6.61–62

We will follow in the footsteps of Alexander the Great so that the geographical description [of India] is intelligible. Diognetus and Baeton, the surveyors of his expeditions, wrote that from the Caspian Gates to Hecatompylos of the Parthians is as many miles as we have said [133 miles]; from there to Alexandria of the Arii, which city the king [Alexander] founded, 575 miles; to Prophthasia of the Drangae 199; to the town of the Arachosii 565; to Hortospanum [Kabul] 175; and then to the Town of Alexander 50 (in some copies differing numbers are found), a city is stated to be sited right below the Caucasus; from it to the river Capheta [Kabul] and the town Peucolatis of the Indians, 237; from there to the Indus river and the town Taxilla 60; to the famous Hydaspes river [Jhelum] 120; to the not less noble Hypasis [Sutledge] 390, which was the end of Alexander's journey....

Itinerarium Antonini 463.3–466.4

Written itineraries, which were more common in the Roman world than itinerary maps, were very important in the development of geographical and marine maps. The best-preserved examples are the Antonine itinerary and the Bordeaux itinerary, which simply provide lists of places along a route with their intervening distance. A short selection of one route in Britain is provided here.

THE ROAD ROUTES OF ANTONINUS AUGUSTUS
THE ROUTE OF THE BRITAINS

From Gesoriacus of the Gauls [Boulogne, France] to the port of Ritupiae [Richborough] of the Britains: in number 450 stades

[Route 1]
From the boundary, that is, from the Wall [Hadrian's], right to Praetorium
[Bridlington?] mpm^2 156
From Bremenium [High Rochester]
To Corstopitum [Corbridge]   *mpm* 20
To Vindomora [Ebchester]   *mpm* 9
To Vinovium [Binchester]   *mpm* 19
To Cataractonium [Catterick]   *mpm* 22
To Isurium [Aldborough]   *mpm* 24
To Eburacum [York],   *mpm* 17
[Home] Sixth Legion Victrix
To Derventio [Stamford Bridge?]   *mpm* 7
To Delgovicia [Malton?]   *mpm* 13
To Praetorium [Bridlington?]   *mpm* 25
                                                   [total: 156]

ROMAN SURVEYING: IT'S ALL IN THE NUMBERS

Agrimensores, the Roman land surveyors, played a prominent role in the distribution of land throughout the Republic and Empire. The measurements and divisions of the land by these professionals were used by the government, military, local communities, and individuals to allocate spaces appropriate for use, whether colonies, military camps or simple allotments. The modern term "centuriation" is used to describe the results of their surveying which employed a rational geometric approach for land division (cf. **8.34, 13.12**) (Dilke 1971 and Campbell 2000).

Frontinus, *Limites* Campbell pp. 10, 16–23 (T 12.5 = L. 29.1)

First of all, they enclosed an area of land by means of four *limites* [boundaries or paths] most commonly of 100 feet in each direction (the Greeks call this a *plethron*, the Oscans and Umbrians a *vorsus*), while our people of 120 feet in each direction since they wanted each of the four sides to consist of 12 ten-foot measurements, just as there are 12 divisions in the day and 12 months in the year. The area first enclosed by four *actus* they say was called a *fundus*. Two of these *fundi* joined together enclose [the area of] a *iugerum*. Then two of these *iugera* joined together create a square portion of land, since its dimensions in every direction consists of two *actus* in this way. Some say that the first unit [like this] was called *sors* [allotment] and, multiplied 100 times, a *centuria*. [Deviations of size follow.]

11.58 Agrippa's map of the world

The reliability of maps of practical application was very much in the hands of the individuals making the surveys. Pliny notes the great variation of shapes and distances on maps depicting the same region, concluding that such inconsistencies are a natural outcome of personal and historical circumstances existing when the survey was done. Lack of control over such elements meant that mistakes were made even by men like Agrippa, whose work was done with painstaking care. This mapping project undertaken by Agrippa may in fact have been begun under the authority not of Augustus, but of Julius Caesar.

Pliny, *Natural History* 3.17

Since Agrippa was a man of such great attentiveness and care, especially in this endeavour, who could believe that he made an error, particularly since he was planning to exhibit his map of the world to the city? And not only him, but even divine Augustus, since he had completed the portico containing the map. The structure had been begun by his sister according to the design and instructions of Marcus Agrippa.

Frontinus (*On the Aqueducts of Rome* 1.17) describes his use of maps of aqueducts for maintenance at the end of the first century AC. The recovery of fragments of the *Forma Urbis*, a marble plan of Rome, provides invaluable evidence regarding the layout of the city at the end of the second century AC.

11.59 The importance of maps to an army on the move

An army is often most exposed to danger when on the march, especially in hostile and unfamiliar territory. To prepare for or lessen the possibility of attacks, commanders could learn about their route by listening to scouts and local inhabitants, but the best generals had maps made.

Vegetius, *Military Science* 3.6

First off, the commander should have a very detailed description for the routes of the entire area in which the war is being waged, so that he knows the spaces between places not only in terms of the number of feet but even the quality of the roads. The description ought faithfully to record shortcuts, side roads, mountains, and rivers. This has even been carried to the point that the more shrewd generals have insisted on having not only the verbal descriptions but even maps of the routes of the provinces in which necessity engaged them, so that they could choose the road for the march not only by mental deliberation but also by the sense of sight.

For details of some surveying tools and techniques, see **8.34**.

B NAVIGATION

The Mediterranean Sea is made to order for sea travel. There are numerous islands, the climate of the surrounding lands does not vary to any great degree, the tidal range is relatively small, and the sailing seasons and winds are largely predictable. Furthermore, most parts of the region are poor in natural resources such as metals, timber, and fertile farmland, forcing dependence on trade with more fortunate areas. Finally, the topography of most of the circum-Mediterranean lands was too rugged, and the technology and infrastructure of land travel too rudimentary or unreliable to foster the movement of goods and people over long distances. Not surprisingly, there is evidence for travel and trade by sea as early as the seventh millennium BC, and by the Early Bronze Age ships were carrying goods and engaging in military action over the whole Mediterranean. With iron tools and a long tradition of ship construction and navigation to build upon, the Greek and Roman cultures each brought the level of maritime expertise and activity to new peaks. In the Hellenistic and Roman periods in particular, major advances were made in the size of ships, the sophistication of their equipment, and the design of the harbours to accommodate them. Canals, too, facilitated trade by water. In fact, throughout antiquity, ships were the only means of efficient long-distance travel or trade, and they remained the largest and most complex machines constructed by any of those cultures (McGrail 2008).

THE SAILING SEASON AND THE DANGERS OF NAVIGATION

11.60 The prominence of the sea in maps

Both the Greeks and the Romans were well aware of the shape of the earth and the relative position of cities and topographical features throughout the known world. By the late Hellenistic period, sophisticated maps were in use. *The Itinerarium Antonini* (**11.57**) also has a section on sea routes.

Strabo, *Geography* 2.5.17

The sea in particular draws the outlines of the land and gives it its shape, producing bays, deep sea, and straits, also isthmuses, peninsulas, and promontories. Both rivers and mountains, however, assist in this. Through such features we gain a clear idea of continents, favourable locations for cities, and all the other striking details a geographic map is filled with. Among these details is the multitude of islands scattered across the open sea and along the whole coastline.

11.61 Sailing season in the Mediterranean

The Mediterranean Sea for the most part is calm throughout the late spring and summer months, but impressive storms are frequent during the winter. As a result, most ships were laid up from at least early November to late March, but some for even longer. Hesiod, writing in the seventh century BC, reflects the distaste for seafaring typical of a largely agricultural society which put to sea to supplement the produce of poor land. Roman literature also frequently reflects this agricultural prejudice; e.g. see Horace, *Epodes* 10. Vegetius, *Military Science* 4.39, provides a simplified account of sailing seasons.

Hesiod, *Works and Days* 618–634

But if you conceive a desire for stormy seafaring – when the Pleiades flee the mighty strength of Orion and dive into the misty sea [late October–early November], then indeed gales blow from all directions. And from that moment no longer keep ships on the wine-dark sea, but think to work the land, as I advise you. Haul your ship out on dry land, surround it with stones to keep out the force of the winds that bring damp, and remove the drain plug so that the rain of heaven may not rot it. Stow all the tackle and fittings in your house, putting in trim the sail – the wings of a sea-roving ship – and hang the well-shaped rudder oar over the hearth smoke. You yourself wait for the proper sailing season to arrive, then drag your ship to the sea and load a suitable cargo in it, that you might bring home profit – just as your father and mine, Perses, the great fool, used to put to sea in ships because he lacked the means for a comfortable life.

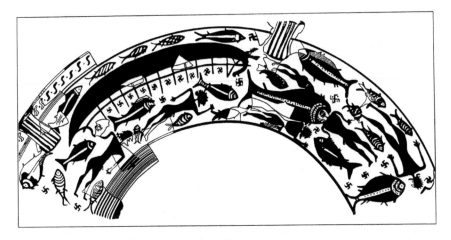

Figure 11.3 Vase painting with drowning sailors.
Source: Krater, eighth century BC, Pithekoussai; adapted from Buchner, 1966.

11.62 Dangers of a winter sea voyage

Andocides, *On the Mysteries* 137–138

What greater danger can befall someone than to make a sea voyage during the winter season. When the gods had my person in such circumstances, my life and my possessions in their power, did they nevertheless save me? Could they not have seen to it that even my corpse was denied burial? Furthermore, it was war time, and triremes and pirates were still at sea; many were taken prisoner by them, lost their goods, and lived the rest of their lives as slaves. And besides there were foreign shores on which many had already been wrecked and had suffered terrible mistreatment and torture and died [fig. 11.3].

11.63 Farmers who seek their fortune in sea trade

Although both Aelian and Philostratus (*Life of Apollonius* 4.32) wrote in the second or third centuries AC, they reflect the same aristocratic suspicion of seafaring that Hesiod voiced long before. The poverty that drove individuals to undergo the perils of the sea is clear, along with a suspicion that an ungentlemanly greed for gain played a part as well.

Aelian, *Letters of Farmers* 18

My neighbour Laches, they say, has given up agriculture and working his farm and sails the Aegean. He measures other seas, too, rides the waves, lives the life of a seagull, and battles unfavourable winds; he plots his course from headland to headland. Keeping an eye out for juicy profits and thinking of striking it rich all at once, he said good-bye to those little goats and his former pastoral life. Unable to live out the shabby, hand-to-mouth existence provided by his fields, and dissatisfied with his state, he dreams of Egyptians and Syrians and examines their bazaars, calculating over and over, by Zeus, the compound interest and counting heaps of money. The profit of a voyage out and back inflames and fires up his imagination, and he does not think of storms, opposing winds, the ever-changing sea, or unseasonable weather. As for us – even if we work hard for little gain, nevertheless the land is much steadier than the sea, and since it is more trustworthy, it offers more certain prospects.

11.64 Piracy and navies in the Bronze Age Mediterranean

Piracy was as old an occupation in the Mediterranean as trade, and many individuals engaged in both during a voyage, depending on circumstances and opportunities. Only central authority was able to suppress piracy by patrolling the seas, a clear priority of the Minoan culture, which depended on sea trade.

Thucydides, *Peloponnesian War* 1.4

Minos is the earliest of those we know from tradition to have acquired a navy. He controlled what is now the Hellenic Sea [Aegean] and ruled the Cycladic islands and was the first coloniser of most of them, driving out the Carians and setting up his sons as governors. Understandably, he cleared piracy from the sea as far as he could, so that his revenues might come to him more easily.

11.65 Roman merchant ships crowd seas and harbours

During the period of Roman domination of the Mediterranean world, an enormous volume of trade moved by sea, from one end of the Mediterranean to the other and along the Atlantic coast of Europe. The familiar rhetorical criticism, however, was still voiced. On the profits of sea trade, see also **11.128**.

Juvenal, *Satires* 14.275–283

Look at our ports and the sea, crowded with great ships! The majority of the human race is now at sea. A fleet will go wherever the hope of profit calls, and it will fly across not just the Carpathian and Gaetulian seas [eastern and southern Mediterranean] but will also leave Calpe [Gibraltar] far behind and hear the setting sun hiss in the Herculean main [Atlantic Ocean]. It is well worthwhile, no doubt, to have seen the sea serpents and mermen of the ocean so that you can return home from there with a tightly packed wallet and boasting of your swollen purse.

SHIP-BUILDING

11.66 The woods used in ship-building

Great care was taken to select wood with the qualities of strength, flexibility, or resistance to rot appropriate to the part of the ship in which it was used. See also **10.14–15**.

Theophrastus, *Enquiry into Plants* 5.7.1–3

Silver fir, fir, and Syrian cedar are in general terms useful for ship-building. They construct triremes and long ships out of silver fir, because of its lightness, round [merchant] ships of fir, because it does not decay – some build triremes of the latter as well, because they are poorly provided with silver fir. In Syria and Phoenicia they build them of Syrian cedar, for they lack even fir; in Cyprus they use Aleppo pine, since the island possesses this and it seems better than their fir. Most parts are made of these woods, but the keel of a trireme is made of oak so that it might stand up to the stress of the ship being hauled out. On merchant ships the keel is made of fir, but they add a false keel of oak to this when they haul it out, or one of beech for smaller ships; the false keel is made entirely of this wood.

Lathe-turned fittings [?] for boats are made from mulberry, manna-ash, elm, or plane, for it must be tough and strong. The worst material is plane wood, for it decays quickly. Some make fittings of this kind for triremes from Aleppo pine, because it is light. The cutwater, to which the false keel is attached, and the forward outrigger frame are made of manna-ash, mulberry, and elm, for these must be strong.

Vegetius, *Military Science* 4.34

Although in building the foundations for houses the quality of the sand and stone is carefully considered, all the materials for building ships must be selected with even more care, because a faulty ship is a greater danger than a faulty house. The *liburna* [a type of warship], therefore, is built of cypress and domestic or wild pine, and especially of silver fir. The use of bronze rather than iron nails for fastening it together is more practical, even though the expense seems somewhat heavier, since bronze lasts longer and is shown to save money. Because of the warmth and moisture, rust quickly consumes iron fastenings, while bronze fastenings keep their integrity even in water.

11.67 Two ways to build a ship

Only a few passages in Greek and Latin literature actually describe the construction of a ship, and these generally provide only scanty details of the tools and procedures involved. In both cultures, shipwrights were humble, anonymous craftsmen who learnt their trade as apprentices and reproduced traditional designs by means of standardised craft procedures and by eye. As a result, no technical handbooks were produced. Many details of construction, however, have survived in literary and historical sources, some of which are given below. The most famous description of ship construction, when Odysseus builds a vessel to continue his voyage home from the island of Ogygia, documents the use of planks joined edge to edge with mortise and tenon joints pegged with treenails [**fig. 11.4**]. This was the typical method of construction in the Mediterranean from the Bronze Age to the late Byzantine period, allowing the assembly of the hull before most of the framing was in place – a sequence implied by Homer. It may have evolved from the Egyptian technique of building ships from large numbers of small, irregular planks connected by numerous mortise and tenon joints, enforced by the absence of large trees in the Nile valley.

Homer, *Odyssey* 5.227–261

As soon as early-born, rosy-fingered Dawn appeared, Odysseus immediately put on a tunic and cloak, and the nymph [Calypso] put on a long white robe.... She gave him a great bronze axe, well fitted to his hands, sharpened on both blades; it had a beautiful olive wood handle, hafted well. She also gave him a well-polished adze. Then she led the way to the farthest part of the island where tall trees stood – alder, poplar, and fir – reaching to the sky, long dry and well-seasoned, which would float for him lightly. But when she had shown him where the tall trees stood, Calypso the radiant goddess went home, and he started cutting planks. The work went quickly for him, and he cut down 20 in all and

TRANSPORT AND TRADE

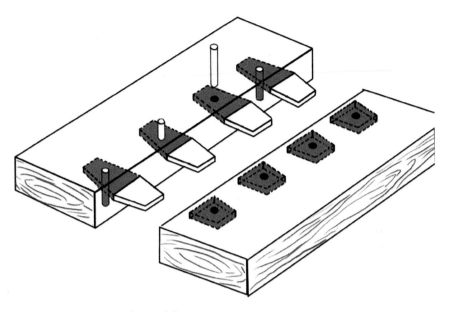

Figure 11.4 Mortise and tenon joint.
Source: Adapted from McGrail and Haad, 2008, fig. 243.

trimmed them with his axe. He smoothed them skilfully and made them true to the line. Meanwhile, Calypso, the radiant goddess, brought him augers, and he shaped the planks to fit one another and bored mortises in them all. Then he hammered the ship together with tenons and dowels. As wide as a man skilled in ship-joinery marks out the lines for the hull of a broad-beamed merchant ship, just so wide did Odysseus make his vessel. He worked on, laying decking planks and fastening them to the close-set frames, then finished the ship with long gunwales. In it he stepped a mast, with a yard fastened to it and made besides a steering oar, to steer it with. Next, he fenced it in completely with a woven wickerwork screen, to serve as a barrier to the waves. Then he piled in much brushwood dunnage. Then Calypso, the radiant goddess, brought him large pieces of cloth to make sails, and he fashioned these well, too. He made fast within the ship braces, brails, and sheets, then he heaved it down to the bright sea on rollers [?].

Herodotus, *Histories* 2.96

The ships in which the Egyptians carry cargoes are constructed from acacia trees, the shape of which is very similar to that of the lotus of Cyrene, but the sap is gum. Cutting planks about two cubits long from these acacia trees, they lay them like courses of bricks [i.e. with alternating joints] and construct the ship in the following fashion. They insert numerous, long, close-set tenons into all the edges of the two-cubit planks [and fasten them together], then they complete construction by stretching thwarts across at the top. They make no use of frames. On the inside they caulk the joints with papyrus. They fashion one steering oar, and mount it in a hole in the sternpost [?]. They use masts of acacia and sails made of papyrus. These boats cannot sail up the river except with a stiff, steady breeze, so they are towed from the bank. For a downstream run it is rigged as follows. A small raft is made of tamarisk wood fastened together with reed matting, and a stone of about two talents weight with a hole through it. The raft is fastened to the ship with a rope and let go to be carried on ahead; the stone is made fast to the stern with another rope. The raft, then, floats along quickly carried by the current, and tows the *baris* (for this is what they call the ship), while the stone is dragged behind and, bumping along the bottom, keeps the boat on a straight course. They have many ships of this type, and some of them carry many thousand talents' weight.

11.68 An account of ship-building

This papyrus, dating to the second half of the third century AC, records work on a substantial boat in an Egyptian boatyard. Two sawyers cut wood for between 3 and 7 shipwrights. They first provide planks of *persea* wood for the hull, then frames of acacia to be inserted in the finished hull – first on one side, then on the other, since scaffolding around the hull is dismantled first on one side. This is the appropriate sequence of work for a boat with a hull of edge-joined planks.

P.Flor. 1.69

(Phaophi) 17: to 6 shipwrights for work on the aforementioned boat, at 7 *drachmai* each, 42 *drachmai*

to 2 sawyers for cutting *persea* wood, at 8 *dr.* each, 16 *dr.*

18: to 5 shipwrights for work on the aforementioned boat, at 7 *dr.* each, 35 *dr.*

to 2 sawyers for cutting *persea* wood, at 8 *dr.* each, 16 *dr.*

19: to 4 shipwrights for work on the aforementioned boat, at 7 *dr.* each, 28 *dr.*

to 2 sawyers for cutting *persea* wood, at 8 *dr.* each, 16 *dr.*

21: to 5 shipwrights for work on the aforementioned boat, at 7 *dr.* each, 35 *dr.*

22: to 4 shipwrights for work on the aforementioned boat, at 7 *dr.* each, 28 *dr.*

23: to 6 shipwrights for work on the aforementioned boat, at 7 *dr.* each, 42 *dr.*

24: to 4 shipwrights for work on the aforementioned boat, at 7 *dr.* each, 28 *dr.*

27: to 7 shipwrights for work on the aforementioned boat, at 7 *dr.* each, 49 *dr.*

Hathyr 1: to 4 shipwrights for work on the aforementioned boat, at 7 *dr.* each, 28 *dr.*
to 2 sawyers for cutting *persea* wood, at 8 *dr.* each, 16 *dr.*
2: to 5 shipwrights for work on the aforementioned boat, at 7 *dr.* each, 35 *dr.*
to 2 sawyers for cutting frames of acacia, at 8 *dr.* each, 16 *dr.*
3: to 4 shipwrights for work on the aforementioned boat, at 7 *dr.* each, 28 *dr.*
to 2 sawyers for cutting frames of acacia, at 8 *dr.* each, 16 *dr.*
4: to 4 shipwrights for work on the aforementioned boat, at 7 *dr.* each, 28 *dr.*
to 2 sawyers for cutting frames of acacia, at 8 *dr.* each, 16 *dr.*
5: to 3 shipwrights for work on the aforementioned boat, at 7 *dr.* each, 21 *dr.*
7: to 4 shipwrights for dismantling the planks of the scaffolding on one side of the aforementioned boat, at 7 *dr.* each, 28 *dr.*
to 2 sawyers for cutting frames of acacia, at 8 *dr.* each, 16 *dr.*
8: to 4 shipwrights for dismantling the planks of the scaffolding on the other side of the aforementioned boat, at 7 *dr.* each, 28 *dr.*

11.69 The Celtic tradition of ship-building

At least since the Bronze Age, shipwrights in the Celtic regions of northern Europe had been constructing ships in a manner very different from those in the Mediterranean. Perhaps because of the heavier seas, great tidal fluctuation, and the enormous regional resources of oak, they built their ships of thick oak planks nailed or tied to heavy frames, with flat bottoms and often no keel. Since the planks of the hull were not edge-joined with mortises, caulking was essential to make the hulls watertight. The importance of the frames to shipwrights working in this tradition may have affected the development of frame-first construction in the Mediterranean in the late Byzantine period. Caesar describes the ships of the local Celtic tribes on the North Sea.

Caesar, *Gallic War* 3.13

The ships of the Veneti were all constructed and fitted out in the same manner. The hulls were somewhat more flat-bottomed than those of our ships so that they might more easily approach shoal water and tidal flats. Their prows, however, were quite high, and their sterns too, well-suited to the magnitude of the waves and storms. The ships were made completely of oak, very resistant to any forceful blow or rough treatment. The thwarts were made of beams one foot high, fixed in place with iron bolts the thickness of a thumb; the anchors were worked with iron chains instead of ropes. They use hides or thin tanned leather for their sails, either because they do not have linen and are ignorant of its use, or – and this seems more likely – they believe that such sails could not hold up against the ocean's great storms and high-gusting winds, and that they could not properly direct such heavy ships. In an encounter with these ships the only advantage ours had was speed, and the use of oars; in all other respects their ships were better suited to the character of the region and the force of its storms.

Strabo, *Geography* 4.4.1

Because of the great tidal variation, they make their ships flat-bottomed and beamy, with high sterns and prows, and they use oak, of which they have an abundance. For this reason, they do not match up [mortise and tenon] joints in the planks but leave open seams which they caulk with tree-moss...³

11.70 Rafts and other atypical water craft

Although we hear the most about large boats built with planked and framed hulls, many other types of craft were found on fresh and salt water in the Graeco-Roman world: rafts, coracles, and even devices similar to modern air-mattresses. These could be makeshift, emergency craft, or vessels remarkably well adapted to the materials and topography of the country. The skin boats Herodotus and Caesar describe are still in use on the Euphrates and in Wales and Ireland. Special boats were also used by Greek and Roman military forces to construct pontoon bridges (see **11.18**).

Xenophon, *Anabasis* 1.5.10

In the course of these desert marches there was a large and prosperous city called Charmande, on the far bank of the Euphrates river, where the men purchased supplies. They crossed the river on floats in the following manner. Taking the hides, they had for tent covers, they heaped up dry grass on them, then drew the edges together and sewed them up to keep the water from reaching the grass. They crossed over on these and obtained provisions....

Livy, *History of Rome* 21.27.5

There, having felled some timber, they [Hannibal's army] quickly built rafts on which to carry across the horses, men, and other baggage. The Spanish, without hesitating, stuffed their clothing into skin bags, put their shields above, and lying on top of them swam across the river on their own.

Herodotus, *Histories* 1.194

I will now describe what seems to me to be the greatest marvel of all in that region, beside the city [Babylon] itself. Their boats which voyage down the river to Babylon are circular and made completely of hides. They build these in Armenia, upstream from Assyria, cutting frames of willow and stretching hides over them as an exterior covering, like the hull of a ship. They neither taper the forward end to a prow nor round off the back end to a stern, but make them round like a shield. Having filled the boat with reeds [as dunnage] they load cargo and let the boat be borne off down the river. The usual cargo they carry along consists of palm wood casks filled with wine. Two men stand up and steer the boat, each with one paddle – one man drawing his paddle towards the boat, the other pushing it away. Some of these boats are built very large, others small; the largest of them can carry a cargo of even 5,000 talents weight [approx. 132 to 186 metric tonnes!]. There's a live ass in each boat, more than one in the larger boats. When they have floated down the river to their destination, Babylon, and disposed of the cargo, they sell off the boat's framework and all the reeds. Then they load the hides on the asses and make the land journey back to Armenia, for – on account of the swift current – it is not possible to sail back up the river by any means. This is the reason they make their boats of hides rather than wood. Once they have driven their asses all the way back to Armenia, they build more boats in the same way.

Caesar, *Civil War* 1.54

Since matters had come to this difficult pass, and all the roads were blocked by the troops and cavalry of Afranius and they could not build bridges, Caesar ordered his soldiers to make boats of a certain type he had become familiar with in Britain some years before. The keel and primary frames were made from light strips of wood; the rest of the hull was woven of flexible branches and covered with hides.

11.71 Some finishing touches

Greek and Roman ships generally did not require caulking, since the mortise and tenon joints held the strakes tightly together as the joints were sealed by swelling of the wood. The exterior of a hull, however, could be pitched to slow the action of marine borers and retard fouling and rot. This process brought the risk of fire, as recorded by Bianor. In the Hellenistic and Roman period painted decoration was sometimes applied to the superstructure with wax-based pigments. Some ships, especially long, lightly built warships, would also be reinforced with heavy cables that ran from stem to stern either inside or outside the hull and could be tightened to take stress off the hull and avoid sagging. Launching a large cargo ship could be a demanding task (see also **11.77–78, 11.104**), but it has been calculated that the crew of a trireme would have been more than sufficient to haul the ship in or out of the water in the manner described by Apollonius.

Bianor, *Greek Anthology* 11.248

It was not the depths of the sea that took this ship (how could they, for she never set sail), nor the south wind, but she perished before encountering south wind and sea. For she was already pegged together all the way up to the thwarts and they were anointing her with fat pine sap, when the pitch, boiling over from the heat of the fire, showed that she, built to be faithful at sea, was less faithful on land.

Pliny, *Natural History* 35.49, 149

Wax is infused with these same colours [purple, indigo, blue, yellow, green, and ceruse] for those paintings that are applied with heat [encaustic], a type inappropriate for walls, but common on warships – now, indeed, also for cargo ships, since we even decorate means of transport with paintings.

Long ago there were two methods of encaustic painting, with wax [on wood and with wax] on ivory, using a *cestrum* [a small pointed rod], until the practice of painting warships began. This added a third method, or using a brush with wax melted by a fire; this type of painting on ships is not spoiled by sun, salt, or wind.

Apollonius of Rhodes, *Argonautica* 1.367–90

First of all, at Argus' command, they girded the ship strongly on the interior with a well-twisted rope, drawing it tight at either end so that the planks might be gripped firmly by the tenons and hold out against the opposing force of the swells. They quickly dug a trench around her as wide as the space the ship covered and from the prow as far into the sea as the ship would run when drawn by their hands. They dug a downward slope in front of the stem and set smooth rollers in place in the trench. They tipped the boat down on the first rollers so that she might slip and be carried along by them. Above, they reversed their oars on either side and bound them, projecting a cubit's length, to the thole pins. The heroes stood on both sides in a line and pushed with hands and chest together, and ... bending to it with all their strength they moved the ship from its stocks and, straining with their feet, forced her onward, and Peleian *Argo* followed swiftly. They cried out from either side as they rushed along. The rollers groaned with the friction beneath the strong keel, and around them thick smoke arose because of the weight, but she slipped down into the sea.

EQUIPMENT, WORKING, AND DESCRIPTIONS OF SHIPS

11.72 State-owned shipyards

In both the Greek and the Roman cultures, the state controlled the production of warships, and as a result more information survives about their design and construction than does for merchant ships. The inscription records regulations for the military harbour at Piraeus around 430 BC. The number of men needed to launch or beach triremes in the Piraeus shipyards was much less than a full crew because winches and greased slipways were used (for ships see Morrison 2000, 1980, Morrison and Coates 1995, McGrail 1981).

Aristotle, *Constitution of Athens* 46.1

The Council also inspects triremes after they have been built, and their gear and ship sheds, and arranges for construction of new triremes or tetraremes, whichever the people vote for, the gear for these, and the ship sheds. But the people select the naval architects by vote.... When it arranges the construction of triremes, the Council selects ten men from itself as overseers of trireme-building.

IG I² 73

It is forbidden for anyone to draw up [a trireme] with fewer than 40 men or to pull one down to the sea with fewer than 20 men, or to apply pitch or work the bracing ropes with fewer than 50 men, or to carry out the anointing [smearing?] with fewer than 100 men, or to take off any piece of the tackle, nor may the trierarch or the captain give orders for the start.

11.73 The first Roman warships

When the Romans found they had to fight large Carthaginian warships during the First Punic War (264–241 BC), they provided themselves with similar ships and trained crews quickly. Although the events related by Polybius seem unrealistic in detail, it is quite possible that the Romans borrowed Carthaginian shipwrights or copied advances in design. Because ships can travel long distances and are conspicuous, innovations in nautical design tend to spread quickly. For another early Roman fleet, see **13.24**.

Polybius, *Histories* 1.20.9–21.3

Seeing that the war was dragging on, they then decided for the first time to build ships – 100 quinqueremes [ships with two banks of oars, with three men on each upper oar and two on each lower one] and 20 triremes [ships with three banks of oars]. The matter caused them much difficulty, since their shipwrights were completely at a loss concerning the construction of quinqueremes – no one in Italy ever having used such ships up to that time. From this affair one might clearly perceive the spirit and daring of the Roman temperament.... When they first undertook to send their troops over to Messene, not only did they not have any decked ships, but no warships whatsoever, or even a single *lembos* [type of ship used for trading or war]. Nevertheless, borrowing pentekonters and triremes from Tarentum, Locri, Elea, and Naples, they recklessly carried their men across on these. On this occasion, when the Carthaginians put out to attack them in the straits, one of their decked ships drew ahead of the others out of eagerness, ran aground, and fell into the hands of the Romans. Using this ship as a model, they then set out to build the whole fleet after its design. Hence it is obvious that, if this event had not occurred, their lack of practical knowledge would have prevented them from carrying out this project to its conclusion.

While those to whom the construction of the ships had been entrusted were occupied with their preparation, those who had assembled the crews were teaching them in the following way on land how to row. They sat the men on rower's benches set up on dry land in the same order as the benches on the ships themselves, stationed the coxswain in their midst, and had them practice falling back all together while bringing their hands back, then to move forward again while pushing out their hands, and to initiate and complete these movements at the coxswain's commands. Once the crews had been trained, they launched the ships immediately upon completion and, after practising real rowing for a short time at sea, they sailed down along the coast of Italy as the commander ordered.

11.74 Tillers and masts as levers

It was a common misconception in antiquity (at least among scientific or technical writers) that high sails provided more power because of the longer lever arm between mast and boat; in fact, the speed of the wind itself increases the farther it gets from the retarding friction with sea or land. Although neither the Greeks nor the Romans seem to have used topmasts, large cargo ships could have topsails rigged on the portion of the mast projecting above the yard. Unlike the sternpost rudder invented in the late medieval period, the steering oar was a balanced rudder that could be moved easily around its long axis by means of a light tiller. Single or double steering oars are very effective at steering, but are more vulnerable to damage at sea than the sternpost rudder.

Vitruvius, *On Architecture* 10.3.5

In just the same way the steersman of a very large merchant ship, holding the tiller of the steering oar, which the Greeks call *oiax*, with one hand and turning it skilfully around the centre and the tight fulcrum joint with a push, steers it, laden as it is with an enormous quantity and weight of cargo and its own mass of wood. And when the sails of this ship are rigged halfway down the mast, it cannot make good speed. But when the yard has been hauled up to the very top of the mast, the ship then makes way with greater power, because the sails take the wind not in proximity to the mast step, which is a fulcrum, but at the mast head, a greater distance away.

Likewise, when oars are tied to their tholes with loops and pushed forward and pulled back by the rowers' hands, the oar blades moving through the seawater far from the fulcrum drive the ship forward with mighty force, breasting the foam in its course as the bow cuts the yielding waters.

11.75 The rigging of small and large ships

In antiquity, as today, sailing ships varied significantly in size and design. The small, only partly decked ship Homer describes had a light-weight mast that could be easily stepped and un-stepped, along with equally simple standing and running rigging. The enormous sailing ships of the Hellenistic period and the Roman Empire, however, were much more complicated to work and maintain. The largest were used to carry bulk foodstuffs, such as grain, wine, and oil over long distances around the Mediterranean. Sails were most often made of linen, sometimes of leather, and in Egypt papyrus could be used.

Homer, *Odyssey* 2.414–428

So, they brought everything and stowed it in the well-benched ship just as the dear son of Odysseus [Telemachus] ordered. Telemachus then boarded the ship, but Athena preceded him and sat down in the stern, and Telemachus sat down close by her. The men cast off the stern cables, then boarded the boat themselves and took their seats on the rowing benches. Flashing-eyed Athena sent them a favourable wind, a strong-blowing west wind singing over the wine-dark sea. Telemachus called out orders to his men to work the rigging, and they listened to his call. They raised the mast of fir, set it in the hollow mast step, and secured it with forestays; they hauled up the white sail with thongs of well-twisted cowhide. Wind filled the belly of the sail and the dark sea sang loudly around the stem of the ship as she went.

Pliny, *Natural History* 19.5

How bold and full of wickedness is life that a plant [flax] is cultivated in order to catch winds and storms, and that it is not enough to be carried by waves alone. Indeed, now it is not even sufficient that sails are larger than ships, but, even though single trees are barely large enough to serve as yards for the sails, nevertheless other sails are added above the yards, and besides these, still others are rigged at the bows and others at the sterns and death is challenged in so many ways.

Seneca, *Letters* 77.1–3

Suddenly today the Alexandrian ships came into our view, those which are customarily sent ahead to announce the arrival of the following fleet; they are called "mail ships". The sight is a welcome one to the people of Campania; the whole population of Puteoli stands on the breakwater and identifies the Alexandrian ships by the special spread of their sails, even in a great crowd of ships. They alone are allowed to keep the masthead sail rigged, which all ships use on the high seas. For nothing helps a ship on its way as much as the upper part of the sail – this is where a ship gets most of its speed. In consequence, whenever the wind has risen and become stronger than is expedient, the yard is lowered, for the wind has less force lower down. When they pass between Capri and the promontory of Sorrento, ... other ships are ordered to make do with the mainsail alone, and the masthead sail of the Alexandrian boats stands out.

11.76 The *Isis*, a great ship from the grain fleet

Lucian provides a thorough, detailed description of one of the ships that carried Egyptian grain from Alexandria to Rome in the second century AC. In this dialogue, the dramatic context is the unexpected arrival of such a large ship in the Piraeus, the port of Athens, by now of little economic importance. The ship had set a course north from Alexandria to sail between Cyprus and the Turkish coast, planning to catch a favouring wind for a course along the south coast of Crete and on to the Straits of Messene and Rome. Prevailing winds and currents often enforced this indirect departure from Egypt and Palestine (cf. **11.91**). A storm forced the ship off course into the Aegean, and she put into Piraeus for shelter.

Lucian, *The Ship* 1.4–6

TIMOLAUS: "What was I supposed to do, then, Lycinus, being at leisure and hearing that such a very large boat, well beyond the norm, had sailed into the Piraeus – one of the ships carrying grain from Egypt to Italy? I believe that you two, you and Samippus, have come down from the city for no other reason than to see the ship".

LYCINUS: "… We stood by the mast for a long time, looking up and counting the rows of hide [on the sail], and marvelling at the sailor going up among the brailing lines and then running quite safely along the yard while holding on to the lifts".

SAMIPPUS: "What a big ship! 120 cubits long, the ship's carpenter said, its beam somewhat more than a quarter of that, and 29 cubits from deck to bottom – where it is deepest along the bilge. Then besides, what a mast, and what a yard it supports! What a forestay keeps it stepped! With what a graceful curve the sternpost rises up and over to a pendant golden goose head! Correspondingly, at the other end, the prow juts out far in front, framed by images of the goddess Isis, after whom the ship is named. And all the rest of the decoration: the paintings, the pennant at the masthead blazing like fire, and even more the anchors and capstans and windlasses and cabins on the poop – they all seemed marvels to me. You might compare the complement of sailors to an army. She was said to carry sufficient grain to feed everyone in Attica for a year. And all this a little old man – a real shrimp! – has just saved [from a storm] by turning the immense steering oars with a little steering bar. He was pointed out to me, a guy with receding curly hair. I think Heron was his name".

11.77 The *Syracusia*, a luxurious grain transport

Athenaeus quotes several long passages from Hellenistic writers that record in great detail the design and decoration of some extraordinary ships (see also **11.78–79**). Although some features of these ships must have been unique, many of the details of design and decoration could be found in a more practical application on smaller contemporary ships. Some of the procedures, however, are recorded out of sequence. The launching of the boat is also recorded by Plutarch, *Marcellus* 14.8, who suggests the use of compound pulleys rather than the screw-gear windlass recorded here.

Athenaeus, *Philosophers at Dinner* 5.206e–209b

Hieron II, King of Syracuse [reigned 269–215 BC], was an ambitious shipbuilder, constructing grain-transports. As raw material he prepared timber from Mt. Aetna, a quantity sufficient for the construction of 60 triremes. In keeping with this he prepared treenails, ground futtocks, frames, and the rest of the necessary materials – some from Italy, other from Sicily; *esparto* grass from Spain for the cables; hemp and pitch from the Rhone valley, and all the other required materials from a wide variety of places. He assembled shipwrights and other craftsmen, and from them selected the Corinthian Archias as the master builder. Hieron urged him to undertake the construction with a will, ... and the ship was half completed in six months [there is a gap in the text at this point]. As each portion was completed, it was overlaid with sheets of lead, for there were 300 craftsmen working the materials, not including their assistants.

Hieron ordered this part of the ship [the hull?] to be dragged down into the sea, to receive the rest of its fittings there. After a great deal of discussion about how to launch it, Archimedes the mechanician alone was able to drag it down, with a few persons to help, He was able to launch such an enormous ship by means of a device involving a screw, for he was the first to invent devices employing the force of the screw. The remaining half of the ship was completed in another six months, and everything was fastened down with bronze rivets, most of which weighed 10 *minai*, while the rest were half as big again. These were fitted into holes drilled by augers and fastened the frames to [the strakes?]. The hull was made watertight by means of lead sheeting laid on over pitched strips of canvas. Once he had completed the outer surface, he went on to complete the interior arrangements.

The ship was designed along the lines of a merchant ship [or possibly "designed for 20 rowers in each bank of oars"], but had three gangways. The lowest, used for handling cargo, could be reached by a number of companionways. The second was designed to provide access to those wishing to reach their cabins. The last and highest was for soldiers. Along the middle gangway, along both sides of the ship, were 30 cabins for the men, each big enough for four couches. The owner's cabin was big enough for 15 couches and had three chambers of three-couch size; the aft kitchen served these. All had their floors laid with mosaic in many different kinds of stones, in which was recounted marvellously the whole story of the *Iliad*. The fittings, ceilings, and doors were all carefully worked. Along the uppermost gangway there was an exercise area and promenades laid out in accordance with the scale of the ship. Along the promenades were gardens of all sorts, wonderfully luxuriant with plants, which were watered by hidden lead pipes. There were also arbours of white ivy and vines planted in large pots filled with earth and irrigated in the same manner as the gardens. These arbours shaded the promenades. Next to these was built a shrine to Aphrodite, of three-couch size, with a floor made of agate and all the other most sought-after stones of Sicily. The walls and ceiling were panelled with cypress, the doors made of ivory and cedar. It was splendidly furnished with paintings, statues, and a service of drinking vessels.

Adjacent was a reading room, of five-couch size, the walls and doors made of boxwood. There was a library there, and in the ceiling there was a hemispherical dome imitating the sundial of Achradina. There was also a bath of three-couch size with three bronze tubs and a wash-basin of variegated Tauromenian stone holding 50 gallons.

There were also other accommodations for passengers and for those keeping watch over the bilge pumps. In addition, there were 10 stables along each side, and nearby were kept their fodder and the belongings of the riders and their slaves. There was a covered water tank at the bow which held 20,000 gallons, made of planks and pitched canvas. Next to it was a fish tank made of planks and lead sheeting; it was filled with seawater, and many fish were kept alive in it. Beams projected from both sides of the boat at equal intervals; on them were mounted wood bins, bread ovens and roasting ovens, grain mills, and other services. A row of supports shaped like giants, nine feet high, ran around the outside of the ship, all spaced at equal intervals to support the superstructure and triglyphs. Furthermore, the whole ship was decorated with appropriate paintings.

There were eight towers equal in height to the ship's deck structures, two at the stern, another two at the prow, and the rest amidships. Each of these was fitted out with two booms terminating in small compartments from which stones could be dropped on any of the enemy that sailed beneath. Four fully-armed young marines were stationed on each of the towers, along with two archers; the whole interior of the towers was full of stones and missiles. A raised platform with parapet and battlements was built across the ship on supports. On it was mounted a catapult capable of throwing a stone weighing 180 pounds or a spear 18 feet long. Archimedes designed the device. It could throw either type of projectile 600 feet. Nearby were protective screens made of thick leather straps and hung from bronze chains. Two stone-carrying booms were hung from each of the three masts, and grappling hooks or masses of lead could be let go on attackers. An iron palisade encircled the ship as a protection against boarders, and all around it were hooked iron projectiles which, fired from catapults, could fasten on to enemy boats and drag them into striking range. Sixty young, fully-armed marines were stationed on either side of the ship and an equal number around the masts with the stone-dropping booms. There were also men in the bronze crow's-nest structure at the masthead: three on the mainmast, then two and one respectively on the other two masts. Slaves kept these men supplied with stones and missiles by hauling them up to the fortified stations in baskets on lines rigged over pulleys.

There were four wooden and eight iron anchors. As for the masts, the timbers for the fore and mizzen were found easily, but that for the mainmast was found with difficulty by a swineherd in the mountains of Bruttium. The engineer Philias of Tauromenium brought it down to the sea. Although the bilge was very deep, it was pumped dry by one man using a water-screw, an invention of Archimedes.

The ship was named *Syracusia*, but when Hieron sent her off, he renamed her *Alexandria*. As ship's boats it had, first, a merchantman-like vessel of 3,000 talents burden [78 tonnes], with a full set of oars, then some fishing-type vessels of 1,500 talents burden, and some small boats besides. The complement of the crew was not less than ... [the number has been lost]. Besides those mentioned above there were 600 at the bow ready for orders.

Sixty thousand measures of grain were loaded on board, 10,000 jars of Sicilian pickled fish, 20,000 talents of wool, and 20,000 talents of other cargo;

besides this there were the provisions for the crew. When Hieron heard that, of all the harbours the ship [was to call at], some could not accommodate it at all and others only with great risk, he decided to send it to Alexandria as a present for King Ptolemy, for there was a shortage of grain in Egypt. He did so, the ship was sailed down to Alexandria, and there it was hauled out on land.

11.78 The largest warship in the ancient world

Another of the passages in the section of Athenaeus dealing with extraordinary ships concerns a very large warship – one that had 40 men to each bank of oars, as opposed to three in the Classical trireme, and six or seven for the standard ship of the line in the Hellenistic period. Several details of this description indicate that the enormous ship was a catamaran, probably with oars projecting both sides of each hull, and 20 men to each bank of oars. In fact, Philopator simply linked two "Twenties", each in itself one of the biggest ships of the line, with a great platform for marines and siege engines. As with the *Syracusia* (**11.77**), the launching of this boat too was a major accomplishment in itself, both because of the weight and the potential for damage. In this case, some portion of the boat was finished on land – perhaps the hull – launched, and floated into a dry-dock for completion (see below, **11.104**). Athenaeus quotes this passage from Callixenus, *On Alexandria*. The passage from Plutarch raised the suspicion that this ship would have been of little practical use but see Murray (2012) for a revisionary understanding.

Athenaeus, *Philosophers at Dinner* 5.203e–204c

[King Ptolemy IV] Philopator [reigned 221–205 BC] built his "Forty" with a length of 280 cubits, a beam of 38 cubits from gangway to gangway, and a height of 48 cubits to the prow ornament. From the stern ornament to the waterline measured 53 cubits. It had four steering oars 30 cubits long, and the oars at the top of each bank, the longest, were 38 cubits long. These, because they had lead in their handles and were quite heavy inboard, were well balanced and very easy to handle in use. The ship was double-prowed and double-sterned and had seven rams. One of these was the principal ram, the rest subordinate, with some on the outriggers. It carried 12 under-girding cables, each one 600 cubits long. It was extremely well proportioned, and the rest of the fittings were remarkable. There was a sculptured figure at bow and stern no less than 12 cubits tall, and every surface was elaborately decorated with encaustic painting.... For the trial voyage it took more than 4,000 men to man the oars and 400 other crewmen, and 2,850 marines on the deck....

It was launched at the beginning from a kind of cradle, which they say was built with the timbers of 50 five-banked ships. The cradle was pulled to the water by a crowd of men, with shouting and the sounding of trumpets.

Plutarch, *Demetrius* 43.4–5

These were the first 15 or 16-banked ships ever seen by mortal eyes [built by Demetrius Poliorcetes of Macedonia, 294 BC]. Later on, of course, Ptolemy Philopator constructed a 40-banked ship, ... but she served merely for show. He intended her for display rather than use, since she was not much different from an immovable building and was put in motion only with danger and difficulty. The beauty of Demetrius' ships, however, did not adversely affect their fighting qualities, ... and their speed and effectiveness were even more notable than their dimensions.

11.79 A palatial riverboat

The passage describing the construction and launching of the "Forty" continues with a description of an equally large riverboat, built by King Ptolemy Philopator as a kind of floating pleasure palace. This ship, too, was a catamaran.

Athenaeus, *Philosophers at Dinner* 5.204d–206c

Philopator also built a river boat, called the "Houseboat", with a length of half a *stadion* [300 feet; 91.5 m], a maximum beam of 45 feet, and a height of slightly under 60 feet, including the collapsible pavilion. Its lines differed from those of both galleys and merchant ships in that its draught had been altered somewhat in order to suit its use on the river. Below the waterline it was broad and flat, while the superstructure was massive, and the periphery of the superstructure projected significantly beyond the hull, particularly at the bow, with very gracefully curving lines. There was a double prow and double stern, all built up high, since the waves in the river often rise quite high.

The hold amidships was fitted out with rooms for dinner parties, berths, and all the other conveniences for elegant living. Two promenade decks encircled three sides of the ship; a single one was no less than 500 feet in length. The lower deck resembled a portico in its design, the upper one a cryptoporticus, built up all round with walls and windows. As you entered by the stern, you met first with a vestibule open in front but with a row of columns at the sides; in the portion facing the prow there was an entrance gate built of ivory and the most costly wood. Entering this you found a sort of presentation stage, constructed with a roof. Matching the forward gate, there was a second gate aft, to one side and a portal with four doors led into it, ... giving access to the largest cabin. It could hold 20 dining couches and was surrounded by a colonnade. The greater part of it was built of Syrian cedar and Milesian cypress. The doors in the peripheral wall, 20 in all, were laminated of fragrant cedar and decorated with ivory. The studs on their front surfaces and the handles were made of copper, which had been fire gilt. The column shafts were made of cypress, while the Corinthian capitals were completely covered in ivory and gold. The whole entablature was of gold, and above it was set a frieze of remarkable figures in ivory more than a foot and a half tall, mediocre in their execution, but remarkable for their rich display. There was a beautiful coffered ceiling of cypress wood, with sculptured decoration covered with gilding. [The rest of the description includes further details of the luxurious furnishings.]

There were many other rooms throughout the whole extent of the area amidships. The mast was 105 feet tall, with a mainsail of linen embellished by a purple topsail.

11.80 The economics of calculating a cargo load

Since most harbours in antiquity were associated with rivers, the point of Aristotle's comment is not that ships loaded to capacity in salt water will sink as they approach a freshwater port, but that ships loaded in a river harbour for a sea voyage should be loaded beyond safe capacity. Otherwise, a portion of the ship's carrying capacity is lost, along with the potential profit. On the other hand, greed could lead to overloading vessels, with consequent loss of property and life.

Aristotle, *Meteorologica* 2.3 (359a)

For the difference in density between salt and freshwater is so great that ships laden with a cargo of the same weight almost sink in rivers but ride in proper trim in salt water and are quite seaworthy. Through an ignorance of this phenomenon, some who load their ships in rivers have lost a chance for profit.

Demosthenes 34, *Against Phormio* 10

After all this, men of Athens, the defendant was left in Bosporus, while Lampis put to sea and was wrecked not far from the port. For even though his ship was already fully loaded, as we hear, he stowed 1,000 steer hides on the deck, and this was the cause of the ship's destruction. He himself was saved in the ship's boat, along with the rest of Dion's slaves, but he lost more than 30 free-men's lives besides the cargo.

11.81 Several ways to load cargo

Techniques of loading and stowing cargo are governed largely by the size, weight, and number of the objects being loaded. Small boats were loaded over the gunwales or up gangplanks by stevedores, as described by Persius, or by a simple crane improvised from the yardarm. Even large cargo vessels, however, such as the one implied by Persius, could be loaded and unloaded in this fashion if the cargo consisted of amphoras of wine or oil, which could be carried by a single person or slung beneath a wooden beam carried by two stevedores. Grain was sometimes shipped in bulk, but it was loaded and unloaded in sacks or baskets. Cranes are documented in use for construction purposes from the Classical period onwards, and they probably appeared at the same time in dockyards as well. Vitruvius may have seen cranes mounted on revolving platforms, but he may also merely be quoting a theoretical Hellenistic source. Obelisks, building stone, or marble sarcophagi, of course, were much more difficult cargoes, requiring special ships and extraordinary techniques.

Persius, *Satires* 5.140–142

All set to go, you load bundles and amphoras on your slaves and shout "Quick aboard the ship!" There's nothing to stop you from a fast passage across the Aegean on your big freighter....

TRANSPORT AND TRADE

Vitruvius, *On Architecture* 10.2.10

This type of device is called a *polyspaston* [compound pulley], because its many small wheels make the work quick and easy. The use of a single crane pole has the advantage that – by inclining it ahead of time as desired – a load can be deposited on the left or right side.

All of these devices mentioned above are suitable for use not only in construction, but also for loading and unloading ships, some being set up vertically, others horizontally on a circular platform that can be turned. In the same manner, but at ground level and without the use of timber uprights, cables and sheave blocks are used to draw ships up on shore.

Pliny, *Natural History* 36.67–70

Ptolemy Philadelphus [283–246 BC] set up an obelisk 80 cubits tall at Alexandria. The Pharaoh Nectanebo [?380–363 BC] had quarried this out and left it uninscribed, but shipping it down the river and setting it up were more impressive achievements than the original quarrying. Some sources report that transport down the river on a raft was arranged by the architect Satyrus, but Callixenus indicates that Phoenix accomplished it. He dug a canal from the Nile up to where the obelisk lay and loaded up two very beamy ships with cubes of the same stone as that of the obelisk, each side one foot long, allowing him to calculate accurately by volume that their weight was twice that of the obelisk. In this way the ships could float in under the obelisk, which was suspended at either end by the banks of the canal. Afterwards, as the blocks were unloaded, the ships rose in the water and lifted the heavy obelisk....

Above all, there came the difficult task of transporting obelisks to Rome by sea, with ships that attracted particular attention. The ship that had carried the first of a pair was dedicated as a testimony of the remarkable achievement in a permanent dry-dock at Puteoli; it was destroyed by fire. The ship in which Caligula brought another obelisk to Italy was kept on display for some years, since it was the most marvellous ship that had ever been seen on the sea. Towers were built on it with pozzolanic mortar at Puteoli and it was towed to Ostia and sunk by order of Claudius as part of the harbour construction. [Pliny has mixed up some of the technical details here; see **11.102**]. Still another concern is the provision of ships to carry obelisks up the Tiber....

INSURANCE, VOYAGES, AND WRECKS

11.82 Contract for a bottomry loan

Since the provision and maintenance of a ship and crew was expensive by itself, the purchase of a cargo in addition often required an outside loan. In the event the ship and cargo were lost at sea, the loan did not have to be repaid, but if the voyage was successful, the significant profit on the cargo was sufficient to provide profits and repayment of the loan with interest. A number of such contracts that were the subject of litigation are preserved in speeches presented to Athenian law courts.

Demosthenes 35, *Against Lacritus* 10–11

Androcles of Sphettus and Nausicrates of Carystus have lent to Artemo and Apollodorus of Phaselis 3,000 silver *drachmai* for a voyage from Athens to Mende or Scione, and from there to the Bosporus – or, if they wish, from there to the western parts of Pontus as far as the Borosthenes – and back again to Athens, at an interest rate of 225 *drachmai* to the thousand – but if they set sail from Pontos to Hieron after the rising of Arcturus, at 300 to the thousand – on the security of 3,000 jars of wine of Mende, which they will carry from Mende or Scione in the twenty-oared ship of which Hyblesius is owner. They offer these jars as security, not owing money on them to anyone else, nor taking any further loans on them, and they will bring back to Athens in the same ship all the merchandise put on board in Pontus as return cargo. And if the merchandise is brought safely to Athens, they will pay back to the lenders within twenty days at Athens the interest due according to the agreement, without deduction except for jettison made by the common agreement of the passengers.... And if the ship in which the merchandise is carried suffers some irremediable damage but the goods on it are saved, let the surviving goods be the common property of the lenders.

11.83 State-funded insurance for a critical cargo

The provision of sufficient quantities of grain to feed the enormous population of the city of Rome was a constant headache for the Roman emperors. Claudius (AD 41–54) initiated various measures to encourage individual shippers.

Suetonius, *Claudius* 18

... [Claudius] employed every possible stratagem to ship grain [to Rome], even in the winter season. For he guaranteed the merchant shippers profits by assuming himself any losses someone might suffer on account of storms, and he instituted great rewards for those who built merchant ships....

11.84 The start of a voyage

When a ship was ready to depart, heralds would announce the fact around the port, so passengers could board.

Philostratus, *Life of Apollonius* 8.14

"Do you know", he said, "of a ship departing for Sicily?" "I do", he replied, "for we have our lodgings at the seaside – and the herald is at the door, since the ship is being made ready this very moment. I gather this from the cries of the crew and the way they work at raising the anchors".

11.85 Early voyages of exploration

Although the Mediterranean, Red Sea, and Persian Gulf were already well travelled in the Late Bronze Age, we have few accounts of the earliest voyages. Phoenician colonists settled Carthage in the ninth century BC, and Greek and Phoenician merchants and pirates were particularly active in exploring the Mediterranean and adjacent seas in the eighth century BC. The one detail Herodotus doubts in the story he recounts about the circumnavigation of Africa actually proves its truth.

Herodotus, *Histories* 1.163

These Phocaeans [from northwestern Asia Minor] were the first Greeks to make long-distance voyages, and it was they who discovered the Adriatic Sea, and Tyrrhenia, Iberia, and Tartessus. They sailed not in round-hulled merchant ships, but in fifty-oared warships.

Herodotus, *Histories* 4.42

When Nechos, the Pharaoh of Egypt [610–595 BC], had finished digging the canal that leads from the Nile to the Red Sea, he sent out Phoenicians in ships, ordering them to sail around Africa until they came to the Pillars of Hercules [Straits of Gibraltar], the northern sea, and thus back to Egypt. So, the Phoenicians set out from the Red Sea into the southern sea. Whenever autumn came they put into land at whatever part of Libya they were sailing past, sowed seed and waited for the harvest. After harvesting the grain they sailed on, and after two years had passed, they passed the Pillars of Hercules in the third year and returned to Egypt. They said something I don't believe, although others might, that the sun was on their right [i.e. to the north] as they sailed around Libya.

11.86 Sea travel to India using the monsoons

By the first century BC Greek and Roman merchants had learnt how to use seasonal trade winds to sail back and forth between the Red Sea and the coast of India. This trade conveyed a wide variety of exotic merchandise to the large Mediterranean market.

Pliny, *Natural History* 6.100–101, 106

Later on [after Alexander's exploration of the east] it was considered the safest route to set out from the Promontory of Syagrus in Arabia with the west wind, which locally they call Hippalus, for Patale [in India], a distance estimated as 1,332 miles. Subsequently it was judged a faster and safer route to depart from the same promontory and head for the Indian harbour of Sigerus. For a long time, this was the course followed, until a merchant found a shortcut and India was brought closer by the greed for profit. In fact, ships set out every year, with troops of archers on board, for pirates were a scourge in those waters.

They sailed back from India at the beginning of the Egyptian month Tybis, our December, or in any case before the sixth day of the Egyptian Mechir, which corresponds to before our 13 January. In this way it is possible to make the return trip in the same year. They set sail from India with a southeast wind [in fact, with the northeast monsoon], and after entering the Red Sea with a southwest or south wind.

11.87 Examples of swift voyages

By the early Empire, very large merchant ships were sailing the Mediterranean, with rigs more complex than the simple square sail typical of the Classical period (see **11.75**). The new rigs and growing familiarity with long-distance routes and trade winds allowed remarkably swift and predictable voyages.

Pliny, *Natural History* 19.3–5

What is more astonishing than that there is a plant [flax, from which linen sails were made] that brings Egypt so close to Italy that two prefects of Egypt, Galerius and Balbillus [AD 55], reached Alexandria from the Straits of Messina in seven and six days, respectively, and 15 years later, during the summer, the praetorian senator Valerius Marianus made Alexandria from Puteoli on the ninth day, with a very light breeze. Or that there is a plant which brings you from Gades and the Straits of Gibraltar to Ostia in seven days, from Nearer Spain in four, from the province of Narbonne in three, from Africa in two....

11.88 A proposal for a ship's log

Until the invention of devices for accurate celestial navigation and the marine chronometer, there was no way for a ship out of sight of land to record accurately the distance travelled in a day. Vitruvius here proposes use of a paddle wheel-driven, geared hodometer to record distances, probably borrowed from a work by a Hellenistic technician. Another nautical version appears in Hero, *Dioptra* 35, and a related land version in **11.38**. For a paddle-wheel arrangement driving a ship, see **2.17**.

Vitruvius, *On Architecture* 10.9.5–7

An axle is passed through the sides of a ship, the ends projecting beyond it. On both ends are mounted wheels four and a half feet in diameter, around the circumference of which are projecting paddles that touch the water. Likewise, the middle of the axle, inside the ship, carries a drum with a single projecting cog. At this point a frame is installed with a drum inside it that has 400 teeth spaced at equal intervals around its circumference, teeth which can mesh with the single tooth on the drum mounted on the axle. In addition, a further, single tooth is mounted on the side of the [second] and projects from it. Above, in another frame fixed to the first, another drum is mounted horizontally and provided with teeth in the same fashion; these teeth mesh with the single tooth fixed to the side of the drum mounted vertically. At each revolution of the vertical drum, its single tooth pushes one of the teeth on the horizontal drum and turns it in a circle. There should be holes in the horizontal drum in which spherical pebbles are to be placed. A small hole is to be drilled in the case holding this drum and its framework, having a channel in which the pebble, when released, can fall against a bronze vessel and announce its passage with a clang.

With this arrangement, when the ship is propelled by oars or the wind, the paddles will touch the water as it slips by, be pushed along by the great force, and will turn the wheels. As these turn, they will move the axle, and the axle the drum. With each revolution, the tooth on this drum will strike one of the teeth of the second drum and move it along slightly. In this way, when the wheels have been turned around 400 times by their paddles, and the vertical drum turned around once, the single tooth mounted on the side of the vertical drum will strike a tooth on the drum mounted on its side. Therefore, every time the rotation of the horizontal drum carries one of the pebbles to the hole, it releases the pebble into the channel. In this way, the sound and the tally of the pebbles will indicate the miles travelled by the ship.

11.89 The sounding-weight as a navigational instrument

Although we tend to think of the sounding-weight, usually a bell-shaped lead weight with a lug to attach a rope, as an instrument intended solely to determine depth, such weights usually had a concave base with studs that held a lump of tallow used to take a sample of the sea bottom. In combination, the knowledge of both the depth and composition of the bottom can allow quite accurate navigation. Herodotus provides the earliest mention of this device, but it must already have been in use for some time. Isidore quotes further information from Lucilius, a Roman poet of the second century BC. See also **11.91**.

Herodotus, *Histories* 2.5

For the nature of the Land of Egypt is this: first, as you sail towards it and are still a day's run from land, if you cast the sounding-weight you will bring up mud and the depth will be eleven fathoms. This shows that the alluvium from the land extends out so far.

Lucilius (Warmington Fragment 1163–64) = Isidore, *Etymologies* 19.4.10

... a sounding-lead: ... a little mass of lead and a rough line of flax.

11.90 Shaping sails, and a landlubber's view of tacking

Although the square rig used by most large ships in antiquity was most efficient when the wind blew from somewhere around the stern, the sail could be turned with the braces or shaped with brailing lines so that some progress could be made by tacking back and forth into a less favourable wind. In particular, the forward end could be bunched up on the yard to form an efficient triangular shape in which the force of the wind is applied aft of the mast. As Aristotle points out, the opposite arrangement could be used to facilitate running with a strong wind. The crew of the ship in the first passage, which set sail from Beirut for Alexandria in the second century AC, at first try to make headway against a rising wind by tacking – much to the consternation of the passengers – but ultimately let the ship run before the storm until it piles up on a reef.

Achilles Tatius, *Leucippe and Clitophon* 2.32, 3.1–4

When the wind turned fair for setting out there was a great commotion throughout the ship, the sailors running in all directions and hauling on ropes as the helmsman called out orders. The yard was pulled round, the sail let go, the ship leaped forward, the anchors were stowed, the harbour was left.... The wind freshened, the sail bellied out and carried the ship along.

On our third day out, a sudden fog poured around us out of a clear sky and took away the daylight. A headwind came up from the sea over the prow of the ship, and the helmsman ordered the yard to be dragged around. The sailors hastened to haul it around and somehow bunched up one end of the sail on the yard by force (for the increasing strength of the gusts of wind impeded their efforts), keeping the other portion fully spread as before to allow tacking for the best wind. The ship lay over on her side at such a steep slope that many of us felt she would roll completely over when the next gust struck. So, we all decamped for the highest point on deck in order to lighten the side of the boat dipped in the sea and by the addition of our weight to force the elevated side back to the level again. We accomplished nothing at all. The high portion of the ship lifted us up rather than our weighing it down. For some time, we struggled to drag into equilibrium the ship, balanced in such a way on the waves. Suddenly the wind shifted to the other side of the ship, which was nearly swamped, and the side that up to now had sloped into the waves was thrown up by a sudden shift of force, while the other upraised side was pushed down into the sea. A great wailing rose from the ship, and with a shout all changed places back again to their former positions, and again a third and fourth time – countless times – we went through the same difficult manoeuvre and followed the heaving of the ship. And even before one crossing was complete, a second, opposite migration was necessary. Carrying our baggage up and down the ship we ran a sort of long-distance race a thousand times, always thinking death was close at hand....

[The storm worsened, and the cargo was jettisoned.] Finally, the helmsman gave up, let go of the tillers, and gave the ship up to the sea. He prepared the skiff and after ordering the sailors to climb on board, started down the ladder. They jumped in right behind him. [At this point a fight broke out as the sailors threatened the passengers trying to join them in the boat. Finally, the skiff was cut loose, while the ship broke up on a reef.]

[Aristotle], *Mechanical Problems* 7 (851b)

Why is it that, when sailors wish to run before an unfavourable wind, they reef that part of the sail near the helmsman and they slacken the sheet that is towards the bow? Is it because the rudder cannot act against a strong wind, but only a moderate one, and for this reason they reef the sail? In this way the wind drives the ship forward and the rudder acts against a favouring wind, working on the seawater like a lever. At the same time the sailors fight against the wind, for they lean over in the opposite direction.

11.91 How to work a doomed ship

The account of the apostle Paul's difficult voyage from Caesarea to Rome includes a storm scene as well, with further details about the measures that could be taken to strengthen and lighten the ship and keep it off the reefs. The ship was driven all the way from the south coast of Crete to Malta.

Acts 27:13–44

When a light south wind began to blow [the sailors] thought they had obtained what they wanted, and raising anchor they coasted along Crete. But not long afterwards the so-called northeaster wind struck us from the shore, and since the ship was caught by the wind and could not make way against it, we gave way to it and were carried along. Running up behind a small island called Cauda, we were able with great difficulty to secure the ship's boat, and after hoisting it up, they employed cables to undergird the ship. Fearing that they might be driven on to the Syrtes, they let down a sea-anchor and so were carried along. Since we were being badly tossed about by the storm, the next day they began jettison of the cargo, and on the third day with their own hands cast overboard the ship's tackle....

When the fourteenth night had come and we were being carried along in the Sea of Adria [Ionian Sea], the sailors suspected that they were nearing land, and casting the sounding-weight, they found 20 fathoms. A short distance along they sounded again and found 15 fathoms. Afraid that we might run up on some shoals, they let out four anchors from the stern and prayed for daylight. The sailors lowered the ship's boat into the sea under the pretence of setting anchors from the prow, scheming to escape from the ship. But Paul said to the centurion and his soldiers, "Unless these sailors remain on the ship, we cannot be saved", and the soldiers cut the rope holding the boat, and let it go.

... We were in all 276 souls on the ship.

When day came, they did not recognise the place, but saw a bay with sandy beach to which they planned to bring the ship, if able to. They cast off the anchors, letting them go in the sea and at the same time untied the lashings on the steering oars; they spread the steering sail to the wind and made for the beach. But they struck some shoals and ran the boat up on them; the bow remained stuck fast, and the stern was broken up by the force of the waves. The centurion ... ordered those who could swim to throw themselves overboard first and head for land, the rest on planks or on some pieces of the ship. And in this way, it happened that all came safely to land.

11.92 Profiting from shipwreck

Marine salvage was an important source of income for individuals who lived near treacherous shoals or headlands; some even went so far as to post false navigational lights to lure ships to their destruction. Xenophon describes a situation on the coastline of the Black Sea west of the Bosporus.

Xenophon, *Anabasis* 7.12–14

Here many of the ships sailing into the Black Sea run aground and are wrecked, for there is shoal water far and wide in the sea. The Thracians who live there have set up boundary markers, and each group plunders the wrecks which wash up in its own territory.... Here many couches were found, and crates, and written books, and a great supply of all the other items shipowners transport in wooden chests.

11.93 What to do when the wind fails

When a large cargo ship was becalmed, little could be done except wait for wind. Warships, however, and small craft could be rowed, especially when a schedule had to be kept or a perishable cargo carried to market. For free sailors, at least, singing was an essential part of this activity.

Longus, *Daphnis and Chloe* 3.21

While they were eating ... a fishing-boat was seen sailing along the coast. There was no wind, but a great calm, and it seemed best to them to row, which they did with a will. They were hastening to bring some fresh fish in good condition to one of the richer citizens in the city, and as they worked the oars, they did what all sailors do to lighten their difficult tasks. One among them, the boatswain, knew some sea-shanties, and the rest, like a kind of chorus, called out all together in response to his song.

11.94 Sea battle with triremes and larger ships

The trireme (Morrison 2000), which had three oars in each bank of oars, every oar manned by a single rower, represented the ultimate in speed and manoeuvrability. The ships were very light for their length and rowing power, so they could move and change course quickly, functioning like a kind of manned torpedo designed to ram enemy ships. After a successful ram, however, the attacker had to back away quickly in order not to be boarded or become a stationary target for another ship. The abridged selection from an account in Herodotus of the famous battle between Athenian and Persian ships and their allies in 480 BC mentions some of the tactics and dangers of this form of warfare. The passage from Polybius underlines the growing importance of marine and hand-to-hand combat in the Hellenistic period, when many of the warships were larger and heavier than triremes. This battle took place in 201 BC. For other techniques used in sea battles, see **13.19–13.22**.

Herodotus, *Histories* 8.84, 86, 89–90, 96

Then the Greeks put all their ships out to sea, and the foreign forces set upon them immediately as they put out.... Since the Greeks fought with order and in proper array, while the foreigners did nothing either in an orderly manner or rationally, it was right that they should come to such an end as befell them.... The majority of the foreigners perished in the sea, since they did not know how to swim. When the [foreign] ships in the front rank were turned to flight, at that moment most of them were destroyed. For the captains in the ranks behind, trying to advance in their ships in order to demonstrate some accomplishment to the king themselves, ran up against their comrades' ships as they fled.... A ship from Samothrace rammed an Athenian ship, but as the Athenian ship was sinking, a ship from Aegina attacked and sank the Samothracian ship. The Samothracians, however, being javelin-throwers, swept the marines from the ship which had sunk theirs, boarded it, and seized it.... When the sea battle broke off, the Greeks towed to Salamis such of the wrecks as still happened to be there.

Polybius, *Histories* 16.3.2–4.14

Attalus [I, King of Pergamon] attacked an eight-bank and ramming her first with an opportune blow below the waterline, he finally sank her in the face of great resistance on the part of the troops on deck. Philip's ship [Philip V, King of Macedon], a ten-bank ship, and the flagship, came into the possession of his enemies by a strange chance. She rammed with great force amidships a *trihemiolia* [a type of small, fast oared ship] that was in her way and stuck beneath the top rowing benches, since Philip's captain was not able to moderate the ship's speed in time. While the ship hung suspended there, she was in a very precarious position and completely incapable of manoeuvring. At this opportune moment two five-bank ships fell upon her, effected severe damage to both sides, and destroyed her and all those on board.... Dionysiodorus mounted a ramming attack at full speed but missed and, passing close alongside, lost all his oars on the starboard side, his towers being carried away at the same time. Once this had happened the enemy crowded about him from all directions and with a shout and cries of excitement destroyed the remainder of the ship's company along with the ship.

For the rest of the ships in the fleet, the contest was equal, for the fact that Philip had a greater number of light, oared vessels was balanced by the greater number of heavy, decked vessels in Attalus' fleet.... If the Macedonians had not deployed their galleys in the battle line among their decked ships, the battle might have come to an easy and quick conclusion, but as it was, these galleys obstructed the use of the Rhodian ships in many ways. For, once the initial order of battle had been disturbed in the first attack, all were mixed up with one another with the result that they could neither easily sail through the enemy line nor turn their ships around, nor in any way employ their own special tactics. The galleys attacked them sometimes about the banks of oars, making coordination of the rowing difficult, sometimes at the prow or at the stern, so that they impeded the activities of the helmsmen and the rowers. But in ramming bow to bow they used a certain technique. Setting the trim of their own ship low at the bow, they received the ram above the waterline, but by inflicting damage on the enemy below the waterline they landed blows that could not be repaired. They seldom, however, used this method of attack. For the most part they avoided close engagement, since the Macedonians put up courageous defence from the deck in fights at close quarters. Generally, they sailed through the line of enemy ships and put their oars out of action, afterwards doubling back and attacking them – sometimes in the stern and sometimes in the side, while they were turning. Driving the ram home, they holed some of the ships and on others destroyed some of the essential tackle.

HARBOURS

11.95 The harbour as refuge and basis for prosperity

Early Bronze Age harbours in the Mediterranean were simply beaches or river mouths selected for proximity to an important settlement and for at least partial protection from storm waves. By the Late Bronze Age, some harbours in the eastern Mediterranean were provided with quays built of blocks, but artificial breakwaters do not seem to have appeared until the eighth century BC and did not become common until the fifth. At all periods, safe arrival at a harbour was the occasion for rejoicing, and the harbour became a metaphor for refuge. Homer describes the double harbour of the sea-loving Phaeacians, ideal because – in an age without artificial breakwaters – sailors could select whichever beach was in the lee of the wind. In the Classical period, the prosperity of Corinth was attributed to her possession of a harbour on either side of the Isthmus of Corinth, allowing traders to trans-ship goods and avoid the dangerous passage around Cape Malea. Boats could even be carried across the Isthmus on dollies, along a stone-paved roadway called the *diolkos* (see **11.14**; Blackman 2008).

Vergil, *Georgics* 1.303–304

[In Winter farmers generally enjoy their gains and rejoice] … as when heavy-laden ships have finally made port, and the happy sailors have crowned the sterns with garlands.

Homer, *Odyssey* 6.262–269

But when we reach the city – there's a wall with high towers around it, and a beautiful harbour on either side of the city, with a narrow entrance passage between them. The seagoing ships are beached along the causeway; everyone has a slipway for himself alone. There, too, is their market place around the beautiful temple of Poseidon, furnished with quarried stone set in the earth. Here they busy themselves with the tackle for their black-pitched ships, cables and sails, and shave down their oars.

Strabo, *Geography* 8.6.20

Corinth is called "wealthy" because of its commerce, situated on the Isthmus of Corinth and controlling two harbours, one of which faces Asia, the other Italy. This facilitates the exchange of goods for peoples so distant from one another. For just as long ago the Straits of Sicily were not easy to navigate, so also the high seas, particularly around Cape Malea, on account of the contrary winds.... For this reason, traders from both Asia and Italy were delighted to bring their merchandise here and avoid the passage around Cape Malea.

11.96 An early Greek breakwater

Although many Greek cities had artificial breakwaters by the time Herodotus wrote (mid-fifth century BC), that at Samos, which had to enclose a large, open bay, was particularly impressive for its length and the depth of water in which it had to be built. Although Herodotus exaggerates the depth of the water, the term he uses for the breakwater – *choma* – indicates a rubble mound, the correct design for this situation.

Herodotus, *Histories* 3.60

I have said a bit more about the people of Samos because they have the three greatest engineering works anywhere in the Greek world. [The first and third are the aqueduct of Eupalinos and the Temple of Hera].... The second is a mole in the sea, enclosing the harbour; built at a depth of 20 fathoms [*c.*37 m; this is inaccurate]; the mound is more than two *stadia* [*c.*366 m] in length.

11.97 The harbour and lighthouse of Alexandria

There was a harbour in the lee of the small island of Pharos – essentially an exposed limestone reef – during the Bronze Age, but, according to Strabo, the Pharaohs chose not to develop the site, probably to preserve Egypt's relative isolation from foreigners. This use of an unimproved natural breakwater is typical of Bronze Age harbours. Alexander the Great, however, recognised the site's potential as a trade centre and commissioned structural improvements. He built the *Heptastadion*, a causeway "Seven *Stadia*" (1,250 m) long connecting Pharos island with the mainland. Ptolemy II (283–246 BC) hired Sostratus of Cnidus to build the lighthouse, one of the marvels of ancient engineering, which served as the model for many Roman lighthouses.

Strabo, *Geography* 17.1.6–10

Pharos is a small, oblong island, quite close to the mainland and with it creating a harbour that has two entrances. The shoreline has the form of a bay framed by two headlands that project seaward, and the islet is located between them closing off the bay – for it lies lengthwise, parallel to the shore. The eastern of the extremities of Pharos is closer to the mainland and the promontory opposite (the promontory is called Lochias) and gives the harbour a narrow entrance. In addition to the narrowness of the passage, there are rocks – some below the water, others projecting above it – which continually break up the swell that rolls in on them from the open sea. This end of the islet is a rock, washed on all sides by the sea, that has on it a tower marvellously constructed of white stone in many stories, and carrying the same name as the island. Sostratus of Cnidus, a friend of the kings, dedicated this, as the inscription says, for the safety of those who sail the sea. Since the coastline was low-lying and harbourless in both directions, and also had reefs and some shoal water, those who sailed in from the open sea had need of some elevated and conspicuous sign in order to set a good course for the harbour entrance.

The western channel also is not easy to enter, but it does not require as much caution. It forms another harbour called Eunostos ["Happy Return"], which lies in front of the closed harbour basin that was artificially excavated. The harbour which has its entrance by the above-mentioned tower of Pharos is the Great Harbour. The innermost basin of the western harbour lies adjacent to the Great Harbour, separated from it by the mole called *Heptastadion*. This mole is a bridge extending from the mainland to the western portion of the island, leaving only two channels into the Eunostos Harbour; these are bridged over. This structure was not only a bridge to the island, but also an aqueduct – at least when Pharos was inhabited....

As for the Great Harbour, in addition to being beautifully enclosed by the mole and by natural topography, it is deep close in-shore, so that the largest ship can tie up by the steps of the quay. It is also divided up into several harbour basins.

The former kings of the Egyptians, enjoying what they possessed and not in need of any imports whatsoever, ... set up a guard at this place ordered to drive away anyone who approached.... But when Alexander came and saw the advantages of the place, he decided to fortify the settlement at the harbour. The advantages are numerous. First, the site is washed by two seas, one on each side: on the north by the so-called Egyptian Sea, on the south by the marshy lake Marea, which is also called Lake Mareotis. The Nile fills this lake through many canals, both on the south and from each side, and the merchandise brought in through these canals is much greater than that brought in by sea, so that the lake port was much richer than the seaport....

Next, after the *Heptastadion*, is the Eunostos Harbour and beyond it the artificial basin, which is called Kibotos and also has ship sheds. Farther along there is a navigable canal which extends as far as Lake Mareotis.

Pliny, *Natural History* 36.83

We must not pass over the great generosity of King Ptolemy, through which he permitted the name of the architect [of Pharos] to be inscribed on the very fabric of the building. Its function, in connection with the navigation of ships at night, is to show a light and provide warning of shoals and the harbour entrance. Structures like it now show their lights at many sites, for example Ostia and Ravenna. There is danger in leaving the beacon fire uninterrupted, in case it should be mistaken for a star, since from a long way off the appearance of the beacons is similar. This same architect is said to have been the first to build a promenade on piers, at Cnidus.

11.98 A famous Phoenician harbour

Although this description of the harbour of Sidon, in present-day Lebanon, dates to the second century AC, the harbour facilities themselves reflect the original Phoenician arrangement. Breakwaters of cut stone blocks springing from sandstone reefs and islets protected an outer basin. Another breakwater marked off an inner basin protected on its seaward side by a sandstone promontory. Channels cut through this promontory allowed seawater to flow into the inner basin and keep it free of silt. The author's somewhat enigmatic description may reflect a knowledge of these flushing channels.

Achilles Tatius, *Leucippe and Clitophon* 1.1

Sidon is a city on the Syrian Sea. It is the mother city of the Phoenicians.... There is a double harbour basin in the bay, wide, and gently closing out the sea. Where the bay forms a hollow on the right side a second channel has been excavated and the water runs in, so there is a second harbour basin within the first. As a result, merchant ships can spend the winter in this calm inner basin, while in summer they enter the harbour's outer basin.

11.99 The harbours of Carthage

Carthage, a Phoenician colony founded near modern Tunis in the ninth century BC, dominated sea trade in the western Mediterranean until the mid-third century BC. Her harbour took the form of basins – called *cothon* – excavated just inside the shoreline, an idiosyncratic arrangement that may have its roots in Phoenicia. The basins were carefully segregated by function.

Appian, *Punic Wars* 8.14.96

The harbour basins were connected by a channel and had a common entrance from the sea 70 feet across, which they closed with iron chains. The first basin was for merchants, and there was a great deal of ship's tackle of all kinds near it. There was an island in the middle of the second basin, and both the island and shore were fringed at intervals with great quays. Shipyards for 220 vessels occupied these quays, and by the shipyards were arsenals with equipment for the warships. Two Ionic columns stood in front of each shipshed, giving the island and the shoreline the appearance of a portico. The admiral's quarters were built on the island, and from there the trumpeter had to give signals, the herald passed on orders, and the admiral kept an eye on things. The island was located near the entrance and rose to a significant height, so that the admiral could observe what was happening at sea, while it was not possible for those approaching by ship to get a clear look at what was going on inside. Not even merchants sailing into the harbour obtained an immediate, clear view of the shipyards, for a double wall surrounded them. There were gates which admitted the merchants from the first basin into the city without passing through the military dockyards.

11.100 How to construct a harbour

Vitruvius provides us with the only detailed information concerning the methods of constructing harbours in the Roman world. Implicit in his discussion is the difference between the natural harbour typical of the Aegean and the artificial harbour typical of the western Mediterranean. The "earth" he refers to in both passages is pozzolana, a volcanic ash that can be substituted for sand to make a mortar that will set in contact with water (see **8.14**). Construction of the type of rubble-mound breakwater Vitruvius only alludes to is described by Pliny, who witnessed the construction of the Great Harbour at Centumcellae, modern Civitavécchia. It is not certain what sort of procedure Procopius refers to, but the most likely explanation for the "coffers" is that they were floating caissons that were filled with pozzolanic concrete until they sank in the desired position (Oleson ed., 2014, Rickmann 1988) [**fig. 11.5**].

Vitruvius, *On Architecture* 5.12.1–7

I must not omit the proper arrangement of harbours but rather explain by what techniques ships are protected in them from stormy weather. Harbours that have an advantageous natural location, with projecting headlands or promontories that shape naturally curved or angled recesses, seem to be the most useful. Colonnades or shipyards are to be constructed around the circumference, or entrances from the colonnades to the markets. Towers are to be built on either side [of the entrance to the harbour], from which chains can be drawn across by means of winches.

If, however, we have no natural harbour situation suitable for protecting ships from storms, we must proceed as follows. If there is an anchorage on one side and no river mouth interferes, then a mole composed of concrete structures or rubble mounds is to be built on either side and the harbour enclosure constructed in this manner. Those concrete structures which are to be in the water must be made in the following fashion. Earth is to be brought from that region which runs from Cumae to the promontory of Minerva and mixed in the mortar used in these structures, in the proportions of two parts earth to one of lime. Next, in the designated spot, formwork enclosed by stout posts and tie beams is to be let down into the water and fixed firmly in position. Then the area within it at the bottom, below the water, is to be levelled and cleared out, [working] from a platform of small crossbeams. The building is to be carried on there with a mixture of aggregate and mortar, as described above, until the space left for the structure within the form has been filled. The places which we have described above, then, have this natural advantage.

But if because of waves or the force of the open sea the anchoring supports cannot hold the forms down, then a platform is to be built out from the shore itself or from the foundations of the mole, made as firm as possible. This platform is to be built out with a level upper surface over less than half its area. The section towards the shore is to have a sloping side. Next, retaining walls one and a half feet wide are to be built towards the sea and on either side of the platform, equal in height to the level surface described above. Then the sloping section is to be filled in with sand and brought up to the level of the retaining walls and platform surface. Next a mass of the appointed size is to be built there, on this levelled surface, and when it has been poured is left at least two months to cure. Then the retaining wall which holds in the sand is cut away, and in this manner erosion of the sand by the waves causes the mass to fall into the sea. By this procedure, repeated as often as necessary, the breakwater can be carried seaward.

But in locations where the earth does not occur naturally, one must use the following procedure. Let double-walled formwork [*arcae duplices*, i.e. cofferdams] be set up in the designated spot, held together by close-set planks and tie beams, and between the anchoring supports have clay packed down in baskets made of swamp reeds. When it has been well tamped down in this manner, and is as compact as possible, then have the area bounded by the cofferdam emptied and dried out by means of water-screw installations and water-wheels with compartmented rims and bodies. The foundations are to be dug there, within the cofferdam. If the foundations are to be on a rocky, solid bottom, the area to be

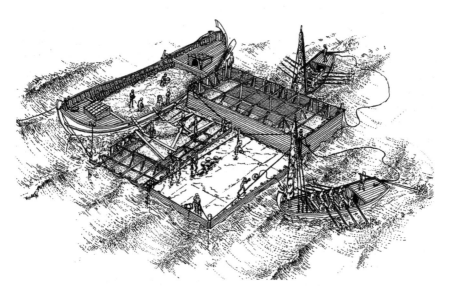

Figure 11.5 Reconstruction of prefabricated *caissons* being loaded with concrete off Caesarea. Brandon *et al.*, 2014, p. 213, fig. 8.52.

Source: Drawing by C. J. Brandon, used by permission.

excavated and drained must be larger than the wall which will stand above, and then filled in with a concrete of aggregate, lime and sand. But if the bottom is soft, the foundations are to be covered with charred alder or olive wood pilings and filled in with charcoal, as described for the foundations of theatres and city walls. Then the wall is to be raised of squared stone with joints as long as possible, so that the stones in the middle may be well tied together by the joints. The space inside the wall is to be filled with rubble packing or concrete. Thus, it may be possible to build a tower upon it.

When all this has been finished, the shipyards must be considered, and in particular that they be laid out facing north – for a southern exposure, on account of its heat, leads to dry-rot, wood worms, ship worms, and other pests, and nourishes and maintains them. Furthermore, because of the danger of fire, these buildings should be constructed with as little wood as possible. There should be no restriction on their size, but they should be built to the dimensions of the largest ships, so that even these will have a roomy berth when they are drawn up on shore.

Pliny the Younger, *Letters* 6.31.15–17

The place itself was just as delightful. The villa is very beautiful; it is fringed by fields of the brightest green and overlooks the seashore and a bay which at this very moment is being turned into a harbour. The breakwater on the left has already been reinforced with construction of the greatest stability, while that on the right is in the process of being built. At the harbour entrance a free-standing mole rises from the sea to serve as a breakwater against seas brought in by the on-shore wind and provide safe entrance to ships on either side. The technique by which the mole is built has got to be seen. A wide barge brings enormous stones right up to it and throws them in one on top of another. Their weight keeps them in position, and little by little a sort of rampart is constructed. A kind of stony hump can already be seen rising above the water which breaks the waves that beat upon it and tosses the spray high in the air with a great roar; the sea all around is white with foam. Masses of concrete will be laid on top of the stones, and as time passes it will come to resemble an island. This will be the harbour, and it will bring safety to many by providing a haven on this very long stretch of harbourless coastline.

Procopius, *On Buildings* 1.11.18–20

There [at Constantinople] he [Justinian] brought to completion with great skill a sheltered harbour where there previously had been none. Finding a shore exposed from both directions to the winds and the force of the breaking waves, he established it in the following way as a refuge for voyagers. He prepared great numbers of a very large, box-shaped formwork – the so-called "coffers", and dropped them in oblique lines on either side [of the basin] for a great distance out from the shore. By repeatedly setting a new course of forms in careful order on top of those laid previously, he constructed two walls angled out towards each other from opposite sides [of the harbour], rising from their foundations deep in the water up to the surface where ships float and manoeuvre. He threw untrimmed boulders on top of them, and when these boulders are pounded by the surf, they toss off the force of the waves. Even when a strong wind rises in the winter, the whole area within the breakwaters remains still, since one entrance into the harbour has been left between them for ships. In that place he also constructed holy shrines, as I recounted above, along with porticoes, markets, public baths, and nearly every other type of building.

11.101 A magnificent early imperial harbour

The harbour facilities of Caesarea Palestinae were built by Herod the Great between 22 and 10/9 BC as a rival to the superb harbour of Alexandria. As a sign of loyalty to the emperor, he named the harbour *Sebastos*, the Greek version of Augustus. Roman engineers and pozzolana sand were brought from Italy, and the resulting harbour was both technologically modern and magnificent in design (see Oleson 1989). Although Josephus did not understand the important role concrete played in the construction, and his calculations of depth are highly inaccurate, his description nevertheless gives a vivid impression of the Roman ability to create harbours at inhospitable locations (Oleson, ed., 2014; Raban 1985) [fig. 11.5].

Josephus, *Jewish War* 1.408–414

Herod noticed a settlement on the coast – it was called Straton's Tower – which, although much decayed, because of its favourable location was capable of benefiting from his generosity. He rebuilt the whole city in white marble, and decorated it with the most splendid palaces, revealing here in particular his natural magnificence. For the whole coastline between Dor and Joppa, midway between which the city lies, happened to lack a harbour, so that every ship coasting along Phoenicia towards Egypt had to ride out southwest head winds riding at anchor in the open sea. Even when this wind blows gently, such great waves are stirred up against the reefs that the backwash of the surge makes the sea wild far off shore. But the king, through a great outlay of money and sustained by his ambition, conquered nature and built a harbour larger than the Piraeus, encompassing deep-water subsidiary anchorages within it.

Although the location was generally unfavourable, he contended with the difficulties so well that the solidity of the construction could not be overcome by the sea, and its beauty seemed finished off without impediment. Having calculated the relative size of the harbour as we have stated, he let down stone blocks into the sea to a depth of 20 fathoms [*c.*37 m]. Most of them were 50 feet long, 9 high, and 10 wide [15.2 × 2.7 × 3.05 m], some even larger [most of these blocks were in fact concrete]. When the submarine foundation was finished, he then laid out the mole above sea level, 200 feet across [61.0 m]. Of this, a 100-foot portion was built out to break the force of the waves, and consequently was called the breakwater. The rest supported the stone wall that encircled the harbour. At intervals along it were great towers, the tallest and most magnificent of which was named Drusion, after Caesar's stepson.

There were numerous vaulted chambers for the reception of those entering the harbour, and the whole curving structure in front of them was a wide promenade for those who disembarked. The entrance channel faced north, for in this region the north wind always brings the clearest skies. At the harbour entrance there were colossal statues, three on either side, set up on columns. A massively built tower supported the columns on the port side of boats entering the harbour; those on the starboard side were supported by two upright blocks of stone yoked together, higher than the tower on the other side.

There were buildings right next to the harbour also built of white marble, and the passageways of the city ran straight towards it, laid out at equal intervals. On a hill directly opposite the harbour entrance channel stood the temple of Caesar, set apart by its scale and beauty. In it there was a colossal statue of Caesar, not inferior to the Zeus at Olympia on which it was modelled, and one of the Goddess Roma just like that of Hera at Argos. He dedicated the city to the province, the harbour to the men who sailed in these waters, and the honour of the foundation to Caesar: he consequently named it Caesarea.

11.102 The harbours of Rome

The site of Rome was 25 km from the coast, but the Tiber river provided an excellent route for travel both down to the sea and upstream into the interior. Until the reign of Claudius (41–54), the city made do with a harbour in the river mouth, at the port city of Ostia, although the harbour at Puteoli in the Bay of Naples – protected by a breakwater – served as an alternative landing place. Claudius' engineers excavated a basin in the soft ground north of the river mouth, protected it with concrete breakwaters, and connected it to the Tiber by means of a canal. A large ship served as the formwork for the foundation for the lighthouse, filled with hydraulic mortar made with ash from Puteoli. Dionysius of Halicarnassus' description of the river port in the period of the kings seems to underestimate the dangers it posed for larger vessels, particularly in bad weather.

Dionysius of Halicarnassus, *Roman Antiquities* 3.44.2–3

The Tiber river descends from the Apennine mountains, flows right past Rome, and empties into the harbourless and exposed coastline of the Tyrrhenian Sea.... Since it is suitable for navigation up to its sources by even rather large river boats, and as far as the site of Rome by great, seagoing merchant ships, he [King Ancus Marcius, later seventh century BC] decided to construct a seaport at the river mouth, using the last stretch of the river as the harbour basin. For the river becomes very broad where it reaches the sea, and encompasses great bays like those of the best maritime harbours. And what is particularly marvellous, its mouth is not obstructed by sandbars thrown up by the sea – something that happens to many large rivers.... Consequently, oared ships – however large they happen to be – and merchant ships of a capacity up to 3,000 (*modii*?) enter at the mouth of the river and arrive at Rome under oar or tow. Larger ships ride at anchor off the entrance and have their cargo off-loaded and new cargo loaded on again by river boats. He founded a walled city on the elbow of land between the river and the sea and named it Ostia ["Opening", "Mouth"], after the location.

Dio Cassius, *Roman History* 60.11.2–5

Nearly all the grain the Romans consumed was imported, but the coastline near the mouth of the Tiber had neither safe landing places nor suitable harbour basins.... Apart from supplies brought in during the summer sailing season and stored in warehouses, there was nothing imported during the winter, and whoever ran the risk [of a winter voyage] suffered disaster. With this in mind, Claudius undertook to construct a harbour basin. He would not be dissuaded, even though his architects replied to him, when he asked how great the cost would be, "You don't want to do it!" But he conceived a project worthy of the dignity and greatness of Rome and brought it to completion. First, he excavated a considerable plot of land near the coast, built quay walls all around it, and let in the sea. Next, in the sea itself he laid down great moles on either side of the basin entrance and thus enclosed a large body of water, and in it he fashioned an island carrying a lighthouse. This Port, as it is still called by the locals, was built by him at this time.

Suetonius, *Claudius* 20.3

At Ostia he constructed a harbour by building breakwaters out from shore to the right and left and placing a mole in front of the entrance, which was in deep water. To give this mole a more stable foundation, he first scuttled the ship in which a large obelisk had been conveyed from Egypt, then laid massive piers above.[4] He topped it off with a very tall tower modelled after the Pharos at Alexandria, so that at night ships might direct their course towards its beacon fire.

Pliny, *Natural History* 16.201–2

It is certain that nothing more remarkable than this ship has been seen on the sea. It carried 120,000 bushels of lentils for ballast, and its length occupied a good portion of the left side of the harbour at Ostia. It was scuttled there by the Emperor Claudius and three great masses the height of towers were built on top of it, made from pozzolana mortar brought here for that purpose.

11.103 A disastrous harbour improvement plan

Artificial breakwaters can interfere with the siltation regime of river or long-shore currents, resulting in the deposit of sand inside the harbour basin or outside the entrance. The harbour at Portus had to be maintained by dredgers, but the "improvement" of the harbour facilities at Ephesus recounted by Strabo resulted in the deposit of silt that ultimately filled the entire basin. An inscription of the

Roman Proconsul L. Antonius Albus has survived (AD 161/2) that prohibits the dumping of rubbish in the harbour, in order to prevent aggravation of the siltation problem.

Strabo, *Geography* 14.1.24

Ephesus has both shipyards and a harbour basin. Architects made the entrance narrow at the orders of King Attalos Philadelphos [reigned 159–138 BC], but all were deceived as to the result. For the king thought that the entrance would be deep enough for large merchant ships, along with the harbour basin itself – this formerly had shallow spots because of silt dropped by the Cayster River – if a mole were thrown up at the entrance, which was very wide. He ordered the mole to be built, but just the opposite happened. Since the silt was hemmed in, it made the whole harbour basin more shallow as far as the entrance. Formerly, the ebb and flow of the tide had been sufficient to take up the silt and carry it outside.

SEG 19 (1963) no. 684

To Good Fortune. The Proconsul L. Antonius Albus proclaims: since it is essential for the greatest metropolis of Asia, if not of the entire world, that the harbour, which receives people who come from all directions, should not be obstructed – when I learnt in what manner individuals damage the harbour, I thought it necessary to prohibit this by means of an edict and to set a suitable penalty for those who disobey. I therefore order those who import wood or marble neither to store the wood nor saw the marble on the quay. For the former by the weight of their cargoes damage the supporting foundations built there for the protection of the harbour basin, and the latter, by throwing in the emery [used to cut the marble] ... diminish the depth and obstruct the flow; both obstruct passage on the quay....

11.104 A dry-dock in the harbour at Alexandria

Ships too large to be dragged up on the beach or on paved ramps inside a harbour to allow repair were sometimes accommodated in dry-docks. This must have been an exceptional procedure, since watertight doors (like lock gates on a canal) to close the compartment apparently were unknown, and it is seldom mentioned. A Phoenician engineer, however, built a dry-dock that could service the 40-banked ship of Ptolemy Philopator (described above, **11.78**). Athenaeus quotes this information from Callixenus, *On Alexandria*.

Athenaeus, *Philosophers at Dinner* 5.204c–d

Subsequently, however, a Phoenician invented a method of launching which involved surrounding the ship with a trench equal to it in length, excavated adjacent to the harbour. He constructed the lower floor for it of regular stone blocks, to a depth of five cubits, and along its whole length he set skids [or "rollers"] crosswise from one side of the trench to the other, leaving a depth of four cubits altogether. Constructing an entrance from the sea, he let in the seawater to fill the trench up and easily drew the ship into it with the assistance of unskilled labour. Closing the opening up again they pumped out the water with pumps, allowing the ship to settle safely on the above-mentioned skids.

DIVING

Diving was not uncommon in antiquity. Sponge and shellfish divers, for example, or salvage divers as well as individuals swimming under water in the context of military activities are well attested. A handful of inscriptions from imperial Rome and Ostia (*CIL* XIV 303; supp. 4620; *CIL* VI 1080, 1872, 29700, 29702) mentions a guild of salvage divers (*corpus urinatorum*), the term *urinator* possibly referring to the physiological reaction of the human body of voiding the bladder in a phenomenon called "cold water diuresis" (Oleson: 1976).

Large bodies of water are dangerous environments for humans for three reasons: humans cannot breathe under water, visibility is extremely limited, and the equalisation of pressure is a challenge. Working under water was, therefore, a risky endeavour, and the people who were engaged in it used a number of, sometimes drastic, techniques to minimise the risks posed by the hostile environment. The use of snorkels appears in ancient literature, as well as rudimentary diving bells – notably in the *Alexander Romance* where Alexander the Great explores the seafloor in an overturned vessel. Other texts mention the perforation of eardrums and nostrils to ease equalisation of pressure under water, the use of earplugs, and some provide anecdotal reports about oil that illuminates the water when it is released from the mouth of a submerged diver. (cf. **3.76–80**)

11.105 The elephant's trunk as a model for snorkels

Aristotle, *On the Parts of Animals* 659a.9–15

Some divers, when they spend a lot of time under the sea, equip themselves with a device, through which they draw air from above the water surface. Nature has made such a device for elephants in the length of their nose. They breathe by holding their nostrils above the surface if ever they have to make their way through the water. For as we have said, the trunk is the elephants' nose.

11.106 Equalising pressure under water

The ear is very sensitive to changes in external pressure. Within two metres under water the external water pressure will depress the eardrum inwards against the air-filled middle ear, which causes intense pain that increases with increasing water depth. The diver has to equalise the pressure periodically, normally by pinching the nose and blowing gently against the closed nostrils, to prevent the eardrum from bursting. A burst eardrum is not only extremely painful, it also means that cold water can suddenly flood the middle ear, leading to impaired senses of balance and hearing. This problem was obviously a concern, as it could lead to life-threatening situations. It is not surprising to read in the Aristotelian corpus that drastic solutions may have been common practice.

[Aristotle], *Problems* 960b.8–17

Why do the eardrums of divers rupture in the sea? Do they burst because they are exposed to pressure by the holding of the breath? But if this is the reason, it should happen also in air. Or is it that when something is unyielding, it is more quickly broken, and more so in a harder medium than a soft one? What is inflated yields less easily. So, the eardrums, as has been said, become inflated under the holding of breath, so that the water, which is harder than air, strikes and ruptures them.

Why do divers tie sponges around their ears? Is it so that the sea, when it enters with force, does not rupture the eardrums? For in this way they do not become filled as they do when the sponges are removed.

[Aristotle], *Problems* 960b.21–34

Why do sponge divers cut their eardrums and nostrils? Is it because in this way they can equalise pressure more easily? For in this way breath seems to escape. They say that they suffer very much from difficulty of breathing by their inability to expel the breath. But when they have vomited forth the breath, as it were, they are relieved. It is strange, therefore, that they cannot breathe for the sake of cooling. This seems to be more necessary. Is it natural that the distress is greater to those holding their breath because they are swollen and strained? There also seems to be an automatic outwards movement of the breath. But if there is also an inward one, we must consider it. For so there seems to be. For they can likewise enable divers to breathe by lowering down cauldrons. These do not fill with water but retain the air. They have to be lowered into the water by force. Whenever they are tilted from the upright position, water flows in.

[Aristotle], *Problems* 961a.24–30

Why do the eardrums of divers rupture less easily if they pour in olive oil? The reason why eardrums rupture has been stated before. The olive oil, however, when it has been poured into the ear, makes the seawater slide off, just as it does on the outside of the body when one is oiled up. Sliding along, it does not strike a blow to the inside of the ear and so does not rupture it.

11.107 A tricky hiding place

Maurice, *Strategikon* 11.4.10

The Slavs are experienced and surpass all people in crossing rivers, and they are excellent at holding out under water. Often in their own country when they are attacked suddenly in a difficult position they dive to the bottom of a body of water, and having made long, hollow reed pipes for this purpose, they hold them in their mouths with the end reaching through the water surface. Lying on the bottom they breathe through the reeds and hold out for many hours without arousing any suspicion. But if by chance the reeds are seen sticking out the water, they look to an inexperienced person like they were growing in the water naturally. But those who are experienced in these matters and recognise the reeds by their cut ends or their position either shove them down into the divers' mouths or pull them out and bring the men to the surface because they can no longer remain under water.

11.108 An unlikely source of light under water

Plutarch, *Causes of Natural Phenomena* (Mor. 915a)

They say that divers, when they blow out oil that they keep in their mouths, have light and clear water in the deep.

Pliny, *Natural History* 2.106.234

In winter the sea is warmer, in autumn more salty; the whole sea is calmed by oil, and because of this, divers disperse it from their mouths, because it calms down harsh nature and carries light with it.

CANALS

11.109 The Nile-to-Red Sea canal

There is much disagreement in ancient sources about the chronology and character of this canal, stimulated in large part by a belief voiced by some that the level of the Red Sea was higher than that of the Nile. There is no unanimity about who finished the canal, and how well it functioned. The allusion by Diodorus to some kind of lock is one of the few places in ancient literature where a solution of this type is assumed, but the design is not entirely clear (cf. Strabo, *Geography* 17.1.25). The emphasis on rapid closure of the barrier may indicate that there was a single door rather than an impoundment lock. The canal was certainly out of service by the late first century BC (cf. Plutarch, *Antony* 69.2–3), and Pliny (*Natural History* 6.165–6) asserts that Ptolemy never finished it at all, because of the difference in water level.

Herodotus, *Histories* 2.158

Psammetichus had a son Nechos who became king over Egypt [610–595 BC], and it was he who first undertook the canal leading from the Nile to the Red Sea. Darius the Persian [521–485 BC] finished digging it. The canal is four days' voyage in length, and its width is sufficient to allow two triremes to be rowed along side-by-side. Water flows from the Nile into it; it leaves the river just above the town of Bubastis, by the Arabian town of Patumus, and runs into the Red Sea. Digging was begun from the part of the Egyptian plain that is closest to Arabia.... During Nechos' reign, 120,000 Egyptians died while excavating it.

Diodorus of Sicily, *History* 1.33.9–12

From the Pelusiac mouth of the Nile there is an artificial canal to the Arabian Gulf and the Red Sea. Nechos, son of Psammetichus, was the first to undertake construction of it. After him, Darius the Persian made progress with the work of excavation for a while, but he finally left it unfinished, for he learnt from certain persons that if he cut through the Isthmus he would carry the responsibility for flooding Egypt. They indicated to him that the Red Sea was at a higher level than Egypt. Later on, Ptolemy II [285–246 BC] brought it to completion and at the most suitable spot fashioned a technically sophisticated barrier. He opened it whenever he wished to sail through, then quickly closed it again, and in practice it functioned very successfully. The stream which flows through this canal is named Ptolemy, after its builder....

11.110 How to dig a canal

The Persian King Xerxes, during his invasion of Greece, decided to cut a canal across the low Isthmus that joined the long, mountainous promontory of Athos in northern Greece with the mainland. The project probably was intended to show the king's power over nature.

Herodotus, *Histories* 7.23

The foreigners, dividing up the ground by ethnic origin, dug in the following way. They made a straight line by the town of Sane, and when the trench became deep, some took position at the bottom and dug, while others handed over the earth that was constantly being excavated to still another group positioned higher up on steps. The latter, in turn, as they received it, handed the earth on to yet another group, until they came to those at the top, who carried it away and dumped it. Everyone except the Phoenicians had double the work when the steep sides of their excavation collapsed. Since they made the trench the same width at top and bottom, this was bound to happen. But the Phoenicians showed their skill in this, as in so many other matters. Taking in hand the portion that was allotted to them, they excavated the upper portion of the canal to double the width that the canal itself was supposed to have, and narrowed it as the work proceeded. At the bottom their excavation was equal in width to that of the others.

11.111 Canals requiring locks

We have no firm evidence that some sort of device similar to the modern canal lock was known in antiquity. Although Diodorus of Sicily suggests the use of such a device on the Nile-Red Sea canal

(**11.109**), other authors do not mention the presence of a lock, or indicate that the canal was left unfinished. Several passages in Roman authors describe officially approved canal projects that would have required extensive systems of locks, but – perhaps significantly – none of them was completed. Pliny the Elder mentions an arrangement of flash-locks on the Tiber, but such an arrangement would have been of little use for travel upstream (Smith 1978, Moore 1950).

Tacitus, *Annales* 13.53

Up to this time [AD 55] there had been peace in Germany.... Vetus was preparing to link the Moselle and Saône rivers by digging a canal between the two. In this way merchandise shipped by the Mediterranean, then on the Rhone and Saône, might by means of this canal then run down the Moselle into the Rhine and so to the Atlantic. This would remove the difficulties of the land journey and make a navigable link between the West and the North. Aelius Gracilis, governor of Belgica, was envious of the project and deterred Vetus....

Pliny the Younger, *Letters* 10.41, 42

Pliny to Trajan:
There is a very large lake in the vicinity of Nicomedia. Marble, farm produce, wood, and lumber are transported across this on boats as far as the road with very little expense or effort, but from that point down to the sea on carts with great effort and greater expense.... <To connect the lake to the sea>[5] is an undertaking requiring a great deal of labour, but there is no lack of it. The countryside is well populated and the city even more so, and it seems very likely that all will gladly assist with a project that will benefit everyone.

The next thing is for you to send a surveyor or architect, if it seems fit to you, to make a careful survey as to whether the lake is above sea level. The local experts assert that it is 40 cubits higher. I have observed a canal in the area dug by one of the kings, but it is not clear whether it was intended to drain the surrounding fields or to connect the lake with the river, for it was left unfinished.

Trajan to Pliny:
We might find it of interest to connect that lake of yours with the sea, but clearly there must be a careful survey of how much water the lake holds and where it comes from. Otherwise it might drain completely once it has been given an outlet to the sea. You might request a surveyor from Calpurnius Macer, and I will send you from here someone skilled in this sort of work.

Pliny, *Natural History* 3.53

In its uppermost reaches the Tiber is a narrow stream, only navigable when its water has been diverted to storage reservoirs and then discharged. The same holds true of its tributaries the Tinia and Glanis, whose flow must be held in this manner for nine days, unless the rain helps.

11.112 Canal boats for night-time travel

The most successful canals in antiquity were those associated with rivers, for the constant flow ensured sufficient depth for boats and scouring of the channel. The canal draining the Pomptine marshes was also used by travellers; there is an amusing account of such a trip taken by the poet Horace in his *Satires* (**11.17**).

Strabo, *Geography* 5.3.6

Near Terracina, as you go towards Rome, a canal runs alongside the Appian Way, fed at numerous points by the flow from marshes and rivers. People navigate it mostly at night, so that embarking in the evening they can disembark at dawn and continue the rest of the way by the road – but it is also navigated by day. A mule tows the boat [see **fig. 11.6**; and cf. the Tiber in **11.102**].

C STANDARDS OF TRADE

COINAGE

The origin and development of coinage, as well as the technology of minting in antiquity, are still only imperfectly understood, largely because the literary evidence is full of errors where it exists at all. What we can say with some certainty is that, in both the Greek East and Italy, minted coins were a relatively late invention, preceded as a mode of exchange for several centuries by unstandardised ingots or bars worth the weight of their constituent metal. This was, of course, an extremely awkward system not only because of the weight of the tokens but because their actual value had to be tested at every exchange. When

Figure 11.6 Towing a river boat on the Durance river.
Source: Lapidary Museum of Avignon, France.

international trade re-emerged in the Aegean region during the Archaic Age, individual states there began to assume responsibility for guaranteeing the purity and weight of the metals used in exchange, issuing the first true coinage bearing symbols attesting to its authenticity. Over time the minting of coins evolved into one of the few examples of mass-production in antiquity, though no literary evidence survives to explain satisfactorily the technology of making blanks, cutting dies, and so on; so our knowledge of how the Greeks struck their coins and how the Romans early on cast their bronze money in moulds is derived principally from a close examination of the coins themselves (Meadows 2008.)

11.113 The barter system of trade

Coins are not mentioned in Homer's epics, where barter of goods and the exchange of unstandardised bits of precious metal are the common means by which items were bought and sold. The large ingots of iron that have been found from this period are shaped rather like miniature ox-hides [**fig. 11.7**], which lends support to Pliny's association of early money with livestock (**11.119**); and the handier spits of metal eventually were to give their name (*obeloi*) to the smallest denomination of Greek coinage.

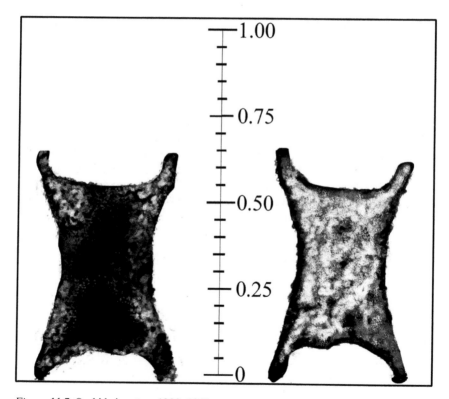

Figure 11.7 Ox-hide ingots, *c.*1200–1050 BC.

Source: Adapted from Bass, 1967, figs. 56–58 and BM 1897, 0401.1535.

Homer, *Iliad* 7.465–475

The sun set, and the work of the Achaeans came to an end. They slaughtered oxen throughout the encampment and took their meal. A fleet of ships had arrived from Lemnos with a cargo of wine, sent by Euneus, ... son of Jason, who had included in the shipment a separate gift of a thousand measures of wine for Agamemnon and Menelaus, the sons of Atreus. From these ships the long-haired Achaeans bought themselves wine by barter, some giving bronze in exchange and others flashing iron, some hides or live cattle and others slaves. And then they laid out a bounteous feast....

11.114 The principles of monetary exchange

In these two passages Aristotle outlines three of the greatest advantages of state-issued money: its portability, the easy reciprocity it makes possible, and its guaranteed value.

Aristotle, *Politics* 1.3.11–15 (1257a–b)

All goods are susceptible to barter, which followed at first from the natural order of things, since men had more than a sufficiency of some objects and less of others. From this it is clear that trade does not naturally belong to the art of profit-making, for the act of exchange was necessary only as a means of providing men with enough to get by on....

Yet it makes sense that barter gave rise to profit-making. For they were compelled to develop the use of money when they began to depend more on foreign trade by importing what they lacked and exporting their surpluses. Since not all of men's natural requirements are easy to transport, they agreed among themselves to give and to receive as a medium of exchange some sort of object that was useful in itself and easy to handle in day-to-day life – iron, for example, and silver and other metals – defined at first informally by their size and weight, but in the end impressed with stamps that certified these values, which no longer had to be measured with each transaction.

Aristotle, *Nicomachaean Ethics* 5.5.8–11 (1133a)

[If there is no proportional equivalence and correlation between products,] the exchange is not equal and falls apart: it is quite possible that the product or service of one party in the exchange is more valuable than the others, which makes it necessary for them to be balanced.... For it is not two doctors who strike an agreement for exchange, but a doctor and a farmer – people, that is, who on the whole are neither alike nor equal, but who must be put on an equal footing. This means that all things subject to exchange must somehow be made to correspond, and it is for this purpose that money has come about, as a kind of intermediary that measures the relative value of all things: how many pairs of sandals, for example, are equivalent to a house or a meal...

So, all things must be measured by some one standard – which is, in fact, demand, the principle that connects everything together: for without demand there will be no exchange, and if demand varies so too will the rate of exchange. Money has, by general agreement, become the exchangeable representation of demand. This explains why the Greek term for money is *nomisma*: it exists not through Nature but by custom (*nomos*).

11.115 The invention of coinage

The traditions related here – that the first Mediterranean coinage originated in Lydia in Asia Minor, and that Aegina was the first Greek city-state to mint coins – are probably not far from the truth. The earliest coins, resembling small water-smoothed pebbles, were of electrum, the alloy of gold and silver that is found naturally in western Asia Minor around Sardis, and in the early sixth century BC, the island of Aegina just south of Athens was active in international trade and so a logical pioneer in this revolutionary development (Kraay 1964, Hill 1922).

Herodotus, *Histories* 1.94

The Lydians were the first people we know who used currency coined from gold and silver, and they were also the first to engage in retail trade.

For an example of the wealth of the Lydians – primarily from the placer gold harvested from the Pactolus River – see Herodotus, *Histories* 1.50–51: King Croesus' gifts to the oracle at Delphi included 4 ingots of pure gold and 117 of electrum, weighing in total 244 talents.

Strabo, *Geography* 8.6.16

According to Ephorus, silver was coined first on the island of Aegina by Pheidon, because the place was a centre of trade, its agricultural poverty having encouraged its inhabitants to earn their livelihood as merchants at sea – whence, Ephorus adds, minor goods are called "Aeginetan merchandise". [See also **11.123**.]

11.116 Persian bullion and coinage

Herodotus here describes the tribute paid to King Darius *c.*500 BC by his dependent satrapies. The passage acknowledges the lack of standardisation typical of this period.

Herodotus, *Histories* 3.89, 95–96

He requested those paying their tribute in silver to use the standard of the Babylonian talent, and those paying in gold to use the Euboic talent – the former being equal to 78 Euboic *minai* [rather than the normal 60]....[6]

Gold coinage is reckoned as thirteen times the value of silver.... This is how the Persian king stores away his tribute: he melts it down, pours it into earthenware jars, and – as each vessel is filled – breaks off the surrounding clay. So, when he needs money he coins [from these ingots] just as much as the occasion demands.

11.117 Standardisation of coinage

The two principal standards accepted for coinage in the classical Greek world were the Aeginetan and the Euboic, the latter eventually adopted by the great trading centres of Athens and Corinth. Still, international trade continued to be hampered by the existence of many more local standards. The following decree was issued by Athens probably in the 420s BC, in an attempt to impose on the cities of her Empire uniform standards of coinage and of weights and measures [**figs. 11.8–9**].

Figure 11.8 Athenian silver tetradrachm, Athena and owl.

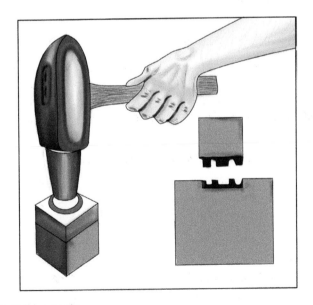

Figure 11.9 Striking a coin.
Source: Adapted from Kleiner, 1975, fig. 2.

R. Meiggs, D. Lewis, *Greek Historical Inscriptions* (Oxford 1969) 45.12–13

The following rider of the Council [of Athens] is added to the oath taken by the Council: "If anyone mints silver coinage in these cities and makes use not of the coinage, weights, and measures of the Athenians, but of foreign coinage, weights, and measures, the punishment is to be in accordance with the previous decree that Clearchus proposed". Private individuals are to surrender foreign silver at their convenience, and their city is to exchange it....

11.118 Spartan resistance to standardisation

The Spartans were well known for their perceived dangers of outside influence upon their austere lifestyle. A famous example of this xenophobic attitude is their retention of the primitive iron spit (the *obolos*) well after others had adopted coins, but they went even further to lessen the real value of their iron currency. Plutarch has described a serious case of corrupt behaviour of a great Spartan, Gylippus, which has led the ephors to reject gold and silver.

Plutarch, *Lysander* 17. 2

[but they ought] to use their hereditary currency. This was iron, which was dipped in vinegar when it first came from the fire so that it might not be worked again, but by the dipping become un-tempered and useless. So really, it was very heavy and difficult to carry, and a great quantity and weight of material had little value.

A very similar event is related for the legendary Lycurgus (Plutarch, *Lycurgus* 19. 1–2).

11.119 The development of Roman coinage

The following passage from Pliny is the *locus classicus* for the historical development of the Roman monetary system. Most scholars question the precision of his dates, but the general outline is reasonable. For striking of coins see **fig. 11.9**.

Pliny, *Natural History* 33.42–48

Not even coined silver was used by the Roman people before the defeat of King Pyrrhus [280–275 BC].... King Servius [traditionally 578–535 BC] was the first to have made bronze into money by stamping a design on it (before which time, according to Timaeus, they used the metal in its unworked state at Rome): they stamped the metal with an image of livestock (*pecus*), whence derives our word for money, *pecunia*....

Silver was coined when Quintus Ogulnius and Gaius Fabius were consuls [269 BC], five years before the start of the First Punic War. It was agreed that a *denarius* should be valued at 10 *librae* of bronze, a *quinarius* at 5 *librae*, and a *sestertius* at 2.5 *librae*. The weight of a *libra* of bronze was, however, reduced during the First Punic War, when the state could not meet its expenses, and it was decreed that the *as* should be struck weighing a sixth of a *libra*, thus saving five-sixths [of the original weight] and paying off the national debt. The design of this bronze coin featured a double-faced Janus on one side and a warship's ram on the other.... The designs of silver coinage were chariots drawn by two horses (*bigae*) or by four horses (*quadrigae*), from which the coins were called *bigati* and *quadrigati*.

Gold coinage, the *aureus*, was struck 51 years after the first silver, a *scripulum* of gold being worth 20 silver *sestertii* (at the standard ratio then of 400 *sestertii* to a *libra* of silver). Later it was agreed to strike 40 *aurei* from a *libra* of gold; little by little the coin's weight was reduced by the emperors, until most recently Nero pegged it at 45 to the *libra*.

But the invention of coinage brought with it the beginning of greed: moneylending was dreamt up, and profit without exertion.

11.120 Debasement and forgery of coinage

Since coins were generally claimed to be worth the value of their bullion, there was great incentive for states to profit by reducing the purity or content of the metal while continuing to peg it at its original value. This happened on and off for centuries, but the beginning of official, continuing, and serious debasement of Roman coinage really occurred under Nero in the mid-first century AC, when a greater proportion of alloy was added to both silver and gold coins, and their weight slightly reduced. As for Pliny's comments about *unofficial* debasement, numismatic forgeries must be as old as coinage itself. The detection technique invented by Gratidianus may have been serration of the edges of the coins, which appears in issues of *denarii* at the appropriate time. Ultimately, it was unsuccessful.

Pliny, *Natural History* 33.132

When he was triumvir [43–31 BC], Antony introduced iron into the silver *denarius*. Some forgers add bronze and others reduce the weight, the standard being 84 *denarii* struck from a *libra* of silver. As a result, a method to test the *denarii* was introduced under a law so well received by the people of Rome that they all – neighbourhood by neighbourhood – voted statues to Marius Gratidianus. It is extraordinary, too, that this is the only one of the arts in which it is the defects that are mastered: it is a *forged* specimen of a silver coin that attracts attention and fetches a higher price than the genuine articles.

WEIGHTS AND MEASURES

According to Herodotus, metrology was invented in Egypt during the Bronze Age, in response to the central government's need to measure and assess agricultural land for the purposes of taxation, from which evolved standard measurements for the area of fields and for the weight and volume of produce. This same motive, not incidentally, was attributed to the Egyptians' development of record-keeping (**11.121**).

Most ancient units of measurement were based on naturally available models: units of length were derived from parts of the human body (digit, palm, cubit, foot, pace), a standard that was always ready to hand; units of area reflected the amount of land a team of oxen could reasonably plough in a day (*plethron* in Greek, *iugerum* to the Romans); many units of weight, like the talent, were based on human capacity (it was considered the load a grown man could carry); and units of volume imitated the vessels used for the commonest dry and liquid produce, grain and wine (hence the Roman *amphora* represented both the vessel and the standard measurement of volume, equal to a cubic Roman foot). Though many states imposed their own local standards on administrative and commercial transactions within their borders, as with coinage there were seldom universally recognised models for these measurements, and even in the relatively unified Mediterranean of the Roman Empire people quantified things using a variety of different standards (Wikander 2008, O'Brien 1981–84, Kisch 1965).

11.121 Egypt and the origins of metrology

Herodotus, *Histories* 2.109

Tradition says that the Pharaoh Sesostris divided up the country among all Egyptians by assigning to each of them a uniform, rectangular plot of land, from which he raised his revenues by imposing payment of an annual tax. If the Nile ever stole part of a man's allotment, he would come to the pharaoh and declare what had happened, and Sesostris would send out inspectors to measure the amount of land eroded so that in the future it might yield the appropriate proportion of the original rent. It seems to me that it was from this that the science of land measurement was invented and then passed into Greece.

11.122 Variations in standards of measurement

It was not so much that systems of measurement differed geographically and chronologically – most states in the Mediterranean used many of the same bases for their units of mensuration – but that those bases had different values in different societies. As one example, consider these three descriptions of ancient Babylon, from authors of varying periods and backgrounds. Many of the variances in their accounts are attributable to variable units of measurement.

Herodotus, *Histories* 1.178

Babylon lies in an extensive plain, and is square in shape with each side measuring 120 *stadia*, the total circuit of the city amounting to 480 *stadia*.[7] Such is the size of the city of Babylon, which was laid out like no other urban centre we know of. A moat encircles it first, deep and wide and filled with water; then a wall 50 royal cubits thick and 200 cubits high (the royal cubit being 3 fingerwidths longer than the standard cubit).[8]

Strabo, *Geography* 16.1.5

Babylon itself lies in a plain. The circuit of its walls is 385 *stadia*; it is 32 feet thick and 50 cubits high between its towers, which are 60 cubits high; the *parodos* or walkway on top of the wall is wide enough that a pair of four-horse chariots can easily pass one another.

Pliny, *Natural History* 6.121

[Babylon] is surrounded by two walls 60 miles in length, 200 feet high, and 50 feet thick (the Babylonian foot being 3 fingerwidths longer than ours).

11.123 The invention of standard weights and measures

Pheidon was considered, with some historical justification, the author not only of Greek coinage, but of widely accepted (though not universally standard) measures.

Herodotus, *Histories* 6.127

... Pheidon, tyrant of the Argives, who devised the system of measures for the Peloponnesians....

Strabo, *Geography* 8.3.33

Pheidon the Argive ... invented the measures called Pheidonian, and weights, and coinage stamped of silver and other metals.

11.124 Standardisation in the marketplace

Most ancient states, like Athens in this example, established uniform local measures and weights to ensure fair market practices. There were usually approved versions of these units housed in the agora or forum, to which customers could refer if suspicious of the measurements used by a local merchant.

Aristotle, *Constitution of Athens* 51

Ten Market Clerks (*agoranomoi*) are chosen by lot, five for the Peiraeus and five for the city proper. To them is assigned legal responsibility for supervising all goods for sale, to ensure the marketing of genuine and dependable merchandise.

Ten Measure Clerks (*metronomoi*) are also chosen by lot, five each for the city and the Peiraeus. They have charge of all measures and weights, to ensure the merchants' use of legitimate ones.

There also used to be ten Grain Inspectors (*sitophylakes*), five for the Peiraeus and five for the city, but now there are 20 and 15 respectively. It is up to them to ensure, first of all, that unmilled grain is sold in the market for fair prices; and secondly, that the millers set the price of barley meal in accordance with the cost of barley, and the bakers likewise match loaves of bread to the cost of wheat, while at the same time keeping to the weights set by the Grain Inspectors, whom the law charges with this responsibility.

They choose by lot ten Superintendents of the Warehouses [in the Peiraeus], charged with supervising the warehouses and with requiring the traders to make available in the city proper two-thirds of whatever grain reaches the harbour warehouses from overseas.

MARKETS, TRADE, AND PRICES

11.125 Local markets in Athens and Rome

The three short excerpts that follow, though meaningless when read out of their dramatic or satirical context, do give an impression of the variety of products available in urban markets in antiquity, and of the busy crowd of motley characters who frequented the Athenian Agora and the Roman Forum. The third passage, by Martial, alludes to an edict from the Emperor Domitian that prevented hawkers and shop owners from blocking traffic by setting up their stalls along the streets and sidewalks – a common sight in Mediterranean countries today, despite similar official injunctions (Tchernia 2011/2016, Horden and Purcell 2000, Rickman 1980, Hopper 1979).

Aristophanes, *Lysistrata* 557–564

LYSISTRATA: Here they are in the crockery stalls, now in the vegetable market. They roam the agora fully armed, like priests of Cybele.
MAGISTRATE: Of course. They have to show they're men.
LYSISTRATA: And yet I can't help laughing when one of them goes shopping for fish carrying his shield decorated with a Gorgon's head!
CALONICE: As god is my witness, I saw a long-haired cavalry commander, still mounted, toss into his bronze helmet some pease-porridge he'd bought from an old woman; and another one, this time a Thracian, shaking his shield and javelin and generally behaving as if he were a mythical king, hailing the shopgirl who sells figs and gobbling down her ripe fruits.

Plautus, *Curculio* 466–482

I'll show you where [in the Forum] you can easily find any type of fellow you want, so no one needs to spend a lot of time trying to meet someone with or without vices, righteous or unrighteous. You want to meet a perjurer? Go to the Assembly. You'll find a liar and blowhard at the Shrine of Venus the Purifier. Look behind the Courthouse for rich and married spendthrifts; the same locale is home to well-used prostitutes and men who make a habit of collecting on promises. Try the Fish-Market for subscribers to the dining clubs. The worthy and wealthy types take their stroll in the lower Forum, and the pure exhibitionists hang out near the drain in the middle Forum. Above the Pool are the self-assured, loquacious, and nasty types who shamelessly malign others without cause – and who themselves are open to enough well-founded criticism. Below the Old Shops you'll find those who lend and borrow with interest. Behind the Temple of Castor are those you don't want to trust too easily. In the Tuscan neighbourhood are men who offer themselves for sale: those who manipulate, and those who rent themselves out to others for manipulation. But I hear the doors creak....

Martial, *Epigrams* 7.61

[Previously,] the inconsiderate pedlar had hijacked the whole city, and no shop's doorstep knew its own limits. Now, Domitian, you have ordered our constricted neighbourhoods to expand, and what recently was a footpath has now become a boulevard. No column now is hung about with wine bottles strung on chains; no praetor is forced to walk in mud down the middle of the street; no hidden razor is drawn in the midst of a crowding throng; no smoke-filled greasy-spoon appropriates every bit of road. Barber, innkeeper, cook, butcher – all stick to their own doorways. Now Rome is alive, when recently it has been one large shopping mall.

11.126 International markets

The following passages illustrate some of the Mediterranean's most important emporia: a generic Bronze Age roadstead, the communally founded trading colony of Naucratis in Egypt, and the huge markets of Alexandria, Athens, and Rome. Corinth is another good example: for a description of that city's unique system of two international harbours separated by an Isthmus, see **11.95**.

Homer, *Odyssey* 15.415–416, 455–456

Then came to the island the Phoenicians, men famed for their ships, greedy rascals, bringing a load of trinkets with them in their black ship.... They stayed among us for an entire year, and stored away in their hollow ship a hoard of provisions they obtained by barter.

Herodotus, *Histories* 2.178–179

Amasis was a philhellene who did a number of favours for some of the Greeks, most notably by presenting to those who came to Egypt the city of Naucratis for their settlement....

In the old days Naucratis was the one and only commercial harbour in Egypt. Its prestige was such that anyone who arrived at any of the other mouths of the Nile had to swear that he had done so involuntarily and then sail his ship to the

Canopic mouth; or, if unable to sail against unfavourable winds, to carry his cargo in flat-bottomed barges all around the Delta until he arrived at Naucratis.

Dio Chrysostom, *Discourses* 32.35–36

In size and location your city [Alexandria] is as distinctive as it could possibly be, and of all the cities under the sun is quite clearly considered second [only to Rome]. Egypt, itself a monumental nation, is like the body of the city – or, better, its appendage – and the qualities of the Nile are beyond expression, its wonder and usefulness distinguishing it from all other rivers. Thanks to the beauty of your harbours, the size of your fleet, and the marketing of an endless supply of goods from every corner of the world, you dominate the entire Mediterranean Sea, and you even hold in your grip the waters that lie beyond: the Red Sea and the Indian Ocean, whose name was hardly heard before your time. So, it happens that the trade and commerce, not just of islands, harbours, and some straits and isthmuses, but essentially of the entire inhabited world, are in your hands. For Alexandria lies, as it were, at the conjunction of the whole world, even of the most remote nations, just like the marketplace of a single city, bringing everyone together into one spot, exposing them to one another, and insofar as possible making them all kinsmen.

Xenophon, *Ways and Means* 3.1–3, 12–13

Now I shall explain how our city [Athens] has all the conveniences and advantages to make it especially suited to commerce. First of all, of course, Athens has the finest and safest harbour facilities for vessels, where it is possible for them to find refuge from storms and ride safely at anchor. And what is more, in most other cities, merchants find it necessary to accept a return cargo, since the local currency of these places is not accepted abroad. But at Athens they have the choice of exchanging their goods for a host of exports in popular demand, or – should they prefer not to take a return cargo – of exporting silver. This latter is a sound business practice, since they will always get more than they paid for it no matter where they sell it.

Still, the number of our trading partners would increase considerably and our commerce would be much more amicable if rewards were offered to any member of the port authority who settled disputes equitably and quickly, so as not to delay the merchant eager to depart....

When capital becomes available, it would be a particularly good idea to add to the existing housing for ship owners around the harbours, and to build for merchants places designed for wholesale trade, and hotels for visitors. And if houses and shops were built for retail merchants in both the Peiraeus and the town proper, they would be at one and the same time an embellishment for the city and a source of much rental income.

Aelius Aristides, *To Rome* 10–13

The sea is drawn like a sort of waistband across the middle of the inhabited world – the middle of your Empire, since they are one and the same. Far and wide around the sea lie vast continents, from each of which you are constantly filled with provisions. From every land and sea are brought the fruits of each season, whatever all the farms and rivers and lakes produce, by Greek or barbarian techniques. It follows that, if anyone wishes to behold all these things, he must travel the known world to gaze on them – or he must be in Rome. For whatever is grown or manufactured in each nation is inevitably found here always in abundance. Every hour, every harvest cycle sees so many merchant ships arriving here with cargoes from all parts that the city is like some communal production centre for the world. You can see so many shipments from India, if you like, or from Arabia Felix that you imagine that the trees there have been left perpetually bare for the inhabitants who, if they need anything, must journey here to Rome to beg for a share of their own produce. Clothes from Babylonia, too, and decorations from the barbarian world beyond arrive here in greater quantities and more easily than whatever cargoes had to be brought from Naxos or Kythnos to Athens. Your farmlands are Egypt, Sicily, and the cultivated parts of Africa....

Everything converges here: commerce, seafaring, agriculture, metallurgy, all crafts present and past, everything that is produced and grown. Whatever one does not see in Rome is not to be counted among things that have existed or now exist.

11.127 Trade beyond the Mediterranean

Long-distance trade beyond the shores of the Mediterranean was common throughout the Classical period, often through intermediaries on the borders of Greek or Roman territory: silk and spices from the Orient are the best examples, though products like amber from northern Europe and wild animals from Africa were also significant imports. The Romans were certainly aware of trade deficits arising from their thirst for luxuries from outside the Empire; Pliny alludes to this in the first passage below, and mentioned it again in *Natural History* 12.84: "The lowest estimate has India, China, and the peninsula of Arabia removing from our Empire each year some 100 million *sestertii*. That is the price our luxuries and our wives cost us". It is extremely fortunate that one coastal pilot survives, *Periplus of the Erythraean Sea*, written by a ship captain in the first century AC. For a discussion of this work, see Casson 1989.

Pliny, *Natural History* 6.101

Now that, for the first time, we have definite knowledge of the route, I have no qualms about laying out the entire voyage from Egypt [to India]. The topic is an appropriate one: not a year goes by without India draining off at least 50 million *sestertii* from our Empire, sending us in payment merchandise that is sold here for a hundred times its original value.

Periplus of the Erythraean Sea 49

Imports to the market of Barygaza[9] include wine (predominantly Italian, also from Laodicaea and Arabia); copper, tin, and lead; coral and topaz; clothing of all sorts, both unlined and hybrid; striped girdles a *cubitus* wide; *storax* [a fragrant gum] and honey clover; raw glass; sulphide of arsenic and powdered antimony; Roman coins of gold and silver, the source of some profit when exchanged for the local currency; and balm, if it is not very costly and in small amounts.

During those times imports for the king included expensive silver plate, musical instruments,[10] beautiful maidens to be used as concubines, fine wine, costly unlined clothing, and special balm.

Exports from the region include nard oil, *kostos* [a spice], *bdellium* [an aromatic gum], onyx, myrrh, and Indian *lukion*; all sorts of fine linen, silk, fabric made from mallow fibre, and [silk] thread; long pepper; and produce brought from the [other] markets.

Those sailing to this port from Egypt in the proper season set out around the month of July....

Pliny, *Natural History* 6.106

They sail back from India at the beginning of the Egyptian month Tybis, our December, ... so it happens that they make the round trip in a single year.

11.128 The profits of maritime trade

Many Roman literary authors express a discomfort with or even a fear of the seas (e.g. Horace, *Epodes* 10), and they sometimes even curse the inventor of the boat for having violated the laws of Nature by turning sea into land. Partly because of this cultural prejudice, and partly because of a law limiting senatorial investment to landed property, the surviving names of ship owners and commercial exporters are more often Greek, Syrian, or African in origin, than Italian. It is no coincidence, then, that Petronius' character Trimalchio, who describes in the passage below how he made his fortune in maritime shipping, was not from an old Roman family but was a *nouveau-riche* freedman.

Petronius, *Satyricon* 76

I took a fancy to business. I won't hold you up with the details: I built five ships, loaded them up with wine (it was worth its weight in gold in those days), and sent them off to Rome. You'd think I'd fixed it in advance: every ship was wrecked – it's the truth, no lie. In a single day Neptune swallowed up 30 million *sestertii*. But do you think I gave up? No siree, I hardly even tasted the loss, it was as if it hadn't happened. I built more ships, bigger, better, and luckier, so no one could say I didn't have guts. You know, a titanic ship is particularly sturdy. I loaded them up with a second cargo of wine, and with bacon, beans, ointments, and slaves. It was at this juncture that Fortunata did a really noble thing: she sold off all her jewellery and all her clothes, and put a hundred gold coins in my hand. This was the yeast that my nest-egg needed. If the gods are on your side, things happen quickly: in one voyage I made a round 10 million *sestertii*.

11.129 The costs of long-distance trade

Despite the very real dangers of sailing, the costs of transporting bulk cargoes in antiquity were many times less by sea than by land, thanks in part to the capacity of merchantmen, their small crew, and the relatively direct routes between major cities, most of which were on or near the Mediterranean coast. The first passage here, dating to AD 301 (see **11.136**), gives us a unique opportunity to compare these shipping costs. The following selection – on the "hidden" costs of transporting frankincense from Arabia to the Mediterranean – was written by Pliny two and a half centuries earlier, before inflation had driven up the price of that fragrant resin some 17-fold.

Diocletian, *Price Edict* (Frank, *ESAR* Vol. 5, p. 369)

Freightage for a wagon-load of 1,200 *librae*	*denarii* 20 per mile
Freightage for a camel-load of 600 *librae*	8 per mile
Freightage for an ass-load	4 per mile

(Graser, *TAPA* 71 [1940] 154–174)

The maximum rates a shipper may charge between certain places and certain provinces:

from Alexandria to Rome	1 military *modius*	*denarii* 16
from Alexandria to Byzantium		12
from Alexandria to Africa		10
from Alexandria to Ephesus		8
from Alexandria to Pamphylia		6
Likewise from Asia to Rome		16
Likewise from Africa to Rome		?
from Africa to Sicily		6
from Africa to the Gauls		4

Pliny, *Natural History* 12.63–65

After the frankincense is collected it is transported by camel to Sabota, where one of the city gates is opened for its arrival; indeed, *not* to follow the highway here was considered a capital offence. At Sabota the priests confiscate a tithe (measured by volume, not by weight) for the god they call Sabis.... Then it can be exported only through the land of the Gebbanitae [in southwestern Arabia], and so a tax is levied on it for their king, too. Their capital, Thomna, lies some 1,500 miles[11] from the Judaean town of Gaza on the Mediterranean coast, a distance divided into 65 daily stages for the camel caravans. There are also fixed percentages of the frankincense that go to the king's priests and scribes as well, and they are not alone: even the guards and their henchmen and the porters and minor functionaries all skim off their shares. And these payments continue all along the route, here for water, there for fodder, or for lodging at the stopping-places or various customs duties, so that the cost amounts to 688 *denarii* for each camel-load that arrives at the Mediterranean coast, even before our Empire's collectors levy their tax. So it is that a *libra* of the best frankincense fetches a price of 6 *denarii*; second quality, 5 *denarii*; and third quality, 3.

Compare Varro *On Agriculture* 1.16.2–4, on the decided advantages of selling to *local* markets.

11.130 The three branches of trade

Aristotle, *Politics* 1.4.2 (1258b)

Within the art of making money, there are three branches that are devoted to the exchange of commodities, or trade. The largest branch is commerce, which itself is composed of three parts: ship owning, transport, and marketing (each of these differing from the others in their relative safety and profits). The second branch is money-lending. The third is wage-earning, of which one part involves the mechanical arts, the other unskilled labour that makes use only of a man's body.

On money-lending as a principal form of investment, see especially Pliny, *Natural History* 33.28; Pliny, *Letters* 10.54.

11.131 Trade guilds

While the ancients had no real equivalent of our modern trade unions, there are many examples of workers in a common profession forming associations for social and funereal, though usually not economic, reasons. These *collegia*, as the Romans called them, were a significant element in the social fabric of antiquity, providing their members with a sense of fellowship and community in an otherwise impersonal culture.

Plutarch, *Numa* 17.1–2

Among King Numa's other administrative measures, it was his partitioning of the population according to trades that is especially acclaimed.... Flute players, goldsmiths, carpenters, dyers, shoemakers, leather finishers, coppersmiths, and potters were all separated into occupational divisions. By grouping the rest of the trades together he formed a single body out of all their members. And he gave them social activities, assemblies, and religious rituals appropriate for each body.

11.132 Agriculture vs. commerce

Common to most ancient societies was a tension between what had traditionally been a predominantly agrarian populace and, as agricultural productivity improved, a growing proportion of workers in commercial and "service" occupations. This passage gives credit to Solon for recognising this tension and acting to resolve it; it is sobering to realise that, centuries later, the Romans were still prohibiting senators from investing in "trade", and Cicero could happily observe that the only decent occupation for a free man was farming.

Plutarch, *Solon* 22.1

Solon realised that the city was teeming with a constant influx into Attica of people from all over in their quest for security, that most of the countryside was unproductive and poor, and that maritime traders were not in the habit of importing produce for those who had nothing to offer in return. So, he directed the citizens towards manual crafts, even having a law passed removing any requirement that a son support a father who had not taught him a trade.... Because he saw that the land characteristically could supply barely enough to sustain those who worked it and was unable to support a large leisured class of agriculturally unproductive people, Solon conferred dignity on the trades, even enjoining the Council of the Areopagus to examine how each man made his living and to reprimand those who were unemployed.

Athens' traditional dependence on imported food, alluded to in this passage, is well illustrated by an observation of Demosthenes 20 (*Against the Leptines* 31–33): "For you know, of course, that we Athenians use far more imported grain than any other peoples. And the amount that comes to us from the Black Sea is equal to all the grain that arrives from all the other areas of supply: ... around 400,000 *medimnoi* come to Athens from the Cimmerian Bosporus [the Crimea] alone".

11.133 Speculative manipulation of the grain market

From time to time in antiquity central governments assumed control over urban food supplies, if only to avoid civil disturbances that might threaten their authority. The state-manipulated *annona* of Rome is the most obvious (and successful) example of this. But in smaller centres, unfettered private enterprise and entrepreneurship were often the norm, despite their potentially deleterious effects on society as a whole. In this example, the first-century AC Pythagorean mystic Apollonius has sworn himself to a period of utter silence, but his self-control is put to the test when he confronts some greedy merchants in Aspendos, a usually prosperous small city on the south coast of Asia Minor.

Philostratus, *Life of Apollonius* 1.15

Apollonius arrived at Aspendos, the third most important city in Pamphylia, situated beside the Eurymedon River. Bitter vetches and other foodstuffs consumed only in dire circumstances were on sale to feed the inhabitants, since the influential merchants were keeping all the grain hoarded away so as to make a profit by selling it outside the province.

A crowd of every age was surrounding the chief magistrate and threatening to burn him alive, even though he was clinging to the imperial images [of Tiberius], which at that time were more venerated and inviolate than the statue of Olympian Zeus.... Apollonius went up to the magistrate and, using sign-language, asked him what was happening. He answered that *he* had done nothing wrong but was suffering as much as the rest of the people, and that if he couldn't make his point, he and his fellow citizens would all perish together. Apollonius turned to those who were standing about and indicated by nodding his head that they should listen to the magistrate. And they not only fell silent out of a feeling of awe towards him, but even laid aside their torches on the nearby altars.

Filled with fresh courage, the magistrate spoke: "This one and that one", he shouted, "are to blame for the famine that plagues us, for they have hoarded their grain and are stockpiling it here and there throughout the province!" The Aspendians began to encourage one another to head for the farms of the accused, but Apollonius convinced them by his head movements not to do that, but rather to summon those held responsible and try to get the grain from them through cooperation.

When the accused arrived, Apollonius came close to breaking his silence and speaking out against them, so affected was he by the tears of the inhabitants: children and women weeping together, and old men lamenting that they were on the point of dying from starvation. But he honoured his vow of silence and instead wrote his rebuke on his tablets, which he gave to the magistrate to read aloud, as follows: "Apollonius to the food-merchants of Aspendos: The earth is Mother of us all, for she is impartial; but you, by pursuing only your own interests, have tried to make her Mother only to yourselves; and unless you stop it, I shall not allow you to stand on her any longer".

The merchants were so frightened by this that they filled the agora with grain, and the city came to life again.

11.134 Food prices in early Rome

Pliny, *Natural History* 18.15

With this kind of traditional custom, not only was Italy agriculturally self-sufficient and in no need of importing foodstuffs from any of the provinces, but the market price of food was incredibly low.... Marcus Varro writes that, in the year Lucius Metellus celebrated his triumph with a long procession of elephants [150 BC], a single *as* could buy a *modius* of emmer wheat, a *congius* of wine, 30 *librae* of dried figs, 10 *librae* of olive oil, or 12 *librae* of meat. And it was not that this low cost was the result of the *latifundia* assembled by individuals who evicted their neighbours, since a law introduced by Licinius Stolo set a limit of 500 *iugera* of fenced land.

11.135 Price variations in the Roman Empire

Pliny, *Natural History* 33.164

I am quite aware that the prices of goods I have quoted in various contexts are not the same from place to place, and change almost every year. It all depends on the costs of shipping, or on the deals each merchant struck, or on some influential wholesaler who can keep the market value inflated. (I still remember a case in Nero's reign when Demetrius was prosecuted before the consuls by [every perfumer on] Seplasia Street [in Capua].) Still, I have had no choice but to quote the prices that were generally current at Rome, to give a sense of the relative worth attached to things.

11.136 An attempted freeze on wages and prices

The remarks of Pliny in the preceding passage are a good introduction to the dilemma that would face Diocletian 250 years later, at the dawn of the fourth century AC: disparate prices for basic necessities throughout the Empire, combined with inflation, devaluation of currency, and chronic unemployment or underemployment. His response was a brave, if misguided, one: to promulgate, in AD 301, a universal edict setting maximum prices that could be charged for particular goods throughout the Roman world, and the highest wages that could be paid for specified work. The strategy was an almost immediate failure – rather than curbing rising costs, it tended to drive newly unprofitable products out of the marketplace – but the edict itself still affords us a remarkable opportunity to understand the relative values of goods and services near the end of antiquity.

Diocletian, *Price Edict* (Frank, *ESAR* Vol. 5, pp. 318–356)

The prices that no one may exceed in the sale of individual goods are indicated here below:

wheat	1 military *modius*	*denarii* 100
barley		60
beans, meal		100
beans, whole		60
lentils		100
oats		30
lupines, raw		60
lupines, cooked	1 Italian *sextarius*	4
rice, cleaned	1 military *modius*	200
sesame		200
mustard		150
mustard, prepared	1 Italian *sextarius*	8
Similarly for wine:		
Tiburtine, Sabine, Falernian, etc.	1 Italian *sextarius*	*denarii* 30
previous year's vintage, first quality		24

previous year's vintage, second quality		16
table wine		8
beer, Celtic or Pannonian		4
beer, Egyptian		2
spiced wine		24

Similarly for oil:		
olive oil, first pressing	1 Italian *sextarius*	*denarii* 40
olive oil, second quality		24
olive oil, table quality		12
liquamen, first quality		16
salt	1 military *modius*	100
honey, best quality	1 Italian *sextarius*	40
Similarly for meat:		
pork	1 Italian *libra*	*denarii* 12
beef		8
goat or mutton		8
sow's udder		20
ham		20
sausage, pork	1 *uncia*	2
sausage, beef	1 Italian *libra*	10
pheasant, fattened		250
pheasant, wild		125
goose, fattened		200
goose, not fattened		100
chicken	1 pair	60
thrushes	10	60
pigeons	2	24
ducks	2	40
hare		150
rabbit		40
sparrows	10	16
dormice	10	40
peacock		300
peahen		200
boar	1 Italian *libra*	16
venison		12
lamb	1 *libra*	12
kid		12
butter	1 Italian *libra*	16

Similarly fish:		
saltwater fish, with rough scales	1 Italian *libra*	*denarii* 24
fish, second quality		16

river fish, best quality		12
river fish, second quality		8
fish, salted		6
oysters	100	100
mussels	100	50
sardines	1 Italian *libra*	16
Similarly:		
artichokes, large	5	*denarii* 10
lettuce, best quality	5	4
cabbage, best quality	5	4
beets, largest	5	4
radishes, largest	10	4
turnips, largest	10	4
green onions, first size	25	4
garlic	1 Italian *modius*	60
capers		100
cucumbers, first size	10	4
melons	4	4
asparagus, from the garden	bunch of 25	6
eggs	4	4
snails, largest	20	4
almonds, cleaned	1 Italian *sextarius*	6
pistachios		16
cherries	4 *librae*	4
apricots	10	4
peaches, largest	10	4
apples, best quality	10	4
plums, yellow, largest	30	4
figs, best quality	25	4
dates	25	4
olives, from Tarsus	20	4
milk, sheep's	1 Italian *sextarius*	8
cheese, fresh	1 Italian *libra*	8
For wages:		
farm labourer, with board	daily	*denarii* 25
stone mason, with board		50
mosaicist, as above		60
wall painter, as above		75
picture painter, as above		150
blacksmith, as above		50
baker, as above		50
shipwright, on a seagoing ship, as above		60

shipwright, on a river boat, as above		50
camel driver, donkey driver, muleteer, with board		25
shepherd, with board		20
veterinary, for clipping and preparing hoofs per head		6
barber	per man	2
sheep shearer, with board	per head	2
water carrier, for a full day's work, with board daily		25
sewer cleaner, for a full day's work, with board		25
polisher, for a used sword		25
scribe, for the best script	100 lines	25
tailor, for cutting and finishing a cloak of larger size		25
tailor, for an ordinary woman's tunic		16
teacher, elementary, per boy	monthly	50
teacher, arithmetic, per boy		75
teacher, Greek and Latin or geometry, per boy		200
advocate or jurist, for pleading a case		1,000
cloakroom attendant	per bather	2
bath attendant in a private facility		2

For hides:

oxhide, untanned, first quality	*denarii* 500
the same, tanned, for shoe leather	750
sheepskin, largest size, untanned	20
the same, tanned	30
seal skin, untanned	1,250
leopard skin, untanned	1,000

For boots:

boots for muleteers or farm workers, first quality, no hobnails	*denarii* 120
shoes, senatorial	100
shoes, equestrian	70
sandals, oxhide, women's, single-soled	30

For leather goods:

military saddle		*denarii* 500
bridle for horse, complete with bit		100
leather sack, first quality		120
leather sack	daily rental	2
packsaddle, for a donkey		250

(Frank, *ESAR* Vol. 5, pp. 363–366)

For wagons [and agricultural gear]:

freight wagon, best quality ...	*denarii* 6,000

passenger wagon ...	3,000
dormitory wagon ...	7,500
threshing sledge, wooden	200
plough with yoke	100
mill, to be powered by donkeys	1,250
mill, to be powered by water	2,000
handmill	250

(Frank, *ESAR* Vol. 5, pp. 382–385, 412)

[Miscellaneous precious commodities:]		
silk, white	1 *libra*	12,000
silk, dyed purple		150,000
wool, from Tarentum, washed		175
wool, dyed purple		50,000
gold, refined, bars or coins		50,000

Notes

1 Cf. *CIL* 9.6075 = *ILS* 5875 for earlier repairs to the Via Appia by Hadrian and local land-owners.
2 *mpm = milia passuum*; a Roman mile is equivalent to 1,000 paces or double-strides, so long distances are given in multiples of 1,000 paces.
3 Reeds were also used; see Pliny, *Natural History* 16.158.
4 For another description of the ship, see **11.81**.
5 The text is corrupt here, but this seems the most likely interpretation.
6 The local standard of talents, which first appear as a measure of gold in Homer, varied in their weights. Consider two examples: the Attic silver talent weighed about 26 kg, and produced 6,000 *drachmai* in coins (see Aristotle, *Athenian Constitution* 10.2); and the Egyptian gold talent of the first century AC. weighed 80 Roman *librae*, more than 27 kg (Pliny, *Natural History* 33.52).
7 The length of a *stadion* was commonly 600 Greek feet, about an eighth of a Roman mile.
8 The standard cubit of 24 *daktyloi* measured about 0.47 m.
9 Modern Broach, some 300 km north of Bombay, where the Narmada River enters the Gulf of Khambhat; Casson 1989: 198–201.
10 Or "slave musicians"?
11 Various distances are given in different versions of the text, but this figure is a reasonable estimate.

Bibliography

Maps, roads, and bridges

Alcock, Susan E., John P. Bodel, and Richard J.A. Talbert, *Highways, Byways, and Road Systems in the Pre-Modern World*. Malden: Wiley-Blackwell, 2012.
Bosanquet, Robert C., "Greek and Roman Towns; I. Streets: The Question of Wheeled Traffic". *Town Planning Review* 5 (1915) 286–293.

Bosio, Luciano, *La Tabula Peutingeriana: una descrizione pittorica del mondo antico.* Rimini: Maggioli, 1983.

Briegleb, Jochen, *Die vorrömischen Steinbrücken des Altertums.* Düsseldorf: VDI Verlag, 1971.

Bunbury, Edward H., *A History of Ancient Geography among the Greeks and Romans from the Earliest Ages till the Fall of the Roman Empire.* 2 Vols. 2nd edn. London: Murray, 1883.

Burford, Alison, "Heavy Transport in Classical Antiquity". *Economic History Review* 13 (1960) 1–18.

Campbell, Brian, *The Writings of the Roman Land Surveyors. Introduction, text, translation and commentary.* Society for the Promotion of Roman Studies, Journal of Roman Studies Monograph No. 9, [London], 2000.

Casson, Lionel, *Travel in the Ancient World.* London: Allen & Unwin, 1974.

Chevallier, Raymond, *Roman Roads.* London: Batsford, 1976.

Cüppers, Heinz, *Die Trierer Römerbrücken.* Mainz: von Zabern, 1969.

Coleman, Kathleen M., "Ptolemy Philadelphus and the Roman Amphitheater". Pp. 49–68 in William J. Slater, ed., *Roman Theater and Society: E. Togo Salmon Papers I.* Ann Arbor, University of Michian Press, 1996.

Dilke, Oswald A.W., *The Roman Land Surveyors. An Introduction to the* Agrimensores. Newton Abbot: David and Charles, 1971.

Dilke, Oswald A.W., *Greek and Roman Maps.* London: Thames & Hudson, 1985.

Forbes, Robert J., "Land Transport and Road-Building". Pp. 131–192 in Robert J. Forbes, *Studies in Ancient Technology*, Vol. 2. 2nd edn. Leiden: Brill, 1965.

French, David H., *Roman Roads and Milestones of Asia Minor Vol. 4. The Roads, Fasc. 4.1 Notes on the Itineraria.* British Institute at Ankara, Electronic Monograph 10, [London], 2016.

Gazzola, Piero, *Ponti romani. Contributo ad un indice sistematico con studio critico bibliografico.* Florence: Olschki, 1963.

Goodchild, Richard G., and Robert J. Forbes, "Roads and Land Travel, with a Section on Harbours, Docks, and Lighthouses". Pp. 493–536 in Charles Singer, Eric J. Holmyard, Alfred R. Hall, and Trevor Williams, eds., *History of Technology*, Vol. 2: *The Mediterranean Civilizations and the Middle Ages.* Oxford: Clarendon Press, 1957.

Hammond, Nicholas G.L., and Lawrence J. Roseman, "The Construction of Xerxes' Bridge over the Hellespont". *Journal of Hellenistic Studies* 116 (1996) 88–107.

Harley, John B. and David Woodward, eds., *The History of Cartography: Cartography in Prehistoric, Ancient, and Medieval Europe and the Mediterranean, Vol. 1.* Chicago: University of Chicago Press, 1987.

Irby, Georgia L., "Greek and Roman Cartography". Pp. 819–835 in Georgia L. Irby, ed., *A Companion to Science, Technology, and Medicine in Ancient Greece and Rome, Vol. II.* Chichester: Wiley-Blackwell, 2016.

Mitchell, Stephen, "Requisitioned Transport in the Roman Empire: A New Inscription from Pisidia". *Journal of Roman Studies* 66 (1976) 106–131.

O'Connor, Colin, *Roman Bridges.* Cambridge: Cambridge University Press, 1993.

Quilici, Lorenzo, "Land Transport, Part 1: Roads and Bridges". Pp. 551–579 in John P. Oleson, ed., *The Oxford Handbook of Engineering and Technology in the Classical World.* Oxford: Oxford University Press, 2008.

Ramsay, A.M., "The Speed of the Roman Imperial Post". *Journal of Roman Studies* 15 (1925) 60–75.

Snodgrass, Anthony M., "Heavy Freight in Archaic Greece". Pp. 16–26 in Peter Garnsey, Keith Hopkins, and Charles R. Whittaker, eds., *Trade in the Ancient Economy*. London: Chatto & Windus, 1983.

Talbert, Richard J.A. *Barrington Atlas of the Greek and Roman World*. Princeton: Princeton University Press, 2000.

Talbert, Richard J.A. *Cartography in Antiquity and the Middle Ages: Fresh Perspectives, New Methods*. Leiden: Brill, 2008.

Talbert, Richard J.A. *Rome's World: The Peutinger Map Reconsidered*. Cambridge: Cambridge University Press, 2010.

Talbert, Richard J.A. and K. Brodersen, eds., *Space in the Roman World*. Münster: LIT, 2004.

Thomson, J. Oliver, *History of Ancient Geography*. Cambridge: Cambridge University Press, 1948.

Ward-Perkins, John B., "Etruscan Engineering: Road Building, Water-Supply, and Drainage". Pp. 1636–1643 in Marcel Renard, ed., *Hommages à A. Grenier*. Collection Latomus, 58.3 (1962).

Werner, Walter, "The largest ship trackway in ancient times: the Diolkos of the Isthmus of Corinth, Greece, and early attempts to build a canal". *The International Journal of Nautical Archaeology* 26.2 (1997) 98–119.

Vehicles and riding

Anderson, John K., *Ancient Greek Horsemanship*. Berkeley and Los Angeles: University of California Press, 1961.

Anderson, John K., "Greek Chariot-Borne and Mounted Infantry". *American Journal of Archaeology* 79 (1975) 175–187.

Anderson, John K., "New evidence on the origin of the spur". *Antike Kunst* 21 (1978) 46–48.

Azzaroli, A., *An Early History of Horsemanship*. Leiden: Brill, 1985.

Bulliet, Richard W., *The Camel and the Wheel*. Cambridge, MA: Harvard University Press, 1975.

Connolly, Peter, and Carol Van Driel-Murray, "The Roman cavalry saddle". *Britannia* 22 (1991) 33–50.

Crouwel, Johan H., *Chariots and other Wheeled Vehicles in Iron Age Greece*. Amsterdam: Allard Pierson Stichting, 1993.

Crouwel, Johan H., *Chariots and other Wheeled Vehicles in Italy before the Roman Empire*. Oxford: Oxbow Books, 2012.

Crouwel, Johan H., and Jaap Morel, *Chariots and Other Means of Land Transport in Bronze Age Greece*. Amsterdam: Allard Pierson Museum, 1981.

Fortemyer, Victoria, "The Dating of the Pompe of Ptolemy II Philadelphus". *Historia* 37 (1988) 90–104.

Greenhalgh, Paul A.L., *Early Greek Warfare: Horsemen and Chariots in the Homeric and Archaic Ages*. Cambridge: Cambridge University Press, 1973.

Harris, Harold A., "Lubrication in Antiquity". *Greece and Rome* 21 (1974) 32–36.

Junkelmann, Marcus, *Die Reiter Roms: Die antike Reitkunst im archäologischen Experiment*. 2 vols. Mainz: von Zabern, 1991–1992.

Leone, Aurora, *Gli animali da trasporto nell' Egitto greco, romano e bizantino.* Rome: Pontifico Istituto Biblico, 1988.
Littauer, Mary A., "Early Stirrups". *Antiquity* 55 (1981) 99–105.
Littauer, Mary A., and Johan H. Crouwel, *Wheeled Vehicles and Ridden Animals in the Ancient Near East.* Handbuch der Orientalistik, Siebente Abteilung, Erster Band, Zweiter Abschnitt, B: Vorderasien, Lieferung 1. Leiden, Cologne: Brill, 1979.
Raepsaet, Georges, "Land Transport, Part 2: Riding, Harnesses, and Vehicles". Pp. 580–605 in John P. Oleson, ed., *The Oxford Handbook of Engineering and Technology in the Classical World.* Oxford: Oxford University Press, 2008.
Raepsaet, Georges, "Land Transport and Vehicles". Pp. 836–853 in Georgia L. Irby, ed., *A Companion to Science, Technology, and Medicine in Ancient Greece and Rome, Vol. II.* Chichester: Wiley-Blackwell, 2016.
Rice, Ellen E., *The Grand Procession of Ptolemy Philadelphus.* Oxford and New York: Oxford University Press, 1983.
Richardson, Nicholas J., and Stuart Piggott, "Hesiod's Wagon: Text and Technology". *Journal of Hellenic Studies* 102 (1982) 225–229.
Röring, Christof W., *Untersuchungen zu römischen Reisewagen.* Koblenz: Forneck, 1983.
Wiesner, Joseph, *Fahren und Reiten.* Archaeologia Homerica, I, F. Göttingen: Vandenhoeck & Ruprecht, 1968.

Navigation: general studies

Arenson, Sarah, *The Encircled Sea: The Mediterranean Maritime Civilization.* London: Constable, 1990.
Ashburner, Walter, *The Rhodian Sea Law.* Repr. Aalen: Scientia, 1976.
Bass, George F., ed., *A History of Seafaring Based on Underwater Archaeology.* New York: Walker, 1972.
Casson, Lionel, *Ships and Seamanship in the Ancient World.* Rev. edn. Princeton: Princeton University Press, 1986.
Casson, Lionel, *The Periplus Maris Erythraei. Text with Introduction, Translation and Commentary.* Princeton: Princeton University Press, 1989.
Casson, Lionel, *The Ancient Mariners. Seafarers and Sea Fighters of the Mediterranean in Ancient Times.* 2nd edn. Princeton: Princeton University Press, 1991.
Casson, Lionel, *Ships and Seafaring in Ancient Times.* London: British Museum Press, 1994.
D'Arms, John H., and E.C. Kopff, eds., *The Seaborn Commerce of Ancient Rome: Studies in Archaeology and History.* Memoirs of the American Academy in Rome 36 (1980).
Gianfrotta, Piero Alfredo, and Patrice Pomey, *Archeologia subacquea: storia, techniche, scoperte e relitti.* Milan: Arnoldo Mondadori Editore, 1981.
Göttlicher, Arvid *Die Schiffe der Antike: Eine Einführung in die Archäologie der Wasserfahrzeuge.* Berlin: Mann, 1985.
Oleson, John P., "A Possible Physiological Basis for the Term *Urinator*, 'Diver'". *American Journal of Philology* 97 (1976) 22–29.
Parker, Anthony J., "Classical Antiquity: The Maritime Dimension". *Antiquity* 64 (1990) 335–346.
Parker, Anthony J., "Cargoes, Containers and Stowage: The Ancient Mediterranean". *International Journal of Nautical Archaeology* 21 (1992) 89–100.

Rougé, Jean, *Ships and Fleets of the Ancient Mediterranean*. Middletown, CT: Wesleyan University Press, 1981.
Schulz, Raimund, *Die Antike und das Meer*. Darmstadt: Primus, 2005.
Starr, Chester G., *The Roman Imperial Navy 3 B.C.–A.D. 324*. 3rd edn. Chicago, 1993.

Navigation: ships and seamanship

Basch, Lucien, "Ancient Wrecks and the Archaeology of Ships". *International Journal of Nautical Archaeology* 1 (1972) 1–58.
Basch, Lucien, "Éléments d'architecture navale dans les lettres grecques". *L'Antiquité Classique* 47 (1978) 5–36.
Basch, Lucien, *Le Musée imaginaire de la marine antique*. Athens: Institut hellénique pour la préservation de la tradition nautique, 1987.
Ben-Eli, Arie L., ed., *Ships and Parts of Ships on Ancient Coins*. Haifa: National Maritime Museum, 1975.
Blackman, David, ed., *Marine Archaeology*. Colston Papers, Vol. 23. Hamden, CT: Archon, 1973.
Casson, Lionel, "Bronze Age Ships. The Evidence of the Thera Wall Paintings". *International Journal of Nautical Archaeology* 4 (1975) 3–10.
Casson, Lionel, "More Evidence for Lead Sheathing on Roman Craft". *Mariner's Mirror* 54 (1978) 139–144.
Casson, Lionel, John R., Steffy, and Elisha Linder, eds., *The Athlit Ram*. College Station: Texas A&M University Press, 1991.
de Graeve, Marie-Christine, *The Ships of the Ancient Near East (c.2000–500 B.C.)*. Leuven: Katholieke Universiteit, 1981.
Gillmer, Thomas C., "The Thera Ships – A Re-Analysis". *Mariner's Mirror* 61 (1975) 321–329; 64 (1978) 125–133.
Göttlicher, Arvid, *Materialien für ein Corpus der Schiffsmodelle im Altertum*. Mainz: von Zabern, 1978.
Göttlicher, Arvid, *Kultschiffe und Schiffskulte im Altertum*. Berlin: Mann, 1993.
Greenhill, Basil, ed., *Archaeology of the Boat. A New Introductory Study*. Middletown: Wesleyan University Press, 1976.
Hagy, James W., "800 Years of Etruscan Ships". *International Journal of Nautical Archaeology* 15 (1986) 221–250.
Irby, Georgia L., "Navigation and the Art of Sailing". Pp. 854–869 in Georgia L. Irby, ed., *A Companion to Science, Technology, and Medicine in Ancient Greece and Rome, Vol. II*. Chichester: Wiley-Blackwell, 2016.
Jenkins, Nancy, *The Boat Beneath the Pyramid. King Cheops' Royal Ship*. New York: Holt, Rinehart & Winston, 1980.
Johnston, Paul F., *Ship and Boat Models in Ancient Greece*. Annapolis: Naval Institute Press, 1984.
Johnstone, Paul, *The Sea-Craft of Prehistory*, 2nd edn. London: Routledge, 1988.
Kapitän, Gerhard, "Ancient Anchors: Technology and Classification". *International Journal of Nautical Archaeology* 13 (1984) 33–44.
Landström, Björn, *Ships of the Pharaohs: 4000 Years of Egyptian Shipbuilding*. London: Allen & Unwin, 1970.
Marsden, Peter, *Ships of the Port of London: First to Eleventh Centuries A.D.* Northampton: English Heritage, 1994.

McGrail, Sean, *The Ship. Rafts, Boats and Ships from Prehistoric Times to the Medieval Era*. London: Her Majesty's Stationery Office, 1981.

McGrail, Sean, "Sea Transport, Part 1: Ships and Navigation". Pp. 606–637 in John P. Oleson, ed., *The Oxford Handbook of Engineering and Technology in the Classical World*. Oxford: Oxford University Press, 2008.

Morrison, John S., *The Ship. Long Ships and Round Ships. Warfare and Trade in The Mediterranean, 3000 B.C.–500 A.D.* London: Her Majesty's Stationery Office, 1980.

Morrison, John S., ed., *The Age of the Galley – Mediterranean Oared Vessels since Pre-Classical Times*. London: Conway, 1995.

Morrison, John S., and Joseph F. Coates, *The Athenian Trireme. The History and Reconstruction of an Ancient Warship*. 2nd edn. Cambridge: Cambridge University Press, 2000.

Morrison, John S., and Joseph F. Coates, *Greek and Roman Oared Warships, 399–30 B.C.* Oxford: Oxbow Books, 1995.

Morrison, John S., and Roderick T. Williams. *Greek Oared Ships, 900–322 B.C.* Cambridge: Cambridge University Press, 1968.

Murray, William M., *The Age of Titans: The Rise and Fall of the Great Hellenistic navies*. Oxford: Oxford University Press, 2012.

Murray, William M., and Photias M. Petsas, *Octavian's Campsite Memorial for the Actian War*. Transactions of the American Philosophical Society, 79.4 (1989).

Oleson, John P., "A Roman Sheave Block from the Harbour of Caesarea Maritima, Israel". *International Journal of Nautical Archaeology* 12 (1983) 155–170.

Paglieri, Sergio, "Origine e diffusione delle navi etrusco-italiche". *Studi Etruschi* 28 (1960) 209–231.

Peretti, Aurelio, *Il periplo di Scilace. Studio sul primo portolano del Mediterraneo*. Pisa: Giardini, 1979.

Pomey, Patrice, and André Tchernia, "Le tonnage maximum des navires de commerce romains". *Archaeonautica* 2 (1978) 233–251.

Rival, M., *La charpenterie navale romaine. Matériaux, méthodes, moyens*. Paris: Centre Camile Julian, 1991.

Steffy, J. Richard, *Wooden Ship Building and the Interpretation of Shipwrecks*. College Station: Texas A&M University Press, 1994.

Whitewright, J., "Ships and Boats". Pp. 870–888 in Georgia L. Irby, ed., *A Companion to Science, Technology, and Medicine in Ancient Greece and Rome, Vol. II*. Chichester: Wiley-Blackwell, 2016.

Navigation, harbours, canals

Blackman, David J., "Ancient Harbours in the Mediterranean". *International Journal of Nautical Archaeology* 11 (1982) 79–104, 185–211.

Blackman, David J., "Sea Transport, Part 2: Harbors". Pp. 638–672 in John P. Oleson, ed., *The Oxford Handbook of Engineering and Technology in the Classical World*. Oxford: Oxford University Press, 2008.

Boyce, Aline A., "The Harbor of Pompeiopolis: A Study in Roman Imperial Ports and Dated Coins". *American Journal of Archaeology* 62 (1958) 67–78.

Houston, George W., "Ports in Perspective: Some Comparative Materials on Roman Merchant Ships and Ports". *American Journal of Archaeology* 92 (1988) 553–564.

Moore, Frank G., "Three Canal Projects: Roman and Byzantine". *American Journal of Archaeology* 54 (1950) 97–111.

Oleson, John P., ed., *Building for Eternity: The History and Technology of Roman Concrete Engineering in the Sea*. Oxford: Oxbow, 2014.

Raban, Avner, ed., *Harbour Archaeology. Proceedings of the First International Workshop on Ancient Mediterranean Harbours, Caesarea Maritima, 24–28.6.83*. Oxford: British Archaeological Reports, 1985.

Reddé, Michel, "La représentation des phares à l'époque romaine". *Mélanges de l'École Française de Rome* 91 (1979) 845–872.

Rickman, Geoffrey E., "The Archaeology and History of Roman Ports". *International Journal of Nautical Archaeology* 17 (1988) 257–267.

Smith, Norman A.F., "Roman Canals". *Transactions of the Newcomen Society* 49 (1978) 75–86.

Uggeri, Giovanni, "La terminologia portuale romana e la documentazione dell' 'Itinerarium Antonini'". *Studi Italiani di Filologia Classica* 40 (1968) 225–254.

Trade, prices, money

Finley, Moses I., *The Ancient Economy*. 2nd edn. Berkeley: University of California Press, 1985.

Frank, Tenney, ed., *An Economic Survey of Ancient Rome*. 6 Vols. Baltimore: Johns Hopkins University Press, 1933–1940.

Garnsey, Peter, Keith Hopkins, and Charles R. Whittaker, eds., *Trade in the Ancient Economy*. Berkeley: University of California Press, 1983.

Hopper, Robert J., *Trade and Industry in Classical Greece*. London: Thames & Hudson, 1979.

Horden, Peregrine and Nicholas Purcell, *The Corrupting Sea*. Hoboken: Wiley-Blackwell, 2000.

Knorringa, H., *Emporos: Data on Trade and Traders in Greek Literature from Homer to Aristotle*. Amsterdam: Paris, 1926.

Rickman, Geoffrey E., "Articles of trade, and problems of production, transport, and diffusion". *Memoirs of the American Academy at Rome* 36 (1980), 261–276.

Tchernia, André, *The Romans and Trade (first published 2011; Translated from the French by James Grieve, with Elizabeth Minchin)*. Oxford Studies on the Roman Economy. Oxford; New York: Oxford University Press, 2016.

Coinage

Hill, George F., "Ancient Methods of Coining". *Numismatic Chronicle* 2 (1922) 1–42.

Jones, John R., *Testimonia Numaria: Greek and Latin Texts Concerning Ancient Greek Coinage*, Vol. 1: *Texts and Translations*. London: Spink, 1993.

Kraay, Colin M., "Hoards, Small Change and the Origin of Coinage". *Journal of Hellenic Studies* 84 (1964) 76–91.

Meadows, A., "Technologies of Calculation, Part 2: Coinage". Pp. 769–776 in John P. Oleson, ed., *The Oxford Handbook of Engineering and Technology in the Classical World*. Oxford: Oxford University Press, 2008.

Reece, Richard, "The Use of Roman Coinage". *Oxford Journal of Archaeology* 3 (1984) 197–210.

Sellwood, David G., "Some Experiments in Greek Minting Technique". *Numismatic Chronicle* 3 (1963) 217–231.

Weights, measures, and numbers

Glautier, Michael W.E., "A Study in the Development of Accounting in Roman Times". *Revue Internationale des Droits de l'Antiquité* 19 (1972) 311–343.

Hecht, Konrad, "Zum römischen Fuss". *Abhandlungen der Braunschweigischen Wissenschaftlichen Gesellschaft* 30 (1979) 1–34.

Hultsch, Frederich, *Griechische und römische Metrologie*. 2nd edn. Berlin: Weidmann, 1882.

Keyser, Paul. "The Origins of the Latin Numerals 1 to 1000". *American Journal of Archaeology* 92 (1988) 529–546.

Kisch, Bruno, *Scales and Weights. A Historical Outline*. New Haven: Yale University Press, 1965.

Lillo, Antonio, *The Ancient Greek Numeral System. A Study of Some Problematic Forms*. Bonn: Habelt, 1990.

McCartney, Eugene S., "Popular Methods of Measuring". *Classical Journal* 22 (1927) 325–344.

Nissen, Heinrich, "Griechische und römische Metrologie". Pp. 833–890 in Iwan von Müller, ed., *Handbuch der klassischen Altertumswissenschaft*, Vol. 1. 2nd edn. Munich: Beck, 1892.

Nowotny, Eduard, "Zur Mechanik der antiken Waage". *Jahreshefte des Österreichischen Archäologischen Instituts*, Wien 16 (1913) Beiblatt.

O'Brien, Denis, *Theories of Weight in the Ancient World*, Vol 1: *Weight and Sizes*. Vol 2: *Weight and Sensation*. Leiden: Brill, 1981–1984.

Wikander, Charlotte, "Technologies of Calculation, Part 1: Weights and Measures". Pp. 759–768 in John P. Oleson, ed., *The Oxford Handbook of Engineering and Technology in the Classical World*. Oxford: Oxford University Press, 2008.

12

RECORD-KEEPING

A TIME-KEEPING

Although time is an artificial and abstract concept (Diogenes Laertius, *Zeno* 7.141), its importance to humans is undeniable. For agriculture, husbandry, and hunting the significance of time is obvious. Crops need to be planted and harvested, and animals move or need to be moved at specific times. These are broad observations of time that were easily observed through changes in the seasons through "time-reckoning". Eventually, however, people required more precise timing, first to celebrate religious festivals in honour of their gods and then to accommodate their increasingly busy, urban life: time had to be measured by using instruments. The change from a rural life – where the general observation of time was adequate – to life in the city – where division of the daylight hours became more and more important – is the basic difference between "time-reckoning" and "time measurement"; only the latter requires the accuracy supplied by instruments.

Recording the passage of solar days, lunar months, and solar years was at first sufficient time-reckoning. This simple and non-technological method for observing time, however, was found to be inadequate for maintaining proper units of time as record-keeping became more important. As a result, adjustments had to be made to calendars (Gawlinski 2016, Brind'Amour 1983, Hannah 2016). Such modifications, although non-technological, are the basis of the modern calendar, and a few examples have been included here (**12.5**). When it became important to divide the day, and later the night, into units of time, instruments began to play a major role in time measurement. The sundial served to divide daytime into appropriate units but was totally inadequate to apportion time at night, thus stimulating the invention and development of the water clock (*clepsydra*). Several problems, however, were never fully resolved, the most important being the division of the day into twelve hours. Since the amount of daylight changes throughout the year, hours could be anywhere from 45 to 75 minutes long. The lack of a set unit of time for the hour in antiquity proved to be a constant source of annoyance for the builders of the clocks (**12.10**).

Vitruvius is our most important written source, providing information about the Hellenistic period that is simultaneously very detailed and confusing. A treatise on a large water clock attributed to Archimedes survives only in Arabic manuscripts and is not treated here. Much archaeological evidence survives to support the written material; two of the best-known examples are the Tower of the Winds (Noble and Price 1968), a monumental water clock in Athens, and the Horologium Augusti (Buchner 1982), a monumental sundial in Rome that used an obelisk as its gnomon. There are, however, no surviving literary references to calendrical computers, such as the first-century BC Antikythera Mechanism found in a shipwreck (Jones, 2017, Price 1974). The closest we come are the descriptions of Archimedes' mechanical model of the universe (**2.45**). (Hannah 2016, 2008)

Note: in the following passages, the term clock is used in the general sense for devices used to keep time, that is, for sundials and water clocks.

12.1 The importance of time

Antiphon, Fragment 77 [137b, 1445] = Plutarch, *Antony* 28.1 (928)

[Antony and Cleopatra are in Alexandria.] There, indulging in the pastimes and sports of a young man at leisure, Antony squandered and consumed with his pastimes the thing which Antiphon calls the most expensive outlay: time.

12.2 Everyman's sundial

Greek Anthology 11.418

By putting your nose opposite to the sun and opening your mouth you will show the hours to everyone passing by.

12.3 The earliest written record of months

The importance of religious celebrations is apparent in the Mycenaean calendar since only tablets concerned with religious offerings have a date given in months (Ventris and Chadwick, *DMG* p. 407). The translations and interpretations are quite tentative: vacant spaces may be a function of the calendar and some phrases have other possible meanings (see *DMG* pp. 475–476 for alternative interpretations).

DMG no. 172 (pp. 284–289, 462–464)

(Reverse)

In the month of Plowistos. Pylos sacrifices at Pa-ki-ja-ne and brings gifts and leads victims.
For the Mistress: one gold cup, one woman.
For Mnasa: one gold bowl, one woman.
For Posidaeia: one gold bowl, one woman.
For the "thrice-hero": one gold cup.

For the "lord of the house": one gold cup.
Pylos ... (blank)

(Obverse):
Pylos sacrifices at the shrine of Poseidon and the city, and brings gifts and leads victims: one gold cup, two women, for Gwowia (and?) Komawenteia....

12.4 Early Greek discoveries

Diogenes Laertius, *Lives of Eminent Philosophers* 2.1

Anaximander was the first man to invent the gnomon and to set it up as a sundial in Lacedaemon to mark the solstices and equinoxes, as Favorinus states in his *Miscellaneous History*. He also built clocks.

Pliny (*Natural History* 2.187) credits Anaximenes, the pupil of Anaximander, with the discovery at Lacedaemon.

Herodotus, *Histories* 2.109

The Hellenes learnt the sundial, the gnomon, and the 12 divisions of the day from the Babylonians.

12.5 Early Roman time-keeping and its refinement

The next passages demonstrate quite well the Roman concept of time and its development at Rome. Pliny clearly indicates the disadvantages of the sundial and the superiority of the water clock. The last passage, which records Julius Caesar's reforms to the calendar, is included because our own calendar is based upon this change.

Livy, *History of Rome* 1.19.6–7

First of all, King Numa [715–673 BC] divided the year into 12 months according to the revolutions of the moon. But since the moon does not completely fill thirty days in each month and eleven days are lacking to form the full year that is designated by the solar revolution, he regulated the year by interposing intercalary months in such a way that, in the twentieth year, the days should coincide with the same position of the sun as when they started; as a result the length of all the years will be full [and no days will be lacking]. He also created the days when public business could or could not be conducted, since sometimes it would be useful if nothing could be proposed to the people.

Pliny, *Natural History* 7.212–215

... In the Twelve Tables [451–450 BC] only sunrise and sunset are listed. After a few years, midday was also added, which the attendant of the consul announced when, from the Curia, he saw the sun between the Rostra and the Graecostasis [an elevated area or structure where foreign ambassadors awaited a summons to address the senate]. He also announced the last hour of the day when the sun retreated from the Maenian Column to the prison, but only on clear days right down to the First Punic War [264–241 BC]. According to Fabius Vestalis, 11 years before war was waged with Pyrrhus [begun in 281 BC], Lucius Papirius Cursor first erected a sundial for the Romans at the Temple of Quirinus when he dedicated the temple that had previously been vowed by his father. But Fabius does not indicate the method of manufacture or the maker of the sundial, nor from where it was brought or from which writer he obtained his information.

Marcus Varro records that the first one set up in public was on a column beside the Rostra by the consul Manius Valerius Messala during the First Punic War after Catania in Sicily was overcome, and was taken from there 30 years after the date given for the Papirian sundial, in the four hundred and ninety-first year of the city [263 BC]. Although its lines did not correspond to the hours, nevertheless they were dependent on it for 99 years until Quintus Marcius Philippus, who was censor with Lucius Paullus, set up a more carefully designed one next to it; and this was accepted as the most welcome of the censor's undertakings. Even then, however, the hours were uncertain in cloudy weather until the next *lustrum* when Scipio Nasica, the colleague of Laenas, first divided the hours of the days and nights equally with a water clock. And he dedicated this clock in the five hundred and ninety-fifth year of the city in a roofed building. For so long a time the day was undistinguished for the Roman people.

Censorinus, *The Natal Day* 20.4–11

Afterwards either by Numa, according to Fulvius, or by Tarquin, according to Junius, the 12 months and 365 days were instituted [as the period for a year].... At last when it had been decided to add an intercalary month of 22 or 23 days in alternate years, so that the civil year corresponded to the natural [solar] year, it was intercalated best in the month of February between Terminalia [the festival on the twenty-third] and Regifugium [the festival on the twenty-fourth]. This was done for a long time before it was perceived that the civil years were somewhat longer than the natural years. The duty for correcting this fault was given to the pontiffs along with the power for them to intercalate at their own discretion. But most of these men further distorted the matter entrusted to them to correct by wantonly intercalating longer or shorter months on account of their hatred or favouritism, by which some men departed from office more quickly or some served longer....

The calendar was so out of step that Gaius Julius Caesar, the *pontifex maximus*, in his third consulship with Marcus Aemilius Lepidus [46 BC], in order to correct the earlier defect inserted two intercalary months, totalling 67 days, between November and December although he had already intercalated 23 days in the month of February. This made the year one of 445 days. At the same time, he provided that the same mistakes would not occur in the future, for once he abolished the intercalary month he regulated the civil year according to the course of the sun. And so, he added 10 days to the 355, which he divided among the seven months with 29 days.... In addition, on account of the quarter day which is known to complete a true year, Caesar decreed that after a cycle of four years a single day, where once there used to be a month, should be intercalated after Terminalia [23 February], which is now called *bissextus* [double-sixth; our leap year day].

CLOCKS

12.6 The affliction of sundials

Plautus, *The Boeotian Woman* (Fragment v.21 Goetz) = Aulus Gellius, *Attic Nights* 3.3.5

May the gods destroy that man who first discovered hours and who first set up a sundial here; who cut up my day piecemeal, wretched me. For when I was a boy, my only sundial was my stomach, by far the best and truest of all clocks. When it advised you, you ate, unless there was no food; now even when there is food it isn't eaten unless the sun allows it. Indeed, now the town is so filled with sundials that the majority of its people crawl about all shrivelled up with hunger.

12.7 Sundials and philosophers all disagree

Philosophers, well known for their inability to agree, were almost as consistently inconsistent as sundials, if we trust the following passage.

Seneca, *The Pumpkinification of Claudius* 2.2

Although I can't tell you the exact hour, which is more easily agreed upon among philosophers than among the sundials, nevertheless it was between the sixth and seventh hour.[1]

12.8 Sundials

Sundials are of great importance in the history of time-keeping, but once created they do not offer much scope for improvement. This does not mean they were not improved or that they could be constructed and used without proper instruction. Before describing the complexities of their construction, Vitruvius relates a precaution for the builder and user of the analemma: the measurement of time at different latitudes required specific knowledge of the length of the shadow cast by the gnomon of the sundial at the equinox (Gibbs 1976).

Vitruvius, *On Architecture* 9.7.1

But we must explain the principles of the shortening and lengthening of the day and separate them from these [previously explained astronomical observations]. For the sun at the equinoctial time, turning through Aries [spring equinox] and Libra [autumn equinox], makes the gnomon cast a shadow eight-ninths of its own length, at [the latitude of] Rome. At Athens the shadows are three-quarters the length of the gnomon, at Rhodes five-sevenths, at Tarentum nine-elevenths, at Alexandria three-fifths, and in all other places the shadows of the equinoctial gnomons are found to be different in varying amounts according to nature.

See Pliny, *Natural History* 2.182 for similar information. After describing the analemma and its applications, Vitruvius provides a partial list of types of sundials and their inventors. The list is important since it demonstrates that even the technology for sundials did not remain static: new types were invented to improve a relatively simple concept. Archaeological evidence supports many of the types listed (Talbert 2017, Price 1969).

Vitruvius, *On Architecture* 9.8.1

Berosus the Chaldaean is reported to have invented the semicircular type, hollowed out of a square block and carved on the lower portion according to the inclination of the equator to the horizon [the polar altitude]; the *Scaphe* or Hemisphere as well as the disc on a level surface by Aristarchus of Samos; the *Arachne* [spider] by the astronomer Eudoxus, or some say Apollon. [Several others are listed.].... Many have also written instructions for making travelling sundials of these types that can be hung up.

Water clocks consisted of four basic types: the simple outflow type, which merely let a measured amount of water escape (**12.9**); the large outflow type with lines marked on the inside to indicate hours (**12.10**); the simple inflow type, which consisted of two vessels: an outflow clepsydra that discharged into a reservoir with hours marked on its interior; and the inflow type with reservoir and overflow pipe that maintained a constant head of pressure (**12.11**). Although other refinements were made in an attempt to account for the lengthening and shortening of the days, the invention of the fourth type is regarded as pivotal for ancient time measurement since this solved the problem of decreasing water pressure as the water emptied from the other types (Hannah 2016, Noble and Price 1968).

12.9 Water clocks in court

Many speeches delivered in court refer to the lack of water or to insufficient water for making appropriate or elaborate charges and accusations; others indicate that the flow of water from the clepsydra was stopped for depositions and other interruptions. The clepsydrae used in the courts were very simple devices from which flowed a measured amount of water; once the water had run out, the time was up.

Plato, *Theaetetus* 172d

Those men, on the other hand, always speak in haste, for the flowing water [of the clock] urges them on.

Demosthenes 45, *Against Stephanus* 1.8

[To the clerk] Take the deposition itself and read it so that out of it I can prove my case. Read; and you [another officer of the court] stop up the water [from the clock].

In the following speech, Demosthenes (although part of the *corpus* of Demosthenes, this speech probably is not by Demosthenes himself) acts on behalf of a woman in a very complicated inheritance case. Several claimants, according to the speaker, have conspired to defraud his client of her inheritance, but he does not have the water (time) to explain all the facts of the case. The claim of insufficient water is quite common, but this passage is unusual since it provides specific amounts, a useful indication that in the courts time was not measured in minutes and hours, but according to the volume of water in the clock.

[Demosthenes] 43, *Against Macartatus* 8–9

And when the Archon took the case to court, and it was to be tried, the claimants had everything readied for the trial. They even had four times as much water [in the clock] as we did for making speeches. For it was necessary that the Archon, men of the jury, pour in [to the clock] an *amphora* [of water] for each of the claimants, and 3 *choes* for the response. The result is that I, acting as the counsel for this woman, cannot explain to the jurors the family relationship and other matters in as fitting a manner as I would have wished, nor am I able to defend myself against the tiniest amount of the charges they have invented about us; for I have only a fifth part of water [a fifth of the time they have].

12.10 A formula for the dimensions of a water clock

The difficulty of trying to build a small water clock with a reservoir of a shape such that the water recedes an equal distance in equal amounts of time is immense. This formula from a papyrus of the third century AC produces a shape "something like a flower-pot". Such measurements for water clocks are very rare and unfortunately the scribe has made some mistakes. Corrections and omissions are in angled brackets. The commentary in *P.Oxy.* 470 is very helpful.

P.Oxy. 470.31–85

The calculation for the construction of clocks [*horologia*] is given thus: make the upper <diameter> of an *olmiskos* [frustrum of a cone] 24 *dactyloi*, the bottom 12 *dactyloi*, and its depth 18 *dactyloi*. If we add the 24 *dactyloi* to the 12 *dactyloi* of the base, there will be 36 *dactyloi*, of which half will be 18, <multiplying this> by 3 because of its rounded shape the product will be 54; of this a third is 18, a quarter is 13.5; the one multiplied by the other will give <a volume of?> 143 <243>; thus it makes 204 [this final number's significance is unknown].... Thus the first line [*grammai*] is 24 *dactyloi* <long>; double this number is 48, of which take away ⅔, the remainder is 47⅓, of this half is 23⅔; multiplied by ⅓ <3> the result is 71; a third <of this> is 23⅔, a quarter is 17⅔+1/12 <17¾>; the product is 310 1/12 <420 1/12>.... [The intervening calculations are here omitted.]

The seventh calculation is <20 [*dactyloi* long], doubled is >40, take away ⅔, the remainder is 39⅓, of which a half is 15⅔ <192/3>.

Aeneas Tacticus provides a refreshingly simple method to regulate the volume of water in a clepsydra.

Aeneas Tacticus, Fragment 48

A water clock is a very useful device for the night-time guards. Since the nights become longer or shorter, it is constructed in the following manner. It is necessary to smear the inside with wax and when the nights become longer to take out some of the wax so that there is room for more water; when the nights become shorter to mould on more so that it holds less water. And it is necessary to make with accuracy the orifice through which the water of a specified period flows out.

12.11 Water clocks

Vitruvius indirectly credits Ctesibius with the invention of the inflow water clock with overflow and float in the reservoir. The second reservoir, with both an opening at the bottom and the overflow at the top, will maintain a constant volume in the tank and thus will provide constant pressure for the outflow to the clock itself. Vitruvius also describes some of Ctesibius' other innovations for water clocks such as the use of special materials to prevent clogging of the openings. This type of water clock is regarded as the forerunner of our mechanical clock.

Vitruvius, *On Architecture* 9.8.2–7

The methods of making water clocks have also been examined.... first of all by Ctesibius of Alexandria.... First, he began by making a hollow opening out of gold or by piercing a gemstone, since these materials are not worn by the action of water nor collect grit so they become clogged. The water flowing in at a regular volume through that opening raises an inverted bowl, which is called the cork or drum by the craftsmen. A bar and revolving drum are attached to this apparatus and both are fitted with regularly spaced teeth, which, when meshing into one another, make measured rotations and movements. Other bars and other drums, toothed in the same manner and driven by the same motion, cause various effects and movements by their turning: figures are moved, cones revolve, pebbles or eggs fall, trumpets sound, and other peripheral actions happen. On these clocks the hours are marked either on a column or a pilaster; a figure emerging from the bottom points them out with a rod throughout the entire day. The rods have to be adjusted by shortening or lengthening them for every day and month by inserting or removing wedges. The shutoffs for controlling the water are made as follows. Two cones are made, one solid and one hollow, turned on a lathe so that one can go into and fit the other. The same rod is used to loosen or to tighten them, producing a powerful or gentle current of water flowing into the vessels. Thus, by these methods and devices, water clocks are constructed for use in winter. But if the shortening or lengthening of the days is not accomplished properly by adding or removing wedges, since the wedges are very often defective, it must be adjusted as follows. The hours are to be marked off transversely on a small column from the analemma, and the lines of the months are also to be marked upon the small column. And this column is to be made to revolve so that as it turns continuously towards the figure and the rod, by which the emerging figure points out the hours, it thus adjusts the shortening and lengthening of the hours according to their months.

12.12 An anaphoric clock coordinated with the stars

The description provided by Vitruvius demonstrates that this device is an ancestor of the astrolabe, although according to this passage it was used principally as a clock.

Vitruvius, *On Architecture* 9.8.8–10

There are also winter clocks of another type, which are called *Anaphorica*, and they make them in the following manner. The hours, radiating from a central point on the front, are indicated by bronze rods on the inscribed analemma. Circles are inscribed on this to mark the limits of the monthly spaces. Behind these rods is a drum, on which the firmament is drawn and painted along with the circle of signs [of the zodiac]. And the drawing is composed of 12 celestial symbols, of which, when moving from the centre, the representation is first greater then it becomes smaller. On the back part of the drum, in the middle, a revolving axle is inserted and on it is wound a flexible bronze chain. At one end hangs a cork, or drum, which is raised by the water, and at the other end a counterpoise of sand, equal in weight to the cork. Thus, as much as the cork is raised by the water, by that much the weight of the sand drags down and turns the axle, and the axle turns the drum. The revolution of this drum causes either a larger or a smaller part of the circle of the signs – by their turning – to indicate the proper length of the hours according to their seasons. For in each of the signs are made holes for the number of days in each month; and a pin, which seems to represent the sun on the clock, indicates the spaces for the hours. This pin as it is moved from hole to hole completes the course of the passing month. Thus, just as the sun passing through the spaces of the constellations lengthens and shortens the days and hours, so the pin on the clock, moving along the holes in the direction opposite to the turning of the drum at the middle, is carried day by day, sometimes over wider and sometimes over narrower spaces, and it produces a representation of the hours and days for the limits of the months.

Vitruvius concludes his discussion by describing a method to control the flow of water in an attempt to make allowance for the varying lengths of daylight hours. Unfortunately, he makes two incorrect assumptions: first that the rate of flow is directly proportional to the head of water, and second that the lengthening and shortening of the days are symmetrically grouped around the equinoxes.

12.13 An ancient alarm clock

The relationship between water clocks and water organs was very close. Vitruvius (**12.11**) states that Ctesibius devised both, combining elements to produce elaborate water clocks that trumpeted. Plato, in fact, is said to have used such a device, perhaps to signal the beginning of his lectures at daybreak. The banqueters in the following passage have just heard a water organ and are debating whether it belongs to wind or stringed instruments.

Athenaeus, *Philosophers at Dinner* 4.174c

But it is said that Plato provided a small notion of its construction by having made a clock for use at night that was similar to a water organ, although it was a very large water clock. And indeed, the water organ does seem to be a water clock.

The banqueters conclude that the water organ belongs to the wind instruments. For a discussion of water organs, see **2.46**.

B HOW TO SELECT: SORTITION VS. ELECTION

In the Greek and Roman worlds, selections for the various offices were made by rulers, sortition/allotment, and election. The last could be done openly in several simple ways: raising hands (Aristotle, *Athenian Constitution* 55.4 and Plutarch, *Phocion* 34–35), walking across the floor (Suetonius, *Tiberius* 31) or acclamation – for example, the loudest shouting by the Spartans (Aristotle, *Politics* 2.6.18 (1271a) and Plutarch, *Lycurgus* 26).

Sortition, basically getting the "short straw", introduced a randomness to the process and became a favourite method of appointment for many offices in the Athenian democracy where all citizens were regarded, ideally, as equally fit and equally responsible for the various duties (for Roman use see Varro, *On Agriculture* 3.17.1 and Cicero, *On Behalf of Gnaeus Plancius* 22.53).

At the same time, however, the tradition of voting with ballots was used in some cases to ensure that the count for a quorum was reached, and eventually became more important with the introduction of secret ballots to safeguard the selection of officials (Staveley 1972 and Rhodes 2004 and 1981).

12.14 Simple techniques of sortition

Allusions to simple allotment procedures appear in the literature from Homer onward. Odysseus (*Odyssey* 9.331–335) states he got the men, whom he would have picked himself, to blind Polyphemus. The texts record a variety of objects used for the selection process: beans, pebbles, seal-stones among others.

Plutarch, *Pericles* 27.2

Pericles has laid siege to Samos in 440/439 BC.

Since it was a troublesome business for him to restrain the Athenians, who were unable to endure the chafing of the delay and were eager to fight, having separated the entire force into eight divisions, he conducted a selection by lot from the groups. He then granted to the division taking the white bean to feast and to take their leisure, while the others kept fighting. And because of this, they say those having enjoyed their comforts call it a "white day", – from the white bean.

Plutarch, *Timoleon* 31.3

Timoleon's mercenary force in western Sicily faces the much larger forces of Hicetas and the Carthaginians at the Crimissus River in 339 BC.

A remarkable strife and eager rivalry fell upon the cavalry officers with Timoleon that created a delay for the battle. For not a single officer was willing to cross the river against the enemy later than another, but each thought himself worthy to be the initiator; thus, their crossing would have no order with them jostling and trying to outrun one another. So Timoleon wishing to decide the leaders by lot, took a signet-ring from each leader, then after casting all the rings into his own military cloak and having mixed them up, he revealed the first one selected, that by chance had as the engraving of its seal a trophy of victory.

12.15 The *kleroterion*: a complex allotment machine

At Athens by the fourth century BC, and probably before, the special needs of the democracy required that the simple lot process be made more secure and open, yet closed when larger numbers of participants were involved for the law courts since selection of jurors received careful consideration first, to ensure a cross-section of the full citizen body and second, to ensure jurors were not known ahead of time in order to ensure as much fairness as possible.

Selection, entrance to court, and the casting of the juror's one vote secretly are part of an elaborate process, which involves an ingenious device, the *kleroterion* [**fig. 12.1**], an allotment machine created to serve the needs of the democratic polis and its tradition of selection by lot.

The following passage relates to the jury courts, which became particularly important in fifth and fourth century BC Athens, both for dispensing justice in as impartial a manner as possible by randomising not only the jurors but the judges and courts and protecting the secrecy of the vote (Dow 1939/2004, Bishop 1979, and Rhodes 1981).

Artistotle, *Constitution of Athens* 63. 1–2; 64–66

The nine Archons appoint by lot [the jurors for 9 of] the law courts, tribe by tribe, the Secretary of the Thesmothetai [Lawgivers] those of the tenth tribe.

There are ten entrances to the law courts, one for each tribe; and 20 *kleroteria* [allotment machines], two for each tribe; and 100 boxes, ten for each tribe, and other boxes into which are thrown the *pinakia* [voting identity cards; **fig. 12.2**] for the jurors selected by lot. And two balloting urns and rods of office,

RECORD-KEEPING

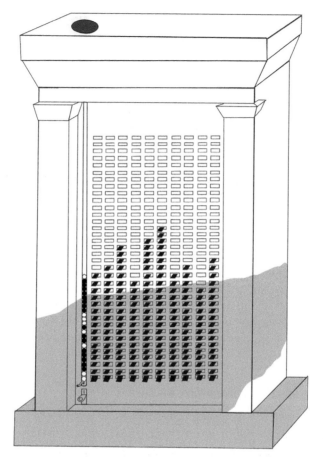

Figure 12.1 Kleroterion: Aristotle's voting machine
Source: Adapted from a broken example, Agora Museum, Athens, Greece; Camp, 1986, fig. 83.

Figure 12.2 Pinakion: voting ballot, Athens, fourth century BC.
Source: Adapted from examples in Agora Museum, Athens, Greece; Camp, 1986, fig. 84.

corresponding to the number of jurors [to be selected from each tribe] are placed at each entrance. Acorn-shaped ballot balls equal to the number of rods are thrown into the urn. On these acorns are engraved the letters of the alphabet starting from the eleventh, the *lambda* ["L"] – used of the thirtieth number –employing as many successive letters as will be courts to be filled ... [qualifications of jurors follows]....

Each juror has a boxwood *pinakion*, inscribed with the names of himself, his father, and his deme, and one letter of the alphabet up to and including *kappa* ["K"; the letter before *lambda*]. For the jurors are divided into ten sections according to tribe, equal numbers of them under each letter.

Once the Thesmothet has selected by lot the letters to be assigned to the law courts, the attendant takes them and affixes to each court the letter obtained by lot.

The ten boxes rest in front of the entrance of each tribe. They are inscribed with the letters of the alphabet up to and including *kappa*. When the jurors have thrown their *pinakia* into the box upon which is engraved the same letter as is inscribed on their own *pinakion*, once the attendant vigorously shakes the box, the Thesmothet draws one *pinakion* from each box.... The man called the "Affixer" [also selected by lot to prevent corruption] sticks the *pinakia* taken from the box into the framed column of slots upon which has the same letter as on the box.... there are five columns of slots on each of the *kleroteria*.

Once the Archon has thrown in the cubical markers, he conducts the allotment for the tribe with the *kleroterion*. The cubes are bronze: black and white. As many white cubes are thrown in [i.e. into the tube] as would be needed to obtain the jurors by lot, one white cube for every five *pinakia;* the black cubes utilise the same method. When the Archon draws out the cubes, the herald summons those selected by lot.... Having heeded the call, the summoned man draws an acorn from the urn, and stretching it out holding up the engraved letter, he first shows it to the appointed Archon. When the Archon has inspected it, he throws the *pinakion* of the man into the box inscribed with the same letter as that on the acorn in order that the man will go into the court to which he is allotted and not into whatever court he may want. Thus, it would not be possible to bring together in a court the jurors someone may wish. Lined up at the side of the Archon are as many boxes as would be courts to be filled, and each has its letter which have been obtained by lot for each court. [a series of checks continue to ensure the juror goes to the proper court to which he has been assigned by lot; officials he encounters have also been selected by lot; once the courts are filled, allotment machines are then used to select the magistrates who are assigned by lot to a particular court; lot is also used to assign jurors certain duties of the court.]

12.16 Simple Spartan election tools

Even with sortition, sometimes the need for confidentiality arose, thus simple methods to ensure secrecy became more elaborate with technological improvements brought in for voting in jury courts. The tradition of a secret vote in Sparta, which was famous for acclamation and openness, is perhaps not anachronistic since camaraderie was essential and, thus, it was best not to 'know' who was against your membership in the mess, a central feature of Spartan society.

Plutarch, *Lycurgus* 12.5–6

It is said that someone wishing to be a member of a mess [common meal that strengthened social, political and military bonds] was approved as fit in the following manner. Each of the messmates, taking a piece of soft bread in his hand, in silence threw it, just like a pebble for voting, into a vessel being carried on the head of a servant. Those thinking him fit throw the bread as is, while those voting for exclusion squeeze it very hard with their hand. The squeezed piece has the effect of a hollow ballot. [a single 'negative' squeezed bread denied entry to the candidate].

12.17 Athenian improvement of the simple voting technique

Plutarch, *Alcibiades* 22.2

Alcibiades has been recalled from Sicily to Athens to face charges but when he arrived at Thurii, he fled, not wishing to have his future decided by a simple secret vote.

When someone recognised him and asked, "Do you have no faith in your country, Alcibiades?" He said, "In everything else; but concerning my own life I lack faith even in my mother lest she mistake a black for a white ballot when she casts her vote".

Aristotle, *Constitution of Athens* 68–69

Aristotle describes the tools and method for voting by jurors; a reflection of the technologically simpler Spartan method (**12.16**) [**fig. 12.3**].

The ballot-discs are bronze and have a stem in the centre; half the stems are hollow, the other half solid. The officials selected by lot for the taking of votes, once the arguments are ended, give to each of the jurors two ballot-discs, a hollow one and a solid one. This is done in full view for the adversaries to see so that the jurors might take neither two solids nor two hollow discs. Then the man appointed to the task hands over tokens; each juror who has voted receives one token made of bronze inscribed with a gamma [the number 3] which when submitted he receives 3 [obols], so that everyone votes since it is not possible for anyone to receive a token unless he has voted.

RECORD-KEEPING

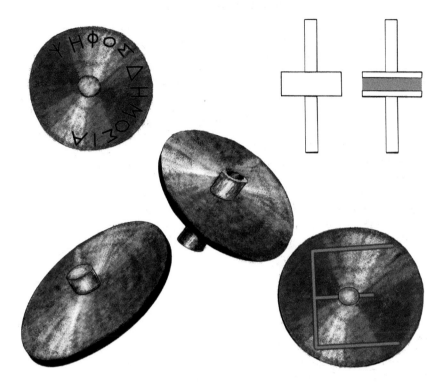

Figure 12.3 Ballot discs.
Source: Adapted from examples in Agora Museum, Athens, Greece, Camp, 1986, fig. 80.

There are two receptacles, one of bronze and the other of wood, standing in the court of justice, which are separated in a manner [from the jurors] so that no one casting his ballots might to so covertly. By casting into these receptacles, the jurors reckon their vote; the bronze for the valid disc, the wooden for the voided disc. And the bronze receptacle has a lid finely worked so as to make room for only one disc, so that the same man may not put in two at a time.

[After some instructions to the courtroom, most importantly that the hollow stem is for the first speaker, the prosecutor, the solid stem for the second speaker, the defendant, the procedure is described].... And the juror, taking his two discs from the ballot stand and pressing on the stem of each disc and not revealing to the litigants neither the hollow nor solid ends, casts the valid one into the bronze receptacle [*amphoreus*, a jar, here], the voided one into the wooden receptacle [Once all discs are cast, officials place the valid discs on a counting-board with holes where the hollow and solid stems are visible to everyone. They are counted by lot-selected officials and the herald delivers the majority-derived verdict; ties go to the defendant.]

12.18 Roman innovation to secure secret ballots

At Rome, protection of the voter was sought by secret ballot during elections via a very visible voting procedure that ensured privacy: the voter was isolated upon a high visible bridge so that the vote could be cast without any immediate pressure. Citizens voted in groups (centuries or tribes), entering a roped-off area and proceeding single file over raised gangways (*pontes*, "bridges"). The voting procedure is depicted on a coin issued by the moneyer P. Licinius Nerva in 113–112 BC [**fig. 12.4**], which with the passages help to explain this innovation.

Plutarch, *Marius* 4.2

In his tribunate Marius introduced a law concerning voting by ballot, supposing it would diminish the power of the influential men in judicial cases. Cotta, the consul, persuaded the senate to fight the law, and to summon Marius to give an account of his action.

Cicero, *On the Laws* 3.38

Cicero expresses his elitist displeasure with secret voting since it prevents fraud from being identified. Here he wishes to revoke the reform introduced by Marius.

... all laws should be annulled which have been proposed in whatever way to conceal a voting tablet in every kind of matter so that one is unable to inspect the tablet, unable to question it, unable to appeal it. And furthermore, the law of Marius which made the bridges [elevated passageways to the balloting boxes] so narrow, also should be abolished.

Figure 12.4 Roman coin with voting procedure.

Source: Republican coin of moneyer P. Licinius Nerva, late second century BC.

12.19 Ostracism: ridding the polis of disruption

Ostracism, a fifth century BC phenomenon in Athens, is more democratic in the modern sense since voting occurs in which a majority decision, rather than random lot, determines who is selected. Archaeological recovery of ostraca, the pottery fragments with the names scratched into them, indicates the potential for cheating and/or influence peddling since the same hand has been identified on deposits of ostraca [**fig. 12.5**] (Lang 1990 and Broneer 1938).

Plutarch, *Aristeides* 7. 4–5

The procedure, to give a general account, happened as follows. Each citizen, having taken an ostracon [a broken piece of pottery] and having scratched on it the name of the citizen he wished to be banished, carried it to an area of the agora which was fenced off in a circle with a railing. Then the Archons first counted out the full number of ostraca together, for if there should be less than 6,000 being cast, the banishment-by-ostracism was null and void. Then [if more than 6,000 ostraca were counted] having sorted the ostraca according to each personal name, they banished by proclamation the man with the name most scratched on the ostraca. He was banished for ten years, with the right to enjoy the income from his estate.

Figure 12.5 Ostraca.

Source: Adapted from examples in Agora Museum, Athens, Greece; Camp, 1986, fig. 39.

RECORD-KEEPING

C WRITING AND COMMUNICATION

More than any other technology, written language makes civilisation possible. Written records can defy time and space, and, by enabling the accumulation of wisdom, they promote permanent knowledge and intellectual progress. In antiquity, the art of writing was much esteemed: many early peoples believed in the divine origin of scripts, and often the few who could write were especially respected – and powerful. There is evidence, for example, that the conservative aristocratic governments of some archaic Greek city-states actively opposed the introduction of the relatively simple alphabet, since the ability to read would give new power and independence to their fellow citizens.

That alphabet is the basis of most modern written scripts, but was certainly not the earliest form. The history of the development of writing is controversial, but the principal evolutionary stages are generally accepted. The early symbolic scripts of the Bronze Age made no connection between the symbol and the spoken name for it in a particular language but instead relied on iconography (pictures to describe an object or event), mnemonic devices (simple symbols to aid in recall), and ideograms (in which symbols represented not just the objects of iconography, but the underlying attributes associated with them): theoretically, at least, all these could be "read" by a speaker of any language, though in practice local custom made many symbols unintelligible to an outsider. The later phonetic scripts that evolved in the late Bronze Age and Archaic period represented the actual sounds of words in a specific spoken tongue – the symbols standing for syllables (consonant and vowel) or single sounds (consonant or vowel, as in our alphabet) – which dramatically reduced the number of symbols that had to be learnt, and so promoted literacy among most levels of society.

Even so, access to written records would always be limited in antiquity. In the absence of multiple copies of books produced on a printing press, the cost of handwritten volumes was extraordinarily high, limiting the average citizen's reading of "literature" to inscriptions engraved into wood, metal and, especially, stone of laws and decrees set up in the agora or forum, and to the volumes housed in the few public libraries that were donated to cities by high-minded citizens (Clarysse and Vandorpe 2008, Dilke 1977).

12.20 Egyptian hieroglyphs

Diodorus here describes the ideographic script of the ancient Egyptians, called *hieroglyphica* ("holy-carvings") because it was used and understood principally by the priests and theocratic bureaucracy and was used mostly for inscriptions on temple walls and in tombs. The two other versions mentioned in his last paragraph – demotic and hieratic – are actually the same script as hieroglyphs, but written in a faster (and so more cursive and conventional) way: because of their simplification of the symbols, both lost the recognisable pictorial character of the original hieroglyphs.

Diodorus of Sicily, *History* 3.4.1–4; 3.3.5

To complete our discussion of the Ethiopians' antiquities, mention must be made of their script, which is called "hieroglyphic" by the Egyptians. It turns out that the shapes of their characters correspond to animals of every sort, to the extremities of the human body, and to tools (especially those used by carpenters). For it is not through a combination of syllabic sounds that their script expresses the underlying meaning, but rather from the allusion latent in the objects represented and their metaphorical meaning that practice has impressed on their memory.

They draw, for example, pictures of a hawk and a crocodile, a snake and elements of the human body: an eye, a hand, a face, and so on. In this case the hawk, because it is pretty well the swiftest winged creature, signifies to them anything that happens swiftly. Then, with apt metaphorical allusions, this sense can be applied to all swift creatures and whatever else conforms closely enough to the nature of those thus identified. The crocodile is symbolic of everything that is evil, the eye is the guardian of right behaviour and protector of the whole body. As for the bodily extremities: the right hand with fingers splayed connotes the acquisition of resources, while the left hand with fingers clenched suggests the defensive protection of property. The same reasoning holds for the other characters, whether parts of the body or tools or anything else. For by tracing closely the allusions that are implicit in any given object and by exercising their minds through long practice and memorisation, they learn to read fluently everything that has been written....

There are two [other] scripts used by the Egyptians: one, labelled "popular" (*demotic*), is learnt by everyone; the other, called "sacred" (*hieratic*), is understood only by the priests of the Egyptians, who learnt it from their fathers as one of the things that should be kept secret.

Herodotus, *Histories* 2.36

When they write or do calculations, Egyptians move their hand from right to left, unlike the Greeks, who go from left to right.[2]

12.21 The mysteries of writing

Though the Bronze-Age Linear B tablets found on Crete and the Greek mainland show that the Mycenaeans knew how to write as early as the fifteenth century BC, literacy disappeared with the subsequent dissolution of the Bronze Age empires of the eastern Mediterranean. The memory of "writing" thus acquired a mystical aura, as seen in this passage from Homer. King Proteus is persuaded by his deceitful wife, Anteia, to do away with the noble Bellerophon, who had rejected her advances. This passage, the only one in either the *Iliad* or *Odyssey* that alludes to literacy, is often taken as proof that the epics assumed their present form before the Greeks once again learnt the art of writing, in perhaps the ninth century BC, from Phoenician traders.

Homer, *Iliad* 6.167–170

Proteus avoided killing Bellerophon outright, being frightened at heart, but sent him off to Lycia with malevolent tokens, having first entered many lethal scratches in a folding tablet, and ordered him to show them to Anteia's father, so that he might be killed.

12.22 The (mythical) origins of the alphabet

Pliny is right here in attributing the Greek alphabet to influences from the Phoenician traders of the early Iron Age, who brought with them, sometime between 950 and 800 BC, the North Semitic alphabet that had evolved in the eleventh century BC in the area of modern Israel and Syria, and vestiges of which persisted in the Greek version (*alpha, beta, gamma=aleph, beth, gimel*). The character Cadmus may well be a mythological invention, and Pliny's other claims here, and their dates, are hardly credible.

Pliny, *Natural History* 7.192–193

It is my belief that the Assyrians have always had written characters, though others assume, with Gellius, that they were invented in Egypt by Mercury, or in Syria. Either way, it is agreed that Cadmus brought to Greece from Phoenicia a 16-letter alphabet, to which four characters – Z, Y, F, and C – were added by Palamedes at the time of the Trojan War, and after him another four – U, X, W, Q – by the lyric poet Simonides. (The value of all these letters is recognised in the Latin alphabet as well.) Aristotle opts for a primitive alphabet of 18 letters, with the additions of Y and Z by Epicharmus rather than Palamedes. Anticlides records a certain Menos as having invented the alphabet in Egypt, 15,000 years before Phoroneus, the primeval Greek king, and uses the monuments to try to prove his attribution. Epigenes, a serious author of the first rank, takes a different view by showing that astronomical observations have been recorded on fired bricks by the Babylonians for the past 730,000 years (Berosus and Critodemus, who record the shortest period of these logs, say 490,000 years). From this it seems that the alphabet has been in use forever. The Pelasgians brought it to Latium.

12.23 The Greek and Latin alphabets

All alphabets used in Europe today are derived from that of ancient Greece. Originally, and inevitably, there were several local Greek alphabets, each with individual peculiarities; but by *c.*350 BC all had disappeared in favour of Ionic Greek, which became the "classical" Greek alphabet of 24 letters. Latin letters were not derived directly from these, but came through the Etruscans in the seventh century BC, when the Romans adopted only those letters that represented sounds in archaic Latin. After their conquest of Greece in 146 BC, they added two Greek letters (*z* and *y*); the first-century AC Emperor Claudius tried in vain to add three more, to make the alphabet more phonetic, but the final two additions to our alphabet – *w* and *j* – were not made until the Middle Ages (Healey 1990).

Herodotus, *Histories* 5.58–59

These Phoenicians who accompanied Cadmus.... and settled in Boeotia taught the Greeks many things, but their greatest lesson was the alphabet, which as far as I can tell was unknown to the Greeks before this time. At first, they used the standard Phoenician script, but with the passage of time they changed both the sound and the shape of the letters.

Most of the land in their neighbourhood at that time was occupied by the Ionian Greeks, who learnt their letters from the Phoenicians and put them to their own use after making a few changes in their forms. And they called the script that they used "Phoenician", which was only right since it had been the Phoenicians who had introduced it to Greece. So, too, the Ionians call papyrus sheets "skins", a holdover from antiquity when a scarcity of papyrus had them using the skins of goats and sheep; and even in my own day many non-Greeks use such skins as a writing surface.

I have personally seen Cadmean letters carved on some tripods in the Temple of Ismenian Apollo at Thebes in Boeotia, and they are essentially the same as the Ionian alphabet.

Tacitus, *Annales* 11.13.3–14.5

Once he learnt that the Greek alphabet, too, had not been invented from beginning to end at a single stroke, the Emperor Claudius added new letter forms [to the Latin alphabet] and promoted their use in public....

In Italy the Etruscans were taught the alphabet by Demaratus from Corinth, while our earliest ancestors learnt it from Evander the Arcadian. And the forms of the Latin characters are the same as the oldest letters of the Greeks. But, like the Greeks, we had a small number of letters to begin with, to which later additions were made. Claudius used this as a precedent for adding three more letters,[3] which remained in use while he was emperor but disappeared afterwards – though they can still be seen in public inscriptions set up in fora and on temples.

12.24 Oral tradition in a literate society

So far we have emphasised the advantages of writing: here Caesar acknowledges that universal literacy has some disadvantages.

Caesar, *Gallic War* 6.14

It is said that those who attend Druid schools learn a great number of verses, which means that some of them remain in training for two decades. They do not think it right and proper to entrust these expressions to writing, although in most other matters, in both public and private circumstances, they use the Greek alphabet. It seems to me that they have instituted this practice for two reasons: because they do not wish the elements of their discipline to become public knowledge, and to prevent those in training from relying on the written word and paying less attention to their memories. And most people do usually experience this: by depending on writing they lose the concentration and retentive mind that come from learning things by heart.

12.25 The benefits of writing

Palamedes, one of many characters credited in antiquity with the invention of the alphabet, here reflects on the two primary benefits of the written word: unlike oral utterance, it defies both space and time.

Euripides, *Palamedes* (Fragment 578 = Nauck p. 542)

I was the one to devise these antidotes to forgetfulness: by my invention of consonants, vowels, and syllables I made it possible for people to learn to read, so that he who is absent overseas can even there know well what is happening back home; so that a dying man can speak through writing to his children, each heir to know his share of the wealth. The written word removes the traps that cause strife among people, because it does not let them lie.

12.26 A negative view of writing

Socrates seems to offer a very negative view of the invention of writing: that it is merely a tool that can be almost useless unless properly used by responsible writers and readers – one might compare similar criticism of the internet today. The criticism is not merely the simple lack of training of one's memory, but much more complex (cf. **12.22–12.25** for other 'creators').

Plato, *Phaedrus* 274e–275b

It is related that Thamus set forth many things both in praise and in blame about each of the arts to Theuth – which would be too long a story to go through – but when he came to the letters he said, "Knowledge of this, o King, will make the Egyptians wiser and will make their memories better; for I have found a blessing of memory and wisdom". But Thamus responded, "Most ingenious Theuth, one man has the ability to bring forth the diverse aspects of art, but another has the ability to judge what portion of harm and of aid it has for those who will use it. And now you, who are the father of letters, and because of that benevolent interest proclaim the opposite for them to what they are capable. For this invention will yield forgetfulness in the minds of those learning it through the lack of practice of their own memory since their trust of writing, which is produced from outside themselves through exotic characters, and they will no longer be the remembrancers, recalling memory from inside themselves. So, you have not discovered a blessing of memory, but of reminding; and you provide the appearance of wisdom to your pupils, not reality. For becoming much-learnt [by reading but] without the discipline of instruction, they will only seem to be very sagacious since they are for the most part without sense and hard to get along with, having become wise in their own mind, rather than truly wise".

12.27 Teaching a child to read and write

Quintilian, *On the Training of an Orator* 1.1.12, 15–16, 24–25, 27, 30–31

There are two reasons why I prefer that a boy begin with the Greek language. Latin, which is more commonly used, he will soak up even if we would rather he not; and it is important that he first be taught Greek culture, from which ours is derived....[4]

Some have expressed the view that reading should not be taught to boys who are younger than seven, since that is in their view the earliest age at which they have a grasp of what education is about and can tolerate the work involved.... The preferable view is held by those like Chrysippus, who want every moment in a boy's life to be attended to: the first three years he concedes to the nurses, but still expects them to play a part in forming the child's mind on the highest principles possible....

At all events I do not approve of what I see to be the common technique of having small children learn the names and order of the letters before they know their shapes. This practice inhibits their recognition of the alphabet, since they rely on what they can recall from memory rather than paying attention to the actual appearance of the letters....

As soon as the child has begun to follow the shapes of the letters, there is some advantage in having them carved as accurately as possible into a wooden board, to act as tracks for guiding the stylus. The pen will thus not wander off course as it does on wax tablets, confined as it is to the groove and unable to stray from the pattern. And by tracing these prescribed impressions faster and more frequently, the fingers will become steadier and the child will not need a helping hand to guide his own....

There is no shortcut for learning syllables: they all have to be learnt by heart. There is no sense in following the common practice of postponing the most difficult ones, which only means that they will be met with surprise when words are written down.... Then let the child start to make words by connecting these syllables and to link them together into a sentence.

For another allusion to bad handwriting, see Plautus, *Pseudolus* 27–30.

WRITING MATERIALS AND BOOK PRODUCTION

12.28 Advice on a composition notebook

Quintilian and Catullus in the next two selections refer to the wax-covered wooden tablets that were the common reusable notebooks of antiquity, their surfaces easily erased by warming them over a flame.

Quintilian, *On the Training of an Orator* 10.3.31–32

It is best to write on wax tablets, which provide the easiest method for erasure, unless it happens that relatively weak eyesight requires the use of parchment instead. But while parchment does aid visual acuity, it also delays the hand and breaks the train of thought because of the repeated removal of the reed pen from the page to refill it with ink.

Whichever format is chosen, pages will have to be left blank for the spontaneous addition of supplementary material, since closely written pages sometimes deter one from making revisions, or in any case obscure the original text when new material is inserted. Please – do not use wax tablets that are excessively wide: I know from experience of an otherwise diligent young man whose compositions were interminable because he would measure their length by counting the number of lines, a foible that no amount of dissuasion could correct, and that was eliminated only with a change of writing tablets.

12.29 Tablets used as notebooks

Catullus, *Poems* 50.1–6

Yesterday, Licinius, when we were at a loose end and both felt frivolous, we had great fun with my waxed tablets. Each of us played with writing light verses, now in one metre, now in another, answering each other's compositions while we joked and drank our wine.

Martial, *Epigrams* 14.6.4 mentions waxed wooden tablets of three and five leaves.

12.30 Writing surfaces, especially papyrus

Papyrus was the principal material for written publication in antiquity: in Greek it is called *byblos*, the same word that means "book" (and, of course, *Bible*). Our "paper", though the word is etymologically related to "papyrus", was a later technological import from China via the Arabs in the eighth century AC (Lewis 1974 for papyrus, Johnson 1970 for parchment).

Pliny, *Natural History* 13.68–72, 74, 77, 81, 89

The character of papyrus will also be discussed, since it is on the use of paper especially that civilised life, or at least its historical memory, depends.

Marcus Varro dates the invention of paper to the foundation of Alexandria in Egypt following the victory of Alexander the Great. Before this time papyrus was not exploited: at first people were in the habit of writing on palm leaves, then on the inner bark of certain trees;[5] later they began to keep public records on sheets of lead,[6] and soon after private accounts on pieces of linen or waxed tablets. We discover in Homer that small writing tablets were employed even before the Trojan era....[7] By and by (also according to Varro), when the two kings Ptolemy and Eumenes were competing with each other over their libraries and Ptolemy laid an embargo on the export of paper, parchment was invented at Pergamum.[8] Thereafter there was no discrimination in the widespread use of paper, the material on which depends the undying record of human life....

Papyrus, then, grows in the swamps of Egypt or where the Nile's still waters have spread beyond their banks and formed pools less than 3 feet deep. Its root grows at a slant and is as thick as an arm; its body is triangular in section and reaches a height of up to 15 feet, ending in a head that resembles a thyrsus but contains no seeds:[9] its sole application is in decorating statues of the gods with wreaths made of its blossoms. The natives treat the roots like lumber, using it not only as firewood but for vessels and other utensils; in fact, it is this very papyrus that they weave into boats, and from the inner bark they make sails and mats as well as clothing and even blankets and ropes. They also chew it either untreated or boiled, though it is only the juice that they actually swallow.

Paper is made from papyrus by splitting it with a needle into strips that are extremely thin but as wide as can be managed. The best papyrus comes from the middle of the plant, the quality declining with each subsequent layer that is split off. The first-quality paper used to be termed "hieratic" and was in antiquity reserved for religious texts. Flattery has now given it the name "Augustus", just as the second-quality paper has been named "Livia" after his wife....

Whatever the quality, paper is woven together on a board that has been moistened with water from the Nile, the sludge providing the binding force of a glue. After both of its ends are trimmed, each strip of papyrus is first plastered onto the table flat and straight, then other strips are laid across them to form a lattice, which is pressed between beams. The sheets are dried in the sun and then joined together, ... though never more than 20 to a roll.... Any unevenness is smoothed off with an ivory tool or a shell, though the text is liable to fading because the polishing makes the page less absorbent and more glossy....[10]

The papyrus plant, too, experiences periods of infertility, and already when Tiberius was emperor a shortage of paper prompted the appointment of members of the Senate to oversee its distribution; on other occasions life became anarchic....

12.31 Saving paper

Because papyrus was relatively inexpensive, to write on the back was the habit of a stingy fellow (Martial) or an enormously productive scholar (Pliny). Concerning the habits of Pliny's prolific uncle, the author of the *Natural History*, see also **14.17**.

Martial, *Epigrams* 8.62

Picens writes his epigrams on the backside of a page, and is distressed because the god Apollo turns *his* backside on him when he does it.

Pliny, *Letters* 3.5.17

He left me 160 notebooks of extracts, written on both sides of the page in a minute hand, a technique that increases the work's actual length.

12.32 Reuse: erasure and starting anew

In **12.41**, Catullus comments on the reuse of paper whose original text has been erased; such a page, whose information can often be recovered, is called a palimpsest. The use of wax tablets (**12.27–29**) and white chalk boards (Aristotle, *Constitution of Athens* 47–48) make reuse of the same surface quite easy; the sponge was also used to allow the reuse of papyrus. Below, Ovid describes a romantic situation where Byblis, who loves her own brother, writes a text in wax to confess her love, but easily changes it when she needs to conceal thoughts and her identity; Suetonius provides more dramatic examples of the use of the sponge.

Ovid, *Metamorphoses* 9.519–531

[After wavering about what to do, Byblis has decided upon her plan]
 ... she said, "Let him see: let me confess my raging desires.
 Alas for me! To what do I fall to ruin? What flame does my mind take up?"
But while reflecting thus, she composes her words with a trembling hand. She takes up an iron stylus in her right and in the other she takes the smooth wax [tablets].
 She begins and she hesitates, she writes and she condemns the tablets,
 And she writes hints and she obliterates, she alters and she censures, and she approves.
 And in her backs-and-forths, having taken up the tablets she puts them down and having put them down she picks them up again.
 She does not know what she desires. Whatever she seems about to do displeases her.
 On her face boldness is mixed with shame. She had written "Sister" but once it was seen, deleted "Sister", and etched other such words on the smoothed-out wax.

Suetonius, *Augustus* 85.2

[Augustus has written numerous prose works and has occasionally composed verse with less success.] For although having commenced his tragedy with great vigour, with the composition not progressing, he destroyed it. To his friends inquiring "What has Ajax been doing?" he replied, "Ajax has blotted himself out on a sponge" [not fallen on his sword as in the myth].

Suetonius, *Caius Caligula* 20

Caligula displays his bizarre sense of humour at an oratory contest.

[Victors received strange praise] ... but those who had displeased the most were ordered to blot out their own writing with a sponge or their tongue, unless they preferred to be chastised with a rod or plunged into the nearest river.

ATTEMPTS TO PROTECT THE INTEGRITY OF INFORMATION AND TO FOIL FORGERS

Technology in antiquity, like today, was often employed to thwart those who tried to profit from others by illegal means (**11.120**). The testing/assaying of metals (**5.4, 6.40–42**), the weighing and measurement of goods (**11.121, 11.123–24**), and regulations to control unsafe technological construction (**8.61, 8.64, 8.75–76**) are some obvious examples. Once the ability to read became more common, safeguards were developed to protect information. They range from seal-stones, codes and invisible inks (**12.35–38, 13.14–15**) to a more elaborate arrangement to stop the proliferation of forged documents during the reign of Nero.

12.33 Seals and seal-stones

First intended to "protect" stores and valuables from tampering or theft, often from servants or slaves according to Cicero (*On the Orator* 2.61), when the "owner" was absent, the use of seal-stones expanded to secure privacy and to guarantee the power behind the owner of the seal-stone; their small size promoted mobility and rulers could "lend" them to their lesser officials to enact their will. This became more common during the Roman Empire when rings with stones would be provided by emperors to officials of lesser status. Mention of seals are widespread in the texts and only a small sampling of their use, power and, appearance are provided below.

Plato, *Laws* 12. 954b–954c

But if the master of the house happens to be away from home, let the ones still at home grant access to the unsealed things to the one doing the search, and let the searcher counter-seal what is already sealed, and let the one searching the house to discover the theft set whomever he might wish as guard over it for five days. And if the master should be away for more than this time, let the searcher with the provision that he has called in the city-magistrates, break open the sealed goods of what is sealed, and let him seal them up again in the same way with the household and city-magistrates attending.

Aristophanes, *Thesmophoriazousae* 414–429

Aristophanes provides evidence of several types of security used to safeguard the virtue of females in the household.

Again, on account of this man [Euripides] the men cast the seals, bars, upon the women's quarters and watch over us. We are incessantly watched and they rear Molossian hounds as monstrosities, a thicker layer of security against the adulterers.... Our husbands now carry some little, very malicious Spartan keys on their persons, fashioned with three secret teeth. Formerly a signet worth three obols with the same sign for three obols was sufficient to secretly open the door for the women procuring them; but now this base-born slave Euripides has taught the men to have seals of worm-eaten wood hanging about them. [The suggestion to poison Euripides follows.]

Plutarch, *Tiberius Gracchus* 10.6

The power of the seal could be symbolic as well as a physical indication of the desire for privacy (cf. Plutarch, *Alexander* 2 and 39.5 and *Demetrius* 22.1). Here Tiberius Gracchus uses his own seal in a state function to frustrate those resisting his own reforms.

He placed his private seal upon the temple of Saturn [the state treasury], so that the quaestors [here the state-treasurers] might take nothing from it nor deposit anything. And he made a proclamation of a penalty for refusal of compliance by the praetors, so that all became fearful to discharge their several duties at hand [plots against Tiberius' life now begin].

Plutarch, *Sulla* 3.4

And since Sulla himself was by nature vainglorious, at that point when for the first time he had emerged from his base and obscure existence and had become a man of some account among his fellow citizens, and was tasting the honour of it, he pushed ahead in this pursuit of ambitious endeavour having an image of his exploit engraved on a seal-ring to wear, and he continued using it always. The device was Bocchus delivering, and Sulla receiving, Jugurtha.

Suetonius, *Augustus* 50

On letters of recommendation, certificates and imperial letters, he first used a sphinx for the sealings, soon an image of Alexander the Great, and lastly his own image, carved by the hand of Dioscurides, which the following emperors continued to use as their seal. He also attached to all letters the particular hour, not only of the day, but even of the night, to indicate when they were rendered.

Pliny, *Natural History* 37.4 provides a longer list of different seals from Polycrates to Augustus, and includes the engravers of the images.

Dio Cassius, *Roman History* 66.2 (Ep. 65.2)

The ring of a Roman Emperor was a kind of state-seal, and the emperor sometimes allowed the use of it to such persons as he wished to be regarded as his representatives.

For he [Mucianus], having delivered up the supreme power to Vespasian, exalted himself greatly on that account and especially because he was addressed as brother by him, and because he possessed the power to transact all business, however great, that he wished, and without the emperor's oversight. And he was able to write [official orders] by merely marking down the [power of] the name of the emperor for which purpose he wore the signet that had been sent to him so that he might possess the Imperator's seal for issuing imperial orders.

Plutarch, *Demetrius* 51

False seals or fraudulent use of seals were potential problems as implied by Demetrius' fear when he was detained by Seleucus.

But Demetrius finding himself in ill-fortuned circumstance such as this, sent word to the friends and commanders who were about his son, and those at Athens and Corinth, ordering them to trust neither the letters nor the seal [allegedly] of himself, but to treat him as dead, and to guard carefully for Antigonus his cities and the rest of his matters.

12.34 A device to prevent falsification

Suetonius, *Nero* 17

Suetonius provides a very cryptic description of the method, but it seems three tablets were used, that had been secured by linen string at three points. The top tablet and the bottom had the contract written upon them and if tampering was suspected on the top tablet, the linen threads would be cut before an official and the two versions compared (cf. Andrews 2013).

A check on forgery was first devised then; that only [sets of three?] tablets were to be sealed and only if they had been drilled through three times with linen string having been passed through the holes [to secure them].

STEGANOGRAPHY AND CRYPTOGRAPHY: OBSCURING THE TEXT

Steganography, concealed writing, has an advantage over cryptography or cipher since it avoids the arousal of curiosity caused by an encrypted message when discovered. Messages were hidden in various ways in antiquity: under clothing, in weapons, beneath the wax of writing tablets (cf. **13.15**). Aeneas Tacticus, *Treatise on the Defence of a Besieged City* 31 provides a wide variety of methods, including many of the following (Leighton 1969 and Reinke 1992).

12.35 Physically hiding the text

Herodotus, *Histories* 5.35.2–3

Aristagoras, the "acting" tyrant of Miletus, in a difficult situation with the Persian overlords, receives unexpected advice from the tyrant Histiaeus, who is under house arrest in Susa.

Dreading each of these things, Aristagoras started to plan a rebellion, for it came about then that the man with the tattooed head came from Susa from the side of Histiaeus, providing the sign for Aristagoras to revolt from the king. For Histiaeus, who wished to signal to Aristagoras to revolt, had no other fail-safe method to prompt him since the roads were being guarded. So, having completely shaved his most trustworthy slave's head, he tattooed it and waited for the hair to grown back. As soon as it was grown in, he sent the man to Miletus having ordered nothing of him other than when he reached Miletus to urge Aristagoras, having shaved his hair, to examine his head. The tattooing on it gave the signal for rebellion, as related by me earlier.

Aulus Gellius, *Attic Nights* 17.9, provides a slightly fuller account.

Herodotus, *The Histories* 7.239.2–4

When Xerxes was decided to make his expedition against Hellas, Demaratus [the deposed king of Sparta], then being present in Susa and having learnt this, wished to make it known to the Lacedaemonians. At risk of detection, he had no other way to alert them than to devise this stratagem: taking a two-leaved, small tablet, he scraped out its wax, and then wrote the intention of the King Xerxes on the wood of the tablet. Next, after completing the message, he poured melted wax back again over the writing, so that the man carrying the tablet would give nothing away to the men guarding the roads. When the tablet arrived at Lacedaemon, the Lacedaemonians were unable to interpret its meaning, until at last, as I have heard, Gorgo, the daughter of Cleomenes' and the wife of Leonidas, having thought about it, made a proposal, bidding them scrape the wax away so that they would find writing on the wood. Being persuaded, they found and read the message, then When they did so, they found and read the message, and then sent it to the other Greeks.

Aulus Gellius, *Attic Nights* 17.9.16–17 records a similar practice by a Carthaginian commander.

12.36 Invisible inks

Invisible inks were used to pass secret messages to recipients who knew the process to make them visible again. These three techniques demonstrate quite different levels of expertise.

Ovid, *Art of Love* 3.627–630

Young women, wishing to communicate with lovers when under strict supervision are advised not to conceal messages under garments and in sandals that are easy to discover. Ovid suggests the skin of the servant and then finally:

Letters composed in new milk are also safe and cheat the eyes; sprinkle them with powdered charcoal, and you will read them. And the writing which is made with a stalk of dampened flax, will deceive so that the undefiled writing surface will bear the invisible words.

Philo, *On Sieges* D77 (102.31–36)

Philo describes a chemical reaction for this next technique.

The communications are written on a new hat or on the skin, with oak-gall crushed and soaked in water. Once the letters have dried, they become invisible, but once flower of copper [copper sulphate] has been pounded [to look] just like black [colouring] in water and a sponge has been soaked in this, whenever they are sponged-off with this, they become visible.

Ausonius, *Letters* 27.22.21–22 (Green 1991)

This letter discusses many methods to keep secrets, but of particular interest is his comment about revealing messages written using invisible ink employing a new(?) technique using heat, rather than cold ashes to cling to the milk.

Trace letters with milk; the paper as it dries will keep them ever invisible. Yet with [the heat of] embers the writing is brought to light.

12.37 Ciphers

When the mysterious nature of writing, a type of cipher in itself (see **12.21**), disappeared as more people became literate, various means to disguise the words were developed. Besides messages literally hidden from view, ciphers were developed or special physical arrangements made that could only be recovered by others with the identical physical arrangements.

THE SPARTAN *SCYTALE* TRANSPOSITION CIPHER

Aulus Gellius, *Attic Nights* 17.9.6–15

But the ancient Lacedaemonians, when they desired to disguise and hide the dispatches done for the state that were sent to their generals to prevent their deliberations from becoming known if they were seized by the enemy, used to send letters written fashioned in the following form. There were two well-turned branches, rather long, of the same thickness and length, smoothed and finished exactly alike. One of these was given to the general setting out to war, the other the magistrates kept at home under their power and their seal. When employment of more secret dispatches arrived, they bound about the staff a leather thong of moderate fineness, but as long as was necessary for purpose, in a simple circular spiral so that the edges of the thong, which coiled around, continually adjoined and clung to each other. Then they inscribed the dispatch on that thong across the transverse edges of the joints, with the lines of writing running from the top to the bottom. Once the letter was written in this way, the thong was unrolled from the branch and sent to the general, who was acquainted with that contrivance. Moreover, the unrolling of the thong returned the letters truncated and mutilated, and their "limbs" and "heads" were strewn about on very diverse parts of the thong. Therefore, if the thong had fallen into the hands of the enemy, nothing at all could be conjectured out of the writing. But when the man, to whom it had been sent, had received it, he wound the thong around the matching staff, which he had, from the top to the bottom, in the same manner as he knew it had to be done, and thus the letters through the same twining of the branch, becoming firmer, came together again, and they offered the dispatch complete and undamaged, and easy to read. This kind of letter the Lacedaemonians call a *scytale*.

Plutarch, *Lysander* 19.4–7, duplicates much of this information specifically applied to the Spartans' recall of Lysander after reports of his abusive behaviour abroad reached them.

SUBSTITUTION CIPHERS

Although simple ciphers could easily be composed, as seen below, for reasons of security more complex codes were developed from an early date (cf. **13.14**, Aulus Gellius, *Attic Nights* 17.9.1–5 and Aeneas Tacticus, *Treatise on the Defence of a Besieged City* 31).

Suetonius, *Caesar* 56.6–7

[Caesar's letters and dispatches are described].... And letters to Cicero and, likewise, to household members concerning private matters exist, in which, if somewhat more secret messages had to be passed, he wrote in cipher. Thus, that is, with the with a construction of the alphabet in its right order one was unable to make out a single word. If anyone is determined to investigate and to follow through on these letters, he must interchange [Caesar's letters] with the fourth letter down of the alphabet, that is, D for A and the rest similarly.

Suetonius, *Augustus* 88

Moreover, when Augustus wrote in cipher, he simply put B for A, C for B and so on, the letters following one after another in the same respect, although for X he used a double A [AA].

12.38 The most secure communication

**Aeneas Tacticus, *Treatise on the Defence of a Besieged City*
31.16–19**

Of all of them, the most secret way to sending messages, but the most laborious, which is done without writing, will now be made clear by me. The process is as follows. Bore 24 holes through a large astragal [a bone joint], six on each side of the astragal. Let the holes of the astragal stand for the letters. Pay attention on which side begins *alpha* [A] and the ciphers indicating [specific letters] that are marked on each side. After this, whenever you wish to communicate any word on them, pass a thread through the holes, such as, for example, if you wish to signify Αἰνείαν [Ainean] in the passing through of a thread, beginning from the side of the astragal on which is *alpha* [A], pass the thread through, and passing by the ciphers next to the *alpha* [A], draw through again when you have come to the side where the *iota* [I] is, and again you pass by the ciphers next to this one, and pass the thread through where *nu* [N] happens to be. And again, you pass by the ciphers next to this one to where *ei* [EI] is, and you pass the thread through. Now, "writing" the remainder of the message in this way, pass the thread into the holes exactly as we just now set down the name. As a result, there will be a ball of thread spun around the astragal, and it will be necessary that the one reconstituting the letters write the letters on a tablet as they are being revealed from the holes. The unthreading happens in the reverse order to that the threading. But it makes no difference that the letters are written in reverse order on the tablet, for the message will be perceived just as clearly. Nevertheless, to understand properly the ciphers is a greater task than the work to create it [simplified variations that use wooden boards follow].

12.39 Reed pens

Martial, *Epigrams* 14.19

You won a writing-case in the raffle. Don't forget to fit it out with reed pens. We've given you everything else; you take care of the minor things.

Martial, *Epigrams* 14.38

The land of [Egyptian] Memphis supplies us with reeds suitable for writing with.

12.40 Copyists

Before the development in fifteenth-century Europe of the printing press with movable type, publication of books was a slow and error-filled procedure in which a reader would recite a work from the original manuscript to be copied down by trained – though often distracted or bored – scribes. As the following two selections from Martial suggest, the competence of these scribes varied.

Martial, *Epigrams* 2.8

If any of the poems in these pages seem to you, my reader, either too obscure or only partly Latin, don't blame me: the copyist botched them while he was hurrying to get through the right number of verses for you.

Martial, *Epigrams* 14.208

The words may run fluently, but the hand is faster than they: the right hand has finished its job while the tongue is still working.

12.41 The finished scroll

The "binding" of the finished scroll involved clipping and smoothing the sides of the roll with pumice stone and tinting them black, attaching the ends to a pair of bone or wooden rollers (*omphaloi*), dipping the whole scroll into cedar oil to protect it from worms and moths, and writing the title in red (*rubrica*) on a projecting label to help identification when the scroll was inserted in its pigeonhole.

Martial, *Epigrams* 3.2.7–11

[To his new book] Now you can go on parade: you've been anointed with cedar oil, both your edges are appropriately distinguished, you revel in the painted ends of your wooden rod, a delicate purple cover protects you, and your label proudly blushes with scarlet dye.

Martial, *Epigrams* 4.10

While my small book is new and the edges of the scroll have not yet been rubbed smooth, while the page is not well dried and panics at the thought of being touched, go take it, lad, as a gift of no great consequence to a dear friend who deserves to be the first to own my trivial creations.[11] Run along – but equip yourself first by taking a Punic sponge along with the book; that suits what I'm giving him. Many corrections, Faustinus, cannot correct my jokes; but a single erasure can!

Catullus, *Poems* 22.3–8

That fellow Suffenus – you know him well, Varus – is an agreeable chap, witty and refined. Still, no one writes nearly as many lines of poetry as he. By my estimate he's got thousands – ten or more – written out in full, not just jotted down in the typical way on pages cleaned off for reuse. No, new books with imperial paper, new rods with decorated ends, red ribbons for the parchment wrapper, the lines marked out with a lead wheel, and the whole thing smoothed by pumice.[12]

12.42 Some texts and volumes of interest

The following selections illustrate some important aspects of the evolution of writing and publishing. Epigraphical texts on wood, metal and stone were obviously meant to last a very long time, as indicated in the first passage by Plutarch. The next text about an abridged edition of Livy's extensive *History of Rome*, reflects the inability of most people to afford complete editions of original works (though Livy's was notoriously long); the third carries completeness to an extreme; the fourth complains about overpriced books; and the fifth is perhaps one of our earliest references to the *codex*, or bound book (Roberts and Skeat 1983, Turner 1977), as a substitute for the relatively clumsy scrolls. The last text from Sicily, in both Greek and Latin, advertises a stone-cutter's workshop. The seemingly short and simple bilingual text has several grammatical irregularities that have resulted in multiple theories about literacy and authorship: Greek, Latin, or Phoenician.

Plutarch, *Solon* 25

Plutarch, writing almost 700 years after Solon, describes remnants of wooden plaques that had been set in special revolving frames to allow reading of laws on both sides.

Solon gave force to all his laws for 100 years, and they were inscribed upon wooden *axones* [tablets] rotating in the frames embracing them. Some small remains of these were still preserved in the Prytaneum in my time.

Martial, *Epigrams* 14.190

The massive Livy is compressed into scanty sheets of parchment: my library can't contain the whole thing.

Pliny, *Natural History* 7.85

There are cases of keen eyesight that are positively beyond belief. According to Cicero, a copy of Homer's epic poem the *Iliad*, written on parchment, was enclosed in a nutshell.

Martial, *Epigrams* 13.3.1–4

The entire horde of offerings in this slender volume of verse will set you back 4 *sestertii*, if you buy it. Is 4 too much? It can set you back 2, and Tryphon the bookseller would still make a profit.

Martial, *Epigrams* 14.192

This bulky object, put together from thick bundles of pages, contains the entire 15 books of Ovid's poem.

IG 14.297 = *CIL* 10.7296

For the different theories regarding levels of literacy and suggestions of Greek, Latin or Phoenician authorship see Tribulato 2011.

Stelai [inscribed monuments] are here inscribed and executed in proper form: for sacred temples and [*syn*?] for public works.	*Tituli* [inscriptions on plaques] are here laid out and cut: for sacred for sacred temples and [*qum*?] for public works.

12.43 The invention of parchment

The earliest codices were made of folded, sewn, and cut quires of papyrus, but their design was greatly improved with the development of vellum, made of the treated skin of a calf, lamb, or kid, a technique perfected in Pergamum (hence our word "parchment"). This material, used much earlier but not common until the fourth century AC, was more durable and economical than papyrus: both its surfaces could be written on without bleeding through. The following three short epigrams allude to early parchment versions of the classic works by Homer, Vergil, and Cicero (Johnson 1970).

Martial, *Epigrams* 14.184

The *Iliad* and Ulysses, enemy of Priam's kingdom, lie buried together in a much-folded skin.

Martial, *Epigrams* 14.186

How short a parchment has taken on the endless Vergil! The first leaf carries the image of the poet himself.

Martial, *Epigrams* 14.188

If that parchment is to be your travelling companion, imagine that you're taking Cicero himself with you to share your long journeys.

BOOKSHOPS AND LIBRARIES

Since publication was so labour-intensive, books were generally priced beyond the means of all but the wealthiest – a class to which, of course, most of the authors of our texts belonged. The first of the selections that follow illustrate some enviable private libraries; the last refer to three of the many public libraries in ancient cities that made literature accessible to many literate if relatively poor citizens.

12.44 A lucky find in a used-book stall

Aulus Gellius, *Attic Nights* 9.4.1–5

When I was returning to Italy from Greece and had arrived at Brundisium, I disembarked and was wandering about that famous port ... when I spied a bundle of books displayed for sale. My hunger aroused, I lost no time in heading for the books. Now all of them were in Greek and filled with amazing stories and legends, incredible things that you'd never heard of, but written by ancients of considerable authority: Aristaeus of Proconnesus and Isigonus of Nicaea, Ctesias and Onesicritus, Philostephanus and Hegesias. The volumes themselves, though, were filthy from long neglect, and looked and felt disgusting. Still, I went up and asked how much they cost, was tempted by a remarkably low price that I had hardly expected, and bought most of them for a pittance. I ran through them all cursorily over the next two nights....

12.45 Private libraries

Seneca, *On Tranquillity of Mind* 9.4–7

One should acquire only as many books as he needs to get by with, and none for show.... What reason do you have for indulging a man whose passion is for bookcases made of citron wood and ivory, who collects the works of obscure or inferior authors and then can't stop yawning from a boredom induced by his thousands of volumes, and whose special delight comes from the books' outer appearance and titles? So it's at the houses of the grossly indolent that you'll see a complete set of orations and history books, their cases piled to the ceiling. It used to be private bathing complexes, small or large; now no house in town is considered complete without a library.

Catullus, *Poems* 68.33–36

It's because I live at Rome that I don't have an extensive collection of texts with me: that's where my house and home is, that's where I spend my time. Only one small book-box[13] of the many I own accompanies me [when I travel to Verona].

Vitruvius, *On Architecture* 6.4.1

Bedrooms and libraries should face east, since their functions require sun in the morning. This will also keep the books in the library from becoming mouldy. Books in rooms facing south or west are damaged by bookworms and dampness because the prevailing, moisture-laden winds, as they spread their humidity over them, foster the growth of the worms and destroy the books with mouldy splotches.

12.46 The peripatetic fate of Aristotle's library
Strabo, *Geography* 13.1.54

Aristotle handed down his library to Theophrastus, to whom he also left his school. He was, as far as we know, the first book collector, and the first to have taught the kings in Egypt a system of cataloguing. Theophrastus handed down the library to Neleus, who took it with him to Skepsis and bequeathed it to his heirs, uneducated people who kept the books in storage without any concern for their condition. When they heard that the Attalids, who controlled their city, were conducting a determined search for volumes with which to stock their library in Pergamum, they buried their collection in some kind of trench. After a long time, their descendants sold off, for a considerable amount of money, the collections of both Aristotle and Theophrastus, which by then had suffered from moisture and moths. The buyer was Apellicon from Teos, who was more a bibliophile than a scholar; in trying to reconstruct the pages that had been eaten through he produced fresh copies of the text by filling in the gaps – incorrectly – and so published editions that were full of errors.

12.47 The earliest public libraries

These three passages give a simple and not entirely accurate account. While Peisistratus did foster the collection and editing of texts, the first Athenian *public* library was probably not established until the third century BC[14] Vitruvius errs in giving the Attalids credit for inspiring the Ptolemies; not *all* the collection of the Alexandrian Museum was destroyed by the Romans; and beware Seneca's customary moralising (Houston 2014, Volker 1981, Platthy 1968).

Aulus Gellius, *Attic Nights* 7.17

Tradition has it that the first person in Athens to have established a public library for the liberal arts was the tyrant Peisistratus. The Athenians themselves took great pains and care continuously to augment the collection. But when Xerxes had occupied Athens and burnt all the city but the Acropolis [480 BC], he stole the entire library and shipped it off to Persia. It was not until a long time later that King Seleucus – the one who was called Nicator[15] – saw to it that every one of the books was returned to Athens.

Later, in Egypt under the Ptolemies, a huge number of books, amounting to almost 700,000 volumes, was partly acquired, partly published on the spot. But in our first war in Alexandria [48–47 BC], while the city was being plundered, all the books were burnt – not wilfully or by premeditation, but accidentally by auxiliary troops.

Vitruvius, *On Architecture* 7, Preface 4

The Attalid dynasty [280–133 BC], inspired by the great pleasure they derived from the pursuit of learning, established an outstanding public library at Pergamum;[16]

then Ptolemy [Philadelphus, 283–246 BC], inspired more by unlimited jealousy and a possessive appetite, expended no less effort in competing with them to build a similarly stocked library at Alexandria.

Seneca, *On Tranquillity of Mind* 9.5

At Alexandria, 40,000 books were burnt. Others may well have praised this library as the most splendid monument to royal wealth, as did Livy, who calls it an outstanding achievement of the Ptolemies' taste and patronage. It was not "taste" or "patronage" that it stood for, but learnt extravagance!

12.48 Library rules

Athenian Agora 3: *Inscriptions* 1.2729 (= Platthy p. 113)

No book is to be removed, since we have so sworn. The library will be open from the first hour till the sixth.

12.49 Neither a borrower nor a lender be

Martial, *Epigrams* 1.117

Whenever you run across me, Lupercus, without waiting you say: "May I send a boy to get from you your book of epigrams? I'll return it to you as soon as I've read it". There's no need to bother your slave-boy, Lupercus: it's a long way for him to get to my neighbourhood and I live at the top of three long flights of stairs. You can find what you're looking for nearer at hand. Everyone knows that you often visit the Argiletum; across from the Forum of Caesar is a shop whose doorposts are covered with blurbs. So, take a minute to scan the list of poets and look for me there. No need to ask Atrectus (that's the shopkeeper's name). From the first or second pigeon-hole he'll offer you a Martial smoothed with pumice and embellished with purple, for 5 *denarii*. "You're not worth that much", you say? You show good sense, Lupercus.

Galen, *Commentary* on Hippocrates' *On Epidemics* 3.239–240[17]

They say that Ptolemy [Euergetes, 247–222 BC] gave the Athenians no small proof of his enthusiasm for collecting old books. Having left with them a security deposit of 15 talents of silver, he borrowed the books of Sophocles, Euripides, and Aeschylus solely to make copies and return them forthwith and undamaged. Once he had finished the copies, at great cost and on the finest paper, he decided to keep the volumes that he had borrowed from the Athenians and sent back to them his copies, with the request that they keep the 15 talents and accept the new volumes as substitutes for the originals that they had lent him. Even if he had not sent the new copies back while keeping the originals, there was nothing the Athenians could have done, since they had accepted the silver on the condition that they should keep it if he kept their books. So, they accepted the new copies and held on to the money as well.

12.50 On plagiarism

Martial, *Epigrams* 1.66

As a thief hungering after my books, you're making a mistake if you think you can become a poet for only the cost of a text of my work and a cheap blank scroll: a standing ovation can't be bought for 6 or 10 *sestertii*. Keep an eye out instead for unpublished poems and unrevised studies known only to one man.... A famous book cannot change its author. But if you can find one whose edges have not yet been smoothed with pumice or embellished with ornamental ends and slip-cover, buy it. And I have just such a one. No one will ever know.

Notes

1 Roughly between midday and 13:00, depending on the time of year.
2 This was not always the Greek style of writing: earlier scripts were written alternately right-left and left-right, like the route of an ox ploughing a field: hence their name, *boustrophedon*.
3 One to stand for the consonantal *u* (our *w*), another for the Greek double consonant *ψ* (*ps*, also Latin *bs*), the third to represent the *y* sound halfway between *i* and *u*.
4 See Quintilian, *On the Education of an Orator* 10.5.2–3, where our educator praises the practice of translating from Greek to Latin.
5 The Latin words for "bark" and "book" are the same: *liber*.
6 In fact, small, rolled pieces of lead seem to have been used exclusively for personal curses. Other metals, though, were used for public documents: bronze, for example, for military discharge papers.
7 See **12.21**.
8 The English word "parchment" derives from the ancient *pergamena*; cf. **12.23**, **12.43**.
9 The thyrsus was the symbolic rod carried by Dionysus, and was capped with a pine cone.
10 Cf. Martial, *Epigrams* 14.209: "Let the rind of the Egyptian papyrus be made smooth by the seashell: the reed pen will run its path without obstruction".
11 Elsewhere (*Epigrams* 8.72.1–3) Martial gives a similar description of an author so eager for friends to read his new work that a copy is sent off even before it is finished: "Though you are not yet spruced up with purple and smoothed off by the rough bite of dry pumice, you're off to follow Arcanus ...".
12 Cf. Catullus, *Poems* 1.1.1–2, the dedication of his book of poems: "To whom am I to present my charming new volume, just now polished smooth with pumice? To you ...".
13 On portable book-carriers, see also Martial, *Epigrams* 14.37, 14.84.
14 See *IG* 2^2 1009; *Hesperia* 20 (1947) 170–172.
15 One of the successors of Alexander the Great, in the early third century BC.
16 To be accurate, it was established by Eumenes II, who ruled from 197 to *c.* 160 BC.
17 Wenckebach pp. 79–80; Platthy pp. 118–119.

Bibliography

Time-keeping

Brind'Amour, Pierre, *Le Calendrier romain. Recherches chronologiques.* Ottawa: University of Ottawa Press, 1983.

Buchner, Edmund, *Die Sonnenuhr des Augustus.* Mainz: von Zabern, 1982.

Cotterell, Brian, Francis P. Dickson, and Johan Kamminga, "Ancient Egyptian Water-Clocks. A Reappraisal". *Journal of Archaeological Science* 13 (1986) 31–50.

Gawlinski, Laura, "Greek Calendars". Pp. 891–905 in Georgia L. Irby, ed., *A Companion to Science, Technology, and Medicine in Ancient Greece and Rome, Vol. II*. Chichester: Wiley-Blackwell, 2016.

Gibbs, Sharon L., *Greek and Roman Sundials.* New Haven: Yale University Press, 1976.

Hannah, Robert, "Timekeeping". Pp. 740–758 in John P. Oleson, ed., *The Oxford Handbook of Engineering and Technology in the Classical World.* Oxford: Oxford University Press, 2008.

Hannah, Robert, "Roman Calendars". Pp. 906–922 in Georgia L. Irby, ed., *A Companion to Science, Technology, and Medicine in Ancient Greece and Rome, Vol. II*. Chichester: Wiley-Blackwell, 2016.

Hannah, Robert, "Time-Telling Devices". Pp. 923–940 in Georgia L. Irby, ed., *A Companion to Science, Technology, and Medicine in Ancient Greece and Rome, Vol. II*. Chichester: Wiley-Blackwell, 2016.

Jones, Alexander, *A Portable Cosmos. Revealing the Antikythera Mechanism, Scientific Wonder of the Ancient World.* New York: Oxford University Press, 2017.

Merritt, Benjamin D., *The Athenian Year.* Berkeley: University of California Press, 1961.

Moraux, Paul, "Le réveille-matin d'Aristote". *Études Classiques* 19 (1951) 305–315.

Noble, Joseph V., and Derek de Solla Price, "The Water Clock in the Tower of the Winds". *American Journal of Archaeology* 72 (1968) 345–355.

Price, Derek de Solla, "Portable Sundials in Antiquity, including an Account of a New Example from Aphrodisias". *Centaurus* 14 (1969) 242–266.

Price, Derek de Solla, *Gears from the Greeks. The Antikythera Mechanism – A Calendar Computer from c.80 B.C.* Transactions of the American Philosophical Society, 64. 7. Philadelphia: American Philosophical Society, 1974.

Samuel, Alan E., *Greek and Roman Chronology. Calendars and Years in Classical Antiquity.* Handbuch der Altertumswissenschaft, Abt. 1, Teil 7. Munich: Beck, 1972.

Talbert, Richard J.A., *Roman Portable Sundials: The Empire in Your Hand.* New York: Oxford University Press, 2017.

How to select: sortition vs. election

Broneer, Oscar. "Excavations on the North Slope of the Acropolis, 1937". *Hesperia* 7 (1938) 228–243.

Bishop, J. David, "The Cleroterium". *The Journal of Hellenic Studies* 90 (1970) 1–14.

Dow, Sterling, "Aristotle, the Kleroteria and the Courts". 62–94 in Rhodes, ed., 2004; originally published in *Harvard Studies in Classical Philology* 50 (1939) 1–34.

Forsdyke, S., *Exile, Ostracism and Democracy. The Politics of Expulsion in Ancient Greece.* Princeton: Princeton University Press, 2005.

Lang, Mabel, *Ostraka. Athenian Agora Vol. 25*. Princeton: American School of Classical Studies, 1990.

Rhodes, Peter J., ed., *Athenian Democracy*. Oxford and New York: Oxford University Press, 2004.

Rhodes, Peter J., *A Commentary on the Artistotelian Athenaion Politeia*. Oxford: Clarendon Press, 1981.

Staveley, Eastland S., *Greek and Roman Voting and Elections*. Ithaca: Cornell University Press, 1972.

Writing and communication

Achard, Guy, *La Communication à Rome*. Paris: Belles Lettres, 1991.

Andrews, Colin, "Are Roman Seal-Boxes Evidence for Literacy?" *Journal of Roman Archaeology* 26 (2013) 423–438.

Beard, Mary, Alan K. Bowman, and Mireille Corbier, "Literacy in the Roman World". *Journal of Roman Archaeology*, Supp. 3. Ann Arbor: Journal of Roman Archaeology, 1991.

Boge, Herbert, *Griechische Tachygraphie und Tironische Noten. Ein Handbuch der antiken und mittelalterlichen Schnellschrift*. Berlin: Akademie, 1973.

Bowman, Alan K., and Greg Woolf, eds., *Literacy and Power in the Ancient World*. Cambridge: Cambridge University Press, 1994.

Boyd, Clarence E., *Public Libraries and Literary Culture in Ancient Rome*. Chicago: University of Chicago Press, 1915.

Bundgard, Jens A., "Why did the Art of Writing Spread to the West? Reflections on the Alphabet of Marsiliana". *Analecta Romana Instituti Danici* 3 (1965) 11–72.

Casson, Lionel, *Libraries in the Ancient World*. New Haven: Yale University Press, 2001.

Chadwick, John, *Linear B and Related Scripts*. Berkeley: University of California Press, 1987.

Clarysse, Willy, and Katelijn Vandorpe, "Information Technologies: Writing, Book Production, and the Role of Literacy". Pp. 715–739 in John P. Oleson, ed., *The Oxford Handbook of Engineering and Technology in the Classical World*. Oxford: Oxford University Press, 2008.

Daly, Lloyd W., *Contributions to a History of Alphabetization in Antiquity and the Middle Ages*. Brussels: Latomus, 1967.

Dilke, Oswald A.W., *Roman Books and their Impact*. Leeds: Elmete Press, 1977.

Driver, Godfrey Rolles, *Semitic Writing from Pictograph to Alphabet*. Rev. edn. Edited by S.A. Hopkins. London: Oxford University Press, 1976.

Harris, William V., *Ancient Literacy*. Cambridge, MA: Harvard University Press, 1989.

Havelock, Eric A., *The Literate Revolution in Greece and its Cultural Consequences*. Princeton: Princeton University Press, 1982.

Havelock, Eric A., *The Muse Learns to Write: Reflection on Orality and Literacy from Antiquity to the Present*. New Haven: Yale University Press, 1986.

Healey, John F., *The Early Alphabet*. Berkeley: University of California Press, 1990.

Heubeck, Alfred, "Schrift". *Archaeologia Homerica*, III, X. Göttingen: Vandenhoeck & Ruprecht, 1979.

Hooker, James T., *Reading the Past. Ancient Writing from Cuneiform to the Alphabet*. Berkeley: University of California Press, 1991.

Houston, George W., *Inside Roman Libraries: Book Collections and Their Management in Antiquity. Studies in the history of Greece and Rome*. Chapel Hill: University of North Carolina Press, 2014.

Hudemann, Ernest E., *Geschichte des römischen Postwesens während der Kaiserzeit*. 2nd edn. Berlin: Calvary, 1905.

Jeffery, Lilian H., *The Local Scripts of Archaic Greece. A Study of the Origins of the Greek Alphabet and its Development from the Eighth to the Fifth Centuries B.C.* Oxford: Clarendon Press, 1961.

Johnson, Richard R., "Ancient and Medieval Accounts of the 'Invention' of Parchment". *California Studies in Classical Antiquity* 3 (1970) 115–122.

Kenney, Edward J., "Books and Readers in the Roman World". Pp. 3–32 in Edward J. Kenney and Wendy V. Clausen, eds., *The Cambridge History of Classical Literature*, II: *Latin Literature*. Cambridge: Cambridge University Press, 1982.

Kenyon, Frederic G., *Books and Readers in Ancient Greece and Rome*. 2nd edn. Oxford: Clarendon Press, 1951.

Leighton, Albert C., "Secret Communication among the Greeks and Romans". *Technology and Culture* 10 (1969) 139–154.

Lewis, Naphtali, *Papyrus in Classical Antiquity*. Oxford: Clarendon Press, 1974.

Millard, Alan R., "The Canaanite Linear Alphabet and its Passage to the Greeks". *Kadmos* 15 (1976) 130–144.

Naveh, Joseph, *Early History of the Alphabet. An Introduction to West Semitic Epigraphy and Palaeography*. Jerusalem: Hebrew University, 1982.

Nissen, Hans J., Peter Damerow, and Robert K. England, *Archaic Bookkeeping: Writing and Techniques of Administration in the Ancient Near East*. Chicago: Chicago University Press, 1993.

Platthy, Jenö, *Sources on the Earliest Greek Libraries, with the Testimonia*. Amsterdam: Hakkert, 1968.

Powell, Barry B., *Writing: Theory and History of the Technology of Civilization*. Chichester/Malden, MA: Wiley-Blackwell, 2009.

Reinke, Edgar C., "Classical Cryptography". *The Classical Journal* 58 (1992) 113–121.

Roberts, Colin H., and T.C. Skeat, *The Birth of the Codex*. London: The British Academy, 1983.

Schlott, Adelheid, *Schrift und Schreiber im alten Ägypten*. Munich: Beck, 1989.

Schubart, Wilhelm, *The Typology of the Early Codex*. Philadelphia: University of Pennsylvania Press, 1977.

Skeat, T.C., "The Length of the Standard Papyrus Roll and the Cost-Advantage of the Codex", *Zeitschrift für Papyrologie und Epigraphik* 45 (1982) 169–175.

Starr, Raymond J., "The Used Book Trade in the Roman World". *Phoenix* 44 (1990) 148–157.

Thomas, Rosalind, *Literacy and Orality in Ancient Greece*. Cambridge: Cambridge University Press, 1992.

Tribulato, Olga, "The Stone-Cutter's Bilingual Inscription from Palero (*IG* XIV 297=*CIL* X 7296): A New Interpretation". *Zeitschrift für Papyrologie und Epigraphik* 177 (2011) 131–140.

Turner, Eric G., *Greek Papyri: An Introduction*. 2nd edn. Oxford: Clarendon Press, 1980.

Turner, Eric G., *The Typology of the Early Codex*. Philadelphia: University of Pennsylvania Press, 1977.

Volker, Michael Strocka, "Römische Bibliotheken". *Gymnasium* 88 (1981) 298–329.

Wendel, Carl, *Kleine Schriften zum antiken Buch- und Bibliothekswesen*. Edited by Werner Krieg. Cologne: Greven, 1974.

13

MILITARY TECHNOLOGY

It is an unfortunate truth that military technology is as old as humankind itself, and that many tools that would prove of benefit to other technologies were in the first instance developed as weapons of war. The clubs and spears designed by Palaeolithic humans to bring down wild game must have proved equally as useful against other groups of hunters. Even the transition to settled life at the beginning of the Neolithic period created new causes for warfare, as the surplus food accumulated by agriculturally productive towns was enough of a temptation to nomadic neighbours that early settlements like Jericho built fortification walls along with houses, shrines, and cisterns. The iron that the Bronze-Age Hittites used for weapons to intimidate their warlike neighbours later proved a revolutionary material for more peaceful functions, like more efficient ploughshares that would ease the cultivation of the stiffer soils of northern Europe and so encourage Mediterranean culture to expand northward.

Little was different in the Graeco-Roman period: most cities, except those founded in the relatively stable period of the first two centuries AC, were protected by defensive fortification walls; and military affairs dominated much of literature, from the epics of Homer to the histories of influential writers like Thucydides, Polybius, and Livy. The military, it seems, was an integral part of ancient life. As a result, we know more about the technology of warfare than any ancient pursuit other than agriculture. It is significant that we have as many manuals for tactics as we do for farming, from Xenophon in the fourth century BC to Frontinus in the first century AC and Vegetius at the end of antiquity. The difficulty here is how to present such a complex and familiar topic within manageable limits, particularly since many of the literary texts are best comprehended with the archaeological evidence at hand as well. With armies and weapons, for example, we have chosen to focus on the ancient connection between social status, armour, and tactics, while the section on fortifications and siege machinery is more typically "technological". (Campbell and Trittle, eds., 2013, Levithan 2013, Erdkamp 2011, Davies 2008, De Souza 2008, Sabin *et al.*, eds., 2007, Keppie 1984).

THE DEVELOPMENT OF ARMIES

13.1 The invention of weapons

It was a common tradition among the ancients to ascribe the invention of technological devices to specific, usually mythological, individuals (cf. **1.6–1.8, 1.22–1.23, 11.123**). Such attributions are hardly historical, but they can be useful as an indication of what the author felt to be worthy of reporting. This list of weapons reveals something of the variety of arms known in the first century AC; notice, for example, the absence of stirrups in the third paragraph (D'Amato 2016).

Pliny, *Natural History* 7.200–2

The Africans were the first to use clubs – they call them "staves" – when they battled the Egyptians. Shields were invented, if not by Chalcus son of Athamas, then by Proetus and Aerisius while they were campaigning against each other. Midias of Messene invented the breastplate. The helmet, sword, and spear were inventions of the Spartans, and greaves and crests for helmets came from the Carians. Some say Jupiter's son Scythes invented the bow and arrow, though others attribute the latter to Perseus' son Perses. Lances were developed by the Aetolians, the spear with a throwing strap by Aetolus, Mars' son, the light skirmishing spears by Tyrrenus, likewise the heavy javelin, the battle-axe by Penthesilea the Amazon. Pisaeus is credited with hunting spears and the version of missile-throwers called the scorpion, while the Cretans invented the catapult and the Phoenicians the ballista and sling.

The bronze war-trumpet came from Tyrrenus' son Pisaeus, the *testudo* [.....] from Artemon of Clazomenae, and, from Epius while at Troy, the style of siege engine called the "horse", now the "ram".[1]

Bellerophon invented horse riding, Pelethronius reins and saddles, and the Centaurs – Thessalians who lived beside Mt. Pelius – cavalry tactics. The Phrygian race was the first to harness two horses to a chariot, and it was Erichthonius who added two more.

During the Trojan War Palamedes invented military formation, the password, tokens for recognition, and sentinels; Sinon in the same war invented signalling from watchtowers.

Truces and treaties were invented by Lycaon and Theseus respectively.

13.2 Bronze age armour

In the first passage Homer describes the new set of armour that the god Hephaestus fashioned for Achilles, to replace the set that the Trojan Hector had stripped from Achilles' friend Patroclus; in the second, a variety of pieces, including the famous boar's-tusk helmet, a noteworthy piece of workmanship whose authenticity has been confirmed by archaeology. Just how much of this armour is actually Mycenaean, and how much reflects developments in the half-millennium between the Trojan War and Homer, is a matter of debate; what is clear from the epic is the importance of the individual warrior rather than the coordinated use of massed troops typical of the later hoplite dispositions.

Homer, *Iliad* 18.478–481, 609–613

First off, he made a large and sturdy shield, cleverly decorating its entire surface, and around it added a shining rim, three layers thick and glittering, and attached a strap for carrying it, made of silver. Five were the layers of the shield itself....

When he had fashioned the large and sturdy shield, he then fashioned for him a breastplate that sparkled brighter than a fire's flame; and he fashioned for him a massive helmet that fit snugly to his temples, well-proportioned and skilfully worked, and attached a crest of gold; and he fashioned for him greaves of flexible tin.

Homer, *Iliad* 10.254–265

With these words one pair clad the other in their dread armour. To Diomedes, son of Tydeus, steadfast Thrasymedes gave his double-edged sword (his own having been left by the ship) and his shield, and on his head, he placed a helmet made of bull's-hide but with no boss or crest, the kind that is called a skullcap and protects the head of strong young men. To Odysseus Meriones gave a bow, a quiver, and a sword, and on his head, he placed a helmet made of hide, stiffened on the inside by a series of taut braces, the outer surface thick with the white teeth of a bright-tusked boar skilfully set on both sides, and furnished with a lining of felt.

13.3 The hoplite army of Greece

As Aristotle notes in the first passage, the evolution of the *hoplites*, the heavy-armed infantryman, occurred in the Archaic Age in conjunction with the evolution of new forms of government. The massed phalanx with long spears revolutionised tactics, and was to be the dominant form of infantry organisation until replaced by the more flexible Roman maniples. It is surprising, then, that literary references to the panoply of hoplite armour are scarce, most of our information coming instead from vase paintings and surviving artefacts. Still, two seventh-century BC poets – Alcaeus and Tyrtaeus, from Mytilene and Sparta respectively – give us a glimpse of the new armament (notice, for example, the larger shield), and of the tactics associated with it, based on their own first-hand experiences in combat (Hanson 1991, Cartledge 1977).

Aristotle, *Politics* 4.10.10 (1297b)

The first pattern of government that evolved among the Greeks after the monarchies was based on military service. In its original form it consisted of the cavalry, from which warfare derived its strength and superiority (heavy-armed infantry being of no use without the sort of systematic organisation of which our ancestors had no tactical experience ...). But as our cities grew in size and those armed as infantrymen became more powerful, a larger number of people took a share in the government.

Alcaeus, Fragment 54 (Diehl) = Fragment 140 (Loeb)

The great hall sparkles with bronze, the entire ceiling decked out for Ares [the god of war] with bright helmets, from which nod crests of white horsehair, ornaments for the heads of men; bright greaves of bronze, defence against strong arrows, conceal the pegs from which they hang; breastplates of new linen and concave shields are cast about, and beside them blades from Chalcis² and a multitude of belts and tunics.

Tyrtaeus, Fragment 8 (Diehl) = Fragment 11 (Loeb), lines 3–4, 11–13, 21–38

Fear not a horde of men, and do not panic, but let every man hold his shield face to face with the enemy's front line.... Of those who stand side by side and dare to advance into a hand-to-hand fight with the front ranks, not many die, while they keep safe their comrades in the rear.... Let each man stand firm with both legs apart and planted in the ground, biting his lip and covering with the belly of his broad shield his thighs and legs below and his chest and shoulders above. Let him shake his hefty spear in his right hand, and flourish his fearful crest upon his head. Let him learn to fight by doing mighty deeds, and not stand holding his shield beyond the missiles' range. Let each man get in close, and in hand-to-hand fighting with his long spear or his sword wound an enemy and take him captive. Placing foot next to foot, leaning shield against shield, let him fight his man standing crest to crest and helmet to helmet and chest to chest, grasping the handle of his sword or the long shaft of his spear. And you, too, the light-armed troops, crouch down beneath the shield on either side and fling your sizeable stones or hurl your smooth spear against the enemy as you stand next to those in full armour.

13.4 The Macedonian war machine

The Macedonians, fighting in the more open terrain of their homeland in northern Greece, relied more on cavalry than the Greeks in the south; and by supplying their close-packed phalanxes with long pikes, they created a front line that was almost impenetrable. But, as Demosthenes observes here, there was more than that to the military success of Macedon under his enemies Philip and Alexander (Markle 1982, 1978).

Demosthenes 9, *Philippics* 3.47–50

In my opinion, nothing has undergone greater change and improvement than the art of war. First off, I understand that the Spartans of a century ago would, like everyone else, spend the summer season of four or five months invading and despoiling the countryside with their hoplites and conscript troops, and then return home again. They were so old-fashioned – rather, so statesmanlike – that ... their manner of waging war was conventional and forthright. But now, you doubtless realise, it is traitors who more often than not are the cause of destruction, and nothing ever results from a regular pitched battle. And yet you hear that Philip roams at will not because of his phalanx of hoplites, but because he has attached to himself an army of lightly armed troops, cavalry, archers, mercenaries, and the like. When he uses these to attack a state suffering from factional discord, and mutual distrust prevents anyone from going out against him to protect their countryside, he brings up his siege machinery and blockades the city. I do not speak of "summer" or "winter", since it makes no difference to him, just as he sets no season apart for a recess from campaigning.

13.5 The early Roman army

These three passages from Livy and Polybius describe the Roman army in the middle of the Republic, the period of Rome's greatest military expansion from the conquest of Italy, through the Punic Wars against Carthage, to the annexation of Macedonia. It will be evident immediately that sources for Roman military technology are much fuller than for their eastern predecessors (Bishop 1993, Robinson 1975, Davies 2008; cf. Greek wafare and arms in Snodgrass 1967, De Souza 2008).

Livy, *History of Rome* 8.8.3–8

The Romans in the past used round shields (*clipei*), but after they began to be paid for military service, they adopted oblong ones (*scuta*) in their place; and what had originally been a phalanx on the Macedonian model later came to be a battle-line drawn up by maniples, with those in the rear arranged in more units.

The *hastati* were the first line, 15 maniples positioned with a narrow gap between them. Each maniple had 20 light-armed soldiers – those equipped with only a thrusting-spear (*hasta*) and javelins – and a second group of soldiers with oblong shields. This front line of battle contained the pick of young men who were at the ideal age for military service. They were followed by an equal number of maniples of the *principes*, more mature soldiers armed with oblong shields and the best of weapons.

This combined body of 30 maniples they call the *antepilani*, because under their standards were stationed another 15 companies, each divided into three parts of which the first in each case was called the *pilus*. A company consisted of 186 men divided among the three *vexilla* or "standards", each of which had 60 soldiers, two centurions, and a standard-bearer. The first standard led the *triarii*, veterans whose mettle had already been proved; the second led the *rorarii*, lesser men in both age and experience; and the third standard the *accensi* [supernumeraries], relegated to the rear because they were the least reliable.

Polybius, *Histories* 6.22–23

They have the least experienced troops (*velites*) carry a sword, javelins, and the *parma*, a circular shield of sturdy construction and large enough to afford protection, measuring as it does 3 feet in diameter. The rest of their uniform includes a simple helmet, over which is sometimes placed a wolf's skin or similar covering, both for protection and as a symbol by which their unit commanders can distinguish them and determine their bravery or cowardice as they bear the brunt of the attack. The javelin's wooden shaft is, on average, 2 cubits long with the diameter of a finger; the point is a hand's span in length, sharpened by being beaten out so finely that it is inevitably bent on first impact and cannot be thrown back by the enemy – otherwise the missile could be used by both sides.

They have the next in age, called the *hastati*, sport the full panoply, of which the Roman version includes first an oblong shield (*scutum*), whose convex outer surface is 2.5 feet wide and 4.0 feet long, and the thickness of a palm's breadth at the rim. It is fashioned of two layers of wood glued together, its outer surface covered with linen and then with calf's skin. Its rim is edged with iron at the top and bottom, giving it protection against a sword's downward cut as well as from being planted hard on the ground. It is also fitted with an iron boss to ward off the direct blows from stones, long pikes, and powerful missiles in general. In addition to the shield they are equipped with a sword, termed "Spanish", that hangs on their right leg; for thrusting it has a sharp point, and for cutting a double edge that owes its effectiveness to the strength and rigidity of the ribbed blade.

Add to these two *pila*, a bronze helmet, and greaves. There are two varieties of *pila*, one heavy and the other light. The heftier ones have either a rounded shaft the diameter of a palm's breadth, or one square in section. Along with these they carry the light *pila*, which resemble similarly sized hunting spears. All these spears have a wooden shaft about 3 cubits long, to which is fitted a barbed iron head of the same length, made effective by an attachment so secure (since it overlaps half the length of the shaft and is clamped on with tightly spaced rivets) that in action the iron head will shatter before the connection gives way....

On top of all this equipment they wear a feathered chaplet with three purple or black plumes rising vertically to a height of about a cubit. The effect of attaching these to the helmet, when taken with the rest of the equipment, is that each man appears twice as tall as he really is, and with a noble presence that strikes fear into his enemies. In addition, the rank and file complete their outfit with a brass breastplate a span square, which they place over their heart and so call a heart-protector (*pectorale*); while those whose property qualification is rated above 10,000 *drachmai* replace the *pectorale* with a coat of chain-mail (*lorica*).

The same armour is common to the *principes* and *triarii* as well, except that the latter carry, instead of the *pila*, long spears for thrusting (*hastae*).

Livy, *History of Rome* 8.8.9–14

Once the army had been drawn up along these divisions, the *hastati* were the first of all the lines to enter battle. If they could not overwhelm the enemy, they would retreat in tight order and be received back into the gaps left between the companies of *principes*, who then took up the fight with the *hastati* following behind. The *triarii* would kneel in their ranks with their left legs outstretched, their shields leaning on their shoulders, the feet of their spears planted in the earth, and their heads at an angle pointing upwards, as if their battle line was bristling with a protective palisade.

If the *principes*, too, had no luck with their attack, they would gradually fall back from the front line towards the *triarii*, ... who, after receiving the *principes* and *hastati* into the gaps between their units, would rise from their kneeling position, "block the lanes" by forming their units into one mass, and then fall on the enemy in a continuous unbroken line, with no expectation of help from the rear. The enemy, in pursuit of ranks they thought they had beaten, were especially intimidated by this tactic when they spied a whole new line of even more troops suddenly rising up before them.

It was usual to raise four legions of 5,000 infantrymen apiece, with 300 cavalry attached to each legion.

13.6 The reforms of Marius

At the end of the second century BC the Roman general Marius, whose support came from the common people rather than the landed aristocracy of the Senate, introduced revolutionary reforms to the Roman military, prompted by political as well as tactical considerations. Three are described here by Plutarch; the fourth – his reorganisation of the legion based on the cohort rather than the maniple, and standardisation of armament – is not explicitly mentioned in our literary sources, but is deduced in part from the disappearance of any reference to maniples and to the *hastati*, *principes*, and *triarii*, after the beginning of the first century BC.

Plutarch, *Marius* 9.1; 13.1; 25.1–2

Immediately following his decisive election [as consul for 107 BC] he conducted a military levy, in which he broke both law and tradition by enrolling a large number of landless citizens of the lowest rank. Such men had not been admitted into the ranks by previous commanders, who allotted arms (like any other benefit) only to those who were eligible by virtue of their property qualifications, the supposition being that each man was pledging his property as a kind of security....

During this campaign he worked constantly on the efficiency of his troops as they marched, drilling them with jogging, long marches, and so on, and requiring each soldier to carry his own equipment and prepare his own meals. So, it came about that men who liked to work and did what they were told in silence and with good humour were subsequently called "Marian mules"....

The story goes that it was for this battle [with the Cimbri] that the design of the *pila* was first modified by Marius. Previously, the end of the wooden shaft that was inserted into the iron head was held in position by a pair of iron nails; but now, leaving one of these nails as it was, Marius removed the second and replaced it with a wooden peg that was easily broken. By this cunning design the *pilum* would no longer be left sticking straight out of the enemy's shield, but rather the force of impact would snap the wooden peg, causing the shaft to swing where it was attached to the iron head and drag along the ground, held fast by the dislocated point.

13.7 Some simple comfort for the soldier

The sound-deadening properties of sponge are mentioned a few times in the sources (Aeneas Tacticus, *Defence of a Besieged City* 19; cf. 20.1–4) and its use in the helmet may help the soldier to keep his orientation. Its use on the legs may be for padding to mitigate disabling blows.

Aristotle, *History of Animals* 5.16.2 (548b1)

There are three types of sponges ... the third, nicknamed "the sponge of Achilles", is exceptionally strong, fine and close-textured. This sponge is placed beneath the helmet and greaves, and it lessens the sound of the blow.

13.8 Necessity is the mother of invention: an early appearance of the piton

When on long campaigns it is impossible to prepare for every eventuality – Alexander the Great's Persian campaign must have presented a huge number of challenges for him and his force. After being taunted with the challenge to produce "winged soldiers" to gain the surrender of the Rock, Alexander offers prizes to the first soldiers who attain the summit. The solution to taking the unassailable Sogdian Rock seems simple in hindsight, but must have been a very creative act considering the total surprise it produced. See Philo, *On Sieges* D73–75 (102.12–24) for special iron pegs used in assaults of fortification walls, as well as unusual leather ladders for stealthed-use at night.

Arrian, *The Anabasis of Alexander* 4.19

The men drawing themselves together, as many as were practised to scale rock in the sieges, numbered 300, and having equipped themselves with the small iron pegs by which their tents were readied to be anchored fast to the snow-clad ground where ever it seemed to be frozen stiff, or into the ground if some areea was bereft of snow. And fastening them to strong cords made from flax, they set out in the night towards the most sheer face of the rock, thus the one least guarded. And fixing fast some of these little pegs into the earth where it showed through and some into the snow least likely to break up, they dragged themselves up the rock, some this way others that way. [Thirty fell and died but] the rest, climbing up and seizing the top of the mountain by dawn, signalled with linen banners towards the Macedonian camp as ordered of them by Alexander. [The barbarians, totally surprised and disoriented, surrender.]

13.9 The imperial Roman war machine

Josephus, *Jewish War* 3.71–97

Anyone who goes out of his way to examine the general organisation of their army will realise that the Romans have acquired an empire of such an extent as a reward for their prowess, not as a gift of fate. For them, military exercises do not coincide just with the beginning of hostilities. Far from sitting by in peacetime with idle hands that they put to work only when the need arises, they act as if they had been born already armed, never taking a holiday from training, never waiting for the critical moment to arrive. Their military exercises are no less vigorous than the real thing: the soldier who treats practice like true combat and puts everything he has into it is remarkably self-confident when he faces action....

They are not easily caught off-guard by a surprise attack. Whenever they invade hostile territory they always build a fortified camp before they engage the enemy in battle.... [The traditional Roman encampment arrangements are here described in detail.].... It is as if a city is created at a single stroke, complete with its market place, artisans' quarter, and council hall where the centurions and military tribunes can pass judgement on whatever disputes are brought before them. The enclosing wall with all the structures inside is finished in less time than it takes to imagine it, thanks to the size and skill of the workforce....

The need to break camp is announced with a trumpet call that sets everyone to work. At this first signal they strike their tents and get everything ready for departure. When the second blast signals them to form up, they are quick to load their equipment onto the mules and donkeys, stand at the ready, ... and then torch the encampment to prevent it from being used by their enemies, knowing that they could easily rebuild it. The third blast from the trumpet signals their departure.... Then they move forward, all marching in silence and in good order, each man keeping his proper place in the ranks as if the enemy was all around them. The infantry are armed with a breastplate and helmet, and carry a blade on each side: the sword on the left is the larger of the two, the dagger on the right measuring no more than a span.[3] The elite foot-soldiers that form the general's bodyguard carry a spear (*hasta*) and round shield (*parma*), rather than the javelin (*pilum*) and oblong shield (*scutum*) of the regular troops, whose equipment also includes a saw, a basket, a spade, and an axe, in addition to a leather strap, a sickle, a length of chain, and three days' worth of rations – the burden carried by an infantryman being not much different from that of a mule.

The cavalryman carries a long sword on his right side, a longish pike in his hand, a large shield held at an angle over his horse's flank, and three or more broad-headed javelins the length of spears, suspended in a quiver alongside; his helmet and breastplate are identical to those of all the infantrymen. The elite horsemen of the general's bodyguard are equipped no differently from the regular mounted troops.

TACTICS

13.10 The elements of strategy

The preceding passages give us much information about the elements of battlefield tactics from the Bronze Age to the Roman Empire; and we have, of course, lengthy literary accounts (of varying accuracy) describing hundreds of ancient battles. By way of summary, consider the following tables of contents, chapter by chapter, of the first three of Frontinus' four books on military strategy for commanders, composed at the end of the first century AC.

Frontinus, *Stratagems* 1 Preface, 2 Preface, 3 Preface

[Book One] Types of stratagems to guide the commander in his necessary preparations for battle:

1 On keeping one's strategy hidden.
2 On scouting out the enemy's strategy.
3 On establishing the character of the war.
4 On leading an army through territory infested by the enemy.
5 On escaping from particularly troublesome positions.
6 On setting ambushes while on the march.
7 How to camouflage or to compensate for things that we lack.
8 On diverting the enemy's attention.
9 On suppressing a military mutiny.
10 How to inhibit an inopportune desire for battle.
11 How to work up an army's passion for battle.
12 On dispelling the fear acquired by soldiers from unfavourable omens....

[Book Two] Types of stratagems that pertain to the conduct of battle:

1 On choosing the right occasion for battle.
2 On choosing the battlefield.
3 On deploying the line of battle.
4 On spreading confusion in the enemy's line.
5 On ambushes.
6 On letting the enemy escape to avoid backing him into a corner where desperation will cause him to renew the attack.
7 On concealing setbacks.
8 On rallying the troops by perseverance.

Types of stratagems that, to my mind, are vital after the battle is over:

9 On tidying up after a victory.
10 On repairing losses after a defeat.
11 On keeping the loyalty of those who waver.

12 What should be done to protect the camp, if we lack confidence in the troops at hand.
13 On retreating....

[Book Three] I have chosen the following types of stratagems for capturing a fortified position by force:

1   On a surprise attack.
2   On bluffing those who are besieged.
3   On securing the betrayal of the position.
4   By what tactics the enemy can be reduced to starvation.
5   How to convince the enemy that the siege will continue uninterrupted.
6   On throwing the enemy garrison off its guard.
7   On diverting streams and contaminating water supplies.
8   On striking fear into the hearts of the besieged.
9   On an assault from an unexpected quarter.
10 On tricks for luring out the besieged.
11 On feigning retreats.

Conversely, types of stratagems for protecting the besieged:

12 On inspiring wariness in one's troops.
13 On sending and receiving dispatches.
14 On bringing in reinforcements and providing rations.
15 How to simulate abundance in a time of deprivation.
16 How to counteract traitors and deserters.
17 On counterattacks.
18 On the perseverance of the besieged.

13.11 A Roman army on the march in hostile territory
Josephus, *Jewish War* 3.115–126

Vespasian, who was eager to invade Galilee himself, struck out from Ptolemais after marshalling up his army in the Romans' usual marching formation. He gave orders that the archers and the lightly armed troops from the ranks of the auxiliaries should set out in the van, to beat off any sudden enemy assaults and to search out suspicious wooded areas that could conceal an ambush. There followed a detachment of heavily armed Roman infantry and cavalry; then ten soldiers from each *centuria* carrying, in addition to their personal kits, tools for laying out the campsite; after them, the "pioneers" (*idiopoioi*), whose job it was to keep the main body of men from being worn out on a difficult march, by straightening curves on the route, smoothing out impassable sections, and cutting down trees that blocked the way. Behind them, Vespasian placed his personal baggage and that of his *legati*, with a substantial cavalry escort for its protection. He himself rode out next, accompanied by his spearmen and the cream of the infantry and cavalry. He was followed by the legions' squads of cavalry, 120 horse to a legion; then came the mules laden with siege machinery and the other engines of war. After them, the *legati*, the prefects of the cohorts, and the military tribunes, escorted by crack troops; then the standards surrounding the eagle, which the Romans have at the front of every legion.... Behind these hallowed symbols came the trumpeters, and then the main body of infantry, six columns abreast, with the customary centurion alongside, keeping an eye on their formation. The attendants from each legion followed as a group behind the infantry, leading mules and yoked animals carrying the soldiers' equipment. After the legions came rank upon rank of auxiliary troops,[4] and finally, the rear guard for protection: light infantry, legionaries, and a strong force of cavalry.

13.12 The military camp

When in the field, the Roman army would construct a fortified encampment at every overnight stop. This description of a camp for two consular legions and auxiliary troops reflects the practice of the second century BC, and has been largely confirmed by archaeological excavation. The modular design of the camp can be compared with the similar procedures for laying out a Greek temple (**8.40**). The simpler first passage is expanded upon in the second (cf. **8.34, 11.57**) (Dilke 1971 and Campbell 2000; Allison 2013, Baatz 1994).

[Hyginus], *On the Fortifications of Camps* 12

In the entrance of the central part of the praetorian at the *via principalis* is so-called *locus gromae* [place of the *groma*], perhaps because the crowd meets there or [because it is the spot] in the regulation of the *metae* [markers for measurement]. Once a *ferramentum* [an iron pole here] has been positioned in that very spot, the *groma* is set up upon it so that the gates of the camp produce a star in accordance with straight [sighting] lines. And the instructors of this art, for the reasons written above, are called *gromatici* [field surveyors].

Polybius, *Histories* 6.27–28, 30–31

This is how the Roman legions pitch their camp. Once the location of the camp has been chosen, the general's tent (*praetorium*) occupies the place in it that is most suitable for both observation and the issuing of orders. First, they place a standard where they intend to pitch the *praetorium*, then a rectangular space is measured off all around the standard so that every side is 100 feet away from it, and the area 4 *plethra* in size. Along whichever side of this central square seems best suited for obtaining water and forage, the Roman legions are arrayed as follows.... They place all the tents of the 12 tribunes in a straight line, parallel to the legions' side of the square and 50 feet away from it, in order to leave room for the horses as well as the pack animals and the tribunes' baggage. These tents are pitched with their backs to the *praetorium* square and facing the outside edge of the camp, which we should consider the front of the whole schema (and be consistent in referring to it that way). The tribunes' tents are placed equally distant from each other, their line stretching the full length of the space occupied by the Roman legions.

They measure off another 100 feet from the front of all these tents. Then, starting from the line drawn and marking out this interval parallel to the tents of the tribunes,[5] they begin constructing the encampments of the legions, employing the following method. They divide the aforementioned line in two, and from this point in a line at right angles to the original line they place the cavalry of either legion facing the other and 50 feet apart (with the second, right-angle line running down the middle of that intervening space). The disposition of the cavalry and the infantry is pretty much the same, the *turmae* and the maniples each being formed into a square. Each of these squares looks onto one of the streets that separate them, along which its length is measured – 100 feet – and for the most part they try to make the depth the same except for the allies.... [There follow extensive details of the disposition of specific units within the scheme outlined.].... Another space of 50 feet separates the *hastati* from the allied cavalry ... [and then] the maniples of the allied infantry....

The square behind the tents of the tribunes, flanking the *praetorium*, is used on the one side for a market and on the other for the quaestor's headquarters and its attendant supplies....

With things laid out in this way, the whole schema of the camp is a perfect square and its general layout, especially its division by streets, gives it an arrangement very similar to that of a city. The rampart (*agger*) stands 200 feet from the tents on all sides, the empty space between serving many useful functions for the army: it is properly designed to allow for the entrance and exit of the legions; ... here, too, they bring the young livestock herded along with the army and the plunder captured from the enemy, and guard them securely overnight; but most important, in attacks by night no fire or missile can reach them, and those few that do are rendered all but harmless thanks to the distance the tents are set back and the area surrounding them.

Vegetius (*Military Science* 1.21–25) adds some different details from the late Empire. [Hyginus], *On the Fortification of Camps*, dating probably from the third century AC, is an entire volume devoted to the subject, but reads more like a theoretical treatise than one based on practical experience.

13.13 The shield or "tortoise" formation

There were two forms of "tortoise" (Greek *chelone*, Latin *testudo*) used by the Roman army: the one described here by Dio Cassius as a useful disposition of the infantry, and the other (see **13.28**), a protective shed built over rams and sappers by the forces besieging a town [**fig. 13.1**].

Dio Cassius, *Roman History* 49.30

The design of this kind of "tortoise" is as follows. The baggage train, the light-armed soldiers, and the cavalry are drawn up in the middle of the army, while the heavy-armed troops [the legionaries] with their long, curved, cylindrical shields are arranged around the outside, facing outward and presenting arms, to form a kind of box in which they surround their comrades. Those others, equipped with flat shields, form up in the middle in close order and raise their shields over their heads and above all the others as well, so that the only things visible throughout the whole armed mass are the shields, which protect everyone in the dense formation from enemy missiles. In fact, the arrangement is so incredibly strong that men can walk on top of it, and even horses and carts can be driven over it when they find themselves faced by a narrow gully. Such is the configuration of this formation, and this explains the term "tortoise", which it derives from its strength and excellent protection.

The Romans make use of the *testudo* in two ways: in approaching and storming a fortified redoubt, often using it even to put soldiers onto the wall itself; or sometimes, when surrounded by enemy archers, to give them the impression of

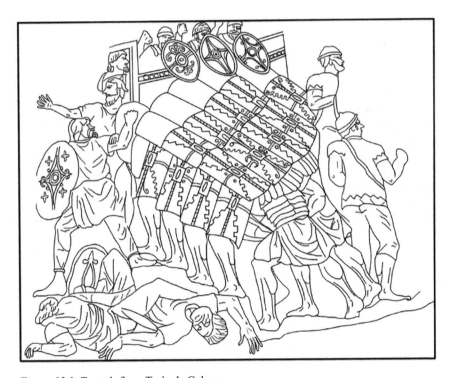

Figure 13.1 *Testudo* from Trajan's Column.
Source: Adapted from Reliefs on Trajan's Column, *in situ*, Rome, Italy; Cichorius, 1896, pl. 51.

extreme fatigue by crouching down *en masse* (even their horses are taught to kneel and lie down), then suddenly leaping up when the enemy approaches and taking them by surprise.

13.14 Communications: fire signals

In antiquity, the passing of messages on the battlefield was as difficult as long-distance communication, as the following two passages reveal. In the first, Aeschylus has the Mycenaean queen Clytemnestra describe her network of fire signals that brought word of the Greeks' conquest of Troy, some 400 km across the Aegean Sea; in the second, Polybius gives details of two complex and agonisingly slow methods of conveying information between units in the field. There are serious difficulties in plotting the route that the signals took in this first passage, principally because some of the locations have not been firmly identified. The distances of the legs that can be determined vary from 24 km to 130 km. Philo, *On Sieges* B55–57 (90.28–45) provides a more direct method.

Aeschylus, *Agamemnon* 280–312

CHORUS: And who among messengers could arrive with such speed?

CLYTEMNESTRA: Hephaestus, sending a bright flame out from Ida. Beacon sent beacon here, their fire like a dispatch rider: Ida to Hermes' crag on Lemnos; then third, lofty Athos, Zeus' own, took up the great torch from the island and, overleaping the strait so as to skim the sea, the energy of the mobile signal fire for sheer delight ... [there is a break here in the text] ... the pine torch transmitting the message of its golden-beamed flame, as another sun, to the peaks [or "signal towers"] of Makistos. Hesitating not at all, nor inattentively overpowered by sleep, he fulfilled his role of messenger. And from afar the light of the beacon reached [or "crossed"] the streams of Euripus and gave the sign to the guards of Messapion. They signalled in turn and sent the message on by setting alight a pile of dry heather. The signal fire, still strong and not yet dimmed like the beaming moon, leapt over the plain of Asopus towards the bare rock of Kithaeron and kindled another relay of courier fire. And there the watchmen did not ignore the light sent from afar, but lit a fire greater than previously mentioned [or "than they had been ordered"]. The light darted over the gorgon-eyed pool and, arriving at the mountain where goats roam, exhorted the fire installation to waste no time. And lighting it up with might aplenty they sent forward a great beard of flame, a blaze that vaulted over the promontory overlooking the Saronic Gulf, until it fell when it reached the watchtowers near our city on the heights of Arachnaion. And then this light, begotten of the Idaean fire, came to rest on this, the roof of the sons of Atreus. Such is the organisation of my torch-bearers.[6]

Polybius, *Histories* 10.43–47

In days gone by fire-signalling was a simple affair and largely useless for those who employed it. For it was necessary that this exercise be carried out through predetermined signals, but since circumstances are of an indefinite nature, the majority of them were not amenable to the use of fire signals.... For it is not possible to devise a prearranged set of signals for matters that are not susceptible to prediction.

Aeneas Tacticus[7] ... says that those who intend to communicate some pressing matter through fire signals to each other should prepare [two] clay jars, of exactly the same size in width and depth.... Then they ought to prepare corks a little narrower than the mouths of the jars, through the middle of which they should stick rods measured off into sections three fingers wide, each section clearly marked off by a line around the rod. On each section the most obvious and common wartime incidents are to be written: for example, on the first, "Cavalry present in the countryside"; on the second, "Heavy-armed infantry"; on the third, "Light infantry"; next, "Infantry with cavalry"; then, "Ships"; after these, "Grain".... With this done [he tells his readers], they should make holes in both jars with exacting care, so that the perforations and outflow are the same for both. Then they are to fill the jars with water and place the corks holding the rods inside, and allow the perforations to release water simultaneously. When this takes place, it will be obvious that (as is necessary if the specifications are identical) the corks descend and the rods are hidden within the jars to the same depth as the amount of water that flows out. In the course of operation, these devices can be made harmonious and of a uniform rate. Then they are to take them to the places where those who will attend to the fire signals are, and position one of those jars in each of those locations. Then (he goes on to say), whenever one of those circumstances written on the rod occurs, the signaller is to raise his torch, and maintain this position until his comrades in on the scheme raise their torch in return. When both torches are visible simultaneously, they are to lower them and in a split second allow the water to start flowing out of the perforations. When the sinking cork and rod brings to the lip of the jar that particular written circumstance that you wish to communicate, the author instructs us to raise our torch. The other party is immediately to stop up the aperture of their jar and look at which one of the circumstances written on the rod is at the lip. This will be the communication, if everything functions at a uniform rate for both devices....

The latest method, invented by Kleoxenos and Demokleitos and perfected by me, is very detailed and capable of accurately transmitting every exigency – though care and accurate observation are necessary when using it. It works like this. The letters of the alphabet are to be divided into five groups of five letters each in alphabetical order (the last group will be missing one letter, but this does not affect the use of the system). Those who plan to signal to each other must now prepare five tablets and write one of the alphabet's five divisions on each tablet, in order.... The signaller will raise the first set of torches on the left, showing which tablet the receiver is to look at: for example, one torch if it is the first, two if the second, and so on. Then he will raise the second set of torches on the right in like fashion, to show which letter of that tablet the receiver of the fire-signal is to write down.... A case in point: If you want to display the message that "About one hundred of the soldiers have gone over to the enemy", first you must pick out from the words those that carry the meaning with the smallest number of letters – for example, in place of the above statement, "Cretan hundred deserted us".... The message will be communicated by torches in the following way. The first letter is *kappa*; this is in the second set, and so on the second tablet. So, the signaller will need to raise two torches on the left (so that the receiver knows he is to look at the second tablet), then five on the right, to indicate *kappa* (for this is the fifth letter of the second set), which the one receiving the fire signals will have to write down on a small tablet. Then the signaller raises four on the left, since *rho* is in the fourth set, then two on the right, for it is the second letter of that fourth set.... Though this system allows any circumstance to be communicated accurately, it requires a lot of work with the torches, since double fire signals have to be made for each letter.

The latest method described by Polybius is now called the Polybius Square or Polybius Checkerboard.

13.15 Communication efforts using divers and swimmers

When the need arose, unusual writing materials, containers, and unusual delivery systems were employed in order to communicate. Two examples are provided here. In the first Hirtius communicates with Decimus, whom Antony has besieged at Mutina [43 BC]; in the second Lucullus contacts the Cyzicenes, who are under siege by Mithridates [73 BC].

Dio Cassius, *Roman History* 46.36.3–5

Because of the river near Mutina and the guard on it they [Hirtius and Octavian] were unable to advance farther. Even so, wishing that their presence be revealed to Decimus so that he might not precipitate any terms, they first tried making fire signals from the loftiest trees. But when he didn't understand, having scratched out some things onto a delicate sheet of lead, they rolled it up like a small piece of papyrus and gave it to a diver to carry across under water by night. Thus Decimus, having learned simultaneously of their presence and of their promise of assistance, replied to them in the same way, and with this method they then continuously revealed all their plans to each other.

Frontinus, *Stratagems* 3.13.6–7

Lucius Lucullus wanted to make the Cyzicines, who were besieged by Mithridates, more confident of his arrival. There was a single narrow entrance to the city, connecting the island to the mainland by a small bridge. Since this was held by the troops of the enemy, he first had letters sewn inside two inflated skins. Then he ordered one of his own soldiers, who was skilled in swimming and maritime matters, to stretch out upon the skins that had been bound together on the lower part by two amply separated straps, and to make the seven-mile-crossing. The soldier accomplished this so skilfully that, with his legs set apart like rudders, he steered his course and beguiled the men patrolling on guard, who were watching from a distance, by appearing to be some marine beast.

The consul Hirtius regularly sent letters inscribed on lead sheets to Decimus Brutus, who was besieged by Antony at Mutina. The letters were bound tight to the arms of soldiers, who swam across the Scultenna river.

13.16 The arming of a cavalryman

Evidence for organised cavalry units exists from the Archaic period, though principally because of the irregular topography they formed a relatively small proportion of the forces of most Greek city-states. Under Philip and Alexander (see **13.4**), in the more open plains of Macedon they were developed as an indispensable protection for the vulnerable wings of the phalanx.

Xenophon, *Art of Horsemanship* 12.1–12

First, then, I say that a cavalryman's breastplate should be made to fit his body.... Since the neck is also a vital part, I note the need for a protective covering for it, attached to the breastplate and shaped to fit the neck.... I think the strongest helmet is surely that of Boeotian design, which again covers especially well everything above the breastplate, while still not obstructing vision. The breastplate should be fashioned in such a way that it does not hinder either sitting or bending over. Around the abdomen, groin, and hips the "wings" or flaps should be of such design and size as to fend off missiles. And since a wounded left hand incapacitates a rider, for it too I recommend the piece of armour devised for it called the gauntlet: this covers the shoulder, upper arm, forearm, and the hand holding the reins; it is amenable to stretching and bending; and in addition to all this, it protects the area under the armpit left exposed by the breastplate. Since it is necessary to raise the right arm if one wants to hurl a javelin or strike [with his sword], the part of the breastplate that inhibits this movement should be removed, and in its place, flaps should be attached at the joints so that, whenever the arm is lifted, they spread out accordingly, and when it is lowered, they close....

Since the rider is exposed to every peril if his mount is wounded, it is necessary to provide the horse as well with armour for the forehead, chest, and along the haunches, this last also protecting the thighs of the rider. Most of all it is important to cover a horse's flanks, for these are at once the most vital and the most vulnerable. The saddle-cloth can serve as some protection for them, but it is equally necessary to arrange this saddle-cloth so that the rider is able to maintain a pretty firm seat, while the horse's back where he is sitting is not chaffed.

As an offensive weapon I recommend a sabre rather than a sword: because of his elevated position, the cavalryman will appreciate the slashing motion of a broad sabre more than the thrust of a sword. Instead of a spear with a [long] reed shaft, which is fragile and awkward to carry, I suggest instead a pair of light spears with cherry wood shafts; for the skilled horseman can throw one and still use the second [as a lance] in all directions forward, to the sides, or behind.

13.17 Cataphracts

The iron-clad horsemen of the fourth-century AC Persian army, described here, were a formidable foe, and foreshadow the mailed knights of medieval Europe.

Ammianus Marcellinus, *History* 24.6.8; 25.1.12–13

On their side the Persians drew up squadrons of cataphract[8] cavalry, which were so massed that the angles of their bodies, seamlessly fitted with metal plates, would dazzle with their sparkle the eyes of those facing them; and the whole crowd of horses was protected with leather coverings.... All of the units [infantry too] were wearing iron armour, and such was the covering of thick plates over every part of their body that the rigid junctures of the armour corresponded to the joints of their limbs; and the reproductions of human features were so carefully fitted to their heads that, as their whole bodies were covered in metal plate, the missiles that fell on them were able to lodge only where there were tiny holes, either placed over the orbs of their eyes to afford them limited vision, or at the ends of their noses to allow them short breaths. Some of them, ready to fight with their long lances, stood without moving, so that you would think they were clamped with bronze bands.

13.18 Elephants

As an engine of war, primarily used to spread panic among the enemy (*ad terrorem*, says Livy), the elephant first made its appearance at the Battle of the Hydaspes against Alexander (Arrian, *History of Alexander* 5.10–11; Quintus Curtius, *Histories of Alexander* 8.13.6; Diodorus of Sicily, *History* 17.87.2), and later the king of Epirus, Pyrrhus, used it in his unsuccessful invasion of Italy in the early third century BC (Plutarch, *Pyrrhus* 21). The Seleucids in the east and the Carthaginians in the west had easiest access to Asian and African elephants, which figured as important elements in their tactics. The following passage describes the famous battle at Zama in North Africa, traditional site of the last important stand of the Carthaginian Hannibal against the Romans (202 BC) (Scullard 1974, Glover 1948).

Polybius, *Histories* 15.9, 11–12, 16

Publius Scipio arranged the lines of his horses in the following way. First the *hastati* with gaps between their maniples, and behind them the *principes*; but rather than stationing their maniples behind the gaps left in the first line (as is the usual habit of the Romans), he arranged them in a line with – and somewhat back from – the front maniples, because of the horde of elephants the enemy had.... He filled the gaps between the first maniples with units of the *velites*, ordering them to bear the brunt of the first onslaught and, if forced from their position by the charge of the elephants, to withdraw – those who could outrun the elephants, in a straight line through the gaps right back to the rear of the whole army, and those who were overtaken to retire to one side or the other along the passages between the lines of maniples.... Hannibal positioned his elephants, more than 80 of them, in front of his entire army and, behind them, the mercenaries, about 12,000 in all.... His wings he secured with cavalry, stationing his Numidian allies on the left and his Carthaginian horsemen to the right....

When both sides were prepared for battle, ... Hannibal ordered the elephant drivers to charge the enemy. But the moment the trumpets and horns blared from all sides, some of the elephants took fright and at once rushed impulsively back at the Numidians who had come to the aid of the Carthaginians.... The rest of the elephants came down on the Roman *velites* in the space between the opposing lines, suffering losses themselves as much as inflicting them on their enemies. Finally, stricken with terror, some of them escaped through the gaps between the Roman maniples – which allowed them through and suffered no injuries, thanks to the foresight of their general – while the others fled along the front line to the right side, where they were pelted with missiles by the cavalry and ended up off the battlefield. At the same time as the elephants were causing such confusion, Laelius attacked the Carthaginian cavalry and drove them from the field at full speed....

Still, Hannibal had shown great skill in preparing at short notice a horde of elephants and then sending them in front of his army to throw the enemy into confusion and breach their formations.

The battle is also described in Livy, *History of Rome* 30.32–35.

The following story of Semiramis, the queen considered by the Greeks as one of the first Assyrian monarchs, was taken by Diodorus from the oriental history of Ctesius of Knidos, whose accuracy (particularly on such early matters) is questionable. As this passage begins, the queen is about to invade India across her eastern border; the supposed date is the late ninth century BC.

Diodorus of Sicily, *History* 2.16.8–9

Realising that she was much weaker because she had no elephants, Semiramis formed a plan to construct mock-ups of these animals, hoping that the Indians would be terrified since they believed that elephants existed only in India. She distributed the carcasses of 300,000 select black cattle among the craftsmen and those who had been levied to work on the contrivances, and made the mock elephants by sewing the hides together and filling them with straw, faithfully reproducing in every regard the outward appearance of these beasts. Each mock-up had a man inside to look after it, and a camel by which it was mobilised so that it appeared, at least to those looking from a considerable distance, to be a real elephant.

13.19 Naval weapons

According to Thucydides (*The Peloponnesian War* 1.13) the first recorded sea battle occurred in the mid-seventh century BC, between Corinth and Kerkyra (Corfu). The design and construction of warships – the early Greek pentekonters, the triremes developed during the Persian Wars, and the titanic quinqueremes of the Hellenistic and Roman periods – have been described earlier (**11.76–77**). These excerpts, from a lengthy and detailed description of the second battle of Salamis, in 307/6 BC, provide some insight into the weaponry and tactics employed by Hellenistic fleets, combining the classical stratagem of ramming with the use of warships rather like floating islands from which were launched assaults more typical of a land campaign.

Diodorus of Sicily, *History* 20.49–51

Demetrius ... assigned full naval complements to all his ships, embarked the best of his troops, and then placed missiles and *ballistae*[9] on board and set up in the bows an adequate number of catapults for bolts three spans long.... When the trumpets sounded the signal and both armies raised their battle cries, all the ships bore down in ramming formation – a terrible sight. First employing bows-and-arrows and stone-throwers, then continuous volleys of javelins, the crews wounded those who came within range; then, once the ships converged and the violent ramming was about to begin, the marines hunched down on the decks and the rowers, urged on by the boatswains, bent to their oars with even more spirit. As the ships were driven together with a violent impact, some sheared off each other's banks of oars, making them useless for retreat and pursuit, and preventing the men on board from joining the fray, though they were spoiling for a fight. Other ships, colliding bow to bow with their rams, rowed backwards for another attempt at ramming, while the men on deck wounded each other, inasmuch as the target for both sides lay close at hand. And when their trierarchs had manoeuvred for a broadside hit and their rams were holding firm, some marines leapt aboard the enemy vessels, inflicting as many serious wounds as they received....

13.20 The corvus

Before the First Punic War (264–241 BC) Rome had never challenged a sea power, but in order to secure her interests in Sicily she had to defeat Carthage, the major naval power of the western Mediterranean. Still, the Romans remained more confident of their abilities in a traditional infantry campaign, and so introduced the *corvus* ("crow") to make a sea battle imitate as closely as possible an action on land. The *corvus* was first used, with great success, off Mylae in 260 BC. For the first portion of this passage, see **11.73**.

Polybius, *Histories* 1.22.3–10

Since the [Roman] ships were poorly built and sluggish, someone proposed to the Romans as a recourse in battle the devices thereafter called *corvi* or "crows". This is how they are built. A round pole four fathoms high and three palms in diameter stood upright on the bow; at the top was a pulley, and around it was fixed a boarding-ladder with steps nailed across it, four feet wide and six fathoms long. There was an oblong hole in the planking that went around the pole at a distance of two fathoms from the near end of the boarding-ladder.[10] Along the sides of the ladder was a railing at the height of a man's knee. At the far end of the device was attached a piece of iron like a pestle shaved to a point and with a ring on its top, so that the whole thing seemed to resemble those machines used for pounding grain. A rope was tied to this ring, by which the *corvus* was raised by the pulley on the pole as the ships made contact, and then released onto the deck of the enemy vessel, sometimes from the prow, and sometimes they swung them around if the ships met broadside. When the ships were bound together with the *corvi* stuck in the deck planking, the marines leapt aboard from all sides if the vessels were lying alongside each other, but if they met bow on they made their assault across the *corvus* itself, proceeding two abreast. In such a case those in the lead would cover the exposed area in front using their large shields as a screen, and those following would protect either side by placing the rims of their shields on top of the railing.

13.21 The grappling iron

A sea battle between Pompey the Younger and Octavian, whose admiral Agrippa – later to distinguish himself at the Battle of Actium in 31 BC – devised a special grapnel for naval combat. This encounter took place off Naulochus in 36 BC.

Appian, *Civil Wars* 5.118, 119

Three hundred ships were rigged out on either side, carrying missiles of every kind, towers, and whatever machines they could conceive of. Agrippa invented the so-called "grip" or grappling iron, which consisted of a piece of wood five cubits long surrounded with iron, and with a ring at either end. One of the rings held the grip – an iron hook – and to the other were attached many small cords that used machines to draw in the grip once it had been released from a catapult and had grabbed hold of an enemy ship....

The grappling hook particularly distinguished itself. Its light weight allowed it to be hurled from a considerable distance, and it stuck fast as soon as it was pulled back by the ropes. Those whom it caught could not easily cut it because of the iron sheathing, and its length made it next to impossible for anyone to cut the ropes (the counter-device of mounting sickles on poles not yet being known). The enemy thought of only one thing in these unexpected circumstances: to back water and pull away. But since the other side was doing the same, the exertion of the crews was matched and the grappling hook did its bit.

13.22 Naval weapons of fire

The nature of ancient ship construction made fire even more efficacious as a naval weapon than a land-based one (for which see **13.31**) (Haldon 2006).

Livy, *History of Rome* 37.11.13

No more than five Rhodian ships got away, along with two vessels from Kos, taking advantage of the panic caused by sparking flames to open an escape route for themselves amid the crush of ships. For they were carrying a blaze of live flames contained in iron buckets suspended on a pair of poles jutting out from the bow.

Polybius, *Histories* 21.7

Pausistratos, the Rhodian admiral, employed a "fire-carrier" that had the shape of a funnel. From both sides of the prow and along the inner surfaces of the hull were laid cords with looped ends, to which were attached poles projecting out to the sea like horns. The funnel-shaped bucket, full of fire, was fastened to the end of these poles with chains, so that when the ship rammed or came alongside an enemy vessel the fire could be shaken out onto it, while the angle of the poles kept it a safe distance from the friendly ships.

Arrian (*Anabasis* 2.19) and Curtius (*Histories* 4.3.2–7) give lengthy descriptions of a fire-ship filled with combustible materials, towed against the Macedonian siege-works, and set aflame by the Tyrians during Alexander's blockade of their city in 332 BC.

Frontinus, *Stratagems* 4.7.9–10

In a naval battle Gnaeus Scipio shot jars full of pitch and pine-resin at the enemy fleet. The point of shooting these projectiles was two-fold: their weight might cause damage [on impact], and by scattering their contents they could provide fuel for a fire.

Hannibal taught King Antiochus to shoot small containers filled with poisonous snakes against an enemy fleet, using fear of the snakes to hamper the marines in fighting and the crew in handling the ship.

What follows is the first mention of Greek fire (in Greek, "sea-fire"), a mysterious chemical mixture that was launched through force pumps and could be extinguished only with vinegar. The naval battle here described, between the Byzantines (called "Romans") and Arabs, really falls beyond the scope of our study, since it took place in the second half of the seventh century AC [**fig. 13.2**] (Mayor 2003).

Theophanes, *Chronographia* 354.14 (de Boor)

Then Kallinikos, an engineer from Syrian Heliopolis, fled to the Romans. After manufacturing sea-fire, he set the ships of the Arabs afire and incinerated all aboard. And in this way the Romans recovered in victory, and invented sea-fire.

MILITARY TECHNOLOGY

Figure 13.2 Greek fire after 12th century illustration from the Madrid Skylitzes.
Source: After twelfth century AC illustration from the Madrid Skylitzes.

13.23 Use of fire in a siege and defensive measures against it

Starting a Fire: A Special Weapon

Aeneas Tacticus, *Treatise on the Defence of a Besieged City* 33.1–2

One must pour pitch and cast tow [a fire-starter of unprepared flax] and sulphur on the protective sheds being brought forward, then throw a blazing fagot fastened with a rush-rope upon the protective shed ... and on the war-machines being moved forward. It is necessary to burn these devices in this manner. Let wooden forms be prepared fashioned like pestles but much larger, and into the ends of the wooden form drive sharp iron pieces, larger and smaller, and around the other parts of the wooden form, above and below; separately, prepare powerful combustibles. The form's appearance should be like the thunderbolt of paintings. Against an approaching siege machine, one must launch this [weapon], fashioned in this way so that it sticks into the machine and so that the fire will persevere once the wooden form has been stuck fast.

DEFENDING AGAINST ATTACKS USING FIRE

Philo, *On Sieges* D34–35 (99.21–28)

So that neither machines nor assault-bridges nor tortoises are set on fire, one must use iron and bronze and lead tiles and wet seaweed [placed] into nets positioned on [the machines] and wetted sponges and fleeces drenched in vinegar or water; or [one must use] birdlime or ash being mixed with blood to smear the beams in the places you think fire is most likely to be attempted.

13.24 Equipping a fleet

In 205 BC, towards the end of the Second Punic War against Carthage, the Roman general Scipio received permission to sign on volunteers and receive allied contributions towards building a new fleet at no cost to the state. Notice the equipment supplied for the large number of marines, who were an essential element in Roman naval strategy.

Livy, *History of Rome* 28.45.14–21

First the communities of Etruria promised the consul assistance, each according to its means. Caere would supply grain for the sailors and all sorts of provisions, Populonia iron, Tarquinia sail cloth, Volaterrae the interior woodwork as well as grain. Arretium promised 3,000 shields, 3,000 helmets, and equal quantities each of javelins, spears, and lances totalling 50,000 pieces, together with enough axes, spades, hooks, basins, and hand mills for a fleet of 40 warships, 120,000 *modii* of wheat, and a contribution towards the maintenance of junior officers (*decuriones*) and oarsmen. The citizens of Perusia, Clusium, and Rusellae would supply fir for shipbuilding (which Scipio supplemented with fir from state-owned forests) and a large quantity of grain. Nursia, Reate, Amiternum, and the entire Sabine territory joined with the communities of Umbria in promising soldiers. Large numbers of Marsi, Paeligni, and Marrucini signed up as volunteers for the fleet. And Camerinum sent a fully equipped cohort of 600 men, even though their treaty with Rome did not require it.

After 30 keels had been laid – 20 quinqueremes and ten quadriremes – Scipio so pressed on with the job that the ships were launched, fully rigged and equipped, on the forty-fifth day after the timber had been hauled in from the forests.

FORTIFICATIONS AND SIEGE ENGINES

13.25 The need for fortification walls

That fortification walls make men soft and complacent is a common literary motif. An unknown poet, for example, is credited in Plato *Laws* 6.778d-e with the line: "Walls should be made, not of earth, but of bronze and iron". And Polyaenus (*Stratagems* 8.16.3) has Scipio concerned that one of his soldiers places greater confidence in the camp's palisade than in his own sword. Aristotle here offers a more realistic assessment (D'Amato 2016, De Souza 2008, Davies 2008, A. Johnson 1983, S. Johnson 1983).

Aristotle, *Politics* 7.10.5–8 (1330b–1331a)

As for fortification walls, those who contend that cities laying claim to valour need not have walls hold a quite outdated opinion, particularly when it is clear that those cities that make such a display of vanity are, in practice, proved wrong. While it is not honourable to use the strength of one's wall to try to protect a city against a foe of equal or only slightly greater numbers, yet it is possible that the superiority of the attackers may happen to prove to be too much for the valour of a few defenders. If in this case the city is to be saved and not suffer harm or humiliation, the greatest possible security and strength of the walls must also be considered the most suitable for warfare, particularly in light of recent inventions that improve the accuracy of the missiles and artillery used in sieges....

It is not enough just to put walls around a city: care must be taken to make them aesthetically pleasing for the city and at the same time appropriate for their military functions – keeping in mind those newly invented machines.

13.26 The walls of Athens and Rome

The existence of fortification walls around cities in antiquity was so much taken for granted that our sources seldom mention them. This is true even of the extensive walls that protected Athens and Rome. In the first selection, Thucydides alludes to the construction of a pair of Long Walls (a third was added later) that joined the fortified city of Athens with its two harbours, at Phaleron and the Piraeus; the second group gives most of the written evidence for the three consecutive walls that protected Rome.

Thucydides, *The Peloponnesian War* 1.107.1, 108.3; 2.13.7

In these times [457 BC] the Athenians began to build their Long Walls, one to Phaleron and the other to the Piraeus, ... and they finished their Long Walls [the next year].... The length of the wall from Phaleron to the circuit wall of the city proper was 35 *stadia*; the part of the circuit wall that was manned was 43 *stadia* in circumference; ... the Long Walls to the Piraeus were 40 *stadia* long; ... and the circuit wall of the Piraeus was 60 *stadia* around....

Livy, *History of Rome* 1.44.3; 6.32.1; *Historia Augusta*, *Aurelian* 39.2

[In the sixth century BC Servius Tullius] surrounded Rome with a rampart [*agger*] and ditches [*fossae*] and a wall [*murus*].... [In 378 BC] new debts were accumulated on account of the tax for building a wall of squared stone, which was arranged by the censors.... [In AD 270] Aurelian expanded the walls of Rome to such an extent that the circuit of his wall was almost 50 miles.[11]

13.27 Construction of a circuit wall and towers

Vitruvius, *On Architecture* 1.5

The foundations of the towers and circuit wall are to be laid in the following manner. The trenches should be dug down to bedrock, if it can be reached, and as extensively along the surface of the bedrock as seems reasonable given the scope of the work. These foundation trenches should be wider than those parts of the walls that will be above ground, and should be filled with structural material that is as solid as possible.

The towers are to project outside the wall, so that the enemy who is determined to approach the wall in an assault will be vulnerable to missiles from the towers, since both his left and right sides will be exposed. Extraordinary care must be taken, it seems, to avoid any easy approach for storming the wall; the roads ought to be led along the contours and be so contrived that the lines of travel do not lead straight to the gates but come from the [defenders'] left – a design that will place nearest the wall the right side of those who are climbing up, the side that is not protected by a shield.

Walled towns should not be laid out in a square shape or with prominent angles, but rather with a circular plan to give a more unobstructed view of the enemy. Those towns with salient angles are hard to defend, because the angle affords greater protection to the enemy than to the population inside.

I do believe that the wall should be made broad enough to allow armed men meeting on the top to pass one another without getting in each other's way. Through the whole thickness, splines of charred olive wood should be set as closely together as possible so that both faces of the wall, bound together by these long thin boards as if by pins, should be permanently stable – for neither rot nor weather nor time can harm this material....

The interval between two towers should be set so that one is no further than a bow-shot from the other, which will ensure that, no matter what section is attacked, the enemy can be repulsed by the scorpions and other missile-shooting devices from the towers on the right and the left. On the inner side of the towers there should be a gap in the wall as wide as the tower, spanned by wooden gangways that give access into the towers along joists with no iron reinforcements. This way, if the enemy captures any part of the wall, the defenders can cut these joists away and (if they work quickly) the enemy will not be given the chance to make his way into any other parts of the towers and wall – unless he is prepared to take a plunge.

The towers should be built round or polygonal, since siege engines very quickly weaken square towers, the constant beating by rams shattering their corners. When it comes to round structures, however, the pounding is directed into the centre, like wedges, and cannot do any damage.

Likewise, the defensive works involving curtain wall and towers are especially safe when used in tandem with earth ramparts [*aggeres*, on the inner face] because they cannot be harmed by rams, tunnelling, or any other machines. Still, it is not reasonable to build an embankment in all places, but only where outside the wall there is a high stretch of ground and level access for attacking the defences. In places like this, first trenches (*fossae*) should be dug....

As for the materials that should be used for the core or outer skin of the wall, it is impossible to make specific prescriptions since not everywhere can we find the supplies that we might want. Where squared stone is available, or flat stone, rubble, fired or unbaked brick,[12] use it....

Philo, *On Sieges* A8 (80.5–11)

Philo provides much information about placement and shapes of fortification walls and towers along the lines of Vitruvius, but also adds some small details about the outer stone skin that appears often in the archaeological record and other written sources (Whitehead 2016).

So, they don't take a shock of any sort from a blow of any sort, the last of the stones [in stone towers] should be bound to each other with lead and iron and mortar, for that occasion when the rock-throwers are advancing so they are unable to knock down the parapets.

Philo, *On Sieges* A13 (80.28–31)

One of our earliest mentions of a technique to conduct repairs to catapult-damaged walls.

Oak beams must be put into the walls and the towers end-on continuously every four cubits, so that if any damage is suffered by stone-throwers we may easily repair it.

13.28 Battering rams

The evolution of fortification walls was matched by the development of weapons designed to penetrate them. The earliest, a device that must be as old as city walls themselves, was the simple battering ram, whose invention is imaginatively recorded here by Vitruvius. Improvements included suspending the ram from slings to increase its force, covering it to protect its operators, and designing a more portable, lightweight version, this last described in the second passage, when the Sabiri and Romans, besieging Petra in the late Empire, find themselves unable to use the traditional wheeled ram because of the rough ground.

Vitruvius, *On Architecture* 10.13.1–2

It is recorded that the ram was first invented for siege operations in the following way. The Carthaginians pitched their camp to besiege Gades. They had already taken a small outpost, and now tried to knock it down. After a while, since they did not have iron demolition tools, they took a timber, raised it up with their hands, and by continually beating its end against the top of the wall they managed to dislodge the upper courses of stones, and gradually they demolished the whole fortification row by row.

Later a Tyrian craftsman by the name of Pephrasmenos, inspired by the principle of this invention, erected an upright pole and hung a crossbeam from it like a set of scales, and by drawing it back and pushing it forward with forceful strokes he razed the wall of Gades.

It was Ceras the Carthaginian, however, who was first to make a base of wood, place wheels underneath, and give it a superstructure of upright poles and crossbeams. Inside he hung the ram, and covered [the framework] with oxhides to protect those who were placed inside the engine to beat on the wall. Because the device had such slow movement, he began to call it a "ram tortoise" (*testudo*).[13]

Procopius, *Gothic War* 8.11.27–31

Now these Sabiri concocted a battering ram, not in the usual fashion, but by introducing a different idea. Rather than building into the engine any beams – uprights or cross-ties – they instead fixed stout saplings everywhere in place of the beams and lashed them to one another, then covered the whole thing with hides, maintaining the shape of the traditional ram. The only beam in it was hung, as is usual, by slack chains along the midline of the engine, its end made sharp and sheathed in iron like the point of an arrow, for repeated battering against the circuit wall. The finished engine was so light that there was no need for it to be dragged forward or nudged along by the men inside; rather, there were 40 men inside it and covered by the hides – the same ones who would pull back the beam and swing it against the wall – who could carry the ram on their shoulders without any difficulty.

13.29 Invention of catapults

Diodorus of Sicily, *History* 14.42.1, 43.3, 50.4

As a matter of fact, the catapult was invented at this time [399 BC] in Syracuse, for the greatest technical minds from all over had been assembled in one place.... Catapults of all sorts were built, as well as a great number of other missiles.... The Syracusans killed many of their enemies by shooting them from the land with catapults that shot sharp-pointed missiles. In fact, this piece of artillery caused great consternation, since it had not been known before this time.

For another version of this story, see **14.19**. Plutarch (*Moralia* 219a) records the comment of Archidamus III, a fourth-century BC Spartan king, when he first saw a missile hurled by a catapult imported from Sicily: "By Herakles, the bravery of an individual no longer matters!" Complex details for the construction of catapults are given in Vitruvius, *On Architecture* 10.10–12 (Schiefsky 2015, Rihll 2007, Marsden 1971, 1969, Hacker 1968).

13.30 The onager

In the section of his work immediately preceding this one, Ammianus has attempted a description of *ballistae*, but the passage is close to indecipherable (he himself acknowledges his "meagre skill"). But for the onager, or "wild ass", he is our principal source, and happily does a slightly better job describing it (though the passage still requires considerable editorial clarification).

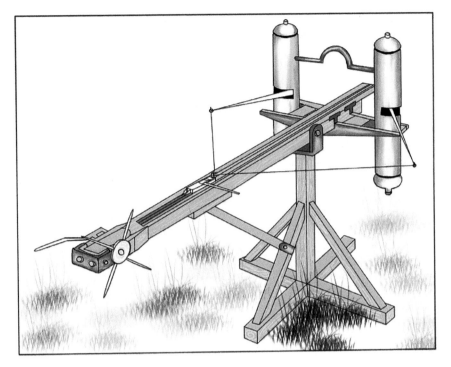

Figure 13.3 Cheiroballistra.
Source: Adapted from Marsden, 1971, pls. 6–8.

Ammianus Marcellinus, *History* 23.4.4–7

On the other hand, the scorpion (which they now call the onager or "wild ass") follows this design. Two beams are hewn out of oak or holm-oak, slightly curved [in the middle] so that they seem to rise up like humps. These are joined together in the manner of a bow-saw: each side piece is pierced by a rather large hole, through which durable ropes are passed and bound between the beams, holding the framework together so that the engine does not break apart.[14] From the middle of the ropes a wooden arm rises obliquely, pointing upwards like the pole of a chariot [unhitched]. The arm can be raised higher or drawn downwards by cords that are wrapped around it [near the top end]. To the tip of this arm are attached iron hooks, from which hangs a sling made of hemp or [?] iron. A large cushion of goat's hair stuffed with fine bits of straw is placed in front of this wooden arm, bound on [a cross-brace] with strong fastenings.

Since a heavy contraption of this sort, placed on a stone wall, dislodges everything beneath it more by its violent concussion than by its weight, it is placed on a pile of turf or a brickwork embankment.

In battle, then, a round stone is placed in the sling and four young men on either side, turning the bars [of the windlass] to which the ropes are fastened, bend the arm nearly horizontal. Then at last the soldier in charge, standing close above the machine, uses a sharp blow from a hefty mallet to strike out the bolt that holds in position the tethered rope of the device. The swift blow releases the arm, which then smacks into the soft cushioning and hurls the stone, which will crush whatever it hits.

The machine is called a *tormentum* because all of the released energy is first created by twisting [for which the Latin stem is *tor-*] the ropes; or a scorpion since it has an upraised stinger; or recently an onager because, when wild asses (*onagri*) are attacked by hunters, they stay at a distance and kick rocks behind them, splitting open the chests of their pursuers or breaking their bones and bursting apart their skulls.

13.31 Fire as an offensive and defensive weapon

The first selection describes the attack on fortified Delium by the Boeotians and their Peloponnesian allies (424 BC). And at the siege of Uxellodunum in Aquitania, Caesar has constructed a ramp and tower to guard access to a spring that was the defenders' only source of water.

Thucydides, *The Peloponnesian War* 4.100

After trying [to storm Delium] by other means, the Boeotians brought up a device that I will now describe, and it did indeed take the city. They cut a great beam in two [lengthwise], hollowed the whole thing out, and carefully fitted it together again, making in effect a pipe. On one end they hung a cauldron on chains, and inserted into it, at an angle from the hollow beam, the iron snout of a set of bellows. Most of the rest of the wood was also clad in iron. Using wagons, they hauled the device forward some distance to a section of the wall that was built principally of vines and logs. When it was within range, they put a great set of bellows up to the end of the beam that faced them, and started working them. When the blast of air went through the sealed tube to the cauldron – which held smouldering coals, sulphur, and pitch – it created a huge flame and set the wall alight. No one could remain there any longer; all were forced to abandon their posts and flee. This is how the fortified city was taken.

Caesar, *Gallic War* 8.42.1

The inhabitants of the town, terrified at this misfortune, filled barrels with hardened fat, pitch, and wooden shingles. They set them on fire and sent them rolling against the siege-works, at the same time attacking with great intensity so that they might use the danger of combat to keep the Romans from extinguishing the fire. Suddenly a great flame shot up, this time from the siege-works, for anything that was hurled down the intervening slope lodged in the protective sheds and the ramp, and ignited the very material that prevented it from reaching the Romans.

13.32 Examples of chemical warfare

Pausanias, *Description of Greece* 10.37.7–8

Solon also devised another stratagem against the Cirrhaeans. The waters of the river Pleistos, which ran into the city in a channel to supply the inhabitants, he diverted in another direction. When the Cirrhaeans still withstood the besiegers by drinking water from the wells and rainfall, he threw roots of hellebore into the Pleistos and then, once he determined that there was enough of the poison in the water, he turned the stream back into its original channel. The Cirrhaeans threw caution to the wind and drank their fill of the water. Stricken with unremitting diarrhoea as a consequence, they abandoned their posts on the wall and let the city be taken....

Philo, *On Sieges* B53 (90.14–23)

Preparation for fending off attacking forces include collection of unusual materials, including poisons for use on weapons (Mayor 2003).

[Have available] also the Arabian potion and the little mollusc, the one that is in the lake 50 stades away from Ake, and mistletoe and salamanders and the poisons of vipers and of asps and naphtha, which is in Babylon, and fish oil – [all are useful] to foul the waters of advancing enemies, and useful if, throughout the ongoing hazards, we drive forward to burn [their] machines and to smear [our own] projectiles, in order to contrive fear and a quick death for the wounded men and for those assaulting the wall.

13.33 Early siegecraft

Thucydides here describes the Spartan siege of Plataea in 430 BC, and in Diodorus we read of the investment of Perinthus by Philip of Macedon, 341/0 BC. Both these sieges show relatively simple techniques on both sides: offensively, circumvallation (the construction of an outer wall to keep the besieged entrapped and to prevent relief from outside), a ramp or towers to give access over the walls, and tunnelling and ramming for weakening the circuit; and, defensively, counter-tunnelling, the construction of inner walls, and the use of fire and dropped weights to incapacitate the enemy's engines.

Thucydides, *The Peloponnesian War* 2.75–76

First the Spartans hedged the Plataeans in with a palisade of stakes made from the trees they had chopped down, so that no one could escape; then they began to raise an earthen ramp towards the city, hoping that the great size of their army engaged in this labour would make this the quickest possible way of taking the place. They cut timbers from Mt. Cithaeron and constructed a lattice-work, which they placed along either side of the ramp instead of a solid wall, to keep the ramp from spreading horizontally too much. They brought wood and stones and earth, and threw them into the ramp together with anything else that might fill up the mound. For 70 days and nights they kept piling up earth without stopping, divided into reliefs so that, while some were carrying earth, others slept or ate, supervised by Spartan commanders of the auxiliary troops from each city, who kept them at their work.

When the Plataeans saw the mound rising higher, they put together a wooden crib and set it on top of their city wall where the enemy was mounding earth against it, then built into it bricks that they removed from the nearby houses. The timbers served as a binding-frame for the bricks so that the construction did not become weak as it rose higher, and they had coverings of skins and hides that kept the workers and framework safe by protecting them from being hit by flaming arrows. Yet while the height of the wall was being raised considerably, the ramp outside was growing at an equal pace.

But the Plataeans had another idea. They breached their wall where the ramp abutted it and set about removing the earth into the city. When the Peloponnesians discovered this, they packed clay into wicker mats and threw this clot into the breach, to prevent the enemy from breaking it up and carrying it off as they had the loose earth.

Frustrated in this ploy, the Plataeans changed their plan and instead dug a tunnel out from the city and, when they calculated that they were under the ramp, began once again to remove the accumulated earth from underneath. And for a long time, the Spartans outside the city, who had no idea what the Plataeans were up to, kept throwing material onto the mound, but it came no closer to completion since it was being hollowed out from underneath and was continually settling into the cavity that was emptied.

But, fearing that even with this strategy their small number would not be able to hold off so many, the Plataeans came up with this additional plan. They stopped working on the high parapet opposite the ramp and, starting on either side of it from the bottom of the wall, began building a crescent-shaped barrier on the inside, curving inward towards the city, so that, if the main wall were taken, this one could hold out, forcing the enemy to build a second ramp against it – not only would they have to repeat their whole effort, but as they advanced into the arc of the new barrier they would be more exposed to attack from both sides.

At the same time as they were raising their ramp, the Peloponnesians brought siege engines up to the city. One of them, moved up to the high defensive parapet opposite the ramp, knocked down a large part of it and struck fear into the Plataeans; but other engines at a different part of the wall were snared in nooses by the defenders and snapped off. The Plataeans also prepared large beams with long iron chains fixed at either end and hanging from two yard-arms tilted forward and jutting over the wall. Whenever an engine was about to attack some section of the wall, they would draw the beam up crosswise [over the enemy ram], and then drop it while letting the chains run loose. The beam would plummet down and break off the head of the ram.

Subsequently the Peloponnesians tried – unsuccessfully – to burn the Plataeans out, and finally resorted to a long siege after completely surrounding the city with a wall to pen in the inhabitants. Some broke out, but those remaining were driven by starvation to surrender. The Spartan victors executed the men, enslaved the women, and destroyed the city.

Diodorus of Sicily, *History* 16.74.2–5

Philip ... organised a siege and, advancing his engines against the city, spent one day after another attacking the walls in shifts. He built towers 80 cubits high, substantially taller than the towers along the wall of Perinthus, and from this superior position kept wearing down the defenders. By shaking the walls with battering rams and undermining them with tunnels, he managed to bring down a good part of the circuit. The Perinthians in turn defended themselves with vigour, and quickly built another wall against the attack....

The king had many bolt-shooting engines of every sort, and with these he devastated the men fighting desperately from the battlements. For their part the Perinthians, while continuing to lose many men every day, received reinforcements, missiles, and catapults from Byzantium.... Still the king stuck to his resolve. He divided his forces into a number of units and day and night made repeated relay attacks on the walls. With 30,000 soldiers and a great mass of ordnance and siege engines (and an unrivalled collection of yet other artillery pieces), he kept wearing down the besieged.

13.34 Hellenistic sieges

The last paragraph of the previous selection alludes to the development of new artillery pieces in the fourth century BC. Advances in engineering techniques under Philip II and Alexander led to significant improvements in siege machinery, especially in the construction of huge towers and extensive protective galleries, all fitted with a variety of newly designed artillery pieces. Here, Diodorus describes the siege of Rhodes in 305 BC by Demetrius I of Macedon who, unsuccessful in his earlier naval assaults on the island, decided to attack by land and won the epithet Poliorcetes (Besieger) for his efforts, though they were ultimately futile; Polybius outlines the siege of Echinus by Demetrius' distant successor, Philip V of Macedon, a century later (211 BC).

Diodorus of Sicily, *History* 20.91.1–8

He prepared a massive quantity of every sort of matériel, and constructed an engine called the *helepolis* [city-taker], bigger by far than any that had come before it; for its base was square, each side measuring nearly 50 cubits, and made of squared timber bound with iron. He divided the space within the frame by beams spaced about a cubit apart so that there would be standing room for the men who were to push the engine forward. The whole mass was quite mobile, placed as it was on eight enormous, solid wheels whose treads were 2 cubits wide and clad in iron plates. To allow progress to the side, they installed pivots [on the wheels], by which the whole engine could easily be moved in any direction.

From the corners rose posts of equal height, just under 100 cubits, inclining towards one another at such an angle that the whole assembly was nine storeys tall, encompassing 4,300 square feet at the first level and 900 square feet at the top. The three sides open to attack he covered by nailing iron plates to their outer surfaces, to protect the machine from firebrands. Along the front of the various levels were windows, the size and shape of each configured to the nature of the missiles that were to be shot out of it. These openings had curtains that could be raised mechanically and protected those on the different levels while they were busy shooting missiles: for they were stitched together out of hides and stuffed with wool to absorb the impact of the *ballistae*. Each of the levels had two wide ladders, one for carrying up supplies and the other for coming down, so that everyone's duties could be performed without confusion.

Those who were to move the engine were chosen from among the whole army: 3,400 men of superior strength, some positioned inside and others behind the engine.[15] As they pushed it forward, the clever design of the structure assisted in the labour of moving it.

Plutarch (*Demetrius* 40) relates that at Thebes the *helepolis* bogged down and moved only 400m in two months. Elsewhere, Diodorus (*History* 20.48) describes a similar engine and its destruction by fire.

Polybius, *Histories* 9.41.1–10

He determined to make his approach on the city against a pair of towers, and so built opposite each of them a shed or tortoise to protect the ram and those filling in the ditches, and between the rams and parallel to the wall, in the space from one tower to the other, he erected a gallery. The appearance of the finished siegeworks was much like that of the defensive wall: the characteristics of the wickerwork gave the superstructure above the sheds the look and arrangement of towers, and the gallery between resembled the curtain wall, since the weaving of the upper wickerwork was divided, as it were, into battlements.

Throughout the lower levels of the [offensive] towers sappers threw on more and more earth as they levelled the unevenness of the ground to allow for the approach of the platforms, and the ram was then driven forward. Water jars and other equipment in case of fires were kept on the second level, and the artillery along with them. A crowd of men stood on the third level at the same height as the towers of the city, ready to engage those defenders who were eager to damage the ram.

From the gallery between the towers, two tunnels were driven forward to the curtain wall. There were also three batteries of *ballistae*, one shooting stones weighing a talent, the others stones of 30 *minai*. Roofed passages had been built underground extending from the encampment to the sappers' sheds, so that those going out from the camp and those returning from the siege-works would not be hit by missiles from the city.

Because the countryside had a plentiful supply of the materials required for this purpose, these siege-works were completely finished in a matter of a few days.

For an attempt to date the development of towers, galleries, and wheeled and tortoise-rams, see Vitruvius, *On Architecture* 10.13.3–6.

13.35 Archimedes defends Syracuse against the Romans

The Roman siege of Syracuse during the Second Punic War is perhaps the most famous example of a blockade by land and sea, though hardly typical: the imaginative counter-siege devices developed by Archimedes make the whole event unique. Some of the defensive engines described here may seem fantastic, but – aside from some hyperbole by Plutarch (and Polybius) – the Hellenistic Greeks had the theoretical knowledge and practical experience to construct them all. It is sobering to realise that Archimedes' remarkable technological inventiveness ended with this incident in Syracuse: he was murdered when the Romans finally took the city.

Plutarch, *Marcellus* 14.9–17.3

The king persuaded Archimedes to build for him machinery for every type of siege, both defensive and offensive. Because he had lived most of his life in peacetime and amid civic affairs, Archimedes had not himself made use of such engines before, but in the present circumstances his apparatus proved especially beneficial to the Syracusans....

When the Romans began their attack from both the land and the sea, the Syracusans were struck dumb with fear, thinking that nothing could hold out against such a powerful assault. But when Archimedes let loose his engines, he launched on the Roman forces missiles of every description and stones of immense mass, which fell with a whiz and speed that one would not believe, mowing down those who stood in their path and throwing the ranks into confusion: nothing could withstand their weight.

As for the Roman navy, yard-arms unexpectedly emerged from the walls, sinking some of the ships by dropping great weights from above, and hoisting others bow-first straight up out of the water in iron claws or beaks shaped like those of cranes, and then plunging them back stern-first. Other ships were turned around and spun about by means of guy ropes and windlasses inside [the city], and then dashed against the steep promontories jutting out just under the wall, causing great loss of life among the crews on board, who were crushed. Again and again some ship would be lifted out of the water up into the air, then whirled back and forth – a horrible sight as it hung there, until its crew fell out and were hurled in all directions and the ship, now empty, would fall onto the walls or slip away once the grip was removed.

There was a siege engine Marcellus was bringing forward on a veritable bridge of ships.[16] While it was being moved towards the wall and still some distance away, a stone weighing 10 talents was shot at it, followed by another and then a third. With a great crash and plume of water, some scored a hit, pulverised the base of the engine, shook the framework violently, and tore it away from the bridge. This confounded Marcellus, who ordered his ships into a full-speed retreat and his foot-soldiers to follow suit.

After deliberating, the Romans decided to come close up to the walls, under cover of darkness if they were able, thinking that the tension cords Archimedes was using imparted such force that the missiles they discharged would fly right over their heads and be thoroughly ineffectual at close quarters, the distance not being right to score a hit. Archimedes, it seems, had long ago prepared for just such an eventuality by making the ranges of his instruments adaptable to any distance and by using compact missiles. There was a line of many small apertures through the wall, which allowed the short-range weapons to be placed so as to hit nearby targets while remaining invisible to the enemy.

So, when the Romans came up close, unnoticed as they thought, they again were met by a barrage of missiles that hit their mark: rocks plummeting almost straight down on them, and arrows shot out of the whole line of the wall. So, they fell back....

In the end, Marcellus saw that the Romans had become so terrified that, if a small cord or bit of wood was seen poking slightly over the wall, they would spin around and flee, shouting "There it is! Archimedes is aiming one of his engines at us!" So, he discontinued all frontal assaults and set up a protracted siege for the duration.

Cf. Polybius, *Histories* 8.3–7 for an even more detailed account of Marcellus' siege engines and Archimedes' counter-siege devices.

Only Diodorus mentions the device that follows, and even he gives an incomplete and confused description of how Archimedes focused the sun's rays into a beam that could ignite ships at a distance. It does not figure in the more detailed – and reliable – accounts of Polybius and Plutarch.

Diodorus of Sicily, *History* 26.18

When Marcellus drew his fleet off the distance of a bow-shot, the old man[17] put together a six-sided mirror, and at equal distances from it he placed small quadrangular mirrors of the same sort, which moved on plates and small hinges. He set this device in the full face of the sun's rays at midday (it works in winter as well as summer). Then, as the rays were reflected onto it, a terrible, fiery combustion was ignited on the ships and left them in ashes, even at the distance of a bow-shot.

Philo, *On Sieges* A76 (85.22–29)

Philo, *On Sieges* D34–40 (99.21–28), briefly describes several types of mechanical tortoises including ones that move on wheels or rollers. Here he provides a defensive measure against them and other mechanical weapons.

Having gathered together ceramic pots both at public expense and from the citizens, and having crammed their mouths tight with seaweed, which is not liable to rot, it is necessary to bury them in the earth, upright and empty, in front of the outermost ditch. After this lay on more earth on top, so that the men walking upon them suffer nothing terrible, but the tortoises being brought forward and machines sink down onto them.

13.36 Use of divers in siege warfare

Divers were deployed in a variety of tactics by island and coastal settlements under siege. Curtius Rufus describes the resistance of the people of Tyre and their attempts to hamper the construction of Alexander's causeway to their island. Philo provides advice for general tactics to frustrate maritime invasion.

Q. Curtius Rufus, *Histories of Alexander the Great* 4.3.10

An extraordinary force [employed against the causeway] were those men who, far from the sight of the enemy, plunged into the sea and penetrated right to the causeway with their stealthed glide. And then with hooks they pulled towards themselves the projecting branches of the trees, which when the trees had come after the branches, they carried many parts of the structure along with themselves into depths. At that point the divers boldly set to work on the logs and tree-trunks, now relieved of their burden, until finally the whole work which had been resting upon the tree-trunks, since its foundation was gone, followed into the depths.

Philo, *On Sieges* C59–61 (95.20–29)

If sections of the walls are on deep water, it is necessary to construct obstacles so that an approach is impossible at this point and that the ram of a large ship cannot punch into the wall, or so that [the enemy], setting assault-bridges, cannot capture a tower. It is also necessary at night, when the weather is stormy, to order divers to cut away the anchors of moored vessels, and to bore out holes in their hulls. In this way, especially, we will stop enemies from remaining at anchor.

13.37 The Romans besiege Jotapata

Josephus, a Jewish general before his defection to the Romans, here describes first-hand the frightening effects of Roman artillery on the defenders of Jotapata in Judaea (AD 67). His lengthy and gripping account of the assault (*Jewish War* 3.141–288, 316–339) is a virtual encyclopedia of siege techniques and defensive counter-measures, including one of the first references to boiling oil being poured on a city's attackers.

MILITARY TECHNOLOGY

Josephus, *Jewish War* 3.235, 240–246

Around late afternoon the Romans set up the ram once more and moved it up to the stretch of wall that had already been weakened by its battering.... Though Josephus' men were falling one upon another under fire from the catapults and *ballistae*, they still were not driven from the wall, but kept pelting with fire and iron and rocks those of the enemy who were working the ram under the wickerwork sheds. Yet they accomplished next to nothing....

The force of the bolt-launchers and the catapults ploughed through whole ranks of men, and the whizzing stones hurled by the siege machinery tore away the battlements and broke off corners of the towers. There was no mass of soldiers, however strong, that could avoid being mowed down from front line to rear by the force and size of those stone projectiles.

Some examples of what happened this night should give an appreciation of the power of the machinery. One of the men standing on the wall with Josephus was struck by a stone that decapitated him and flung his skull, like a shot from a sling, three *stadia*. And that same day a pregnant woman was just stepping out of her house when she was hit in the belly, and the foetus was hurled half a *stadion* away.

EPILOGUE

13.38 Roman legions in the late Empire

This account of the Roman army in the late fourth or fifth century AC represents its state at the end of the Roman Empire and dawn of the Middle Ages in the western Mediterranean. Yet it is not so very different from the legions described by Polybius and Plutarch half a millennium earlier.

Vegetius, *Military Science* 2.25

Not only the number of soldiers, but also the sort of iron implements they use has made it routine for the Roman legion to emerge victorious. First of all, it is provided with javelins that no armour or shield can withstand. Each century regularly has its own *ballista* mounted on a wagon, with mules assigned for pulling it and an attachment of 11 men for loading and aiming it [**fig. 13.4**; cf. **fig. 13.3**].
]. The larger these engines are, the further and more forcefully are their missiles shot. They are not only used to protect the camp, but are also placed in the field behind the front line of heavy-armed troops. Neither armoured cavalry nor infantry with shields can withstand their impact. There are usually 55 wheeled *ballistae* in one legion. Likewise, there are ten onagers – that is, one for every cohort – carried on two-wheeled ox-carts and fully loaded, so if it happens that the enemy comes to attack the palisade, the camp could be defended by arrows and stone projectiles.

The legion also carries along dugouts made from individual logs, together with long ropes and sometimes even iron chains to bind together these *monoxili*, as they call them [Greek for "single piece of wood"]. Once flooring is laid over them, the foot-soldiers and cavalry can cross in safety rivers that otherwise have no bridges and cannot be forded.

The legion has iron hooks, which they call "wolves", and curved iron blades fixed to very long poles; as well, two-pronged mattocks, spades, and hoes [for digging trenches], and troughs and buckets for carrying the earth; it also has picks, hatchets, axes, and saws for felling and sawing timber and pales. In addition, there are specialists with every kind of iron tool, who construct large and small siege sheds (*testudines*), and *musculi* ["mussels"], rams, cribs (which they call *vinea* [after vine trellises]), and even mobile towers – all designed for assaulting enemy cities.

So then, to keep from saying too much in listing off every last item: the legion ought to carry along everything deemed necessary in every mode of warfare, so that it can create a fortified city wherever it pitches camp.

Figure 13.4 *Carroballista* from Trajan's Column.

Source: Adapted from reliefs on Trajan's Column, *in situ*, Rome, Italy; Cichorius, 1896, pl. 51.

Notes

1 This is one of the more plausible ancient explanations of the famous Trojan Horse.
2 This city on the island of Euboea was thought to derive its name from the nearby copper mines, *chalk* – being the Greek root for copper and bronze.
3 Josephus here reverses the standard practice of wearing the sword (*gladius*) on the right (compare **13.5** and **13.9**).
4 The Greek says "mercenaries", but it was traditional to station auxiliary troops at the rear of an army on the march (e.g. Polybius, *Histories* 6.40.6–8).
5 This space in front of the tribunes' tents was used as the camp's principal road (*via principalis*).
6 Such a system of communication was not a fictional creation by the playwright: Homer (*Iliad* 18.210–213) refers to beacon-fires at the siege of Troy, and a version was being used in Aeschylus' own time (see Thucydides, *The Peloponnesian War* 2.94.1).
7 In his surviving fourth-century BC treatise *On Resisting a Siege* (7.4), Aeneas Tacticus briefly refers his reader to a passage on this subject in one of his lost works, which was presumably the source for this long summary by Polybius.
8 The Greek word *kataphraktes* refers to a coat of mail.
9 *Ballistae* is a term normally used for artillery that hurls stones, while catapults were loaded with bolts.
10 Clearly the boarding-ladder must have been hinged at this point, the nearer two fathoms remaining fixed horizontally from the deck, the outer four fathoms being held vertical until dropped onto the enemy vessel.
11 Or "10 miles" (the Latin is ambiguous). The actual circuit of the wall, which survives in large part, is 12 miles.
12 Pausanias (*Description of Greece* 8.8.7–8), in recording the capture of Mantinea in the fourth century BC when its besiegers diverted a river against its unfired brick city wall and "melted it like wax in the sun", observes that the real advantage of brick over stone in fortifications is its resiliency when pounded by artillery.
13 But see the more plausible explanation of the name in **13.13**.
14 Ammianus neglects to mention here that these ropes are twisted and provide the tension to the throwing arm. The side-beams were in fact held together by wooden cross-pieces.
15 It is clear that not all these men could be employed at one time: Diodorus either errs in the number, or includes several teams that would work in relays.
16 Mentioned in *Marcellus* 14.3: eight ships lashed together to form a platform for the huge engine.
17 Archimedes was at this time over 75, according to Diodorus.

Bibliography

General studies and surveys

Adcock, Frank E., *The Roman Art of War under the Republic*. Cambridge, MA: Harvard University Press, 1940.

Adcock, Frank E., *The Greek and Macedonian Art of War*. Berkeley: University of California Press, 1957.

Anderson, John K., *Military Theory and Practice in the Age of Xenophon*. Berkeley: University of California Press, 1970.

Bertholet, Florence, and Christopher Schmidt Heidenreich, eds., *Entre archéologie et épigraphie: nouvelles perspectives sur l'armée romaine*. Echo, 10. Frankfurt am Main; Bern: Peter Lang, 2013.

Brélaz, Cédric, and Sylvian Fachard, eds., *Pratiques militaires et art de la guerre dans le monde grec antique: études offertes à Pierre Ducrey à l'occasion de son 75e anniversaire. Revue des Études Militaires Anciennes, 6–2013*. Paris: Éditions A. et J. Picard, 2013.

Campbell, Brian, *Greek and Roman Military Writers: Selected Readings*. Oxford and New York: Routledge, 2004.

Campbell, Brian, and Lawrence A. Trittle, eds., *The Oxford Handbook of Warfare in the Classical World*. Oxford: Oxford University Press, 2013.

Campbell, Brian, *The Roman Army, 31 BC–AD 337: A Sourcebook*. London: Routledge, 1994.

Davies, Roy W., ed., *Service in the Roman Army*. New York; Columbia University Press, 1989.

Devoto, James G., *Philon and Heron: Artillery and Siegecraft in Antiquity*. Chicago: Ares, 1996.

Dixon, Karen, and Pat Southern, *The Roman Cavalry from the First to the Third Century AD*. London: Batsford, 1992.

Ducrey, Pierre, *Warfare in Ancient Greece*. New York: Schocken, 1986.

Erdkamp, Paul, ed., *A Companion to the Roman Army*. New York: Wiley-Blackwell, 2011.

Everson, Tim, *Warfare in Ancient Greece*. Stroud: Sutton, 2004.

Faulkner, Raymond O., "Egyptian Military Organization". *Journal of Egyptian Archaeology* 39 (1953) 32–47.

Garlan, Yvon, *War in the Ancient World. A Social History*. London: Chatto & Windus, 1975.

Griffith, Guy T., *The Mercenaries of the Hellenistic World*. London: Cambridge University Press, 1935.

Hanson, Victor Davis, *The Western Way of War. Infantry Battle in Classical Greece*. Oxford: Oxford University Press, 1990.

Humble, Richard, *Warfare in the Ancient World*. London: Cassell, 1980.

Junkelmann, Marcus, *Die Legionen des Augustus. Die Römische Soldat im Archäologischen Experiment*. Mainz: von Zabern, 1986.

Junkelmann, Marcus, *Die Reiter Roms*, Vol. 1: *Reise, Jagd, Triumph und Circusrennen*. Mainz: von Zabern, 1991.

Junkelmann, Marcus, *Die Reiter Roms*, Vol. 2: *Zubehör, Reitweise, Bewaffnung*. Mainz: von Zabern, 1992.

Keppie, Lawrence J.F., *The Making of the Roman Army: From Republic to Empire*. London: Batsford, 1984.

Kromayer, Johannes, ed., *Antike Schlachtfelder in Griechenland. Bausteine zu einer antiken Kriegsgeschichte*. 4 vols. Berlin: Weidmann, 1903–1931.

Kromayer, Johannes, and Georg Veith, *Heerwesen und Kriegführung der Griechen und Römer*. Handbuch der Altertumswissenschaft, IV.3.2. Munich: C.H. Beck, 1928.

Le Bohec, Yann, *The Imperial Roman Army*. London: Batsford, 1994.

Lee, Geoff, Helene Whittaker, and Graham Wrightson, eds., *Ancient Warfare: Introducing Current Research, Vol. 1*. Newcastle upon Tyne: Cambridge Scholars Publishing, 2015.

Levithan, Josh, *Roman Siege Warfare*. Ann Arbor: University of Michigan Press, 2013.

Rihll, Tracey E., "Technology in Aineias Tacticus: Simple and Complex". Pp. 265–289 in Maria Pretzler and Nick Barley, eds., *Brill's Companion to Aineias Tacticus*. Leiden and Boston: Brill, 2017.

Sabin, Philip, Hans van Wees, and Michael Whitby, eds., *The Cambridge History of Greek and Roman Warfare, I: Greece, the Hellenistic World and the Rise of Rome*. Cambridge: Cambridge University Press, 2007.
Sarantis, Alexander, and Neil Christie, eds., *War and Warfare in Late Antiquity (2 Vols.)*. Late antique archaeology, 8. Leiden; Boston: Brill, 2013.
Saulnier, Christiane, *L'armée et la guerre dans le monde étrusco-romain*. Paris: De Boccard, 1980.
Southern, Pat, *The Roman Army*. Oxford: Oxford University Press, 2007.
Spence, Iain G., *Cavalry of Classical Athens*. Oxford: Oxford University Press, 1993.
Sullivan, Denis F., *Siegecraft: two tenth-century instructional manuals by "Heron of Byzantium"*. Dumbarton Oaks Studies 36. Washington: Harvard University Press, 2000.
Webster, Graham, *The Roman Imperial Army of the First and Second Centuries AD*. 3rd edn. Totowa, NJ: Barnes & Noble, 1985.
Whitehead, David, *Philo Mechanicus: On Sieges. Translated with Introduction and Commentary*. Stuttgart, 2016.
Yadin, Yigael, *The Art of Warfare in Biblical Lands in the Light of Archaeological Study*. 2 Vols. New York: McGraw-Hill, 1963.

Equipment and tactics

Ahlberg, Gudrun, *Fighting on Land and Sea in Greek Geometric Art*. Skrifter Utgivna av Svenska Institutet i Athen, 16. Stockholm: Svenska Institutet i Athen, 1971.
Best, Jan G.P., *Thracian Peltasts and their Influence on Greek Warfare*. Groningen: Wolters-Noordhoff, 1969.
Bishop, Michael C., *Roman Military Equipment from the Punic Wars to the Fall of Rome*. London: Batsford, 1993.
Bishop, Michael C., and Jonathan C.N. Coulston, *Roman Military Equipment from the Punic Wars to the Fall of Rome*. London: Batsford, 1993.
Buchholz, Hans-Günter, and Joseph Wiesner, *Kriegswesen, 1: Schutzwaffen und Wehrbauten*. Archaeologia Homerica I, E. Göttingen: Vandenhoeck & Ruprecht, 1977.
Buchholz, Hans-Günter, Joseph Wiesner, Stephan Foltiny, and Olaf Höckmann, *Kriegswesen, 2: Angriffswaffen: Schwert, Dolch, Lanze, Speer, Keule*. Archaeologia Homerica I, E. Göttingen: Vandenhoeck & Ruprecht, 1980.
Campbell, Brian, "Teach Yourself How to Be a General". *Journal of Roman Studies* 77 (1987) 13–29.
Cartledge, Paul, "Hoplites and Heroes: Sparta's Contribution to the Technique of Ancient Warfare". *Journal of Hellenic Studies* 97 (1977) 11–27.
D'Amato, Raffaele, "Arms and Weapons". Pp. 801–816 in Georgia L. Irby, ed., *A Companion to Science, Technology, and Medicine in Ancient Greece and Rome, Vol. II*. Chichester: Wiley-Blackwell, 2016.
Engels, Donald W., *Alexander the Great and the Logistics of the Macedonian Army*. Berkeley: University of California Press, 1978.
Garbsch, Jochen, et al., *Römische Paraderüstungen*. Munich: C.H. Beck, 1978.
Glover, Richard F., "The Tactical Handling of the Elephant". *Greece and Rome* 17 (1948) 1–11.
Glover, Richard F., "Some Curiosities of Ancient Warfare". *Greece and Rome* 19 (1950) 1–9.

Goldsworthy, Adrian K., *The Roman Army at War 100 BC–AD 200*. Oxford: Clarendon Press, 1996.
Greenhalgh, Peter A.L., *Early Greek Warfare: Horsemen and Chariots in the Homeric and Archaic Ages*. Cambridge University Press, 1973.
Hanson, Victor Davis, ed., *Hoplites: The Classical Battle Experience*. London: Routledge, 1991.
James, Simon, "Archaeological Evidence for Roman Incendiary Projectiles". *Saalburg Jahrbuch* 39 (1983) 142–143. 4 illus.
Kolias, Toxiarchis, *Byzantinische Waffen*. Vienna: Österreichische Akademie der Wissenschaften, 1988.
Littauer, Mary Aiken, "The Military Use of the Chariot in the Aegean in the Late Bronze Age". *American Journal of Archaeology* 76 (1972) 145–157.
Markle, Minor M., III, "Use of the Sarissa by Philip and Alexander of Macedon". *American Journal of Archaeology* 82 (1978) 483–497.
Markle, Minor M., III, "Macedonian Arms and Tactics under Alexander The Great". Pp. 87–111 in B. Barr-Sharrar, Eugene N. Borza, eds., *Macedonia and Greece in Late Classical and Early Hellenistic Times*. Washington: National Gallery, 1982.
Mayor, Adrienne, *Greek Fire, Poison Arrows & Scorpion Bombs: Biological and Chemical Warfare in the Ancient World*. Woodstock: Overlook Duckworth, 2003.
McLeod, William, "The Range of the Ancient Bow". *Phoenix* 19 (1965) 1–14; "Addenda". 26 (1972) 78–82.
Richmond, Ian, *Trajan's Army on Trajan's Column*. Edited by Mark Hassall. London: British School at Rome, 1982.
Robinson, H. Russell, *The Armour of Imperial Rome*. New York: Scribner, 1975.
Snodgrass, Anthony, *Arms and Armour of the Greeks*. Ithaca: Cornell University Press, 1967.
Wheeler, Everett, *Stratagem and the Vocabulary of Military Trickery*. Mnemosyne, Supp. 108. Leiden: Brill, 1988.

Fortifications

Adam, Jean-Pierre, *L'architecture militaire grecque*. Paris: Picard, 1981.
Allison, Penelope M., *People and Spaces in Roman Military Bases*. Cambridge: Cambridge University Press, 2013.
Baatz, Dietwulf, *Bauten und Katapulte des römischen Heers*. Stuttgart: Steiner, 1994.
Breeze, David J., Brian Dobson, *Hadrian's Wall*. London: Allen Lane, 1976.
D'Amato, Raffaele, "Siegeworks and Fortifications". Pp. 784–800 in Georgia L. Irby, ed., *A Companion to Science, Technology, and Medicine in Ancient Greece and Rome, Vol. II*. Chichester: Wiley-Blackwell, 2016.
Davies, Gwyn, "Roman Warfare and Fortifications". Pp. 691–714 in John P. Oleson, ed., *The Oxford Handbook of Engineering and Technology in the Classical World*. Oxford: Oxford University Press, 2008.
De Souza, Philip, "Greek Warfare and Fortifications". Pp. 673–690 in John P. Oleson, ed., *The Oxford Handbook of Engineering and Technology in the Classical World*. Oxford: Oxford University Press, 2008.
Foss, Clive, David Winfield, *Byzantine Fortifications: An Introduction*. Pretoria: University of South Africa Press, 1986.

Garlan, Yvon, *Recherches de poliorcétique grecque*. Bibliothèque des Écoles Françaises d'Athènes et de Rome, 223. Paris: de Boccard, 1974.
Johnson, Anne, *Roman Forts of the 1st and 2nd Centuries AD. in Britain and the German Provinces*. London: Adam & Charles Black, 1983.
Johnson, Stephen, *Late Roman Fortifications*. London: Batsford, 1983.
Lawrence, Arnold Walter, *Greek Aims in Fortification*. Oxford: Clarendon Press, 1979.
Miller, Martin C.J., and James G. De Voto, eds., *Polybius and Pseudo-Hyginus, Fortification of the Roman Camp*. Amsterdam: Gieben, 1994.
Scoufopoulos, Niki C., *Mycenaean Citadels*. Studies in Mediterranean Archaeology, 22. Göteborg: Studies in Mediterranean Archaeology, 1971.
Scullard, Howard H. *The Elephant in the Greek and Roman World*, London: Thames and Hudson, 1974.
Todd, Malcolm, *The Walls of Rome*. London: Paul Elek, 1978.
Winter, Frederick E., *Greek Fortifications*. Toronto: University of Toronto Press, 1971.

Siege engines

Baatz, Dietwulf, "Recent Finds of Ancient Artillery". *Britannia* 9 (1978) 1–17.
Baatz, Dietwulf, "Hellenistische Katapulte aus Ephyra (Epirus)". *Athenische Mitteilungen* 97 (1982) 211–233.
Hacker, Barton C., "Greek Catapults and Catapult Technology: Science, Technology, and War in the Ancient World". *Technology and Culture* 9 (1968) 34–50.
James, Simon, "Archaeological Evidence for Roman Incendiary Projectiles". *Saalburg Jahrbuch* 39 (1983) 142–143.
Marsden, Eric W., *Greek and Roman Artillery: Technical Treatises*. Oxford: Clarendon Press, 1971.
Marsden, Eric W., *Greek and Roman Artillery: Historical Development*. Oxford: Clarendon Press, 1969.
Rihll, Tracey E., *The Catapult: A History.* Yardley, PA: Westholme Publishing, 2007.
Schiefsky, Mark J., "Technê and method in ancient artillery construction: the *Belopoieca* of Philo of Byzantium". Pp. 613–651 in Brooke Holmes and Klaus-Dietrich Fischer, eds., *The Frontiers of Ancient Science: Essays in Honor of Heinrich von Staden.* Berlin, Munich, Boston: Walter de Gruyter, 2015.

14

ATTITUDES TOWARDS LABOUR, INNOVATION, AND TECHNOLOGY

Most of the material included in this chapter is directly concerned with attitudes towards labour, technology, and innovation in the Greek and Roman cultures, but the connection sometimes is implicit rather than explicit. The subject is so large that naturally only a selection of the most important passages could be included. Furthermore, many passages in earlier chapters are relevant to this topic. Nevertheless, the overall evolution of attitudes from the eighth century BC through the late Roman period should be clear. The literary sources show a general disdain for manual labour, the so-called banausic prejudice. Ample evidence in the form of funerary monuments or other types of material and pictorial remains – particularly from the Roman period – demonstrates, however, that artisans and craftspeople were proud of their work. They prominently displayed their accomplishments, and some managed to accumulate by their manual labour and business acumen wealth in magnitudes that rivalled that of the senatorial classes. The Tomb of the Haterii [**fig. 8.8**] or the Tomb of Eurysaces the Baker [**fig. 4.2**] are striking examples. Cuomo (2007) gives an excellent analysis of the interaction between humans and the technology that surrounded them in Greek and Roman antiquity.

The sequence of topics in this chapter is, roughly, early attitudes towards labour and innovation, the appearance of banausic prejudice and its expression in the Greek and Roman cultures, positive and negative attitudes towards experiment, technology, and innovation, and visions of future progress in human knowledge (Greene 2008, Cuomo 2007).

ATTITUDES TOWARDS LABOUR AND THE PROFESSIONS

14.1 The high status of craftsmen

The Greek term for craftsman – *demiourgos* or *demioergos* – literally means "one who serves the community". In the eighth century BC the title was applied to practitioners of respected professions, but by the fifth century BC it had been transferred to the humble class of craftsmen.

Homer, *Odyssey* 17.382–386

For who of his own accord ever approaches and summons a stranger from elsewhere unless it be one of those who practise a craft of public benefit (*demioergoi*): a seer, or a healer of illnesses, or a master at building with timbers, or one inspired by song who gives pleasure with his singing? For these men are summoned throughout the boundless earth.

14.2 The dignity of work

Although the immediate context of Hesiod's discussion here is agricultural labour, which continued to be held in high regard throughout antiquity, the enthusiastic tone of his comments nevertheless is striking. See also Hesiod's comments on the role of competition in technological advance (**1.29**).

Hesiod, *Works and Days* 303–311, 410–413

Both gods and humans are angry with one who lives an idle life, for by nature he is like the stingless drones who consume the labour of the honey bees, eating without working. Let it be your care to arrange your tasks in the right order, that your storehouses might be full of the season's food-stuffs. Through work men become wealthy in flocks and property, and in labouring hard they are much better loved by the immortal gods. Work is not shameful; it is idleness that brings shame....

Do not procrastinate until tomorrow or the day after tomorrow, for a shiftless worker does not fill his granary, nor does a procrastinator. Careful attention helps the work along, but the dilatory man is always wrestling with destruction.

14.3 Origins of the Greek prejudice against craftsmen

By the fifth century BC the prejudice against manual labour, even in the skilled crafts, was well embedded in Greek society. Herodotus, however, seems to have enough perspective on the issue to realise that this social attitude is not inevitable and that there were variations in its degree even in Greek society. The city of Corinth, for example, depended heavily on manufacture and commerce, and perhaps as a result this prejudice is not expressed as clearly there. The term Plutarch uses for handicraft – *techne banausos* – incorporates the root of *banausia*, a pejorative term for manual labour current by the fifth century BC. The term may have originated from *baunos*, "furnace", an installation essential to most craft activities (for an older view see Mondolfo 1954).

Herodotus, *Histories* 2.167

I observe that the Thracians, Scythians, Persians, Lydians, and nearly all foreigners hold those who learn trades, and their descendants in less regard than the rest of society. They consider those who have nothing to do with the handicrafts more noble, particularly those who practise the art of war. This much is certain, that the Greeks, and especially the Spartans, learned this custom somewhere. The people of Corinth feel the least prejudice for those who practise the handicrafts.

Plutarch, *Agesilaus* 26.4–5

Agesilaus ... ordered the allied soldiers to sit down in mixed groups and the Spartan soldiers apart by themselves. Then he had his herald call on the potters to stand up first, and after they rose, next the smiths, then in turn the carpenters, and the builders and the rest of the crafts in order. So all the allies stood up except for a few, but not one of the Spartans, for they were forbidden to learn or practise a handicraft. Then Agesilaus said with a laugh, "You see, Gentlemen, how many more soldiers we are sending out than you are".

14.4 Money stimulates the crafts and trades

The comic playwright Aristophanes based his play *Wealth* on the premise that no one would be willing to work if wealth were equitably divided up among the population. It is need that drives the worker rather than any love of the craft (Greene 2000, Wilson 2002, Finley 1965).

Aristophanes, *Wealth* 160–164, 166–167, 510–516

Every craft and every skill among humans was invented on your account [money]. For this one person sits and cobbles shoes, someone else works in bronze, and another in wood. This one casts gold – gold derived from you.... One fulls cloth, another washes sheep skins. One is a tanner, another sells onions.

For if Wealth regained his sight and parcelled himself out equally, no human being would practise a trade or a science. And if both of these disappeared from the earth, who would have a care to work metal for you, or build a ship, or sew, or make a wheel, work with leather, make bricks, do the wash, tan hides, or break the field with a plough to reap the fruits of Demeter – if it were possible to live in idleness without a thought for any of these things?

Hellenistic kings also used their tremendous resources to stimulate technological creativity and to create immense machines for war (**11.77–79, 13.14–35**).

14.5 The worker or slave as a "living tool"

The detached attitude of Aristotle towards free and slave labourers is in part a result of the low status of labour among upper-class Greeks, an offshoot of the social attitude that devalued the practice of trades and handicrafts. For an agricultural context, see **3.10**.

Aristotle, *Politics* 1.2.4–5 (1253b–1254a)

As for tools, some are inanimate, others animate; for example, to the helmsman a steering oar is an inanimate tool, the lookout man an animate tool – for an assistant in the crafts is in the category of a tool. So also an article of property is a tool for living, and property in general is a collection of tools, and the slave is an animate article of property. Every assistant is a kind of tool that takes the place of several tools. For if every tool were able to complete its own task when ordered – or even anticipate the need – just as the statues of Daidalos supposedly did, or the [wheeled] tripods of Hephaistos which Homer says "entered of their own accord the assembly of the gods"[1] – if shuttles could pass through the web by themselves or *plectra* play the harp, master craftsmen would have no need of assistants, and masters no need of slaves.

14.6 The drawbacks of banausic occupations

Xenophon puts in the mouth of Socrates a concise statement of the ostensible reasons the Greek elite were prejudiced against those who practised trades and occupations. Since Socrates himself is said to have been trained as a stone-cutter, it is uncertain whether the sentiments are actually his, or those of Xenophon. This prejudice is based in part on philosophical rather than social or economic objections, since the major drawback of a banausic occupation is the obstacles it puts in the way of service to the mind or soul and to the state, which is a metaphor for the mind. The author of a similar work in the *corpus* of Aristotle provides a contrast between banausic occupations and agriculture. The pragmatic Roman author Columella emphasises the value of a well-trained farm manager.

Xenophon, *Estate Management* 4.2–3

To be sure, the so-called banausic arts are spoken against and quite rightly held in contempt in our states, for they ruin the bodies of those practising them and those who supervise, forcing them to sit still and pass their time indoors – some even to spend the day at a fire. As their bodies are softened, so too their minds become much more sickly. In addition, the so-called banausic arts leave no leisure time for paying attention to one's friends or state, so that the persons who practise them have the reputation of treating their friends badly and being poor defenders of their homeland. In some states, particularly those with a warlike reputation, it is forbidden for any citizen to practise the banausic arts.

[Aristotle], *Household Management* 1.2.2–3

Of the acquisitive occupations the foremost are those that are natural, and first of these is cultivation of the land; second are the extractive occupations, such as mining and the like. Agriculture has a special justice about it, for it does not derive gain from other men – whether willing, as with the mercantile or salaried occupations, or unwilling, as with the occupation of war. Rather it is a natural occupation, since all creatures receive sustenance from their mother, and in the same fashion all humans from their mother the earth.

In addition, agriculture contributes greatly to manly character. Unlike the banausic arts, which make the body weaker, agriculture accustoms the body to outdoor living and hard physical labour, and makes it all the more capable of undergoing the dangers of war. For only the possessions of the farmers are outside a city's defences.

Columella, *On Agriculture* 11.1.9–11

All these tasks [moving heavy loads; digging; ploughing; mowing; pruning; grafting; feeding of livestock] the manager, as I said before, cannot evaluate unless he is also an expert in them so that he can correct whatever has been done wrong in any of them. For it is not enough to reprimand the person who made the mistake if one does not teach how to do it right. I therefore enthusiastically say again that the future manager must be trained just like a future potter or craftsman. I cannot really say whether these trades can be learned faster as they are not as broad in scope. Agriculture, however, is a big and diverse subject, and if we wished to list all its parts, it would be hard to grasp them all. This is why I cannot wonder enough, as I justifiably complained in the beginning of my work, that masters can be found in all the other arts, which are less necessary for life, but in agriculture neither students nor teachers. Unless the magnitude of the subject has created a sense of awe of either learning or professing this almost limitless science it should not be neglected out of shameful despair. For the art of oratory is not abandoned either just because a perfect orator is nowhere to be found, nor philosophy because no one of accomplished wisdom is to be found. On the contrary, many encourage themselves to learn at least some parts of them even though they cannot learn them in their entirety.

14.7 The role of craftsmen in the "ideal" state

In discussing their "ideal" states, both Plato and Aristotle naturally give the craftsmen a much lower status than the rulers, who are essentially philosophers. The basis for this discrimination, however, is not so much the degrading character of banausic occupations, as the fact that they draw attention away from meditation and dialectical enquiry. Plato goes further in prescribing that even craftsmen should follow just their own single trade, which presumably they are best at, and not meddle in others.

Plato, *Laws* 8.846d–847a

For the other craftsmen we have to arrange matters as follows. First of all, no resident citizen – nor any servant of a resident citizen – shall be among those who practise the technical crafts. For a citizen has taken on a sufficient craft, and one that requires much practice along with a great deal of study: acquiring and preserving the common public system of the state – something, which he ought not to attend to as an avocation. Scarcely any human is capable by nature of attending to two pursuits or crafts properly, nor, again, to practise one properly himself and supervise another practising a different one. This, then, must become a primary rule in the state, that no smith shall act as a carpenter, nor shall a carpenter supervise anyone working as a smith rather than at carpentry, ... but each individual in the state, having found his craft, shall earn his living from that alone.

Aristotle, *Politics* 3.2.8–9 (1277a–b)

We assert that there are several categories of slave, as there are several kinds of employment for them. One category belongs to the handicraftsmen, that is, as their name implies, those who live by the work of their hands. The banausic craftsman is one of them. For this reason, some states long ago, before the appearance of extreme democracy, did not allow them to hold office.

Aristotle, *Politics* 8.2.1 (1337b)

It is clear, therefore, that the young must be taught those of the useful arts that are absolutely necessary, but not all of them. It is obvious that the liberal arts should be kept separate from those that are not liberal and that they must participate in such of the useful arts as will not make the participant vulgar (*banausos*). A task or an art or a science must be considered banausic if it makes the body or soul or mind of free men useless for the practice and application of virtue. For this reason, we term "banausic" those crafts that make the condition of the body worse, and the workshops where wages are earned, for they leave the mind preoccupied and debased.

14.8 Writing about the banausic crafts is degrading

It is interesting that Aristotle and other ancient authors, such as Archimedes, who obviously saw the practical value of technology and technological ideas, could not free themselves – in their writing at least – from the social prejudice directed at these subjects. Aristotle clearly knows of technical handbooks but does not care to reproduce their information, while Archimedes allegedly did not want to write about the splendid practical applications of his theoretical research (see also **14.15**). Although Aristotle seems to have a low opinion of the access craftsmen could have to philosophical knowledge, in a passage of the *Metaphysics* (here, **1.27**) he nevertheless recognises degrees of knowledge or skill among craftsmen themselves (cf. Oleson 2004).

Aristotle, *Politics* 1.4.3–4 (1258b–1259a)

The most scientific of these productive industries are those where chance plays the least role; the most banausic those in which the body suffers the most damage; the most servile those in which the most use is made of the body; the most degrading those in which there is the least call for virtue. A general account of each of these industries has been given above, but – while it would be useful for their practical application to give an account of each one in detail – it would be vulgar to spend much time on them. Accounts have been written about these industries by certain authors, ... so that anyone who cares to can learn about them from their books.

Pappus, *Mathematical Collection* 8.3

Some say that Archimedes of Syracuse worked out the causes and systems of all these [mechanical and scientific devices].... Carpus of Antioch has written somewhere that Archimedes composed only a single book concerned with the mechanical arts, *On the Construction of an Orrery*, and that he did not think it worthwhile to write about his other inventions.

14.9 The cultural duty of Rome

As the climax of a famous passage in which Aeneas, the mythical ancestor of the founders of Rome, reviews the images of his still-to-be-born descendants, these verses have long been felt to sum up the purpose of Roman culture – at least as far as the elite were concerned. Given this attitude, it is no wonder that most Roman aristocrats with any pretensions to advancement followed some military career and paid at least lip service to the notion that war was the most noble occupation.

Vergil, *Aeneid* 6.847–853

Others, no doubt, will shape with more gentle touch bronze statues that breathe with life or draw living portraits from the marble block; they will plead legal cases better, or with a rod will trace the circling path of the heavens and predict the rising of the constellations. You, Roman, remember to govern the nations with your rule – these will be your arts: to crown peace with our way of life, to spare the conquered, and to defeat the proud in war.

Vegetius, *Military Science* 3.10

Who, however, would doubt that the art of war is superior to everything else, for through it liberty and prestige are retained, provinces are enlarged, and Empire is preserved.

14.10 Roman aristocratic attitudes towards occupations

Although Cicero did not descend from a noble Roman family, he nevertheless embodies the typical Roman aristocrat's attitudes towards class, wealth, and occupations. The classification he proposes resembles some of the opinions of Aristotle quoted above. In the end, leisure (*otium*) was the ultimate goal, since it gave time for the pursuit of philosophy. The word for work, the opposite of leisure, was *negotium*, literally "not-leisure", reflecting the social priorities. In Cicero's mind it was leisure that made the traditional gods particularly blessed, while the Stoic god had to engage in a tedious *negotium*.

Cicero, *On Duty* 1.42

Now, as to which crafts and other means of earning a living are suitable for a gentleman to practise and which are degrading, we have been taught more or less the following. First of all, those occupations that stir up people's ill will are condemned – such as tax-gatherer or money-lender. Also vulgar and unbecoming to a gentleman are all the jobs hired workers take on, whose labour is purchased rather than their skill, for their very salary is the remuneration of their servitude. Those must also be considered vulgar who buy from wholesale merchants in order to sell immediately, for they do not get any profit without *significant* deception – and there is nothing more base than misrepresentation. All craftsmen spend their time in vulgar occupations, for no workshop can have anything liberal about it. The lowest esteem is accorded those occupations that service the sensual pleasures – "fishmongers, butchers, cooks, poulterers", as Terence writes....

But the professions that require a greater degree of intelligence or from which a significant social benefit is derived – such as medicine, or architecture, or the teaching of liberal subjects – these are honourable for those to whose social rank they are appropriate. Trade, on the one hand, if it is on a small scale, must be considered vulgar, but if it is wholesale and on a large scale, importing great quantities from all directions and redistributing it to a large clientele without misrepresentation, it should not be spoken ill of.... However, of all the gainful occupations, none is better, none more profitable, none more pleasant, none more worthy of a free man than agriculture.

Cicero, *Brutus* 257

Therefore, one should consider not how useful someone is, but what his real worth is. For there are very few who excel at painting or sculpting, but there can never be a lack of labourers or porters.

Cicero, *On the Nature of the Gods* 1.19.50–53

You Stoics, Balbus, also like to ask us what sort of life the gods lead, and how they pass their time. Their life, in fact, is the happiest one could conceive of, a life bountiful in all good things. God does nothing; he is not involved in any occupations, he does not undertake any tasks. He simply finds joy in his wisdom and virtue and knows with absolute certainty that he will forever enjoy pleasures both consummate and eternal.

This is the god we should call properly happy; that god of yours seems truly overworked. For if the world itself is god – what can be less restful than to turn around the axis of the heavens with amazing speed, without any rest? Nothing can be happy unless it is at rest. But if some god or other is present within the world who governs it and steers it, who regulates the courses of the stars, the change of the seasons, and the deviations and patterns of everything that exists, and who watches over land and sea to protect human life and human interests – what a tedious and laborious business (*negotia*) he is involved in. For we judge a life truly happy that has tranquillity of mind and a complete freedom from all duties.

14.11 The "gentleman" architect

Although Cicero ranks the profession of architect honourable in the previous passage, he clearly does not regard it as an aristocratic occupation, but one "for those to whose social rank it is appropriate".

In his handbook of architectural practices, Vitruvius, a man of middle rank, is careful to indicate that he does not seek out commissions and thus – not surprisingly! – has not become wealthy or famous from the profession. He even voices approval of the amateur gentleman architect who designs his own projects.

Vitruvius, *On Architecture* 6, Preface 5–6

But I, Caesar, have not given myself over to study for the purpose of making money from my profession; rather, I am of the opinion that a slender fortune accompanied by a good reputation is better than wealth accompanied by infamy. As a result, I have found little celebrity. I hope, however, that by publishing these volumes I will be known even to posterity. Nor is it surprising that I am so little known to the general public. Other architects beg and wrangle for commissions. My teachers, however, passed on to me the rule that one ought to be asked to take on a job rather than ask for it oneself.... Therefore our forebears used to entrust commissions first of all to architects of good family, enquiring next if they had been properly brought up....

But when I note that a profession of such great magnificence is practised by individuals without training or experience, who are ignorant not only of architecture, but even of construction, I cannot help praising those propertied gentlemen who find strength in their own self-instruction and build for themselves. They conclude that – if the job must be given to individuals without experience – they are quite entitled to spend the sum of money according to their own wishes rather than those of someone else.

14.12 The four classes of occupations and arts

In this passage, Seneca makes explicit the derivation of some of these Roman banausic attitudes from Greek philosophy, in this case the Hellenistic philosopher Posidonius. Occupations rise in status as they become more visible to the elite and less directly associated with physical labour.

Seneca, *Letters* 88.21–23

Posidonius asserts that the arts fall into four classes: first the vulgar and degrading, then those concerned with entertainment, those concerned with education of the youth, and finally the liberal arts. The vulgar, which depend on manual labour, are staffed by workmen and are concerned with satisfying everyday needs. They have not even a pretence of beauty or honour. The entertainment arts are those which are focused on pleasing the eye and ear. You might assign to this category the stage technicians who construct scaffolding that rises by itself or floors that rise silently upwards, and other surprising novelties, such as objects that seem whole but fall apart, that seem fragmentary but join together of their own accord, or that stand erect then gradually collapse.... The arts concerned with the education of the youth, which have some similarity to the liberal arts, are those that the Greeks called the cycle of studies. We, however, call them the liberal arts. Those alone are the liberal arts, however – and, in fact, to speak more accurately, the "free" arts – whose concern is excellence of mind and character.

14.13 The choice of a "proper" profession

In this amusing short work, Lucian recounts his family's decision to send him out to an uncle's sculpture workshop as an apprentice. After a disappointing first day and a consequent thrashing, the boy has a dream in which Sculpture and Education fight over him. Sculpture promises Lucian a strong body and fame as an artist, while Education wins his loyalties by promising him fame and financial success as a rhetorician, diplomat, and virtuous and wise man. The prejudice against physical work is explicit.

Lucian, *The Dream* 6–9,13

Two women, taking me by the hands, were dragging me each to herself with all their strength.... They were shouting at each other, too, one of them calling out "He belongs to me, and you want to take him!" While the other said "It's useless to claim what belongs to someone else!" One of them was made up like a worker, masculine, with greasy hair, calloused hands, her clothing tucked up under her belt, covered with marble dust – just like my uncle when he cut stones. The other one had a pleasing face, was comely in appearance, and nicely dressed. Finally, they agreed to award me to whichever one of them I wanted to be with. The rough, masculine one spoke first.

"Dear boy, I am the profession of Sculpture, which you began to learn yesterday. If you are willing ... to follow and stay with me, first of all you will be royally kept, and will have strong shoulders and will be free of any sort of envy.... Don't feel disgust at my humble appearance and soiled clothing. From such a start the famous Pheidias interpreted the figure of Zeus, and Polykleitos made the Hera, and Myron found praise and Praxiteles admiration.... If you become one of them, how could you miss being famous yourself ...?"

Sculpture said all this and more, stumbling over her words and making a lot of grammatical mistakes.... But when she finished, the other one began as follows.

"I, my child, am Education, already familiar and known to you, even though you still have more to learn. This woman has already told you what benefits she will provide if you become a stone-cutter. Indeed, you will be nothing more than a labourer, working away with your body and placing all your hope of a livelihood in it, without any reputation yourself, receiving scanty and ignominious pay.... And even if you should turn into a Pheidias or Polykleitos and create many wonderful sculptures, everyone would praise your art, but no one in his right mind who caught a glimpse of you would pray to become like you. For whatever you might become, you would be considered a worker (*banausos*), a handicraftsman, one who lives by manual labour....

"But if you turn away from such great and noble men [philosophers and literary figures], from glorious deeds and revered sayings, from a fine appearance, from honour and glory and praise and privilege and power and office, ... then you will put on a filthy tunic, take on the appearance of a slave, and you will hold crowbars, chisels, hammers, and gravers in your hands, bending over your work. You will be a groveller and have a groveller's ambitions, humble through and through. You will never hold your head up high and conceive manly or liberal thoughts, and although you will plan out your creations to be properly proportioned and well shaped, you will not take any care that you yourself will be well proportioned and well ordered. On the contrary, you will make yourself more worthless than a block of stone".

ATTITUDES TOWARDS EXPERIMENT AND INNOVATION

14.14 Some defects in Greek higher education

Even if someone like Lucian chose higher education over a trade, his problems were not necessarily over. Reliable teachers had to be found, and wealth and leisure were necessary to allow a course of study. In the end, much of the subject matter taught was selected because it was amenable to proof by dialectic, rather than by experiment. As a result of this approach, the Greek and Roman elites cut themselves off from adequate investigation of the experimental and life sciences, and the applied technologies that might have grown out of them. Diodorus suggests as well that the need for teachers to support themselves by attracting students hampered real research. He contrasts this approach with the more institutionalised methods of teaching in Mesopotamia (Greene 2008).

Plato, *Republic* 7.528b–c

"But this subject [solid geometry], Socrates, seems not yet to have been investigated".

"The cause of that", I said, "is two-fold. First, because no state holds it in high esteem, the subject is investigated without energy, since it is difficult. Second, the researchers need a director, without whom they would not discover anything. Such a person is difficult to find, and if he could be found, as things stand now, researchers in this field are too arrogant to listen to him. But if the state were to become a joint supervisor and treat these subjects with some respect, the researchers would listen and the truth would be revealed through continuous and strenuous investigation".

Diodorus of Sicily, *History* 2.29.5–6

Among the Greeks, however, the student who has approached a number of subjects without preparation then takes up philosophy late. But after working willingly on it up to a certain point, distracted by the need to earn a living, he leaves. Only a few, focusing intently on philosophy, remain in the course of study for the sake of earning a living from it, always trying out something new concerning the major issues rather than following those who went before them. As a result, the foreigners, who always stick to their own subject matter, have a firm control of detail, while the Greeks, who aim at the profit to be made from contract work, constantly found new schools of thought ... leaving their students wandering in confusion.

14.15 The insufficiency of the experimental method

A related impediment to technological advance in the Graeco-Roman world was the philosophical "insufficiency" of the experimental method. Since experiment departed from the dialectical approach to research by involving the senses and physical apparatus, it was felt to be defective in its apprehension of the "truth". As a result, even very skilled inventors such as Archimedes seem to have been reluctant to write proper treatises about their inventions, or even about the applied sciences (cf. Reece 1969 with Fögen 2016).

Plato, *Republic* 7.530e–531a

"And in all this we will keep a lookout for what concerns us", I said. "What is that?" "That none of our pupils undertakes to learn something incomplete and defective, something that does not always lead to the point that everything must lead to, just as we have already stated about astronomy. Or do you not know that they employ a similar procedure with regard to harmony, transferring consideration to hearing and measuring audible concords and sounds against one another, wasting their effort just as the astronomers do?"

Plutarch, *Moralia* 8.2.718e

Now in all the so-called mathematical subjects, as in smooth and undistorted mirrors, traces and images appear of the truth about what can be known. But Geometry in particular ... leads comprehension upward and diverts it, as if purified and freed completely from sense perception. For this reason Plato himself criticised Eudoxus, Archytas, and Menaechmus for undertaking to make the problem of doubling the cube the object of instruments and mechanical devices, as if they were trying to find two mean proportionals not by reason, but by any way practicable. For in this way, he thought, the innate benefit of Geometry was lost and corrupted, slipping back again into the realm of sense perception....

Plutarch, *Life of Marcellus* 17.3–4

Archimedes' intellect, however, was so great and his mind so deep, and he possessed such a wealth of scientific theory that – although from inventions such as these [especially siege engines] he earned the reputation and fame for more than human sagacity – he had no desire to leave behind treatises about these matters. He considered the occupation of an engineer and, in short, every art that applied itself to practical needs, as ignoble and vulgar (*banauson*), and he focused his intellectual ambition only on those subjects the beauty and remarkable character of which were intrinsically free of necessity.

14.16 Theory is not everything

Some engineers appreciated that theoretical methodology alone did not always result in proper constructions. Philo offers a positive evaluation of theoretical approach and care, but ultimately concludes that experimental, hands-on, work is also needed, not only for future work but that it was likely the original stimulus for the much appreciated later theoretical work itself. His subtle criticism of theoretical work is echoed also by Hero, *Belopoeica* W 72.

Philo of Byzantium, *Artillery Manual* 49.12–51.4

I understand you are fully aware that the [artillery-maker] trade contains something unintelligible and baffling to many people; at any rate, many who have undertaken the building of engines of the same size using the same construction, similar wood and identical metal without even changing its weight, have made some with long range and powerful impact and others which fall short of these. Asked why this happened, they had no reason to give; thus, the remark made by Polyclitus, the sculptor, is appropriate for what I am going to say. He maintained that perfection was achieved gradually in the course of many calculations ... [since without calculation many small errors together make a large total error].... In the old days some engineers were on the way to discovering that the fundamental basis and unit of measure for the construction of engines was the diameter of the hole.... It was impossible to obtain it except by experimentally increasing and decreasing the size of the hole ... [later engineers also experimented to arrive at the basic principle of construction].... Not everything can be accomplished by the theoretical methods of pure mechanics, but much is to be found by experiment.

For instance, the correct proportions of buildings could not possibly have been determined right from the start and without the benefit of previous experience as is clear from the fact that the old builders were extremely unskilful, not only in general building, but also in shaping the individual parts. The progress to proper building was not the result of one chance experiment.... By experimentally adding to the bulk here and subtracting there, by tapering, and by conducting every possible test, they made them appear regular to the sight and quite symmetrical, for this was aim in that craft.

14.17 Greek learning and Roman research

Although Cicero appreciates the remarkable achievements of Greek philosophers and scientific or technical writers, he boasts that the Romans had made many independent discoveries in fields of interest to them, or had improved on many of the Greek accomplishments. In contrast, Pliny the Elder, who a century later wrote a long, highly derivative compendium of learning about the natural world, frequently expresses regret at the absence of "pure" scientific research among his contemporaries, and even neglect of the published Greek works. His *Natural History* was intended to remedy this neglect of past discoveries, but reveals a mind with little originality. An unintentionally amusing account of the elder Pliny's incessant reading and note-taking is given by his nephew Pliny the Younger, *Letters* 3.5 (Fögen 2016, DeLaine 2002).

Cicero, *Tusculan Disputations* 1.1.2

... I have always believed that the Romans have constantly shown themselves more clever at invention than the Greeks, or have improved upon what they received from the Greeks – at least in those fields they judged worthy of their effort. For surely we adhere to better and more elegant customs and rules of life, household and family arrangements, and without question our ancestors determined the course of the Republic with better regulations and laws. What should I say about the art of war, in which our men have proven much superior not only in bravery, but also even more so in discipline? As for those qualities that come from nature rather than study, neither the Greeks nor any other people can be compared. For where has such seriousness of purpose, such steadfastness, greatness of spirit, honesty, loyalty, such superior merit in every aspect been found in any group, that it might stand comparison with our ancestors?

Pliny, *Natural History* 14.2–4

For who would not agree that the establishment of intercommunications throughout the whole world through the majesty of the Roman Empire has improved life by the exchange of merchandise and by alliance in a blessed peace, and that even those things formerly hidden have been brought into common use? Nevertheless, by Hercules, no one is found who knows the wealth of information handed down by writers of former times. The research of the men of long ago was so much more productive or their industry so much more fortunate when, 1,000 years ago at the very beginnings of literacy, Hesiod began to publish his instructions to farmers and numerous others followed his line of research. For this reason, our task is greater, since now we have to investigate not only what was found out later, but also the discoveries made by the pioneers, since a general disregard has brought about complete destruction of the record. Who can find another cause for this failing besides the state of the world? Indeed, other customs have crept to the fore and human attention is fixed on other things—only the arts of profit-making find favour.

Pliny, *Natural History* 2.117–118

More than 20 ancient Greek authors have published their observations about this subject [meteorology]. For this reason I am all the more surprised that when the world was in conflict and divided up into kingdoms, that is, torn limb from limb, so many men were interested in devoting themselves to a subject so difficult to investigate.... But now, during such a happy time of peace, under the rule of an emperor so pleased at the advance of literature and the arts, nothing whatsoever is being added to the sum of knowledge through original research, and in fact not even the discoveries made by our predecessors are carefully studied.

Pliny, *Natural History* 1, Preface 17–18

Through a reading of about 2,000 books, very few of which students touch on account of the abstruseness of their contents, we have collected into 36 volumes 20,000 noteworthy facts abstracted from 100 authors, with the addition of a great many facts, which either our predecessors were unaware of or experience discovered later. I do not doubt that many facts escaped even me, since I am only human, distracted by duty, and I pursue these interests only in my spare time, that is at night....

14.18 The genesis of a Greek inventor

Inventors and their innovations were not totally neglected in the ancient world. In certain technologies, such as warfare or water-lifting, and among entrepreneurial groups, innovations were actively developed and even fostered by support from the elite. Although from a humble family, Ctesibius of Alexandria managed through his own inventive talents to gain the support of the Ptolemies (Greene 2008).

Vitruvius, *On Architecture* 9.8.2–4

Ctesibius was born at Alexandria, the son of a barber. He stood out from the crowd by his intelligence and great industry and had the reputation of delighting in mechanical contrivances. Once, he wanted to hang a mirror in his father's shop in such a manner that, when it was pulled down and pulled back up again, a hidden cord would move a counterweight accordingly. He installed the following device. He attached a wooden conduit beneath a ceiling beam and mounted pulley wheels there, then he strung a cord through the conduit to a corner of the room, where he set up vertical pipe sections. In these he arranged a lead weight to be let down by the cord. Thus, when the weight ran down into the narrow pipe sections and compressed the air, and the volume of pressurised air was pushed forcibly down through the opening of the pipe out into the atmosphere, meeting with a barrier it produced a clear sound.

Therefore Ctesibius, by noticing that noise and clear sounds were generated by the air as it was pushed along and the wind forced out, became the first to design water-organs based on these principles. He also explained water pumps and automata and many kinds of gadgets, among them the construction of water clocks.

14.19 Hellenistic "think-tanks" for military technology

In addition to fostering the work of solitary inventors and technicians, the Ptolemies and other Hellenistic kings, such as Dionysius I of Syracuse (c.430–367 BC), also assembled groups of specialists in technologies of particular interest to them, especially the design and improvement of catapults and other siege engines. The sources suggest that these rulers understood the benefits of collaborative effort. For another version of the story, see **13.29**.

Philo of Byzantium, *Artillery Manual* 50.3

Alexandrian craftsmen achieved this first [advances in the theory and construction of siege engines], being heavily subsidised because they had ambitious kings who fostered craftsmanship. Not everything can be accomplished by the theoretical methods of pure mechanics, but much is to be found by experiment.

Diodorus of Sicily, *History* 14.41.3–4, 42.1

Dionysius, therefore, immediately assembled technicians, commanding them to come from the cities he ruled, and luring them from Italy and Greece – and even from Carthaginian territory – with high wages. For he intended to manufacture weapons in great numbers and projectiles of every sort, and in addition tetraremes and quinqueremes – no quinqueremes yet having been constructed at that time. After assembling a great number of technicians, he divided them into work-groups according to each one's own talents....

In fact, the catapult was invented in Syracuse on this occasion, since the most able technicians were gathered together from all over into one place. The high wages stimulated their enthusiasm, along with the numerous prizes offered to those judged the best....

14.20 Roman attitudes towards innovation in warfare

The Roman elite, too, had the reputation of being open to innovation in warfare, one technology in which the advantages of "labour-saving" technology were immediately and graphically apparent. But by the early Empire, this openness had been replaced by the typical imperial complacency and conservatism, exemplified by Frontinus. The terrifying character of this technology was well understood, and Pliny provides a strong rhetorical criticism of the use of iron in military technology.

Polybius, *Histories* 6.25.10–11

The same description applies to the Greek shields, which are useful for both attack and defence, being solid and firm. Noticing these advantages, the Romans quickly copied [the Greek armament]. For this too is one of their virtues, to adopt new customs and emulate what is better.

Frontinus, *Stratagems* 3, Preface

I will relate first the stratagems of use for besieging cities, then those that might help the besieged. Leaving aside engineering works and catapults, fields where innovation has long since reached its limit and in which I see no further scope for the applied arts, we recognise the following types of stratagems connected with siege operations....

Pliny, *Natural History* 34.138–139

Next, an account must be given of the mines producing iron, a substance that serves the best and worst part of the apparatus for living. Indeed, with iron we plough the earth, plant trees, trim the living vine props, compel the vines to renew their youth yearly by trimming them of the spent growth. With iron we construct buildings, quarry stone, and we use it for every other useful application. But we employ iron as well for war, slaughter, and banditry, not only in hand-to-hand combat, but also on a winged missile, now fired from catapults, now thrown by the arm, now actually fletched with feathers. I think this last to be the most criminal artifice of human ingenuity, in as much as we have taught iron to fly and given it wings so that death might reach a man more quickly.

14.21 Innovative talent among non-Roman peoples

By the fourth century AC, many of the ideals and institutions of the Roman Empire had become tired, or stultified by a growing bureaucracy. It then became a kind of rhetorical trope that the non-Roman peoples pressing on the borders of the Empire were more innovative and adaptable than the citizens of the Empire themselves.

Anonymous, *On Matters of War*, Preface, 4

Although the barbarian peoples neither exert influence through eloquence nor find distinction in high office, they are, however, intimately involved with inventions, their very nature helping them along.

Procopius, *Gothic War* 8.11.27–28

When they saw that the Romans were in despair and did not know how to deal with the situation, they worked out a device of a type that no Roman or Persian had ever thought of since humans came into existence – although there are now and always have been great numbers of engineers in both states. Throughout history both have often had the need for such a device when storming fortifications in ground that is rough and difficult of access, but not one of them had the inspiration that came now to the barbarians. In this way, as time passes, human nature always feels the impulse to keep up with it through innovations in practical procedures.

14.22 The conflict between philosophy and technology

In this very long, rhetorical letter, Seneca develops at length the argument (against the Hellenistic philosopher Posidonius) that technological inventions are not the product of wisdom (*sapientia*), but of ingenuity (*sagacitas*), a lower form of knowledge. Like the banausic crafts themselves, in Seneca's opinion, practical ingenuity is of little value relative to philosophical wisdom. He mentions many inventions and technologies in the course of the letter.

Seneca, *Letters* 90.10–13

In this also I differ from Posidonius, when he concludes that tools for the crafts were invented by wise men.... It was human ingenuity, not human wisdom, that invented those things. In this also I differ from him, that it was the wise who devised mines for iron and copper when the earth, scorched by forest fires, poured out metal from liquefied surface veins of ore. The sort of men who discovered these mines are the same sort who work them today. Nor does the question of whether the hammer or the tongs came first seem as subtle to me as it did to Posidonius. Both were invented by someone with a mind that was nimble and sharp, but not great or elevated – along with anything else that must be sought with a bent body and a mind focused on the ground.

14.23 Hostile responses to innovations

Although the interpretation of both these stories remains controversial, they seem to reveal reluctance on the part of the emperor – the greatest potential patron in the Roman system – to allow innovations that would disturb the economic system. Vespasian may be referring to the intentional use of labour-intensive procedures as a sort of make-work project, while the story in Petronius was concocted in response to the spread of glassblowing, to explain why the major remaining drawback of the material – brittleness – could not be overcome. See **10.86** for another version of this story.

Suetonius, *Life of Vespasian* 18

To an engineer who promised to transport some heavy columns to the Capitoline Hill at a low cost, he gave a significant reward for his scheme, but refused to put it into operation, saying "You must let me feed the poor folk".

Petronius, *Satyricon* 50–51

Pardon me if I say that I personally prefer glassware, since glass vessels do not give off an odour. If they were not breakable, I would prefer them to gold. And now they are very cheap. In fact, there was once a craftsman who made a glass bowl that was unbreakable. He was given an audience with the emperor [Tiberius], bringing along his gift. He had the emperor hand it back and threw it to the floor. The emperor was as frightened as could be, but the man picked the bowl up from the ground – it was dented just like a vessel made of bronze! He took a little hammer from his shirt and fixed it perfectly without any problem. By doing so he thought he had made his fortune, especially after the emperor said to him "No one else knows how to temper glassware like this, do they?" Just see what happened. After he said "No", the emperor had his head chopped off, because if this invention were to become known, we would treat gold like dirt.

ATTITUDES TOWARDS TECHNOLOGY

14.24 A sophist's pride in craftsmanship

Although Greek intellectuals looked down on the life of craftsmen, Hippias boasted of his achievements in various arts and crafts. In fact, he practised the trade of sophist – a paid lecturer on practical wisdom – and the all-round talents he boasted of were simply a way of bolstering this image. He, too, would have found it shameful to engage in craft activities for pay.

Plato, *Lesser Hippias* 368b–d

Certainly, of all men you have the greatest familiarity with the largest array of arts, as I once heard you boast.... You said that you once arrived at Olympia wearing only items that you had made yourself. First of all – for you began with that – the ring you wore was your own work, showing that you knew how to engrave rings, and a seal stone was your work besides, and a strigil, and an oil flask – all of which you made. Then you said that you had yourself cobbled the sandals that you wore, and you had woven your cloak and tunic. But what seemed to everyone the most unusual thing and a sign of the greatest knowledge, was when you said the sash you were wearing around your tunic was like the Persian sashes and very valuable, and that you had made it. In addition to these things you said that you had come with poetic works – epics, tragedies, and satires – and many writings of various genres composed in prose.

14.25 The Christian approach to work and the crafts

Although the elite class of both Greek and Roman society was contemptuous of craftsmen and salaried workers in general, individuals who had worked at some trade often were proud of their skills and accomplishments. Their voices, however, are seldom heard in the ancient sources (but see **10.19**). Both because of its dogma and because of its early appeal to the lower class, Christianity placed a high value on work and on human skills, and many Christian sources elevate work to a virtue (Ovitt 1986).

St. Paul, *Second Letter to the Thessalonians* 3: 6–10

We command you, Brothers, in the name of Jesus Christ, to keep away from any brother who is living in idleness and not according to the tradition which you received from us. For you know yourselves how you ought to imitate us. We were not idle when we were with you; we did not eat anyone's bread without paying for it, but with toil and labour we worked day and night, in order not to be a burden to any of you. It was not because we do not have that right, but to give you in our conduct an example to imitate. For even when we were with you, we gave you this command: if anyone will not work, let him not eat.

Didache 12

If someone wishes to settle among you and has a craft, let him work and eat. But if he does not have a craft, use your own judgement in providing for him, so that no one shall live with you in idleness because he is a Christian.

14.26 The marvellous variety of human technologies

In this chapter St. Augustine describes and praises the "good things with which God has filled this life doomed to condemnation". The products of human technology take a prominent place.

St. Augustine, *City of God* 22.24

For beside the arts of living well and gaining eternal happiness, ... has not the human intellect discovered and put to use so many and such great technologies, in part to serve real needs, in part for the sake of pleasure ...? What marvellous, stupendous accomplishments human effort has achieved in the fields of construction and textile production! How far it has progressed in agriculture and navigation! What accomplishments of imagination and application in the production of all kinds of vessels, various types of statues and paintings.... What great inventions for capturing, killing, taming irrational animals! And against humans themselves, how many types of poisons, weapons, devices! What medicines and remedies humans have discovered for protecting and restoring their health! How many sauces and appetisers they have found to make eating a pleasure! What a great number and variety of signs for conveying thought and for persuasion, the most important of which are words and letters! ... What skill in measuring and counting! With what sharp minds humans grasp the paths and arrangement of the heavenly bodies! How great the knowledge of this world humans have filled themselves with! Who could describe it?

14.27 The imperfect state of human technology

Many Greek and Roman intellectuals understood that all human knowledge is imperfect, and that much remained to be discovered or invented.

Archimedes, Fragment (*Hermes* 42 [1907] 246, lines 13–17)

For I assume that through the method I have demonstrated some persons who now exist or who will exist in the future will arrive at other theorems that have not yet occurred to me.

Seneca, *Speculations about Nature* 7.25.4–5

A time will come when careful research over a longer period of time will bring to the light of day matters [about the heavens] that now lie hidden. A single lifetime is not sufficient for the investigation of such great matters, even if it were focused full-time on the heavens. And besides, we divide these few years of ours unequally between study and enjoyment. As a result, this knowledge will be revealed through long, successive ages. A time will come when our descendants will be astonished that we were ignorant of things that are so apparent.

Note

1 See **2.48**.

Bibliography

Bedini, Silvio, "The Role of Automata in the History of Technology". *Technology and Culture* 5 (1964) 24–41.

Blundell, Susan, *The Origins of Civilization in Greek and Roman Thought*. London: Croom Helm, 1986.

Borger, Theo, and Michael Erler, *Proklos Diadochos. Über die Vorsehung, das Schicksal and den freien Willen an Theodorus, den Ingenieur (Mechaniker)*. Beiträge zur klassischen Philologie, 121. Meisenheim am Glan: Anton Hain, 1980.

Cole, Thomas, *Democritus and the Sources of Greek Anthropology*. Cleveland: American Philological Association, 1967.

Cuomo, Serafina, *Technology and Culture in Greek and Roman Antiquity*. Cambridge: Cambridge University Press, 2007.

Dalley, Stephanie, and John P. Oleson, "Sennacherib, Archimedes, and the Water Screw". *Technology and Culture* 44.1 (2003) 1–26.

DeLaine, Janet, "The Temple of Hadrian at Cyzicus and Roman Attitudes to Exceptional Construction". *Papers of the British School at Rome* 70 (2002) 205–230.

Dodds, Eric R., *The Ancient Concept of Progress, and Other Ideas on Greek Literature and Belief*. Oxford: Clarendon Press, 1973.

Edelstein, Ludwig, *The Idea of Progress in Classical Antiquity*. Baltimore: Johns Hopkins University Press, 1967.

Finley, Moses L, "Technical Innovation and Economic Progress in the Ancient World". *Economic History Review* 18 (1965) 29–45.

Fögen, Thorsten, "Roman Responses to Greek Science and Scholarship as a Cultural and Political Phenomenon". Pp. 958–972 in Georgia L. Irby, ed., *A Companion to Science, Technology, and Medicine in Ancient Greece and Rome, Vol. II*. Chichester: Wiley-Blackwell, 2016.

Frontisi-Ducroux, Françoise, *Dédale: Mythologie de l'artisan en Grèce ancienne*. Paris: Maspero, 1975.

Gabba, Emilio, ed., *Tecnologia, economia, e società nel mondo romano. Atti del convegno di Como, 27–29 settembre, 1979*. Como: Banca Popolare, 1980.

Geoghegan, Arthur T., "The Attitude Towards Labor in Early Christianity and Ancient Culture". *Studies in Christian Antiquity*, 6. Washington: 1945.

Greene, Kevin, "Perspectives on Roman Technology". *Oxford Journal of Archaeology* 9 (1990) 209–219.

Greene, Kevin, "How was technology transferred in the western provinces?" Pp. 101–105 in M. Wood, F. Queiroga, eds., *Current Research in the Romanization of the Western Provinces*. Oxford: Tempus Reparatum, 1992.
Greene, Kevin, "Technological Innovation and Economic Progress in the Ancient World: M. I. Finley Reconsidered". *Economic History Review* 53 (2000) 29–59.
Greene, Kevin, "Inventors, Invention, and Attitudes Toward Innovation". Pp. 800–820 in John P. Oleson, ed., *The Oxford Handbook of Engineering and Technology in the Classical World*. Oxford: Oxford University Press, 2008.
Kiechle, Franz, *Sklavenarbeit und technischer Fortschritt im römischen Reich*. Wiesbaden: Franz Steiner, 1969.
Mondolfo, Rodolfo, "The Greek Attitude to Manual Labour". *Past and Present* 6 (1954) 1–5.
Mondolfo, Rodolfo, *Polis, lavoro, e tecnica*. Edited by M.V. Ferriolo. Milan: Feltrinelli, 1982.
Moseé, Claude, *The Ancient World at Work*. London: Chatto & Windus, 1969.
Mueller, Reimar, "Die Bewertung der Technik in der Kulturtheorie der Antike". Pp. 415–426 in Joachim von Herrmann, and Irmgard Sellnow, eds., *Produktivkräfte und Gesellschaftsformationen in vorkapitalistischer Zeit*. Berlin: Akademie Verlag, 1982.
Mund-Dopchie, Monique, "La Notion de progres chez les grecs. Mise au point préliminaire". *Les Études Classiques* 51 (1983) 201–218.
Oleson, John P., "Well-Pumps for Dummies: Was There a Roman Tradition of Popular Sub-Literary Engineering Manuals?" Pp. 65–86 in Franco Minonzio, ed., *Problemi di macchinismo in ambito romano*. Como: Comune di Como, 2004.
Ovitt, George, Jr., "The Cultural Context of Western Technology: Early Christian Attitudes Towards Manual Labor". *Technology and Culture* 27 (1986) 477–500.
Pasoli, Elio, "Scienza e tecnica nella considerazione prevalente del mondo antico: Vitruvio e l'architettura". Pp. 63–80 in *Scienza e tecnica nelle letterature classiche. Seste giornate filologiche genovesi, 23–24 febbraio 1978*. Genoa: Istituto di Filologia classica e medievale, 1980.
Pavlovskis, Zoja, *Man in an Artificial Landscape: The Marvels of Civilization in Imperial Roman Literature*. Leiden: Brill, 1973.
Phillips, Eustace D., "The Greek Vision of Prehistory". *Antiquity* 38 (1964) 171–178.
Pleket, Henri W., "Technology and Society in the Graeco-Roman World". *Acta Historiae Neerlandica* 2 (1967) 1–25.
Pleket, Henri W., "Technology in the Greco-Roman World: A General Report". *Talanta* 5 (1973) 6–47.
Pot, Johan H.J. van der, *Die Bewertung des technischen Fortschritts: Eine systematische Übersicht der Theorien*. 2 vols. Assen: Van Gorcum, 1985.
Price, Derek J. de Solla, "Automata and the Origins of Mechanism and Mechanistic Philosophy". *Technology and Culture* 5 (1962) 9–23.
Reece, David W., "The Technological Weakness of the Ancient World". *Greece and Rome* 16 (1969) 32–47.
Schneider, Helmuth, ed., *Das griechische Technikverständnis: Von den Epen Homers bis zu den Anfängen der technologischen Fachliteratur*. Impulse der Forschung, 54. Darmstadt: Wissenschaftliche Buchgesellschaft, 1989.
Vemant, Jean Pierre, *Myth and Thought among the Greeks*. London: Routledge & Kegan Paul, 1983.

White, Kenneth D., "Technology and Industry in the Roman Empire". *Acta Classica* 2 (1959) 78–89.
White, Kenneth D., "Technology in Classical Antiquity. Some Problems". *Museum Africum* 5 (1976) 23–35.
White, Lynn, Jr., "Cultural Climates and Technological Advance in the Middle Ages". *Viator* 2 (1971) 171–201.
Wilson, Andrew, "Machines, Power and the Ancient Economy". *Journal of Roman Studies* 92 (2002) 1–32.

INDEXES

A INDEX OF PASSAGES TRANSLATED

This index lists only the citations translated in the text of the book. The selection numbers in the book are given in parentheses after each citation. A brief characterisation and date are given for each author; all dates are AC unless BC is indicated. An author's name appears in square brackets, e.g. [Apollodorus], when a work has been falsely attributed to an author by ancient sources, or in cases of uncertain authorship.

Achilles Tatius, Greek novelist, second century: *Leucippe and Clitophon* 1.1 (11.98), 2.32 (11.90), 3.1–4 (11.90).
Acts, a New Testament book composed c.60: 27:13–44 (11.91).
Aelian, rhetorician, c.170–235: *Letters of Farmers* 18 (11.63); *On the Nature of Animals* 15.1 (3.75), 15.11 (3.76).
Aelius Aristides, man of letters, c.117–181: *To Rome* 10–13 (11.126).
Aeneas Tacticus, Greek general and military writer, mid-fourth century BC: Fragment 48 (12.10), *Treatise on the Defence of a Besieged City* 31.16–19 (12.38), 33.1–2 (13.23), 34 (10.89), 34.1 (2.25).
Aeschines, Athenian orator, c.397–322 BC: *Against Timarchus* 1.97 (10.128).
Aeschylus, Greek tragedian, 525/4–456 BC: *Agamemnon* 280–312 (13.14); *The Persians* 293–294 (5.9); *Prometheus Bound* 442–506 (1.6).
Aëtius of Amida, Byzantine Greek physician and medical writer, mid-fifth to mid-sixth century: *Medical Books* 16.89.1–18 (10.117).
Agathias, Alexandrian poet and historian, c.531–580: *Histories* 5.9.2–5 (8.32).
Alcaeus, Greek lyric poet, flor. c.600 BC: Fragment 54 (13.3).
Ammianus Marcellinus, Roman historian, c.330–395: *History* 14.1.9 (8.73), 14.4.13–15 (8.47), 23.4.4–7 (13.30), 24.6.8 (13.17), 25.1.12–13 (13.17).
Anaxagoras, early Greek philosopher, c.500–428 BC: Diels-Kranz Fragment 59.A.102 (1.18).
Andocides, Athenian orator, c.440–390 BC: *Mysteries* 38 (5.14), 137–138 (11.62).
Anonymous, *On Matters of War*, imperial bureaucrat and military technologist, c.350: Preface, 4 (14.21), 17.1–3 (2.17).
Antipater of Thessalonica, Greek epigrammatist, late first century BC to early first century: *Greek Anthology* 9.418 (2.12).
Antiphon, Attic orator, c.480–411 BC: Fragment 77 [137b, 1445] (12.1).
Apicius, Roman gourmet, fourth century: *On Cooking* (selections) (4.26).

INDEXES

[Apollodorus], a valuable compendium of Greek myth, first or second century: *Library* 1.4.2 (10.119).

Apollonius of Rhodes, Greek epic poet, third century BC: *Argonautica* 1.367–90 (11.71), 2.1001–7 (6.19).

Appian, Roman historian, *c.*160: *Punic Wars* 8.14.96 (11.99); *Civil Wars* 1.1.7 (3.3), 2.20.147 (2.52), 5.118, 119 (13.21).

Apuleius, Roman novelist and rhetorician, *c.*155: *The Golden Ass* 7.15 (4.3), 9.10–13 (2.19).

Archimedes, Greek mathematician and inventor, *c.*287–212 BC: Fragment (*Hermes* 42 [1907] p. 246, lines 13–17) (14.27).

Aristophanes, poet of Old Attic Comedy, *c.*457–385 BC: *Clouds* 771–773 (2.4); *Ecclesiazusae* 1–16 (2.31); *Frogs* 108–115 (11.30); *Knights* 315–321 (10.69), 792–793 (10.78); *Lysistrata* 557–564 (11.125), 567–586 (10.34); *Thesmophoriazousae* 414–430 (12.33); *Wealth* 160–167, 510–516 (14.4).

Aristotle, Greek philosopher, 384–322 BC: *Constitution of Athens* 46.1 (11.72), 47.2 (5.10), 51 (11.124), 63.1–2, 64–66 (12.15), 68–69 (12.17); *History of Animals* 5.16.2.548b1 (13.7), 5.19.551b (10.29); *Metaphysics* 1.1.11–17.981a-982a (1.27), 1.3.3–5.983b (9.1); *Meteorologica* 2.3.358b (4.21), 2.3.359a (11.80), 4.6.383a-b (6.23); *Nicomachaean Ethics* 5.5.8–11.1133a (11.114); *On Marvellous Things Heard* 25–26.832a (6.2), 49.834a (6.38), 62.835a (6.32); *On the Parts of Animals* 659a.9–15 (11.105), 3.4.667b (10.119), 3.5.668a (9.9), 4.2.677a (10.119); *On the Soul* 1.3.406b (2.49), 3.8.432a (1.18); *Politics* 1.1.5.1252b (10.5), 1.2.4–5.1253b-1254a (14.5), 1.3.11–15.1257a-b (11.114), 1.4.2.1258b (11.130), 1.4.3–4.1258b-1259a (14.8), 2.5.1.1267b (11.47), 2.21–2, 29–34.1267b (8.17), 3.2.8–9.1277a-b (14.7), 4.10.10.1297b (13.3), 7.6.1.1327b (1.15), 7.10.5–8.1330b-1331a (13.25), 8.2.1.1337b (14.7).

[Aristotle], anonymous authors of works in Aristotelian *corpus*, fourth century BC to fifth century: *Household Management* 1.2.2–3 (14.6); *Mechanical Problems* Preface, 847a.10–25 (1.20), Preface, 848a.20–38 (2.44), 1.849b-850a (2.38), 7.851b (11.90), 18.853a-b (2.36), 20.853b-854a (2.38), 22.854a (2.39), 28.857a-b (9.26), 32.3 and 5 (960b15 and 21) (3.78); *Problems* 960b.8–17 (11.106), 960b.21–34 (11.106), 961a.24–30 (11.106).

Arrian, Greek historian and military commander, late first to mid-second century: *The Anabasis of Alexander* 4.19 (13.8).

Artemidorus, Greek writer on interpretation of dreams, second century: *Interpretation of Dreams* 1.48 (2.20), 1.51 (10.64), 2.20 (10.64).

Athenaeus, Greek writer, *c.*200: *Philosophers at Dinner* 1.28c (10.77), 3.124c-f (2.32), 4.174c (12.13), 5.198e-f (2.51), 5.197e-202f (11.36), 5.203e-204c (11.78), 5.204c-d (11.104), 5.204d-206c (11.79), 5.206e-209b (11.77), 7.102 (316f) (3.80).

Athenian Agora, Inscriptions, a collection of various dates: 3.1 2729 (12.48).

St. Augustine, early church father, 354–430: *City of God* 7.4 (10.129), 22.24 (14.26); *Letters* 185.4.15 (2.18).

Aulus Gellius, Roman writer, *c.*130–180: *Attic Nights* 3.3.5 (12.6), 3.3.14 (2.18), 7.17 (12.47), 9.4.1–5 (12.44), 10.12.9–10 (2.50), 15.1.4–6 (8.77), 17.9.6–15 (12.37), 19.10.1–4 (8.39).

Aurelius Victor, historian and politician of the Roman Empire (*c.*320–390): *[Epitome] On the Caesars* 13.12–13 (8.75).

Ausonius, Latin poet, died *c.*395; ed. Green: *Letters* 27.22.21–22 (12.36); *Moselle* 240–49 (3.74), 267–9 (6.29), 359–64 (2.16).

Bianor, Greek writer of epigrams, first century BC – first century AC: *Greek Anthology* 11.248 (11.71).

733

INDEXES

Bruns, *Fontes Iuris Romanae*, 1909, a collection of Roman legal documents from various periods: no. 113 (5.12).

Caesar, Roman politician, general, and historian, 100–44 BC: *Gallic War* 3.13 (11.69), 4.17–18 (11.19), 5.12 (5.6), 6.14 (12.24), 8.42.1 (13.31); *Civil War* 1.54 (11.70).

Cassiodorus, Roman politician and writer, *c.*490–583: *Letters* 3.53.1, 6 (9.3).

Cato, Roman politician and writer, 234–149 BC: *On Agriculture* 1 (3.5), 2 (3.12), 10–11 (3.7), 22.3 (11.45), 38 (8.58), 39.1–2 (3.25) 40 (3.52), 48 (3.51), 56–59 (3.11), 74 (4.6), 88.1–2 (5.31), 92 (3.27), 95 (3.27), 98 (10.10), 135.1–2 (3.9), 135.3–5 (10.71), 155 (3.18), 162 (4.25).

Catullus, Roman poet, *c.*87–57 BC: *Poems* 17.23–6 (11.42), 22.3–8 (12.41), 50.1–6 (12.29), 64.310–19 (10.51), 68.33–36 (12.45).

Celsus, Roman encyclopaedist, known for his extant medical work, *c.*25 BC – AD 50: *On Medicine* 1 Proem (10.109), 1 Proem 23–26 (10.121), 1 Proem 74–75 (10.121), 2.32.1 (10.117), 3.18.12 (10.117), 5.2 (10.120), 5.19 (10.120), 5.23 (10.117), 5.25.1 (10.117), 6.5 (10.114), 7.26 (10.119), 7.5 (10.120), 8.3 (10.119), 8.5 (10.119), 25.1 (10.117).

Censorinus, Roman grammarian, third century: *Natal Day* 20.4–11 (12.5).

Cicero, Roman orator, 106–43 BC: *Against Verres* 2.5.11.27 (11.33); *Brutus* 257 (14.10); *Letters to Atticus* 14.9 (8.63); *Letters to his Brother Quintus* 3.1.1–2 (8.36); *On Behalf of Murena* 51 (8.65); *On Behalf of Sextus Roscius Amerinus* 7.19 (11.44); *On Duty* 1.42 (14.10); *On the* Laws 2.24 (10.114), 3.38 (12.18); *On the Nature of the Gods* 1.19.50–53 (14.10); *On the Orator* 2.86.353 (8.79); *Republic* 1.14.21–22 (2.45); *Tusculan Disputations* 1.1.2 (14.17), 1.45 (10.122).

Claudian, Latin poet, ca.400: *Shorter Poems* 51 (2.45).

Clement of Alexandria, Greek theologian, *c.*150–215: *Protrepticus* 4.48 (7.8).

Columella, Latin agricultural writer, *c.*65: *On Agriculture* 1.6 (3.6), 1.7.6–7 (3.13), 1.9.2–3 (3.31), 2.2.1–7 (3.16), 2.2.22–8 (3.31), 2.12.1–7 (3.28), 2.20.1–3 (3.33), 2.20.5 (3.35), 2.9.9 (3.27), 3.1.3–4 (3.42), 5.8–9 (3.49), 6 Preface 1–3 (3.54), 6.2.1–7 (3.56), 6.19.1–3 (3.56), 6.37.10–11 (3.58), 7.1 (3.57), 7.2.1–2 (3.59), 7.8.1–5 (4.22), 8.15.1–6 (3.64), 8.17.1–3,6 (3.67), 9.15.1–16.1 (3.66), 11.1.9–11 (14.6), 11.3.52–53 (3.39), 12.3.6 (10.53), 12.16.5 (4.25), 12.19–21, 39, 41 (4.24), 12.47.2–4 (4.25), 12.50.1–3 (4.18), 12.52.1, 6–7 (4.19), 12.52.10–11 (4.19); *On Trees* 2.1–3.5 (3.43), 18 (3.51).

CIG, *Corpus Inscriptionum Graecarum*, collection of Greek inscriptions from various periods: 1688 (11.15).

CIL, *Corpus Inscriptionum Latinarum*, collection of Latin inscriptions from various periods: 1.593.20–55 (11.6), 1.593.56–67 (11.49), 1.638 (11.9), 1.698 (8.59); 2.759 (11.25), 2.760 (11.25), 2.5181, 19–31 (9.43), 2.5181, 46–56 (6.16), 2.5181, 58–60 (5.12); 3.199 (11.9), 3.7251 (11.28), 3.8267 (11.9), 3 p. 948, no. 10 (5.15); 4.138 (8.62); 6.2305 (3.16); 8.2728 (9.13), 8.22371 (11.9); 10.6854 (11.9), 10.7296 (12.42), 15.184 (10.140), 15.213 (10.140), 15.415–19 (10.140), 15.650a (10.140), 15.651a (10.140), 15.761 (10.140).

Crosby, M., *Hesperia* 10 [1941] 40–49 (5.10).

Q. Curtius Rufus, Roman historian, first or second century: *Histories of Alexander the Great* 4.3.10 (13.36), 10.10.13 (10.122).

St. Cyprian, Christian writer, *c.*200–258: *Letters* 77.2.4 (5.37).

Demosthenes, Athenian orator, 384–322 BC: 9, *Philippics* 3.47–50 (13.4); 27, *Against Aphobus* 1.9–10 (10.136); 34, *Against Phormio* 10 (11.80); 35, *Against Lacritus* 10–11 (11.82); 37, *Against Pantaenetus* 38 (5.11); 42, *Against Phaenippus* 5–7, 20 (3.1); 43, *Against Macartatus* 8–9 (12.9); 45, *Against Stephanus* 1.8 (12.9).

INDEXES

Deuteronomy, an Old Testament book composed *c.* seventh century BC: 11:8–11 (3.20).

Didache, collection of early Christian Greek saying, second century: 12 (14.25).

Digest, Justinian's codification of Roman law in 50 books, *c.*533: 8.2.17 (2.3), 9.2.27.29 (9.88), 50.16.235 (9.21).

Dio Cassius, Graeco-Roman historian, *c.*164–229: *Roman History* 46.36.3–5 (13.15), 48.51.1–2 (2.5), 49.30 (13.13), 51.165 (10.122), 56.18.1–3 (1.9), 60.11.2–5 (11.102), 66.1–2 (12.33), 68.13.1–6 (11.21), 69.4.1–5 (8.39), 71.3.1 (11.22).

Dio Chrysostom, Greek orator and popular philosopher, ca. 40–115: *Discourses* 32.35–6 (11.126).

Diocletian, Roman Emperor, 284–305: *Price Edict* selections (11.129, 11.136).

Diodorus of Sicily, Graeco-Roman historian, ca. 40 BC: *History* 1.8.7–9 (1.17), 1.33.9–12 (11.109), 1.34.2 (9.32), 1.98 (7.2), 2.16.8–9 (13.18), 2.29.5–6 (14.14), 3.3.5 (12.20), 3.4.1–4 (12.20), 3.12.1–13.1 (5.18), 3.13.2–14.4 (6.5), 4.76.1–3 (7.1), 4.76.4–6 (8.39), 4.80.5–6 (8.19), 5.13.1–2 (6.21), 5.22.2 (6.13), 5.26.2–3 (4.20), 5.27.1–3 (5.26), 5.36–38 (5.20), 14.41.3–4 (14.19), 14.42.1 (13.29, 14.19), 14.43.3 (13.29), 14.50.4 (13.29), 16.74.2–5 (12.29), 18,26.3 (10.122), 18.27.3–4 (11.40), 20.91.1–8 (12.30), 20.49–51 (13.19), 26.18 (12.31).

Diogenes Laertius, Greek writer on lives and doctrines of ancient philosophers, third century: *Anaximander* 2.1.2 (11.51); *Aristippus* 2.103 (8.43); *Lives of Eminent Philosophers* 2.1 (12.4).

Dionysius of Halicarnassus, Graeco-Roman historian and rhetorician, first century BC: *Roman Antiquities* 3.44.2–3 (11.102), 3.45.2 (11.24).

Dioscorides, Greek physician and writer about medicine, first century: *Pharmacology* 1.106–111 (10.65), 4.75.2–7 (10.115), 5.75 (6.30), 5.115 (8.12), 5.134 (10.37).

DMG, J. Chadwick, M. Ventris, *Documents in Mycenaen Greek*, a collection and translation of Linear B documents, Late Bronze Age: no. 1 (10.125), 4 (10.125), 21 (10.125), 26 (10.125), 28 (10.125), 41 (8.7), 47 (8.7), 172 (12.3), 211, 217, 219, 224, 226–27 (10.32), 252 (10.126), 253.1–2 (6.27), 254 (6.27), 265 (10.126), 273 (10.126), 278 (10.126), 282 (10.126), 317 (10.32).

Euripides, Greek tragedian, *c.*485–406 BC: *Palamedes* Fragment 578 (12.25).

Ezra, An Old Testament book composed *c.* fifth century BC: 6:3–4 (8.45).

Feissel, D., *Syria* 62 (1985) 79–83 (10.35).

Festus, Latin grammarian and scholar, late second century: *On the Significance of Words* 9.78 (2.23), (Excerpts of Paulus 276 M.; 381 Th.) (10.55).

Frontinus, Roman military and political figure, administrator, and writer, *c.*30–104: *On the Aqueducts of Rome* 1.16 (9.18), 1.17–19 (9.19), 1.23–25, 29, 31, 33–36 (9.20), 2.75–76 (9.21), 2.77–78, 87–88 (9.22), 2.92 (9.23), 2.103, 105, 107, 109 (9.24), 2.112–115 (9.21), 2.116–125, 127 (9.25); *Limites* Campbell p. 10, 16–23 (11.57); *The Science of Land Measurement* Campbell p. 12, 19–29 (8.34); *Stratagems* 1 Preface (13.10), 2 Preface (13.10), 3 Preface (13.10, 14.20), 3.13.-7 (13.15), 4.7.9–10 (13.22).

Gaius, Roman jurist, second century, quoted in *Digest* 50.16.235 (10.20).

Galen, Greek physician, philosopher, and writer, *c.*129–?199/216: *Commentary on Hippocrates' "On Epidemics"* 3.239–240 (12.49); *On the Composition of Drugs According to Places* 12.966 14K (10.115); *On the Method of Curing* Vol. 10 p. 445 (10.116); *On the Usefulness of the Parts of the Body* 1.2–3 (1.18).

Geoponika, a Byzantine Greek farming manual of the tenth century that preserves excerpts of many earlier, now-lost, agricultural writers: 20.46.1–2 (4.24).

Greek Anthology, an ancient collection of Greek poems of various periods: 6.204, 205 (9.19), 9.418 (2.11), 11.408 (9.107), 11.418 (12.2), 16.221 (5.36), 16.323 (9.84).

Greek Historical Inscriptions (Meiggs and Lewis), a collection of Greek inscriptions from the eighth to late-fifth centuries BC: 45.12–13 (10.107).

Gregory of Nyssa, erudite theologian and writer, c.335–c.395: *On Ecclesiastes* 3.656A Migne (2.16).

Harpocration, Greek grammarian in Alexandria, probably second century: *s.v. Lampas* (1.7)

Hero of Alexandria, Greek mechanical engineer, c.65: *Dioptra* 34 (11.38), 37 (2.40); *Mechanics* 3.1 (11.35), 3.21 (2.42); *Pneumatics* 1.15–16 (2.53), 1.16–17 (2.54), 1.21 (2.55), 1.28 (9.35), 1.34 (2.56), 1.38 (2.57), 1.43 (2.7), 2.11 (2.9).

Herodian, Greek writer of Roman history, mid-second to mid-third century: *History of the Empire after Marcus* 4.2 (10.124).

Herodotus, Greek historian, c.480–425 BC: *Histories* 1.68 (10.2), 1.94 (11.115), 1.163 (11.85), 1.178 (11.122), 1.193.1–4 (3.22), 1.194 (11.70), 2.5 (11.89), 2.35 (10.54), 2.36 (12.14), 2.62.1 (2.24), 2.86–88 (10.122), 2.96 (11.67), 2.109 (11.121, 12.4), 2.124–25 (8.4), 2.158 (11.109), 2.167 (14.3), 2.178–79 (11.126), 3.60 (9.13, 11.96), 3.89 (11.116), 3.95–96 (11.116), 3.96 (10.8), 4.36 (11.52), 4.42 (11.85), 4.64 (10.62), 4.74 (10.30), 4.152 (7.12), 4.195 (5.29), 5.12 (10.52), 5.35.2–3 (12.35), 5.52–53 (11.1), 5.58–59 (12.23), 6.46–47 (5.13), 6.119 (5.29), 6.127 (11.123), 7.23 (11.110), 7.26.3 (10.120), 7.36 (11.18), 7.239.2–4 (12.35), 8.84, 86, 88–90, 96 (11.94), 8.98 (11.26), 9.37.1–4 (10.114).

[**Herodotus**], biography of Homer of uncertain authorship, c. second century?: *Life of Homer* 32 (10.74).

Hesiod, early Greek poet, c.700 BC: *Works and Days* 20–26 (1.29), 50–52 (2.23), 107–178 (1.1), 303–311, 410–413 (14.2), 427–436 (3.29), 536–546 (10.60), 618–634 (11.61).

St. Hilary of Poitiers, bishop in Gaul, opponent of Arianism, c.315–367: *On the Trinity* 10.14.329 (10.115).

Hippocrates (and the Hippocratic Corpus), medical writer regarded as the father of western medicine, c.460–375 BC (and anonymous followers/writers, sixth to fourth centuries BC): *On Diseases* 4.55.29–35 (6.20); *On Joints* 72 (10.119); *On Regimen* 1.13.1–4 (6.24); *On Regimen in Acute Diseases* 18 (10.110); *The Physician* 2.14–16 (10.118), 2.19–21 (10.118), 7 (10.118).

Historia Augusta, collection of biographies of Roman Emperors, composed c.350: *Hadrian* 17.5–7 (10.108); *Marcus Antoninus* 23.8 (11.50), *Pertinax* 8.6–7 (11.39), *Caracalla* 9.4–5 (8.72), *Elagabalus* 25.2–3 (10.72), *Severus Alexander* 24.5–6 (9.42), *The Three Gordians* 33.1 (3.71), *Aurelian* 39.2 (13.26).

Homer, Greek epic poet, c.750 BC: *Iliad* 6.167–70 (12.21), 7.465–75 (11.113), 10.254–65 (13.2), 15.410–13 (8.33), 17.389–93 (10.66), 18.414 (10.108), 18.369–79 (2.48), 18.410–11 (10.1), 18.468–82 (6.26), 18.478–81 (13.2), 18.483–565 (10.7), 18.599–601 (10.73), 18.609–13 (12.2), 19. 23–39 (10.121); *Odyssey* 2.414–28 (11.75), 3.432–34 (10.1), 3.475–86 (11.2), 4,440–446 (10.121), 5.227–61 (11.67), 5.243–45 (8.33), 6.69–70 (11.32), 6.232–235 (10.6), 6.262–69 (11.95), 7.107 (10.36), 7.112–33 (3.40), 9.105–31 (1.2), 9.218–23, 237–39, 244–49 (4.22), 9.383–90 (10.3), 9.391–94 (6.24), 14.23–4 (10.68), 15.415–16 (11.126), 15.455–56 (11.126), 17.382–86 (14.1), 24.226–31 (10.68).

Horace, Roman poet, 65–8 BC: *Epodes* 10 (10.113); *Satires* 1.5.1–8 (4.21), 1.5.9–23 (11.17).

INDEXES

[Hyginus], military writer, probably from the third century: *On the Fortifications of Camps* 12 (13.12).

IG, Inscriptiones Graecae, a collection of Greek inscriptions from various periods: 1^2 73 (11.72), 1^2 374.248–56 (7.10), 1^2 1084 (10.19), 1^3 449.1–41 (8.52), 1^3 475.1–20 (8.53), 1^3 475.54–71 (8.55), 2^2 1534, fr. A, r. 61, 2^2 1668 (8.57), 2^2 1673.11–43 (11.35), *IG* 2/32 1666.38ff. (8.54), 7.3073 (8.50), 14.297 (12.42).

IGR, Inscriptiones Graecae ad res Romanas pertinentes, a collection of Greek inscriptions of various periods relevant to Roman subjects: 1.1142 (11.9).

Isidore, Bishop of Seville and Latin writer, 602–636: 19.4.10 (10.83), 19.19.1–2 (10.20).

Isocrates, Athenian orator, 436–338 BC: 29–40 (1.8).

Itinerarium Antonini, Parthey and Pinder, eds., Berlin: 1848. A collection of 225 route-lists of places on Roman roads across the Empire, written in the late first to late third century, perhaps largely in Caracalla's reign (198–212): 463.3–466.4 (11.57).

Josephus, Jewish soldier, statesman, and historian, c.37–100: *Jewish Antiquities* 3.120–21 (2.43); *Jewish War* 1.408–14 (11.101), 2.189–90 (10.80), 3.71–97 (13.9), 3.115–26 (13.11), 3.235, 240–46 (12.32).

Jeremiah, prophet in the Old Testament: 10:4 (7.11).

Juvenal, Roman satirist, c.?50/65–127: *Satires* 3.190–202 (8.63), 3.283–88 (2.31), 5.67–75 (4.7), 10.56–64 (7.9), 14.33–37 (11.16), 14.275–83 (11.65).

Lactantius, rhetorician and erudite Christian writer who became an adviser to Constantine I, c.250–c.325: *On the Deaths of the Persecutors* 5 (10.120).

Latin Anthology, a collection of Latin poems of various periods: 103 (2.18), 284 (2.12).

Leonidas of Tarentum, Greek epigrammatist, third century BC: *Greek Anthology* 6.204, 205 (10.18).

Livy, Roman historian, c.50 BC–AD. 15: *History of Rome* 1.19.6–7 (12.5), 1.44.3 (13.26), 6.32.1 (13.26), 8.8.3–8 (13.5), 8.8.9–14 (13.5), 21.27.5 (11.70), 21.37.2–3 (11.11), 28.45.14–21 (13.24), 37.11.13 (13.22), 41.27.5 (11.6).

Longus, author of Greek romances, late second-early third century: *Daphnis and Chloe* 3.21 (11.93).

Lucan, Roman writer of epic and minor works, 39–65: *Pharsalia* 10.141–43 (10.31).

Lucian, Greek satirical writer, c.120–180: *The Dream* 6–9, 13 (14.13); *Greek Anthology* 11.408 (10.106); *Hippias* or *The Bath* 4–8 (9.41); *The Ignorant Book Collector* 6 (10.114); *The Ship* 1.4–6 (11.76).

Lucilius, Roman satiric poet, c.160–102/1 BC: Warmington Fragment no. 1163–64 (11.89).

Lucretius, Roman poet, c.94–55 BC: *On the Nature of Things* 4.513–19 (8.35), 5.517 (2.11), 5.925–1025 (1.4), 5.1091–1104 (1.16), 5.1241–1265 (6.1), 5.1350–60 (10.24), 5.1361–69 (1.13), 5.1370–78 (3.81), 5.1448–57 (1.24), 6.160–63 (2.22), 6.808–15 (5.23).

Lysias, Greek orator, c.459–380 BC: *Against Eratosthenes* 12.19 (10.136).

Macrobius, Roman grammarian and critic, c.400: *Saturnalia* 3.16.15 (9.45), 7.12.13–16 (4.13).

Martial, Roman poet, c.40–104: *Epigrams* 1.41.2–5 (2.22), 1.66 (12.50), 1.117 (12.49), 2.8 (12.40), 3.2.7–11 (12.41), 4.10 (12.41), 6.93 (9.45), 7.61 (11.125), 8.62 (12.31), 9.37 (10.114), 11.77 (9.46), 12.48 (8.47), 12.48.5–8 (10.110), 12.74 (10.85), 13.3.1–4 (12.42), 14.19 (12.39), 14.38 (12.39), 14.94 (10.85), 14.130 (10.70), 14.184, 186, 188 (12.43), 14.190 (12.42), 14.192 (12.42), 14.208 (12.40).

Maurice, successful Byzantine general, who was emperor from 582–602, credited with a praiseworthy military manual, 539–602: *Strategikon* 11.4.10 (11.107).

Meiggs, R. and D. Lewis, *Greek Historical Inscriptions* 45.12–13 (11.117).
Mesomedes, Greek lyric poet, second century: *Greek Anthology* 16.323 (10.83).
Mitchell, S., *Journal of Roman Studies* 66 [1976] 107–108 (11.29).
Moschion, Greek tragic poet, third century BC: Fragment 6 (1.30).
Nicander, Greek didactic poet, second century BC: *Theriaca* 367–371 (11.16).
On Ancient Medicine, a medical treatise, fifth century BC: 3.5–32 (1.25).
Oppian, *Hunting* 1.147–157 (3.68), 4.85–211 (3.70); *Fishing* 3.72–91 (3.73), 5.612–74 (3.79).
Oribasius, Greek medical writer, c.320–400: *Compendium of Medicine* 49.4.52–58, 5.1–5,7–9 (2.41).
Ovid, Roman poet, 43 BC–AD 17: *Amores* 1.14.25–50 (10.114); *Art of Love* 3.26 (12.36); *Fasti* 3.815–24 (10.25); *Metamorphoses* 4.121–24 (9.25), 6.53–60 (10.56), 6.383–391 (10.119), 9.522–531 (12.32).
P.Flor., a collection of Greek papyri of various periods from Egypt, now in Florence: 1.69 (11.69).
P.Holm., a Greek papyrus from Roman Egypt, now in Stockholm, (Halleux 1981) (6.42, 10.93, 10.94).
P.Leid., a collection of Greek papyri of various periods from Egypt, now in Leiden, (Halleux 1981) (6.42, 10.50, 10.93).
P.Mich., a collection of Greek and Latin papyri of various periods from Egypt, now in Michigan: VIII.7.29.30. (10.110).
P.Oxy., a collection of Greek papyri of various periods from Oxyrhynchus in Egypt: 470.31–85 (12.10), 3595 (10.79), 3536 (9.85).
P.Petr., a collection of Greek papyri of various periods from Egypt: 3.43 (11.23).
P.Tebt., a collection of Greek papyri of various periods from Tebtunis in Egypt: 703.87–117 (10.38).
Palladius, Roman agricultural writer, fourth century: *On Agriculture* 1.18 (4.11), 1.41 (2.14), 1.42 (3.8), 7.2.2–4 (3.33).
Pappus, mathematician of Alexandria, c.320: *Mathematical Collection* 8.1–2 (2.33), 8.3 (14.8), 8.52 (2.35).
St. Paul, apostle of early Christian church, first century: *Second Letter to the Thessalonians* 3.6–10 (14.25).
Paul, distinguished Roman jurist, late second to third century, quoted in *Digest* 1.15.3–4 (8.64).
Paul of Aegina, Byzantine physician famous for his compilation of earlier medical writers, c.625–c.690: *Medical Compendium in Seven Books* 6.48.1.
Paulinus of Nola, Christian bishop and writer, 353/4–431: *Letters* 49.1, 2, 3, 12 (9.36).
Pausanias, Greek traveller and geographer, c.150: *Description of Greece* 1.26.6–7 (2.31), 1.30.2 (2.23), 1.40.1 (9.13), 1.40.4 (7.6), 1.44.6 (11.3), 2.3.3 (6.31), 2.25.8 (8.5), 3.12.10 (6.25), 3.17.6 (7.3), 5.16.1 (8.18), 8.14.8 (7.3), 8.16 (10.123), 8.54.5 (11.3), 9.3.2 (7.1), 9.4.1 (7.5), 9.10.2 (7.5), 9.32.3 (9.10), 9.36.5 (8.5), 9.37.5–7 (8.48), 9.38.2 (8.5), 9.38.6–8 (9.11), 10.4.1 (8.86), 10.11.2 (5.13), 10.16.1 (10.4), 10.32.8 (11.3), 10.37.7–8 (12.28), 10.38.3 (10.63), 10.38.6 (7.3); *Description of Arcadia* 8.17.2 (7.5).
Periplus of the Erythraean Sea, description of sea routes from Egypt to India, first century: 49 (11.127).
Persius, Roman satirist, 34–62: *Satires* 5.140–42 (11.81).
Petronius, Roman novelist, first century: *Satyricon* 50–51 (14.23), 76 (11.128), 79 (11.48).

INDEXES

Philo of Alexandria, Hellenistic Jewish philosopher, c.30 BC–AD 45: *On the Confusion of Tongues* 38 (9.33).
Philo of Byzantium, technological writer, probably late third century BC: *Artillery Manual* 49.12–51.4 (14.16), 50.3 (14.19); *On Sieges* (ed. Whitehead) A8 (13.27), A13 (13.27), A76 (13.35), B53 (13.32), C59–61 (13.36), D34–35 (13.23); *Pneumatics* 61 (2.10), 65 (9.29), appendix 1, chapter 2 (9.34).
Philostratus, Greek philosophical writer, c.170–248: *Imagines* 1.1.294.5–12 (1.30), 1.9.308.23–35 (1.12); *Life of Apollonius* 1.15 (11.133), 3.57 (3.77), 8.7.5 (10.27), 8.14 (11.84).
Philostratus the Younger, Greek writer, third century: *Imagines* 3.1.395 (1.12).
Pindar, Greek lyric poet, 518–438 BC: *Nemean Odes* 6.10–12 (3.26); *Olympian Odes* 1,26–27 (10.114).
Plato, Greek philosopher, c.429–347 BC: *Critias* 111b-c (10.11); *Laws* 1.643b-c (10.132), 3.677a-679b (1.5), 6.761a-b (9.9), 8.846d-847a (14.7), 12.954b-954c (12.33); *Lesser Hippias* 368b-d (14.24); *Republic* 2.369b-370b (1.26), 4.421d-e (10.75), 7.528b-c (14.14), 7.530e-531a (14.15); *Statesman* 303d-e (6.8); *Theaetetus* 172d (12.9); *Phaedrus* 274e (12.26).
Plautus, Roman comic poet, c.254–184 BC: *Boeotian Woman* Fragment v.21 (12.6); *Comedy of Asses* 707–9 (2.18); *Curculio* 466–82 (11.125); *The Pot of Gold* 505–22 (10.59).
Pliny the Elder, Roman soldier, administrator, and writer, 23–79: *Natural History* 1, Preface 17–18 (14.17), 2.106.234 (11.108), 2.117–18 (14.17), 2.239 (2.4), 3.17 (11.58), 3.53 (11.111), 3.138 (5.2), 4.10 (11.14), 5.57–58 (3.21), 6.61–62 (11.57), 6.100–1 (11.86), 6.101 (11.127), 6.106 (11.86, 11.127), 6.121 (11.122), 7.1–4 (1.19), 7.29 (10.114), 7.55 (10.124), 7.84 (11.44), 7.85 (12.42), 7.192–93 (12.22), 7.194 (8.2), 7.196 (10.23), 7.200–2 (13.1), 7.212–15 (12.5), 8.167 (3.58), 8.171–73 (3.58), 8.190–93 (10.26), 8.196 (10.58), 8.197 (10.43), 8.207–9 (3.61), 8.217–218 (3.82), 9.120–21 (10.95), 9.125–41 (10.46), 10.52–54 (3.63), 10.152–3 (3.62), 11.104–6 (3.82), 12.1–2 (1.11), 12.1–4 (3.41), 12.32 (4.23), 12.38–39 (10.28), 12.59 (10.141), 12.63–65 (11.129), 13.4–5 (10.100), 13.7 (10.103), 13.8 (10.104), 13.19 (10.105), 13.20 (10.101), 13.24 (10.101), 13.28–44 (3.53), 13.68–72 (12.30), 13.74 (12.30), 13.75 (10.134), 13.77 (12.30), 13.81 (12.30), 13.89 (12.30), 13.91–101 (10.17), 14.2–4 (14.17), 14.10–12 (3.44), 14.54–5 (4.16), 14.86 (4.16), 14.124 (4.14), 14.126–30 (4.14), 14.133–35 (4.15), 14.149 (4.20), 15.1 (3.48), 15.5 (4.19), 15.23 (4.19), 15.33–34 (10.67), 15.57 (3.52), 15.62–65 (4.25), 15.87 (10.48), 16.4 (2.29), 16.23 (2.27), 16.34–42 (10.14), 16.38 (10.22), 16.52–60 (10.22), 16.156–78 (10.16), 16.184–92 (10.13), 16.195 (10.12), 16.200 (10.12), 16.200–1 (8.26), 16.201–2 (11.102), 16.206–32 (10.15), 16.207–8 (2.22), 17.50–57 (3.23), 17.101 (3.52), 17.115–17 (3.45), 17.120 (3.52), 17.164–66 (3.44), 17.191–93 (3.47), 17.199–202 (3.44), 17.250 (3.19), 18.5–21 (3.2), 18.15 (11.134), 18.35 (3.3), 18.39–43 (3.14), 18.71 (4.9), 18.86–90 (4.4), 18.92 (4.4), 18.97–98 (4.1), 18.102–4 (4.5), 18.105–6 (4.6), 18.107–8 (4.8), 18.167–70 (3.22), 18.171–73 (3.30), 18.193–94 (3.23), 18.195–200 (3.32), 18.258–63 (3.37), 18.317 (4.12), 19.3–5 (11.87), 19.5 (11.75), 19.10–12 (10.61), 19.14–15 (10.28), 19.16–18 (10.37), 19.19–20 (10.31), 19.23–4 (8.70), 19.27–30 (10.31), 19.38–39 (3.82), 19.48 (10.41), 19.55 (2.32, 4.21), 19.60 (3.19, 3.38, 9.38), 19.173–74 (10.30), 21.83–84 (10.90), 22.2–4 (10.44), 23.140 (10.63), 24.96 (10.41), 28.66 (10.35), 28.91 (10.35), 28.191 (10.92), 29.11 (2.25), 31.38–39 (9.4), 31.49 (5.24), 31.57–58 (9.15), 31.73–85 (5.31), 31.106 (10.81),

31.109–10 (10.81), 33.1–3 (5.1), 33.42–48 (11.119), 33.59 (6.40), 33.60–63 (6.3), 33.62 (6.2), 33.66–78 (5.21), 33.69 (6.6), 33.93–94 (6.36), 33.95–98 (5.3), 33.99–100 (6.7), 33.103–4 (6.10), 33.106–9 (6.11), 33.122 (6.18), 33.123 (6.9), 33.126 (5.4), 33.127 (6.40), 33.132 (11.120), 33.139 (10.9, 10.131), 33.145 (10.9), 33.164 (11.135), 34.4 (6.33), 34.11 (10.130), 34.13 (8.72), 34.15 (7.3), 34.36–37 (7.13), 34.41 (7.14), 34.69 (7.7), 34.95 (6.12), 34.97–99 (6.34), 34.99 (10.10), 34.123–25 (10.99), 34.138–39 (14.20), 34.141 (6.35), 34.143–45 (6.20), 34.146 (6.22), 34.164–165 (5.8), 34.148 (2.6), 34.149 (6.22), 34.150 (6.35), 34.156–59 (6.14), 34.160–62 (6.39), 34.164–65 (5.8), 35.6 (7.15), 35.41–43 (10.99), 35.46 (10.47), 35.49 (11.71), 35.50 (10.96), 35.149 (11.71), 35.150 (10.39), 35.151 (7.4), 35.153 (7.15), 35.159–61, 163 (10.76), 35.166 (8.14), 35.169–73 (8.8), 35.175 (10.35), 35.178 (5.28), 35.179 (2.30), 35.182 (10.10), 35.183–85 (10.40), 35.196–98 (10.35), 35.199 (10.10), 36.1–3 (5.32), 36.14 (5.33), 36.20 (7.7), 36.27 (10.123), 36.47–50 (8.25), 36.51–53 (5.38), 36.55–57 (5.35), 36.67–70 (11.81), 36.83 (11.97), 36.90 (8.46), 36.95–97 (8.44), 36.100 (8.30), 36.101 (8.85), 36.104–6 (9.17), 36.106 (8.6), 36.117 (8.71), 36.121–23 (9.17), 36.124–25 (9.12), 36.125 (5.36), 36.159–62 (5.40), 36.171–72 (8.11), 36.173 (9.5), 36.175–77 (8.13), 36.184–89 (8.21), 36.190–94 (10.80), 36.195 (10.86), 36.198–99 (10.82), 36.200–1 (2.26), 37.29 (10.82).

Pliny the Younger, Roman statesman and writer, c.61–112: *Letters* 1.6 (3.69), 2.17.4–5, 7–9, 11, 23 (8.67), 3.5.17 (12.31), 3.19 (3.4), 6.31.15–17 (11.100), 8.17.1–2,6 (8.75), 10.39.1–6 (8.38), 10.40 (8.38), 10.41, 42 (11.111), 10.120 (11.28).

Plutarch, Greek philosopher and biographer, c.50–120: *Agesilaus* 26.4–5 (14.3), 40.3 (10.122); *Alcibiades* 22.2 (12.17); *Alexander* 18.4.4–6 (8.82), 24.7–8 (2.23), 26.2–4 (8.17), 35.1–4 (2.30), 35.14 (10.72), 36.1–2 (10.49), 63.3–6 (10.115); *Antony* 28.1 (12.1); *Aristeides* 7.4–5 (12.19); *Brutus* 12 (10.111); *Cato* 5.3 (8.56), 9 (10.117); *Causes of Natural Phenomena/Moralia* 915a (11.108); *Comparison of Nicias and Crassus* 1.1–2 (5.17); *Concerning Curiosity* 12 (10.113); *Demetrius* 43.4–5 (11.78), 51 (12.33); *Demosthenes* 11 (10.113); *Dion* 6 (10.115), 9.3 (10.111); *Gaius Gracchus* 7.1–2 (11.8), 11.1 (8.17); *Julius Caesar* 58.4–5 (8.82); *Lycurgus* 12.5–6 (12.16); *Lysander* 17.2 (11.118); *Marcellus* 14.9–17.3 (12.31), 14.7–9 (2.37), 17.3–4 (14.15); *Marius* 4.2 (12.18), 9,1, 13.1 (13.6), 21.3 (3.24), 25.1–2 (13.6); *Moralia* 1.2.619a (6.37), 4.6.672b (4.20), 8.2.718e (14.15), 8.8.730e (1.10), 36.73c-d (6.24), 636c (7.6), 674a (7.8), 843d (5.11), 20.588f (10.73), 21.974e (9.27), 915a (3.76, 11.108), 950b-c (3.77); *Nicias* 3.5 (11.20); *Numa* 17.1–2 (11.131); *Pelopidas* 21.2 (10.120); *Pericles* 12.5–7 (8.49), 13.5 (8.28), 27.2 (12.14); *Solon* 21.4 (2.24), 22.1 (11.132), 24 (10.114), 25 (12.42); *Sulla* 3.4 (12.33); *Tiberius Gracchus* 10.5–6 (12.33); *Timoleon* 31.3 (12.14).

Pollux, Greek scholar and grammarian, second century: *Lexicon* 7.108 (6.28), 10,124-125 (10.53).

Polybius, Greek historian, c.200–115 BC: *Histories* 1.20.9–21.3 (11.73), 1.22.3–10 (13.20), 3.4.10–11 (1.28), 6.22–23 (13.5), 6.25.10–11 (14.20), 6.27–28, 30–31 (13.12), 6.53 (7.15), 9.41.1–10 (12.30), 10.28 (9.4), 10.43–47 (13.14), 12.13.11 (2.51), 15.9, 11–12, 16 (13.18), 16.3.2–4.14 (11.94), 21.7 (13.22), 34.9.10–11 (6.4).

Procopius of Caesarea, Greek historian, c.500–?: *History of the Wars* 5.14.6–11 (11.10), 5.19.8–9 (2.14), 5.19.19–22 (2.15), 8.11.27–31 (12.24, 13.20), 8.17 (10.29); *On Buildings* 1.11.18–20 (11.100); *Gothic War* 8.11.27–28 (14.29).

Psalms, a collection of Hebrew poems composed from the tenth to the sixth centuries BC: 12:6 (6.6).

INDEXES

Ptolemy, astronomer, mathematician, and geographer, *c.*135: *Geography* 1.20.1–22.6 (11.55).
Quintilian, Roman rhetorician, *c.*30–100: *On the Training of an Orator* 1.1.12, 15–16, 24–25, 27, 30–31 (12.27), 2.13.16 (11.16), 7 Preface 2 (7.9), 10.3.31–32 (12.28).
Sallust, Roman historian and politician, 86–*c.*35 BC: *Catiline* 31.9 (8.65).
Scholiast on Demosthenes: *Against Androtion 13* (7.5).
Seneca, Roman statesman and philosopher, *c.*4 BC–AD 65: *Letters* 57.1–2 (11.13), 70.20 (9.47), 71.15 (1.31), 77.1–3 (11.75), 86.8–12 (9.40), 88.21–23 (14.12), 90.10–13 (14.22), 90.20 (10.57), 90.25 (8.69), 90.32 (8.3), 115.8–9 (5.35); *On Providence* 4.9 (8.66); *On Tranquillity of Mind* 9.4–7 (12.45), 9.5 (12.47); *Pumpkinification of Claudius* 2.2 (12.7); *Speculations about Nature* 1.3.2 (10.35), 3.24.2–3 (9.44), 4B.13.3 (2.32), 7.25.4–5 (14.27).
Servius, Latin grammarian and commentator, fourth century: *On the Aeneid* 7.14 (10.55).
Sophocles, Greek tragedian, *c.*496–406 BC: *Antigone* 332–372 (1.22); *Nauplius* Fragment 432 (1.23).
Statius, Roman poet, *c.*45–96: *Silvae* 1.5.45–46 (8.68), 1.5.57–59 (8.68), 4.3.20–55 (11.12); *Thebaid* 6.880–85 (5.22).
Strabo, Graeco-Roman historian and geographer, *c.*64 BC–AD 21: *Geography* 1.1.20 (11.53), 2.5.9 (11.54), 2.5.10 (11.56), 2.5.17 (11.60), 3.2.8 (6.18), 3.2.9 (5.26), 4.4.1 (11.69), 4.6.12 (5.27), 5.1.12 (10.2, 10.135), 5.2.5 (5.34), 5.2.6 (5.7), 5.3.6 (11.112), 5.3.8 (9.16, 11.5), 5.4.7 (11.13), 7 Fragment 34 (6.2), 8.2.1 (11.14), 8.3.33 (11.123), 8.6.16 (11.115), 8.6.20 (11.95), 9.1.23 (5.33, 6.15), 9.2.18 (9.11), 9.2.40 (9.11), 11.2.19 (5.26), 11.5.6 (11.43), 12.3.30 (2.11), 12.3.40 (5.23), 12.8.14 (5.35), 13.1.54 (12.46), 13.4.14 (10.42), 14.1.5 (8.27), 14.1.24 (11.103), 14.1.37 (11.46), 14.2.23 (11.4), 14.6.5 (6.17), 16.1.5 (11.122), 16.1.9–10 (9.8), 16.1.15 (2.30), 16.1.27 (11.30), 16.2.13 (9.6), 16.2.23 (10.45), 16.2.25 (10.80), 17.1.6–10 (11.97), 17.1.30 (9.37), 17.1.48 (9.7).
Suetonius, Roman biographer, *c.*69–160: *Caesar* 44.1 and 3 (8.82), 56.6 (12.37); *Augustus* 18 (10.124), 43.5 (8.80), 49.3 (11.27), 50 (12.33), 80 (10.110 and 10.116), 85 (12.32), 88 (12.37); *Tiberius* 40 (8.81); *Caius Caligula* 20 (12.32), 39.1 (2.18), 53.2 (8.78); *Claudius* 18 (11.83), 20.1–2 (9.12), 20.3 (11.102), 25.2 (11.50), 32 (9.12); *Nero* 17 (12.34), 20 (10.115), 38.1 (8.65); *Vespasian* 18 (13.23); *Domitian* 14.4 (8.74).
Sulpicius Severus, Latin historian, *c.*360–420: *Dialogues* 1.13 (9.30).
Supplementum Epigraphicum Graecum (SEG), yearly updates on Greek epigraphical scholarship, 19 [1963] 684 (11.103).
Sylloge Inscriptionum Graecarum*[3] *(SIG), a collection of Greek inscriptions from various periods: 86 (11.25).
Synesius, Christian neoplatonist, pupil of Hypatia, bishop of Ptolemais and writer of various texts, *c.*370–*c.*413: *Letter* 15 (9.2).
Tacitus, Roman historian, *c.*56–117: *Agricola* 21 (1.9); *Annales* 1.79 (8.83), 3.31 (11.7), 11.13.3–14.5 (12.23), 11.20 (5.16), 12.24 (8.16), 13.53 (11.111), 15.40 (8.65), 15.43 (8.76), 16.6 (10.122); *Histories* 5.6 (5.30).
Tertullian, Christian writer, *c.*160–240: *Against the Valentinians* 7 (8.62); *On the Soul* 10.4 (10.119), 30.3 (1.9).
Theaetetus Scholasticus, Greek epigrammatist, third century BC: *Greek Anthology* 16.221 (5.36).
Theocritus, Greek bucolic poet, *c.*300–260 BC: *Idylls* 15.29–32 (10.91), 21.6–14 (3.72).
Theophanes, Byzantine Greek historian, *c.*760–818: *Chronographia* 354.14 (13.22).
Theophrastus, Greek philosopher, *c.*370–285 BC: *Concerning Fire* 1–2 (2.21), 5 (2.1),

741

21–22, 57 (2.31), 24 (5.24), 28–29, 37 (2.28), 59 (2.25), 63 (2.22), 70 (5.25) 73 (2.22); *Concerning Odours* 14–19 (10.102), 21–22 (10.103), 25 (10.104), 40 (10.105); *Enquiry into Plants* 5.2.1 (10.11), 5.5.1 (10.13), 5.5.6 (10.13), 5.7.1–3 (11.66), 5.9.1–4, 6 (2.27), 5.9.6–7 (2.22), 9.3.1 (10.88); *On Stones* 16 (2.29), 41–43 (5.39), 45–47 (6.41), 49 (5.5, 6.32), 51–52 (5.5), 56 (10.97), 57 (10.98), 60 (6.9), 63–64 (5.19), 65–66 (8.12).

Thucydides, Greek historian, *c.*455–400 BC: *The Peloponnesian War* 1.4 (11.64), 1.10.2 (8.84), 1.107.1 (13.26), 1.108.3 (13.26), 2.13.7 (13.26), 2.75–76 (12.29), 3.15.1 (11.14), 4.100 (13.31).

Tyrtaeus, Greek elegiac poet, *flor. c.*684 BC: Fragment 8 (13.3).

Ulpian, Roman jurist, mid-second century, quoted extensively in *Digest* 8.2.17 (2.3), 9.2.27.29 (10.87), 33.7.12.16–18 (8.64).

Varro, Roman agricultural writer, 116–27 BC: *On Agriculture* 1.2.22–23 (5.43), 1.17–22 (3.10), 1.50 (3.33), 1.51 (3.34), 1.52 (3.35), 1.54.2 (3.46), 1.54.3 (4.10), 1.55 (3.50), 1.55.4–6 (4.17), 1.57 (3.36), 1.59.2 (4.25), 1.63 (3.36), 2.1.11–24 (3.55), 2.2.18 (10.33), 2.3.6–10 (3.60), 2.4.7–8 (3.61), 2.4.11–12 (3.61), 2.4.22 (3.68), 2.6.2–3 (3.57), 2.7.18 (3.60), 2.10.10–11 (3.59), 2.11.1 (3.66), 2.11.5–9 (3.66), 3.4.2–5.6 (3.65), 3.9.1–7 (3.62), 3.9.11–12 (3.62), 3.16.12–17 (3.66), 3.17.2–4 (3.67); *On the Latin Language* 5.140 (31) (11.31), 5.143 (8.16).

Vegetius, Roman military historian, *c.*383–450: *Military Science* 2.11 (10.137), 2.25 (13.38), 3.6 (11.59), 3.7 (11.22), 3.10 (14.9), 4.34 (11.66).

Vergil, Roman poet, 70–19 BC: *Aeneid* 6.847–53 (14.9); *Georgics* 1.71–83 (3.26), 1.121–46 (1.3), 1.160–75 (3.29), 1.303–4 (11.95), 2.226–58 (3.17), 3.156–65 (3.56), 4.228–35 (3.66), 4.484 (2.8); *Moretum* 16–29 (4.2), 39–51 (4.2).

Vitruvius, Roman architect and military engineer, *c.*50–26 BC: *On Architecture* 1.2–10 (8.37), 1.2.8 (8.15), 1.5 (13.27), 2.1.1–2 (1.16), 2.1.2–3 (8.1), 2.1.7 (8.1), 2.3.4 (8.8), 2.4.1–5.1 (8.13), 2.6.1 (8.14), 2.7.5 (5.41), 2.8.1–2 (8.22), 2.8.4 (7.22), 2.8.7 (8.11), 2.8.16–17 (8.61), 2.8.18–19 (8.9), 2.8.20 (8.10), 3.3.11–13 (8.42), 3.4.1–2 (8.43), 3.4.5 (7.42), 4.1.9–10 (8.41), 4.3.3–6 (8.40), 5.10 (9.39), 5.12.1–7 (11.100), 6 Preface 5–6 (14.11), 6.1.2, 12 (8.20), 6.3.1–2 (8.60), 6.4.1 (12.45), 7 Preface 4 (12.47), 7 Preface 16 (8.29), 7.2.1–2 (8.23), 7.3.1–2 (8.31), 7.3.3–6 (8.24), 7.8.1–4 (6.7), 7.9.4 (10.138), 7.9.5 (6.40), 7.11.1 (10.139), 8.1.1–2, 4–6 (9.2), 8.5.1 (8.34), 8.6.1–11 (9.14), 8.6.12–15 (9.5), 9.7.1 (12.8), 9.8.1 (12.8), 9.8.2–4 (14.18), 9.8.2–7 (12.11), 9.8.8–10 (12.12), 10.1.1–4 (2.34), 10.1.4–6 (1.14), 10.2.10 (8.51, 11.81), 10.2.11–12 (11.34), 10.3.5 (11.74), 10.4.1–4, 5.1 (9.28), 10.5.1–2 (2.13), 10.6.1–4 (9.31), 10.7.1–4 (2.47), 10.8.1–6 (2.46), 10.9.1–4 (11.37), 10.9.5–7 (11.88), 10.13.1–2 (13.28).

Xenophanes, Greek poet and philosopher, *c.*570–478 BC: Fragment 18 (1.20).

Xenophon, Greek historian, *c.*428–354 BC: *Anabasis* 1.5.5 (5.42), 1.5.10 (11.70), 7.12–14 (11.92); *Art of Horsemanship* 7.1–2 (11.41), 12.1–12 (13.16); *Cyropaedia* 6.2.33–34 (11.30), 6.2.36 (11.11), 8.2.5 (10.127), 8.6.17–18 (11.26); *Estate Management* 4.2–3 (14.6); *Memorabilia* 1.1.7–8 (1.21), 2.7.5–6 (10.133), 2.7.12 (10.133), 2.7.14 (10.133), 3.6.12 (6.18), 3.8.8–10 (2.2); *On Hunting* 1.1, 18 (3.69), 6.5–8 (3.69), 11.1 (3.69); *Ways and Means* 3.1–3 (101.126), 3.12–13 (111.126), 4.2–3 (5.9), 4.14–17 (5.14).

INDEXES

B SUBJECT INDEX

This is a selective index of proper names, substances, procedures, and topics. Names of individuals and geographical sites have been included here only if especially significant. Readers interested in a particular subject can find relevant passages easily by examining the Table of Contents. References are to page numbers; **bold** type indicates a detailed or especially significant description

abrasives 214, 389, 402, 435
adjutages **353–356**, 358; *see also quinaria*
Aegina 490, 574, 594
Aeneas Tacticus 442, 628, 679
aeölipile **39–40**
Aesculapius **451–452**
Ages of Humankind **13–14**
agriculture: agricultural workers 101–102; animal husbandry 100–101, **131–144**; arboriculture **123–131**; calendars **105**; *vs.* commerce 607; crops and cultivation **114–117**; early Roman 90; economics **101**, 106, 121, 124; equipment 101; farms and estates **89–93**; the farmstead 93–105; the harvest 117–122; horticulture 122; origins 15, 22, 27; status of **712**, 716; tools **96–98**
Agrippa 281, 285, 323, **350–351**, 354, 359, **545**, 685
alcohol **172–173**; *see also* beer; mead; wine
Alexander the Great **59**, 243, 278, 309, **328**, 342, 345, **460–461**, **483–484**, 535, 543, 568, 576, 648
Alexandria 37, **40**, 75–76, **278**, **363**, **425**, **477–478**, **492**, 497, 532–533, 566, 569, **576–578**, 584–585, **601–602**, 605, 644, 658–659
alloys 189, 221, 231, **241–243**
alphabets 638, **640–641**
alum 188, 204, 340; uses of 416
amalgam 227, 229
amphitheatre 285, 286, **322**, 327
amurca 113, 119–120, 165, 396, 428
anaesthesia 460, 462–463
Anaximander 540, 623
anchors 520, 523, 553, 560, 562, 571–572
animal husbandry 100–101, **131–144**
animal power *see* energy, animal
animals: classification 132; honouring of 309; wild 27, **145–147**
annona 607
antimony **230**
Antioch 323

apiaries **142–143**
Apollodorus 295–296, 522
Apollonius 555, 607
Appian Way 173, **514**
apprenticeship **491**
aqueducts 43–44, 202, **345–361**, 510; *see also* pipelines and aqueducts; siphon; tunnels
Arachne 406, 423, 626
arboriculture **123–131**
archaeology, as historical source 7–8
arches 268
Archimedean screw *see* screw, water
Archimedes 61, 63, 70, 561–562, **699–700**, 714, 720
architects 276, **292–296**; status of **716–717**
Archytas 75
Argo 556
Aristotle 658
armies: Greek **666–668**; Roman **668–678**, **701–703**
arms and armour 239–240, **665–672**
asbestos 58, 405, 408, **410**
asem 247
Aspendos 607
assaying **189**, **244–248**
assembly lines *see* specialisation
asses *see* donkeys
Athens 19–20, 89, 164, **191–192**, 211, 214, **298**, 307–308, 322, 330, **434**, 567, **595–596**, **599–602**, 607, **631–635**, 637, **658**, **689**
Attalus I Soter 574
Attalus II Philadelphus 585
auger 126, 403, 561
Augustus Caesar *see* Caesar
Aurelian 689
automata 61, **74–82**
aviary **141**
awnings 321–322

Babylon/Babylonia 56–57, 108–109, 554–555, 598–599
Baiae 36, 276, 381, 515

743

bakers **164–165**, 600, 611
balance **63–64**
Balearic Islands 152
ballistae 293, **684**, 692, **697–698**, **702**
ballots **630–636**
banausia **710–715**, **717–718**, 720, 726
Barbegal 43
barbers **454**
barrels **404–405**
barter **592–593**, 601
barylkos 65–66
baths 36, 56, 488, 561, 612; Roman 323, **376–381**; *see also* hygiene; hypocausts
battles: land **667–681**, **694–702**; naval **573–575**, 684–685
beer 159, **172–173**, 610
bees and bee-keeping **142–145**
beeswax 141, **143**, 145, **442**, 555
Belisarius 44
Bellerophon 639, 665
bellows 55, 74, 229–230, **240–241**
Bench of Hippocrates 451, **474–475**
birdlime 15, **442**
bitumen 57, 113, 169, **207–208**, 248, 396
blacksmiths *see* smiths
boars 139, 145–146
bodies: cremation, inhumation vs **485–487**; dead, preservation **480–485**
bones, setting 66–68, **474–475**
books 643, **653–660**
boustrophedon 660n2
bow drill 392, 403
branding 133
brass **241–243**, 390
bread **159–165**
breakwaters 559, **576–585**, 586
breeding: cattle **133–134**; goats **137–138**; mules **135–136**; pigs **138–139**; sheep **136–137**
bricks 267; fired 288, 433, **495–496**, 640; unfired **270–272**, **315**
bridges 243, 346, 508, 511–513, 518, **520–525**
Britain 190, 200, 231, 544, 555
bronze 13, 190, 233, 239–240, **242**, 245, 262, 394, 396, 592; casting and recycling **262–263**; Corinthian **241**; sculpting 254; statues **256–257**
Bronze Age engineering 269, **344–345**
bucket chains, for water-lifting 363, **365–369**
buildings, origins **266–267**; *see also* construction

burning-glass **36**, 51

Cadmus 640–641
Caesar, Augustus 152, 318, 325, 582; sealstones of **648**
Caesar, Julius 76, **328–329**, 454, 517, **521**, **539**, 545, 555, **623–625**
Caesarea Palestinae **582**
caissons **579–582**
calendars: agricultural **104–105**, 122, 131; origin **622–625**
Caligula 47, 326–327, **345**, **566**, **647**; *see also* Gaius Caesar
camp, military 357, **675–676**
canals 109, 152, **342**, 345, 517, **519**, 566, 568, **588–591**; *cf. diolkos*
Cappadocia 120
Capua 98, 231
cargo 510, 554, 558, 562, **565–566**, 584, 593, 602–604
carroballista **703**
Carthage/Carthaginians 278, 557, 578–579, 668, 682–683, 688, 691
casting: bronze 239, **261–263**; economics **262**; iron 236, **239**; plaster and wax 264
catamaran 563–564
cataphracts **682**
catapults 60–61, 319, 562, 690, **692**, 694, 697–698, 702, 723–724; *see also carroballista; cheiroballistra*; onager
catheter **470**
Cato 93, 98–101, 110, 121–122, 125, 309
cattle, training and treatment **133–135**
cavalry **681–682**
Celts 553
ceramics **431–435**, 441
cerd 362–363
chalk 112, 130, 271, 278, 284, **289**, **382**, 391, 403, 415, **455**, **538**, 646
chalk line **289**, 403, 538
charcoal **53–55**, 107, 131, 163, 230–231, 233, 238, 281, 299, 312, 441, 446, 580, 650
chariots 488, **509–510**, 516, 529, 536, 599
Chauci 56
cheese 159, **174–175**
cheiroballistra **693**
chemical warfare **695**
chemistry, applied **440–450**
China 22, 409
chorobates **290**
Christianity **727–728**
ciphers 649, **651–653**

INDEXES

Circus Maximus 301–302
cisterns **340–341**, 351, 353
cities, fire-fighting **318–319**; *see also* construction; fire
civilisation: the rise of **31–32**; written language and 638
clamps, in construction 283, 308, 313, 323
Claudius 345, 566–567, 583–584, 640–641
clay **217**: construction 119, 130, 142–143, 167, 313, 360, 377, 428, 696; mines **198**; vessels 112, 163, 177–178, 337, 388, 431–432, 434, 442, 696
cleansers **442**
Cleopatra 444
clepsydra 621, **626–629**
climata 293, 543
climate: and agriculture 93, 124, 142; and construction 276, 280; and technological development **23**, 280
clocks **622–630**; anaphoric **629–630**; water **623–624**, **626–628**, 630, 723; *see also* sundials
clothes/clothing 23, 26, 426, **429**, 491–493
Clytemnestra 678
coal **56**, 419
cogwheel 42–43, 569–570
coinage 9, **241–242**, **591–597**; debasement 597; Roman **596–597**; Spartan iron spits 596; standardisation **596**
collegia **606**
colonists 91
Colossus at Rhodes 263
Columella 94
communication: military **678–681**; secure **649–653**; writing and **638–660**
compost **110–111**; *see also* manure
competitive spirit as motivator **31**, **295–296**, 347, 644, **657–659**; *see also* rivalry, as driver of technological progress
concrete **274–276**, **579–582**
Constantinople 56, 301, 581
construction: building codes, fire and flood protection **324–326**; city planning **277–285**; contracts 307, 314; costs **307–309**; domestic **314**, 318; earthquake and 263, 288, **299–300**; economics **307–309**, 313; high-rise living **317–318**; lime kiln **312**; machines **305–307**; metal in **322–323**; origins **267–270**; site selection **276**; structural disasters **326–327**; techniques **302–303**; tools **289–292**, **303**; walls and wall-facings **282–285**

cooking utensils **179–180**
Copaic basin drainage **344–345**
copper 189, 192, 194–195, 199, **231**, 233, **244–245**
corbelling **269**
Corinth 517, 575–576
Corinthian order **297–298**
cork 149, **399**
corvus **684–685**
Cos 409
cosmetics **447–450**
cothon 578
cotton **408**
craftsmen, attitudes towards 30–31, **709–715**, **727–729**
crampons **536**
cranes 210, 301, **305–307**, 565–566
crank **63**, 371
Crassus 197, 318
cremation, inhumation vs **485–487**
Crete 121, 255
Cronus 13
crop rotation **111–113**
Ctesibius 73–74, 371–372, 628, **723**
cubits 598–599; *see* weights and measures
cultivation 99–100, **114–117**; labour required for **114–115**
cupellation **225–226**
Cyclopes 14, 174, 238–239, 269
Cyrenaica 153

Dacia 196
Daedalus 75, 136, **255**, **295**
Dalmatia 201
Danube 520, 522
Darius 300, 394, 421, 518, 525, 588, 594
Dead Sea 207–208
defrutum 95, **169–170**, 176
Delphi 195, 392, 518
Delphic knife 393
Demetrius I of Macedon 697
Demetrius of Phaleron 75
demiourgos 709
dentures **457**; *see also* prosthetics
Diana, Temple of, at Ephesus **299**, **530**
Didyma 285
Diocletian 604, 609
diolkos **517**, 575
Dionysius I of Syracuse 723–724
Dionysius II of Syracuse 324
dioptra **290**
diploë 464
dissection, and vivisection **476–479**

distillation 173
diving **586–588**; as military tactic **680–681**, **701**; for sponges **149–152**, **586–587**
division of labour *see* specialisation
domes 286, **288**, 295, 561
Domitian 324, 515, 600–601
domus 314
donkeys **135**, **160–161**, 612
doorbell **78–79**
Doric order **296**, 298
drainage **107**, 187, 210, **342–346**, 351–352, 371
Druids 641
dry-dock **585**
ducks **141**
dunnage 551, 554
dyes and dyeing **416–427**, 444, 446, 458

earthquake *see* construction, earthquake and
Echinus 697
ecology, environmental damage **152–153**, **233**, 235, **397**, 418; *see* hushing, pollution
education: defects of **719–720**; process of 146, 295, **491–492**, 643, 717–719
eggs 53, **139–140**
Egypt **108–109**, 162, 173, **197–198**, 213, **225**, **268–269**, 278, 301–302, **342**, 368, 371, **375**, **408**, 416, **422**, 435, **437**, **478**, 523, 566, **568**, **588–589**, **597–598**, 638–639, **644**
Elba 190, 236
election, sortition vs **630**
electrum 247, 594
elephants, in warfare **682–683**
embroidery **425**
emmer 163, 165
encaustic painting 556, 563
energy: animal **46–47**; geothermal **36–37**; human **46–49**; hydraulic **107**; magnetic 37; solar **34–36**; steam power **39–40**; wind **37–38**
engineering: disasters 272, 317, **326–327**; grand projects **328–329**; mechanical **60–81**
envy, as driver of technological progress **31**; *see* competitive spirit as motivator
Ephesus 299, 530, 584
Erechtheum 262, **308–309**
esparto grass **410**
Etruscans 641

Euphrates River 108–109, 342
experimentation 221, **719–727**

fallowing **112–113**
Fates, the Three **421**
feathers **140**
fermentation 163, **168–169**, 172
fertiliser *see* compost; manure
fertility *see* soil, fertility
fire: defending against **687–688**; extinguishing **53**, **373**, **442**; fighting **318–319**; making 36, **50–51**; nature of **49–50**; origins of 15–16, 18–19, **24**; preserving **52**; prevention, in construction **324–326**
fire, uses of **53–54**; in metallurgy **222–223**, 229, **245–248**; in mining 201, 204–205; in warfare **678–680**, **685–688**, **693–695**
fire drills **50–51**, 55
fire signals **678–680**
fishing 27, 145, **148–149**
flax 101, 111, **414–415**, 426, 559, 569–570
flaying 427, 451, **477–480**
flight/flying **75**
flood 342, 352; destruction of humankind by **16–17**; Egypt and Mesopotamia **108–110**, 342; protection from in construction **324–326**, 329, 344–345
flooring 134, 178, **281**, **303–305**, **311**, 317, **320–321**, 367, **376–377**, 438, 561, 585
flour **162–164**; *see also* grain; mills
food, prices 607–613
food preparation: origins of 29–30; specialisation 489
forgery: of coins **597**; foiling **647–649**
formwork **579–581**, 583
fortifications **688–701**, 702
forum 277, 600–601
foundations, in construction 216, 267, 272, 294, **299–300**, 310, 312, 366–367, 550, **579–584**, 689
fountains 347, 350–351
frankincense 497, **605–606**
fruit trees and fruit 105–106, 123–124, 128–129, 131, 152, 177, 611
Fucine Lake 345–346
fuel **49–57**
fulling **413–414**
furnaces, smelting 54–55, **225–229**, 232–233, 236, **240**
furniture **401–402**, 430, **493–494**

Gades 569, 691

gadgets, mechanical **70–81**
Gaius Caesar 351, 517; *see also* Caligula
Galen 461
galena 188, 190, 232
gardens 108, **122**, **375–376**, 561
garum **176**
Gaul 116, 118, 121, 173, 205
gears 42, **65–66**, **70–71**, 569–570
geese **140–141**
gems, artificial **444–445**
geothermal energy *see* energy, geothermal
Germany 20–21, 197
glass **435–440**
glastum 417
globes **541–543**
glue/gluing **400**, 446, **474–475**, **476**, 480, 645
goats 110, 132–133, **136–138**, 146, 153, 174–175, 396, 610, 641
gods and humans **13**, **18–19**, **26–27**, **710**
gold: mines *see* mines and mining; native 18, 205–206, 223–224; refining 207, 223–224, **225–229**, 245, 248; solder 247; working 394
grafting 22, **125–126**, **129–130**
grain 13, 31, 101, 109, 111, 114, 120, **159–162**; grain ships **559–562**; granaries 120; mills **46–47**; speculative manipulation of market 607–608; storage 120–121; threshing and winnowing **119–120**
grapes 96–97, 100–101, **123–126**, 130, 177
grappling hook 562, **685**
Greek fire **686**
guilds 433, **606**
gypsum 122, **273**, 284

Hadrian 295–296, 452, 510, 513, 522, 544
Hagia Sophia **288**
ham 177
hand, as tool 16, **25–26**
Hannibal 188, 211, 515, 554, 682–683, 686
harbours **575–586**
hares **145–146**, 152
harvesting **117**, **119**, 128
haymaking 117, 120–122
health and well-being **451–487**
heating: passive solar **320–321**, 376; water 381
Hellespont 520
hemp 101, 405, **410**

hens 139–140
Hephaestus 19, 74, 239, 391
Hera, Temple of, at Samos 285, **301**, 576
herbs and spices, for preservation **480–484**, 486
Herod the Great 582
Hero of Alexandria 60, 61, 62, 68
Hierapolis 417
hieroglyphs **638–639**
Hieron II of Syracuse 560–562
Hippocrates 451
Hippodamus 277–278, **538**
hodometer **533–534**, 540
hoe 98
hoists, cranes and 61, **305–307**
honey 108, 140, **142–143**, 159, 168, 172, 175, 455–456, 475; as preservative **178–179**, 420, 484
hoplites **666–668**
horses 27, 509, 513, 520, **525–526**, 599, 612; mounting **536**; vehicles and **529–537**
horseshoe **536**
horticulture **122–123**
human energy *see* energy, human
hunting 15–16, 27, **145–147**
hushing **200–203**
hydraulic technology 5, 108, **344–346**, 583
hydrometer **338**
hygiene: personal 382, **452**; sponges **453–454**; strigil, oil and **452–453**; *see also* baths
hypocausts **319–321**, 376–377

ice **58–59**, 173–174, 272, 537
ignitabulum 52
incertum 282
India 14, 209, 541, 543, 568, 602–604, 683
indigo **419**
inhumation, cremation vs **485–487**
ink **446–447**
insulae 314, **315–317**, 325–326
insurance, for sea voyages **566–567**
inventiveness 200, 295, **723–725**
Ionic order 298
iron: in construction **322–323**; ore **190**, 593; production **235–239**, **391–392**; uses of 13, 15, 724–725; welding **244**, **392–393**
irrigation **107–109**, 122, **342–346**, **362–363**, **375–376**
iugerum 90

Jerusalem 69
Jews 173
Jocasta 261
Jotapata 701
Julius Caesar *see* Caesar, Julius
Jupiter *see* Zeus/Jupiter
Justinian 409, 581

kilns: for brick 266–267; for ceramics 388, **431–432**, 435; for lime 283, **312**, **495–496**; for metallurgy 225–226; for sculpture 253
kleroterion **631–633**

labour: attitudes towards *see banausia*; division of *see* specialisation; slave *see* slave labour
ladders 311, 318, 441, 571, **684–685**, 698; leather 671
lamps/lighting **52–53**, **57–58**, **80–81**, 160, 204, **323–324**, 340, **490**
landlord, absentee **102**
land transport, economics **537**; *see also* chariots; donkeys; horses; mules; oxen; postal services; roads; tunnels; vehicles
languages 25, 198, 638
large-scale production **487–497**
laserpicium 153; *see also silphium*
latifundia **91**, 121, 608
latrines, communal **382**; *see also* hygiene; toilets, Roman; urinals, Roman
Laurion 191–192, 196, 232
laws: coinage 595–597; mining and metallurgy **192–195**, **233**; public baths **380–381**; shipbuilding 556–557; urban transport 510–512, **539**; *see also* construction; wages and prices
lead **190**, **230–232**, 445, 450, 561; in pipes **348–349**
learning, human **25**; *see also* education, process of
leather 411, **426–430**, 612
leaven **162–163**
levers 23, **62–65**, 72–73, 79–80, 159, **165**, **167–168**, 373, 400, **558**, 572
libraries **656–660**
lighthouses **576–578**, 583–584
lime, slaked **283**
linen 322, **408**, 410–411, 414–415, 426
liquamen 176
literacy, disadvantages **641–642**
literary sources, for ancient technology **4–6**

litharge 248n2, 476
litter, for transport **529**
livestock 94–95
locks, canal 585, **589–591**
locusts 153
log, ship's **569–570**
looms 23, **423–423**, 424
lora 167, 170
Lycurgus 596
Lydia 594

Macedon/ia 149, 667–668
machines: basic **59–69**; in construction **305–307**; siege *see* sieges/siege machinery
magnetic energy *see* energy, magnetic
Mago 92, 159, 176
manual labour, attitudes towards *see banausia*
manure 95–96, 103, **110–113**, 129, 131; *see also* compost; fertiliser
maps/mapping **540–546**
marble and hard stone: moving 280; powdered 284; quarries **211–214**; tools **303–305**; trade **213**; veneers **214–215**, **284–285**; walls 284, **308–312**, 315; floors **303–305**
Marcellus 699–700
Marius, Gaius 111–112, 635–636, **670**
markets 98, **599–608**
masonry, stone **272–273**, 279
mead 172
meadows 92, 94, 103, 121–122, 152
measures, weights and 28, **597–600**
mechanician 60–61, 71
mechanics 23, **26–27**, **60–61**
medicine: anaesthesia 460, **462–463**; antidote to poison **463**; drug recipes **461**; origins **451**; physicians' equipment (materials, sizes, varieties) **463–476**; wound care **475–476**
Mediterranean Sea 547
meningophylax 473
merchantmen **549**, **558–562**, 605
mercury **227–229**
Mesopotamia **109**, 207, 342, 344
metallurgy: environmental impact **233–235**; fire, uses of in **222–223**, 229, **245–248**; origins of 19; recipes **246–248**
metalwork, divine instruction **394**
metalworking **390–396**
metrology, origins of **598**
mica 122

748

milk 133, 137, 140, 159, 174
mills 46–49, 96, 97, 98, 135, 159–161; rotary 160–161, 172; water 42–45
minerals 189
mines and mining 185–209; administration 191–197; clay 198; dangers 203–205; gold 195–199, 200–202, 207; hushing 202–203; laws 192–194, 232–233; moral consequences 187–188; placer deposits 205–206; prospecting 188–189; Roman soldiers 197; silver 54, 191, 196–197, 199–200; techniques 197–207; waterwheels 336
minium 228–229, 245–246, 495
Minos 255, 549
Mithridates 463
molae 47, 172
moles *see* breakwaters
monsoons 568
mordants and mordanting 416–417, 420–421
mortar and pestle 159
mortise and tenon 550, 551, 554–555
mosaic 281, 437, 561
Moselle 45, 148–149
mules 48, 135–136, 309, 519, 591, 612
mulsum 176
mummification 481–484
murex 418–419
Mycenae/Mycenaean 231, 269, 344, 411, 487–489, 509, 622–623, 665–666, 678
Mylasa 410, 510

naphtha 56–57
nature: human interference 152–153; as inspiration for technology 21–24, 26–27, 61–62
Naucratis 601
naval weapons 684–686
navies 548, 688
navigation 546–591
Nero 90, 92, 153, 318, 325–326, 455, 517
nets: for fishing 146, 149, 426; for hunting 146, 426
Nicaea 294
Nicomedia 590
Nile 108, 342, 371, 375, 566, 588–589, 601–602, 645
Nilometer 342
Numa 433, 606, 623–624
numbers, origin of 19, 28
nutcracker 64
nuts: and bolts 69; threaded 66–69

obelisks 301–302, 565–566, 584
occupations, classes and status 717–719
octopus 152
Odysseus 14, 123, 160, 174, 223–224, 238, 289, 392, 394, 429, 508, 550–551, 589, 630, 666
office, selecting for 630–637
olive oil 57–58, 95, 170–172, 414, 428, 608, 610
olives and olive orchards 96–97, 99–101, 127–130
onager 135, 692–694
optical refinements, in construction 298–299
oral tradition vs writing 641–642
orchards 96, 123, 128–129
ore: iron 235–236; milling 226; washing 227
origin of technologies *see* sources of technological innovation
organ, water *see* water, organ
Ostia 316, 345, 566, 569, 578, 583–584
ostracism 637
oxen 23, 46, 90, 94, 116, 362, 530, 537, 539, 598, 612; training 133–134

Palamedes 28, 640–642, 665
palm trees 130–131
Pantheon 287–288, 322
papyrus 130, 492, 552, 641, 644–645
parchment 644–645, 654–656
Parthenon 298–299, 302–303, 307–309
Parthia 163
passum 170, 176
pâté 140
pearls 445–446
peat 56
pecunia 90, 132, 596
pens *see* stylus
pentekonters 557, 568
pepper 209
perfumes 447–450
Pergamum 645, 656, 658
Perinthus 695, 697
Persia 509, 514, 525, 594, 658
personal appearance, skin care 455–456
personal improvement, body and mind 456–457
pesticides 113; *see also amurca*; threshing and threshing floors
petroleum 56–57, 207–208
Phaeacia 123, 575

Pharos **576–577**, 584
Pheidon 594, 599
Philip II of Macedon 667, 695
Philip V of Macedon 574, 697
Philon 309
Phoenicians 549, 568, 578, 585, 589, 601, 639–641
pigments **445–447**, 495
pigs **138–139**
pilum 90
pipelines and aqueducts **346–361**; *see also* aqueducts; siphon; tunnels
pipes **353–357**; lead **348–349**; terracotta 177–178, 253, 257–258, 262, **348–350**, 496–497
Piraeus 309–310, 559, 602, 689
pirates 548, 568–569
pisé 270
pistons 73, **371–373**, 374
pitch **103**, 105, **112**, 126, 130, **137**, **168–169**, 177–178, 208, 242–243, 370, **404–405**, 435, **441**, 446, 476, 555, **557**, **686–687**, 694
pithoi 434
pitons 671
placer deposits **205–206**
plagiarism 660
planetarium 70–71
plaster of Paris 273–274
Plataea 695–697
plaumoratum 116
ploughs and ploughing 14–15, 23, 27, 96–98, **114–117**
plumb bob 291–292
pneumatic pressure 73, **365**, **371–373**
poison 462–463, 648, **686**, **695**, 728; antidote recipes **463**
pollution, air **234**, 418, **427**
pomerium 277
Pompeii 518
Pons Sublicius 524
porridge **600**
Portugal 192–195
postal services **525–528**, 537
potter's wheel 57, **431**
pottery **431–435**
poultry **139–141**
pozzolana **275–276**, 579, 582, 584
preservation: dead bodies **480–485**; food **168–171**, **177–179**; trophies and symbols **479–480**
presses 61, 66, 96–97, 101, **165–168**, **171–172**, 176; *see also* olive oil; wine

processions 75–76, 264, 522, 532–533, 539
projectile extraction *see* Spoon of Diocles
Prometheus 12, **18–19**, 31, **52**
prospecting 185, **188–189**
prosthetics **457–460**
protective coating/finish 103, 113, **178**, **272**, 275, **395–396**, 400, **435**, 441
pruning **126–127**
Ptolemy II Philadelphus 37, 75, 566, 576–577, 589, 659
Ptolemy III Euergetes 659
Ptolemy IV Philopator 563–564, 585
public speaking 456
pulleys 59, **62–63**, 77–79, 81, **305–307**, **367–368**, 562, **566**
pumice 215, 654
pumps 34, 71, **73–74**, 361, 369, **371–376**, 586, 723
Puteoli 230, 275–276, 313–314, 516, 559, 566, 569
pyramids **268–269**, 352

qanats **339**
quarrying 22, **209–217**; moralistic evaluation 211
quenching, iron **238**
quicklime **273–274**, 414
quinaria **354**, **356–357**
quinqueremes 557, 574, 688, 724; *see also* triremes

rafts 552, **554–555**, 566
rams: in sieges **690–692**, 695, 699, 701–702; on warships 563, **573–575**, 684
raspatories 464
reasoning 28
reclamation, land **344–345**
Red Sea 22, 568, 588–589
reeds and rushes 401
refining, ore **224–233**
refrigeration **58–59**
repair and recycling **50–51**, 53, **112**, **232–233**, 262, **344–345**, **358–361**, 411, 437, 494–495, **513**, 518, **523–524**, **528–529**, 585–586, **691**; for writing materials **645–647**; *see also* wax tablets
reservoirs, urban **348–349**, 351–360, 364
reticulatum 273, **282–283**; *see also incertum*
revetment **44–45**, **214–215**; *see also* veneers

Rhine 521
Rhodes 697
rice **165**
rivalry, as driver of technological progress **31**; *see* competitive spirit as motivator
riverboat **564**
roads 28–29, 93, 107, 268, 279, 313, 351, 421, 430, 508, 509–519, 525, 528, 530, 534–535, 538–539, 545–546, 555, 590, 601, 690; roadside services 528–529
robots **74–81**
Rome 43–44, 46, **164**, 216, **270–274**, 285, **286–287**, **295–296**, **322–323**, 330, **350–352**, 358–360, **524**, **583–584**, 600, **689**, **715**; domestic construction **314–319**
Romulus 90
roofing **285–288**
rope 430
rotary mill *see* mills, rotary
Royal Road (Persia) **509**
rust, iron **243**

Sabiri 691–692
sailing season 546–548
sails 19, 551, 553, 558–559, **570–572**
Salamis 525, 574
salt 52, 102, 134, 143, 152, 169, **171–173**, 175, **177**, 190, **208–209**, 225, 230–231, 247–248, 341, 364, 419, 428, 430, 437, 449, 455, 565, 610
salvage divers 586
Samos 198, 301, 346, 576
sand, in construction 274–275
saqiya **369**
Sardis 406, 509, 594
Sasema 217
sauces **175–177**
saws **98**, **215–216**, 400, 403; water-powered **44–45**, **214–215**
scalping **426**
Scipio, Publius 682
screw 61–62, **66–69**, 168, **560–561**; nut and 66–68; water 200, **369–371**, 374, 562, 580
sculpturae 252
sculpture: masters 261; origins **255–259**; process 259, 261; *chryselephantine* 258; terracotta 257–258; *sphyrelaton* 256–257; wooden 256; *see also* statues
scythes 96, **121**
Scythia 410, 426
sea-fire 686

seals and seal-stones **647–649**
secret ballot *see* voting
security 322, **324**, 454, **463**, **497**, **647–653**, **688–689**
seeds and sowing 14–15, 94, 97, 103, **117**
Semiramis 683
Servius Tullius 689
sewers **270**, 330, **350–351**, 510
shadufs **109**, 122, 208, **361–362**, 376
sheep 110, **136–137**, 174, 407, 412, 417, 641
shields 394, 493–494, **666–667**, 724
shingles 399
ship-building **549–556**; Celtic tradition 553–554; economics 552–553
ships: grain ships **559–563**; launching 585–586; logs **569–570**; rigging **558–559**; warships 46, 550, 555, 556, **557–558**, **563–564**, 568, **573–576**, 578, **684–688**
shipwrecks 548, **573**, 604
shipwrights 14, 550, **552–553**, 557, 560, 611–612
shipyards **556**, 578–581, 585
shock absorbers 535–536
shoes 489–490, 612
sickles 15, 96–98, 109, **117–118**, 236, 672
Sidon 578
sieges/siege machinery 665, 668, 675, 687, 690, **691–703**
sieves 160, 163
sign-language 608
silk **408–409**, 613
silphium **153**; *see also laserpicium*
silver: coinage 594, 596–597, 599; mines *see* mines and mining, silver; ore **188**, 191, 205, **224–225**; polish 395; refining 225–226, 232–233, 245, 246–248; working 395, 490
siphon 77, 81, **349**, 373
skin care **455–456**
slave labour 47, 48–49, 588–589, **711–712**; in agriculture **91–93**, **99–100**, 102–104, 375; in construction **268**; vs free labour **103–104**, 196; in industry 161; in mining **196–200**, 204, **225**; in quarrying 214
sleeping potions **461–462**
smelters and smelting 54–55, 193, **225–229**, **232–233**, 236, 240
smiths 54, 56, **238–240**, **391–392**, 611
Smyrna 538
snorkels 586

751

snowshoes 536
soap **443**; cf. 452–453
soapwort 417
socialisation 18, 25, 29
Socrates **27**, **29–30**, 36, **229**, 234, 431–432, 462, 491–492, **642–643**, **712**, 719
soda 437
Sogdian Rock 671
soil: fertility **105–113**; types **105–107**
solar energy *see* energy, solar
solder **243–244**
Solon 607, 695
sophists 727
sortition, as method of appointment to office **630**
sounding-weight **570**, 572
sources of technological innovation: deities and heroes **13–15**, **18–19**, **27–31**, 39, 42, 52, 71, 74, 146, 224, 289, 394, 406–407, **451–452**, **640–641**, 665; nature **15–18**, **21–24**, **26–27**, **31–34**, 35, 42, 50–51, **53–54**, **61–62**, 73, 107, 129, 140, 149, 174, **222–223**, 243, 266–267, 280, **297–298**, 350–351, 381, 402, 407, 410, 523, 586, 589, 593, **597**, 604–605, 677, 710, 722; humans **15–21**, **25–30**, **37–41**, 44, 60–61, 63, 65, 70–71, **73–82**, 136, 255, 267–268, **295–296**, 321, 354, 406, 425, 434, **474–476**, 540, 594, **597–599**, 623, 628–630, **639–642**, 649–655, 664–665, 671, 685, 689
Spain 120, 188, **199–202**, 205, 224, 228, **233–234**, 236, 336, 380, 569
Sparta 239, 330, 509, 666
Spartans 479, **484**, 517, **596**, **634**, 648, **651–652**, **695–696**, **710–711**
specialisation 29–30, 487–488, **489–497**
specula **470–471**
speech 18, **25**, **27–28**
speed, of ships **569**
spindles and spinning 406, 412–413, **421–423**
sponges: diving for **149–152**, **586–587**; for hygiene 453–454; for protection in battle 671
Spoon of Diocles **475–476**
statuaria 252
statues **263–264**; cast-bronze **256–257**; colour of 261
steel: iron and **235–241**; production **238–239**
steelyard **63–64**
steering-oars 558, 560

steganography **649**
stevedores 565
Stoics 716
stone, quarried: transport **279–280**; types of 215–216; uses **216**
streets: lighting 57, **323–324**, 538; urban **538–539**
strigil, oil and **452–453**
stucco 274–275, **281–284**
stylus 643
sugar **175**
sundials 621–622, **623–625**, 626
surveying **289–290**, **544–546**
Syracuse 71, **699–700**
syringe 482

tactics, military **673–688**
talent (weight) 594, 613n6
tanning, leather 406, **427–428**
tattooing, as means of secure communication 650
tawing, leather **428**
technology: archaeology as historical source 7–8; attitudes towards **727–729**; drivers of 31 (cf. sources of); goals of 31; literary sources for ancient technology 4–6; philosophy vs **725–726**; and society in antiquity 1–4
tempering, iron **238–239**
temples 279, **296–308**
testudo **665**, **676–677**, 692
textiles **405–426**
Thales 71, 335–336
Theophrastus 658
theoretical work, attitudes towards 721
thermae **376–381**; *see also* baths
Thisbe 344
thong drill 473
Thrace 120, 201–202
threshing and threshing floors 95, **119–120**
Tiberius Caesar 122, 439, 645
Tigris River 108–109, 342
tiles 107, 137, 141, 160, 179, 215, 262, 267–268, 271, 298, 311, 314, 318, 376–377, 433, **495–497**
time-keeping **621–630**
tin 190, 200, **231–232**, 247–248
Tiryns 269
toilets, Roman **381–383**; *see* hygiene; latrines, communal; urinals, Roman
tortoise: military **676–677**, **691–692**, 698–699, 702; for transport 531
touchstones **189**, 244, **246**

town planning **276–279**; *see* Hippodamus; *pomerium*
traction bench *see* Bench of Hippocrates
trade: long-distance **603–606**; standards of **591–613**; three branches of 606
Trajan 294–295, 522, 590
trapetum 159, 165, **171–172**
trees 21–22, 54, **123–125**, 128, **129–131**, 397–399, 401
trepanation 463, **472–474**
triremes 549–550, 555–557, 560, 573–574, 588
Trojan Horse 704n1
tunnels: for aqueducts **346–347**; highway **516–517**; in mining 198, 201–202; in sieges 696, 698–699
turbines **39**
Turkey 214
tympanum 363–364

urbanisation **20–21**, 277–285, **328–331**
urinals, Roman 381; *see also* latrines, communal; toilets, Roman
urine 414, 416, 420, 427, 444
utensils, cooking **179–180**

valves 72–73, 371–373, 374
vaults **269–270**, **283–288**, 321, 323, 325, 377
vegetables **122–123**, 369, 611
vehicles **529–536**
veneers 213, **284–285**, 400–402; *see* revetment
venter 348–349
verdigris 233, 445–446
Vespasian 414, 674
vessels, bronze 263
vinegar 229, 404, 407, 421, 428, 435, 442, 444–445; in mining 201
vineyards and viticulture 14, **97**, 99, 101, **124–127**
vintage **126–127**
vivisection **477–479**
voice, techniques for improvement 456
voting: sortition **630–633**; secret ballot **634–637**

wages and prices **307–309**, 530–532, **552–553**, **608–613**
walls: construction of **283**; painting 316
warfare, nobility of 714–715
warships 46, 550, 555, 556, **557–558**, **563–564**, 568, 573, 578, 684, 688

water: for agriculture 93, 107, 143; as beverage 59, **173–174**; clocks *see* clocks, water; dispenser 79–80; divining **337–339**; heater 381; hydraulic-engineering projects **345–346**; importance of **336**; mills *see* mills, water; organ **71–73**, 630, 723; pipelines and aqueducts 346–361; pressure, diving and 586–587; screw *see* screw, water; sources of **339–341**; wheel **40–45**, 159, 336, **362–365**, 368, 580
water-lifting devices 46, 107, **361–376**
wattle and daub **272**
wax 76, 112, 137, **143**, 169, **178**, 252, 262, **264**, **281**, 364, 396, 402, 405, **442–444**, 481, **484–486**, **556**, 628
wax tablets **643–646**, 649–650
weapons: invention of 665; large-scale production **493–495**; *see also* arms and armour; naval weapons; siege machinery
weaving 19, 406, **414–415**, **422–425**
wedge 62
weights and measures 8–9, **597–600**; origins 28
welding: iron 244; solders and **243–244**
wells 336, **338–340**
wheel: and axle 38, 62, 80; potter's 57, 431; water *see* water, wheel
whetstone 121
willow, uses 402, 554
windmills **37–39**
wine 58–59, 95–97, **124–127**, 170, 173, 609–610; making **167–170**
winnowing **119–120**
woad 417
wood: in construction 279, **285–287**; in ship-building **549–555**
woodworking **396–405**
wool **407**, **411–413**, 492–493, 613
worship 18
wound care 475–476
writing: benefits of **642**; and communication **638–660**; concealment and encryption **649–653**; forgery 647–649; materials **643–647**; negative view **642–643**; origins 19, 639

Xerxes 520, 525, 589

yokes **116–117**, 509, 516, 613

Zama 682
Zeus/Jupiter 13–15, 18, 255

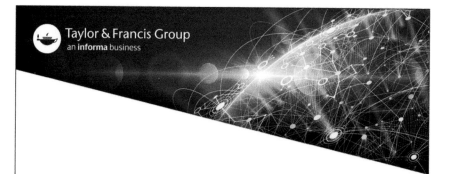